D0992127

MOLECULAR MECHANISMS OF OXYGEN ACTIVATION

MOLECULAR BIOLOGY

An International Series of Monographs and Textbooks

Editors: BERNARD HORECKER, NATHAN O. KAPLAN, JULIUS MARMUR, AND HAROLD A. SCHERAGA

A complete list of titles in this series appears at the end of this volume.

MOLECULAR MECHANISMS OF OXYGEN ACTIVATION

Edited by

OSAMU HAYAISHI

Department of Medical Chemistry
Kyoto University Faculty of Medicine
Kyoto, Japan

ACADEMIC PRESS   New York San Francisco London 1974
A Subsidiary of Harcourt Brace Jovanovich, Publishers

ACADEMIC PRESS, INC.
111 Fifth Avenue, New York, New York 10003

United Kingdom Edition published by
ACADEMIC PRESS, INC. (LONDON) LTD.
24/28 Oval Road, London NW1

Library of Congress Cataloging in Publication Data

Hayaishi, Osamu, Date
    Molecular mechanisms of oxygen activation.

    (Molecular biology; an international series of mono-
graphs and textbooks)
    Includes bibliographical references.
    1.   Oxygenases.   I.   Title.   II.   Series.
QP603.085H39   574.1'9258   72-9987
ISBN 0–12–333640–6

CONTENTS

3   Heme-Containing Dioxygenases

Philip Feigelson and Frank O. Brady

4   Nonheme Iron Dioxygenase

Mitsuhiro Nozaki

5   α-Ketoglutarate-Coupled Dioxygenases

Mitchel T. Abbott and Sidney Udenfriend

6   Microsomal Cytochrome P-450-Linked Monooxygenase Systems in Mammalian Tissue

Sten Orrenius and Lars Ernster

7   Flavoprotein Oxygenases

Marcia S. Flashner and Vincent Massey

8   Pterin-Requiring Aromatic Amino Acid Hydroxylases

Seymour Kaufman and Daniel B. Fisher

9   Copper-Containing Oxygenases

Walter H. Vanneste and Andreas Zuberbühler

10   Chemical Models and Mechanisms for Oxygenases

Gordon A. Hamilton

11   Superoxide Dismutase

Irwin Fridovich

12   Cytochrome c Oxidase

Peter Nicholls and Britton Chance

13   Peroxidase

Isao Yamazaki

LIST OF CONTRIBUTORS

Numbers in parentheses indicate the pages on which the authors' contributions begin.

MICHEL T. ABBOTT (167), Department of Chemistry, San Diego State University, San Diego, California.

INGEMAR BJÖRKHEM (29), Department of Chemistry, Karolinska Institutet, Stockholm, Sweden

FRANK O. BRADY (87), The Institute of Cancer Research, and Department of Biochemistry, College of Physicians and Surgeons, Columbia University, New York, New York

BRITTON CHANCE (479), Johnson Research Foundation, School of Medicine, University of Pennsylvania, Philadelphia, Pennsylvania

HENRY DANIELSSON (29), Department of Chemistry, Karolinska Institutet, Stockholm, Sweden

P. DEBRUNNER (559), Department of Physics, University of Illinois, Urbana, Illinois

LARS ERNSTER (215), Department of Biochemistry, University of Stockholm, Stockholm, Sweden

PHILIP FEIGELSON (87), The Institute of Cancer Research, and Department of Biochemistry, College of Physicians and Surgeons, Columbia University, New York, New York

DANIEL B. FISHER (285), Laboratory of Neurochemistry, National Institute of Mental Health, Bethesda, Maryland

MARCIA S. FLASHNER* (245), Department of Biological Chemistry, The University of Michigan, Ann Arbor, Michigan

IRWIN FRIDOVICH (453), Department of Biochemistry, Duke University Medical Center, Durham, North Carolina

I. C. GUNSALUS (559), Department of Biochemistry, University of Illinois, Urbana, Illinois

MATS HAMBERG (29), Department of Chemistry, Karolinska Institutet, Stockholm, Sweden

GORDON A. HAMILTON (405), Department of Chemistry, The Pennsylvania State University, University Park, Pennsylvania

OSAMU HAYAISHI (1), Department of Medical Chemistry, Kyoto University Faculty of Medicine, Kyoto, Japan

SEYMOUR KAUFMAN (285), Laboratory of Neurochemistry, National Institute of Mental Health, Bethesda, Maryland

J. D. LIPSCOMB (559), Department of Biochemistry, University of Illinois, Urbana, Illinois

VINCENT MASSEY (245), Department of Biological Chemistry, The University of Michigan, Ann Arbor, Michigan

J. R. MEEKS† (559), Department of Biochemistry, University of Illinois, Urbana, Illinois

E. MÜNCK (559), Department of Physics, University of Illinois, Urbana, Illinois

PETER NICHOLLS (479), Biophysics Unit, ARC Institute of Animal Physiology, Babraham, Cambridge, England

MITSUHIRO NOZAKI (135), Department of Medical Chemistry, Kyoto University Faculty of Medicine, Kyoto, Japan

STEN ORRENIUS (215), Department of Forensic Medicine, Karolinska Institutet, Stockholm, Sweden

BENGT SAMUELSSON (29), Department of Chemistry, Karolinska Institutet, Stockholm, Sweden

* Present address: Department of Biology, Syracuse University, Syracuse, New York.
† Present address: Diamond Shamrock Corporation, T. R. Evans Research Center, Painesville, Ohio.

SIDNEY UDENFRIEND (167), Roche Institute of Molecular Biology, Nutley, New Jersey

WALTER H. VANNESTE (371), Laboratorium voor Fysiologische Scheikunde, Faculteit van de Diergeneeskunde, Rijksuniversiteit Gent, Gent, Belgium

ISAO YAMAZAKI (535), Biophysics Division, Research Institute of Applied Electricity, Hokkaido University, Sapporo, Japan

ANDREAS ZUBERBÜHLER (371), Institut für Anorganische Chemie, Universität Basel, Basel, Switzerland

PREFACE

"Oxygenases," a comprehensive treatise on these and related enzymes, was first published in 1962. During the past decade, considerable progress has been made regarding the basic mechanisms involved in activation of molecular oxygen and the physiological functions of various oxygenases. Information and data acquired from the detailed analyses of a large number of isolated and highly purified enzyme systems, together with results and conclusions derived from biological and medical investigations *in vivo*, constitute a body of knowledge from which generalized concepts and basic principles of oxygenases and related enzymes have emerged. Advances in this field of research have indeed been so rapid and diverse that keeping abreast of the progress has presented difficulties even for the experts. In view of these considerations, I have undertaken the editing of "Molecular Mechanisms of Oxygen Activation" covering the major new advances that have been made in our understanding of this vital process.

In the previous volume, the emphasis was primarily on the nature of the reactions rather than on the enzymes and mechanisms of catalysis, and in a sense this side-stepped the most interesting questions which relate to the activation process itself. However, the availability of new physical and chemical techniques has made possible remarkable advances in our knowledge of the structure of many of the enzymes and the development of sophisticated models to account for this activity. One particularly notable advance was the recognition of ternary complexes containing enzyme, substrate, and oxygen as obligatory intermediates in many of the di- or monooxygenase-catalyzed reactions. In addition, entirely new areas of research have been opened up with the discovery of important new biological products such as the prostoglandins and new classes of oxygen-activating enzymes such as the cytochromes P-450, the flavoprotein oxygenases, and the α-ketoglutarate-requiring oxygenases. These oxygenases were unknown or unrecognized in 1962. Superoxide, superoxide dismutase, and super-oxide-utilizing enzymes provided additional insight into the mechanism by which molecular oxygen is activated and reduced.

The new discoveries have highlighted the important and widespread role of oxygen activation in contemporary biological processes. It is my hope that a review of this current knowledge of various aspects of oxygenases will be of value to investigators already working in the field and will provide a foundation for other researchers who wish to become involved in aspects of this important and exciting problem.

I express my appreciation to Dr. B. L. Horecker for his assistance with the editorial work and to the staff of Academic Press for their cooperation. The writing and editing of this volume were carried out during my tenure as a Fogarty International Scholar in the summers of 1972 and 1973.

OSAMU HAYAISHI

1

GENERAL PROPERTIES AND BIOLOGICAL FUNCTIONS OF OXYGENASES

OSAMU HAYAISHI

I. HISTORICAL BACKGROUND

A. Early History

Oxygen, one of the most abundant elements on the earth and directly or indirectly essential for almost all forms of life, has been the subject of

1

intensive studies by biochemists and physiologists ever since Lavoisier initiated the study of biological oxidation processes some 200 years ago. Since that time, the mechanisms by which various nutrients are oxidized by living organisms have remained among the most important and interesting problems in biological science.

Lavoisier and his contemporaries defined the term "oxidation" as the addition of oxygen atoms to a substrate, (X), while the opposite process, that of reduction, was regarded as the removal of oxygen from an oxide [Eq. (1)].

$$X + O \underset{\text{reduction}}{\overset{\text{oxidation}}{\rightleftarrows}} XO \tag{1}$$

As early as 1896, Bertrand observed that living organisms contain a number of enzymes which catalyze the oxidation of various biological compounds, and these were designated "oxydases." The early workers generally assumed that oxygen was affected and modified by "oxydases" in such a way that stable oxygen molecules were activated then bound to substrates. The nature of the so-called activated oxygen, however, was unknown for many years, although organic peroxides and ozonides were postulated as active forms by a number of investigators. At the turn of the twentieth century, Bach and his co-workers proposed that oxygen reacted with an acceptor, A, in the primary reaction to produce an organic peroxide, which then reacted with a substrate, (X), to form an oxide (1).

$$A + O_2 \xrightarrow{\text{oxygenase}} A\overset{O}{\underset{O}{\diagdown\diagup}} \tag{2}$$

$$A\overset{O}{\underset{O}{\diagdown\diagup}} + X \xrightarrow{\text{peroxidase}} AO + XO \tag{3}$$

The enzymes which catalyzed reactions (2) and (3) were named "oxygenase" and "peroxidase," respectively. This hypothetical mechanism, however, failed to gain general acceptance and was eventually abandoned, chiefly because experimental evidence for the formation of organic peroxides during general oxidative processes was lacking.

About 20 years later, Otto Warburg proposed a theory of cell respiration which was to considerably influence students of biological oxidation in subsequent years. According to this theory, the essential process in cell respiration is the activation of oxygen catalyzed by heme-containing enzymes, "the respiratory enzyme" (das Atmungsferment) (2). In many ways this theory was reminiscent of that of Bach, since Warburg assumed that the primary reaction in cell respiration is between molecular oxygen and heme iron (Fe) as follows:

$$\text{Enz--(Fe)} + O_2 \rightarrow \text{Enz--(Fe)--}O_2 \tag{4}$$

The oxidation of the organic molecule then follows according to Eq. (5).

$$\text{Enz--(Fe)--}O_2 + 2\,X \rightarrow \text{Enz--(Fe)} + 2\,XO \tag{5}$$

This role of oxygen molecules per se in biological oxidation processes was vigorously challenged and questioned following the discovery by Schardinger of an enzyme in milk which catalyzed the conversion of aldehydes to acids in the presence of methylene blue but in the total absence of oxygen. The oxidation of aldehyde was accompanied by the concomitant reduction of methylene blue under anaerobic conditions. This finding prompted Wieland to investigate the nature of biological oxidation processes and eventually to propose a generalized mechanism which could work out in the total absence of oxygen (3). According to this scheme the essential characteristic of biological oxidation processes is the removal or transfer of electrons from the substrate molecule, (XH_2), to an appropriate acceptor, A, [Eq. (6)].

$$XH_2 + A \underset{\text{reduction}}{\overset{\text{oxidation}}{\rightleftharpoons}} X + AH_2 \tag{6}$$

This "dehydrogenation" theory was supported by the ingenious experiments of Thunberg, who was able to demonstrate enzymic oxidation of a variety of substrates in the presence of methylene blue under completely anaerobic conditions. Furthermore, in the last several decades many dehydrogenases, which do not utilize oxygen, have been isolated, purified, and crystallized from animal and plant tissues as well as from microorganisms. Pyridine nucleotides, flavin nucleotides, and cytochromes have been found to act as electron acceptors for various dehydrogenases. In the occasional case when oxygen molecules serve as the immediate electron acceptor, the enzymes have been called "oxidases."

According to our present knowledge, classic oxidases can be divided into three categories. In the first category, the enzyme catalyzes the transfer of one electron to one molecule of oxygen forming superoxide anion as shown in Eq. (7). Xanthine oxidase is an example of this type of enzyme.

$$XH + O_2 \rightarrow X + O_2^- + H^+ \tag{7}$$

In the second group, two electrons are transferred to one molecule of oxygen to produce hydrogen peroxide as a product [Eq. (8)]. D-Amino acid oxidase is a classic example of this type of enzyme.

$$XH_2 + O_2 \rightarrow X + H_2O_2 \tag{8}$$

The enzymes of the third class catalyze the transfer of four electrons to a molecule of oxygen producing water as a product [Eq. (9)]. In this case the bond between the two oxygen atoms must be cleaved, whereas in the first and second cases the bond between the two oxygen atoms is preserved. Cytochrome oxidase and ascorbate oxidase are examples of the third class of oxidases.

$$XH_4 + O_2 \rightarrow X + 2H_2O \tag{9}$$

Thus, the early concept of oxygen incorporation by Lavoisier was almost completely replaced and molecular oxygen was considered to serve as a terminal electron acceptor in cellular respiration. In fact, in 1932, Wieland made the following statement in his famous book, "On the Mechanism of Oxidation" (3): "Limiting ourselves to the chief energy-supply foods, we have in this class carbohydrates, amino acids, the higher fatty acids, and glycerol. There is no known example among them of an unsaturated compound in the case of which it is necessary to assume direct addition of oxygen, that is, additive oxidation."

According to this theory, when the overall reaction can be formulated as an addition of oxygen, it was assumed that hydration or hydrolysis is involved, and that the oxygen atoms are derived from the water molecule rather than from atmospheric oxygen. An example of this mechanism is shown in Eqs. (10) and (11), where the substrate is hydrated in the primary reaction, followed by dehydrogenation in a second reaction. The sum [Eq. (12)] is an addition of oxygen to the substrate, but the oxygen is derived from the water molecule rather than from atmospheric oxygen.

$$Sub + H_2O \rightarrow SubH_2O \tag{10}$$
$$SubH_2O + A \rightarrow SubO + AH_2 \tag{11}$$
$$Sum: \quad Sub + H_2O + A \rightarrow SubO + AH_2 \tag{12}$$

The enzymic transformation of aldehydes to acids catalyzed by aldehyde dehydrogenases is an example of an oxidation in which the oxygen atom is derived from the water molecule. In this case acyl thioesters formed in the oxidation step are converted to acids by hydrolysis [Eq. (13)].

$$\tag{13}$$

B. Discovery of Oxygenases

In 1950, during a study of tryptophan metabolism, an enzyme was isolated from cells of a pseudomonad which catalyzed oxidative ring cleavage of the benzene ring of catechol forming *cis,cis*-muconic acid as the reaction product [Eq. (14)] (4). This enzyme, which was termed "pyro-

$$\text{(structure)} + O_2 \longrightarrow \text{(structure)} \tag{14}$$

catechase," exhibited properties unlike those of an "oxidase" or "dehydrogenase," for it was not associated with any of the previously known coenzymes or electron carriers, and none of the dyes or artificial electron acceptors tested could replace oxygen as an oxidant. According to the then generally accepted belief, biological oxidation proceeded exclusively by the removal of electrons or hydrogen atoms from substrates, and direct addition of molecular oxygen was excluded from consideration. Oxygen might still be incorporated into substrates by hydration reactions involving water, but prior or subsequent dehydrogenation process would remove hydrogen or electrons as in the case of aldehyde oxidases. In 1955, however, when a heavy oxygen isotope was used as a tracer in $^{18}O_2$ and $H_2^{18}O$, it was demonstrated that the two oxygen atoms incorporated into the product of the above reaction were derived exclusively from molecular oxygen rather than from water (5). Concurrently and independently, Mason and his collaborators, using the same isotope, found that during the oxidation of 3,4-dimethylphenol to 4,5-dimethylcatechol catalyzed by a phenolase complex, the oxygen atom incorporated into the substrate molecule was derived exclusively from molecular oxygen but not from the oxygen of water [Eq. (15)] (6).

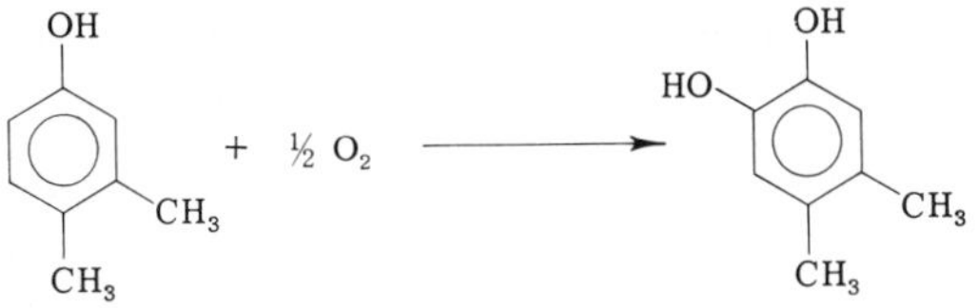

$$\text{(structure)} + \tfrac{1}{2} O_2 \longrightarrow \text{(structure)} \tag{15}$$

These findings conflicted sharply with the then current concept that oxygen could act only as an ultimate electron acceptor in biological oxidation and that all oxygen atoms incorporated into substrates are derived from the oxygen atoms of water. These two newly discovered reactions may be schematically represented by Eqs. (16) and (17).

$$X + O_2 \rightarrow XO_2 \tag{16}$$

$$X + \tfrac{1}{2} O_2 \rightarrow XO \tag{17}$$

The overall reaction in Eq. (16) may be visualized as the addition of both atoms of an oxygen molecule to a molecule of substrate, (X). It was soon acknowledged that in the case of Eq. (17), one of the atoms of molecular oxygen is incorporated into a substrate molecule and the other reduced to H_2O in the presence of an appropriate electron donor, (DH_2), such as NADH, NADPH, tetrahydrofolic acid, or ascorbic acid [Eq. (18)].

$$X + O_2 + DH_2 \rightarrow XO + H_2O + D \qquad (18)$$

These two types of reactions both involve "oxygen fixation" into a substrate molecule, and they are therefore different from the classic oxidase reactions shown in Eqs. (7), (8), and (9). They are similar to the oxygenation reactions known to occur in chemical or photochemical processes, and we therefore proposed a new term "oxygenase" to designate enzymes which catalyze such oxygen fixation reactions (7).

II. NOMENCLATURE, CLASSIFICATION, AND GENERAL PROPERTIES OF OXYGENASES

From the historical point of view as well as from the preceding discussion, it is evident that the term "oxygenase" may be appropriately assigned to a group of enzymes presumably catalyzing the activation of oxygen and the subsequent incorporation of either one or two atoms of oxygen per mole of various substrates. The terms "mono" and "di" oxygenases are generally assigned, respectively, to the enzymes catalyzing these two types of reactions (8).

When the substrate, the acceptor of oxygen, is hydrogen, the enzyme has been called an oxidase. In that sense, oxidases may be envisaged as a special class of oxygenases for reactions in which hydrogen atoms serve as the oxygen acceptor. These enzymes will, however, not be considered extensively here, as evidence indicates that enzymic reduction of oxygen to HO_2, H_2O, or H_2O_2 would involve the activation of hydrogen rather than oxygen, and in addition this subject has been extensively covered in a number of recent reviews.

Because information concerning the mechanism of action of oxygenases is still limited even though the field is progressing rapidly, any classification scheme would necessarily be arbitrary and perhaps temporary. Two subclasses, di- and monooxygenases, have been employed by the majority of workers in this field, although the distinction between the di- and monooxygenase subclasses, discussed below, is phenomenological and historical and may not be so permanent and meaningful.

A. Dioxygenases

Dioxygenases are defined as enzymes catalyzing reactions in which both atoms of molecular oxygen are incorporated into substrates. In the many instances where one substrate can act as the oxygen acceptor [Eq. (16)], the term *"intra*molecular dioxygenases" may be used. The dioxygenases acting upon two acceptor substrates, which have recently been reported from a number of laboratories (9), may be referred to as *"inter*molecular dioxygenases."* One of the two substrates for the latter type has so far been invariably α-ketoglutarate, and the overall reaction may be schematically shown by Eq. (19). More detailed discussions of this type of enzyme will be found in a later chapter by Abbott and Udenfriend (9).

$$\alpha\text{-Ketoglutarate} + O_2 + X \rightarrow \text{succinate} + CO_2 + XO \qquad (19)$$

A third class of dioxygenases include various enzymes which require NADH or NADPH as an electron donor. Although it is quite possible that reactions of this class may involve a simple dioxygenation reaction followed by a reductive step, these two processes may be coupled in such a way as to justify a separate category. The formation of catechol from anthranilate (10) is an example of this type of reaction. The overall reaction may be represented as follows:

$$X + O_2 + NADH + H^+ \rightarrow X \overset{\displaystyle OH}{\underset{\displaystyle OH}{\diagup \diagdown}} + NAD \qquad (20)$$

Some dioxygenases, such as tryptophan 2,3-dioxygenase, contain heme as the sole prosthetic group, while others, such as pyrocatechase, contain nonheme iron, or like quercetinase, contain copper as the prosthetic group.

B. Monooxygenases

Monooxygenases are defined as a group of enzymes which catalyze the incorporation of one atom of molecular oxygen into a substrate while the other is reduced to water. These enzymes are sometimes referred to as mixed function oxygenases or mixed function oxidases since they are apparently bifunctional, carrying out oxidase activity on one site and oxygenase activity on the other. Recently, the term "hydroxylase" has been used by some investigators; however, this term is misleading and would be better avoided because not all hydroxylation reactions are catalyzed by monooxygenases. In some cases, dihydroxy compounds are formed by the introduction of two oxygen atoms derived from the same oxygen molecule. The reaction is presumably catalyzed by a dioxygenase. Alternatively, the oxygen atom of a newly formed hydroxyl group may be derived from water rather than from molecular oxygen. The formation of 6-hydroxynicotinic acid from nicotinic acid (11) and that of barbituric acid from uracil (12)

TABLE I

CLASSIFICATION OF OXYGENASES

Oxygenase	EC No.[a]
A. Dioxygenase	
1. Intramolecular dioxygenase	1.13.11
a. Hemoprotein	
b. Iron-sulfur protein	
c. Copper protein	
2. Intramolecular dioxygenase	1.14.11
3. Miscellaneous	1.13.99
B. Monooxygenase	
1. Internal monooxygenase	1.13.12
2. External monooxygenase	1.14
a. Pyridine nucleotide-linked flavoprotein	1.14.13
b. Flavin-linked hemoprotein	1.14.14
c. Iron-sulfur protein-linked hemoprotein	1.14.15
d. Pteridine-linked monooxygenase	1.14.16
e. Ascorbate-linked cupper protein	1.14.17
f. With another substrate as reductant	1.14.18

[a] EC number refers to the new numbering system introduced in 1972 by the International Union of Biochemistry, Enzyme Nomenclature Commission.

and 6-hydroxylation of the pteridine ring (13) are examples of the latter type of reaction.

Monooxygenases may be classified on the basis of the electron donor involved. Since in some cases the primary reductant is unknown, these may eventually be reclassified.

1. Internal Monooxygenases

The simplest type of monooxygenase catalyzes the incorporation of a single atom of molecular oxygen concomitant with the reduction of the other oxygen atom by electrons derived from the substrate. Thus, the overall reaction may be expressed by the following equation:

$$XH_2 + O_2 \rightarrow XO + H_2O \tag{21}$$

Since the reducing agent is internally supplied, these enzymes may be referred to as internal monooxygenases. The first of these to be crystallized was the lactate oxidative decarboxylase from *Mycobacterium phlei* (14). This enzyme catalyzes the conversion of lactate to acetate with the incorporation of one atom of oxygen into acetate, the evolution of one mole of CO_2, and the reduction of one atom of oxygen to water as follows (15):

$$CH_3CHOHCOOH + O_2 \rightarrow CH_3COOH + CO_2 + H_2O \qquad (22)$$

2. External Monooxygenases

While the internal monooxygenases do not require external reducing agents, more common types of monooxygenases require various kinds of electron donors. The overall reactions are schematically represented by Eq. (18). The electron donor (DH_2) serves as a basis for the subclassification of external monooxygenases. Some examples are shown below.

a. *Flavoprotein Monooxygenases with Reduced Pyridine Nucleotides as* DH_2. Salicylate 1-monooxygenase (16), a flavoprotein, is an example of this type of enzyme and catalyzes the reaction shown below.

$$+ \; O_2 \; + \; NADH \; + \; H^+ \longrightarrow \qquad + \; H_2O \; + \; NAD \; + \; CO_2 \qquad (23)$$

b. *Heme-Containing Monooxygenases.* Aryl 4-monooxygenase (liver microsomal cytochrome P-450) (17, 18) catalyzes hydroxylation of a variety of substrates with reduced flavin as DH_2.

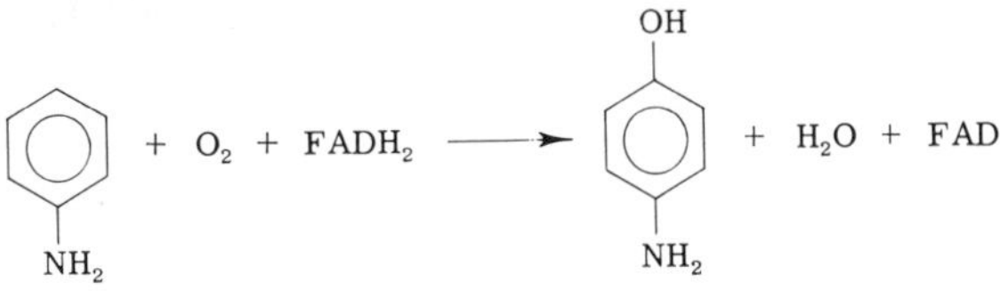

$$+ \; O_2 \; + \; FADH_2 \longrightarrow \qquad + \; H_2O \; + \; FAD \qquad (24)$$

c. *Heme-Containing Monooxygenases with a Reduced Iron-Sulfur Protein as* DH_2. Camphor 5-monooxygenase (19, 20) is a typical example of this type of enzyme.

Camphor $+ O_2 + 2$ reduced putidaredoxin $\xrightarrow{+2\,H^+}$ 5-*exo*-hydroxycamphor

$$+ H_2O + 2 \text{ oxidized putidaredoxin} \qquad (25)$$

d. *Pteridine-Linked Monooxygenases.* Phenylalanine-4-monooxygenase (21) catalyzes the formation of tyrosine from phenylalanine as follows:

L-Phenylalanine $+ O_2 +$ tetrahydropteridine $\rightarrow$ L-tyrosine $+ H_2O +$ dihydropteridine

$$(26)$$

This group of enzymes will be discussed in detail by Kaufman in a separate chapter (22).

e. With Ascorbate as DH_2. Dopamine β-monooxygenase (23, 24) is an example of this type of enzyme.

3,4-Dihydroxyphenylethylamine $+ O_2 +$ ascorbate

$$\rightarrow \text{norepinephrine} + H_2O + \text{dehydroascorbate} \quad (27)$$

f. With Another "Substrate" as DH_2. Monophenol monooxygenase (25) is an example of this type. In this case dopa may be considered as electron donor in the reaction.

$$\text{Tyrosine} + O_2 + \text{dopa} \rightarrow \text{dopa} + H_2O + \text{dopa quinone} \quad (28)$$

A brief summary of the above classification scheme is shown in Table I.

III. CHEMICAL ASPECTS OF OXYGEN FIXATION REACTIONS

A. General Comments

As mentioned above, biological oxidation processes may occur in three apparently different ways: (1) by the removal of an electron, as when a ferrous ion (Fe^{2+}) is converted to a ferric ion (Fe^{3+}); (2) by the removal of hydrogen, as when alcohol is oxidized to aldehyde; or (3) by the addition of oxygen to a molecule, as when an aromatic compound (Ar) is oxygenated to a hydroxy derivative.

$$Fe^{2+} \rightarrow Fe^{3+} + e^- \quad (29)$$

$$CH_3CH_2OH \rightarrow CH_3CHO + H_2 \quad (30)$$

$$ArH + \tfrac{1}{2}O_2 \rightarrow Ar\text{--}OH \quad (31)$$

Although these three types of reactions appear to be dissimilar, they all involve a transfer of electrons. Oxidation is defined as removal of electrons with an increase in valency, whereas reduction is defined as a gain in electrons with a decrease in valency. When carbon and oxygen interact to form covalent C–O bonds [Eq. (31)], electrons are shared between both atoms. Since these electrons are somewhat closer to the oxygen nucleus than to the carbon nucleus, the oxygen has partially gained electrons and is reduced while the carbon has partially lost electrons and is oxidized.

Therefore, the fundamental principle governing the above three types of reaction is basically the same; however, they are catalyzed by different types of enzymes in the cell. The enzymes which catalyze dehydrogenation of primary substrates [Eqs. (29) and (30)] are designated "dehydrogenases" and the enzymes which catalyze the addition of oxygen have been called "oxygenases."

B. Molecular Oxygen as a Substrate

Oxygenases utilize two different species of substrate, namely, molecular oxygen and the oxygen acceptor, which is called the substrate and may be either an organic or an inorganic compound. It has not yet been clarified whether gaseous oxygen or oxygen dissolved in water is utilized by these enzymes. It is generally believed that the latter type of oxygen is different from the former. In water, oxygen is considered to exist largely in a dimerized form, O_2–O_2, because of its biradical nature, and probably forms a charge transfer complex with a water molecule (26). Whatever form the oxygen dissolved in water takes, the type of oxygen that serves as substrate for oxygenases is in rapid equilibrium with gaseous oxygen under normal experimental conditions and can be clearly distinguished through the use of $^{18}O_2$ from the oxygen in water molecules or other compounds in the reaction mixture.

For the details of the isotope technique involving $^{18}O_2$ and $H_2^{18}O$, readers are referred to a previous article in "Oxygenases" (27). It is not always possible, however, to demonstrate $^{18}O_2$ incorporation into products after reaction in an $^{18}O_2$-containing atmosphere for the following reasons. (1) Oxygen atoms in carbonyl groups are easily exchanged with the oxygen of water molecules; therefore, analytical results are not always stoichiometric. In order to avoid this difficulty, carbonyl groups in the reaction product may be further reduced, oxidized, or conjugated in some way to stabilize the oxygen. (2) The oxygenated intermediate may be labile or transient and not susceptible to trapping in a stable form. Desaturation of fatty acids (28) and the side chain of heme (29), which will be discussed in detail in separate chapters, are examples of this case. In these reactions, the oxygenated intermediate appears to be extremely labile or enzyme-bound and is presumed to dehydrate immediately. Therefore a demonstration of the incorporation of molecular oxygen in this transient intermediate is impossible [Eqs. (32)–(34)].

$$\text{R—CH}_2\text{—CH}_2\text{—R}' + O_2 + H_2X \rightarrow [\text{R—}\overset{\overset{\text{OH}}{|}}{\underset{\underset{\text{H}}{|}}{\text{C}}}\text{—CH}_2\text{—R}'] + H_2O + X \quad (32)$$

$$[\text{R—}\overset{\overset{\text{OH}}{|}}{\underset{\underset{\text{H}}{|}}{\text{C}}}\text{—CH}_2\text{—R}'] \rightarrow \text{R—CH=CH—R}' + H_2O \quad (33)$$

$$\text{Sum:} \quad \text{R—CH}_2\text{—CH}_2\text{—R}' + O_2 + H_2X \rightarrow \text{R—CH=CH—R}' + 2H_2O + X \quad (34)$$

The overall reaction [Eq. (34)], therefore, represents desaturation of the substrate with the concomitant reduction of both atoms of molecular oxygen to water.

C. The Role of Substrate Which Acts as Oxygen Acceptor

As for the acceptor for molecular oxygen, a great variety of both organic and inorganic compounds can be oxygenated. In general, oxygen-rich compounds, such as carbohydrates, are not favorable substrates for oxygenases since these usually have many reactive groups containing oxygen such as the hydroxyl, carbonyl, or formyl, and their biochemical function does not require further oxygenation. On the other hand, lipids and aromatic compounds are often metabolized by oxygenases. These oxygen-deficient compounds generally require oxygenation in order to become biologically active or more soluble in water. Because of the hydrophobic nature of lipids and aromatic compounds, molecular oxygen is the preferred hydroxylating agent rather than water. In contrast, purines and pyrimidines with their hydrophilic ring systems are usually hydroxylated by the addition of water, followed by dehydrogenation.

In addition to its role as an oxygen acceptor, the substrate usually acts also as an allosteric regulator. Only in the presence of the substrate does molecular oxygen bind to the active center of the enzyme and become catalytically active. Because this phenomenon appears to be a common and unique feature of many oxygenases, it will be discussed further in a following section.

D. Reaction Mechanisms and the Nature of Active Oxygen

One of the most intriguing and challenging problems in the field of oxygenases has been the mechanism whereby the two substrates interact with the enzyme and become activated; e.g., does molecular oxygen bind to the enzyme first to form EO_2 or does the acceptor first bind to the enzyme to form an ES complex? Is it molecular oxygen, or the substrate, or both, which are activated by the oxygenase? Is there any evidence for the activation of molecular oxygen, and, if so, what is the nature of the so-called active oxygen?

These mechanisms will be dealt with in detail in separate chapters; however, essentially speaking, recent work from a number of laboratories has indicated rather conclusively that the enzyme binds oxygen only in the presence of substrate to form the ternary complex, ESO_2, in which oxygen and substrate interact to form a product. Such a ternary complex was postulated in 1964 on the basis of binding experiments (8), but more direct experimental evidence was not available until 1967 when a ternary complex of tryptophan 2,3-dioxygenase–tryptophan-O_2 was demonstrated by spectrophotometric experiments. This oxygenated form of the enzyme was later shown by the stopped-flow technique to be the so-called obligatory intermediate of the reaction (30–33). Since that time similar oxygenated intermediates have also been observed with protocatechuate 3,4-dioxygenase (an iron-sulfur protein dioxygenase) (34, 35), lysine monooxygenase (a flavoprotein) (36), and also with cytochrome P-450 (37–39). In each case, the enzyme has to bind the organic substrate first before it can be oxygenated, in contrast to oxygen-carrying pigments such as hemoglobin and hemoerythrin which are freely and reversibly oxygenated in the absence of any effector or substrate. In fact, several lines of evidence indicate that the substrate, tryptophan, binds specifically to the heme coenzyme of tryptophan 2,3-dioxygenase; as a consequence, the state of the heme is altered in such a way that its reactivity toward ligands is increased by several orders of magnitude (31–33).

In contrast, Tai and Sih (40) have reported that in the case of steroid dioxygenase molecular oxygen is bound first to the enzyme and the EO_2 complex then reacts with the organic substrate. Their conclusion is based on steady-state kinetics data, and further studies are necessary to elucidate possible differences between this enzyme and others in the class.

The nature of so-called active form of oxygen in the above-mentioned ternary complex has been one of the most important as well as the most difficult questions in this field. All oxygenase-catalyzed reactions are exothermic and are therefore irreversible. Nevertheless, molecular oxygen is a rather inert compound and at room temperature reacts slowly with the substrate compounds in the absence of enzymes. This low kinetic reactivity is usually explained on the basis that molecular oxygen is in a triplet ground state. The direct reaction of a triplet molecule with organic molecules in the singlet state is electronically spin forbidden so that a substantial activation energy is required. For this reason singlet oxygen has been suggested as a likely intermediate in many oxygenase-catalyzed reactions (41, 42); however, definitive evidence is so far unavailable. It has recently been discovered that during enzymic hydroxylation of aromatic substrates the substituent (deuterium, tritium, chlorine, bromine, etc.) displaced by the entering hydroxyl group migrates to an adjacent position in the aromatic ring (43). On the basis of extensive studies of this phenomenon, called the "NIH shift," the active oxygen species involved in certain monooxygenases was postulated to be oxenoid (44). These reactions will be discussed in greater detail in a separate chapter by Hamilton (45).

On the other hand, evidence has appeared indicating that O_2^-, superoxide anion, may be the active form of oxygen in the case of intestinal tryptophan 2,3-dioxygenase (46), and similar experimental results reported by Coon and co-workers indicated that superoxide anion may be involved in the hydroxylation reactions catalyzed by hepatic cytochrome P-450 (47). It is, however, uncertain whether or not superoxide anion is the general form of active oxygen in all oxygenase-catalyzed reactions or whether these enzymes represent a new class of enzyme which utilizes superoxide anion rather than molecular oxygen as an oxygenating agent. It is still quite feasible that intestinal tryptophan 2,3-dioxygenase will prove to be the first example of such a new group of enzymes which may be referred to as "superoxygenases."

E. Uncoupling of Oxygenase Activity

Certain oxygenases, particularly monooxygenases, have exhibited oxidase rather than oxygenase activity under certain conditions. Such a phenomenon has often been referred to as "uncoupling." In 1968, Okamoto *et al.* (48) reported that when imidazoleacetate monooxygenase was treated with varying amounts of mercurials or silver nitrate, the enzyme was almost completely inactivated with respect to oxygenase activity at a ratio of approximately 2 moles of reagent per mole of enzyme. The oxidase activity of the enzyme, however, was unaffected by this treatment. It is of particular interest that some mercurials even enhanced oxidase activity. The reaction product of oxidase activity was identified as hydrogen peroxide, while that of the oxygenase reaction was water. Other oxygenases are known to exhibit oxidase activity when allowed to react with certain substrate analogs. Thus, salicylate monooxygenase was shown to catalyze the oxidation of NADH forming H_2O_2 when benzoate was substituted for salicylate (49). Lysine monooxygenase could act as an amino acid oxidase and produce α-ketoacid analogs of certain substrate analogs such as ornithine and 2,8-diaminooctanoate with the stoichiometric formation of ammonia and H_2O_2 (50).

One possible interpretation of the above findings is that only the substrates with an appropriate size or structure fit the active center of the enzyme molecule forming an enzyme–substrate complex, after which the substrate can be oxygenated by the attack of molecular oxygen. On the other hand, substrates which do not exactly fit the active center of the enzyme may form another type of enzyme–substrate complex, in which the dehydrogenation of the substrate by FAD does not couple with oxygen activation, and the reduced FAD is then merely oxidized by oxygen to form H_2O_2. Such an interpretation is substantiated by a more recent observation by Yamamoto *et al.* (51) that the "fragmented substrate" such as L-alanine plus propylamine is not oxygenated but oxidized. L-Alanine alone is completely inactive as substrate; however, when propylamine is present, oxidation, but not oxygenation, of alanine was demonstrated in the presence of a crystalline preparation of L-lysine monooxygenase. Other α-monoamino acids are also oxidized in the presence of alkylamines with various carbon chain length, but the highest oxidase activity is observed when the total chain length of both amino acid and amine is nearly identical with that of lysine.

IV. BIOLOGICAL FUNCTION OF OXYGENASES

A. General Comments

The significance of biological oxygen fixation in medicine, agriculture, microbiology, and also in food technology, cosmobiology, public health problems, and biochemistry in general has now been well established. Oxygenases play important roles in biosynthesis, transformation, and degradation of essential metabolites such as amino acids, lipids, sugars, porphyrins, vitamins, and hormones. They also play a crucial role in the metabolic disposal of foreign compounds such as drugs, insecticides, and carcinogens. Furthermore, they participate in the degradation of various natural and synthetic compounds by soil and airborne microorganisms in nature and are therefore of great significance in environmental science.

Many monooxygenases catalyze hydroxylation of both aromatic and aliphatic compounds. Monooxygenases also catalyze a seemingly diverse group of reactions including epoxide formation, dealkylation, decarboxylation, deamination, and N- or S-oxide formation. Although the overall reactions catalyzed by various monooxygenases appear grossly unlike each other, the primary chemical event is identical since these processes are all initiated by the incorporation of one atom of molecular oxygen into the substrate. After the initial monooxygenation reaction the compounds become more soluble in water or become biologically more reactive in the sense that they are susceptible to the action of various dioxygenases. The major reaction catalyzed by dioxygenases is the cleavage of an aromatic double bond, which may be located (a) between two hydroxylated carbon atoms, (b) adjacent to a hydroxylated carbon atom, or (c) in an indole ring. However, a similar type of reaction also occurs with aliphatic substrates such as β-carotene, which yields vitamin A as the product (52).

B. Oxygenases in Amino Acid Metabolism

The important role played by oxygenases in the metabolism of aromatic amino acids such as tryptophan, phenylalanine, tyrosine, and various phenolic compounds has long been recognized; for example, tryptophan undergoes a variety of metabolic transformations. Among these, two of the most important are initiated by a monooxygenase and a dioxygenase (Fig. 1). The black arrows in this figure denote oxygenase reactions, and the white arrows indicate reactions catalyzed by enzymes other than oxygenases. The oxygen atoms shown in heavy print represent molecular oxygen which has been incorporated into these substrates by the action of various oxygenases. It can be seen that essentially all the oxidative steps are catalyzed by oxygenases rather than by oxidases and dehydrogenases.

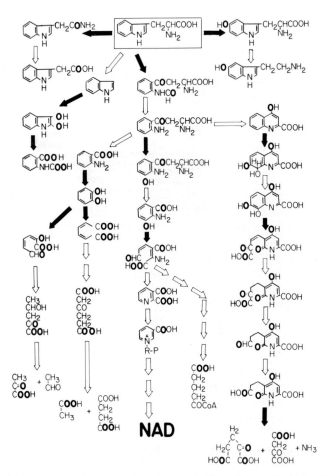

FIG. 1. Oxygenase reactions involved in the pathway of tryptophan metabolism. The black arrows denote oxygenase reactions and the white arrows represent reactions catalyzed by enzymes other than oxygenases.

Tryptophan serves as a precursor for the biosynthesis of serotonin, a potent vasoconstrictor substance and a neurohumoral agent. The enzyme tryptophan 5-monooxygenase catalyzes the initial step of this metabolic transformation and has been purified from the brain stem of various species of mammals (53–55). One atom of molecular oxygen is introduced into tryptophan to form 5-hydroxytryptophan, which is then decarboxylated to serotonin. Tryptophan 2,3-dioxygenase (56, 57) catalyzes the insertion of two atoms of molecular oxygen into the pyrrole moiety of tryptophan, forming formylkynurenine as a product; hence, it is sometimes referred to as tryptophan pyrrolase. This reaction is the initial step in a metabolic sequence which leads to the biosynthesis of a coenzyme, nicotinamide-adenine dinucleotide (NAD), from tryptophan in mammals and in some microorganisms and is probably a rate-limiting step and a likely site of regulation of this pathway. Formylkynurenine thus formed is converted to kynurenine which is then hydroxylated to 3-hydroxykynurenine by the specific action of kynurenine 3-monooxygenase. This enzyme is localized in the outer membrane of rat liver mitochondria (58) and contains flavin adenine dinucleotide (FAD) as its prosthetic group (59). It is induced by oxygen in *Saccharomyces cerevisiae*; the biosynthesis of pyridine nucleotides therefore starts mainly from tryptophan in aerobic cultures but begins from aspartate and a C_3 fragment under anaerobic conditions (60). In mammals, administration of L-thyroxine causes a decrease of about 50% in kynurenine 3-monooxygenase activity, presumably because of decreased biosynthesis of this enzyme, thus providing a reasonable biochemical explanation for the low NAD(P) level in hyperthyroid animals (61).

3-Hydroxykynurenine is transformed to 3-hydroxyanthranilic acid, which is then cleaved by a specific dioxygenase. This reaction is physiologically of great importance since the resulting compound can either be converted to acetyl-coenzyme A through various intermediates including glutaryl-coenzyme A, or it can form a new pyridine ring yielding picolinic acid and quinolinic acid as reaction products. Quinolinic acid has been shown to yield nicotinic acid ribonucleotide, the precursor of NAD (62).

Similarly, the metabolism of phenylalanine and tyrosine is catalyzed by a number of oxygenases; for example, the enzymic formation of tyrosine from phenylalanine and the formation of epinephrine, norepinephrine, melanin, and thyroxine are catalyzed by a number of consecutive reactions involving various mono- and dioxygenation reactions. Regarding many hereditary metabolic disorders as studied in humans, the biochemical anomaly has been traced to the absence of activity of specific oxygenases in these pathways. In normal individuals, almost all of phenylalanine metabolism is channeled through tyrosine by the action of phenylalanine 4-monooxygenase. Kaufman and co-workers have shown that tetrahydropteridine derivatives serve as the direct hydrogen donor in this process (21, 22). Hereditary deficiency of phenylalanine 4-monooxygenase results in phenyl-ketonuria, an inborn error of metabolism and a common cause of mental retardation. Similarly, alkaptonuria is the result of the hereditary deficiency of homogentisate 3,4-dioxygenase, which catalyzes the conversion of homogentisic acid to 4-maleylacetoacetate.

Hydroxyprolines, primarily 4-hydroxyproline, together with a small amount of the 3-hydroxy isomer, are unique constituents of collagen, a major component of cartilage and other connective tissues, and the most abundant protein in the body. Early isotopic studies indicated that free hydroxyproline is not incorporated into collagen but that hydroxylation of peptidylproline yields collagen hydroxyproline. This reaction is catalyzed by a dioxygenase which has been partially purified and requires ferrous ion, ascorbate, and α-ketoglutarate (63–65). The hydroxylation of peptidyl-lysine appears to proceed in a similar manner (65, 66). These reactions will be reviewed in more detail in a separate chapter (9).

A novel type of amino acid decarboxylase, lysine monooxygenase, which yields the corresponding acid amide instead of the amine, was initially isolated from a pseudomonad and crystallized in our laboratory (67). Although the physiological significance of this type of reaction is not yet clear, two different enzymes have been discovered since then which act specifically on L-arginine (68) and L-tryptophan (69).

The formation of hypotaurine and cystine sulfinate from cysteamine and L-cysteine, respectively, are catalyzed by two distinct dioxygenases (70, 71), which will be discussed in detail later.

C. Oxygenases in Lipid Metabolism

Oxygenases play a versatile and ubiquitous role in the metabolism of lipid and related compounds. They also play a major role in the bio-synthesis, transformation, and degradation of steroids, fatty acids, prosta-glandins, vitamin A, bile acids, etc. As early as 1867, Pasteur made the classic observation that anaerobically growing yeast cells degenerated in structure unless periods of aerobiosis were interspersed, although yeast had been known to be capable of maintaining life under strictly anaerobic conditions (72). This observation suggested that molecular oxygen is required for the biosynthesis of some compounds essential to morphogenesis. In fact, subsequent work from a number of laboratories indicated that molecular oxygen is vital for the biosynthesis of some essential lipid com-ponents of cell membrane (73, 74). Oxygenases in lipid and steroid metabo-lism will be dealt with in a separate chapter by Hamberg et al. (75).

D. Oxygenases in Carbohydrate Metabolism

As previously mentioned, sugars and polyalcohols are, in general, not favorable substrates for oxygenases. However, myoinositol is converted to D-glucuronate (76) and ascorbate is cleaved to oxalate and threonate by a dioxygenase (77). Although both of these enzymes are dioxygenases requiring iron as a cofactor, further studies are required for their characterization.

E. Oxygenases in Nucleic Acid Metabolism

Only three oxygenases which act upon nucleic acid and its derivatives have been characterized. All three enzymes require α-ketoglutarate (α-KG) as cosubstrate. These are D-thymidine α-KG dioxygenase (78), 5-hydroxymethyluracil, α-KG dioxygenase (79), and thymine, α-KG dioxygenase (80), all present in *Neurospora*. A more detailed account of these enzymes is given in the chapter by Abbot and Udenfriend (9).

F. Oxygenases in the Metabolism of Various Aromatic Compounds

Investigations of the microbial oxidation of various aromatic compounds have contributed not only to the knowledge of general pathways of their metabolism and regulation but also to the development of molecular genetics (81). These oxidative enzymes also play a major role in the degradation of various compounds in nature by soil microorganisms, a phenomenon which is of vital importance in the control of pollution.

The liver microsomal enzyme, cytochrome P-450, plays a crucial role in the metabolism of a variety of foreign compounds, xenobiotics, including drugs, carcinogenic and carcinostatic substances, antibiotics, insecticides, etc. Because of the physiological and pharmacological importance, it has been the subject of intensive investigation by enzymologists, pharmacologists, and toxicologists. Details of the reaction mechanism and the properties of cytochrome P-450 will be described in a separate chapter by Orrenius and Ernster (82). Some physiological aspects are briefly described here.

It has been known for some time that some of the carcinogenic hydrocarbons are not independently active but become carcinogenic when oxygenated; for example, liver microsomal enzyme converts dibenzanthracene to the corresponding epoxide (83), which then becomes bound to DNA or histones (84). The epoxides of the carcinogenic hydrocarbons, i.e., benzanthracene, dibenzanthracene, and methylcholanthrene, are more active in producing malignant transformations than the parent hydrocarbons (85). These results taken together indicate that the monooxygenation of these hydrocarbons is a prerequisite for their oncogenic activity.

Other examples for the metabolic conversion of noncarcinogenic compounds to carcinogens are known; for example, some aromatic amines are oxygenated to become carcinogenic hydroxylamine or N-oxide derivatives (86).

Similarly, Brodie and co-workers have recently shown that bromobenzene is metabolized to a more reactive metabolite, presumably an arene oxide, by cytochrome P-450 in rats. They speculated that the latter, by forming a covalent bond to the liver proteins, can lead to extensive liver necrosis (87). The incidence of necrosis was higher when P-450 was induced by pretreatment with phenobarbital and lower when P-450 was inhibited by SKF-525A, indicating clearly the involvement of the microsomal aryl monooxygenase system in the hepatotoxic effects of aromatic compounds.

An interesting property of P-450 is the fact that it is one of the few enzymes in animal tissues which can be induced by a substrate. This raises the possibility of enhancing the activity of drug metabolizing enzyme systems by various drug analogs, thus allowing the pharmaceutical chemists to design a variety of potentially useful therapeutic agents. On the other hand, inhibitors of this system might prolong the activity of a drug or decrease the effect of carcinogenic agents. One such inhibitor, SKF-525A, has been formulated but as yet is not widely used for clinical purposes.

It has been proposed that one of the causes for the decline in the population of pigeons, hawks, and other birds is the result of accumulation of DDT in these animals. One of the effects of DDT is induction of higher P-450 levels, and because P-450 is involved in steroid metabolism the balance of steroid hormones is altered. As a consequence the mating habits and reproductive processes of the birds changed resulting in a decrease in numbers (88, 89).

Long-range effects of DDT on humans and other mammals are as yet unknown; however, it has been established that DDT does stimulate the enhanced production of microsomal P-450. As the assimilation of foreign substances from our surroundings increases the levels of this enzyme could also increase above normal levels. Since knowledge of long-range effects of high levels of the enzymes is lacking, it is imperative that the induction and mechanism of action of these enzymes be investigated so that an evaluation of the biological effects of pollutants can be made (90, 91).

The presence of cytochrome P-450 has also been demonstrated in insects, and the resistance to insecticides, which insects develop, is partly because of induction of higher levels of this enzyme system (92, 93). Knowledge of this induction process, as well as the oxygenation mechanisms of insecticides, will assist in formulating compounds for pest control.

Finally, an enzyme similar to P-450 has been found in plants. The leaves of the cotton plant (*Gosypium hirsutum* L.) contain an enzyme which catalyzes the *N*-dealkylation of a pesticide (94). Such an enzyme may be responsible for oxidative destruction of pesticide residues in plants.

V. NATURAL DISTRIBUTION OF OXYGENASES

A. Microbial Oxygenases

Oxygenases are ubiquitously distributed in nature. They have been isolated from animals, plants, and microorganisms. Aerobic microorganisms such as *Pseudomonas*, *Mycobacteria*, and *Nocardia* are rich sources of oxygenases, whereas strictly anaerobic microorganisms appear to be lacking these enzymes.

Quantitative studies of the oxygen fixation reaction have revealed that a considerable amount of molecular oxygen is incorporated into cell constituents of aerobic microorganisms (95); for example, when *Pseudomonas*, a strictly aerobic bacterium, was grown in the presence of ^{18}O-containing air, harvested, dried, and analyzed by a mass spectrometer, almost 10% of the oxygen of the cell material under certain conditions of growth was found to be derived from gaseous oxygen. When similar experiments were conducted in the presence of $H_2^{18}O$, approximately 50–60% of the oxygen of cell materials was found to be derived from water; the rest presumably came from organic substrates, phosphates, and other ingredients in the medium.

The presence of physiologically important oxygenases in animals is generally confined to specialized organs and tissues, and the amounts are rather limited. For this reason mammalian oxygenases in large quantities are difficult to obtain. On the other hand, a number of oxygenases in bacteria are inducible, obtained easily in large quantities, and therefore practical for initiating enzyme studies. Almost all the oxygenases which have been crystallized are of microbial origin. These enzymes can be produced in large quantities and have potential application in environmental control since a variety of chemicals and natural compounds are degraded efficiently by oxygenases. Unstable enzymes can be stabilized by bonding them to a variety of solid materials, and such "immobilized" enzymes offer promise in large-scale operations in industry as well as in environmental control.

B. Animal Oxygenases

In animals, oxygenases are widely distributed in almost all types of cells and participate in the general metabolism of a variety of nutrients as well as foreign compounds. As previously stated, oxygenases play a major role in the biosynthesis of a number of hormones, including steroid hormones, prostaglandins, epinephrine, norepinephrine, thyroxine, and serotonin.

A great variety of oxygenases have been isolated from liver, kidney, brain, lung, adrenals, thyroid gland, gonad, and other tissues. Of special interest is the recent finding of a tryptophan 2,3-dioxygenase, which has a limited distribution in the lower one-third of the small intestine of the rabbit (96). This enzyme has the unique property of utilizing superoxide anion O_2^- as substrate rather than O_2. It participates in the degradation of D- as well as L-tryptophan and also of serotonin and its precursor, 5-hydroxytryptophan (97); thus, it may play a major role in controlling serotonin levels in the intestine.

In animal cells, oxygenases are not always associated with one specific cell compartment but are distributed in various cellular fractions; for example, homogentisate 1,2-dioxygenase (98) and 3-hydroxyanthranilate 3,4-dioxygenase (99) are both located in the soluble, supernatant fraction of liver cells, whereas kynurenine 3-monooxygenase is found in the outer membrane of rat liver mitochondria (100, 101). Aryl 4-monooxygenase, commonly known as cytochrome P-450, is associated with the endoplasmic reticulum. A number of steroid monooxygenases are associated with adrenal mitochondria.

C. Plant Oxygenases

In plants, a great variety of oxygenases which participate in the biosynthesis, transformation, and degradation of cell constituents have been found. Of particular significance to medicinal chemists are oxygenases involved in alkaloid biosynthesis.

Hydroxylation and oxygenative cleavage of aromatic rings have been suggested as being steps in alkaloid biosynthesis; however, many of these suggestions are admittedly speculative and tentative. Among these, Senoh *et al.* (102) suggested the possibility that the biosynthesis of pyridine, α-pyrone, and α-tetronic acid nuclei originate from metapyrocatechase-type oxidations, followed by some types of isomerization reactions; for example, stizolobic acid and stizolobinic acid, new amino acids isolated from the etiolated seedlings of *Stizolobium hassjoo*, are presumably derived from 3,4-dihydroxyphenylalanine *in vivo* by a metapyrocatechase-type cleavage reaction and subsequent cyclization and dehydrogenation of the resultant α-hydroxymuconic semialdehyde derivatives (Fig. 2).

Lipooxygenase (103) is prevalent in plants and is important in food processing and preservation. The enzyme catalyzes the conversion of lipids containing a *cis,cis*-1,4-pentadiene system to the corresponding hydroperoxides. Details will be discussed in a separate chapter (75).

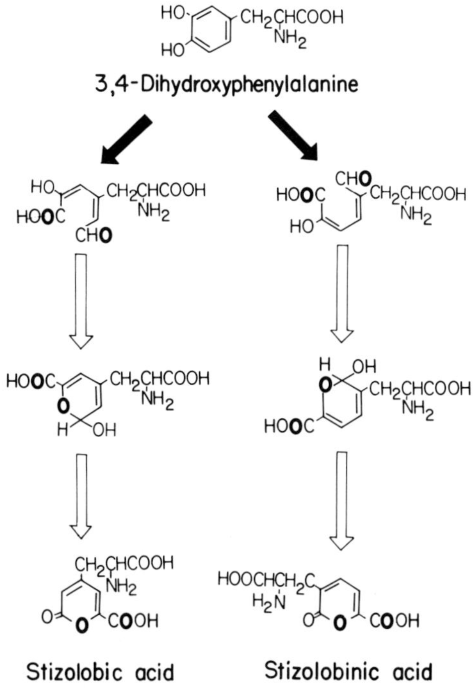

FIG. 2. Proposed pathways of biosynthesis of stizolobic and stizolobinic acids in plants.

By analogy with mammalian and bacterial systems, many oxygenase reactions are expected to be involved in aromatic metabolism in plants since phenolic compounds in higher plants are second only to carbohydrates in abundance. These compounds have great structural variety, ranging from derivatives of simple phenols, such as quinol, to complex polymeric materials, such as lignin. Overall pathways of biosynthesis of these compounds have been investigated in many laboratories, but very little is known about the detailed mechanisms of the enzymic reactions. Nevertheless, the accumulation of positive evidence from both tracer and enzymological investigations suggests that oxygenation reactions participate in the biosynthesis of these complex phenols and alkaloids.

Since phenolase was found to catalyze the incorporation of one atom of oxygen into phenol compounds, participation of this enzyme in the hydroxylation reactions of aromatic compounds during the biosynthesis of phenolic compounds has been suggested by a number of investigators. Many questions, including the precise role of the enzyme in *in vivo* systems, are as yet unanswered. Likewise, peroxidase, abundant in plant tissues, was reported to catalyze various oxygenation reactions besides peroxidatic ones. These are hydroxylation of aromatic compounds in the presence of di-hydroxyfumarate and oxygen (105) and monooxygenation of amino acids, such as methionine (104) and tryptophan (106), to form the corresponding acid amides in the presence of pyridoxal phosphate and molecular oxygen. The physiological significance of these reactions of peroxidase is obscure, but overall reactions of the latter type are identical to those catalyzed by lysine monooxygenase (67) and other similar flavin-containing mono-oxygenases.

The inhibition of photosynthetic carbon dioxide fixation by oxygen was initially observed by Warburg in 1920. The nature of this inhibition was explained on the basis of an inhibition of photosynthesis and a stimulation of photorespiration. When spinach leaves were exposed to an atmosphere containing $^{18}O_2$, the label was rapidly incorporated into the carboxyl group of glycine and serine (107). An oxygen-dependent production of phospho-glycolate is catalyzed by purified soybean ribulose diphosphate carboxylase and by crude extracts of soybean and corn leaves (108). These lines of evidence strongly suggest that ribulose diphosphate carboxylase may act as an oxygenase under certain conditions.

In conclusion, speculations can be made concerning the appearance and function of oxygenases in the evolution of life on earth. It is generally agreed that the earliest primitive forms of life on earth existed in an anaer-obic environment. Therefore, the minimum manifestation of life is capable of occurring in the absence of molecular oxygen. Certainly the existence of various bacteria, which survive and grow solely under anaerobic conditions, is a present-day reminder that molecular oxygen is not essential to all forms of life. With the appearance of oxygen in the earth's atmosphere, however, advanced forms of life began to emerge. Most probably the initial advan-tage of primitive oxygen-utilizing organisms was the ability to produce and utilize energy more efficiently in the presence of oxygen than in its absence. Thus, the role of oxygen as the terminal electron acceptor in respiration, which was early linked to the formation of energy donors such as ATP, became a prominent part of the metabolic apparatus of most living or-ganisms. It is worthy of mention at this point that oxygenases are involved both in the biosynthesis (29) and degradation of heme coenzyme (109). Heme oxygenase catalyzes the oxidation of heme at the α-methene bridge to form biliverdin and carbon monoxide in equivalent amounts.

Oxygenases probably appeared for the first time with the subsequent development of extensive biosynthetic pathways for essential compounds, of intracellular structures with defined function, and of regulatory mecha-nisms for metabolism and growth. Molecular oxygen has always been required in oxidative formation of vitamins essential for coenzyme synthesis, of unsaturated fatty acids and sterols present in membrane structures, and of hormones and mediators which nerve impulses presently utilize. Since oxygen and heavy metals were a ubiquitous part of the environment of these primitive organisms, why could not these be used for these important reactions in preference to other more complex or less common oxidizing agents? In an evolutionary sense, therefore, oxidases preceded oxygenases, and the latter appear to be associated with the most advanced and specialized biochemical processes.

The appearance of oxygen on earth not only affected the energy-producing faculty of living things but also the nature of compounds which play indispensable roles in most forms of life. In contrast to the eons of time required for the development of oxygenase enzymes, only for a decade has man known of their existence.

ACKNOWLEDGMENTS

A part of this writing was carried out during the tenure of Fogarty Scholar-in-Residence at the National Institutes of Health, Bethesda, Maryland, in the summer of 1972. The author wishes to express his appreciation to the following persons for critical reading of the manuscript and providing useful suggestions: Drs. B. L. Horecker, Simon Black, M. I. Harris, M. Nozaki, and S. Yamamoto. Thanks are also due Miss Mariko Ohara for assistance in the preparation and edition of the manuscript.

REFERENCES

1. Bach, A., and Chodat, R. (1903). *Ber. Deut. Chem. Ges.* **36**, 600–605.
2. Warburg, O. (1949). "Heavy Metal Prosthetic Groups and Enzyme Actions." Oxford Univ. Press, London and New York.
3. Wieland, H. (1932). "On the Mechanism of Oxidation," p. 26. Yale Univ. Press, New Haven, Connecticut.
4. Hayaishi, O., and Hashimoto, K. (1950). *J. Biochem.* **37**, 371–374.
5. Hayaishi, O., Katagiri, M., and Rothberg, S. (1955). *J. Amer. Chem. Soc.* **77**, 5450–5451.
6. Mason, H. S., Fowlks, W. L., and Peterson, E. (1955). *J. Amer. Chem. Soc.* **77**, 2914–2915.
7. Hayaishi, O., Rothberg, S., and Mehler, A. H. (1956). *Abstr. 130th Meet. Amer. Chem. Soc.* 53C.
8. Hayaishi, O. (1964). *Proc. Plenary Sessions, Int. Congr. Biochem. 6th, New York* 31–43.
9. Abbott, M. T., and Udenfriend, S. (1973). Chapter 5, this volume.
10. Kobayashi, S., Kuno, S., Itada, N., Hayaishi, O., Kozuka, S., and Oae, S. (1964). *Biochem. Biophys. Res. Commun.* **16**, 556–561.
11. Hunt, A. L., Hughes, D. E., and Lowenstein, J. M. (1957). *Biochem. J.* **66**, 2P.
12. Hayaishi, O., and Kornberg, A. (1952). *J. Biol. Chem.* **197**, 717–732.
13. Breuer, H., and Knuppen, R. (1961). *Biochim. Biophys. Acta* **49**, 620–621.

14. Sutton, W. B. (1957). *J. Biol. Chem.* **226**, 395–405.
15. Hayaishi, O., and Sutton, W. B. (1957). *J. Amer. Chem. Soc.* **79**, 4809–4810.
16. Yamamoto, S., Katagiri, M., Maeno, H., and Hayaishi, O. (1965). *J. Biol. Chem.* **240**, 3408–3413.
17. Mitoma, C., Posner, H. S., Reitz, H. C., and Udenfriend, S. (1956). *Arch. Biochem. Biophys.* **61**, 431–441.
18. Omura, T., Sato, R., Cooper, D. Y., Rosenthal, O., and Estabrook, R. W. (1965). *Fed. Proc. Fed. Amer. Soc. Exp. Biol.* **24**, 1181–1189.
19. Yu, C. A., and Gunsalus, I. C. (1969). *J. Biol. Chem.* **244**, 6149–6152.
20. Katagiri, M., Ganguli, B. N., and Gunsalus, I. C. (1968). *J. Biol. Chem.* **243**, 3543–3546.
21. Kaufman, S. (1963). *Proc. Nat. Acad. Sci. U.S.* **50**, 1085–1093.
22. Kaufman, S. Chapter 9, this volume.
23. Friedman, S., and Kaufman, S. (1965). *J. Biol. Chem.* **240**, 4763–4773.
24. Friedman, S., and Kaufman, S. (1966). *J. Biol. Chem.* **241**, 2256–2259.
25. Brown, F. C., and Ward, D. N. (1957). *J. Amer. Chem. Soc.* **79**, 2647–2648.
26. Diner, S. (1964). *In* "Electronic Aspects of Biochemistry" (B. Pullman, ed.), pp. 237–281. Academic Press, New York.
27. Samuel, D. (1962). *In* "Oxygenases" (O. Hayaishi, ed.), pp. 31–86. Academic Press, New York.
28. Bloch, K. (1969). *Accounts Chem. Res.* **2**, 193.
29. Sano, S. (1966). *J. Biol. Chem.* **241**, 5276–5283.
30. Ishimura, Y., Nozaki, M., Hayaishi, O., Tamura, M., and Yamazaki, I. (1967). *J. Biol. Chem.* **242**, 2574–2576.
31. Hayaishi, O., Ishimura, Y., Nakazawa, T., and Nozaki, M. (1968). *Colloq. Ges. Biol. Chem.* **19**, 196–216.
32. Hayaishi, O. (1969). *Ann. N.Y. Acad. Sci.* **158**, 318–335.
33. Ishimura, Y., Nozaki, M., Hayaishi, O., Nakamura, T., Tamura, M., and Yamazaki, I. (1970). *J. Biol. Chem.* **245**, 3593–3602.
34. Fujisawa, H., Hiromi, K., Uyeda, M., Nozaki, M., and Hayaishi, O. (1971). *J. Biol. Chem.* **246**, 2320–2321.
35. Fujisawa, H., Hiromi, K., Uyeda, M., Okuno, S., Nozaki, M., and Hayaishi, O. (1972). *J. Biol. Chem.* **247**, 4422–4428.
36. Yamamoto, S., Hirata, F., Yamauchi, T., Nozaki, M., Hiromi, K., and Hayaishi, O. (1971). *J. Biol. Chem.* **246**, 5540–5542.
37. Ishimura, Y., Ullrich, V., and Peterson, J. A. (1971). *Biochem. Biophys. Res. Commun.* **42**, 140–146.
38. Estabrook, R. W., Hildebrandt, A. G., Baron, J., Netter, K. J., and Leibman, K. (1971). *Biochem. Biophys. Res. Commun.* **42**, 132–139.
39. Peterson, J. A., Ishimura, Y., and Griffin, B. W. (1972). *Arch. Biochem. Biophys.* **149**, 197–208.
40. Tai, H. H., and Sih, C. J. (1970). *J. Biol. Chem.* **245**, 5072–5078.
41. Foote, C. S. (1968). *Accounts Chem. Res.* **1**, 136.
42. Wasserman, H. H., Scheffer, J. R., and Cooper, J. L. (1972). *J. Amer. Chem. Soc.* **94**, 4991–4996.
43. Guroff, G., Daly, J. W., Jerina, D. M., Renson, J., Witkop, B., and Udenfriend, S. (1967). *Science* **157**, 1524–1530.
44. Jerina, D. M. (1973). *Chem. Technol.* **4**, 120–127.
45. Hamilton, G. A. Chapter 11, this volume.
46. Hirata, F., and Hayaishi, O. (1971). *J. Biol. Chem.* **246**, 7825–7826.

47. Strobel, H. W., and Coon, M. J. (1971). *J. Biol. Chem.* **246**, 7826–7829.
48. Okamoto, H., Nozaki, M., and Hayaishi, O. (1968). *Biochem. Biophys. Res. Commun.* **32**, 30–36.
49. White-Stevens, R. H., and Kamin, H. (1970). *Biochem. Biophys. Res. Commun.* **38**, 882–889.
50. Nakazawa, T., Hori, K., and Hayaishi, O. (1972). *J. Biol. Chem.* **247**, 3439–3444.
51. Yamamoto, S., Yamauchi, T., and Hayaishi, O. (1972). *Proc. Nat. Acad. Sci. U.S.* **69**, 3723–3726.
52. Olson, J. A., and Hayaishi, O. (1965). *Proc. Nat. Acad. Sci. U.S.* **54**, 1364–1370.
53. Nakamura, S., Ichiyama, A., and Hayaishi, O. (1965). *Fed. Proc., Fed. Amer. Soc. Exp. Biol.* **24**, 604.
54. Grahame-Smith, D. G. (1967). *Biochem. J.* **105**, 351–360.
55. Ichiyama, A., Nakamura, S., Nishizuka, Y., and Hayaishi, O. (1970). *J. Biol. Chem.* **245**, 1699–1709.
56. Knox, W. E., and Mehler, A. H. (1950). *J. Biol. Chem.* **187**, 419–430.
57. Hayaishi, O., Rothberg, S., Mehler, A. H., and Saito, Y. (1957). *J. Biol. Chem.* **229**, 889–896.
58. Okamoto, H., Yamamoto, S., Nozaki, M., and Hayaishi, O. (1967). *Biochem. Biophys. Res. Commun.* **26**, 309–314.
59. Okamoto, H., and Hayaishi, O. (1967). *Biochem. Biophys. Res. Commun.* **29**, 394–399.
60. Schott, H., Staudinger, H., and Ullrich, V. (1971). *Hoppe-Seyler's Z. Physiol. Chem.* **352**, 1654.
61. Okamoto, H., Okada, F., and Hayaishi, O. (1971). *J. Biol. Chem.* **246**, 7759–7763.
62. Nishizuka, Y., and Hayaishi, O. (1963). *J. Biol. Chem.* **238**, PC483–485.
63. Hutton, J. J., Jr., Tappel, A. L., and Udenfriend, S. (1966). *Biochem. Biophys. Res. Commun.* **24**, 179–184.
64. Kivirikko, K. I., and Prockop, D. J. (1967). *Arch. Biochem. Biophys.* **118**, 611–618.
65. Kivirikko, K. I., and Prockop, D. J. (1967). *Proc. Nat. Acad. Sci. U.S.* **57**, 782–789.
66. Hausmann, E. (1967). *Biochim. Biophys. Acta* **133**, 591–593.
67. Takeda, H., and Hayaishi, O. (1966). *J. Biol. Chem.* **241**, 2733–2736.
68. Thoai, N. V., and Olomucki, A. (1962). *Biochim. Biophys. Acta* **59**, 533–544.
69. Kosuge, T., Haskett, M. G., and Wilson, E. E. (1966). *J. Biol. Chem.* **241**, 3738–3744.
70. Cavallini, D., De Marco, C., Scandurra, R., Dupré, S., and Graziani, M. T. (1966). *J. Biol. Chem.* **241**, 3189–3196.
71. Lombardini, J. B., Singer, T. P., and Boyer, P. D. (1969). *J. Biol. Chem.* **244**, 1172–1175.
72. Warburg, O. (1966). *In* "Current Aspects of Biochemical Energetics" (N. O. Kaplan and E. P. Kennedy, eds.), p. 103. Academic Press, New York.
73. Bloch, K., Borek, E., and Rittenberg, D. (1946). *J. Biol. Chem.* **162**, 441–449.
74. Andreasen, A. A., and Stier, T. J. B. (1953). *J. Cell. Comp. Physiol.* **41**, 23.
75. Hamberg, M., Samuelsson, B., Björkhem, I., and Danielsson, H., Chapter 2, this volume.
76. Charalampous, F. C. (1959). *J. Biol. Chem.* **234**, 220–229.
77. White, G. A., and Krupa, R. M. (1965). *Arch. Biochem. Biophys.* **110**, 448–461.
78. Shaffer, P. M., McCroskey, R. P., Palmatier, R. D., Midgett, R. J., and Abbott, M. T. (1968). *Biochem. Biophys. Res. Commun.* **33**, 806–811.
79. Abbott, M. T., Dragila, T. A., and McCroskey, R. P. (1968). *Biochim. Biophys. Acta* **169**, 1–6.

80. Abbott, M. T., Kadner, R. J., and Fink, R. M. (1964). *J. Biol. Chem.* **239,** 156–159.
81. Ribbons, D. W. (1965). *Annu. Rep. Chem. Soc.* **62,** 445.
82. Orrenius, S., and Ernster, L. Chapter 6, this volume.
83. Selkirk, J. K., Huberman, E., and Heidelberger, C. (1971). *Biochem. Biophys. Res. Commun.* **43,** 1010–1016.
84. Grover, P. L., and Sims, P. (1970). *Biochem. Pharmacol.* **19,** 2251–2259.
85. Grover, P. L., Sims, P., Huberman, E., Marquardt, H., Kuroki, T., and Heidelberger, C. (1971). *Proc. Nat. Acad. Sci. U.S.* **68,** 1098–1101.
86. Cramer, J. W., Miller, J. A., and Miller, E. C. (1960). *J. Biol. Chem.* **235,** 885–888.
87. Brodie, B. B., Reid, W. D., Cho, A. K., Sipes, G., Krishna, G., and Gillette, J. R. (1971). *Proc. Nat. Acad. Sci. U.S.* **68,** 160–164.
88. Peakall, D. B. (1967). *Nature (London)* **216,** 505–506.
89. Peakall, D. B. (1970). *Sci. Amer.* **222,** 73–97.
90. Street, J. C. (1969). *Ann. N.Y. Acad. Sci.* **160,** 274–290.
91. Gielen, J. E., and Nebert, D. W. (1971). *Science* **172,** 167–169.
92. Krieger, R. I., Feeny, P. P., and Wilkinson, C. F. (1971). *Science* **172,** 579–580.
93. Terriere, L. C., Yu, S. J., and Hoyer, R. F. (1971). *Science* **171,** 581–583.
94. Frear, D. S. (1968). *Science* **162,** 674–675.
95. Hayaishi, O. (1967). *J. Amer. Chem. Soc.* **79,** 5576–5577.
96. Yamamoto, S., and Hayaishi, O. (1967). *J. Biol. Chem.* **242,** 5260–5266.
97. Hirata, F., and Hayaishi, O. (1972). *Biochem. Biophys. Res. Commun.* **47,** 1112–1119.
98. Flamm, W. G., and Crandall, D. I. (1963). *J. Biol. Chem.* **238,** 389–396.
99. Stevens, C. O., and Henderson, L. M. (1959). *J. Biol. Chem.* **234,** 1188–1190.
100. Okamoto, H., and Hayaishi, O. (1970). *J. Biol. Chem.* **245,** 3603–3605.
101. Okamoto, H., and Hayaishi, O. (1971). *Amer. J. Clin. Nutr.* **24,** 805–806.
102. Senoh, S., and Sakan, T. (1966). *In* "Biological and Chemical Aspects of Oxygenases" (K. Bloch and O. Hayaishi, eds.), pp. 93–99, Maruzen, Tokyo.
103. Theorell, H., Holman, R. T., and Åkeson, Å. (1947). *Acta Chem. Scand.* **1,** 571–576.
104. Buhler, D. R., and Mason, H. S. (1961). *Arch. Biochem. Biophys.* **92,** 424–437.
105. Mazelis, M., and Ingraham, L. L. (1962). *J. Biol. Chem.* **237,** 109–112.
106. Riddle, V. M., and Mazelis, M. (1964). *Nature (London)* **202,** 391–392.
107. Andrews, T. J., Lorimer, G. H., Tolbert, N. E. (1971). *Biochemistry* **10,** 4777–4782.
108. Bowes, G., Ogren, W. L., and Hageman, R. H. (1971). *Biochem. Biophys. Res. Commun.* **45,** 716–722.
109. Tenhunen, R., Marver, H. S., Schmid, R. (1969). *J. Biol. Chem.* **244,** 6388–6394.

2

OXYGENASES IN FATTY ACID AND STEROID METABOLISM

MATS HAMBERG, BENGT SAMUELSSON, INGEMAR BJÖRKHEM,
and HENRY DANIELSSON

I. INTRODUCTION

Introduction of oxygen functions is important in the biosynthesis and metabolism of fatty acids, prostaglandins, and steroids. The mechanism of the oxygenation has been extensively studied during the last decade and significant progress has been made concerning transformation of fatty acids by lipoxygenases, desaturases, α- and ω-oxidizing systems and concerning transformation of steroids by various hydroxylases and lyases. Our knowledge of the biosynthesis of prostaglandins from polyunsaturated fatty acids stems entirely from research during the last decade. Detailed knowledge concerning the mechanism of oxygenation of fatty acids and steroids has been obtained in studies with highly purified enzyme systems such as lipoxygenase and some cytochrome P-450-dependent oxygenases from adrenals and liver.

This chapter attempts to summarize work concerning fatty acid and steroid oxygenases which has appeared in the very vast literature since 1962. The discussion will be limited to such aspects of the subject which seem most relevant to the question of the mechanism of oxygenation and which seem to offer the greatest promise for future lines of research. In the case of the biosynthesis and metabolism of steroids, the discussion is limited to mammalian oxygenases.

II. LIPOXYGENASE

Some properties of pure soybean lipoxygenase (EC 1.13.1.13) have been recently reported by Stevens *et al.* (1). The molecular weight was found to be 108,000. Evidence was presented that the enzyme consists of two subunits of 54,000 molecular weight. The dissociation could be effected by treatment with guanidine hydrochloride or sodium dodecyl sulfate. The amino acid composition was determined, and it was shown that the protein contained four residues of free sulfhydryl groups and four residues of half-cystine per molecule. No unusual amino acids were found. Ca^{2+} ion has been found to activate soybean lipoxygenase (2).

A. Specificity of Soybean Lipoxygenase

Soybean lipoxygenase catalyzes the aerobic formation of conjugated *cis,trans*-diene hydroperoxides from unsaturated fatty acids containing methylene group-interrupted *cis* double bonds (3, 4). Hamberg and Samuelsson demonstrated that soybean lipoxygenase has a high degree of specificity with respect to fatty acids serving as substrates as well as the position oxygenated in the fatty acids (5, 6). As seen in Fig. 1 only fatty

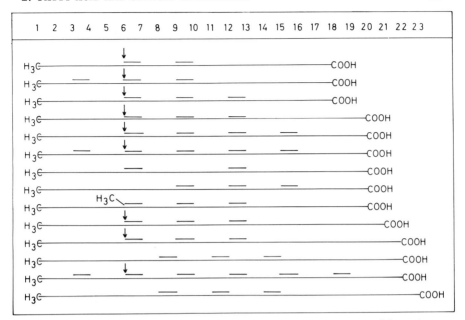

FIG. 1. Positional specificity of soybean lipoxygenase. From Hamberg and Samuelsson (6).

acids having a pair of methylene group-interrupted double bonds located between ω6 and ω10 served as substrates for the enzyme. It could further be shown that in fatty acids serving as substrates only the ω6 position was oxygenated. Dolev *et al.* obtained the same result with linoleic acid, i.e., 13-hydroperoxy-9,11-octadecadienoic acid was the only isomer formed (7). These findings have been confirmed and extended by Holman *et al.* (8), who tested a series of 13 isomeric octadecadienoates as substrates for soybean lipoxygenase. Natural linoleic acid (9,12-octadecadienoic acid) was the best substrate in the series. A shift of the pair of double bonds by only one carbon atom (8,11- and 10,13-octadecadienoic acids) strongly suppressed the rate of oxygenation. Interestingly, 13,16-octadecadienoic acid formed a second peak of reactivity (Fig. 2). This finding provides further support for the concept that the specificity of the enzyme is directed toward the ω6 carbon atom of unsaturated fatty acids. The factor(s) determining this specificity is not known. It may consist of hydrophobic bonding of the terminal part of the fatty acid molecule on the enzyme surface which might in some way favor oxygenation of the ω6 carbon. However, conformational changes in the fatty acid molecule brought about by changes in the positions of the double bonds may also be of importance. It is interesting to note that

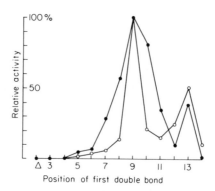

FIG. 2. Substrate specificity of soybean lipoxygenase upon positional isomers of octadecadienoic acid: (○) −Ca²⁺ and (●) +Ca²⁺. From Holman *et al.* (8)

of the CoA-esters of the above-mentioned series of isomeric octadecadienoates, the 9,12- and 13,16-octadecadienoyl derivatives were the best substrates also for hepatic acyl CoA:phospholipid transferases from a number of species (9). Several other workers have confirmed recently that 13-hydroperoxy-9,11-octadecadienoic acid is the major compound formed from linoleic acid by soybean lipoxygenase (10–12). However, small and varying amounts of the 9-hydroxy isomer were also isolated. This was also found in the original papers by Hamberg and Samuelsson (5, 6). Conflicting reports as to the origin of this acid have appeared. Veldink *et al.* obtained 9D-hydroxystearate by reduction and catalytic hydrogenation of reaction mixtures of linoleic acid and soybean lipoxygenase and suggested that an alternate orientation of linoleic acid on the enzyme surface might result in formation of the 9D-hydroperoxide (11). On the other hand, using a method for assignment of the optical configurations of microgram amounts of 9- and 13-hydroperoxyoctadecadienoates, Hamberg found that the 9-hydroperoxide was largely racemic, indicating that it is formed by a nonenzymic reaction (13). It may be mentioned that the recently discovered isoenzyme of soybean lipoxygenase has been reported to catalyze the formation of equal amounts of 9- and 13-hydroperoxyoctadecadienoates from linoleic acid (14). However, the configuration of the products was not established.

B. Mechanism

A mechanism for the formation of conjugated *cis,trans*-diene hydroperoxides from unsaturated fatty acids by soybean lipoxygenase was proposed by Siddiqi and Tappel in 1957 (15). The enzyme was considered as an electron sink which abstracted an electron or a hydrogen atom from the methylene group between the double bonds. The oxygen molecule was introduced in the second step. Support for the free radical intermediates postulated came from the observed inhibitory effects of various anti-oxidants, notably nordihydroguaiaretic acid (4). Since then, further support for the formation of free radicals has accumulated, e.g., initiation of sulfite oxidation during lipoxygenase oxygenation (16) and appearance of an ESR signal when lipoxygenase and linoleic acid are mixed (17).

Hamberg and Samuelsson established the absolute configuration at ω6 of hydroperoxides formed from linoleic acid and 8,11,14-eicosatrienoic acid (6). Both hydroperoxides were found to have the L configuration. In the case of 13-hydroperoxy-9,11-octadecadienoic acid this configuration has been confirmed by an independent method based on gas–liquid chromatographic separation of diastereoisomeric derivatives of methyl esters of 2-hydroxy acids. In this investigation the 13-hydroperoxide was found to be 97–98% 13L and 2–3% 13D (13).

In the formation of ω6-hydroperoxy acids by action of soybean lipoxygenase the ω6 double bond of the substrate is isomerized and one hydrogen is lost from the methylene group at ω8. The stereochemistry of this hydrogen removal has been determined by the use of stereospecifically tritium-labeled 8,11,14-eicosatrienoic acids (6). It was found that 15-hydroperoxy-8,11,13-eicosatrienoic acid formed from [13D-^{3}H, 3-^{14}C] 8,11,14-eicosatrienoic acid retained 93% of the tritium label, whereas the hydroperoxide formed from [13L-^{3}H, 3-^{14}C]8,11,14-eicosatrienoic acid lost 90% of the tritium. This shows that the hydrogen elimination from the ω8-methylene group is in fact stereospecific and that the 13L-hydrogen (pro-S) is selectively removed. Furthermore, analysis of the labeled 8,11,14-eicosatrienoic acid remaining after different times of incubation revealed that there was a significant enrichment of tritium in the precursor when the 13L-tritio acid was used. The presence of an isotope effect indicates that elimination of the ω8L-hydrogen occurs in the first step of the transformation, at least before covalent bond formation between oxygen and substrate has taken place. The observed stereochemistry may be the result of elimination of hydrogen and attack by oxygen from the same side (I) or from different sides (II) of the enzyme-bound fatty acid:

(I)   (II)

Similar results have also been obtained with linoleic acid (18).

Several investigators suggest, on the basis of different types of evidence, that a reactive enzyme–fatty acid–oxygen intermediate is formed in the conversion of fatty acids into hydroperoxides by lipoxygenases.

Direct support for such a reactive intermediate has been provided by Chan (19) in a recent work on coupled oxygenation of 1,3-dienes by soybean lipoxygenase. In one experiment tetraphenylcyclopentadienone (III) was converted into its endoperoxide (IV) by treatment with soybean lipoxygenase and ethyl linoleate:

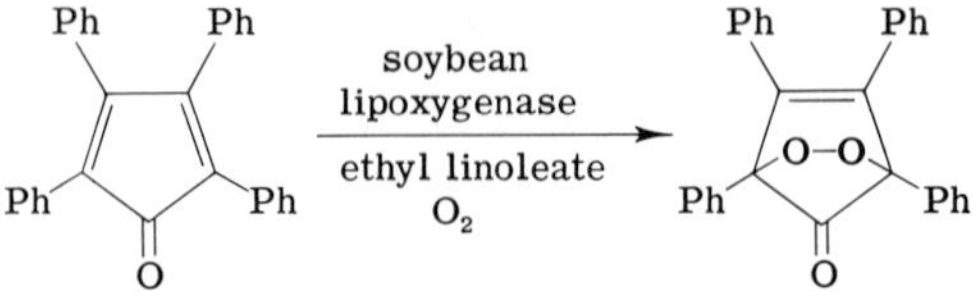

(III) (IV)

This reaction did not take place when boiled enzyme was used, nor when the 1,3-diene was added to a preincubated mixture of enzyme and ethyl linoleate. Therefore, the endoperoxide was probably formed from the 1,3-diene by an oxygenated intermediate formed by lipoxygenase, ethyl linoleate, and O_2. This intermediate was suggested to be a close analog of singlet oxygen. The coupled oxygenation by lipoxygenase may be formulated.

$$E + S + O_2 \rightleftharpoons [E{-}S{-}O_2] \xrightarrow[R \;\curvearrowright\; RO_2]{} E + X$$

An enzyme–fatty acid–oxygen intermediate has also been proposed by Smith and Lands (20) on the basis of studies of the inactivation of the enzyme during incubations with unsaturated fatty acids. Kinetic data indicated that the intermediate either dissociated into active enzyme and product or rearranged into inactive enzyme:

$$E + S + O_2 \rightleftharpoons [E{-}S{-}O_2] \begin{cases} \nearrow E + P \\ \searrow \text{Inactive E} \end{cases}$$

It may be mentioned that as early as 1946 Bergström (21) observed that lipoxygenase is inactivated rapidly in the presence of linoleic acid and oxygen (Fig. 3).

The formation of an oxygenated enzyme-bound intermediate has also been suggested by Graveland (22) to explain a number of oxygenated fatty acids isolated after incubations of linoleic acid with wheat flour doughs:

$$CH_3-(CH_2)_4-\underset{\underset{OOH}{|}}{CH}-CH=CH-CH=CH-(CH_2)_7-COOH \qquad L_1$$

$$CH_3-(CH_2)_4-\underset{\underset{OH}{|}}{CH}-CH=CH-CH=CH-(CH_2)_7-COOH \qquad L_2$$

$$CH_3-(CH_2)_4-\underset{O}{\underset{\diagdown\diagup}{CH-CH}}-CH=CH-\underset{\underset{OH}{|}}{CH}-(CH_2)_7-COOH \qquad L_3$$

$$CH_3-(CH_2)_4-\underset{\underset{OH}{|}}{CH}-\underset{\underset{OH}{|}}{CH}-CH=CH-\underset{\underset{OH}{|}}{CH}-(CH_2)_7-COOH \qquad L_4$$

A minor isomer of each of these compounds was also isolated. L_2 was probably formed by reduction of L_1 by thiol groups or other reducing agents present in the flour. L_4 was formed from L_3 by hydrolytic opening of the oxirane ring. These two reactions may not be enzymic. L_3 was not formed from L_1 or L_2 but probably from an oxygenated intermediate occurring in the formation of L_1 from linoleic acid. Purified wheat flour lipoxygenase

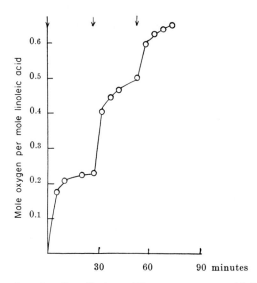

FIG. 3. Oxygenation of sodium linoleate. Lipoxygenase was added at arrows. From Bergström (21).

did not catalyze the formation of L_3 or L_4 from linoleic acid (23). When the enzyme was adsorbed to glutenin, however, it appeared to catalyze the formation of L_3 and L_4 from linoleic acid (23).

Garssen *et al.* (24) have isolated 13-oxo-9,11-octadecadienoic acid, 13-oxo-9,11-tridecadienoic acid, and *n*-pentane in anaerobic incubations of 13-hydroperoxy-9,11-octadecadienoic acid and linoleic acid with soybean lipoxygenase. The same compounds were formed when arachidonic acid was used instead of linoleic acid, indicating that 13-hydroperoxyoctadecadienoic acid was the precursor. The authors suggested that the linoleic acid free radical formed in the initial reaction between linoleic acid and lipoxygenase (6) abstracted a hydrogen atom from 13-hydroperoxyoctadecadienoic acid. The resulting 13-peroxy-9,11-octadecadienoic acid was then converted into the products by a series of rearrangements.

C. Other Lipoxygenases

Lipoxygenases occur widely in the plant kingdom. Urd bean, mung bean, pea, peanut, potato tuber, wheat, barley, flaxseed, alfalfa, and corn have all been reported to contain lipoxygenase activity (cf. ref. 4). In more recent years lipoxygenases have been partially purified from a number of these sources and the products formed on incubation with unsaturated fatty acids have been characterized (Table I).

Single plants may contain more than one lipoxygenase. On disc-gel electrophoresis of soybean extracts three or four bands of lipoxygenase activity appeared (25). Recently, one of these isoenzymes was isolated in a high state of purity (26). This enzyme had a pH optimum at 6.6 in contrast to that of the major lipoxygenase in soybeans (pH 9.5). On incubation with linoleic acid equal amounts of 9- and 13-hydroperoxyoctadecadienoates were produced (14).

TABLE I

MAJOR HYDROPEROXYOCTADECADIENOATES FROM LINOLEIC ACID

Plant	Hydroperoxyoctadeca-dienoate(s)	Ref.
Soybean (lipoxygenase-1)	13L-Hydroperoxy	6, 7
Soybean (lipoxygenase-2)	9- and 13-Hydroperoxy	14
Flaxseed	13-Hydroperoxy	10
Wheat	13-Hydroperoxy	22, 23
Alfalfa	9- and 13-Hydroperoxy	12
Corn	9D-Hydroperoxy	27
Potato	9D-Hydroperoxy	28

III. BIOSYNTHESIS OF PROSTAGLANDINS

The conversion of arachidonic acid into prostaglandin E_2 (PGE$_2$) by homogenates of vesicular gland from sheep was reported in 1964 by van Dorp et al. (29) and by Bergström et al. (30). The transformation of 8,11,14-eicosatrienoic acid and 5,8,11,14,17-eicosapentaenoic acid into PGE$_1$ and PGE$_3$, respectively, was also demonstrated (31, 32) (Fig. 4). Subsequently, PGF$_{1\alpha}$ (33–35) as well as a number of monohydroxy acids (34–36) were isolated after incubation of 8,11,14-eicosatrienoic acid with preparations of sheep vesicular gland. Homogenates of guinea pig lung catalyzed the formation of PGE$_2$ and PGF$_{2\alpha}$ from arachidonic acid (37). This transformation could also be effected by homogenates of rabbit kidney medulla (38). A large number of other tissues has also been found to contain prostaglandin synthetase activity, although the yields of prostaglandins from added precursors are generally very low (35). In the case of rat stomach homogenates it has been shown that large amounts of endogenous substrates are released from phospholipids before and during incubation (39). A similar phenomenon may explain the low percentage conversions of incubated labeled precursors in most tissues (39). The vesicular gland of sheep is a rich and convenient source of prostaglandin synthetase and most studies on the enzyme system, substrate specificity, and reaction mechanism have been carried out with this tissue.

A. Enzyme System

The conversion of 8,11,14-eicosatrienoic acid into PGE$_1$ requires the microsomal fraction and the high-speed supernatant of a homogenate of sheep vesicular gland as well as molecular oxygen. The high-speed supernatant is probably needed to furnish a thermostable reducing cofactor. It can be replaced by reduced glutathione or tetrahydrofolate but not NADH or NADPH (40, 41).

An additional reducing equivalent is required in the formation of PGF compounds. Apparently, glutathione cannot serve this purpose since this cofactor in fact decreased the yield of PGF$_{1\alpha}$ from 8,11,14-eicosatrienoic acid (41). Recently, it was found that the formation of PGF$_{2\alpha}$ from arachidonic acid was stimulated by the addition of Cu^{2+} and the dithiol, dihydrolipoamide (42) or by L-epinephrine (43). However, the identity of the reducing cofactor that stimulates formation of PGF compounds under physiological conditions is unknown.

The heat-labile microsomal component of the biosynthetic system has been obtained in soluble form by treatment of the microsomal fraction

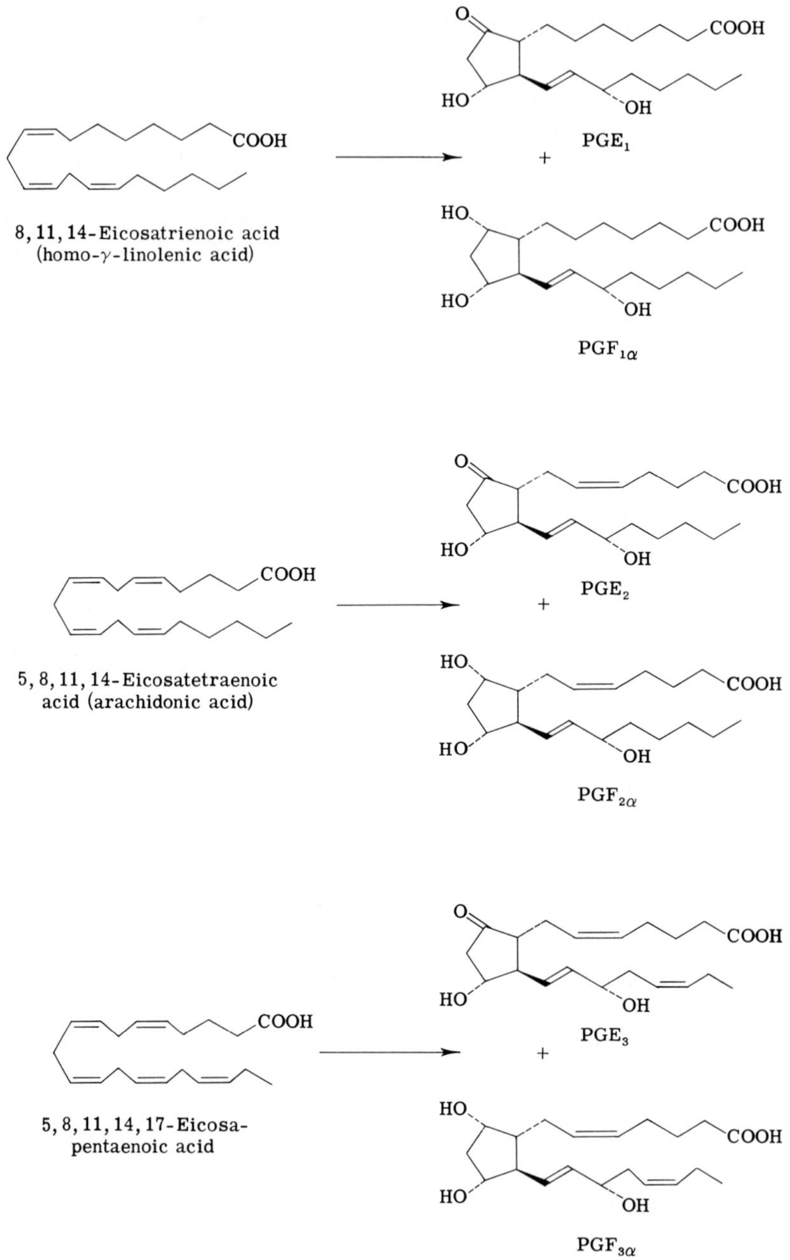

FIG. 4. Structures of the primary prostaglandins and their precursors.

THE
UNIVERSITY OF WINNIPEG
PORTAGE & BALMORAL
WINNIPEG, MAN.   R3B 2E9
CANADA

2. FATTY ACID AND STEROID METABOLISM   39

with the nonionic detergent, Cutscum (44, 45). A tenfold purification of the microsomal component was accomplished by ammonium sulfate precipitation (40–60% saturation) and chromatography on DEAE-cellulose.

B. Substrate Specificity

It was shown early that in addition to 8,11,14-eicosatrienoic, arachidonic, and 5,8,11,14-17-eicosapentaenoic acids, 10,13,16-docosatrienoic acid (31) and 7,10,13-nonadecatrienoic acid (32) were substrates for prostaglandin synthetase. The principal products were identified as di-homo-PGE$_1$ and nor-PGE$_1$, respectively. Later, Struijk et al. (46) synthesized a number of analogs and homologs of the known precursors and tested their conversion into prostaglandins. Trienoic acids of the ω6 type were efficiently converted if the chain lengths were C$_{19}$ to C$_{21}$. In the case of tetraenoic acids of the ω6 type also the C$_{18}$ and C$_{22}$ homologs were converted into prostaglandins. The positions of the double bonds relative to the methyl end appeared to be critical since C$_{20}$ trienoic acids of the ω5, ω8, and ω9 types did not serve as substrates. A small conversion, however, was noted in the case of the ω7 isomer. Additional synthetic analogs have been prepared by the Dutch workers. They found that fatty acids 21:4 ω7, 19:4 ω5, 21:3 ω7, and 19:3 ω5 as well as 8,11,14-eicosatrienoic acids (ω6) containing an extra double bond in positions 2, 3, or 4 were all precursors of PGE compounds (47, 48).

C. Mechanism of Reaction

Three oxygens (two hydroxyl groups and one keto group) are introduced in the conversion of 8,11,14-eicosatrienoic acid into PGE$_1$. By mass spectrometric analysis of PGE$_1$ biosynthesized from 8,11,14-eicosatrienoic acid in the presence of $^{18}O_2$ it was shown that the two hydroxyl groups were derived from molecular oxygen (49, 50). Later it was shown that also the keto group of the five-membered ring was derived from O$_2$ (50, 51). In these studies exchange of the keto group was minimized and one of the epimeric PGF$_1$ compounds formed was analyzed by mass spectrometry. These studies were extended by Samuelsson (51) in experiments with a mixture of $^{16}O_2$ and $^{18}O_2$. On mass spectrometric analysis of a derivative of PGF$_1\beta$ containing the two oxygens of the five-membered ring but not the oxygen of the side chain, only two species of molecules were found, viz., $^{16}O-^{16}O$ and $^{18}O-^{18}O$ derivatives. This showed that the two oxygens of the five-membered ring were derived from the same molecule of oxygen (cf. Fig. 5).

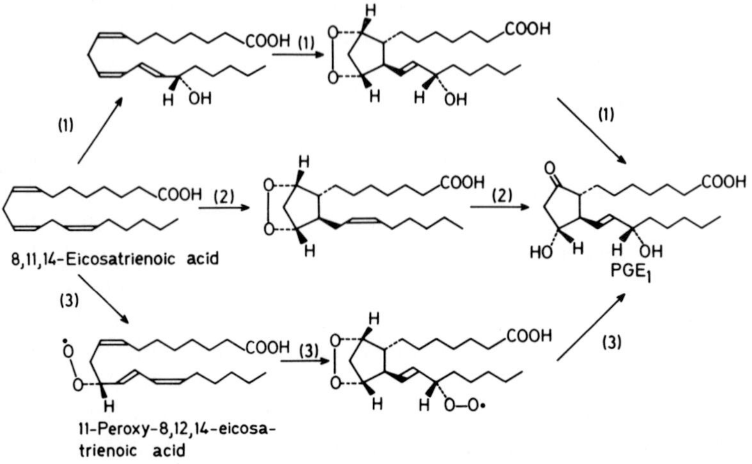

FIG. 5. Hypothetical pathways in the biosynthesis of PGE$_1$ from 8,11,14-eicosatrienoic acid.

The fate of the hydrogens at C-8, C-11, and C-12 of 8,11,14-eicosatrienoic acid has also been studied (52). The corresponding tritium-labeled acids were prepared from [6-^{3}H]-, [9-^{3}H]-, and [10-^{3}H]stearic acids which were desaturated by the *Tetrahymena pyriformis* into γ-linolenic acids. These acids were then elongated to the desired tritium-labeled 8,11,14-eicosatrienoic acids. PGE$_1$ retained the tritium label in each case. Later it was shown that [10-^{3}H]- and [15-^{3}H]8,11,14-eicosatrienoic acids were converted into PGE$_1$ with retention of the label, and that [9-^{3}H]8,11,14-eicosatrienoic acid yielded PGE$_1$ devoid of tritium (36).

These experiments were compatible with the possibility that the first oxygenated intermediate in the biosynthesis of PGE$_1$ contained either one or two oxygens at C-15 [Fig. 5, pathway (1)], or two oxygens between C-9 and C-11 in an endoperoxide [Fig. 5, pathway (2)], or two oxygens at C-11 [Fig. 5, pathway (3)]. Pathway (1) could be excluded by incubation of 15L-hydroperoxy- and 15L-hydroxy-8,11,13-eicosatrienoic acids. No conversion into PGE$_1$ could be detected (53). In pathway (2) the first oxygenated intermediate retains both hydrogens at C-13, whereas in pathway (3) the first intermediate has lost one hydrogen atom from C-13. Substitution of the hydrogen which is removed from C-13 with tritium would result in a kinetic isotope effect in the reaction where tritium is removed. This would lead to enrichment of tritium in the compound undergoing elimination of tritium from C-13. Accordingly, it should be possible to distinguish between pathways (2) and (3) by noting whether or not enrichment of tritium occurred in 8,11,14-eicosatrienoic acid remaining after incubation of [13-³H]-8,11,14-eicosatrienoic acid. It was found that the 13D-tritio acid was converted into PGE₁ with retention of most of the tritium, whereas the 13L-tritio acid was converted into PGE₁ with loss of most of the tritium (53). In a separate experiment it was proved that the 13D-tritium atom was retained at C-13 of PGE₁. The hydrogen eliminated from C-13 of 8,11,14-eicosatrienoic acid is thus the pro-S hydrogen, the same hydrogen as that eliminated in the conversion of 8,11,14-eicosatrienoic acid into 15L-hydroperoxy-8,11,13-eicosatrienoic acid by soybean lipoxygenase (54). Of particular significance was the finding that 8,11,14-eicosatrienoic acid remaining after incubation of the 13L-³H acid was enriched in tritium (53), favoring pathway (3) (Fig. 5) in which the hydrogen elimination from C-13 occurs in the first step.

The structures of a number of by-products formed together with PGE₁ and PGF₁α from 8,11,14-eicosatrienoic acid have been determined (Fig. 6). The formation of one of these, 11-hydroxy-8,12,14-eicosatrienoic acid, lends support to the concept that the oxygenation starts at C-11 in a lipoxygenase type of reaction (34, 36). Furthermore, 11L-hydroxy-12,14-eicosadienoic acid is formed from 11,14-eicosadienoic acid (35). Another monohydroxy acid formed from 8,11,14-eicosatrienoic acid is 12-hydroxy-8,10-heptadecadienoic acid (34–36). It could be concluded from isotope experiments that the three carbon atoms lost in the formation of this acid were not derived from the carboxyl end or the pentyl side chain of 8,11,14-eicosatrienoic acid. The compound was devoid of tritium when biosynthesized from [9-³H]-, [10-³H]-, and [11-³H]8,11,14-eicosatrienoic acids, indicating that C-9 to C-11 of the precursor was eliminated in its formation (36). The identity of the three-carbon fragment with malonaldehyde was established by chromatography on Sephadex G-10 and subsequent formation of δ-N-2(pyrimidinyl)-L-ornithine and 2-hydroxypyrimidine on acid treatment with L-arginine and urea, respectively. The finding that [9-³H]- and [11-³H]8,11,14-eicosatrienoic acids yielded tritium-labeled malonaldehyde shows that carbons C-9 to C-11 are eliminated in form of malonaldehyde in the formation of 12-hydroxy-8,10-heptadecadienoic acid. Most probably this occurs by rearrangement of the postulated endoperoxide (Figs. 5 and 6). The demonstrated *trans* configuration of the Δ⁸ double bond of 12-hydroxy-8,10-heptadecadienoic acid is in agreement with this hypothesis since the side chains of the endoperoxide should be *trans* oriented as they are in the prostaglandins.

Additional prostanoic acid derivatives have been identified among the products formed from 8,11,14-eicosatrienoic acid with the microsomal fraction of sheep vesicular gland homogenates. One of these compounds

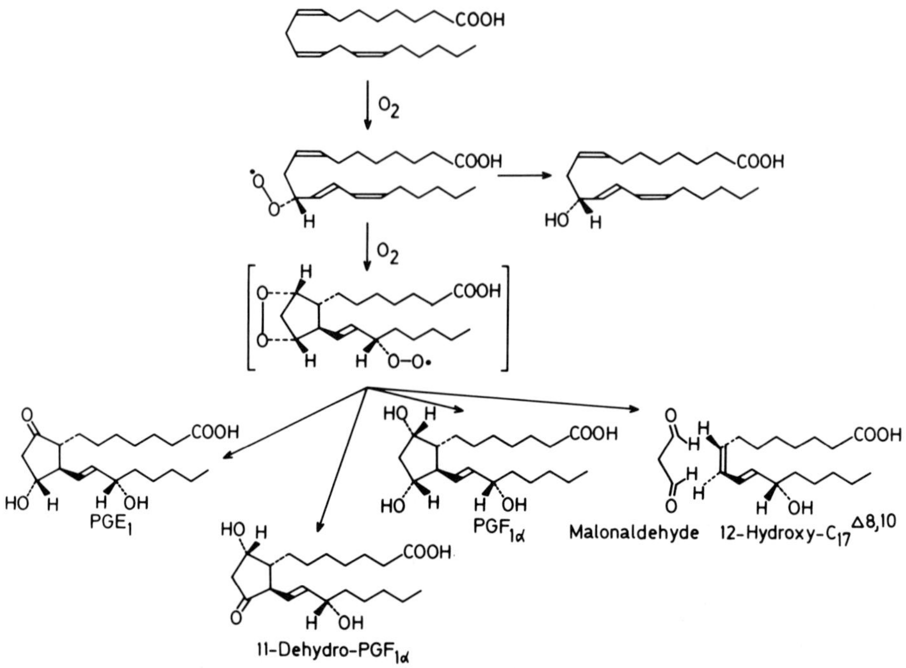

Fig. 6. Mechanism in the formations of PGE₁, PGF₁α, 11-dehydro-PGF₁α, 11-hydroxy-8,12,14-eicosatrienoic acid, 12-hydroxy-8,10-heptadecadienoic acid, and malonaldehyde from 8,11,14-eicosatrienoic acid.

was identified as PGF$_{1\alpha}$ (33–35, 53). Interestingly, no interconversion between PGE$_1$ and PGF$_{1\alpha}$ could be shown. A similar situation has been found in the conversion of arachidonic acid into PGE$_2$ and PGF$_{2\alpha}$ by homogenates of guinea pig lung (37). These findings lend further support to the contention that the PGE and PGF compounds are formed from a common intermediate, *viz.*, the endoperoxide (cf. ref. 56). Another derivative formed on incubation of 8,11,14-eicosatrienoic acid with the microsomal fraction has been identified as 11-dehydro-PGF$_{1\alpha}$ (35, 53, 55). This compound is probably also derived from the endoperoxide (Fig. 6).

Pace-Asciak and Wolfe (57, 58) have identified a number of cyclic ether derivatives as by-products in incubations of arachidonic acid with homogenates of rat stomach and sheep vesicular gland. The results of these authors lend further support for the postulated mechanism of prostaglandin biosynthesis.

D. Sheep Vesicular Gland Lipoxygenase

The first reaction catalyzed by prostaglandin synthetase is closely related to the reaction by which conjugated fatty acid hydroperoxides are formed by plant lipoxygenases (Fig. 7). That the enzyme system responsible for prostaglandin formation in fact can catalyze a lipoxygenase type of reaction is apparent from the results of incubations of linoleic acid and 11,14-eicosadienoic acid with sheep vesicular gland. The products formed were identified as 9L-hydroxy-10,12-octadecatrienoic acid (about 80%), 13L-hydroxy-9,11-octadecatrienoic acid (about 20%) (34, 36, 59), and 11L-hydroxy-12,14-eicosadienoic acid (35), respectively. The stereochemistry in the hydrogen removal from C-11 occurring in the transformation of linoleic acid (ref. 59, Fig. 7) was found to be the same as that in the lipoxygenase-catalyzed oxygenation of linoleic acid (18) and 8,11,14-eicosatrienoic acid (54).

E. Inhibitors

The first data on inhibitors of prostaglandin synthetase were provided by Nugteren *et al.* (35) who observed an inhibitory effect by certain metal ions, *viz.*, Cu^{2+}, Zn^{2+}, and Cd^{2+}. Reduced glutathione, but not other SH compounds, partly abolished this effect. No inhibition was observed with *p*-chloromercuribenzoate, EDTA, or KCN. That fatty acids may serve as inhibitors was first shown by Pace-Asciak and Wolfe (60), who found that α-linolenic, linoleic, and oleic acids in concentrations of 0.9–3.6 $\times$ 10^{-3} *M* partially inhibited the conversion of arachidonic acid into PGE_2 in homogenates of sheep vesicular gland and rat stomach. Nugteren later reported

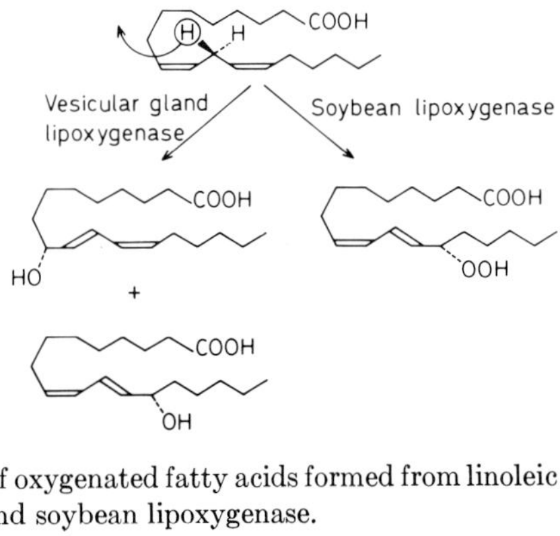

FIG. 7. Structures of oxygenated fatty acids formed from linoleic acid by sheep vesicular gland lipoxygenase and soybean lipoxygenase.

that two conjugated fatty acids, 5,8,12,14-eicosatetraenoic and 8,12,14-eicosatrienoic acids, were strong inhibitors of prostaglandin synthetase (61). These acids are structurally related to 11-peroxy-8,12,14-eicosatrienoic acid which has been postulated as an intermediate in the conversion of 8,11,14-eicosatrienoic acid into PGE_1 and $PGF_{1\alpha}$ (Figs. 5 and 6). Interestingly, the conjugated fatty acids had stronger affinity to the enzyme than had the precursor acids. Recently, decanoic acid was found to be an inhibitor of the enzyme (62). The inhibition was quite dependent upon a chain length of 10 carbons since nonanoic and undecanoic acids did not give appreciable inhibition. This finding is of interest since the distance between the methyl group and the oxygen function is similar in the inhibitor and the proposed intermediate 11-peroxy-8,12,14-eicosatrienoic acid. An acetylenic acid, 5,8,11,14-eicosatetraynoic acid, strongly inhibits prostaglandin synthetase (63) as well as soybean lipoxygenase (64, 65). In this case it has been suggested (65) that the inhibitory action results from the formation of an allene by enzyme-catalyzed abstraction of a hydrogen from C-13. Selective inhibition of PGE_1 but not $PGF_{1\alpha}$ formation from 8,11,14-eicosatrienoic acid by a bicyclo [2.2.1] heptene derivative was reported by Wlodawer *et al.* (66). This derivative is structurally related to the endoperoxide intermediate postulated in prostaglandin biosynthesis (Fig. 6). Indomethacin [1-(p-chlorobenzoyl)-5-methoxy-2-methylindole-3-acetic acid] and aspirin were recently shown to inhibit prostaglandin synthetase (67–69). The concentrations required for 50% inhibition were 0.75 μM for the former and 35 μM for the latter compound. Later another analgesic, naproxen [d-2-(6'-methoxy-2'-naphthyl)-propionic acid], was also found to be an inhibitor (70).

Inhibitors of prostaglandin biosynthesis seem to be of considerable value in assessing the physiological roles of the prostaglandins (cf. refs. 71, 72).

IV. FATTY ACID HYDROXYLATIONS

A. α-Oxidation

α-Oxidation may be defined as a series of reactions by which a fatty acid is converted into its next lower homolog with explusion of CO_2. Such degradation of fatty acids was first observed in extracts of peanut cotelydons by Martin and Stumpf (73). Later, Hitchcock and James (74) found that homogenates of young leaf tissue from castor oil and pea plants catalyzed a similar reaction. Also mammalian brain (75) and liver (76, 77) appear to catalyze α-oxidation of fatty acids.

The peanut system requires an H_2O_2-generating system and NAD. The only intermediates detected were long-chain fatty aldehydes. When imidazole was added to the incubation mixture the formation of aldehydes was prevented, apparently because of inhibition of the "fatty acid peroxidase." The mechanism of the peroxidase reaction has been discussed by Martin and Stumpf (73).

The leaf system uses molecular oxygen instead of H_2O_2. α-Hydroxy fatty acids have been detected as intermediates. Interestingly, both D-α-hydroxy and L-α-hydroxy acids were formed during α-oxidation of fatty acids by acetone powders of pea leaves (78, 79). The L-α-hydroxy acid is further metabolized to a fatty acid containing one carbon less, whereas the D-α-hydroxy acid accumulates. In the presence of imidazole the α-oxidative degradation of palmitic acid but not 2-hydroxypalmitic acid is prevented, indicating that imidazole inhibits α-hydroxylation. The mechanism in the formation of α-hydroxy acids in the leaf system has been studied with palmitic acids stereospecifically labeled with tritium at C-2 and C-3 (80, 81). It was found that D-α-hydroxypalmitic acid biosynthesized from [2D-^{3}H]palmitic acid lost about 76% of the label but retained the label when formed from [2L-^{3}H]palmitic acid. Furthermore, D-α-hydroxypalmitic acid formed from [3D-^{3}H]- and [3L-^{3}H]palmitic acids retained the tritium label. Therefore, this α-hydroxy acid was formed by direct hydroxylation at C-2 and not by hydration of an α,β-unsaturated derivative. The hydroxylation occurred with retention of the absolute configuration as is the case with most "mixed function oxidases":

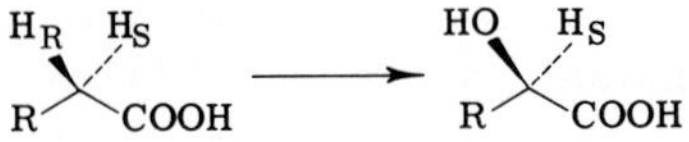

Later experiments showed that [2L-^{3}H]palmitic acid incubated with a particulate fraction from young pea leaves yielded pentadecanal with retention of the tritium label whereas [2D-^{3}H]palmitic acid gave pentadecanal with loss of tritium (82). The precursor of the aldehyde was concluded therefore to be D-α-hydroxypalmitic acid and not the L-isomer. To explain these results it was suggested that α-oxidation in leaves may occur by two routes: (a) via D-α-hydroxy acids and aldehydes, both of which can accumulate, and (b) via L-α-hydroxy acids which do not accumulate (Fig. 8) (82).

Very recent experiments have demonstrated that α-hydroxy acids are formed also by the α-oxidation system of peanut cotelydons (83). When palmitic acid was used as the substrate D-α-hydroxypalmitic acid accumulated whereas L-α-hydroxypalmitic acid was probably further degraded.

Evidence for the existence of an α-oxidation pathway for fatty acids in brain tissue is so far incomplete (cf. ref. 77). Fulco and Mead (84) isolated and degraded brain tetracosanoic and 2-hydroxytetracosanoic acids from rats injected with [1-^{14}C]acetate. They found a similar ^{14}C-distribution

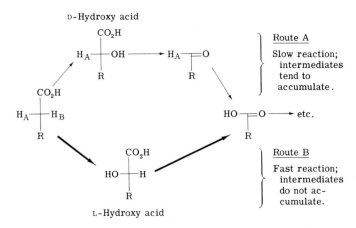

FIG. 8. Possible mechanisms of α-oxidation in leaves. Hitchcock and Morris (82).

pattern in the two acids and suggested that in brain 2-hydroxy fatty acids are formed from the corresponding nonhydroxy fatty acids. Similar results were later obtained by Hajra and Radin (85, 86). However, attempts to carry out hydroxylation of long-chain fatty acids *in vitro* with brain preparations have been unsuccessful. Conversion of 2-hydroxy fatty acids into nonhydroxy fatty acids containing one carbon less have been demonstrated both *in vivo* and *in vitro* (cf. 77). Rat brain microsomes as well as the high-speed supernatant, NAD, and ATP were required to convert 2-hydroxytetracosanoic and 2-hydroxyoctadecanoic acids into tricosanoic and heptadecanoic acids, respectively. There was also evidence that the 2-keto fatty acids were intermediates in the conversion.

Degradation of phytanic acid (3,7,11,15-tetramethylhexadecanoic acid) in rat liver mitochondria has been studied by Tsai *et al.* (76). The conversion was found to be initiated by α-oxidation leading to the formation of of pristanic acid (2,6,10,14-tetramethylpentadecanoic acid). This process required NADPH and O_2. Fe^{3+} greatly stimulated the reaction whereas Fe^{2+} inhibited it. Small amounts of α-hydroxyphytanic acid accumulated during the process suggesting that the first reaction was α-hydroxylation. In later reactions pristanic acid was degraded into pristenic acid (2,6,10,14-tetramethyl-2-pentadecenoic acid) and 4,8,12-trimethyltridecanoic acid by β-oxidation. Stokke (77) independently reported on the α-oxidation of a similar β-methyl substituted compound, 3,9,9-trimethyldecanoic acid, in guinea pig liver mitochondria. The liver mitochondrial system for α-oxidation of fatty acids differs from the other α-oxidation systems. Thus, the brain system for conversion of α-hydroxy acids into lower fatty acids is effective on long-chain fatty acids but not phytanic acid. Furthermore, this system is associated with the microsomal rather than the mitochondrial fraction and is stimulated and not inhibited by Fe^{2+}. The α-oxidation systems of peanut cotelydons and leaf are both inhibited by imidazole which is not the case with the liver mitochondrial system. It will be interesting to learn more of the stereochemistry and mechanism of the mitochondrial system, particularly whether the α-hydroxy-β-methyl fatty acids that accumulate are true intermediates or arise by incomplete stereospecificity of the α-hydroxylating enzyme as in the leaf system.

B. Ricinoleic Acid Biosynthesis

At least two pathways occur in the formation of ricinoleic acid. In immature sclerotia and in mycelial cultures of *Claviceps purpurea*, the ergot fungus, the nearest precursor for ricinoleic acid formation appears to be linoleic acid (87). This reaction does not require O_2 and may consist of addition to the Δ^{12} double bond of H_2O (yielding ricinoleic acid) or of carboxyl groups of long-chain fatty acids, directly yielding acylated derivatives of ricinoleic acid which occur in ergot oil. In the castor bean, on the other hand, biosynthesis of ricinoleic acid occurs by hydroxylation of oleyl-CoA in a reaction requiring O_2 and NADH (88–90). The hydroxylase is highly specific. Stearic, elaidic, palmitoleic, and linoleic acids and their CoA derivatives, are inactive.

By use of oleates stereospecifically labeled with tritium at C-12, Morris (91) demonstrated that hydroxylation at C-12 occurs with loss of the 12D-hydrogen, i.e., with retention of the absolute configuration. The finding that ricinoleic acid biosynthesized from *erythro*-[12,13-^{3}H$_2$]oleate retained 75% of the tritium (Fig. 9) shows that linoleate is not an intermediate.

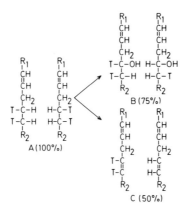

FIG. 9. Formation in castor bean of ricinoleic acid (B) and linoleic acid (C) from *erythro*-[12,13-^{3}H$_2$]oleic acid (A).

Interestingly, linoleic acid was also formed from oleic acid on incubation with homogenates of castor bean. By using [12D-³H]-, [12L-³H]-, as well as *erythro*-[12,13-³H₂]oleates it was shown that the 12D- and 13D-hydrogens were removed from oleic acid in the formation of linoleic acid (91) (Fig. 9).

C. ω-Oxidation

Verkade *et al.* (92) discovered the biological oxidation of fatty acids at the ω carbon atom. Feeding humans and animals with medium-chain fatty acids as triglycerides led to the urinary excretion of dicarboxylic acids of the same chain length. It was proposed that certain fatty acids were oxidized at the ω position and further metabolized by β-oxidation (93). ω-Oxidation was later encountered in studies on the *in vivo* metabolism of α-dimethyl substituted fatty acids (94–96), hydrocarbons (97), and prostaglandins (98).

Wakabayashi and Shimazono (99, 100) and Robbins (101, 102) independently showed that in mammalian liver ω-oxidation of a fatty acid consists of an initial hydroxylation into the ω-hydroxy acid followed by successive oxidations to the aldehyde acid and the dicarboxylic acid. The hydroxylation occurs in the presence of liver microsomes and requires O_2 and NADPH, whereas the oxidation steps are catalyzed by soluble enzymes and NAD. Medium-chain fatty acids (C_8 to C_{12}) are the best substrates for ω-oxidation *in vitro* although longer chain acids can also be oxidized (103, 104). Other substrates for *in vitro* ω-oxidation include fatty acid amides (99, 100), branched-chain fatty acids (105), hydrocarbons (106), and prostaglandins (107).

The mechanism in the hydroxylation of decanoic acid by rat liver microsomes has been recently studied (108). The hydroxy acids isolated consisted of about 92% 10-hydroxydecanoic acid and about 8% 9-hydroxydecanoic acid, predominantly the 9L-hydroxy isomer. Biosynthesis of 10-hydroxydecanoate occurred with loss of one hydrogen from C-10 of the precursor acid and with retention of both hydrogens at C-9. Similarly, in the formation of 9-hydroxydecanoic acid one hydrogen was lost from C-9 and the three hydrogens at C-10 were retained. These results showed that unsaturated derivatives, e.g., 9-decenoic acid, or carbonyl derivatives were not intermediates in the formation of the two hydroxy acids. By use of [9L-²H]decanoic acid it was demonstrated that hydroxylation at C-9 occurred with retention of the absolute configuration. An isotope effect was noted in the formation of 9-hydroxydecanoate from [9-²H₂]decanoate but not in the formation of 10-hydroxydecanoate from [10-²H₃]decanoate.

ω-Hydroxylation of fatty acids and alkanes is inhibited by carbon monoxide (109–111, 104) as well as by NADP and oxidized cytochrome c

(110), indicating that in these hydroxylations NADPH–cytochrome c reductase and cytochrome P-450 are involved. Evidence for the role of NADPH–cytochrome c reductase as electron carrier in ω-hydroxylations has also been provided by Wada et al. (109) who observed an inhibitory effect on the hydroxylation when an antibody against NADPH–cytochrome c reductase was added to the microsomes. Different results have been reported on the effect of phenobarbital treatment on ω-hydroxylating activity. Das et al. (110) and Lu et al. (111) found that treatment with phenobarbital stimulated hydroxylation of laurate. On the other hand, Wada (109) found that treatment with phenobarbital decreased stearate hydroxylation. Björkhem and Danielsson (104) separately assayed ω1- and ω2-hydroxylations of a number of fatty acids and found that ω2-hydroxylations, but not ω1-hydroxylations, were stimulated by phenobarbital treatment. Interestingly, carbon monoxide also influenced these hydroxylations differently, ω2-hydroxylations being inhibited to a lesser extent than ω1-hydroxylations (104). ω2-Hydroxylation of $[9\text{-}^2H_2]$deca-noate, but not ω1-hydroxylation of $[10\text{-}^2H_3]$decanoate, was accompanied by a significant isotope effect suggesting that differences exist between the reactions by which ω1- and ω2-hydroxy acids are formed (108).

The microsomal components required for ω-hydroxylations of fatty acids have been recently studied by Lu et al. (111, 112) and by Ichihara et al. (113) and Kusunose et al. (114). These workers solubilized the enzyme system by treatment with deoxycholate (111, 112) or Triton X-100 (113, 114). Lu et al. subjected the solubilized preparation to DEAE-cellulose chromatography. Stepwise elution with increasing concentrations of KCl yielded three fractions containing cytochrome P-450, NADPH–cytochrome c reductase, and a heat-stable factor, respectively. The soluble preparation of cytochrome P-450 exhibited a carbon monoxide difference spectrum similar to that found for the microsome-bound form. The reductase was partially purified by adsorption on alumina Cγ gel and chromatography on hydroxylapatite and was found to catalyze electron transfer from NADPH to cytochrome P-450 under anaerobic conditions. The NADPH–cytochrome c reductase activity of the preparations was found to be unrelated to their ability to reduce cytochrome P-450 and therefore to function in the hydroxylating system. Therefore, the name "NADPH–cytochrome P-450 reductase" was suggested as being more suitable for the reductase component. The activity of the heat-stable factor resulted from lipid–soluble material, later shown to be replaceable by phosphatidyl-choline (115). Molecules containing two oleoyl residues or one or two lauroyl residues appeared to be the most active. Data were presented to show that the lipid factor was essential for the enzymic reduction of cyto-chrome P-450.

ω-Hydroxylation of fatty acids and hydrocarbons has also been observed in yeast (116) and bacterial systems (117–120). Tulloch *et al.* reported that a yeast, *Torulopsis* sp., fermented long-chain fatty acids and hydrocarbons to produce ω1- and ω2-hydroxy fatty acids. These were then combined with sophorose to give extracellular hydroxy acid sophorosides (116, 121). The relative proportions of the ω1- and ω2-hydroxy acids produced depended on the chain length and degree of unsaturation of the substrate. Stearic and oleic acids were hydroxylated mainly at ω2 to give 17L-hydroxy acids. Experiments with whole cells and $^{18}O_2$ showed that the hydroxyl oxygen was derived from O_2 (122). Further studies with [17D-^{2}H]-, [17L-^{2}H]-, [17-^{2}H$_2$]-, [18-^{2}H$_3$]-, and [16,18-^{2}H$_5$]stearates showed that the 17L-hydroxy-stearate was produced by loss of the 17L-hydrogen and retention of the 17D-hydrogen as well as the five hydrogens at C-16 and C-18 (122, 123). 16- and 17-Octadecenoic acids as well as 17-ketostearic acid could be excluded as intermediates and the reaction could be visualized as a substitution of H with OH occurring with retention of the absolute configuration as in most aliphatic hydroxylations. A significant isotope effect was observed in the hydroxylation of [17L-^{2}H]stearate. Recently, the hydroxylation of oleic acid by cell-free extracts of a species of *Torulopsis* was described (124). The reaction required the particulate fraction sedimenting at 48,000 *g*, O_2, and NADPH. Strong inhibition was observed with CO but not with CN^- or N_3^-.

An enzyme system in *Pseudomonas oleovorans* that catalyzes ω-hydroxylation of fatty acids and alkanes in the presence of O_2 and NADH has been studied in detail (118, 125). C_8 and C_{12} were found to be the optimal chain lengths for alkanes and fatty acids, respectively. Three protein components necessary for the reaction were separated and partially purified (126, 127). These were identified as a rubredoxin, a NADH-rubredoxin reductase, and an ω-hydroxylase. Rubredoxin was bleached into a colorless form on reduction by dithionite or by NADPH in the presence of NADPH-ferredoxin reductase from spinach. The reduction of cytochrome c by rubredoxin in the presence of NADPH and the spinach reductase or of NADH and the bacterial reductase further indicated that rubredoxin could act as an electron carrier (Fig. 10). *Pseudomonas* rubredoxin was later obtained in a highly purified form (128). The apparent molecular weight was 12,800. Two atoms of iron were present per molecule, but no labile sulfide was present. Oxidized rubredoxin gave electron paramagnetic resonance signals at $g = 4.3$ and 9.4 indicating the presence of high spin ferric ion in a rhombic field. One electron per atom of iron was accepted upon reduction by NADPH and spinach NADPH–ferredoxin reductase. Rubredoxin gave intense and detailed optical rotatory dispersion and circular dichroism spectra, suggesting that the iron-ligand chromophores are in a

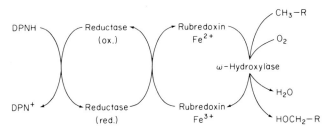

FIG. 10. Electron transport chain in ω-hydroxylation of fatty acids and alkanes by the soluble system of *Pseudomonas oleovorans*. From Peterson *et al.* (127).

highly asymmetric environment. Recently, the interesting observation was made that alkyl hydroperoxides are reduced to alcohols in the presence of NADH, rubredoxin, and NADH–rubredoxin reductase. However, this finding does not necessary implicate alkyl hydroperoxides as intermediates in the formation of 1-alkanols from hydrocarbons (129).

V. FATTY ACID DESATURATION

Bloomfield and Bloch (130) were the first to show that biosynthesis of unsaturated fatty acids may occur by a reaction requiring molecular oxygen and reduced pyridine nucleotide. They found that anaerobically grown *Saccharomyces* yeast cells were capable of biosynthesis of saturated fatty acids only, whereas under aerobic conditions unsaturated fatty acids also appeared. The particulate fraction sedimenting at 105,000 g catalyzed the conversion of palmityl-CoA into palmitoleyl-CoA in the presence of O_2 and NADPH. On the basis of experiments with 9- and 10-hydroxystearic acids (131) it was initially suggested that oxygenated fatty acids were intermediates in the desaturation process. However, when 9- and 10-hydroxystearates as well as their CoA esters were incubated with oleate-forming systems no or very small conversion was found (132, 133).

Certain animal tissues also have the enzymes necessary for introduction of double bonds in fatty acids. This was first shown by Bernhard *et al.* (134) who found that the microsomal fraction plus the high-speed supernatant of a rat liver homogenate catalyzed formation of oleic acid from stearic acid. Marsh and James (135, 136) studied this system further and demonstrated that the desaturation required the microsomal fraction, O_2, and NADPH when stearyl-CoA was used as substrate. Small amounts of hydroxystearates were also obtained. However, it was not believed that free hydroxystearates or hydroxystearyl-CoA esters were intermediates in the desaturation process, although formation of enzyme-bound oxygenated derivatives could not be ruled out. In the light of later work on ω-hydroxylation of fatty acids it seems probable that the hydroxy acids isolated by Marsh and James were predominantly 17- and 18-hydroxystearates which are formed from stearic acid in the presence of liver microsomes, O_2, and NADPH (103). A cyanide-sensitive factor and cytochrome b_5, but not cytochrome P-450, appear to be involved in the transport of electrons from reduced pyridine nucleotide in the liver desaturase system (137). This system apparently requires lipids for activity (138). Extraction of the microsomes with aqueous acetone was found to result in inactivation and the activity was restored by addition of a mixture of phospholipids, triglycerides, and fatty acids (138). A similar lipid requirement has been found for the ω-hydroxylation system of rabbit liver (111). The role of lipids in these systems is not clear. One possibility is that the enzymes require a nonaqueous medium for catalytic activity. Another possibility is that the fatty acids form acyl-lipids which may be the true substrates, as has been suggested by Gurr *et al.* (139, cf. also ref. 140). The desaturation of monounsaturated acyl-CoA esters into diunsaturated derivatives also requires the microsomal fraction, reduced pyridine nucleotide, and O_2 (141). It has been repeatedly observed that mammalian tissues are unable to form linoleic acid. In agreement with this 6,9-octadecadienoic acid is the product formed by desaturation of oleyl-CoA (141).

Desaturation of fatty acids has also been described in bacterial systems, e.g., *Mycobacterium phlei* (142) and *Corynebacterium diphteriae* (143), and in plant systems, e.g., *Chlorella vulgaris* (144), safflower seeds (145), *Euglena gracilis*, and spinach (146). From the two last-mentioned sources a soluble stearyl-ACP (acyl carrier protein) desaturase has been isolated and separated into three components, a reduced pyridine nucleotide oxidase, a ferredoxin, and the desaturase. All components were required for desaturation of stearyl-ACP. The chain for transport of electrons from reduced pyridine nucleotide to O_2 (147) is shown in Fig. 11. The soluble *Euglena* desaturase system would seem to offer a better chance than the particulate desaturases for detecting hypothetical oxygenated fatty acid

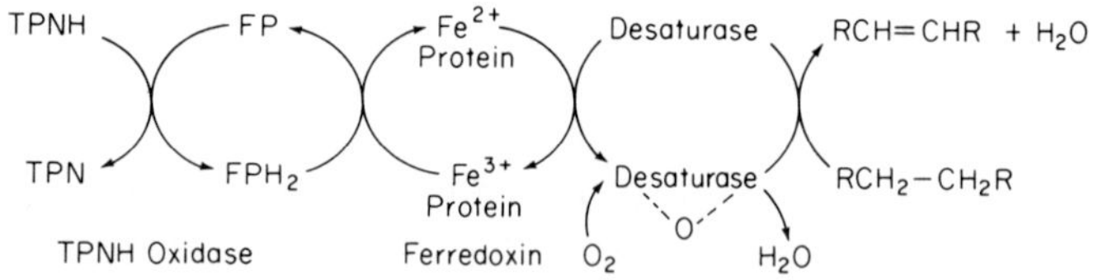

Fig. 11. Electron transport chain for oxygen activation in the soluble system for fatty acid desaturation in *Euglena gracilis*. From Bloch (147).

derivatives as intermediates in the desaturation. However, no transformation of stearyl-ACP into polar derivatives was found, nor was 9- or 10-hydroxystearyl-ACP converted to oleic acid, providing strong evidence against the formation of free oxygenated intermediates in fatty acid desaturation.

Schroepfer and Bloch (143) showed that introduction of a double bond in a fatty acid occurs stereospecifically, i.e., only two out of the four possible hydrogens are removed. Growing cultures of *Corynebacterium diphteriae*, a microorganism capable of converting stearate into oleate, were incubated with stereospecifically tritium-labeled stearates. The oleates isolated after incubation of [9D-³H]- and [9L-³H]stearates lost and retained, respectively, most of the original tritium label, showing that the 9D-hydrogen (pro-R) is removed selectively from stearate in oleate formation. These results and the finding that oleate biosynthesized from *erythro*-[9,10-²H₂]stearate was enriched with respect to dideuterated molecules, but not monodeuterated molecules, show that the hydrogen removed from C-10 also has the D configuration (pro-R) (Fig. 12). Even if the process is most simply visualized as a *cis* removal of hydrogens, a *trans* elimination cannot be ruled out since the conformation of the fatty acid during the dehydrogenation is not known (143). Interestingly, the stearate remaining after incubation of [9D-³H]stearate was enriched significantly with tritium, suggesting that removal of the 9D-hydrogen is the initial step of the reaction.

Using a similar technique, Morris *et al.* (148, 149) studied the formation of oleate, linoleate, and α-linolenate in *Chlorella vulgaris*. Experiments with the *erythro* and *threo* isomers of [9,10-²H₂]stearate and of [12,13-²H₂]- and [15,16-²H₂]oleates proved that introduction of double bonds at Δ⁹, Δ¹², and Δ¹⁵ all occur by removal of hydrogens with the same configuration. By using [9D-³H]-, [9L-³H]-, [12D-³H]-, and [12L-³H]stearates it could be shown that the absolute configuration of the hydrogens removed in for-

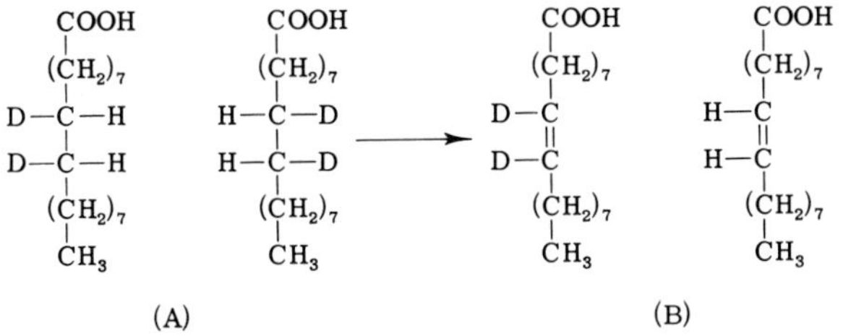

FIG. 12. Formation in *Corynebacterium diphteriae* of oleic acid (B) from *erythro*-[9,10-²H₂]stearic acid (A).

mation of the Δ^9 and Δ^{12} double bonds were D (pro-R). Morris (150) reported that in oleate formation from stearate in goat mammary gland and hen liver also the 9D- and 10D-hydrogens were removed. The stereochemistry in these desaturations was thus the same as that found by Schroepfer and Bloch.

VI. BIOSYNTHESIS OF CHOLESTEROL

A comprehensive review of the biosynthesis of cholesterol has been recently published (151). The reactions from acetate to lanosterol are well defined, whereas the sequence of some steps in the conversion of lanosterol into cholesterol has not been definitely established. Oxygenases are involved in the cyclization of squalene and in the removal of the methyl groups at C-4 and C-14 of lanosterol. It is possible that an oxygenase is involved also in the introduction of the Δ^5 double bond. All the oxygenases are located in the microsomal fraction. Wada (152) has suggested that one or more of the oxygenases contains cytochrome P-450 since the overall biosynthesis of cholesterol from mevalonate is inhibited by carbon monoxide. However, there is no definite evidence that any of the oxygenases so far studied contains cytochrome P-450. The oxygenases in cholesterol biosynthesis have not been extensively purified, and in the case of the demethylation at C-14 it has not been possible to study the oxygenation as a separate step.

A. Cyclization of Squalene

In the previous edition of "Oxygenases" the conversion of squalene into lanosterol was presumed to be initiated by OH^+, formed in a reaction between NADPH and oxygen. Since no stable intermediate could be isolated, it was believed that the reaction was a single rapid concerted move. This contention was consistent with all experimental data until 1966. In this year, Corey *et al.* (153) and van Tamelen *et al.* (154) showed independently that the cyclization of squalene is a two-step reaction with 2,3-oxidosqualene as intermediate (see Scheme 1).

2,3-Oxidosqualene was found to be converted into lanosterol by liver homogenate, and radioactivity of squalene could be trapped in 2,3-oxidosqualene when this compound was added as trapping agent. Furthermore, 2,3-oxidosqualene, labeled with ^{18}O, was converted into [^{18}O]lanosterol (155). The conversion of squalene into lanosterol involves two different enzymes, squalene epoxidase and 2,3-oxidosqualene cyclase. The squalene epoxidase requires oxygen and NADPH and is thus a mixed function oxidase, whereas the cyclase is independent of oxygen (156, 157). The

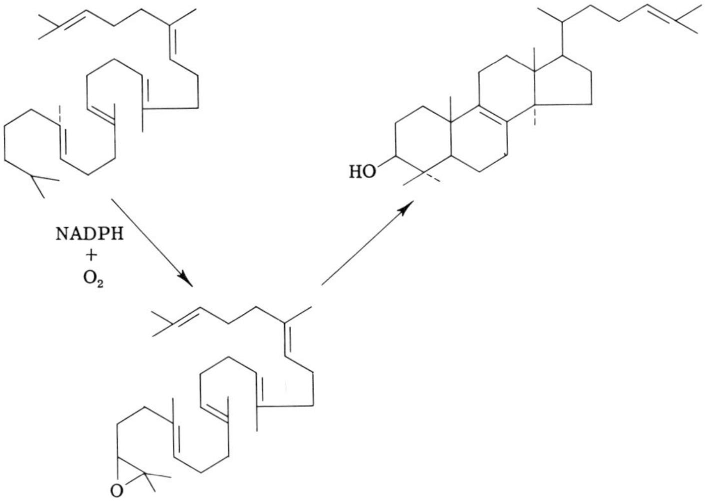

SCHEME 1

epoxidase is located in the microsomal fraction of rat liver homogenate, whereas the main part of the cyclase activity is located in the soluble fraction (156, 157). The epoxidase can be studied separate from the cyclase activity by the use of 10,11-dihydrosqualene as substrate, since the product formed is not a substrate for the cyclase (157, 158). The epoxidase is more resistant to heat treatment than the cyclase, and treatment of rat liver microsomes at 50° for 5 minutes destroys all the cyclase activity (156, 157). The epoxidase is probably independent of cytochrome P-450 since carbon monoxide does not reduce the activity (157). The epoxidase is stimulated severalfold by a heat stable protein in the soluble fraction of rat liver homogenate (156, 157). This protein has been purified about 300-fold and was found to have a molecular weight of about 16,000 (159). The protein aggregates to a molecular weight of about 150,000 during binding of squalene and some sterols to yield a noncovalent squalene– or sterol–protein complex. It has been suggested that the protein (squalene and sterol carrier protein, SCP) plays a role in reactions that squalene and other precursors of cholesterol undergo in the microsomes (159).

B. Demethylation at C-4 and C-14

During the conversion of lanosterol into cholesterol, the methyl groups at C-4 and C-14 are eliminated as carbon dioxide (160). The methyl group at C-14 is removed prior to those at C-4 (161) and the methyl group at

C-4α prior to that at C-4β (162–164). During the last few years considerable information has accumulated concerning the demethylation at C-4, mainly through the work of Gaylor *et al.* The elimination of the methyl groups in the 4 position is considered to proceed by stepwise oxidation of the methyl group to an alcohol, an aldehyde, and a carboxylic acid which undergoes decarboxylation. Miller *et al.* (165) have shown that the demethylation of 4-methylated steroids by rat liver microsomes can be separated into one (or several) step which requires oxygen and NADPH and one (or several) step which requires oxidized cofactors and is independent of oxygen. It was first believed that the 4-hydroxymethyl derivative was the final product in the aerobic step and that the 4-hydroxymethyl derivative was oxidized by microsomal alcohol and aldehyde dehydrogenases in anaerobic steps. The presence in the microsomal fraction of alcohol and aldehyde dehydrogenases active on 4-hydroxymethyl derivatives has been conclusively shown (166). It has also been shown that horse liver alcohol dehydrogenase can replace the microsomal alcohol dehydrogenase (166). However, the main pathway in the conversion of 4-methylated steroids into the corresponding carboxylic acids does not appear to involve alcohol and aldehyde dehydrogenases but instead three consecutive hydroxylations (167–169).

Aerobic incubation of 4α-methyl-5α-cholest-7-en-3β-ol with a microsomal, NAD-depleted preparation together with a NADPH-generating system was found to yield 3β-hydroxy-5α-cholest-7-ene-4α-carboxylic acid as the final product (167–169). A similar type of hydroxylation has been suggested for the oxidation of 19-hydroxy-4-androstene-3,17-dione into 3,17-diketo-4-androsten-19-al (170) and for the oxidation of ethanol into acetaldehyde by microsomes and NADPH (171). However, in the case of oxidation of ethanol, recent work indicates that the mechanism is a peroxidation of ethanol as a result of NADPH-dependent formation of hydrogen peroxide in the microsomes (172). The possibility has not been excluded that the oxidative formation of a 4-carboxylic acid from the 4-hydroxymethyl derivative is also the result of a peroxidation. The rate-limiting oxidase involved in the demethylation of 4-methylated steroids is probably independent of cytochrome P-450 since the reaction is not inhibited by carbon monoxide, ethyl isocyanide, or nicotinamide (173). After 100-fold purification of the enzymes involved in the demethylation of 4-methylated steroids, the preparation was almost devoid of cytochrome P-450 (173). The demethylation is inhibited by cyanide (173). A cyanide-sensitive protein has been extracted from liver microsomes, and this protein may be involved in demethylation of lanosterol (174) as well as desaturation of fatty acids (137, 174). The decarboxylation of 4-carboxylated steroids formed in the microsomes is dependent on a 3β-hydroxy-

steroid dehydrogenase which catalyzes the rate-limiting step in the decarboxylation (175).

It seems probable that the reactions in the removal of the 14α-methyl
group in lanosterol are analogous to those in the removal of the 4-methyl
groups. The carboxylic acid intermediate has not been isolated. It has
been suggested that mixed function oxidases are involved in the removal of
the 14-carboxyl group. It has been shown that a Δ^7 or a Δ^8 double bond
is a structural requirement for removal of the 14-methyl group (176–177),
that the 15α-hydrogen is removed during the conversion of lanosterol into
cholesterol (178–180), and that a $\Delta^{8,14}$-sterol is a probable intermediate
in the reaction (151). On the basis of these findings, two mechanisms have
been postulated which both may involve mixed function oxidases (151)
(Scheme 2).

In the first mechanism the C-14 carboxyl group is eliminated directly to
yield a $\Delta^{8(14)}$-steroid and this $\Delta^{8(14)}$-steroid is dehydrogenated to yield a
$\Delta^{8,14}$-steroid. The conversion of the $\Delta^{8(14)}$-steroid into the $\Delta^{8,14}$-steroid may
involve a mixed function oxidase, and it has been reported that the for-
mation of 5α-cholesta-8,14-dien-3β-ol from 5α-cholest-8(14)-en-3β-ol cata-

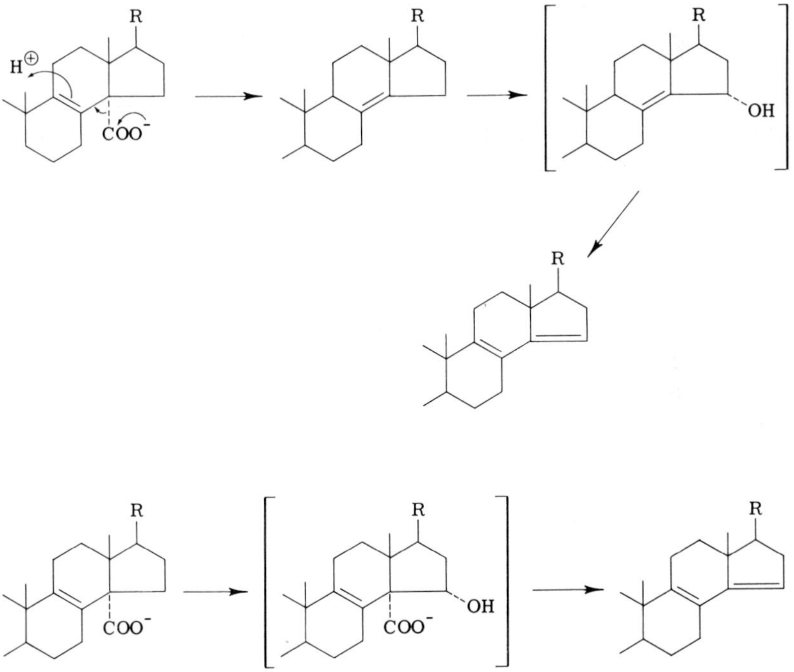

SCHEME 2

lyzed by the microsomal fraction is dependent on oxygen (181, 182). A 15α-hydroxylated intermediate may be involved, and it has been shown that 5α-cholest-8(14)-en-3β,15α-diol is converted rapidly into 5α-cholesta-8,14-dien-3β-ol by the microsomal fraction (182, 183).

In the second mechanism, decarboxylation at C-14 is preceded by loss of the 15α-hydrogen, either as a result of introduction of a 15α-hydroxyl group by a mixed function oxidase or of a direct abstraction of a hydride ion from the 15α position. The carboxyl group is eliminated either together with the hydride ion or with the hydroxyl group.

C. Introduction of the Δ^5 Double Bond

The conversion of 5α-cholest-7-en-3β-ol into, 5,7-cholestadien-3β-ol seems to be an obligatory step in the biosynthesis of cholesterol. Oxygen is required in the reaction, and it has been suggested that a mixed function oxidase is involved (151). It has been shown with specifically tritium-labeled steroids that the reaction involves *cis* removal of the 5α- and 6α-hydrogens (184). However, 5α- or 6α-hydroxylated steroids do not appear to be intermediates (185, 186). NADPH or NADH does not seem to be involved, and Scallen and Schuster have presented evidence that NADP or NAD is the cofactor required, NADP being slightly preferred (187). They have suggested that there might be a direct transfer of hydrogen to the pyridine nucleotide, which is then oxidized by an electron chain with oxygen as the final acceptor. On the other hand, Aberhart and Caspi (188) have shown that if NAD is involved as a cofactor, it is not likely that there is a direct transfer of a hydride ion from the 6α position of the substrate to the 4 position of NAD. Conversion of 6α-^{3}H-labeled 5α-cholest-7-en-3β-ol into 5,7-cholestadien-3β-ol occurred without significant transfer of ^{3}H to NAD; 99% of the liberated tritium was obtained as ^{3}H$_2$O. This seems to suggest the 6α-hydrogen is lost as a proton during the reaction.

Evidence has been recently presented that the SCP (sterol carrier protein) is involved in the conversion of 5α-cholest-7-en-3β-ol into 5,7-cholestadien-3β-ol (159).

VII. CONVERSION OF CHOLESTEROL INTO STEROID HORMONES

Biosynthesis of steroid hormones occurs in adrenal cortex, corpus luteum, ovary, placenta, and testis and the following hydroxylases participate: 11β-, 17α-, 18-, 19-, 21-hydroxylases and a 20α,22R-dihydroxylating system which possibly consists of one 20α- and one 22R-hydroxylase. The C-17 to C-20 and C-20 to C-22 lyases are also mixed function oxidases. A simplified scheme of the steps in the biosynthesis of steroid hormones is shown in Fig. 13. There are several pathways in addition to those shown in Fig. 13, and many of the hydroxylations may occur with 3β-hydroxy-Δ⁵-steroids instead of 3-keto-Δ⁴-steroids as substrates. The adrenal cortex contains all the different hydroxylases. Corpus luteum, ovary, and placenta contain almost exclusively the hydroxylases required for biosynthesis of estrogens, and testis contains mainly the hydroxylases required for biosynthesis of androgens. The 17α- and the 21-hydroxylases are located in the microsomal fraction, the 19-hydroxylase in the microsomal as well as the mitochondrial fraction, and the other hydroxylases in the mitochondrial fraction. The C-17 to C-20 lyase is located in the microsomal fraction and the C-20 to C-22 lyase in the mitochondrial fraction.

A. 11β-Hydroxylase

The enzyme system is most active toward 11-deoxycorticosterone and 11-deoxycortisol and requires oxygen and NADPH. The hydroxylase is located in the inner membrane of adrenal mitochondria (189). Already in 1958, evidence was presented to indicate that the 11β-hydroxylating system consisted of several components (cf. preceding edition of "Oxygenases"). Since then, all the components have been isolated and purified from adrenal mitochondria. The different components can be combined to form an active 11β-hydroxylating system. The transfer of electrons from NADPH to oxygen during the hydroxylation occurs as follows:

NADPH → flavoprotein → nonheme iron protein → cytochrome P-450 → oxygen

The flavoprotein, NADPH–adrenodoxin reductase, has been isolated and purified from bovine adrenal mitochondria by Omura *et al.* (190). The flavoprotein is immunologically different from the NADPH–cytochrome P-450 reductase in liver microsomes (191). The nonheme iron protein (adrenodoxin) was isolated from bovine adrenals independently by Suzuki and Kimura (192) and by Omura *et al.* (193). The molecular weight is about 15,000, and the protein contains two iron atoms and two labile sulfides. The molecule accepts one electron per molecule in the redox reaction. It has a relatively high oxidation–reduction potential (E_0 + 0.164 mV at pH 7.4). The nonheme iron protein cannot be replaced by spinach ferredoxin in the reconstituted 11β-hydroxylating system.

Cooper *et al.* (194) showed that cytochrome P-450 is involved in the 11β-hydroxylation. They found that the hydroxylating activity of bovine adrenal mitochondria is inhibited by carbon monoxide and that this inhibition is reversed by light at 450 nm. The cytochrome P-450 can be pre-

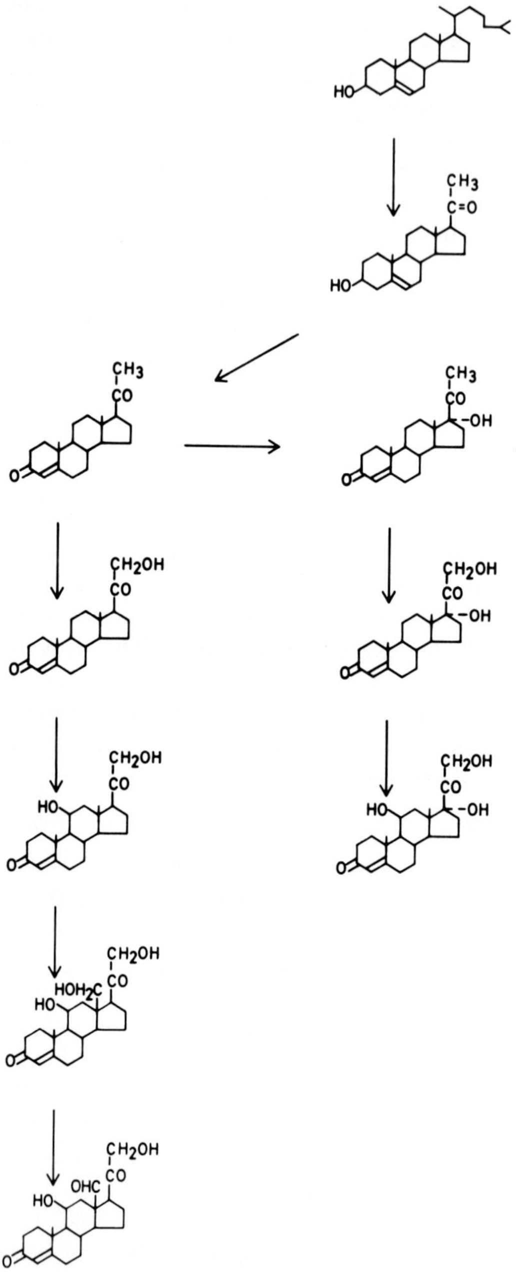

FIG. 13. Pathways in the biosynthesis of steroid hormones.

pared from adrenal mitochondria by extraction with Triton N-101 (195, 196). The cytochrome P-450 isolated by this procedure is in the ferric form, and optical and paramagnetic resonance spectroscopy indicates that it is a low spin system ($S = \frac{1}{2}$) (197). The E_0 value is -400 mV. The cytochrome P-450 involved in 11β-hydroxylation differs from that involved in the conversion of cholesterol into pregnenolone with respect to carbon monoxide sensitivity (198), substrate-induced difference spectra (199), and solubility properties (200). The two types of cytochrome P-450 have been recently separated from each other (199, 201).

In the scheme given on p. 59, the role of NADPH is only to reduce the flavoprotein. A dual role of NADPH in the 11β-hydroxylation has been proposed by Sih et al. (202). These authors have suggested that NADPH might be necessary to reduce a hypothetical perferryl ion ($Fe^{3+}-O-O^+$) to a hydroperoxo complex ($Fe^{2+}-O-OH^-$). In consonance with the concept of a dual role of NADPH, the isolated flavoprotein removes preferentially the 4B-hydrogen of NADPH, whereas in a reconstituted 11β-hydroxylating system, catalyzing the deoxycortocosterone-dependent oxidation of NADPH, the 4A-hydrogen of NADPH is removed preferentially (202). Reduction of the flavoprotein can be achieved by a large amount of NADH, whereas NADPH cannot be replaced by NADH in the 11β-hydroxylating system (202). In contrast, Huang and Kimura (203) have reported recently that only the reduced nonheme iron protein is necessary as reductant for the cytochrome P-450 in 11β-hydroxylation and that NADPH and the flavoprotein can be omitted. It is then clear that if NADPH is capable of direct reduction of the cytochrome P-450 complex this cannot be an essential step in the 11β-hydroxylation.

The properties of the reconstituted 11β-hydroxylating system have been studied by several authors (190, 192, 193, 203–206). Cooper et al. (205) have shown that the reconstituted system has optimal activity when the molar ratio flavoprotein:nonheme iron protein:cytochrome P-450 is about 1:50:1. The nature of the binding of the substrate to the cytochrome P-450 has been studied by substrate-induced difference spectra. As is discussed in more detail in Chapter 6, addition of a substrate to cytochrome P-450 induces either Type I spectral changes (peak at 390 nm and through at 420 nm) or Type II spectral changes (peak at 420 nm and through at 390 nm) in the difference spectra. Deoxycorticosterone gives a Type I difference spectrum (204, 207). It seems probable that this spectral change is the result of a direct complexing with the ferric cytochrome (207). In the steady state the cytochrome P-450 in the reconstituted system is almost completely in the ferric form, and reaction of cytochrome P-450 in the reduced (Fe^{2+}) form with oxygen leads rapidly to the formation of oxidized (Fe^{3+}) cytochrome P-450. When cytochrome P-450 (Fe^{3+}) alone is reduced and subsequently equilibrated with gas mixtures containing oxygen, 11β-hydroxylation occurs only in the presence of reduced nonheme iron protein (197). It is evident that oxygenated cytochrome P-450 (Fe^{2+}) is capable of carrying out a fast and efficient 11β-hydroxylation only if the nonheme iron protein is supplied as a specific electron donor.

The effect of steroids on the reaction of cytochrome P-450 with oxygen has been recently studied with the 11β-hydroxylating system (206). The K_m value for oxygen was about 9 μM, which is higher than that reported (208) for cytochrome c oxidase (about 4 μM). It was also shown that the reduction of cytochrome P-450 was essentially biphasic with an initial rapid phase and a later slow phase. The rate of the initial rapid phase was accelerated by Type I substrates and reduced by pregnenolone which is a Type II substrate. It might be mentioned that a similar biphasic reduction has been described for hepatic microsomal cytochrome P-450 (115). The finding that deoxycorticosterone combines with ferric cytochrome P-450 and that this facilitates reduction seems to indicate that the reduction of cytochrome P-450 occurs in the form of a cytochrome P-450 (Fe^{3+})–substrate complex as has been proposed for hydroxylations by liver microsomes. However, reduced cytochrome P-450 from adrenal mitochondria can also combine with deoxycorticosterone and an alternative pathway is possible in which the cytochrome P-450 is reduced prior to combination with the substrate. Scheme 3 (206) has been suggested.

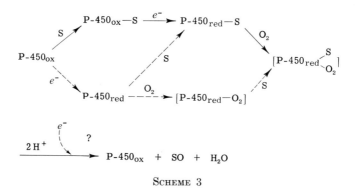

SCHEME 3

In 11β-hydroxylation, two electrons are required and these appear to be transported by the same electron transport chain. In contrast, hydroxylation by liver microsomes may require two electron transport chains (cf. below).

B. Conversion of Cholesterol into Pregnenolone

The mechanism of side chain cleavage of cholesterol in the biosynthesis of steroid hormones has been extensively studied by several groups. Most authors have used partially purified preparations from adrenal mitochondria. Some studies have been performed with preparations from corpus luteum (209), ovaries (210, 211), and placenta (212), and these preparations have the advantage of not being contaminated with 11β-hydroxylating activity. Until recently, side chain cleavage of cholesterol was considered to start with a 20α-hydroxylation followed by a 22R-hydroxylation yielding 20α,22R-dihydroxycholesterol (213–215). The 20α,22R-dihydroxycholesterol was then attacked by a C-20 to C-22 lyase with formation of isocaproic aldehyde and pregnenolone (216, 217). The 20α-hydroxylation was suggested to be rate limiting in the conversion of cholesterol into pregnenolone and the site of action of ACTH (218, 219). Some authors have been able to trap 20α-hydroxycholesterol as an intermediate (220, 221). However, others (210, 222–224) have failed and the role of 20α-hydroxycholesterol as an obligatory intermediate is not established. On the other hand, the role of 20α,22R-dihydroxycholesterol as an intermediate seems well established, and this compound accumulates to a significant extent when labeled cholesterol is incubated with an acetone powder of adrenal mitochondria (217, 225). The compound has also been isolated in crystalline form from bovine adrenal glands (226).

In some recent publications from the Worcester group (225), the individual rate constants for formation and disappearance of each of the hypothetical intermediates, 20α-hydroxycholesterol, 22R-hydroxycholesterol, and 20α,22R-dihydroxycholesterol, were determined with acetone powders of adrenal mitochondria. Evidence was obtained that 20α,22R-dihydroxycholesterol was a much more important intermediate than either 22R-hydroxycholesterol or 20α-hydroxycholesterol. It was suggested that the major pathway to 20α,22R-dihydroxycholesterol did not involve any monohydroxylated cholesterol as intermediate and that the formation of 20α,22R-dihydroxycholesterol was a concerted attack of oxygen. Lier and Smith (227) have suggested that cholesterol 20α-hydroperoxide might be an intermediate. When cholesterol 20α-hydroperoxide was incubated with bovine adrenal mitochondria, 20α,22R-dihydroxycholesterol was obtained as a product together with 20α,21-dihydroxycholesterol. The reaction did not require molecular oxygen or added cofactors, and the enzyme could be solubilized from sonicated acetone-dried mitochondria. The partially purified enzyme was found to contain cytochrome P-450. It may be mentioned that microsomal cytochrome P-450 from liver can act as a peroxidase for several steroid hydroperoxides (228).

Van Lier and Smith (229) have also reported some trapping experiments which support the contention that cholesterol 20α-hydroperoxide might be an intermediate in the conversion of cholesterol into pregnenolone. Recently, Lutrell *et al.* (230) reported that (20R)-20-*t*-butyl-5-pregnene-3β,20-diol is converted into pregnenolone when incubated with sonicated mitochondria. Since this compound is completely substituted at C-22, it is evident that this compound is converted into pregnenolone without a 22-hydroxylated intermediate. Two general mechanisms were postulated for the conversion of (20R)-20-*t*-butyl-5-pregnene-3β,20-diol into pregnenolone, one of radical type and one of ionic type. It was pointed out that the actual cleavage process may be some hybrid type between these two mechanisms.

The mechanism of the C-20 to C-22 lyase has not been definitely established. When 20α,22R-dihydroxycholesterol was incubated in an ^{18}O atmosphere, it was found that no ^{18}O was incorporated into the metabolites (231), suggesting that oxygen in the 20-keto group of pregnenolone is derived from the oxygen of the 20α-hydroxyl group. This mechanism is analogous to the one suggested for the C-17 to C-20 lyase (cf. below). However, it was not excluded that the oxygen at C-20 could be exchanged under the conditions employed.

In most studies with intact adrenal mitochondria or purified systems the side chain cleavage of cholesterol has been assayed as the overall conversion of cholesterol into pregnenolone. It seems probable that the 20α,22R-dihydroxycholesterol synthetase activity and the C-20 to C-22 lyase activity utilize the same electron transport chain and perhaps also the same cytochrome P-450. Bryson and Sweat (232) have shown that the cholesterol side chain cleavage is dependent upon a flavoprotein, a nonheme iron protein, and cytochrome P-450. Furthermore, the reconstituted 11β-hydroxylating system used by several authors also shows side chain cleavage activity (203, 204). Thus, it seems probable that the electron transport chain in side chain cleavage is the same as that in 11β-hydroxylation. However, the species of cytochrome P-450 differs (198–201).

Several reports have appeared dealing with the spectral changes which occur when cholesterol as well as different hypothetical intermediates are added to partially purified side chain cleavage systems (233–235). Recently, the significance of these spectral changes has been questioned (236), and it was shown that different preparations gave different types of spectra with the same substrates. It was concluded that great caution must be exercised in drawing conclusions concerning mechanisms from substrate-induced spectra when crude enzyme preparations are used.

NADPH is required for cholesterol side chain cleavage as well as other hydroxylations in adrenal mitochondria. It has been shown that with intact mitochondria NADPH can be substituted with some Krebs cycle intermediates, and in many cases side chain cleavage as well as 11β-hydroxylation proceed much better with succinate than with extramitochondrial NADPH (209, 237–239). It is evident that succinate can generate intramitochondrial NADPH and that the mitochondria are more or less impermeable to extramitochondrial NADPH. It has been shown that aging or treatment with salt solutions, which cause the mitochondria to swell, increases the stimulation by extramitochondrial NADPH (209, 237, 239). Addition of succinate not only stimulates steroid hydroxylating activity but also the rate of reduction of NAD in the mitochondria (237, 240). It has been suggested that the reduction of NAD in the mitochondria is coupled with the generation of intramitochondrial NADPH and steroid hydroxylating activity. Dinitrophenol, hyperbaric oxygen, arsenate, and amytal inhibit mitochondrial reduction of NAD as well as steroid hydroxylating activity (239, 240). Scheme 4 shows a postulated pathway (240) for electrons in the side chain cleavage of cholesterol where F_T, F_D, and F_S are flavoproteins involved in the oxidation of NADPH, NADH, and succinate, respectively, and NH Fe is nonheme iron.

There might also be other possibilities for generation of NADPH in the mitochondria such as generation by isocitric acid dehydrogenase (240). Although the 11β-hydroxylating system and the side chain cleavage system appear to use the same electron transport chain from NADPH to the specific cytochrome P-450, it is possible that the two systems use NADPH generated by different mechanisms, perhaps in different compartments within the mitochondria (240). It has been suggested that the side chain cleavage system has its own pool of NADPH, generated by reversed electron transport at the expense of high energy compounds synthesized in conjunction with the oxidation of succinate (240).

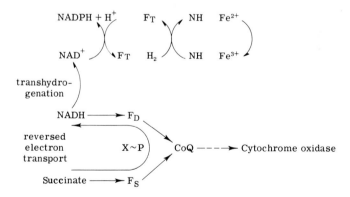

SCHEME 4

C. 21-Hydroxylase

The 21-hydroxylation is catalyzed by the microsomal fraction of adrenal cortex. The enzyme is active toward pregnenolone, progesterone, and 17α-hydroxyprogesterone. The finding that carbon monoxide inhibits 21-hydroxylation and this inhibition is reversed by light indicates that cytochrome P-450 participates (241). However, Matthijssen and Mandel (242, 243) have reported the presence of 21-hydroxylase activity in a partially purified preparation from bovine adrenals which was practically devoid of cytochrome P-450. The activity was not stimulated by the addition of cytochrome P-450. A cytochrome P-450-containing 21-hydroxylating system has been recently purified about 10-fold from steer adrenal microsomes by Mackler *et al.* (244). The enzyme was found to catalyze 21-hydroxylation of all substrates to a similar extent at an early stage of the purification procedure but was then found to lose the ability to 21-hydroxylate pregenenolone and progesterone. It was suggested that this might result from the presence of more than one microsomal 21-hydroxylase or from different reaction mechanisms for different substrates.

The substrate-induced difference spectra obtained with different substrates for the 21-hydroxylase have been studied by several authors (245–248). It has been shown that only such steroids which induce Type I spectral changes are hydroxylated by the system, that the spectral changes are very rapid and precede the hydroxylation step, and that the spectral changes disappear when the substrate is transformed into its hydroxylated product. These observations have led to the assumption that steroid-induced spectral changes result from formation of a primary enzyme–substrate complex. However, recent studies with a partially purified 21-hydroxylating system seem to suggest that the Type I spectral changes are associated with the transformation of cytochrome P-450 from an inactive to an active state which permits enhanced electron flow from the cytochrome (246). Thus, the Type I spectral changes may be independent of the oxygen activation step and the hydroxylation of the compound (246).

D. 17α-Hydroxylase and Biosynthesis of Androgens

17α-Hydroxylation has been studied in testis and adrenals. Attempts to purify the 17α-hydroxylating system from testis have failed. A partially purified system can be obtained by ammonium sulfate fractionation of extracts of adrenals, and an active 17α-hydroxylating system can be precipitated at 40% saturation with ammonium sulfate (249). However, when this protein fraction is combined with the precipitate obtained at 80% saturation with ammonium sulfate, only 11β-hydroxylation occurs. It is evident that the presence of the 11β-hydroxylating system inhibits the 17α-hydroxylase *in vitro*. The inhibiting factor in the 11β-hydroxylating system is probably the nonheme iron protein (249).

17α-Hydroxylated steroids are precursors of androgens. The conversion of pregnenolone into testosterone may follow several different pathways in which the order of the changes in the A ring and the 17α-hydroxylation differs (250). The conversion of 17α-hydroxylated C_{21} steroids into the corresponding C_{19}-steroids requires a C-17 to C-20 lyase which is localized in the microsomal fraction of testis. The enzyme requires oxygen and NADPH. A possible mechanism was suggested in the previous edition of "Oxygenases," and the reader is referred to this book for detailed information. In consonance with the proposed mechanism, the oxygen which appears in the 17α-hydroxyl group has been shown to be transferred to the 17-keto group in the C_{19}-steroid formed (231). It is interesting that 17α,20α-dihydroxy-4-pregnen-3-one is a competitive inhibitor of the C-17 to C-20 lyase (251), and this compound may play a role in the regulation of testicular steroidogenesis.

Certain microorganisms can convert progesterone into testosterone without the intermediate formation of a 17α-hydroxysteroid (252). Although this pathway has been described to occur in the ovary (253), it is probably of little importance, quantitatively, at least in the dog and the rat (250).

E. 19-Hydroxylation and Biosynthesis of Estrogens

19-Hydroxylated intermediates are obligatory intermediates in the biosynthesis of estrogens from C_{19}-steroids (254, 255). The 19-hydroxylation has been mainly studied with microsomes from human placenta but is known to occur also in adrenal mitochondria (256). The 19-hydroxylating system is dependent upon NADPH. It does not appear to involve cytochrome P-450 since the conversion of 4-androstene-3,17,dione into 19-hydroxy-4-androstene-3,17-dione as well as into estrone is not inhibited by carbon monoxide under conditions of limited oxygen supply (257, 258). Evidence has been presented that 3,17-diketo-4-androsten-19-al is an intermediate in the biosynthesis of estrogens (259, 260). It is interesting that the conversion of 19-hydroxy-4-androstene-3,17-dione into the corresponding aldehyde requires oxygen and NADPH, and it has been suggested that the oxidation is in fact a second hydroxylation of the C-19 methyl group in analogy with the reactions involved in the oxidation of the methyl groups in lanosterol (260). Concerning the mechanism of the further conversion of 3,17-diketo-4-androsten-19-al into estrone, the following facts are known: (1) NADPH- and oxygen-dependent oxygenase(s) is involved which is not sensitive to carbon monoxide (255); (2) the 1β- and 2β-hydrogens are removed during the reaction (261–263); and (3) the C-19 group is eliminated as formaldehyde or formic acid (264, 265). On the basis of these findings the Worcester group (261) already in 1962 suggested the possible mechanisms shown in Scheme 5. In an alternative pathway the C-19 group is eliminated in monohydroxylated form which gives formaldehyde as elimination product.

In the first mechanism there is a 1,2 elimination followed by spontaneous elimination of the C-19 group. In the second mechanism the 1β-hydrogen is eliminated by replacement with an activating group such as a hydroxyl group which is eliminated more or less simultaneously as the C-19 group. In the third mechanism the loss of the 2β-hydrogen is explained as an enolization of the 3-keto group. The hydrogen at C-1 is then eliminated as a hydride ion and this elimination is concerted with elimination of the C-19 group. As yet, none of the three pathways postulated has been definitely excluded. It might be mentioned that the aromatization step has been studied with some C_{18}-steroids such as 19-nor-4-androstene-3,17-dione (266, 267). This compound can be hydroxylated to yield 1β-hydroxy-19-nor-4-androstene-3,17-dione under incubation conditions used for aromatization. Subsequent dehydration catalyzed by enzymes, base, or acid gives rise to estrone. This seems to support the contention that there might be 1β-hydroxylated intermediates in the biosynthesis of estrogens. However, the enzymic aromatization of 19-nor-4-androstene-

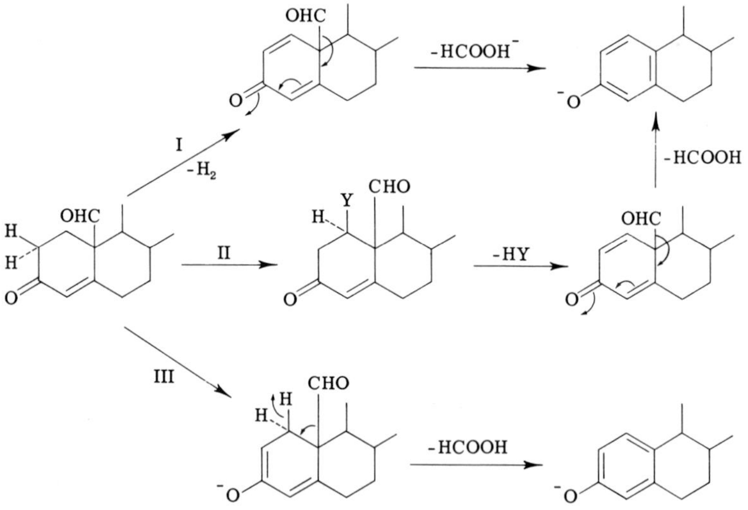

SCHEME 5

3,17-dione was found to be sensitive to carbon monoxide in contrast to the overall conversion of 4-androstene-3,17-dione and 3,17-diketo-4-androsten-19-al into estrone (258). This finding does not exclude completely the possibility that there are 1β-hydroxylated intermediates in the biosynthesis of estrone since the 1β-hydroxylase active on C_{18}-steroids might be different from that active on C_{19}-steroids. It is also possible that 1β-hydroxylation is not the rate-limiting step in the aromatization of 3,17-diketo-4-androsten-19-al in which case a carbon monoxide sensitivity of the 1β-hydroxylase might go undetected (258). It has been recently shown that the oxygenase(s) involved in the aromatization of 3,17-diketo-4-androsten-19-al is less sensitive to reduced tension of oxygen than the oxygenases involved in the conversion of 4-androstene-3,17-dione into 19-hydroxy-4-androstene-3,17-dione and the corresponding aldehyde (258).

VIII. METABOLISM OF STEROID HORMONES IN LIVER

Mixed function oxidases in the liver play an important role in the metabolism of steroid hormones into more polar compounds which can be eliminated from the body. Different C_{18}-, C_{19}-, and C_{21}-steroids can be hydroxylated in a number of different positions in the steroid nucleus. The combined action of these hydroxylases and 5α- and 5β-reductases, various oxidoreductases, and conjugating enzymes is responsible for the complicated pattern of steroid hormone metabolites excreted.

In Table II the microsomal hydroxylations of the major steroid hormones and their metabolites have been listed. In all cases studied, these hydroxy-

TABLE II

Hydroxylations of Steroid Hormones and Metabolites of Steroid Hormones Catalyzed by Mammalian Liver Microsomes

Substrate	Position hydroxylated	References
Estradiol	2, 15α, 16α, 6α, 6β, 7α	268–271
Dehydroepiandrosterone	7α, 7β, 16α	268, 272–274
Androstenedione	6α, 6β, 7α, 16α, 18	268, 275
Testosterone	1β, 2β, 6α, 6β, 7α, 15β, 16α, 18	268, 276–281
5α-Dihydrotestosterone	2β, 7α, 16α	277
5α-Androstane-3α,17β-diol	2β	277
Pregnenolone	2α, 2β, 7α, 7β, 16α	274, 282–283
Progesterone	2α, 6α, 6β, 7α, 15α, 15β, 16α	268, 280, 284
Cortisol	2α, 6β	268, 285

lations seem to involve cytochrome P-450. They are inhibited by carbon monoxide and with a few exceptions also stimulated by treatment with phenobarbital and some other drugs (274, 276, 285–290). In the case of 7α-hydroxylation of testosterone cytochrome P-448 may be involved since this hydroxylation is stimulated by treatment with methylcholanthrene (286) (cf. Chapter 6).

It has been suggested that the microsomal fraction of liver homogenate contains a hydroxylating system which is common for hydroxylation of steroid hormones, fatty acids, and drugs, and which involves NADPH–cytochrome P-450 reductase and cytochrome P-450 (110, 291, 292). The physiological substrates for the system would be steroid hormones and fatty acids since the K_m values for these hydroxylations are much lower than those for drug hydroxylations. Some steroids are known to produce the same difference spectra (Type I) as various drugs when mixed with liver microsomes, indicating a similar binding to the same or similar species of cytochrome P-450 (293). Furthermore, hydroxylation of various steroids is inhibited by certain drugs, indicating competition for the same site on the cytochrome (286, 292, 294, 295). As discussed in Section IV,c and in Chapter 6, a procedure has been described for the partial purification of cytochrome P-450, NADPH–cytochrome P-450 reductase, and a lipid factor, which has been identified as phosphatidylcholine. When added together the three components catalyze ω-hydroxylation of lauric acid, N-demethylation of benzphetamine and chlorocyclizine, and hydroxylation of testosterone and some alkanes (296).

According to current concepts, mainly based on studies with drugs as substrates for the partially purified system, NADPH reduces the NADPH–cytochrome P-450 reductase which in turn reduces the substrate–cytochrome P-450 complex formed in a reaction between oxidized cytochrome P-450 and substrate (286, 297). The reduced substrate–cytochrome P-450 complex combines with molecular oxygen to form a complex which breaks down to give cytochrome P-450 and hydroxylated substrate. Recent work by Estabrook and collaborators (298, 299) indicates that cytochrome P-450 (Fe^{2+})–substrate–O$_2^-$ may be an intermediate in the reaction. These authors detected the formation of a new spectral species during the steady state of oxidation of a Type I drug by hepatic microsomes from pheno-barbital-treated rats in the presence of oxygen and NADPH. The new spectral species exhibited absorption maxima at about 440 and 590 nm in the difference spectrum which is very similar to the difference spectrum calculated from studies of the oxygenated intermediate of reduced cyto-chrome P-450 from *P. putida* (300) (cf. Chapter 7).

The magnitude of absorption of the spectral species depended upon the Type I hydroxylatable substrate, the concentration of cytochrome P-450 and cytochrome b_5, the oxygen concentration, and the use of NADPH as source of reducing equivalents (298). The spectral change was found to decrease with time as oxygen was consumed. The presence of reduced cytochrome P-450 (Fe^{2+})–substrate–O_2^- as an intermediate requires acceptance of two electrons, and it was suggested that the second electron arises from a separate electron transport chain involving cytochrome b_5. Support for this contention was provided by the finding that there are changes in the NADPH-dependent reduction of cytochrome b_5 during the steady state hydroxylation of drugs (299). Furthermore, cytochrome b_5 can be reduced by NADH and addition of NADH stimulates hydroxylation of drugs (299) as well as steroids and fatty acids (301) under conditions of limited supply of NADPH. The contention that a superoxide ion is an intermediate in the reaction is supported by the finding that superoxide dismutase inhibits drug hydroxylation by a reconstituted system from liver, and addition of a superoxide-generating system supports the hydroxylation of drugs in the absence of NADPH and NADPH–cytochrome P-450 reductase (302) (cf. Chapters 11 and 12).

A tentative scheme for the hydroxylation of drugs by the microsomal hydroxylating system as proposed by Estabrook and collaborators (298) is shown in Scheme 6.

Kinetic studies indicate that the rate-limiting step in the overall hydroxylation of several drugs is the reduction of the substrate–cytochrome P-450 complex (297). However, the presence of a marked isotope effect in the hydroxylation of some specifically tritium- or deuterium-labeled steroids and fatty acids seems to suggest that the rate-limiting step in some hydroxylations might be the abstraction of hydrogen from the substrate in the ternary complex (303–305).

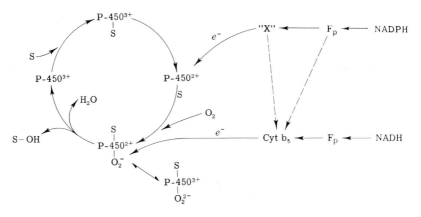

SCHEME 6

There is as yet no definite answer to the question of the possible multiplicity of the microsomal hydroxylating system catalyzing hydroxylation of steroid hormones, fatty acids, and drugs. As mentioned above, administration of phenobarbital leads to an increase in the level of microsomal NADPH–cytochrome P-450 reductase and cytochrome P-450 parallel with an increase in hydroxylation of most steroid hormones (286, 297). Administration of methylcholanthrene leads to an increase in the level of cytochrome P-448 parallel with an increase in hydroxylation of benzpyrene and 7α-hydroxylation of testosterone (297). There is little information concerning the structural differences between cytochrome P-450 and cytochrome P-448. In a recent study, hydroxylation of testosterone was performed with soluble systems containing cytochrome P-450 or cytochrome P-448 (296). Definite differences were observed in the relative rates of 6β-, 7α-, and 16α-hydroxylation of testosterone by the two systems. It was concluded that the specificity was determined mainly by the cytochrome.

Even if the only structural heterogeneity as yet observed for the microsomal hydroxylation of steroid hormones, fatty acids, and drugs is the participation of either cytochrome P-450 or cytochrome P-448, it seems difficult to believe that only two unique systems are involved in hydroxylation of steroid hormones. The following facts seem to suggest additional heterogeneity: (1) the same steroid molecule is hydroxylated in several different positions; (2) the degree of inhibition by carbon monoxide as well as the degree of stimulation by various drugs differ with different hydroxylations; (3) the effects of various inhibitors and ions differ with different hydroxylations; (4) the effects of age, sex, diet, starvation, and various hormones differ with different hydroxylations; and (5) abstraction of hydrogen from the substrate is rate limiting in some hydroxylations but not in others. It is improbable that one or two hydroxylating systems is responsible for the great number of hydroxylations occurring, and it seems equally improbable that every single hydroxylation has its own electron transport chain and its own type of cytochrome P-450. The substrate specificity might result from a specific type of NADPH–cytochrome P-450 reductase, a specific type of cytochrome P-450, or a specific type of protein interaction with the cytochrome and/or the reductase.

In the present review, no attempt will be made to summarize the numerous studies concerning different influences on different steroid hydroxylations catalyzed by the microsomal fraction of mammalian liver. The most detailed studies have been performed with testosterone (276, 286), dehydroepiandrosterone (274), and pregnenolone (274). Most of the information concerns 6β-, 7α-, and 16α-hydroxylation of testosterone, and these studies illustrate well that the different microsomal, cytochrome P-450-dependent hydroxylations respond differently to the same factor

(286). All the testosterone hydroxylations are inhibited by carbon monoxide but to a different degree. The ratio of CO to O_2 needed for 50% inhibition of 6β-, 7α-, and 16α-hydroxylation is 1.54, 2.36, and 0.93, respectively. Chronic treatment of rats with phenobarbital, phenylbutazone, or DDT stimulates the hydroxylation of testosterone in the 16α position to a greater extent than in the 6β and 7α positions. In contrast, chronic treatment with methylcholanthrene increases the 7α-hydroxylation and decreases the 16α-hydroxylation. In dogs, the effect of phenobarbital is different from that in rats. Studies on testosterone hydroxylation as a function of age indicate that the 6β- and 16α-hydroxylations increase as the animal matures, whereas the 7α-hydroxylation decreases. The ability of microsomes to hydroxylate testosterone in the 6β and 16α positions decreases after storage of the enzyme preparation, whereas the 7α-hydroxylation is unchanged. Administration of chlorothione or addition of this compound *in vitro* inhibits the 16α-hydroxylation but not the 7α-hydroxylation.

IX. BIOSYNTHESIS AND METABOLISM OF BILE ACIDS

Figure 14 shows the probable sequence of reactions in the main pathway for biosynthesis of bile acids in mammals (306). Hydroxyl groups are introduced into the 6β, 7α, 12α, 24α, and 26 positions. In certain mammalian species, hydroxylation might also occur in the 6α and 23 positions and in certain vertebrates also in the 16α and 27 positions. With the exception of the 24α- and 26-hydroxylases, which might be present in the mitochondria as well as in the microsomes, all the hydroxylases are located in the microsomal fraction of liver homogenate. It is probable that cytochrome P-450 participates in all hydroxylations since they are inhibited by carbon monoxide although to varying extent (306). The 7α-, 12α-, and 26-hydroxylases show certain characteristics which indicate that these hydroxylases differ from those involved in hydroxylation of steroid hormones, fatty acids, and various drugs. It is noteworthy that side chain cleavage in the main pathway for the conversion of cholesterol into bile acids does not appear to involve a C-24 to C-25 lyase of mixed function oxidase type. Instead, the cleavage seems to be thiolytic as in the β-oxidation of fatty acids (306).

A. 7α-Hydroxylase

The 7α-hydroxylase catalyzes the rate-limiting step in the biosynthesis of bile acids from cholesterol. The hydroxylation is inhibited by carbon monoxide (274) and by antibody against NADPH–cytochrome P-450

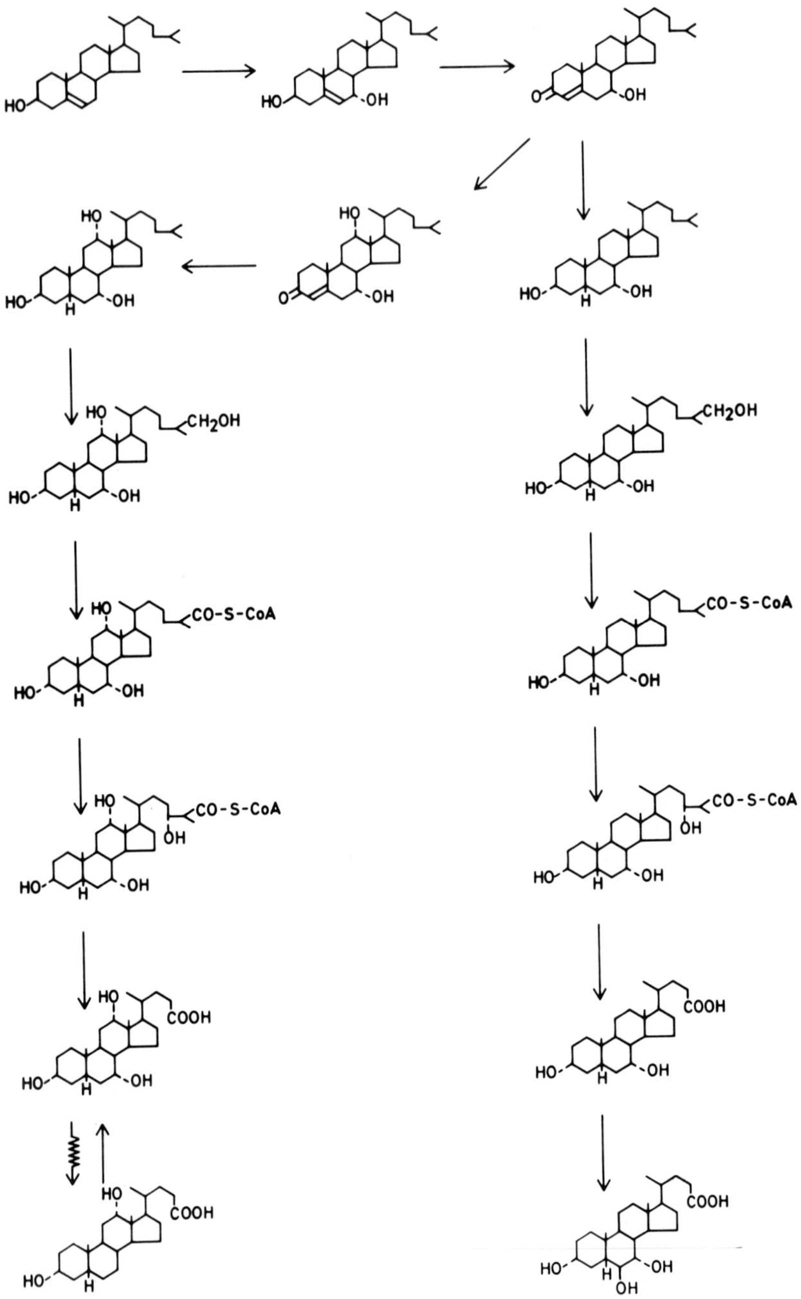

FIG. 14. Pathways in the biosynthesis of bile acids.

reductase (307) indicating the participation of NADPH–cytochrome P-450 reductase and cytochrome P-450 in the hydroxylation. The cytochrome P-450 may be different from that involved in most hydroxylations of steroid hormones and drugs. Treatment with phenobarbital does not increase the 7α-hydroxylation, at least not in some strains of rats (274, 308). Biliary drainage leads to a several-fold stimulation of the 7α-hydroxylase (274, 309), whereas the content of cytochrome P-450 in liver decreases (307). The 7α-hydroxylating system is inhibited by phosphate buffer in concentrations higher than 0.2 M which contrasts with several hydroxylations involved in the metabolism of steroid hormones and drugs by liver microsomes (310). To assay 7α-hydroxylation of cholesterol in liver microsomes it is necessary to inhibit the NADPH-dependent lipid peroxidation by the addition of EDTA or β-mercaptoethylamine (274, 311). In the absence of these reagents, 7α- as well as 7β-hydroxycholesterol and 7-keto-cholesterol will be formed as autoxidation products of cholesterol by the lipid peroxides formed. Some preliminary reports on the solubilization of the 7α-hydroxylase have appeared (312, 313), but there is no definite evidence that a solubilized 7α-hydroxylating system has been obtained. The 7α-hydroxylase active on cholesterol differs from the 7α-hydroxylases active on dehydroepiandrosterone, pregnenolone, testosterone, and tauro-deoxycholic acid since the effects of phenobarbital, methylcholanthrene, and biliary drainage differ (274, 286, 314, 315). The rate-limiting steps in the hydroxylation of taurodeoxycholic acid and cholesterol differ since substitution of the 7α-hydrogen with tritium leads to an isotope effect in the 7α-hydroxylation of taurodeoxycholic acid but not in the 7α-hydroxylation of cholesterol (303). Interestingly, the taurodeoxycholic acid 7α-hydroxylase shows a relatively low sensitivity toward carbon monoxide as do several other hydroxylases with which breakage of the C–H bond is a rate-limiting step in the hydroxylation (304, 305).

B. 12α-Hydroxylase

According to one report, the 12α-hydroxylase is insensitive to carbon monoxide (316) and according to another (317) it is inhibited to some extent by carbon monoxide. If a cytochrome P-450 is involved, it seems to be different from the cytochrome P-450 involved in the hydroxylation of steroid hormones and drugs since phenobarbital inhibits 12α-hydroxylation (316, 317). Cytochrome P-448 is probably not involved since the 12α-hydroxylase activity is unaffected by methylcholanthrene treatment (314). Since cytochrome c inhibits the reaction, NADPH–cytochrome P-450 reductase might participate. However, if it participates, it is not rate limiting since the activity of this enzyme increases upon treatment with thyroxine, whereas the activity of the 12α-hydroxylase decreases (318).

C. 26-Hydroxylase

There is one 26-hydroxylating system in the mitochondrial fraction and one in the microsomal fraction of rat liver homogenate. Both may depend on cytochrome P-450. It is noteworthy that liver mitochondria contain only small amounts of cytochrome P-450. The mitochondrial system is stimulated to a greater extent than the microsomal system by treatment with phenobarbital (319). The mitochondrial system has a broad substrate specificity, whereas the microsomal system appears to be specific for 5β-cholestane-3α,7α-diol and 5β-cholestane-3α,7α,12α-triol (319). Both the mitochondrial and the microsomal systems are partially stereospecific (320–322). It might be mentioned that in addition to 26-hydroxylation, the microsomal, but not the mitochondrial, fraction catalyzes 23ξ-, 24α-, 24β-, and 25-hydroxylations of 5β-cholestane-3α,7α,12-triol (323). These hydroxylations differ from the 26-hydroxylation with respect to sensitivity toward carbon monoxide and treatment with phenobarbital. The characteristics of these hydroxylations are similar to those of hydroxylations of steroid hormones and drugs. It has been shown with specifically deuterium-labeled steroids that the abstraction of the hydrogen may be rate limiting in 24-hydroxylation but not in 26-hydroxylation of 5β-cholestane-3α,7α,12α-triol (304). The microsomal 26-hydroxylation, but not the 24-hydroxylation, is inhibited when the reaction is carried out in deuterated water, supporting the contention that the rate-limiting steps in the two hydroxylations differ (304).

D. 6β-Hydroxylase

The 6β-hydroxylase is active on lithocholic acid and taurochenodeoxycholic acid. This enzyme system involves cytochrome P-450 and NADPH–cytochrome P-450 reductase and is stimulated by treatment with phenobarbital (315, 324, 325). A purification of the 6β-hydroxylase system can be obtained by suspending the microsomal fraction of rat liver homogenate in 1 M phosphate buffer and centrifuging the mixture at 100,000 g (324). The 6β-hydroxylating activity is retained in the supernatant. However, the procedure does not lead to a true solubilization since subsequent dilution with water and centrifugation precipitates most of the activity (310).

REFERENCES

1. Stevens, F. C., Brown, D. M., and Smith, E. L. (1970). *Arch. Biochem. Biophys.* **136,** 413.
2. Koch, R. B. (1968). *Arch. Biochem. Biophys.* **125,** 303.

3. Tappel, A. L. (1963). *In* "The Enzymes" (P. D. Boyer, H. Lardy, and K. Myrbäck (eds.), 2nd ed., Vol. 8, p. 275. Academic Press, New York.
4. Tappel, A. L. (1961). *In* "Autoxidation and Antioxidants" (W. O. Lundberg, ed.), Vol. 1, p. 325. Wiley (Interscience), New York.
5. Hamberg, M., and Samuelsson, B. (1965). *Biochem. Biophys. Res. Commun.* **21**, 531.
6. Hamberg, M., and Samuelsson, B. (1967). *J. Biol. Chem.* **242**, 5329.
7. Dolev, A., Rohwedder, W. K., and Dutton, H. J. (1967). *Lipids* **2**, 28.
8. Holman, R. T., Egwim, P. O., and Christie, W. W. (1969). *J. Biol. Chem.* **244**, 1149.
9. Reitz, R. C., Lands, W. E. M., Christie, W. W., and Holman, R. T. (1968). *J. Biol. Chem.* **243**, 2241.
10. Zimmerman, D. C., and Vick, B. A. (1970). *Lipids* **5**, 392.
11. Veldink, G. A., Vliegenthart, J. F. G., and Boldingh, J. (1970). *Biochim. Biophys. Acta* **202**, 198.
12. Chang, C. C., Esselman, W. J., and Clagett, D. O. (1971). *Lipids* **6**, 100.
13. Hamberg, M. (1971). *Anal. Biochem.* **43**, 515.
14. Christopher, J., and Axelrod, B. (1971). *Biochem. Biophys. Res. Commun.* **44**, 731.
15. Siddiqi, A. M., and Tappel, A. L. (1957). *J. Amer. Oil Chem. Soc.* **34**, 529.
16. Fridovich, I., and Handler, P. (1961). *J. Biol. Chem.* **236**, 1836.
17. Walker, G. C. (1963). *Biochem. Biophys. Res. Commun.* **13**, 431.
18. Hamberg, M., and Samuelsson, B., unpublished observations.
19. Chan, H. W. S. (1971). *J. Amer. Chem. Soc.* **93**, 2357.
20. Smith, W. L., and Lands, W. E. M. (1970). *Biochem. Biophys. Res. Commun.* **41**, 846.
21. Bergström, S. (1946). *Ark. Kemi* **21**, 15.
22. Graveland, A. (1970). *J. Amer. Oil Chem. Soc.* **47**, 352.
23. Graveland, A. (1970). *Biochem. Biophys. Res. Commun.* **41**, 427.
24. Garssen, G. J., Vliegenthart, J. F. G., and Boldingh, J. (1971). *Biochem. J.* **122**, 327.
25. Guss, P. L., Richardson, T., and Stahlmann, M. A. (1968). *J. Amer. Oil Chem. Soc.* **45**, 272.
26. Christopher, J., Pistorius, E., and Axelrod, B. (1970). *Biochim. Biophys. Acta* **198**, 12.
27. Gardner, H. W., and Weisleder, D. (1970). *Lipids* **5**, 678.
28. Galliard, T., and Phillips, D. R. (1971). *Biochem. J.* **124**, 431.
29. van Dorp, D. A., Beerthuis, R. K., Nugteren, D. H., and Vonkeman, H. (1964) *Biochim. Biophys. Acta* **90**, 204.
30. Bergström, S., Danielsson, H., and Samuelsson, B. (1964). *Biochim. Biophys. Acta* **90**, 207.
31. Bergström, S., Danielsson, H., Klenberg, D., and Samuelsson, B. (1964). *J. Biol. Chem.* **239**, PC4006.
32. van Dorp, D. A., Beerthuis, R. K., Nugteren, D. H., and Vonkeman, H. (1964). *Nature (London)* **203**, 839.
33. Kupiecki, F. P. (1965). *Life Sci.* **4**, 1811.
34. Hamberg, M., and Samuelsson, B. (1966). *J. Amer. Chem. Soc.* **88**, 2349.
35. Nugteren, D. H., Beerthuis, R. K., and van Dorp, D. A. (1966). *Rec. Trav. Chim. Pays-Bas* **85**, 405.
36. Hamberg, M., and Samuelsson, B. (1967). *J. Biol. Chem.* **242**, 5344.
37. Änggård, E., and Samuelsson, B. (1965). *J. Biol. Chem.* **240**, 3518.
38. Hamberg, M. (1969). *FEBS Lett.* **5**, 127.

39. Pace-Asciak, C., Morawska, K., Coceani, F., and Wolfe, L. S. (1967). *Prostaglandin Symp. Worcester Foundation Exp. Biol.* p. 371. Wiley (Interscience), New York.
40. Samuelsson, B. (1967). *Progr. Biochem. Pharmacol.* **3,** 59–70.
41. van Dorp, D. A. (1967). *Progr. Biochem. Pharmacol.* **3,** 71–82.
42. Lee, R. E., and Lands, W. E. M. (1972). *Biochim. Biophys. Acta* **260,** 203.
43. Sih, C. J., Takeguchi, C., and Foss, P. (1970). *J. Amer. Chem. Soc.* **92,** 6670.
44. Samuelsson, B., Granström, E., and Hamberg, M. (1967). *in Prostaglandins, Proc. Nobel Symp. 2nd* (S. Bergström and B. Samuelsson, eds.), pp. 31–44. Almqvist and Wiksell, Stockholm.
45. Granström, E., unpublished results.
46. Struijk, C. B., Beerthuis, R. K., and van Dorp, D. A. (1967). *in Prostaglandins, Proc. Nobel Symp. 2nd* (S. Bergström and B. Samuelsson, eds.), pp. 51–56, Almqvist and Wiksell, Stockholm.
47. Beerthuis, R. K., Nugteren, D. H., Pabon, H. J. J., and van Dorp, D. A. (1968). *Rec. Trav. Chim.* **87,** 461.
48. Beerthuis, R. K., Nugteren, D. H., Pabon, H. J. J., Steenhoek, A., and van Dorp, D. A. (1971). *Rec. Trav. Chim. Pays-Bas* **90,** 943.
49. Ryhage, R., and Samuelsson, B. (1965). *Biochem. Biophys. Res. Commun.* **19,** 279.
50. Nugteren, D. H., and van Dorp, D. A. (1965). *Biochim. Biophys. Acta* **98,** 654.
51. Samuelsson, B. (1965). *J. Amer. Chem. Soc.* **87,** 3011.
52. Klenberg, D., and Samuelsson, B. (1965). *Acta Chem. Scand.* **19,** 534.
53. Hamberg, M., and Samuelsson, B. (1967). *J. Biol. Chem.* **242,** 5336.
54. Hamberg, M., and Samuelsson, B. (1967). *J. Biol. Chem.* **242,** 5329.
55. Granström, E., Lands, W. E. M., and Samuelsson, B. (1968). *J. Biol. Chem.* **243,** 4104.
56. Wlodawer, P., and Samuelsson, B. (1973). *J. Biol. Chem.* (in press).
57. Pace-Asciak, C., and Wolfe, L. S. (1971). *Biochemistry* **10,** 3657.
58. Pace-Asciak, C. (1971). *Biochemistry* **10,** 3664.
59. Hamberg, M., unpublished observations.
60. Pace-Asciak, C., and Wolfe, L. S. (1968). *Biochim. Biophys. Acta* **152,** 787.
61. Nugteren, D. H. (1970). *Biochim. Biophys. Acta* **210,** 171.
62. Wallach, D. P., and Daniels, E. G. (1971). *Biochim. Biophys. Acta* **231,** 445.
63. Ahern, D. G., and Downing, D. T. (1970). *Biochim. Biophys. Acta* **210,** 456.
64. Blain, J. A., and Shearer, G. (1965). *J. Sci. Food Agr.* **16,** 373.
65. Downing, D. T., Ahern, D. G., and Bachta, M. (1970). *Biochem. Biophys. Res. Commun.* **40,** 218.
66. Wlodawer, P., Samuelsson, B., Albonico, S. M., and Corey, E. J. (1971). *J. Amer. Chem. Soc.* **93,** 2815.
67. Vane, J. R. (1971). *Nature (London) New Biol.* **231,** 232.
68. Smith, J. B., and Willis, A. L. (1971). *Nature (London) New Biol.* **231,** 235.
69. Ferreira, S. H., Moncada, S., and Vane, J. R. (1971). *Nature (London) New Biol.* **231,** 237.
70. Tomlinson, R. V., Ringold, H. J., Qureshi, M. C., and Forchielli, E. (1972). *Biochem. Biophys. Res. Commun.* **46,** 552.
71. Samuelsson, B., and Wennmalm, Å. (1971). *Acta Physiol. Scand.* **83,** 163.
72. Hamberg, M., and Samuelsson, B. (1972). *J. Biol. Chem.* **247,** 3495.
73. Martin, R. O., and Stumpf, P. K. (1959). *J. Biol. Chem.* **234,** 2548.
74. Hitchcock, C., and James, A. T. (1964). *J. Lipid Res.* **5,** 593.
75. Bowen, D. M., and Radin, N. S. (1968). *Advan. Lipid Res.* **6,** 255.
76. Tsai, S.-C., Avigan, J., and Steinberg, D. (1969). *J. Biol. Chem.* **244,** 2682.
77. Stokke, O. (1968). *Biochim. Biophys. Acta* **152,** 213.

78. Hitchcock, C., Morris, L. J., and James, A. T. (1968). *Eur. J. Biochem.* **3**, 419.
79. Hitchcock, C., Morris, L. J., and James, A. T. (1968). *Eur. J. Biochem.* **3**, 473.
80. Morris, L. J., and Hitchcock, C. (1968). *Eur. J. Biochem.* **4**, 146.
81. Morris, L. J. (1970). *Biochem. J.* **118**, 681.
82. Hitchcock, C., and Morris, L. J. (1970). *Eur. J. Biochem.* **17**, 39.
83. Markovetz, A. J., Stumpf, P. K., and Hammarström, S. (1972). *Lipids* **7**, 159.
84. Fulco, A. J., and Mead, J. F. (1961). *J. Biol. Chem.* **236**, 2416.
85. Hajra, A. K., and Radin, N. S. (1963). *J. Lipid Res.* **4**, 270.
86. Hajra, A. K., and Radin, N. S. (1963). *J. Lipid Res.* **4**, 448.
87. Morris, L. J., Hall, S. W., and James, A. T. (1966). *Biochem. J.* **100**, 29C.
88. James, A. T., Hadaway, H. C., and Webb, J. P. W. (1965). *Biochem. J.* **95**, 448.
89. Yamada, M., and Stumpf, P. K. (1964). *Biochem. Biophys. Res. Commun.* **14**, 165.
90. Galliard, T., and Stumpf, P. K. (1966). *J. Biol. Chem.* **241**, 5806.
91. Morris, L. J. (1967). *Biochem. Biophys. Res. Commun.* **29**, 311.
92. Verkade, P. E., Elzas, M., van der Lee, J., de Volff, H. H., Verkade-Sandbergen, A., and van der Sande, D. (1932). *Koninklijke Ned. Akad. Wetensch. Proc.* **35**, 251; (1933). *Z. Physiol. Chem.* **215**, 225.
93. Verkade, P. E. (1938). *Chem. Ind. (London)* **57**, 704.
94. Bergström, S., Borgström, B., Tryding, N., and Westöö, G. (1954). *Biochem. J.* **58**, 604.
95. Den, H. (1965). *Biochim. Biophys. Acta* **98**, 462.
96. Stokke, O., Try, K., and Eldjarn, L. (1967). *Biochim. Biophys. Acta* **144**, 271.
97. McCarthy, R. D. (1964). *Biochim. Biophys. Acta* **84**, 74.
98. Hamberg, M., and Samuelsson, B. (1969). *J. Amer. Chem. Soc.* **91**, 2177.
99. Wakabayashi, K., and Shimazono, N. (1961). *Biochim. Biophys. Acta* **48**, 615.
100. Wakabayashi, K., and Shimazono, N. (1963). *Biochim. Biophys. Acta* **70**, 132.
101. Robbins, K. C. (1961). *Fed. Proc., Fed. Amer. Soc. Exp. Biol.* **20**, 272.
102. Robbins, K. C. (1968). *Arch. Biochem. Biophys.* **123**, 531.
103. Preiss, B., and Bloch, K. (1964). *J. Biol. Chem.* **239**, 85.
104. Björkhem, I., and Danielsson, H. (1970). *Eur. J. Biochem.* **14**, 473.
105. Björkhem, I., and Danielsson, H. (1970). *Eur. J. Biochem.* **17**, 450.
106. Kusunose, M., Ichihara, K., and Kusunose, E. (1969). *Biochim. Biophys. Acta* **176**, 679.
107. Israelsson, U., Hamberg, M., and Samuelsson, B. (1969). *Eur. J. Biochem.* **11**, 390.
108. Hamberg, M., and Björkhem, I. (1971). *J. Biol. Chem.* **246**, 7411.
109. Wada, F., Shibata, H., Goto, M., and Sakamoto, Y. (1968). *Biochim. Biophys. Acta* **162**, 518.
110. Das, M. L., Orrenius, S., and Ernster, L. (1968). *Eur. J. Biochem.* **4**, 519.
111. Lu, A. Y. H., Junk, K. W., and Coon, M. J. (1969). *J. Biol. Chem.* **244**, 3714.
112. Lu, A. Y. H., and Coon, M. J. (1968). *J. Biol. Chem.* **243**, 1331.
113. Ichihara, K., Kusunose, E., and Kusunose, M. (1970). *Biochim. Biophys. Acta* **202**, 560.
114. Kusunose, E., Ichihara, K., and Kusunose, M. (1970). *FEBS Lett.* **11**, 23.
115. Strobel, H. W., Lu, A. Y. H., Heidema, J., and Coon, M. J. (1970). *J. Biol. Chem.* **245**, 4851.
116. Tulloch, A. P., Spencer, J. F. T., and Gorin, P. A. J. (1962). *Can. J. Chem.* **40**, 1326.
117. Kester, A. S., and Foster, J. W. (1963). *J. Bacteriol.* **85**, 859.
118. Kusunose, M., Kusunose, E., and Coon, M. J. (1964). *J. Biol. Chem.* **239**, 1374.
119. McKenna, E. J., and Kallio, R. E. (1965). *Annu. Rev. Microbiol.* **19**, 183.
120. Cardini, G., and Jurtshuk, P. (1968). *J. Biol. Chem.* **243**, 6070.

121. Tulloch, A. P., Hill, A., and Spencer, J. F. T. (1968). *Can. J. Chem.* **46**, 3337.
122. Heinz, E., Tulloch, A. P., and Spencer, J. F. T. (1969). *J. Biol. Chem.* **244**, 882.
123. Jones, D. F. (1968). *J. Chem. Soc. C.* 2827.
124. Heinz, E., Tulloch, A. P., and Spencer, J. F. T. (1970). *Biochim. Biophys. Acta* **202**, 49.
125. Kusunose, M., Kusunose, E., and Coon, M. J. (1964). *J. Biol. Chem.* **239**, 2135.
126. Peterson, J. A., Basu, D., and Coon, M. J. (1966). *J. Biol. Chem.* **241**, 5162.
127. Peterson, J. A., Kusunose, M., Kusunose, E., and Coon, M. J. (1967). *J. Biol. Chem.* **242**, 4334.
128. Peterson, J. A., and Coon, M. J. (1968). *J. Biol. Chem.* **243**, 329.
129. Boyer, R. F., Lode, E. T., and Coon, M. J. (1971). *Biochem. Biophys. Res. Commun.* **44**, 925.
130. Bloomfield, D. K., and Bloch, K. (1960). *J. Biol. Chem.* **235**, 337.
131. Lennarz, W. J., and Bloch, K. (1960). *J. Biol. Chem.* **235**, PC26.
132. Light, R. J., Lennarz, W. J., and Bloch, K. (1962). *J. Biol. Chem.* **237**, 1793.
133. Gurr, M. I., and Bloch, K. (1966). *Biochem. J.* **99**, 16C.
134. Bernhard, K., von Bülow-Köster, J., and Wagner, H. (1959). *Helv. Chim. Acta* **42**, 152.
135. Marsh, J. B., and James, A. T. (1962). *Biochim. Biophys. Acta* **60**, 320.
136. James, A. T., and Marsh, J. B. (1962). *Biochim. Biophys. Acta* **57**, 170.
137. Oshino, N., Imai, Y., and Sato, R. (1966). *Biochim. Biophys. Acta* **128**, 13.
138. Jones, P. D., Holloway, P. W., Peluffo, R. O., and Wakil, S. J. (1969). *J. Biol. Chem.* **244**, 744.
139. Gurr, M. I., Robinson, M. P., and James, A. T. (1969). *Eur. J. Biochem.* **9**, 70.
140. Harris, R. V., James, A. T., and Harris, P. (1967). *In* "Biochemistry of Chloroplasts" (T. W. Goodwin, ed.), Vol. 2, p. 241. Academic Press, New York.
141. Holloway, P. W., Peluffo, R. O., and Wakil, S. J. (1963). *Biochem. Biophys. Res. Commun.* **12**, 300.
142. Fulco, A. J., and Bloch, K. (1964). *J. Biol. Chem.* **239**, 994.
143. Shroepfer, G. J., and Bloch, K. (1965). *J. Biol. Chem.* **240**, 54.
144. Harris, R. V., and James, A. T. (1965). *Biochim. Biophys. Acta* **106**, 456.
145. McMahon, V., and Stumpf, P. K. (1964). *Biochim. Biophys. Acta* **84**, 359.
146. Nagai, J., and Bloch, K. (1968). *J. Biol. Chem.* **243**, 4626.
147. Bloch, K. (1969). *Accounts Chem. Res.* **2**, 193.
148. Morris, L. J., Harris, R. V., Kelly, W., and James, A. T. (1967). *Biochem. Biophys. Res. Commun.* **28**, 904.
149. Morris, L. J., Harris, R. V., Kelly, W., and James, A. T. (1968). *Biochem. J.* **109**, 673.
150. Morris, L. J. (1970). *Biochem. J.* **118**, 681.
151. Goad, L. J. (1970). *In* "Natural Substances Formed Biologically from Mevalonic Acid" (T. W. Goodwin, ed.), p. 45. Academic Press, New York.
152. Wada, F., Hirat, K., and Sakamoto, Y. (1969). *J. Biochem. (Tokyo)* **65**, 171.
153. Corey, E. J., Russey, W. E., and Ortiz de Montanello, P. R. (1966). *J. Amer. Chem. Soc.* **88**, 4750.
154. van Tamelen, E. E., Wilett, J. D., Clayton, R. B., and Lord, K. E. (1966). *J. Amer. Chem. Soc.* **88**, 4752.
155. van Tamelen, E. E., Wilett, J. D., and Clayton, R. B. (1967). *J. Amer. Chem. Soc.* **89**, 3371.
156. Yamoto, S., and Bloch, K. (1970). *In* "Natural Substances Formed Biologically from Mevalonic Acid" (T. W. Goodwin, ed.), p. 35. Academic Press, New York.

157. Yamoto, S., and Bloch, K. (1970). *J. Biol. Chem.* **245,** 1670.
158. Corey, E. J., and Russey, W. E. (1966). *J. Amer. Chem. Soc.* **88,** 4751.
159. Ritter, M. C., and Dempsey, M. E. (1971). *J. Biol. Chem.* **246,** 1536.
160. Olson, J. A., Lindberg, M., and Bloch, K. (1957). *J. Biol. Chem.* **226,** 941.
161. Gautschi, F., and Bloch, K. (1958). *J. Biol. Chem.* **233,** 1343.
162. Sharpless, K. B., Snyder, T. E., Spencer, T. A., Maheshwari, K. K., Guhn, G., and Clayton, R. B. (1968). *J. Amer. Chem. Soc.* **90,** 6874.
163. Miller, W. L., and Gaylor, J. L. (1970). *J. Biol. Chem.* **245,** 5375.
164. Rahman, R., Sharpless, K. B., Spencer, T. A., and Clayton, R. B. (1970). *J. Biol. Chem.* **245,** 2667.
165. Miller, W. L., Kalafer, M. E., Gaylor, J. L., and Delwiche, C. V. (1967). *Biochemistry* **6,** 2673.
166. Moir, N. J., Miller, W. L., and Gaylor, J. L. (1968). *Biochem. Biophys. Res. Commun.* **33,** 916.
167. Miller, W. L., and Gaylor, J. L. (1970). *J. Biol. Chem.* **245,** 5369.
168. Miller, W. L., and Gaylor, J. L. (1970). *J. Biol. Chem.* **245,** 5375.
169. Miller, W. L., Brody, D. R., and Gaylor, J. L. (1971). *J. Biol. Chem.* **246,** 5147.
170. Akhtar, M., and Skinner, S. J. M. (1968). *Biochem. J.* **109,** 318.
171. Lieber, C. S., and De Carli, L. M. (1970). *J. Biol. Chem.* **245,** 2505.
172. Thurman, R. G., Ley, H. G., and Scholz, R. (1972). *Eur. J. Biochem.* **25,** 420.
173. Gaylor, J. L., and Mason, H. S. (1968). *J. Biol. Chem.* **243,** 4966.
174. Gaylor, J. L., Moir, N. J., Siefried, H. E., and Jefcoate, C. R. E. (1970). *J. Biol. Chem.* **245,** 5511.
175. Rahimtula, A. D., and Gaylor, J. L. (1972). *J. Biol. Chem.* **247,** 9.
176. Richards, L., and Hendrickson, J. B. (1964). *In* "The Biosynthesis of Sterols, Terpenes and Acetogenins," p. 274. Benjamin, New York.
177. Knight, J. C., Klein, P. D., and Szczepanik, P. A. (1966). *J. Biol. Chem.* **241,** 1502.
178. Canonica, L., Fiecchi, A., Kienle, M. G., Scala, A., Galli, G., Paoletti, E. G., and Paoletti, R. (1968). *J. Amer. Chem. Soc.* **90,** 3597.
179. Gibbons, G. F., Goad, L. J., and Goodwin, T. W. (1968). *Chem. Commun.* 1458.
180. Akhtar, M., Watkinson, I. A., Rahimtula, A. D., Wilton, D. C., and Munday, K. A. (1969). *Biochem. J.* **111,** 757.
181. Lee, W.-H., Lutsky, B. N., and Schroepfer, G. J., Jr. (1969). *J. Biol. Chem.* **244,** 5440.
182. Schroepfer, G. J., Jr., Fourcans, B., Huntoon, S., Lutsky, B. N., and Vermillion, J. (1971). *Fed. Proc., Fed. Amer. Soc. Exp. Biol.* **30,** 1105.
183. Huntoon, S., and Schroepfer, G. J., Jr. (1970). *Biochem. Biophys. Res. Commun.* **40,** 476.
184. Paliokas, A. M., and Schroepfer, G. J., Jr. (1968). *J. Biol. Chem.* **243,** 453.
185. Slaytor, M., and Bloch, K. (1965). *J. Biol. Chem.* **240,** 4598.
186. Dewhurst, S. M., and Akhtar, M. (1967). *Biochem. J.* **105,** 1187.
187. Scallen, T. J., and Schuster, M. W. (1968). *Steroids* **12,** 683.
188. Aberhart, D. J., and Caspi, E. (1971). *J. Biol. Chem.* **246,** 1387.
189. Satre, M., Vignais, P. J., and Idelman, S. (1969). *FEBS-Lett.*, **5,** 135.
190. Omura, T., Saunders, E., Estabrook, R. W., Cooper, D. Y., and Rosenthal, O. (1966). *Arch. Biochem. Biophys.* **117,** 660.
191. Masters, B. S. S., Baron, J., Taylor, W. E., Isaacson, E. L., and LoSpalluto, J. (1971). *J. Biol. Chem.* **246,** 4143.
192. Suzuki, K., and Kimura, T. (1965). *Biochem. Biophys. Res. Commun.* **19,** 340.

193. Omura, T., Sato, R., Cooper, D. Y., Rosenthal, O., and Estabrook, R. W. (1965). *Fed. Proc.* **24,** 1181.
194. Cooper, D. Y., Novack, B., Foroff, O., Slade, A., Saunders, E., Narasimhulu, S., and Rosenthal, O. (1967). *Fed. Proc., Fed. Amer. Soc. Exp. Biol.* **26,** 341.
195. Cooper, D. Y., Schleyer, H., Estabrook, R. W., and Rosenthal, O. (1969). *Excerpta Med. Congr. Ser. no.* **184,** *Progr. Endocrinol.* 784.
196. Cooper, D. Y., Schleyer, H., and Rosenthal, O. (1968). *Hoppe-Seyler's Z. Physiol. Chem.* **349,** 1592.
197. Schleyer, H., Cooper, D. Y., Levin, S. S., and Rosenthal, O. (1971). *Biochem. J.* **125,** 10p.
198. Wilson, L. D., and Harding, B. W. (1970). *Biochemistry* **9,** 1621.
199. Jefcoat, C. R., Hume, R., and Boyd, G. S. (1970). *FEBS Lett.* **9,** 41.
200. Young, D. G., Holroyd, J. D., and Hall, P. F. (1970). *Biochem. Biophys. Res. Commun.* 184.
201. Isaka, S., and Hall, P. F. (1971). *Biochem. Biophys. Res. Commun.* **43,** 747.
202. Sih, C. J., Tsong, Y. Y., and Stein, B. (1968). *J. Amer. Chem. Soc.* **90,** 5300.
203. Huang, J. J., and Kimura, T. (1971). *Biochem. Biophys. Res. Commun.* **43,** 737.
204. Mitani, F., and Horie, S. (1970). *J. Biochem. (Tokyo)* **68,** 529.
205. Cooper, D. Y., Narashimulu, S., and Rosenthal, O. (1968). *Advan. Chem. Ser.* **77,** 220.
206. Ando, N., and Horie, S. (1971). *J. Biochem. (Tokyo)* **70,** 557.
207. Sweat, M. L., Young, R. B., and Bryson, M. J. (1969). *Arch. Biochem. Biophys.* **130,** 66.
208. Laser, H. (1952). *Proc. Roy. Soc. (London) Ser. B.* **140,** 230.
209. Yago, N., Dorfman, R. I., and Forchielli, E. (1967). *J. Biochem. (Tokyo)* **62,** 345.
210. Sulimovici, S., and Boyd, G. S. (1968). *Eur. J. Biochem.* **3,** 332.
211. Robinson, J., and Stevenson, P. M. (1971). *Eur. J. Biochem.* **24,** 18.
212. Mason, J. I., and Boyd, G. S. (1971). *Eur. J. Biochem.* **21,** 308.
213. Shimizu, K., Hayano, M., Gut, M., and Dorfman, R. I. (1961). *J. Biol. Chem.* **236,** 695.
214. Constantopoulos, G., and Tchen, T. T. (1961). *J. Biol. Chem.* **236,** 65.
215. Shimizu, K., Gut, M., and Dorfman, R. I. (1962). *J. Biol. Chem.* **237,** 699.
216. Staple, E., Lynn, W. S., Jr., and Gurin, S. (1956). *J. Biol. Chem.* **219,** 845.
217. Constantopoulos, G., Satoh, P. S., and Tchen, T. T. (1962). *Biochem. Biophys. Res. Commun.* **8,** 50.
218. Koritz, S. B. (1963). *Biochem. Biophys. Acta* **56,** 63.
219. Hall, P. F., and Young, D. G. (1968). *Endocrinology* **82,** 559.
220. Solomon, S., Levitan, P., and Lieberman, S. (1956). *Rev. Can. Biol.* **15,** 282.
221. Ichii, S., Omata, S., and Kobayashi, S. (1967). *Biochim. Biophys. Acta* **139,** 308.
222. Koritz, S. B., and Hall, P. F. (1964). *Biochemistry* **3,** 1298.
223. Hall, P. F., and Koritz, S. B. (1964). *Biochim. Biophys. Acta* **93,** 441.
224. Simpson, E. R., and Boyd, G. S. (1967). *Eur. J. Biochem.* **2,** 275.
225. Burstein, S., Kimball, H. L., and Gut, M. (1970). *Steroids* **15,** 809.
226. Dixon, R., Furutachi, T., and Lieberman, S. (1970). *Biochem. Biophys. Res. Commun.* **40,** 161.
227. van Lier, E., and Smith, L. L. (1970). *Biochem. Biophys. Acta* **210,** 153.
228. Hrycay, E. G., and O'Brien, P. J. (1971). *Biochem. J.* **125,** 12p.
229. van Lier, E., and Smith, L. L. (1970). *Biochem. Biophys. Res. Commun.* **40,** 510.
230. Lutrell, B., Hochberg, R. B., Dixon, W. R., and Donald, P. D. (1972). *J. Biol. Chem.* **247,** 1462.

231. Nakano, H., Inano, H., Sato, H., Shikita, M., Tamaoki, B. (1967). *Biochim. Biophys. Acta* **137,** 335.
232. Bryson, M. J., and Sweat, M. L. (1968). *J. Biol. Chem.* **243,** 2799.
233. Mitani, F., and Horie, J. (1969). *J. Biochem. (Tokyo)* **65,** 269.
234. Whyner, J. A., and Harding, B. W. (1968). *Biochem. Biophys. Res. Commun.* **32,** 921.
235. Van Lier, E., and Smith, L. L. (1970). *Biochim. Biophys. Acta* **218,** 320.
236. Burstein, S., Nama, C. O., Gut, M., Schleyer, H., Cooper, D. Y., and Rosenthal, O. (1972). *Biochemistry* **11,** 4.
237. Hall, P. F. (1967). *Biochemistry* **6,** 2794.
238. Hall, P. F. (1967). *Biochem. Biophys. Res. Commun.* **26,** 320.
239. Hall, P. F. (1968). *In* "Biogenesis and Action of Steroid Hormones" (R. I. Dorfman, K. Yamasaki, and M. Dorfman, eds.), Geron-X, Los Altos, California, p. 93.
240. Hall, P. F. (1970). *In* "The Androgens of the Testis" (K. B. Eik-Nes, ed.). Dekker, New York.
241. Ryan, K., and Engel, L. L. (1957). *J. Biol. Chem.* **225,** 103.
242. Matthijssen, C., and Mandel, J. E. (1967). *Biochim. Biophys. Acta* **146,** 613.
243. Matthijssen, C., and Mandel, J. E. (1970). *Steroids* **15,** 541.
244. Mackler, B., Haynes, B., Tattanoi, D. S., Tippit, D. F., and Kelley, V.-C. (1971). *Arch. Biochem. Biophys.* **145,** 194.
245. Lewis, A. M., and Bryan, G. T. (1971). *Life Sci.* **10,** 901.
246. Narasimhulu, S. (1971). *Arch. Biochem. Biophys.* **147,** 391.
247. Rosenthal, O., and Narasimhulu, S. (1969). *Methods Enzymol.* **15,** 597.
248. Cooper, D. Y., Narashimhulu, S., Rosenthal, O., and Estabrook, R. W. (1968). *In* "Functions of the Adrenal Cortex" (K. McKerns, ed.), Vol. 2, p. 897. Appleton, New York.
249. Young, R. B., and Sweat, M. L. (1967). *Arch. Biochem. Biophys.* **121,** 576.
250. Eik-Nes, K. B. (1970). *In* "The Androgens of the Testis" (K. B. Eik-Nes, ed.), p. 1. Dekker, New York.
251. Inano, H., Nakano, H., Shikita, M., and Tamaoki, B.-I. (1967). *Biochem. Biophys. Acta* **137,** 540.
252. Fonken, G. S., Murray, H. C., and Reineke, L. M. (1960). *J. Amer. Chem. Soc.* **82,** 5507.
253. Dorfman, R. I., and Ungar, F. (1965). *In* "Metabolism of Steroid Hormones," p. 411. Academic Press, New York.
254. Longchamp, J. E., Gual, C., Ehrenstein, M., and Dorfman, R. I. (1960). *Endocrinology* **66,** 416.
255. Wilcox, B. R., and Engel, L. L. (1964). *Steroids, Suppl.* **I,** 49.
256. Medhi, A. Z., and Snador, T. (1971). *Steroids* 143.
257. Meigs, R. A., and Ryan, K. J. (1968). *Biochim. Biophys. Acta* **165,** 476.
258. Meigs, R. A., and Ryan, K. J. (1971). *J. Biol. Chem.* **246,** 83.
259. Morato, T., Hayano, M., Dorfman, R. I., and Axelrod, L. R. (1961). *Biochem. Biophys. Res. Commun.* **6,** 334.
260. ₁Akhtar, M., and Skinner, S. J. M. (1968). *Biochem. J.* **109,** 318.
261. Morato, T., Raab, K., Brodie, H. J., Hayano, M., and Dorfman, R. I. (1962). *J. Amer. Chem. Soc.* **84,** 3764.
262. Fishman, J., Guzik, H., and Dixon, D. (1969). *Biochemistry* **8,** 4304.
263. Osawa, Y., and Spaeth, D. G. (1971). *Biochemistry* **10,** 66.
264. Morato, T., Hayano, M., Dorfman, R. I., and Axelrod, L. R. (1961). *Biochem. Biophys. Res. Commun.* **6,** 334.

265. Dorfman, R. I., Gual, C., Morato, T., Hayano, M., and Gut, M. (1962). *Abstr. Int. Congr. Hormonal Steroids, Milano Italy, May* p. 270.
266. Townsley, J. D., and Brodie, H. J. (1966). *Biochem. J.* **101**, 25c.
267. Townsley, J. D., and Brodie, H. J. (1968). *Biochemistry* **7**, 33.
268. Dorfman, R. I., and Ungar, F. (1965). "Metabolism of Steroid Hormones." Academic Press, New York.
269. Breuer, H., Knuppen, R., and Haupt, M. (1966). *Nature (London)* **212**, 76.
270. Breuer, H., Nocke, L., and Pangels, G. (1960). *Acta Endocrinol.* **34**, 359.
271. Breuer, H., Knuppen, R., Ortlepp, R., Pangels, G., and Puck, A. (1960). *Biochim. Biophys. Acta* **40**, 560.
272. Heinrichs, W. L., and Colas, A. (1968). *Biochemistry* **7**, 2273.
273. Sulcová, J., and Starka, L. (1968). *Steroids* **12**, 113.
274. Johansson, G. (1971). *Eur. J. Biochem.* **21**, 68.
275. Gustafsson, J.-Å., and Lisboa, B. P. (1970). *Acta Endocrinol.* **65**, 84.
276. Conney, A. H., and Klutch, A. (1963). *J. Biol. Chem.* **238**, 1611.
277. Lisboa, B. P., and Gustafsson, J.-Å. (1968). *Eur. J. Biochem.* **4**, 496.
278. Lisboa, B. P., and Gustafsson, J.-Å. (1968). *Eur. J. Biochem.* **6**, 419.
279. Lisboa, B. P., and Gustafsson, J.-Å. (1969). *Biochem. J.* **115**, 583.
280. Björkhem, I., Einarsson, K., Gustafsson, J.-Å., and Somell, A. (1972). *Acta Endocrinol.* **71**, 569.
281. Lisboa, B. P., and Gustafsson, J.-Å. (1969). *Eur. J. Biochem.* **9**, 402.
282. Starka, L., Sulcová, J., Dahm, K., Döllefeld, E., and Breuer, H. (1966). *Biochim. Biophys. Acta* **115**, 228.
283. Danielsson, H., and Johansson, G. (1972). *FEBS Lett.* **25**, 329.
284. Gustafsson, J.-Å., and Lisboa, B. P. (1970). *Eur. J. Biochem.* **15**, 525.
285. Burstein, S. (1968). *Endocrinology* **82**, 547.
286. Kuntzman, R. (1969). *Ann. Rev. Pharmacol.* **9**, 21.
287. Burstein, S., and Bhavnani, B. R. (1967). *Endocrinology* **80**, 351.
288. Conney, A. H., Levin, W., Ikeda, M., and Kuntzman, R. (1968). *J. Biol. Chem.* **243**, 3912.
289. Welsh, R. M., Levin, W., and Conney, A. H. (1968). *J. Pharm. Exp. Ther.* **160**, 171.
290. Kuntzman, R., Jacobson, M., Levin, W., and Conney, A. H. (1968). *Biochem. Pharm.* **17**, 565.
291. Kuntzman, R., Jacobson, M., Schneidman, K., and Conney, A. H. (1964). *J. Pharmacol. Exp. Ther.* **146**, 280.
292. Tephly, T. R., and Mannering, G. J. (1968). *Mol. Pharmacol.* **4**, 10.
293. Orrenius, S., Kupfer, D., and Ernster, L. (1970). *FEBS Lett.* **6**, 249.
294. Wada, F., Shimakawa, H., Takasugi, M., Kotake, T., and Sakamoto, Y. (1968). *J. Biochem. (Tokyo)* **64**, 109.
295. Kupfer, D., and Orrenius, S. (1970). *Eur. J. Biochem.* **14**, 317.
296. Lu, A. Y. H., Kuntzman, R., West, S., Jacobson, M., and Conney, A. H. (1972). *J. Biol. Chem.* **247**, 1727.
297. Gillette, J. R. (1971). *Metabolism* **20**, 215.
298. Estabrook, R. W., Hildebrandt, A. G., Baron, J., Netter, K. J., and Leibman, K. (1971). *Biochem. Biophys. Res. Commun.* **42**, 132.
299. Hildebrandt, A. G., and Estabrook, R. W. (1971). *Arch. Biochem. Biophys.* **143**, 66.
300. Ishimura, Y., Ullrich, V., and Peterson, A. (1971). *Biochem. Biophys. Res. Commun.* **42**, 140.
301. Björkhem, I. (1973). Unpublished observation.

302. Strobel, H. W., and Coon, M. J. (1971). *J. Biol. Chem.* **246,** 7826.
303. Björkhem, I. (1971). *Eur. J. Biochem.* **18,** 299.
304. Björkhem, I. (1972). *Eur. J. Biochem.* **27,** 354.
305. Björkhem, I., and Hamberg, M. (1972). *Biochem. Biophys. Res. Commun.* **47,** 333.
306. Danielsson, H. (1973). Mechanisms of Bile Acid Biosynthesis, *In* "The Bile Acids" (P. P. Nair and D. Kritchevsky, eds.). Plenum, New York, Vol.2., p. 1.
307. Wada, F., Hiraka, K., Nakano, K., and Sakamoto, Y. (1969). *J. Biochem.* (*Tokyo*) **66,** 699.
308. Shefer, S., Hauser, S., and Mosbach, E. H. (1972). *J. Lipid Res.* **13,** 69.
309. Danielsson, H., Einarsson, K., and Johansson, G. (1967). *Eur. J. Biochem.* **2,** 44.
310. Björkhem, I., and Danielsson, H. (1973). *Biochem. Biophys. Res. Commun.* **51,** 766.
311. Mitton, J. R., Scholan, N. A., and Boyd, G. S. (1971). *Eur. J. Biochem.* **20,** 569.
312. Scholan, N. A., and Boyd, G. S. (1968). *Biochem. J.* **108,** 27.
313. Boyd, G. S., Scholan, N. A., and Mitton, J. R. (1969). *Advan. Exp. Med. Biol.* **4,** 443.
314. Johansson, G. (1970). *Biochem. Pharmacol.* **19,** 2817.
315. Einarsson, K., and Johansson, G. (1969). *FEBS Lett.* **4,** 177.
316. Suzuki, M., Mitropoulos, K. A., and Myant, N. B. (1968). *Biochem. Biophys. Res. Commun.* **30,** 516.
317. Einarsson, K. (1968). *Eur. J. Biochem.* **5,** 101.
318. Mitropoulos, K. A., Suzuki, M., Myant, N. B., and Danielsson, H. (1968). *FEBS Lett.* **1,** 13.
319. Björkhem, I., and Gustafsson, J. (1973). *Eur. J. Biochem.* **36,** 201.
320. Mendelsohn, D., and Mendelsohn, L. (1969). *Biochem. J.* **114,** 1.
321. Berséus, O. (1965). *Acta Chem. Scand.* **19,** 325.
322. Mitropoulos, K. A., and Myant, N. B. (1965). *Biochem. J.* **97,** 25c.
323. Cronholm, T., and Johansson, G. (1970). *Eur. J. Biochem.* **16,** 373.
324. Voigt, W., Thomas, P. J., and Hsia, S. L. (1968). *J. Biol. Chem.* **243,** 3493.
325. Voigt, W., Fernandez, E. P., and Hsia, S. L. (1970). *Proc. Soc. Exp. Biol. Med.* **133,** 1158.

3

HEME-CONTAINING DIOXYGENASES

PHILIP FEIGELSON AND FRANK O. BRADY

I. INTRODUCTION

Relatively few species of heme proteins interact directly with molecular oxygen. These are limited to heme proteins whose iron is in the divalent state and possesses an available sixth coordination position. The electron transporting cytochromes satisfy the first criterion but not the second (1), and the heme-containing peroxidases satisfy the second but not the first (2). Four types of interaction of oxygen with heme proteins are known to occur in nature, exemplified by hemoglobin (carrier), cytochrome oxidase (reduction), cytochrome P-450 (monooxygenase), and tryptophan oxygenase (dioxygenase). Hemoglobin reversibly binds O_2 and serves its well-

87

known function as a carrier between the lungs and the tissues but does not physiologically or chemically activate oxygen. Cytochrome oxidase, which exists in mitochondria as a membrane-bound copper–heme complex (3) and in *Pseudomonas aeruginosa* as a soluble heme protein (3), acts as a terminal electron acceptor catalyzing electron flow to molecular oxygen and its consequent reduction to water. In contrast, the heme-containing mono- and dioxygenases are enzymes which catalyze the activation and insertion of one or both atoms of molecular oxygen into an organic substrate. Mono-oxygenases, e.g., cytochrome P-450, disrupt the oxygen molecule with one atom being incorporated into the organic substrate and the other being reduced to water by two reducing equivalents initially derived from reduced pyridine nucleotides. Hepatic and microbial L- and intestinal DL-tryptophan oxygenases are the only heme proteins known to catalyze a dioxygenase reaction, i.e., the insertion of both atoms of molecular oxygen into an organic substrate. This chapter, consequently, will be focused upon tryptophan oxygenases (EC 1.13.1.12).

II. HISTORY OF L-TRYPTOPHAN OXYGENASE

In 1931, Kotake and collaborators in Japan demonstrated the presence of a new compound in the urine of rabbits fed tryptophan which they identified as kynurenine (4). They were also the first to describe the conversion of L-tryptophan to L-kynurenine *in vitro* by extracts from rabbit liver. Since this involved rupture of the pyrrole ring of tryptophan they named the responsible enzyme system "tryptophan pyrrolase" (5). Knox and Mehler, in 1950, determined that two separable enzymes were involved in this reaction with the first enzyme catalyzing the conversion of trypto-phan to formylkynurenine and the second enzyme hydrolyzing formyl-kynurenine to kynurenine and formic acid. Since their studies suggested a requirement for H_2O_2 and a sensitivity to catalase, they renamed the enzyme "tryptophan peroxidase-oxidase" (6). Later, the inability of catalase to inhibit the ongoing catalytic reaction and the classic demonstration that $^{18}O_2$ was directly incorporated into formylkynurenine indicated that H_2O_2 served no catalytic function and the enzyme was renamed "tryptophan pyrrolase" (7, 8). With the growing appreciation that this was one of a class of enzymes that catalyze insertion of molecular oxygen into organic substrates (9, 10) the enzyme was rechristened "L-tryptophan oxygenase," the name it holds today; during all these etymologic transformations the unsuspecting enzyme was still catalyzing the oxidation of tryptophan to formylkynurenine (Fig. 1).

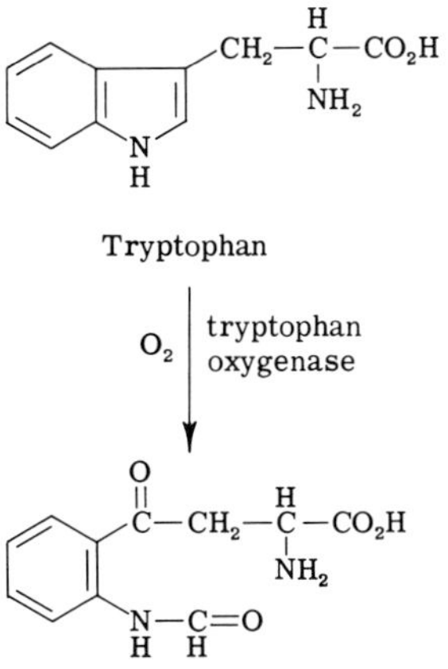

FIG. 1. Reaction catalyzed by tryptophan-2,3-dioxygenase.

The presence of an iron porphyrin component in partially purified prepa-rations of pseudomonad and hepatic tryptophan oxygenases was adduced by Tanaka and Knox from spectral and inhibitor studies (8). This was confirmed and extended by Feigelson and Greengard who purified catalyti-cally inert, hepatic apotryptophan oxygenase whose full catalytic activity was restorable by the addition of exogenous ferriprotoporphyrin IX (11–13). The hepatic and pseudomonad L-tryptophan oxygenases were purified to homogeneity and their spectral, chemical, and physical properties described (14–18). In 1965, Maeno and Feigelson reported metal analyses and ligand inhibitor studies which indicated the presence and catalytic participation of copper in purified tryptophan oxygenase (19). Following certain hesi-tations (15), this has been recently confirmed and extended indicating that the enzymes isolated from *Pseudomonas acidovorans* and rat liver are tetramers, containing two hemes and two copper moieties, which are catalytically essential (20, 21).

Considerable interest in tryptophan oxygenase was generated by a series of findings concerning the rapid augmentation of its catalytic activity in livers of animals, subsequent to parenteral administration of tryptophan (6, 22) or adrenal cortical hormones (23). The biochemical mechanisms underlying substrate and hormonal induction have been explored in depth by a variety of investigators. The substrate-induced increase in hepatic enzymic activity reflects a decrease in the rate of degradation *in vivo* caused by a tryptophan-mediated saturation of apoenzyme with its heme cofactor (13, 24–26), leading to its stabilization (27–29) and accumulation (30). Steroid mediated enzyme induction was found to result from a puromycin-sensitive (31) and actinomycin D-sensitive (32) increase in rate of enzyme synthesis (28), resulting largely in elevated levels of apoprotein (30). Other interesting aspects of the regulation of the catalytic activity of this enzyme involve its neonatal appearance and development (33, 34) and its genetically controlled role in invertebrate eye pigment formation (35–37).

In addition to physiological and developmental regulation in the cellular levels of tryptophan oxygenase, the existence of allosteric properties which determine the catalytic efficiency of existent enzyme molecules has become increasingly evident (15, 16, 29, 38–49). These positive and negative allosteric alterations determine the catalytic efficiency, conformational state, and the reactivity of the enzymic heme for its physiological substrate, oxygen, and for other heme ligands. The degree of protonation of the enzyme, a type of Bohr effect, determines the degree of allosteric interaction between the subunits of hepatic tryptophan oxygenase (45). Furthermore, the ability of another enzyme, i.e., xanthine oxidase (50–54) and chemical reductants (8, 13, 29, 55–58) to activate the enzyme and its allosteric inhibition by metabolites generated in the conversion of tryptophan to niacin all illustrate the multitudinous regulatory influences impinging upon this enzyme. Tryptophan oxygenase has thus evolved as an enzyme of considerable biochemical interest from the viewpoints of catalytic mechanism, allosteric and feedback regulation, substrate induction via stabilization, hormonal induction via control of gene expression, and conformational changes in the enzyme during catalysis and under the influence of allosteric effectors.

III. BIOLOGICAL ROLE AND DISTRIBUTION OF TRYPTOPHAN OXYGENASE

The classic nutrition studies from the laboratory of Elvehjem (59, 60) demonstrated that tryptophan could replace nicotinic acid as a cure for pellagra in man and black tongue in dogs. This was confirmed with the demonstration that [14C]tryptophan was metabolically converted to [14C]-nicotinic acid in vivo (61, 62) and that tryptophan administration to rodents led to augmented hepatic NAD levels (63, 64). Tryptophan oxygenase catalyzes the first committed reaction in the metabolic conversion of tryptophan to nicotinic acid and the pyridine nucleotides. This enzyme is subject to elaborate regulatory controls which will be reviewed below. Liver is the only mammalian organ known to contain L-tryptophan oxygenase activity, and this enzyme may be considered as a specific marker for hepatic tissue (65). That the presence of tryptophan oxygenase is a subtle reflection of the normal, fully differentiated liver cell is testified by the progressively decreasing levels of this enzyme in a series of Morris hepatomas of increasing malignancy (66–68). A variety of hepatoma cells in tissue culture seem to have lost the capability of synthesizing tryptophan oxygenase as a consequence of their dedifferentiation.

In addition to being present in cells which catalyze the conversion of L-tryptophan to the pyridine moiety of NAD and NADP, a tryptophan oxygenase capable of acting on D-tryptophan as well as L-tryptophan has been isolated from rabbit intestine (69, 70). L-Tryptophan oxygenase is present in a variety of microorganisms which catabolize tryptophan with the generation of energy and biologically available nitrogen (71, 72). Thus, various microorganisms may utilize L-tryptophan as the primary carbon and nitrogen source for growth. *Pseudomonas acidovorans* (ATCC 11299b) develops the capability of metabolizing L-tryptophan by inducing the synthesis of L-tryptophan permease, L-tryptophan oxygenase, formyl-kynurenine formamidase and subsequent enzymes in this metabolic pathway (73, 74). It is of interest to note that L-kynurenine is capable of the feedback induction of L-tryptophan permease and of L-tryptophan oxygenase. It is uncertain whether L-tryptophan itself or its metabolite L-kynurenine is the physiological inducer for the synthesis of these enzymes in *Pseudomonas acidovorans* (73). The ability of L-kynurenine to feedback induce the synthesis of the specific L-tryptophan permease may be the first reported instance of metabolite feedback induction of a permease-transport system (74).

Tryptophan oxygenase is also present in invertebrates where its genetic control influences the proportion of the eye pigments derived from kynurenine (35–37).

IV. PHYSICOCHEMICAL PROPERTIES OF TRYPTOPHAN OXYGENASE

A. Purification

Pseudomonas acidovorans, grown on tryptophan, and rat liver of animals previously injected with tryptophan alone or tryptophan in concert with glucocorticoid hormones possess elevated (induced) levels of tryptophan oxygenase and have routinely served as sources for its purification. Tryptophan oxygenase is readily extracted with neutral salt solutions from sonicated, L-tryptophan-induced cells of *Pseudomonas acidovorans* (ATCC 11299b) and is found entirely in the high-speed supernatant (100,000 g) fraction of the hepatic cytosol. The enzymes from both sources appear to be "soluble" proteins and can be purified by classic procedures. In the absence of its substrate, L-tryptophan, the enzyme is very labile during purification, but in the presence of saturation levels of L-tryptophan the enzyme is stabilized and even withstands a thermal denaturation step during its purification (15, 20). Orthodox procedures have been used for the purification of tryptophan oxygenase involving, for the pseudomonad enzyme, sonication of the cells, high-speed centrifugation, treatment with streptomycin sulfate to remove nucleic acids, followed by ammonium sulfate fractionation, a heat step, and DEAE-cellulose chromatography. Preparative acrylamide disc-gel electrophoresis followed by physical excision of and extraction of the brown, cuproheme enzyme protein band yields homogeneous preparations of the pseudomonad enzyme (15). A somewhat modified procedure involving isoelectric precipitation, thermal activation (75) of the enzyme and denaturation of contaminent proteins, ion-exchange chromatography, and preparative acrylamide disc-gel electrophoresis yields homogeneous hepatic tryptophan oxygenase (20) with congruity of catalytic activity and the single protein band obtained upon disc acrylamide electrophoresis (Fig. 2). Tryptophan oxygenase is isolated entirely as holoenzyme only if hematin (5 μM) is added to the enzyme solution prior to the thermal denaturation step; omission of exogenous hematin results in loss of enzymic heme to other heme-binding denatured proteins and

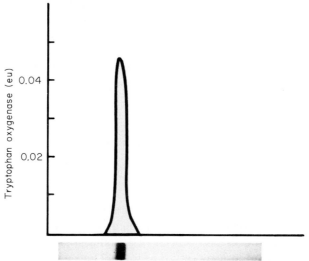

FIG. 2. Polyacrylamide disc-gel electrophoresis of hepatic tryptophan oxygenase. (Bottom) Gel stained for protein. (Top) Replicate gel stained for enzyme activity and scanned spectrophotometrically at 321 nm (20).

subsequent isolation of tryptophan oxygenase as a mixture of apo- and holoenzyme (12, 13). The enzymes from both sources are quite stable when stored anaerobically in 0.1 M K phosphate buffer, pH 7.0, containing L-tryptophan.

B. Spectral and Physicochemical Properties

L-Tryptophan oxygenases of microbial and mammalian origin are heme-copper proteins. Concentrated solutions of the enzyme are brown and exhibit absorption spectra in the visible and the ultraviolet typical of ferriheme proteins (15, 20). These spectra with high Soret to α,β-absorption ratios are typical of high spin ($d_{5/2}^5$) ferriheme proteins (76) such as cytochrome c peroxidase, horseradish peroxidase, and ferrimyoglobin (77). Table I lists the absorption maxima (λ_{max}, nm) and extinction coefficients (ϵ_{mM}) for the spectra of the pseudomonad and rat liver enzymes and some of their derivatives. As yet, no absorption bands are assignable to the copper moieties of the enzymes; this may be a result of the expected low absorption intensities of this cofactor as compared with the heme absorption.

As can be seen in Table I, the pseudomonad (Fig. 3) and the rat liver (Fig. 4) L-tryptophan oxygenases have distinctly different absorption spectra. This is especially obvious for the ferriheme hepatic enzyme (Fig. 4) in which the α and β bands are confluent, exhibiting a broad absorption

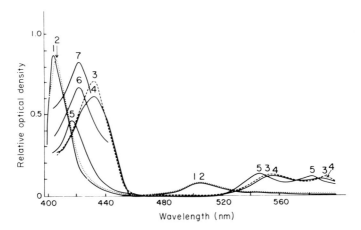

FIG. 3. Absorption spectra of pseudomonad tryptophan oxygenase (93). 1 (——) ferriheme, 2 (···) ferriheme plus L-tryptophan, 3 (- - -) ferroheme, 4 (——) ferroheme plus L-tryptophan, 5 (——) ferroheme plus L-tryptophan plus oxygen (80), 6 (——) ferroheme plus CO, and 7 (——) ferroheme plus L-tryptophan plus CO.

TABLE I

SPECTRAL AND PHYSICOCHEMICAL PROPERTIES OF L-TRYPTOPHAN-2,3-DIOXYGENASE

Absorption spectra $\dfrac{\lambda_{max}\,(nm)}{(\epsilon_{mM})}$		Pseudomonad				Ref.
Ferriheme	405 (340)	502 (20.5)	534s	580 (6.5)	632 (8.7)	15
Ferriheme + Trp	407 (320)					15
Ferriheme + CN	419 (235)	537 (18)	570s			43
Ferroheme	432 (295)	510 (14)	554 (19.5)	588 (14.3)		44
Ferroheme + Trp	432 (190)	510	556	585		82
Ferroheme + CN	432 (280)	535s	564 (19.5)	590 (14.5)		43
Ferroheme + CO + Trp	421 (350)	537 (27)	564 (28)			44
Ferroheme + O_2 + Trp	418 (300)	545 (22)	580 (16)			82

Absorption spectra $\dfrac{\lambda_{max}\,(nm)}{(\epsilon_{mM})}$		Rat liver			Ref.
Ferriheme	406 (353)	500 (20)	540s	630 (8.5)	20
Ferriheme + Trp	408 (290)	500 (16)	530 (12)	590 (10.5)	20
Ferroheme	420 (140)	555 (12)			20
Ferroheme + Trp	423 (120)	540 (10)			20

centered at 555 nm. Addition of the substrate, L-tryptophan (3 mM), causes slight hypochromic red shifts in the Soret peak (405 nm → 407 nm for the pseudomonad enzyme; 406 nm → 408 nm for the hepatic enzyme) and more subtle changes in the α,β region for the ferriheme enzymes. Reduction of the enzymes with dithionite yields spectra with a Soret maximum at 432 nm for the pseudomonad enzyme and 420 nm for the hepatic enzyme. The absorption spectra of the cyanide derivative of ferri- and ferro-pseudomonad tryptophan oxygenase are shown in Fig. 5. These

TABLE I (Continued)

	Pseudomonad	Ref.	Rat liver	Ref.
Molecular weight	122,000	15	167,000	20
Subunits	4	15	4	20
Subunit molecular weight	31,000	15	43,000	20
Subunit structure	(?)	15	$\alpha_2\beta_2$	20
$s_{20,w}$ (+Trp)	6.4	15	7.6	20
Cofactor content				
Heme	2	21	2	20
Copper	2	21	2	21
Specific activity (μmoles formylkynurenine formed per minute per mg protein at 20°)	17.0	15	2.5	20
Turnover number (moles formylkynurenine per minute per mole enzyme)	2070		420	
K_m (L-Trp), with 0.25 mM O_2	300 μM (pH 7.0)		250 μM (pH 8.0)	
K_m (O_2), with 10 mM L-Trp	38 μM (pH 7.0)		37 μM (pH 8.0)	
Reactivity to antipseudomonad tryptophan oxygenase antibodies	+		−	
Reactivity to antihepatic tryptophan oxygenase antibodies	−		+	

Amino acid composition	Pseudomonad	Ref.	Rat liver	Ref.
Acidic residues	20%	15	25%	20
Basic residues	6%		13%	
Aliphatic residues	63%		49%	
Aromatic residues	11%		13%	
Half-cystine residues	0		6	
Tyrosine residues	26		56	
Phenylalanine residues	29		75	
Tryptophan residues	16		17	
Histidine residues	27		41	
Methionine residues	24		9	
Total number residues/mole tetramer	1060		1488	

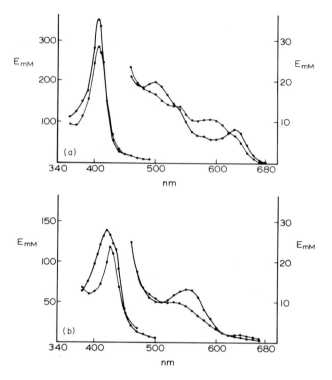

Fig. 4. Absorption spectra of rat liver tryptophan oxygenase (20). (a) Ferriheme minus (●—●) and plus (●- - -●) L-tryptophan. (b) Ferroheme minus (●—●) and plus (●- - -●) L-tryptophan.

spectra are reminiscent of the cyano-horseradish peroxidase complex first reported by Keilin and Hartree (78, 79). Figure 6 depicts the spectral changes accompanying the conversion of pseudomonad tryptophan oxygenase to its carbon monoxide derivative. As is well known, CO, like O_2, combines only with divalent and not with trivalent heme. Reduction of the enzyme with dithionite yields the ferroheme form of the enzyme, which, when equilibrated with gas mixtures of varying proportions of CO, results in the progressive conversion of increasing proportions of the enzyme to the carbon monoxide derivative. The presence of L-tryptophan markedly increases the affinity of the enzymic ferroheme for CO (39).

Hayaishi and co-workers have reported the spectrum of an oxygenated form of pseudomonad tryptophan oxygenase (80–82). This spectrum possesses maxima (418, 545, and 580 nm) similar to those of the oxygenated ferroheme forms of hemoglobin, myoglobin, peroxidase, and cytochrome P-450 (83). It was obtainable in the presence of saturating amounts

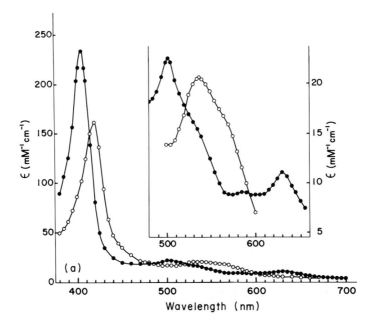

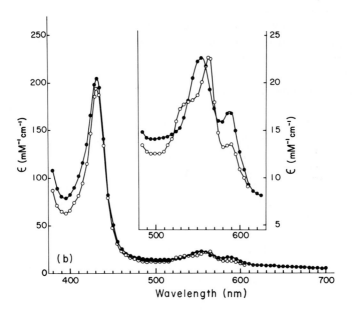

Fig. 5. Absorption spectra of cyanide derivative of pseudomonad tryptophan oxygenase (43). (a) Ferriheme and (b) ferroheme, minus (●–●) and plus (○–○) KCN.

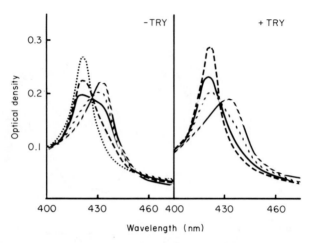

FIG. 6. Effect of L-tryptophan upon the formation of the carbon monoxide complex of pseudomonad tryptophan oxygenase (39). Left, minus L-tryptophan plus CO, 0% (···—), 8% (- - -), 30% (——), 100% (— —), and 100% plus L-tryptophan (···). Right, plus L-tryptophan plus CO, 0% (···—), 1% (- - -), 4% (——), and 100% (— —).

of L-tryptophan, 100% O_2 (continuously bubbled) at low temperature (5°) after reduction with dithionite.

The molecular weights of the pseudomonad and hepatic enzymes were evaluated by sedimentation equilibrium and velocity analytic ultracentrifugation, Sephadex chromatography, and sucrose gradient centrifugation and found to be 122,000 (15) and 167,000 (20), respectively. Dissociation of the native enzyme molecules with sodium dodecyl sulfate, guanidine hydrochloride, urea, or high pH yields subunits of molecular weight of 31,000 for the pseudomonad and 43,000 for the hepatic enzymes (15, 20). Thus, each of the enzymes are tetramers of subunits of the same size. Recent studies indicate that the subunits of hepatic tryptophan oxygenase are nonidentical and separable into two distinct species by virtue of their differential mobilities on a polyacrylamide gel at pH 12.5. The subunit composition of the hepatic enzyme is $\alpha_2\beta_2$, composed of α and β subunits of approximately 43,000 molecular weight (20). Whether the subunits of the pseudomonad enzyme are identical or distinguishable into two distinct species remains uncertain at the present time. The amino acid compositions of the two enzymes (Table I) are distinctly different emphasizing that although the reaction catalyzed may be the same the proteins are quite different. Particularly, note the differences in half-cystine and methionine residues in each enzyme and the absence of cross reactivity with their respective antibodies.

Analyses of wet ashed samples of the enzyme for iron content employing bathophenanthroline, atomic absorption spectra, or spectrophotometric estimation of the heme concentration using the dipyridine hemochromogen method are compatible with the presence of 2 moles of heme per tetramer of the pseudomonad and hepatic enzymes (20, 21). The dipyridine hemochromogen derivative of the heme of tryptophan oxygenase is spectrally indistinguishable and apparently identical with the dipyridine hemochromogen derivative of authentic ferroprotoporphyrin IX (82). The functional role of the heme in catalytic processes is exemplified by (a) restoration of catalytic activity of hepatic apotryptophan oxygenase preparations upon the addition of hematin (13, 57); (b) the fact that various heme ligands, e.g., CO, CN, azide, inhibit the catalytic activity (8) and that the inhibition by carbon monoxide is competitive with oxygen (39); and (c) the photochemical action spectrum of the carbon monoxide tryptophan oxygenase complex possesses a maximum at 421 nm which is indistinguishable from the absorption spectrum of the L-tryptophan–carbon monoxide–ferroheme enzyme ternary complex (39). The competitive reversal by oxygen of the carbon monoxide-inhibited tryptophan oxygenase activity in conjunction with the photochemical action spectrum implicate enzymic heme iron as a site of binding of carbon monoxide and oxygen and demonstrate the requirement that the sixth ligand position of enzymic heme must be available for catalytic function, presumably being a site of combination and activation of molecular oxygen prior to its insertion into tryptophan (39).

The existence of two forms of pseudomonad tryptophan oxygenase both with ferriheme absorption spectra, only one of which was catalytically active in the absence of exogenous reductants, indicated the existence of an enzymic component other than heme which when oxidized yielded catalytically inactive enzyme (19). By analogy with cytochrome oxidase we suspected the possibility that L-tryptophan oxygenase might contain copper as well as heme. In 1965, Maeno and Feigelson reported that during purification of pseudomonad tryptophan oxygenase enrichment with respect to copper content paralleled purification of the enzyme (19). Furthermore, the copper complexing agents bathocuproinesulfonate, cuprizone, salicylaldoxime, and diethyldithiocarbamate selectively inhibited tryptophan oxygenase activity. More recent studies with homogeneous preparations of pseudomonad and rat liver L-tryptophan oxygenases confirm the presence and functional role of copper in tryptophan oxygenase (21). As shown in Fig. 7, Sephadex G-200 column chromatography of a homogeneous preparation of the pseudomonad enzyme indicates a relatively constant ratio of copper content to enzymic activity across the elution peak. Using the molecular weight of 122,000, the data in Fig. 7 indicate

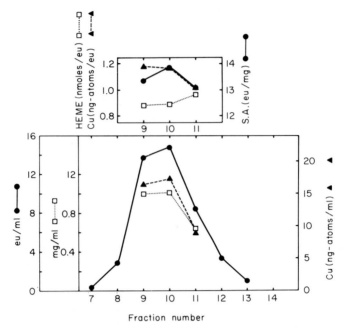

FIG. 7. Sephadex G-200 chromatography of homogeneous pseudomonad tryptophan oxygenase (21). Assayed for protein (□···□). copper (▲—▲), and enzymic activity (●—●).

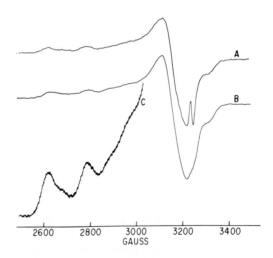

FIG. 8. EPR spectra of Cu^{2+} signal of pseudomonad tryptophan oxygenase (91). Temperature, $-182°$. A, modulation amplitude, 10 gauss; B, 40 gauss; and C, $\times5$.

the presence of 1.97 gm-atoms of copper per tetramer of pseudomonad tryptophan oxygenase. Parallel studies with hepatic tryptophan oxygenase also indicate 2 moles of copper per mole tetrameric enzyme.

The kinetic characteristics of pseudomonad and hepatic tryptophan oxygenase are somewhat similar, the most conspicuous difference being the fivefold difference in turnover number (Table I). The Michaelis constant for the substrates, L-tryptophan (250–300 μM) and O_2 (37–38 μM), are essentially identical for the two enzymes. As will be discussed later, both enzymes are affected by allosteric modifiers, each enzyme possessing its own characteristics.

Electron paramagnetic resonance (EPR) studies with homogeneous pseudomonad tryptophan oxygenase, which had been treated with ferricyanide to oxidize all the copper present to the paramagnetic Cu^{2+}, yielded the EPR spectra shown in Fig. 8. These EPR spectra indicate, in addition to a high spin ferric heme ($g = 6.2$) signal (not shown), the presence of a Cu^{2+} signal with $g_\perp = 2.065$, $g_{||} = 2.265$, and $A = 170$ gauss, typical of

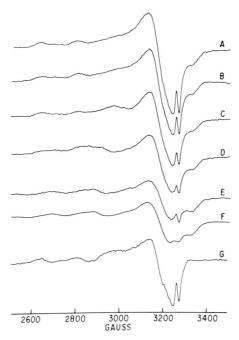

FIG. 9. Stoichiometric cyanide titration of Cu^{2+} EPR signal of pseudomonad tryptophan oxygenase (91). Copper concentration, 0.6 mM; cyanide concentration (A to E), 0, 0.1 mM, 0.3 mM, 0.6 mM, 1.1 mM; G, spectrum A minus spectrum E. Under the conditions of this experiment there was no change in the high spin ferriheme EPR signal, i.e., no cyanoheme complex was formed.

the so-called nonblue copper proteins (84). The chemically oxidized enzyme depicted in Fig. 8 possessed a 100% requirement for an exogenous reductant such as ascorbate to elicit maximal enzymic activity. It is also of interest to note that stoichiometric levels of cyanide will combine with and titrate the copper moieties of tryptophan oxygenase (Fig. 9), whereas higher cyanide levels are necessary to yield EPR or optical spectral changes of the heme moieties of the enzyme.

The activation studies to be subsequently described in this chapter will indicate that catalytically active enzyme contains Cu^+ moieties and that the degree to which they have undergone oxidation to the cupric (Cu^{2+}) state is the degree to which the enzyme has developed a dependence upon exogenous reductants to realize full catalytic activity.

The oxidation reduction midpoint potential of the heme of pseudomonad tryptophan oxygenase was determined by the method of Wilson and Dutton (85–87). This potential (E_M, 7.0) was determined by observing the oxidation–reduction potential (E_H) dependence of the absorbance change from 410 to 430 nm which occurred when the heme of tryptophan oxygenase was reduced at pH 7.0. A plot of E_H (mV) versus the logarithm of the ratio

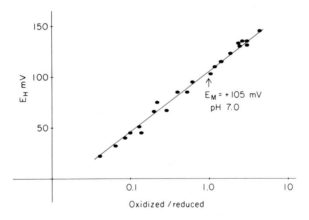

FIG. 10. Redox potential of the heme of pseudomonad tryptophan oxygenase. Enzyme (2 μM, specific activity, 11.0) in 0.2 M KP$_i$, pH 7.0 was titrated anaerobically (argon) by the method of Wilson and Dutton (85–88). Mediators used were pyocyanine (4 μM), phenazine methosulfate (20 μM), phenazine ethosulfate (20 μM), N,N,N',N'-tetra-methyl-p-phenylenediamine (10 μM). Ferricyanide was used as an oxidant and dithionite as a reductant. Titration was at 20° and took 1 hour to complete. Absorbances at 410 and 430 nm were measured to calculate ratio of oxidized to reduced heme. Sequence of the experiment was (1) reduction with dithionite, (2) stepwise reoxidation with ferricyanide, and (3) stepwise rereduction with dithionite. Midpoint potential of the heme of tryptophan oxygenase was +105 mV. This experiment was performed by Dr. David F. Wilson of the Johnson Foundation, University of Pennsylvania.

of the oxidized form to reduced form is shown in Fig. 10. The midpoint potential, E_M, 7.0, for the heme in tryptophan oxygenase (with $n = 1$) is $+105$ mV. This potential is more negative than the potentials reported for the hemes of cytochrome oxidase [$a = +375$ mV; $a_3 = +230$ mV (88)], cytochrome c ($+260$ to $+290$ mV), hemoglobin ($+170$ mV), and cytochrome b_2 [$+120$ mV (89)], but more positive than cytochrome b [$+35$ mV (87)]. Since the binding site of O_2 on tryptophan oxygenase is presumably ferroheme, the E_M, 7.0, of $+105$ mV indicates that the ferroheme is potentially capable of donating an electron to a bound O_2 during catalysis.

V. CATALYTIC MECHANISM

A. Reductive Activation and the Role of Copper

During early studies on tryptophan oxygenase the catalytic activities of crude extracts, as well as partially purified preparations of both the hepatic (6) and microbial (72) enzymes, were found to be stimulated by hydrogen peroxide generating systems. After discovering that H_2O_2 was not a substrate in the catalytic reaction, it was proposed by Knox that its role was in the reduction of the catalytically inactive ferriheme cofactor to the catalytically active ferroheme state (8). Subsequent reports documented the ability of a variety of reductants to activate tryptophan oxygenase: notably, sodium ascorbate (8, 13, 29, 55–58), NaBH₄ (56), sodium dithionite (13, 56), dithiothreitol (14), xanthine and xanthine oxidase (54), and liver mitochondria and microsomes (90). The assumption that all these systems functioned exclusively by converting ferriheme to ferroheme was shown to be unwarranted by Maeno and Feigelson (14). They demonstrated that although the pseudomonad enzyme could be converted to the ferroheme form by treatment with ascorbate in the presence of L-tryptophan, after removal of these agents by Sephadex G-25 gel filtration, the heme of the enzyme autoxidized to the trivalent state and the resulting enzyme was catalytically active and manifested no lag phase. It was concluded that reduction of a component of the pseudomonad enzyme other than the heme prosthetic group was involved in its reductive activation. The complete absence of cysteine and cystine in the pseudomonad enzyme precluded disulfide reduction as the responsible parameter (15) (Table I).

A recent study in this laboratory (74) demonstrated that the variety of reductants which had been reported to be activators of pseudomonad and rat liver tryptophan oxygenases functioned by any combination of three mechanisms depending on the reductant: direct reduction and/or indirect reduction via superoxide (O_2^-), and/or indirect reduction via hydrogen peroxide. The relative contribution of each process was determined by observing whether or not superoxide dismutase and catalase (separately or in concert) were capable of preventing activation by a given reductant. The nonspecific nature of the mechanism of reductive activation led us to infer that the requirement for a reducing agent to manifest maximal enzymic activity was the result of an oxidative inactivation occurring during purification and handling of the enzyme. It was clear that enzyme kept continually in the presence of saturating amounts of L-tryptophan did not manifest a requirement for reductive activation.

These considerations motivated the search for another cofactor in tryptophan oxygenase; the subsequent discovery (19), "undiscovery" (15), and rediscovery [Fig. 7, Table I (21)] of copper in both pseudomonad and rat liver tryptophan oxygenases provided an appropriately oxidizable–reducible candidate. Recent evidence suggests that the catalytically active species of tryptophan oxygenase possess a particular protein configuration and oxidation reduction state depicted as $E(Cu^+, Fe^{3+})$ or its valence isomer $E(Cu^{2+}, Fe^{2+})$. When an enzyme sample is allowed to become oxidatively inactivated, it is oxidized to $E(Cu^{2+}, Fe^{3+})$, the species of the enzyme which requires exogenous reductants for maximal catalytic activity. Use of the metal complexing agents bathocuproinesulfonate (BCS) (Cu^+ specific), diethyldithiocarbamate (Cu^{2+} specific), and bathophenanthrolinesulfonate (BPS) (Fe^{2+} and Cu^+ chelator) helped determine the valence state of copper in active and inactive tryptophan oxygenase (21).

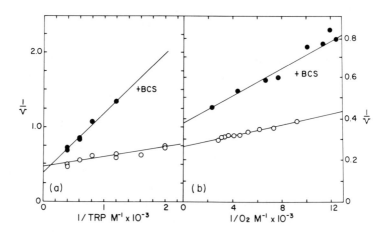

FIG. 11. Lineweaver-Burk plots of the inhibition of enzyme activity of pseudomonad tryptophan oxygenase by bathocuproinesulfonate (BSC) (21). (a) BCS (0.21 mM) and (b) BCS (0.42 mM). K_i (BCS) calculated from (a) was 31 μM.

Bathocuproinesulfonate inhibits the ongoing catalytic reaction, competitively with L-tryptophan and noncompetitively with O_2, indicating the presence of Cu^+ in active enzyme molecules (Fig. 11). Bathophenanthrolinesulfonate behaves similarly. Diethyldithiocarbamate does not inhibit an ongoing catalytic reaction, but it does prevent reductive activation, indicating the presence of Cu^{2+} in inactive enzyme molecules. The enzyme–BCS (Fig. 12) and enzyme–BPS complexes are spectrally indistinguishable from authentic BCS–Cu^+ and BPS–Cu^+ complexes in solution, possessing absorption maxima at 475 and 460 nm, respectively. Quantitative determination of the proportions of Cu^+ and Cu^{2+} in partially oxidized samples of pseudomonad tryptophan oxygenase by EPR spectroscopy (91) and enzyme–BCS complex formation (Fig. 13) confirmed that the ratio of reductant-dependent to reductant-independent catalytic activity is identical to the ratio of Cu^{2+} to Cu^+ forms of the enzyme (21).

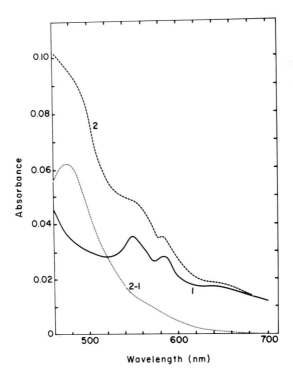

FIG. 12. Difference spectrum (curve 2-1) of pseudomonad tryptophan oxygenase (reduced with dithionite) (curve 1) and its complex with BCS (curve 2) (21).

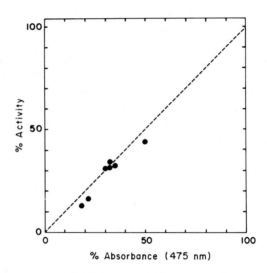

FIG. 13. Plot of percent activity (minus ascorbate) as a function of the amount of BCS difference spectrum (minus dithionite) for pseudomonad tryptophan oxygenase (21).

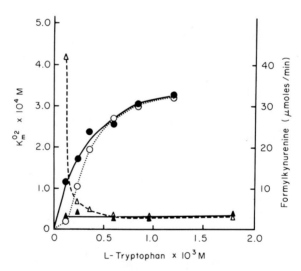

FIG. 14. A plot of velocity versus L-tryptophan concentration for pseudomonad tryptophan oxygenase, plus ($\bullet$–$\bullet$) and minus ($\bigcirc$···$\bigcirc$) α-methyltryptophan. Also, a plot of $K_m^{O_2}$ as a function of L-tryptophan concentration, plus ($\blacktriangle$–$\blacktriangle$) and minus ($\triangle$- - -$\triangle$) α-methyltryptophan (38).

B. Kinetic Properties

Pseudomonad tryptophan oxygenase manifests allosteric behavior (see below) when examined kinetically. As shown in Fig. 14, a plot of catalytic velocity as a function of L-tryptophan concentration results in a non-hyperbolic, sigmoidal curve (38, 41). The addition of DL-α-methyltrypto-phan (1.0 mM) to this system shifts the curve to the left, changing the curve to a hyperbolic shape and lowering the concentration of L-tryptophan required for half-maximal velocity. α-Methyltryptophan is not a substrate and at this concentration is not an inhibitor of the catalytic reaction. These findings are compatible with the existence of two types of tryptophan binding sites, a catalytic site(s) and a regulatory site(s). At all levels of L-tryptophan plots of catalytic velocity versus O_2 concentration are hyperbolic.

Three alternative sequences exist by which L-tryptophan and O_2 may combine with the catalytic site(s) of the enzyme [Eqs. (1)–(3)].

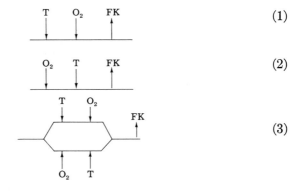

Each of these schemes was evaluated kinetically by making use of α-methyl-tryptophan to saturate the allosteric site(s) and of CO as an inhibitor of the catalytic reaction. The results shown in Fig. 15 indicate that CO is a competitive inhibitor with respect to O_2 and an uncompetitive inhibitor with respect to L-tryptophan for the pseudomonad enzyme (39, 92). These results are kinetically consistent only with Eq. (1), a bi-uni, ordered mechanism in which L-tryptophan binds before O_2 (92).

C. Binding Sites for L-Tryptophan and O_2

By analogy with other copper and heme proteins and chemical models, both L-tryptophan and O_2 may potentially combine with either or both the

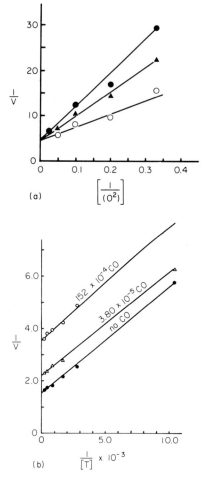

FIG. 15. Lineweaver-Burk plots of the inhibition by carbon monoxide of pseudomonad tryptophan oxygenase. (a) Oxygen varied (39); (b) L-tryptophan varied (92). K_i^{CO} calculated from (a) was 70 μM.

heme and copper moieties of tryptophan oxygenase. The following lines of evidence indicate that L-tryptophan may interact with both the copper and heme moieties:

1. The Cu^{2+} complexing agent, BCS, is a competitive inhibitor with respect to L-tryptophan and a noncompetitive inhibitor with respect to O_2 of the catalytic reaction (Fig. 11).

2. L-Tryptophan is capable of reductive activation (55).

3. L-Tryptophan causes a hypochromic shift of the heme Soret peak (Table I, Figs. 2 and 3).

4. L-Tryptophan causes a transition of one of the two high spin ferrihemes to a low spin ferriheme (91).

Evidence that O_2 binds to heme is as follows.

1. CO is a competitive inhibitor with respect to O_2 of the catalytic reaction, generating an enzyme–CO complex with a typical heme-CO absorption spectrum (Fig. 15a). The photodissociation action spectrum of CO-inhibited enzyme with concomitant recovery of catalysis corresponds to the enzyme–CO absorption spectrum [Fig. 16 (39)].

2. An "oxygenated" ferroheme catalytic intermediate (See Table I) has been reported (80–82).

Evidence that O_2 may interact with copper:

1. Oxidative inactivation of tryptophan oxygenase results in the conversion of Cu^+ to Cu^{2+}, suggesting that a molecule of O_2 oxidizes the copper and escapes as O_2^-.

2. Other copper enzymes and proteins are known to interact with O_2 (84).

It is apparent from the preceding statements that the binding sites of L-tryptophan and O_2 on the enzyme have not yet been unambiguously established. The preponderant existent evidence would suggest that at the active site one substrate, tryptophan, binds to the cuprous moiety and the

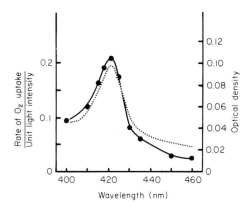

FIG. 16. Photochemical action spectrum of the carbon monoxide complex of pseudomonad tryptophan oxygenase (39). Absorbance spectrum (- - -) and action spectrum (●—●). Performed with dithionite-reduced enzyme in the presence of L-tryptophan.

other substrate, oxygen, binds to the heme iron. It seems quite possible that in the activated transition complex either or both substrates may serve as bridging ligands between the heme and copper moieties, bringing the entire system into electronic poise. Interaction of this nature may generate enzyme-bound electrophilic tryptophan and nucleophilic O_2 species which would interact to form the reaction product, formylkynurenine.

D. Stereochemical Requirements for Substrate and Inhibitor Specificity

In order to determine the stereochemical requirements at the catalytic site(s) of pseudomonad and rat liver tryptophan oxygenases for binding the substrate, L-tryptophan, analogs of tryptophan were examined for their ability to serve as substrates or inhibitors of the two enzymes. D-Tryptophan is neither a substrate nor an inhibitor of either enzyme. Most analogs were used as DL mixtures; it was assumed that only the L form bound, and all calculations shown in Table II were based on this assumption.

As can be seen in Table II, only those analogs which have a ring substitution at the 6 position can serve as substrates; a ring substitution at the 4 or 5 position renders the analog an inhibitor. This may be because of a change in the electron densities at N-1, C-2, and C-3 caused by the type of substitution. In certain instances steric hindrance may be caused by the introduction of a bulky substituent on the indole ring. However, steric considerations seem unlikely with respect to fluorine substitutions for hydrogen. Side chain substitutions result in analogs which are only inhibitors. It is apparent from the data in Table II that the electron density environment of the indole ring of L-tryptophan, as well as the stereochemistry and constitution of the side chain, are important in determining binding and catalysis by pseudomonad and rat liver tryptophan oxygenases (93).

E. Valence State during Catalysis

Two alternate hypotheses have been promulgated concerning the valence state of the heme of tryptophan oxygenase during catalysis. The first was originally proposed by Tanaka and Knox (8) and later supported by Hayaishi (80–82, 94); it considered that only the ferroheme form of the enzyme was catalytically active and that it remained as ferroheme during catalysis. Hayaishi stated: "Ferrous heme in the enzyme is first activated by tryptophan and then reacts with oxygen to form an intermediary ternary complex. Both substrates, tryptophan and oxygen, are activated in the

TABLE II

STEREOSPECIFICITY OF CATALYTIC SITE[a]

K_m for Substrates (mM)		
	Pseudomonad	Rat liver
L-Tryptophan	0.30	0.25
DL-6-Fluorotryptophan	0.72	0.10
DL-6-Methyltryptophan	b	0.24
D-Tryptophan	b	b

K_i for Inhibitors (mM)		
	Pseudomonad	Rat liver
Ring substitution		
DL-4-Fluorotryptophan	1.0	0.33
DL-4-Methyltryptophan	0.67	0.28
DL-5-Fluorotryptophan	0.39	0.12
DL-5-Methyltryptophan	0.55	0.19
L-5-Hydroxytryptophan	0.001	0.01
D-5-Hydroxytryptophan	c	c
DL-5-Methoxytryptophan	>5.0	>5.0
Side chain substitution		
Indole	2.7	0.14
Indolepropionic acid	0.21	0.08
Tryptamine	1.8	0.10

[a] Data from Forman (93).
[b] Not a substrate.
[c] Not an inhibitor.

complex and interact, yielding the product, formylkynurenine" (94). Knox's view was: "The role of peroxide in the reaction was demonstrated to be its conversion of the inactive ferric enzyme in the presence of substrate into the active ferrous form of the enzyme" (8). This may be formulated as

$$E(Fe^{3+}) \xrightarrow{\text{ }} E(Fe^{2+}) \xrightarrow{+Trp} E(Fe^{2+})-Trp \xrightarrow{+O_2} E(Fe^{2+})-Trp-O_2 \rightarrow E(Fe^{2+}) + FK \quad (4)$$

Inactive (red) Active

Our studies on reductive activation indicated that a ferriheme form of the enzyme could be active depending on the valence state of the copper cofactor [E(Cu$^+$, Fe^{3+}) is active and E(Cu^{2+}, Fe^{3+}) is inactive (14, 39, 56,

91)]. Appreciating the fact that O_2 can only bind to ferroheme and not to ferriheme, this laboratory proposed that the heme of tryptophan oxygenase underwent an oscillation in valence during the catalytic cycle. "During catalysis the heme iron atom of the enzyme undergoes oscillation in charge, being sequentially reduced by tryptophan and reoxidized by oxygen. The oxidation is the more rapid process, and during the steady state reaction the heme prosthetic group is maintained largely as ferriprotoporphyrin. The ferroprotoporphyrin form accumulates during anaerobiosis or can be detected during the steady state reaction with the use of CO" (14). This may be formulated as [Eq. (5)]:

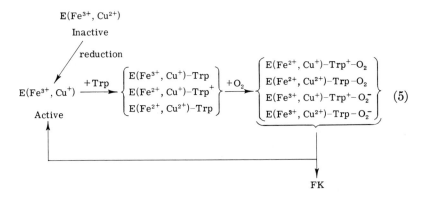

After L-tryptophan combines with the catalytically active ferriheme form of the enzyme, the ET complex may exist in unknown proportion as the depicted valence isomeric species, certain of which contain ferroheme. Oxygen may combine with the ET complex, generating the depicted ETO_2 valence isomeric species, again in unknown proportion. Subsequent interaction on the enzyme of activated L-tryptophan and oxygen generates the product, formylkynurenine, and regenerates the ferriheme form of the enzyme.

This reaction sequence is consistent with the CO inhibition study of Forman and Feigelson (92) (Fig. 15B) which indicated that CO inhibition of pseudomonad tryptophan oxygenase was uncompetitive with respect to L-tryptophan and competitive with respect to O_2. These findings are not compatible with the existence during the steady state reaction of any tryptophan-free form of ferroheme tryptophan oxygenase, which is capable of binding CO and by inference O_2. These data seem readily compatible with our mechanism [Eq. (5)], wherein no free ferroheme form of the enzyme exists, and seem kinetically incompatible with Eq. (4), which postulates continuous regeneration of free ferroheme enzyme. Furthermore, in our studies, during the steady state reaction tryptophan oxygenase exhibited an absorption spectrum typical of the ferriheme enzyme–tryptophan complex (λ_{max} = 407 nm) for the pseudomonad enzyme. After consumption of O_2 in the reaction mixture, only one-half of the ferriheme rapidly reduced to ferroheme (λ_{max} = 432 nm) (Fig. 25). Addition of CO to the steady state reaction likewise resulted in the trapping of only half of the heme as the carbon monoxide complex (14). We interpreted these observations to indicate that one of the enzymic heme iron moieties underwent valence change and that the two heme molecules per enzyme molecule were not functionally equivalent.

Using dithionite reduced enzyme at low temperature, and with O_2 being continuously supplied, Hayaishi's laboratory has made the important discovery of the existence during steady state catalysis of a spectral species with λ_{max} = 418 nm (Table I). Its spectral properties and conditions for its appearance and decay are all compatible with the view that this may be the enzyme–tryptophan–oxygen ternary complex (80–82, 94). The major hesitation in unequivocally accepting this as the normal catalytic intermediate is the fact that 100% of their dithionite-reduced enzyme [E(Cu$^+$, Fe^{2+})] entered the oxygenated state in contrast to only 50% of ascorbate-activated enzyme [E(Cu$^+$, Fe^{3+})], doing so under different conditions (14). It remains for future studies to definitively ascertain the relationship between fully reduced [E(Cu$^+$, Fe^{2+})] and partially reduced [E(Cu$^+$, Fe^{3+})] active enzyme species, their oxygenated intermediates, and the valence states throughout catalysis of the 2 moles of heme and copper which exist within each tetrameric tryptophan oxygenase molecule.

VI. REGULATION OF L-TRYPTOPHAN OXYGENASE ACTIVITY

A. Control of the Level of Enzyme Protein

1. Microbial Substrate Induction

The genus *Pseudomonas* possesses marked nutritional versatility in its ability to utilize a large variety of organic compounds as carbon and nitrogen sources. In many of these instances it has been established that the microorganisms develop their ability to metabolize these compounds by the induction of high levels of the enzymes involved in their metabolism. In 1951, Stanier *et al.* (71) demonstrated that *Pseudomonas* grown upon L-tryptophan developed the ability to metabolize it to kynurenine, anthranilic acid, and catechol. Extracts from these tryptophan-induced cells were shown to contain the enzyme system which catalyzed the conversion of tryptophan to kynurenine (72). These authors recognized the similarities in properties of this catalytic activity in their microbial extracts with those reported earlier for liver tryptophan oxygenase (then known as pyrrolase and/or peroxidase) by Kotake and Masayama (5) and Knox and Mehler (6).

The biochemical events underlying the induction of elevated levels of tryptophan oxygenase and the other enzymes involved in the adaption by these microorganisms to the metabolism of tryptophan as sole carbon source have not been as intensively explored as, for example, has β-gatacto-sidase induction in *Escherichia coli.* By analogy, one expects that enzyme induction in *Pseudomonas* enabling tryptophan catabolism also involves inducer regulation of gene transcription, presumably involving derepression with consequent elevated levels of the mRNA species coding for the inducible enzymes which are then translated cytoplasmically into the appropriate inducible enzymes. However, the biochemical details of induction of this pathway in *Pseudomonas* remain largely unexplored. Some information, descriptive and indirect, does exist. Thus, Palleroni and Stanier found in a strain of *Pseudomonas fluorescens* that L-kynurenine was the coordinate feedback inducer of tryptophan oxygenase and of kynurenine formamidase (95). Tremblay *et al.* (96), studying *P. fluorescens,* reported the induction of these enzymes to be sequential and not coordinate. Working with *Pseudomonas acidovorans* (ATCC 11299b) we found that when unin-duced cells were grown in a medium containing small amounts of yeast extract and high levels of tryptophan, a typical diauxic growth ensued. After consumption of the yeast extract, the *Pseudomonas* strains developed high induced levels of tryptophan oxygenase and resumed growth (Fig. 17)

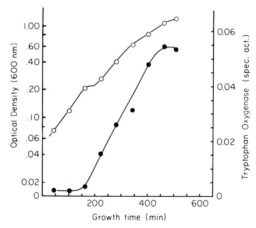

FIG. 17. Bacterial growth (O—O) of *P. acidovorans* and specific activity (●—●) of tryptophan oxygenase when cultured on inducing medium, containing 0.05% yeast extract and 5 mM L-tryptophan (74).

(73). Other biochemical alterations relative to adaptation to growth upon L-tryptophan are summarized in Fig. 18 (74) which indicates: (1) bacteria grown on either L-tryptophan or L-kynurenine contain induced high levels of tryptophan oxygenase and kynurenine formamidase, confirming feedback metabolite induction of these enzymes; (2) the ratio of the *in vivo* induced activities of tryptophan oxygenase and the formamidase is alterable under various conditions, indicating that these two enzymes are either noncoordinately transcriptionally regulated or are each subject to separate secondary regulation at the translational or post-translational levels; and (3) 7-azatryptophan which lacks inducing ability per se, when added to media containing tryptophan, results in a synergistically augmented induction by tryptophan (73).

As depicted in Fig. 18, one of the adaptive events by which *P. acidovorans* accommodates to growth upon L-tryptophan is the appearance of a specific L-tryptophan transport system. This permease activity is sensitive to azide and dinitrophenol and thus seems to implement active transport. Although this permease is absolutely specific for L-tryptophan, it nevertheless is feedback inducible by L-kynurenine; this may be the first example of metabolite induction of a specific permease. Comparative studies with various tryptophan analogs indicate that permease induction is not strictly coordinate with that of tryptophan oxygenase (73, 74).

2. Mammalian Substrate Induction

In 1951, Knox and collaborators reported increased tryptophan oxygenase activities in livers of animals that had received tryptophan or glucocorticoids parenterally (23). The phenomenon of substrate and hor-

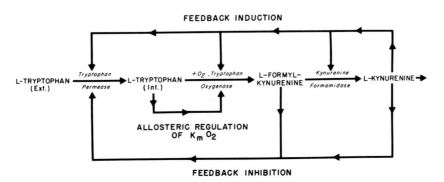

FIG. 18. Induction and regulation of the early enzymes of tryptophan metabolism in *P. acidovorans*. Extracellular solute is designated "Ext.," intracellular is "Int." (73).

monal induction is depicted in Fig. 19 (24). These findings stimulated investigations in numerous laboratories, and a certain degree of insight now exists concerning the biochemical processes underlying both substrate and hormonal control of the tissue levels of tryptophan oxygenase.

It was important to determine whether the increases in activity reflected altered catalytic efficiency (removal of an inhibitor, activation, etc.) or elevated levels of enzyme protein. Antibodies prepared against purified hepatic tryptophan oxygenase were employed in immunochemical titrations of the enzyme protein. It was found that increased levels of enzyme protein, as measured immunochemically, paralleled the hormonal- and substrate-mediated induced increases in catalytic activity (24, 28, 30, 97). Thus, both hormonal and substrate induction resulted in elevated tissue levels of tryptophan oxygenase. Tryptophan oxygenase, which is exclusively present in the soluble cytosol fraction of the liver, was found to be an enzyme with a readily dissociable heme prosthetic group and to exist in the hepatic cytosol only one-third saturated with respect to its heme cofactor (11, 13, 24, 26, 97–101). Competition for heme between apo-tryptophan oxygenase and other hepatic heme binding proteins was demonstrated. With the finding that microsomes were functioning as heme donating organelles, an appreciation developed that a complex intracellular competition for heme between various heme-donating and heme-binding proteins occurred (11, 24–26). This became of acute interest when it was found that the earliest detectable event during substrate induction by tryptophan was an increase in the proportion of tryptophan oxygenase

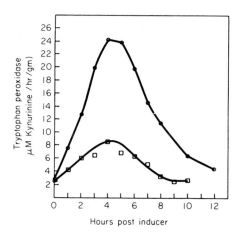

Fig. 19. The tryptophan (O—O) and cortisone (□—□) induced elevations of rat liver tryptophan oxygenase (24).

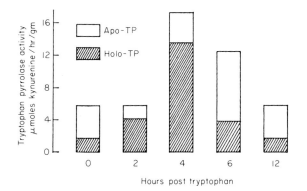

FIG. 20. Variations of hepatic tryptophan oxygenase activity of cell sap and its degree of saturation with hematin during the course of tryptophan induction (24).

which existed as holoenzyme and that this preceded the rise in total amount of enzyme activity present (Figs. 20 and 21) (24). Thus, substrate induction consisted of two discrete phases of elevation of enzymic activity, the first being the result of conversion of catalytically inactive apoenzyme to catalytically active holoenzyme and a second, later, phase being a rise in the total enzyme protein present. It was recognized at that time that "saturation of the enzyme with (heme) activator, with concomitant lowering of free apoenzyme levels, precedes and may be responsible for the subsequent appearance of additional enzyme" (24). Further investigations of this phenomenon were continued in two discrete directions: into enzyme chemistry to ascertain the molecular basis by which tryptophan altered

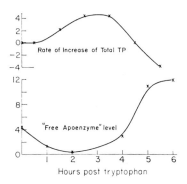

FIG. 21. Kinetic relationship between the rate of accumulation of hepatic tryptophan oxygenase and the apoenzyme level in cell sap during substrate induction (24).

the apo- to holoenzyme ratio and into physiological systems to explore the significance of altering apo- to holoenzyme ratios in controlling the enzyme levels.

The apoenzyme of rat liver tryptophan oxygenase was purified. The catalytically inactive apoenzyme, upon saturation with its heme cofactor, was fully catalytically active (13). When the affinity of apoenzyme for hematin was measured as a function of tryptophan concentration, it was found that the K_m for heme decreased as the enzyme became progressively saturated with tryptophan (Fig. 22) (13). Furthermore, heme binding proteins or competitive inhibitors of heme could not remove or displace heme from tryptophan oxygenase in the presence of tryptophan (26). Thus, conversion of E to the ET complex converts it to a form which has higher affinity for H than does the free E. The kinetic data are compatible with the sequence shown in Eq. (6) (13).

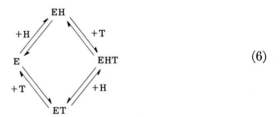

$$(6)$$

Dubnoff and Dimick, in 1959, found that the presence of tryptophan decreased the rate of inactivation of tryptophan oxygenase *in vitro* and concluded: "The changes in level of this basically unstable enzyme are consistent with the view that the enzyme protein is being continually synthesized and broken down. When the cellular level of tryptophan rises more of the newly synthesized enzyme is stabilized. It is not necessary to assume in the case of tryptophan peroxidase that the substrate initiates the synthesis of new enzyme protein" (102).

This remarkable insight was subsequently supported by the demonstration that tryptophan oxygenase was one of the most rapidly turning over, and, hence, one of the most responsive of the hepatic proteins ($t_{1/2}$ of 2.3 hours) (103). The elegant experiments of Schimke demonstrated that the decay in enzymic activity, which follows inhibition of protein synthesis *in vivo*, was markedly diminished in the presence of the substrate, tryptophan, and that the immunochemically demonstrable rise in the level of tryptophan oxygenase in tryptophan-induced livers (30) was not accompanied by elevated rates of enzyme synthesis, as measured by [14C]-amino acid incorporation into the enzyme protein (27–29).

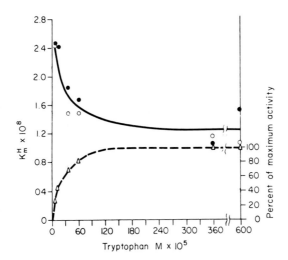

Fig. 22. The effect of hematin concentration on K_m^{Trp} (solid line) and percent maximum activity (dashed line) for hepatic tryptophan oxygenase (13).

Furthermore, it was established that inhibition of genetic transcription by actinomycin D did little to influence substrate induction, whereas inhibition of ongoing protein synthesis by puromycin and cycloheximide did prevent substrate induction (31, 32). Thus, the mechanism of mammalian substrate induction of tryptophan oxygenase is both largely post-transcriptional and post-translational and seems to occur in the following manner. Rat liver tryptophan oxygenase exists in the cytosol only one-third saturated with its heme cofactor and two-thirds as apoenzyme. In the presence of tryptophan, which combines with the apoenzyme, the apoenzyme–tryptophan complex has higher affinity for heme than the apoenzyme alone, and thus competes more successfully with other proteins for hepatic heme. In the presence of high *in vivo* levels of tryptophan, essentially all the enzyme becomes converted to the enzyme–heme–tryptophan complex. This enzyme–heme–tryptophan (EHT) complex seems to be more slowly degraded *in vivo* than is the apoenzyme alone; thus, with a continued normal rate of enzyme protein synthesis and a decreased rate of its breakdown, increased levels of enzyme protein accumulate. The increased catalytic activity observed during substrate induction thus reflects first an increasing proportion of the apoenzyme converted to catalytically active holoenzyme, which is less subject to metabolic turnover, leading to increased amounts of enzyme protein in the liver.

3. Mammalian Hormonal Induction

Hepatic tryptophan oxygenase is representative of the hepatic enzymes whose levels are augmented subsequent to glucocorticoidal steroid administration. This is one of the changes in gluconeogenic and amino acid metabolizing enzymes that characterizes and is to some degree responsible for the gluconeogenesis which ensues following glucocorticoid administration to animals (104, 105). Glucocorticoid hormones such as cortisol, corticosterone, cortisone, and dexamethasone act directly upon the isolated perfused liver (106), as well as upon fetal liver cells in organ culture (107), to form elevated levels of tryptophan oxygenase, thus indicating the hepatocyte to be a direct target cell for these hormones. The detailed biochemical processes by which hormonal enzyme induction occurs is still moot, but certain facts are available and reasonable mechanisms can be considered (108, 109). It has been shown that the hormonally induced synthesis of hepatic tryptophan oxygenase is accompanied by an increased rate of [14C]amino acid incorporation into the enzyme protein (28). This results in increased levels of enzyme protein (30) in a process distinct from substrate induction and which does not involve prior saturation of apo-tryptophan oxygenase with its heme cofactor (compare Fig. 23 with Fig. 20) (24, 97, 99). Furthermore, hormonal enzyme induction is sensitive not only to inhibitors of protein synthesis but also to actinomycin D which inhibits gene transcription (32). These studies indicate involvement of hormonal control over gene function and presumably indicate hormonally enhanced synthesis of the species of mRNA which codes for tryptophan oxygenase and other inducible proteins. Figure 24 indicates the time course

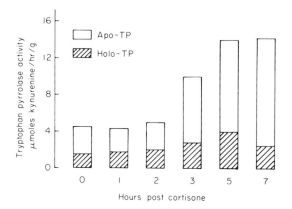

Fig. 23. Variations of hepatic tryptophan oxygenase activity of cell sap and its degree of saturation with hematin during the course of cortisone induction (24).

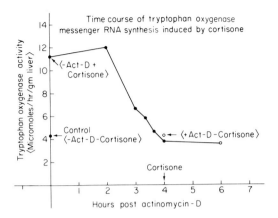

FIG. 24. The time course for hormonal induction of the gene product (mRNA) responsible for the synthesis of hepatic tryptophan oxygenase. (Performed by medical students in the Class of 1974, College of Physicians and Surgeons, Columbia University.)

for the appearance of hormone-evoked gene products which determine the level of hepatic tryptophan oxygenase. These findings may be interpreted to indicate that essentially, immediately following parenteral hormone administration, the increased synthesis of the mRNA coding for tryptophan oxygenase ensues and that this process is completed within 2 hours. Ultimately, this elevates in turn the proportion of the polysomal population synthesizing tryptophan oxygenase with consequent increased levels of this enzyme protein in the cytosol of the hormonally induced hepatocytes. Hypotheses concerning the role of tissue specific hormonal receptor proteins, their transport into the nucleus, their possible interaction with chromatin, the selective activation of RNA polymerases, augmentation in ribosomal RNA synthesis, the possible stabilization of certain species of mRNA, and theories of translational and post-translational hormonal regulation are exciting and provocative aspects of contemporary regulatory biochemistry, which albeit involves tryptophan oxygenase as a prototype of the inducible enzyme but is a large subject which has been adequately reviewed elsewhere (105, 108–110).

In summary, both substrate induction by tryptophan and hormonal induction by glucocorticoids have the same ultimate net effect upon tryptophan oxygenase, i.e., elevated hepatic levels of this enzyme protein. The underlying biochemical processes are quite distinct for these two types of induction. As summarized in Table III, substrate induction is a consequence of an unmodified rate of enzyme synthesis with a decreased rate of enzyme degradation in vivo (metabolic stabilization), whereas hormonal

TABLE III

COMPARISON OF SUBSTRATE AND HORMONAL INDUCTION OF HEPATIC
TRYPTOPHAN OXYGENASE

	Type of induction	
	Substrate	Hormonal
Increase of enzymic activity/gm liver	+	+
Increase in immunochemically titratable enzyme protein/gm liver	+	+
Conversion of apo- to holotryptophan oxygenase	+	−
Stabilization of tryptophan oxygenase	+	−
Increased incorporation of [^{14}C]amino acids into tryptophan oxygenase	−	+
Puromycin	Inhibited	Inhibited
Cycloheximide	Inhibited	Inhibited
Actinomycin D	No effect	Inhibited

enzyme induction involves a gene-mediated enhancement in the rate of synthesis of the inducible enzyme, tryptophan oxygenase.

B. Control of Catalytic Efficiency—Allosteric Modulation

It is evident from the aforementioned studies that the rate of synthesis of tryptophan oxygenase is subject to substrate induction in *Pseudomonas* and hormonal induction in mammals. Furthermore, the rate of enzyme degradation *in vivo* is regulated by the cellular tryptophan level. These processes, acting over a time scale of several hours, serve to determine the intracellular level of this enzyme protein. In addition to this type of regulation, tryptophan oxygenases of both microbial and mammalian origin contain regulatory site(s) which influence the catalytic efficiency with which existent enzyme molecules function. Inductive regulation of the amount of enzyme and allosteric modulation of its catalytic efficiency may serve for long-term and rapid regulation of catalytic function, respectively.

When pseudomonad tryptophan oxygenase was purified to homogeneity, studies of its kinetic and physicochemical properties became possible (Fig. 25). When the rate of formation of formylkynurenine as a function of L-tryptophan concentration was carefully measured, a nonhyperbolic saturation curve was observed (Fig. 14) (38). That this did not result from simple homotropic interaction between identical catalytic sites was

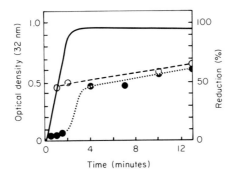

FIG. 25. Relationship between pseudomonad tryptophan oxygenase activity and the valence state of the heme prosthetic group (14). Formation of formylkynurenine (——), rate of reduction of heme (●- - -●), rate of formation of carbon monoxide complex (○- - -○). Replicate cuvettes were used for each determination.

shown by the finding that α-methyltryptophan, which at the levels used was neither a substrate nor an inhibitor of the enzyme, eliminated sigmoidicity; in its presence catalytic activity was a hyperbolic function of L-tryptophan concentration. The simplest interpretation of these data was to postulate the existence of two types of L-tryptophan binding sites on the enzyme: one catalytically productive and the other noncatalytic, but which had to be occupied for catalysis to proceed. L-Tryptophan may bind to either site, whereas α-methyltryptophan binds only to the noncatalytic site. Since this noncatalytic site is physically distinct from the catalytic site and serves to determine whether catalysis proceeds, it fulfills the operational definition of an allosteric site. From this viewpoint, α-methyltryptophan is a positive allosteric effector and L-tryptophan is an atypical homotropic–heterotropic, substrate-effector. Tryptophan is homotropic because it binds to more than one site on the enzyme and heterotropic in the sense that the enzymic sites to which it binds are not identical.

The functional significance of the allosteric site became evident upon discovery that the $K_m^{O_2}$ decreases as tryptophan oxygenase becomes progressively saturated with L-tryptophan and that if all the allosteric sites were first saturated by α-methyltryptophan (effector), then the enzyme manifested maximum affinity for oxygen even at low L-tryptophan levels (Fig. 14). Thus, saturation of the allosteric site with either the normal substrate-effector, L-tryptophan, or, in the presence of low levels of tryptophan, with the analog, α-methyltryptophan, brought about an alteration in the protein which increased the affinity of the heme at the active site for the second substrate, O_2 (38).

One may consider the possible functional significance of the dependence of the $K_m^{O_2}$ upon the environmental tryptophan level (Fig. 22) (13). In *Pseudomonas* grown under nutritional conditions wherein tryptophan was limiting, the organism would benefit by preserving its biosynthesized tryptophan for protein synthesis and diminish its catabolic flow via tryptophan oxygenase. Low intracellular tryptophan levels would not only lead to progressive decrease in the saturation of the catalytic site by tryptophan but also to a decreased saturation of the allosteric site of tryptophan oxygenase, resulting in lowered affinity of the enzyme for its other substrate, oxygen. This tripartite synergy would result in a sharp decrease in the proportion of intracellular tryptophan which would be catabolized by tryptophan oxygenase. Thus, an allosteric intramolecular homeostatic mechanism provides for the preservation of tryptophan for anabolic utilization under conditions of tryptophan insufficiency.

The aforementioned studies concerning alterations in enzymic behavior were based upon measurements of catalytic rate (38). The existence of homogeneous preparations of pseudomonad tryptophan oxygenase in conjunction with the high Soret extinction coefficients of this copper-heme protein enabled direct measurements to be made upon the role of the allosteric site in modulating the reactivity of enzymic heme with ligands such as carbon monoxide (39, 41, 44) and cyanide (40, 41, 43). As described earlier, carbon monoxide is an analog of oxygen which binds to heme iron at the catalytic site with the formation of a new spectral species (Fig. 6). A saturating level of tryptophan enhances by almost two orders of magnitude the equilibrium binding of CO to the enzyme (Fig. 26) (41).

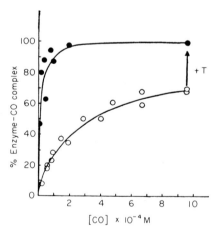

Fig. 26. The effect of L-tryptophan on the affinity for carbon monoxide of the ferro-heme form of pseudomonad tryptophan oxygenase (41).

Whereas α-methyltryptophan alone has no such effect in the presence of a nonsaturating level of tryptophan, saturation of the allosteric site with α-methyltryptophan synergises with tryptophan to bring about the allosteric transition to the form of the enzyme with higher affinity for CO (Fig. 27) (44). What remains undetermined in these experiments is the degree to which the allosteric transition is directly to a form with higher affinity for CO or to a form which has higher affinity for tryptophan. Is it the progressive saturation of the catalytic site by tryptophan under these conditions which is ultimately responsible for the enhanced affinity of the enzymic heme for CO?

Hepatic tryptophan oxygenase differs from the microbial enzyme in numerous respects that have been discussed earlier (Table I). The hepatic enzyme may be purified as the heme-free apoenzyme. Tryptophan both *in vivo* and *in vitro* enhances the affinity of the apoenzyme for heme. It is as yet undetermined whether the binding of tryptophan to the catalytic or to the allosteric sites of hepatic tryptophan oxygenase is responsible for increased affinity for its heme prosthetic group. α-Methyltryptophan, administered parenterally, also increases the proportion of the enzyme as the enzyme–heme complex (EH); hence, saturation of the allosteric site of the enzyme is involved in the transition to the enzymic configuration with high affinity for heme. Furthermore, the fact that α-methyltryptophan induces increases in enzyme levels *in vivo* in a pattern similar to that ob-

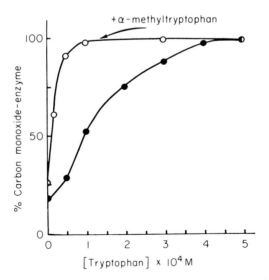

Fig. 27. The effect of α-methyltryptophan upon the equilibrium formation of the carbon monoxide complex of the ferroheme form of tryptophan oxygenase (44).

tained with tryptophan (111), involving enzyme stabilization (29), suggests the possibility that both tryptophan and α-methyltryptophan evoke a conformational change in apotryptophan oxygenase which imparts high affinity for heme and resistance to proteolytic digestion. However, definitive studies to verify this hypothesis remain to be done.

Early clues concerning the allosteric nature of hepatic tryptophan oxygenase were the observations of Schimke that certain frozen and thawed preparations of hepatic tryptophan oxygenase showed sigmoidal saturation kinetics with respect to tryptophan and that α-methyltryptophan antagonized this sigmoidicity (29). α-Methyltryptophan was also found to stabilize the enzyme against thermal inactivation and proteolysis (29). Recent studies with homogeneous preparations of the copper-heme holoenzyme of hepatic tryptophan oxygenase (20) have shown it to undergo interesting and complex allosteric transitions (45). Catalytic activity as a function of tryptophan concentration at pH 6.2 and 8.0, in the absence and presence of α-methyltryptophan, is depicted in Fig. 28, reciprocal $1/v$ versus $1/T$ in Fig. 29, and as Hill plots in Fig. 30 (45). It is evident that at pH 8.0 the saturation of the enzyme by tryptophan is hyperbolic with linear double reciprocal plots and a Hill coefficient of $n = 1.0$, all of which are uninfluenced by the presence or absence of α-methyltryptophan. Therefore, at pH 8.0 there is no detectable interaction

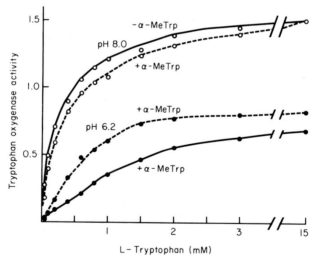

Fig. 28. Plot of velocity versus L-tryptophan concentration for hepatic tryptophan oxygenase at pH 6.2 (●) and pH 8.0 (○), minus (——) and plus (- - -) α-methyl-tryptophan (45).

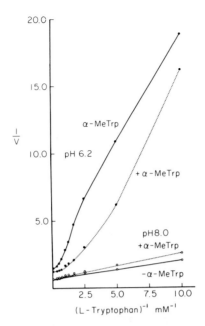

FIG. 29. Lineweaver-Burk plots of the data in Fig. 28 (45).

between the tryptophan and/or α-methyltryptophan binding sites, i.e., no allosteric behavior is evidenced. At pH 6.2 nonhyperbolic saturation is evident with nonlinear double reciprocal plots and Hill coefficients of 0.8 and 1.6 at low and high tryptophan levels, respectively. At pH 6.2 the effector, α-methyltryptophan, lowers the [tryptophan]$_{0.5}$ and results in a linear Hill plot with a Hill coefficient for tryptophan of 1.6. These data

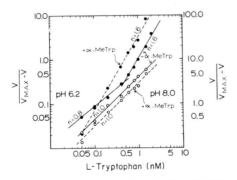

FIG. 30. Hill plots of the data in Fig. 28 (45).

indicate that allosteric interaction between the subunits of the enzyme occurs only when an as-yet unidentified component (histidine?) of the enzyme with a pK near neutrality is protonated. Raising the pH to 8.0 converts the enzyme to a form without functional allosteric interaction between its subunits.

The feedback inhibition of hepatic tryptophan oxygenase by NAD derivatives was reported by Wagner (112) and by Cho-Chung and Pitot (47). Tryptophan oxygenase derived from *Xanthomonas pruni* is inhibited by intermediates of the biosynthetic pathway wherein tryptophan is converted to NAD (46, 113). It has been recently established that 3-hydroxy anthranilate is a negative feedback modulator of hepatic tryptophan oxygenase, which converts hyperbolic saturation kinetics by tryptophan (pH 7.9) to a markedly sigmoidal kinetics (45). Thus, allosteric control of the catalytic activity of microbial and hepatic tryptophan oxygenases is dependent upon the ionization state of the enzyme and upon complex interaction and competition of positive and negative modulators for the allosteric site(s).

VII. DL-TRYPTOPHAN OXYGENASE OF RABBIT INTESTINE

Kotake and Ito, in 1937, isolated D-kynurenine from the urine of rabbits which had been fed a diet containing D-tryptophan (114). Recently, Hayaishi and co-workers partially purified from the ileum of rabbit intestine the enzyme responsible for this conversion (69, 70). The enzyme, as isolated, has been established as a heme protein (λ_{max} = 406, 499, 535, and 560 nm) although the stoichiometry of heme and protein has not been determined since the preparation is not homogeneous (69). Nothing is known as to its copper content. This enzyme, commonly referred to as D-tryptophan oxygenase, may not be properly named since L-tryptophan also serves as an excellent substrate, although with a different pH optimum than evidenced with D-tryptophan (L-tryptophan, pH optimum 6.5, K_m = 20 μM; D-tryptophan, pH optimum 7.5, K_m = 300 μM) (69, 70). This enzyme is perhaps more suitably referred to as intestinal DL-tryptophan oxygenase.

Intestinal DL-tryptophan oxygenase possesses many characteristics similar to pseudomonad and hepatic tryptophan oxygenases, as well as some unique differences. L-Tryptophan, but not D-tryptophan, causes a hypochromic shift of the Soret peak from 406 to 409 nm, this shift being more pronounced at pH 8.0 than at pH 6.0. Ligands such as cyanide, azide, and CO inhibit catalysis by the intestinal enzyme and form spectrophotometrically identifiable heme derivatives (69). The substrate specificity of intestinal tryptophan oxygenase is quite broad. Besides being able to catalyze the dioxygenative cleavage of the pyrrole ring of both D- and L-tryptophan (69), this enzyme can utilize as substrates D- and L-5-hydroxy-tryptophan, tryptamine, and serotonin (115), all of which are inhibitors and not substrates of the pseudomonad and hepatic enzymes (Table II).

As presently isolated, intestinal tryptophan oxygenase possesses a requirement for a reducing agent (in this case, sodium ascorbate, 5 μM, in the presence of methylene blue, 5 μM) in order to elicit maximal enzymic activity (69, 70). This reductive activation is reminiscent of that occurring with oxidized preparations of the pseudomonad and hepatic enzymes and which has been shown to involve Cu^{2+} to Cu^{+} reduction (see above) (21, 74). The requirement for exogenous reductant may be a consequence of the intestinal enzyme being purified in the absence of its substrates, DL-tryptophan. A recent report has suggested the involvement of superoxide anion, O_2^{-}, in the reaction catalyzed by intestinal tryptophan oxygenase (116). The experiments reported to date do not rigorously differentiate between a requirement for O_2^{-} in reductive activation or in catalysis. The above-cited similarities between the intestinal and hepatic tryptophan oxygenases lead one to suspect the possibility that the intestinal enzyme may also be a copper-heme protein and that the requirement for superoxide is similar to that seen in the activation of oxidized preparations of the pseudomonad and hepatic L-tryptophan oxygenases (74).

REFERENCES

1. Keilin, D. (1966). "The History of Cell Respiration and Cytochrome." Cambridge Univ. Press, London and New York.
2. Chance, B., Yonetani, T., and Mildvan, A. S. (eds.) (1971). "Probes of Structure and Function of Macromolecules and Membranes," Vol. 2. Academic Press, New York.
3. Chance, B., Nicholls, P., and Erecinska, M., this volume.
4. Kotake, Y., Iwao, J., Kiyokawa, M., Shichiri, G., Ichihara, K., Otani, S., Tsujimoto, J., and Sakata, H. (1931). Z. Physiol. Chem. 195, 139–192.
5. Kotake, Y., and Masayama, T. (1936). Z. Physiol. Chem. 243, 237–244.
6. Knox, W. E., and Mehler, A. H. (1950). J. Biol. Chem. 187, 419–430.
7. Hayaishi, O., Rothberg, S., Mehler, A. H., and Saito, Y. (1957). J. Biol. Chem. 229, 889–896.
8. Tanaka, T., and Knox, W. E. (1959). J. Biol. Chem. 234, 1162–1170.
9. Mason, H. S., Fowlks, W. L., and Peterson, E. (1955). J. Amer. Chem. Soc. 77, 2914–2915.
10. Hayaishi, O., Katagiri, M., and Rothberg, S. (1955). J. Amer. Chem. Soc. 77, 5450–5451.
11. Feigelson, P., and Greengard, O. (1961). J. Biol. Chem. 236, 153–157.
12. Feigelson, P., and Greengard, O. (1961). Biochim. Biophys. Acta 50, 200–202.

13. Greengard, O., and Feigelson, P. (1962). *J. Biol. Chem.* **237**, 1903–1907.
14. Maeno, H., and Feigelson, P. (1967). *J. Biol. Chem.* **242**, 596–601.
15. Poillon, W. N., Maeno, H., Koike, K., and Feigelson, P. (1969). *J. Biol. Chem.* **244**, 3447–3456.
16. Koike, K., Poillon, W. N., and Feigelson, P. (1969). *J. Biol. Chem.* **244**, 3457–3462.
17. Ishimura, Y., Okazaki, T., Nakazawa, T., Ono, K., Nozaki, M., and Hayaishi, O. (1966). *In* "Biological and Chemical Aspects of Oxygenases" (K. Bloch and O. Hayaishi, eds.), pp. 416–422. Maruzen, Tokyo.
18. Tan Yip, A., and Knox, W. E. (1969). *Enzym. Biol. Clin.* **10**, 233–242.
19. Maeno, H., and Feigelson, P. (1965). *Biochem. Biophys. Res. Commun.* **21**, 297–302.
20. Schutz, G., and Feigelson, P. (1972). *J. Biol. Chem.* **247**, 5327–5332.
21. Brady, F. O., Monaco, M. E., Forman, H. J., Schutz, G., and Feigelson, P. (1972). *J. Biol. Chem.* **247**, 7915–7922.
22. Knox, W. E., and Mehler, A. H. (1951). *Science* **113**, 237–238.
23. Knox, W. E. (1951). *J. Exp. Pathol.* **32**, 462–469.
24. Feigelson, P., Feigelson, M., and Greengard, O. (1962). *Recent Progr. Horm. Res.* **18**, 491–512.
25. Greengard, O., and Feigelson, P. (1961). *J. Biol. Chem.* **236**, 158–161.
26. Feigelson, P., and Greengard, O. (1962). *J. Biol. Chem.* **237**, 1908–1913.
27. Schimke, R. T., Sweeney, E. W., and Berlin, C. M. (1964). *Biochem. Biophys. Res. Commun.* **15**, 214–219.
28. Schimke, R. T., Sweeney, E. W., and Berlin, C. M. (1965). *J. Biol. Chem.* **240**, 322–331.
29. Schimke, R. T., Sweeney, E. W., and Berlin, C. M. (1965). *J. Biol. Chem.* **240**, 4609–4620.
30. Feigelson, P., and Greengard, O. (1962). *J. Biol. Chem.* **237**, 3714–3717.
31. Nemeth, A. M., and de la Haba, G. (1962). *J. Biol. Chem.* **237**, 1190–1193.
32. Greengard, O., Smith, M. A., and Acs, G. (1963). *J. Biol. Chem.* **238**, 1548–1551.
33. Nemeth, A. M., and Nachmias, V. T. (1958). *Science* **128**, 1085–1086.
34. Nemeth, A. M. (1959). *J. Biol. Chem.* **234**, 2921–2924.
35. Egelhaaf, A. (1958). *Z. Naturforsch.* **13b**, 275–279.
36. Egelhaaf, A., and Caspari, E. Z. (1960). *Z. Vererbungslehre* **91**, 373–379.
37. Egelhaaf, A. (1963). *Z. Vererbungslehre* **94**, 349–384.
38. Feigelson, P., and Maeno, H. (1967). *Biochem. Biophys. Res. Commun.* **28**, 289–293.
39. Maeno, H., and Feigelson, P. (1968). *J. Biol. Chem.* **243**, 301–305.
40. Maeno, H., and Feigelson, P. (1968). *Biochemistry* **7**, 968–970.
41. Feigelson, P. (1969). *Advan. Enzyme Regul.* **7**, 119–127.
42. Poillon, W. N., and Feigelson, P. (1971). *Biochemistry* **10**, 753–760.
43. Koike, K., and Feigelson, P. (1971). *Biochemistry* **10**, 3378–3384.
44. Koike, K., and Feigelson, P. (1971). *Biochemistry* **10**, 3385–3390.
45. Schutz, G., and Feigelson, P. (1972). *J. Biol. Chem.* **247**, 5333–5337.
46. Wagner, C., and Brown, A. T. (1970). *J. Bacteriol.* **104**, 90–97.
47. Cho-Chung, Y. S., and Pitot, H. C. (1967). *J. Biol. Chem.* **242**, 1192–1198.
48. Cho-Chung, Y. S., and Pitot, H. C. (1968). *Eur. J. Biochem.* **3**, 401–406.
49. Forman, H. J., and Feigelson, P. (1972). *J. Biol. Chem.* **247**, 256–259.
50. Chytil, F. (1961). *Biochim. Biophys. Acta* **48**, 217–218.
51. Chytil, F., Skrivanova, J., and Brana, H. (1966). *Can. J. Biochem.* **44**, 283–286.
52. Chytil, F. (1968). *J. Biol. Chem.* **243**, 893–899.
53. Julian, J. A., and Chytil, F. (1969). *Biochem. Biophys. Res. Commun.* **35**, 734–740.
54. Julian, J., and Chytil, F. (1970). *J. Biol. Chem.* **245**, 1161–1168.

55. Feigelson, P., Ishimura, Y., and Hayaishi, O. (1964). *Biochem. Biophys. Res. Commun.* **14**, 96–101.
56. Brady, F. O., Forman, H. J., and Feigelson, P. (1971). *J. Biol. Chem.* **246**, 7119–7124.
57. Tokuyama, K., and Knox, W. E. (1964). *Biochim. Biophys. Acta* **81**, 201–204.
58. Knox, W. E., and Ogata, M. (1965). *J. Biol. Chem.* **240**, 2216–2221.
59. Elvehjem, C. A., Madden, R. J., Strong, F. M., and Wooley, D. W. (1937). *J. Amer. Chem. Soc.* **59**, 1767–1768.
60. Elvehjem, C. A., Madden, R. J., Strong, F. M., and Wooley, D. W. (1938). *J. Biol. Chem.* **123**, 137–149.
61. Heidelberger, C., Gullberg, M. E., Morgan, A. F., and Lepkovsky, S. (1948). *J. Biol. Chem.* **175**, 471–472.
62. Heidelberger, C., Abraham, E. P., and Lepkovsky, S. (1949). *J. Biol. Chem.* **179**, 151–155.
63. Feigelson, P., Williams, J. N., Jr., and Elvehjem, C. A. (1951). *J. Biol. Chem.* **193**, 737–741.
64. Feigelson, P., Williams, J. N., Jr., and Elvehjem, C. A. (1951). *Proc. Soc. Exp. Biol. Med.* **78**, 34–36.
65. Knox, W. E. (1955). *Methods Enzymol.* **2**, 242–253.
66. Cho, Y. S., Pitot, H. C., and Morris, H. P. (1964). *Cancer Res.* **24**, 52–58.
67. Dyer, H. M., Gullino, P. M., and Morris, H. P. (1964). *Cancer Res.* **24**, 97–104.
68. Cho-Chung, Y. S., and Pitot, H. C. (1968). *Cancer Res.* **28**, 66–70.
69. Yamamoto, S., and Hayaishi, O. (1967). *J. Biol. Chem.* **242**, 5260–5266.
70. Higuchi, K., and Hayaishi, O. (1967). *Arch. Biochem. Biophys.* **120**, 397–403.
71. Stanier, R. Y., Hayaishi, O., and Tsuchida, M. (1951). *J. Bacteriol.* **62**, 355–356.
72. Hayaishi, O., and Stanier, R. Y. (1951). *J. Bacteriol.* **62**, 691–709.
73. Rosenfeld, H., and Feigelson, P. (1969). *J. Bacteriol.* **97**, 697–704.
74. Rosenfeld, H., and Feigelson, P. (1969). *J. Bacteriol.* **97**, 705–714.
75. Schutz, G., and Feigelson, P. (1972). *Anal. Biochem.* **46**, 149–155.
76. Peisach, J., Blumberg, W. E., Wittenberg, B. A., and Wittenberg, J. B. (1968). *J. Biol. Chem.* **243**, 1871–1880.
77. Smith, D. W., and Williams, R. J. P. (1970). *Struct. Bonding (Berlin)* **7**, 1–45.
78. Keilin, D., and Hartree, E. F. (1951). *Biochem. J.* **49**, 88–104.
79. Keilin, D., and Hartree, E. F. (1955). *Biochem. J.* **61**, 153–171.
80. Ishimura, Y., Nozaki, M., Hayaishi, O., Tamura, M., and Yamazaki, I. (1967). *J. Biol. Chem.* **242**, 2574–2576.
81. Hayaishi, O. (1969). *Ann. N.Y. Acad. Sci.* **158**, 318–335.
82. Ishimura, Y., Nozaki, M., Hayaishi, O., Nakamura, T., Tamura, M., and Yamazaki, I. (1970). *J. Biol. Chem.* **245**, 3593–3602.
83. Peterson, J. A., Ishimura, Y., and Griffin, B. W. (1972). *Arch. Biochem. Biophys.* **149**, 197–208.
84. Vallee, B. L., and Wacker, W. E. C. (1970). *Proteins* **5**, 98–99.
85. Wilson, D. F., and Dutton, P. L. (1970). *Arch. Biochem. Biophys.* **136**, 583–584.
86. Dutton, P. L. (1971). *Biochim. Biophys. Acta* **226**, 63–80.
87. Wilson, D. F., and Dutton, P. L. (1970). *Biochem. Biophys. Res. Commun.* **39**, 59–64.
88. Wilson, D. F., Lindsay, J. G., and Brocklehurst, E. S. (1972). *Biochim. Biophys. Acta* **256**, 277–286.
89. White, A., Handler, P., and Smith, E. L. (1964). "Principles of Biochemistry," p. 302. McGraw-Hill, New York.

90. Greengard, O., Mendelsohn, N., and Acs, G. (1966). *J. Biol. Chem.* **241,** 304–308.
91 Brady, F. O., Feigelson, P., and Rajagopalan, K. V. (1973). *Arch. Biochem. Biophys.* **157,** 63–72.
92. Forman, H. J., and Feigelson, P. (1971). *Biochemistry* **10,** 760–763.
93. Forman, H. J. (1971). Ph.D. Thesis, Columbia Univ., New York.
94. Ishimura, Y., Nozaki, M., Hayaishi, O., Tamura, M., and Yamazaki, I. (1968). *In* "Structure and Function of Cytochromes" (K. Okunuki, M. D. Kamen, and I. Sekuzu, eds.), pp. 188–195. Univ. of Tokyo Press, Tokyo.
95. Palleroni, N. J., and Stanier, R. Y. (1964). *J. Gen. Microbiol.* **35,** 319–334.
96. Tremblay, G. C., Gottlieb, J. A., and Knox, W. E. (1967). *J. Bacteriol.* **93,** 168–176.
97. Feigelson, P., and Greengard, O. (1963). *Ann. N.Y. Acad. Sci.* **103,** 1075–1082.
98. Greengard, O., and Feigelson, P. (1960). *Biochim. Biophys. Acta* **39,** 191–192.
99. Greengard, O., and Feigelson, P. (1961). *Nature (London)* **190,** 446–447.
100. Feigelson, P., and Greengard, O. (1961). *Biochim. Biophys. Acta* **52,** 509–516.
101. Greengard, O., and Feigelson, P. (1963). *Ann. N.Y. Acad. Sci.* **111,** 227–232.
102. Dubnoff, J. W., and Dimick, M. (1959). *Biochim. Biophys. Acta* **31,** 541–542.
103. Feigelson, P., Dashman, T., and Margolis, F. (1959). *Arch. Biochem. Biophys.* **85,** 478–482.
104. Ashmore, J., and Morgan, D. (1967). *In* "The Adrenal Cortex" (A. B. Eisenstein, ed.), pp. 249–292. Little, Brown, Boston, Massachusetts.
105. Feigelson, M., and Feigelson, P. (1965). *Advan. Enzyme Regul.* **3,** 11–28.
106. Goldstein, L., Stella, E. J., and Knox, W. E. (1962). *J. Biol. Chem.* **237,** 1723–1726.
107. Wicks, W. D. (1968). *J. Biol. Chem.* **243,** 900–906.
108. Tomkins, G. M., and Gelehrter, T. D. (1972). *In* "Biochemical Actions of Hormones" (G. Litwack, ed.), Vol. 2, pp. 1–17. Academic Press, New York.
109. Feigelson, P., Yu, F.-L., and Hanoune, J. (1971). *In* "The Human Adrenal Cortex" (N. P. Christy, ed.), pp. 257–272. Harper, New York.
110. Kenney, F. T., Reel, J. R., Hager, C. B., and Witliff, J. T. (1968). *In* "Regulatory Mechanisms for Protein Synthesis in Mammalian Cells" (A. San Pietro, M. R. Lamborg, and F. T. Kenney, eds.), pp. 119–142. Academic Press, New York.
111. Greengard, O. (1964). *Biochim. Biophys. Acta* **85,** 492–494.
112. Wagner, C. (1964). *Biochem. Biophys. Res. Commun.* **17,** 668–673.
113. Brown, A. T., and Wagner, C. J. (1970). *J. Bacteriol.* **101,** 456–463.
114. Kotake, Y., and Ito, N. (1937). *J. Biochem.* **25,** 71–77.
115. Hirata, F., and Hayaishi, O. (1972). *Biochem. Biophys. Res. Commun.* **47,** 1112–1119.
116. Hirata, F., and Hayaishi, O. (1971). *J. Biol. Chem.* **246,** 7825–7826.

Supplementary References

Beadle, G. W., Mitchell, H. K., and Nyc, J. F. (1947). *Proc. Nat. Acad. Sci. U.S.* **33,** 155–158.
Civen, M., and Knox, W. E. (1959). *J. Biol. Chem.* **234,** 1787–1790.
Civen, M., and Knox, W. E. (1960). *J. Biol. Chem.* **235,** 1716–1718.
Feigelson, P. (1964). *Biochim. Biophys. Acta* **92,** 187–190.
Feigelson, P., Ishimura, Y., and Hayaishi, O. (1965). *Biochim. Biophys. Acta* **96,** 283–293.
Feigelson, P., and Maeno, H. (1966). *In* "Biological and Chemical Aspects of Oxygenases" (K. Bloch and O. Hayaishi, eds.), pp. 411–415. Maruzen, Tokyo.
Ghosh, D., and Forrest, H. S. (1967). *Arch. Biochem. Biophys.* **120,** 578–582.

Gray, G. D. (1966). *Arch. Biochem. Biophys.* **113,** 502–504.
Greengard, O., and Dewey, H. K. (1971). *Proc. Nat. Acad. Sci. U.S.* **68,** 1698–1701.
Hayaishi, O., and Nozaki, M. (1969). *Science* **164,** 389–396.
Knox, W. E. (1954). *Biochim. Biophys. Acta* **14,** 117–126.
Knox, W. E. (1966). *Advan. Enzyme Regulat.* **4,** 287–297.
Knox, W. E., and Auerbach, V. H. (1955). *J. Biol. Chem.* **214,** 307–313.
Knox, W. E., and Piras, M. M. (1966). *J. Biol. Chem.* **241,** 765–767.
Knox, W. E., and Piras, M. M. (1967). *J. Biol. Chem.* **242,** 2959–2965.
Knox, W. E., Piras, M., and Tokuyama, K. (1966). *J. Biol. Chem.* **241,** 297–303.
Koike, K., and Okui, S. (1964). *Biochim. Biophys. Acta* **81,** 602–604.
Nozaki, M., Okuno, S., and Fujisawa, M. (1971). *Biochem. Biophys. Res. Commun.* **44,** 1109–1116.
Piras, M. M., and Knox, W. E. (1967). *J. Biol. Chem.* **242,** 2952–2958.
Pitot, H. C., and Cho, Y. S. (1961). *Biochim. Biophys. Acta* **50,** 197–199.
Seglen, P. O., and Jervell, K. F. (1969). *Biochim. Biophys. Acta* **171,** 47–57.
Tokuyama, K. (1968). *Biochim. Biophys. Acta* **151,** 76–87.

4

NONHEME IRON DIOXYGENASE

MITSUHIRO NOZAKI

I. INTRODUCTION

Dioxygenases are enzymes that catalyze the incorporation of two atoms of molecular oxygen into various substrates as shown in Eq. (1),

$$S + O_2 \rightarrow SO_2 \qquad (1)$$

where S represents a typical substrate. This type of reaction was first demonstrated by Hayaishi and his colleagues in 1955 (32) with the pyrocatechase reaction [Eq. (2)]. By using the stable oxygen isotope, $^{18}O_2$,

$$\qquad (2)$$

these authors found that the two atoms of oxygen incorporated into the product by the action of pyrocatechase were both derived exclusively from molecular oxygen (32).

Since then, new dioxygenases have been discovered in all types of living organisms and shown to perform a variety of functions (38). Among these, cleavage of the aromatic ring is one function that appears to depend largely, perhaps entirely, upon this type of enzyme. Thus, extensive studies on the reaction mechanism of dioxygenases have been carried out primarily with phenolic dioxygenases that cleave the phenol or catechol ring. Detailed study of the properties and mechanisms of dioxygenase have been hampered greatly by the difficulty in obtaining a highly purified preparation. In 1963, Nozaki et al. (61, 62) succeeded in obtaining the first crystalline dioxygenase. Since then, several dioxygenases have been obtained in crystalline form (36, 38) so that critical analysis of these enzymes is now possible. All of the phenolic dioxygenases that have thus far been purified and characterized contain nonheme iron as the sole cofactor. In this chapter, the discussion will be focused on recently developed concepts concerned with the reaction mechanism of these nonheme iron-containing dioxygenases.

II. CATECHOL DIOXYGENASES

A. Mode of Ring Fission

Among a number of o-dihydroxyphenyl compounds that are cleaved by microbial dioxygenases, three modes of ring fission have so far been demonstrated (67): (1) oxygenative cleavage of the bond between carbon atoms bearing the hydroxyl groups of an o-dihydroxyphenyl compound (intradiol cleavage, A in Fig. 1); (2) cleavege of the bond between carbon atoms 2 and 3 (proximal extradiol cleavage, B in Fig. 1); and (3) cleavage of the bond between carbon atoms 4 and 5 (distal extradiol cleavage, C in Fig. 1). Pyrocatechase, which catalyzes the conversion of catechol to cis,cis-muconic acid, is a typical example of type A. Metapyrocatechase, the first dioxygenase obtained in crystalline form (61, 62), is a typical example of the extradiol type (16). Protocatechuic acid can be cleaved by the action

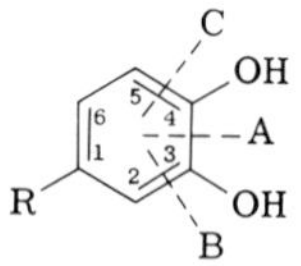

FIG. 1. Three modes of ring fission of o-dihydroxyphenyl compounds by dioxygenases.

of protocatechuate 3,4-dioxygenase (76) as well as protocatechuate 4,5-dioxygenase (15) to form *cis,cis*-β-carboxymuconic acid and α-hydroxy-γ-carboxymuconic-ε-semialdehyde, respectively. The former is another example of intradiol cleavage whereas the latter is an example of distal extradiol cleavage. 3,4-Dihydroxyphenylacetate 2,3-oxygenase has also been obtained in crystalline form (46, 47) and catalyzes the ring fission between carbon atoms 2 and 3 of 3,4-dihydroxyphenylacetate to form α-hydroxy-δ-carboxymethylmuconic-ε-semialdehyde (1) (proximal extradiol cleavage). Likewise, metapyrocatechase catalyzes the cleavage of the carbon bond between positions 2 and 3 when the 1- or 2-substituted catechol is used as substrate. This is another example of proximal extradiol cleavage (67).

Differences between intradiol and extradiol dioxygenases exist not only in their function but also in their physical appearance: The former are red in color whereas the latter are colorless. All of these differences appear to be related to the state of iron bound to the enzyme (63). Various properties of these dioxygenases acting on *o*-dihydroxyphenyl compounds are summarized in Table I. Both intradiol and extradiol enzymes contain nonheme iron as the sole cofactor, but the intradiol enzymes contain trivalent iron whereas the extradiol enzymes contain iron in the divalent form.

B. Intradiol Dioxygenases

1. Pyrocatechase

Pyrocatechase (catechol:oxygen 1,2-oxidoreductase, EC 1.13.1.1) is an enzyme that catalyzes the cleavage of the aromatic ring of catechol to *cis,cis*-muconic acid with the consumption of two atoms of molecular oxygen [Eq. (2)]. The enzyme was first isolated in 1950 (31) and characterized to be a dioxygenase in 1955 (32). Since then, various methods of enzyme purification from different strains of bacteria have been described and several reaction mechanisms for this enzyme have been proposed.

The most purified preparation of pyrocatechase obtained from extracts of benzoate-induced cells of *Pseudomonas arvilla* C-1 (ATCC 23974) has a specific activity of 29.6 μmoles/minute/mg of protein and is homogeneous as judged by ultracentrifugal and electrophoretic criteria. The molecular weight is estimated to be approximately 90,000. This enzyme contains 2 gm atoms of iron per mole of enzyme protein (51). A similar pyrocatechase was also purified from *Pseudomonas fluorescence* (ATCC 11250) (33, 57). Another pyrocatechase purified from phenol-induced *Brevibacterium fuscum* P-13 has a molecular weight of 64,000. Ferrous ion and reduced glutathione are required for maximal activity (44). The substrate specificity for this

TABLE I

PROPERTIES OF DIOXYGENASES ACTING ON o-DIHYDROXYPHENYL COMPOUNDS

Enzymes	Source materials	Substrate	Mode of reaction	Cofactors	Iron gm atoms/mole of enzyme	Molecular weight	References
Pyrocatechase	Pseudomonas arvilla C-1[a]	Catechol	Intradiol	Fe^{3+}	2	95,000	51, 60
Pyrocatechase	Brevibacterium fuscum P-13	Catechol	Intradiol	Fe^{3+}	1	64,000	44, 48, 54
Protocatechuate 3,4-dioxygenase	Pseudomonas aeruginosa	Protocatechuic acid	Intradiol	Fe^{3+}	8	700,000	21, 23
Metapyrocatechase	Pseudomonas arvilla	Catechol	Extradiol (proximal)[b]	Fe^{2+}	3	140,000	62, 65
3,4-Dihydroxyphenylacetate 2,3-dioxygenase	Pseudomonas ovalis	3,4-Dihydroxyphenylacetic acid	Proximal extradiol	Fe^{2+}	4-5	100,000	46, 48
Protocatechuate 4,5-dioxygenase	Pseudomonas sp.	Protocatechuic acid	Distal extradiol	Fe^{2+}	1	150,000	69
3,4-Dihydroxy-9,10-secoandrosta 1,3,5(10)-triene-9,17-dione-4,5-dioxygenase	Nocardia restrictus	3,4-Dihydroxy-9,10-secoandrosta 1,3,5(10)-triene-9,17-dione	Proximal extradiol	Fe^{2+}	1	280,000	80

[a] A similar pyrocatechase is also purified from Pseudomonas fluorescens (57).

[b] When 4- or 3-substituted catechol derivatives are used as substrate, they are exclusively cleaved at proximal site by the action of metapyrocatechase.

enzyme is somewhat different from the one described above (56). This enzyme contains 1 gm atom of iron per mole of enzyme and the iron is easily removed from the protein by incubating the enzyme with a non-metabolizable substrate analog, ethyl protocatechuic acid. The apoenzyme retains less than 5% of the original activity and shows no absorption maxima in the visible region. Reconstitution of the enzyme concomitant with full recovery of the activity is achieved by treating the apoenzyme with a stoichiometric amount of ferrous ion in the presence of oxygen (54, 55). Although the holoenzyme has no sulfhydryl group which reacts with 5,5'-dithiobis(2-nitrobenzoate) (DTNB), the apoenzyme has one DTNB-reactive sulfhydryl group. By treatment with the reagent, apo-enzyme loses its ability to reconstitute the holoenzyme unless it is treated with cysteine or β-mercaptoethanol. These results suggest that one cysteinyl residue in the protein is involved in the binding of ferric ion (55).

a. Spectral Properties. The concentrated solution of highly purified pyrocatechase has a pronounced red color with a broad absorption between 390 and 650 nm. The peak is at about 440 nm and the molecular absorption at 440 nm is estimated to be 4670. The absorbance decreases when the enzyme is heated or treated with acid, and this decrease parallels the loss of enzymic activity suggesting that the absorbance property is intimately related to enzymic activity (51). The red color also decreases when sodium dithionite is added and is restored when the solution is exposed to air. *o*-Phenanthroline removes iron from pyrocatechase when it is added in the presence of sodium dithionite. The decrease in enzymic activity occurs in association with the increase in absorbance at 508 nm caused by the formation of a Fe(II)-*o*-phenanthroline complex. This complex could be separated from the enzyme protein by exhaustive dialysis. The apoprotein thus obtained contains about 10% of total iron and is almost colorless. Upon incubating this preparation with ferrous ion in the presence of oxygen, the red color is partially restored with simultaneous reactivation of the enzyme. These results suggest that the trivalent iron bound to the enzyme is responsible for the red color and is an integral part of the enzyme reactivity (60).

The absorption spectrum of the native enzyme is not altered by the addition of oxygen nor is it changed by exhaustive evacuation. However, when the substrate, catechol, is added under anaerobic conditions, the color of the enzyme solution changes to grayish blue with a concomitant increase in the absorbance at 710 nm, indicating the possible formation of an enzyme–substrate complex. The absorption spectrum is restored to the original level after catechol is degraded to *cis,cis*-muconic acid by the addition of oxygen (51). When the enzyme is titrated with catechol under anaerobic conditions, 2 moles of the substrate per mole of enzyme are required to bring about the maximal absorption change demonstrating that 2 moles of substrate can combine with the enzyme (66).

b. Electron Spin Resonance Properties. Pyrocatechase shows a sharp electron spin resonance (ESR) signal at $g = 4.28$ which is known to result from the spin state of ferric ion (A in Fig. 2). This signal markedly decreases upon the addition of sodium dithionite and reappears when the solution is exposed to air (59, 60). A similar signal change has also been reported with *Brevibacterium* pyrocatechase (48). A decrease in the signal also occurs concomitant with loss in enzymic activity when the enzyme is heated or treated with acid. When the substrate, catechol, is added to the

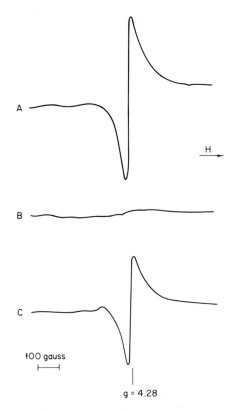

FIG. 2. ESR spectra of pyrocatechase. Pyrocatechase (15.4 mg per 0.2 ml) was incubated with 10 μl of 0.1 M catechol and then shaken in air. The curves represent ESR spectra of pyrocatechase: A, before the addition of catechol; B, after the addition of catechol, and C, after shaking in air. [Reproduced with permission from Nakazawa (60).]

enzyme under anaerobic conditions, the signal disappears instantaneously (B in Fig. 2). The signal is restored to the original level after all of the substrate has been degraded to *cis*,*cis*-muconic acid by the addition of air (C in Fig. 2) (59, 60).

c. Circular Dichroism. As shown in Fig. 3, the circular dichroism (CD) of pyrocatechase exhibits several strong positive bands between 250 and 300 nm, a moderate negative band at 327 nm, and a weak but broad negative one around 500 nm in addition to a strong negative band at 222 nm, which is characteristic of an α-helix in proteins. The negative circular dichroic bands at 327 and 500 nm and a part of the positive bands between 250 and 300 nm are associated with enzymic activity. When the bound iron is removed, the bands at 327 and 500 nm completely disappear and those between 250 and 300 nm are partially diminished. Upon incubation with ferrous ion under aerobic conditions these bands are restored to almost the original level. Thus, the iron bound to the enzyme may be responsible for the CD bands as well as the enzyme activity. The disappearance of the bands above 300 nm also occurs on addition of reducing agents, indicating that the iron in the enzyme is in the trivalent state, which is consistent with the spectral and ESR results (58).

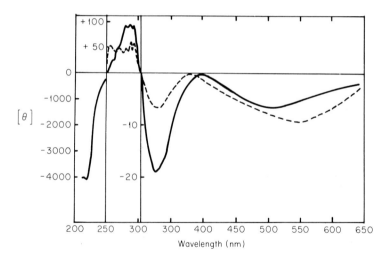

FIG. 3. CD spectra of pyrocatechase. The measurement of CD spectra was performed with enzyme solutions of 11.0 and 1.25 mg/ml for the regions 300–650 and 245–300 nm, respectively, in a 10-mm light path quartz cell. For the determination of the lower wavelength region, an enzyme solution (0.25 mg/ml) was used with a 2-mm light path cell. Solid line represents CD spectrum of the native pyrocatechase and dotted line that of the enzyme after addition of 0.5 μmole of catechol under anaerobic conditions. [Reproduced with permission from Nakazawa (58).]

In the presence of the substrate, catechol, under anaerobic conditions, changes in the magnitude and the position of the circular dichroic bands related to enzymic activity occur, indicating an alteration in the state of the ligands around the iron following the binding of substrate (58).

2. Protocatechuate 3,4-Dioxygenase

Protocatechuate 3,4-dioxygenase (protocatechuate:oxygen 3,4-oxido-reductase, EC 1.13.1.3) catalyzes the conversion of protocatechuic acid to β-carboxymuconic acid with the insertion of two atoms of molecular oxygen [Eq. (3)].

$$
\text{protocatechuic acid} + O_2 \longrightarrow \text{β-carboxymuconic acid} \tag{3}
$$

This is another example of an intradiol dioxygenase, and catalytic properties as well as appearance of the enzyme are very similar to those of pyrocatechase. This enzyme was first described by Stanier and Ingraham in 1954 (76). Since then, a number of investigators have described various methods of purifying the enzyme from *Pseudomonas* (70), *Nocardia* (5), and *Neurospora* (29). The enzyme was finally obtained in crystalline form from *p*-hydroxybenzoate-induced cells of *Pseudomonas aeruginosa* (ATCC 23975) (21). The crystalline enzyme is homogeneous as judged by ultra-centrifugation, starch gel electrophoresis, and immunoelectrophoresis. The molecular weight is estimated to be approximately 700,000 and the molecular activity is calculated to be 45,500 at 24°. The enzyme contains about 8 gm atoms of iron per mole of enzyme (21) and appears to consist of 8 subunits (23).

a. Spectral Properties. A concentrated solution of protocatechuate 3,4-dioxygenase has a deep red color with a broad absorption band between 400 and 650 nm (A in Fig. 4). The red color disappears upon addition of sodium dithionite and reappears either when the solution is exposed to air or when potassium ferricyanide is added to the solution under anaerobic conditions. These results indicate that the trivalent iron bound to the enzyme is responsible for the visible absorption spectrum of the enzyme (23).

When the substrate, protocatechuic acid, is added to the enzyme under anaerobic conditions, the visible absorption spectrum shows an increase in absorbance with a slight red shift of the peak (B in Fig. 4). The spectrum is restored to the original one after the substrate is depleted by the addition of oxygen. Similar, but not identical, spectral changes are observed when various substrate analogs or competitive inhibitors are added to the en-

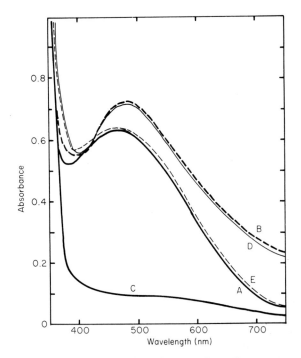

Fig. 4. Effect of reducing agent and oxidant on absorption spectrum of enzyme–substrate complex of protocatechuate 3,4-dioxygenase. The enzyme, 46.3 mg in 2.9 ml of 50 mM tris-Cl buffer, pH 8.5 was placed in a cuvette (A). After air in the cuvette was evacuated by gassing with argon, 0.10 ml of 10 mM protocatechuic acid solution which had been bubbled with argon was added (B), and then 1.1 mg of solid sodium dithionite were further added under a stream of argon (C). After the enzyme was fully reduced, 2.88 mg of solid potassium ferricyanide were added in a similar manner (D) and then the cuvette was shaken strongly in air (E). [Reproduced with permission from Fujisawa et al. (23).]

zyme. When the enzyme is titrated with protocatechualdehyde, a competitive inhibitor, approximately 7.5 moles of the compound per mole of enzyme are required to cause a maximal change in absorption at 450 nm, suggesting that 8 moles of substrate can combine with the enzyme. The spectrum of the enzyme–substrate complex decreases markedly when sodium dithionite is added to the complex (C in Fig. 4) and is restored to that of the original complex upon the addition of ferricyanide (D in Fig. 4). Further addition of oxygen converts the spectrum to that of the original enzyme (E in Fig. 4). These results suggest that the iron in the enzyme–substrate complex is also in the trivalent state (23).

b. Electron Spin Resonance. Like pyrocatechase, protocatechuate 3,4-dioxygenase shows a sharp ESR signal at $g = 4.31$. Since this signal decreases markedly upon the addition of sodium dithionite and is restored by exposure to air, the signal appears to be attributable to the ferric ion bound to the enzyme. The signal height at $g = 4.31$ instantaneously diminishes upon addition of substrate under anaerobic conditions but, differing from the case with pyrocatechase, the signal never completely disappears. These results suggest that the addition of substrate causes a modification in the ligand field of ferric ion in the enzyme rather than a change in valency from the ferric to the ferrous state. This explanation is consistent with the spectrophotometric observations mentioned above. The decreased signal is restored to the original level when the substrate is completely exhausted by the introduction of air. Decrease in ESR signal is also observed when various substrate analogs or competitive inhibitors are used. However, changes in ESR signal caused by these agents are somewhat different from the one caused by the substrate in that they show a marked anisotropy (23).

From a temperature dependence study of the ESR property, Peisach *et al.* (71) suggested that the ligands of iron are sulfur atoms that are arranged in a tetrahedron around the metal. Upon addition of substrate or substrate analogs under anaerobic conditions, new resonances at $g = 6.4$ and 5.6 are observed which arise from Fe^{3+} in a nearly tetragonal environment, analogous to the geometric environment of ligands that exists in heme. It is concluded, therefore, that the binding of substrate to the enzyme causes a change in ligand symmetry, which makes the iron accessible for O_2 binding.

C. Extradiol Dioxygenases

1. Metapyrocatechase

Metapyrocatechase (catechol:oxygen 2,3-oxidoreductase, EC 1.13.1.2) catalyzes the conversion of catechol to α-hydroxymuconic ϵ-semialdehyde with the insertion of two atoms of molecular oxygen [Eq. (4)]. The enzyme

$$\text{(4)}$$

has been purified from extracts of *Pseudomonas arvilla* (ATCC 23973) that had been grown with benzoate as the major carbon source and has been obtained in crystalline form. This enzyme, first described by Dagley and Stopher in 1959 (16), was found to be extremely unstable in the presence of air (50, 85). However, a low concentration of organic solvent such as acetone or ethanol was found to protect the enzyme from inactivation so that all purification procedures have been carried out in a buffer solution containing 10% acetone; thus, the first crystallization of a dioxygenase was achieved with metapyrocatechase (61, 62). Although the mechanism of the protective action of organic solvent is not understood, the enzyme is protected not only from inactivation by oxidizing agents but also from denaturation caused by urea (65). As discussed below, most extradiol enzymes as well as some enzymes other than oxygenase are stabilized in the presence of a low concentration of organic solvent (82).

The crystallized metapyrocatechase has a specific activity of about 110 μmoles/minute/mg of protein at 24° and contains 1 gm-atom of iron per mole of enzyme based on a molecular weight of 140,000 (65). The enzyme is, however, further activated about 2.5-fold by treatment with both cysteine and ferrous ion under anaerobic conditions. The fully activated preparation has a specific activity of about 270 μmoles/minute/mg of protein and appears to contain 3 gm atoms of iron per mole of enzyme (63). This result has recently been confirmed by Takemori et al. (84). In fact, the iron in the enzyme is easily removed from the protein when the enzyme is dialyzed against a buffer which does not contain acetone, and the specific activity of various preparations of enzyme is in association with their iron content. The iron in the enzyme appears to be of the ferrous form and oxidation of the iron by air or H_2O_2 leads to inactivation of the enzyme. The inactivated enzyme can be fully reactivated by incubation with ferrous ion and a reducing agent under anaerobic conditions (65).

a. Spectral and ESR Properties. Unlike intradiol dioxygenases, metapyrocatechase is colorless and shows no significant absorption in the visible range. The enzyme shows an absorption peak at 280 nm with a shoulder at around 290 nm which is characteristic of a simple protein. Native metapyrocatechase shows no significant ESR signal at around $g = 4.0$. However, inactivated enzyme prepared by treatment with H_2O_2 shows a broad signal around $g = 4.2$ which is characteristic of ferric ion. These results suggest that the iron in the native enzyme is in the divalent state and that oxidation of the iron results in inactivation of the enzyme (35).

b. Circular Dichroism. Circular dichroism spectra of metapyrocatechase are shown in Fig. 5. The enzyme shows a strong negative CD band at 225 nm in the region corresponding to the absorption of the peptide backbone and has an ordered structure with a relatively low content of α-helix. This ordered structure is partially destroyed by treatment of the enzyme with urea or alkali but not by resolution of the bound iron. In the vicinity of side chain absorption bands (250–300 nm), positive dichroic bands are

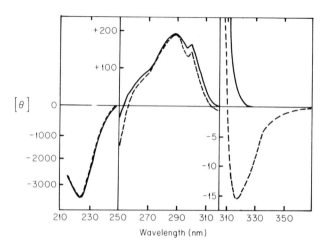

Fig. 5. CD spectra of metapyrocatechase. The measurement of CD spectra was performed under anaerobic conditions with 10 mm light path quartz cuvette. Enzyme solutions used were 16, 0.86, and 0.05 mg/ml and catechol added to the enzyme solution were 1, 0.05, and 0.01 mM for the regions 300–600, 250–300, and 220–250 nm, respectively. (——) Before the addition of catechol; (- - -) after the addition of catechol. [Reproduced with permission from Hirata *et al.* (39).]

detected at 265, 288, and 300 nm. However, no significant CD bands are observed in the visible range. Following catechol addition under anaerobic conditions, a new negative circular dichroic band appears at 317 nm with a concomitant decrease at 300 nm. A similar band is observed when various enzymically active substrate analogs are employed. On the other hand, the addition of nonmetabolizable competitive inhibitors do not alter the circular dichloic spectra of the enzyme. When the enzyme is titrated by substrate under anaerobic conditions, 3 moles of catechol per mole of enzyme are required to bring about maximal change of the CD band at 317 nm, suggesting that 3 moles of catechol can combine with 1 mole of enzyme. However, with the apoenzyme, no such change in the CD band is observed. Addition of iron to the complex bring about the appearance of the CD band. Moreover, the magnitude of this CD band was stoichiometrically related to the amount of catechol as well as of the iron bound to the enzyme (39). In many cases, optical activity near 300 nm can be attributable to tyrosyl residues, tryptophanyl residues, or disulfide linkages, or to all three (4, 28). The negative CD band at 317 nm of metapyrocatechase is suggested to result from the asymmetric configuration of the tyrosyl residue resulting from the interaction between substrate and the iron atom (39).

   c. *Interaction of the Enzyme with Substrates.* Both *o*-phenanthroline and
α,α'-dipyridyl, which are well-known chelators of ferrous ion, exert a
protective effect against the action of oxidizing agents or sulfhydryl inhibitors on metapyrocatechase. On the other hand, when such chelating
agents are present in the reaction mixture, a competitive type of inhibition
for the catechol is observed. From these results, we once proposed that the
substrate combines with iron, forming a chelate complex, during catalysis
(34). However, other nitrogen bases which are extremely poor chelators
such as *m*-phenanthroline, α-naphthoquinoline, and quinoline also show
the protective and inhibitory effects. The effects of *m*-phenanthroline or
α-naphthoquinoline are even more pronounced than that of *o*-phen-
anthroline (Fig. 6). These observations lead us to postulate that the binding
of substrate is stabilized through hydrophobic interaction rather than by
iron chelation (65). On the other hand, since these nitrogen bases prevent
enzymic reactivation and since catechol protects the enzyme from H_2O_2
inactivation with half-maximal protection at a concentration of the order

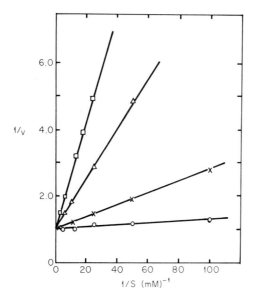

FIG. 6. Competitive inhibition of metapyrocatechase activity by nitrogen bases.
The reaction mixture contained in a final volume of 3.0 ml, 50 mM potassium phosphate
buffer, pH 7.5, inhibitors, catechol, and 1 μg of metapyrocatechase. The inhibitors
added to the reaction mixture were ($\bigcirc$) none, ($\times$) 2 mM *o*-phenanthroline, ($\triangle$) 0.1 mM
m-phenanthroline, and ($\square$) 0.1 mM α-naphthoquinoline. [Reproduced with permission
from M. Nozaki *et al.* (65).]

of the K_m value, the substrate may combine with the enzyme at a site near the iron and may interact with the iron (65).

Equilibrium dialyses of the enzyme against different catechol concentrations under anaerobic conditions reveal that 3 moles of catechol are found to combine with 1 mole of enzyme (63). Bovine serum albumin does not combine with catechol under these conditions, but the apoenzyme is also found to combine with 3 moles of catechol per mole of enzyme (37). These findings are consistent with the above explanation that the substrate combines primarily with the protein moiety of the enzyme. On the other hand, all attempts to indicate the interaction between the enzyme and molecular oxygen, the other substrate, have so far been unsuccessful. From these binding experiments, a sequential reaction mechanism is proposed where an organic substrate first combines with the enzyme and then reacts with oxygen to form a ternary complex, which is then followed by release of the product (36, 37, 66).

Sedimentation analysis of the enzyme at an alkaline pH reveals that the enzyme consists of 2–3 subunits. From the fact that the iron content, the number of substrate molecule bound to the enzyme, and the number of subunits of metapyrocatechase all coincide roughly, it is plausible to assume that each subunit contains one atom of iron and that all of the iron in the total molecule participates in the enzyme reaction as a part of the substrate binding sites (64).

2. 3,4-Dihydroxyphenylacetate 2,3-Dioxygenase

3,4-Dihydroxyphenylacetate 2,3-dioxygenase catalyzes the reaction shown in Eq. (5) (1). The enzyme has been purified from *Pseudomonas*

$$+ \ O_2 \longrightarrow \qquad\qquad (5)$$

ovalis that had been grown with *p*-hydroxyphenylacetate as the major carbon source and has also been obtained in crystalline form (46, 47). Like metapyrocatechase, the enzyme is also stabilized by the presence of a low concentration of organic solvent such as acetone or ethanol (46). The enzyme contains 4–5 gm atoms of iron per mole of enzyme, based on a molecular weight of approximately 100,000. The enzyme is dissociated into small subunits with a molecular weight of 30,000 concomitant with the release of iron either by prolonged storage in air or by treatment with *p*-chloromercuribenzoate. It is, therefore, proposed that the most of the iron in the enzyme is involved in the association of the subunits, although at least 1 gm atom of iron is at the active site (75).

a. Electron Spin Resonance Studies. The iron atoms in the native enzyme are in the state which is undetectable by ESR, but a signal at $g = 4.3$ characteristic of ferric ion develops when excess substrate is added under aerobic conditions. Incubation of this reaction mixture for 3 minutes at room temperature results in the disappearance of the signal. During this period, oxygen in the solution is used up with concomitant formation of the product. When oxygen is reintroduced to the solution, the signal reappears. These reversible changes of the ESR signal can be repeated several times by appropriate addition of oxygen in the presence of excess substrate. Since the signal does not appear at all under strictly anaerobic conditions, the enzymic activity is maintained almost unchanged throughout these experiments, and since both oxygen and substrate are required for the appearance of the ESR signal, the enzyme with the iron in the ferric state is suggested to be the intermediate form of the enzyme (48). When catechol, a slowly reacting substrate, is added to the enzyme solution under aerobic conditions, a sharp signal is observed at $g = 4.3$, in association with the appearance of a deep violet color. The signal and the color disappear when the reaction has been completed. By using 4-bromocatechol, a non-metabolizable substrate analog, a sharp ESR signal and the violet color appear which is irreversible. These results indicate that the ESR signal and the violet color may result from a ternary complex of the enzyme (Fe III)–substrate–oxygen complex which could be an intermediate during the enzymic reaction (48).

The ESR spectrum of the enzyme shows signals at around the $g = 2$ region, $S_m = 1.99$, 2.01, and 2.02, following the evacuation of the enzyme solution and the introduction of nitric oxide gas, an oxygen analog, into the ESR tube. On addition of catechol, these signals disappear and a signal at $g = 4.3$ appears. These results appear to support a reaction mechanism proposed for metapyrocatechase (59), where the enzyme reacts first with oxygen and then with substrate concomitant with the valency change in the bound iron, although this explanation has been reevaluated by recent experiments (40, 66).

3. Protocatechuate 4,5-Dioxygenase

Protocatechuate 4,5-dioxygenase (EC 1.13.1.8) catalyzes the reaction shown in Eq. (6). This enzyme was first described by Dagley and Patel in 1957 (15), and the partial purification of the enzyme and some of its properties were described by Cain (6). The enzyme was further purified from

$$\tag{6}$$

p-hydroxybenzoate-induced *Pseudomonas testeroni* (17) and to an almost homogeneous state from a pseudomonad grown with protocatechuic acid as the major carbon source (69). The most purified preparation has a specific activity of 160 μmoles/minute/mg of protein at 24° and a molecular weight of approximately 150,000. The enzyme contains 1 gm atom of iron per mole of enzyme. Although the enzyme is very unstable and is easily inactivated during storage either at 0° or at room temperature, the enzymic inactivation can be counteracted by the presence of 10% ethanol. Rapid inactivation of the enzyme is also observed during the catalysis (17, 69). Although this inactivation is partially prevented by L-cysteine (17), it appears to result simply from the removal of iron from the enzyme protein, since the inactivated enzyme is fully activated by addition of ferrous ion (69).

4. Steroid Oxygenase

As shown in Eq. (7), 3,4-dihydroxy-9,10-secoandrosta-1,3,5(10)-triene-9,17-dione-4,5-dioxygenase (steroid oxygenase) catalyzes the conversion of 3,4-dihydroxy-9,10-secoandrosta-1,3,5(10)-triene-9,17-dione [(I) in Eq. (7)] into 4,9-diseco-3-hydroxyandrosta-1(10),2-diene-5,9,17-trion-4-oic acid [(II) in Eq. (7)] (27). The steroid oxygenase is capable of catalyzing

$$+ \ O_2 \longrightarrow \tag{7}$$

(I)   (II)

the oxidative cleavage of several substituted catechols in addition to its natural substrate, 3,4-dihydroxy-9,10-secoandrosta-1,3,5(10)-triene-9,17-dione. Among these, 3-isopropylcatechol is the most reactive artificial substrate.

The enzyme was purified to a state of apparent homogeneity from *Nocardia restrictus* induced with progesterone. Like other extradiol dioxygenases, this enzyme is also stabilized by acetone. The molecular weight of the enzyme is estimated to be around 280,000. The enzyme has an ultraviolet absorption maximum at 280 nm and contains 1.13 gm atoms of ferrous ion per mole of enzyme (80). Metal chelating agents, including o-phenanthroline, 8-hydroxyquinoline, and α,α'-dipyridyl, inhibit the enzyme and the inhibition is noncompetitive with respect to both organic substrate and molecular oxygen. Sulfhydryl inhibitors also inhibit the enzyme at a concentration of 1 mM. In 0.05 N NaOH, the enzyme can be dissociated into identical subunits having a sedimentation coefficient of 2.6 S, suggesting that the enzyme consists of several subunits (80).

  a. Kinetic Studies. Kinetic studies have been conducted with the steroid dioxygenase in which the concentration of one substrate was varied while the level of the other substrate was fixed at one of several concentrations in different experiments. In double reciprocal plots of initial velocity against substrate concentration, intersecting lines were obtained which is consistent with sequential mechanism. Studies on the influence of the product on reaction rate indicate that the product seems to give rise both to simple product inhibition and dead end inhibition. Experiments with 4-isopropyl-catechol, a structural analog of the organic substrate, show that the enzyme is inhibited competitively with respect to the organic substrate as expected but uncompetitively with respect to molecular oxygen. These results are consistent with an ordered bi-uni mechanism where molecular oxygen is added first, organic substrate combines next, and, finally, the product is released (81).

III. OTHER DIOXYGENASES REQUIRING NONHEME IRON

A. 3-Hydroxyanthranilate Oxygenase

  3-Hydroxyanthranilate oxygenase (EC 1.13.1.6) catalyzes the cleavage of the benzene ring of 3-hydroxyanthranilic acid between carbon atoms 3 and 4 to yield α-amino-β-carboxymuconic semialdehyde as shown in Eq. (8) (52). This enzyme is found in the liver and kidney of a number of

$$\text{(8)}$$

species (72). Like the extradiol dioxygenases, this enzyme displays a specific requirement for ferrous ion, is rather unstable with respect to storage, and is inactivated rapidly during catalysis (77). This inactivation appears to be the result of oxidation or loss of bound ferrous ion (53). The inactivated enzyme is activated in the presence of ferrous salts by acidification to about pH 4 or by a high urea concentration (6.8 M). Metal binding agents, including o-phenanthroline, remove the bound iron to inactivate the oxygenase. The substrate, 3-hydroxyanthranilate prevents this removal of iron from the active enzyme by o-phenanthroline and also blocks the complete reconstitution of the active enzyme from iron-free enzyme (68).

The Michaelis constants for oxygen and 3-hydroxyanthranilate are 3.1×10^{-4} M and 4×10^{-5} M, respectively. Lineweaver-Burk plots at various fixed levels of the second substrate yield a set of parallel lines, suggesting that the binding of the first substrate is separated from the binding of the second substrate by a very slow irreversible reaction. Picolinic and quinolinic acids, but not the other pyridine compounds tested, inhibit the reaction competitively with respect to 3-hydroxyanthranilate but uncompetitively with respect to oxygen. These results are consistent with the view that a slow process separates the binding of the two substrates (68).

B. Homogentisate Oxygenase

Homogentisate oxygenase (EC 1.13.1.5) is an enzyme contained in mammalian tissues as well as in microorganisms and catalyzes the conversion of homogentisate to maleylacetoacetic acid with incorporation of two atoms of molecular oxygen [Eq. (9)]. Requirement of ferrous ion for

$$\tag{9}$$

enzymic activity was demonstrated by Suda and Takeda (78, 79) and confirmed by subsequent studies (13, 49, 74). Utilization of molecular oxygen in this reaction was verified by the use of oxygen isotope (14). The enzyme has been obtained in crystalline form from *Pseudomonas fluorescens* adapted to tyrosine (2). The crystallized enzyme is homogeneous on ultracentrifugation and the molecular weight of the enzyme is estimated to be about 380,000. Like the mammalian liver enzyme, bacterial homogentisate oxygenase requires ferrous ion as a cofactor. For maximal activity, the enzyme requires at least 10 minutes preincubation with ferrous ion at pH 6.0. Both glutathione and ascorbate are also required for maximal activity at pH 6.0, but only the former is essential at pH 5.4. The K_m values are 6×10^{-4} M for homogentisate and 1×10^{-4} M for ferrous ion at pH 6.0. Unlike the mammalian liver enzyme, the bacterial enzyme is fairly stable on aging or during storage. p-Chloromercuribenzoate inhibits the enzyme competitively with respect to ferrous ion but noncompetitively with respect to homogentisate (2). These results are in agreement with the observations of Flamm and Crandall (20) but not with those of Tokuyama (86) with mammalian liver enzymes.

The homogentisate oxidase in the soluble phase of calf liver has also been partially purified (20, 83). Like other extradiol dioxygenase, this enzyme is also stabilized by the presence of 10% acetone (83). For the activation of this enzyme, a combination of ascorbate, reduced glutathione, and ferrous ion is required at pH 7.0 in phosphate buffer. At pH 5.3 in acetate buffer, neither ascorbate nor reduced glutathione is required; ferrous ion alone fully activates the enzyme. These results indicate that activation is related to the soluble level of ferrous ion. Kinetic studies show that p-chloromercuribenzoate, methylmercuric bromide, and lewisite inhibit the enzyme by competing with ferrous ion but not with homogentisate for a common enzymic binding site. Ferrous ion, organic mercurials, and reducing agents are found to protect the enzyme from irreversible aerobic oxidation. From these results, the existence of ferrous mercaptans at the active center and their combination with oxygen during the enzymic reaction are suggested (20).

The reconstitution process to holoenzyme from apoenzyme has recently been studied by Takemori et al. (83). The activation is appreciably affected by variations in medium pH. The rate of activation of the apoenzyme increases greatly by decreasing pH. Activation at pH 5.6 reaches nearly the maximum level within 2 hours, whereas at pH 6.8, more than 24 hours are required for full activation. Reducing agents affect neither the rate nor extent of the activation. The activity of homogentisicase is strongly inhibited by various diphenolic compounds as well as chelating agents specific for ferrous ion. However, reconstitution of homogentisicase from the apoenzyme with ferrous ion is strongly prevented only by the substrate, homogentisate, or by 2,5-dihydroxybenzoate, a compound structurally similar to the substrate (83). A comparable type of substrate inhibition has been reported in the activation of other oxygenases containing metal ions (30, 65), suggesting that the substrate binding site is intimately related to the metal binding site.

C. 2,5-Dihydroxypyridine Oxygenase

2,5-Dihydroxypyridine oxygenase (EC·1.13.1.9) was first described by Behrman and Stanier (3) with cell-free extracts of *Pseudomonas putida*. A partially purified enzyme catalyzes the oxidation of 1 mole of 2,5-dihydroxypyridine to equivalent amounts of maleamic acid and formic acid with the consumption of 1 mole of oxygen. Crystallization of the enzyme was recently achieved by Gauthier and Rittenberg (25) from *P. putida* grown with sodium nicotinate as the major carbon source. The purified enzyme is labile and the half-life is approximately 2–3 days. The presence of dithiothreitol in the buffer extends the half-life to about 2 weeks. Addition of ferrous sulfate does not protect the enzyme, whereas the combination of dithiothreitol and ferrous sulfate is actually detrimental during long-term incubation at $0°$. The enzyme is sensitive to the sulfhydryl reagents such as p-chloromercuribenzoate and N-ethylmaleimide when these compounds are added directly to the reaction mixture prior to the addition of substrate. The enzyme is also inhibited to some extent by EDTA and KCN, and it is very sensitive to α,α'-dipyridyl, o-phenanthroline, and H_2O_2.

The crystalline enzyme yields a single absorption peak at 280 nm and a single major band with minor diffuse bands on polyacrylamide electrophoresis in the presence of dithiothreitol. In the absence of dithiothreitol, the intensity of the minor bands increases. On acrylamide gels containing sodium dodecyl sulfate, a single band with mobility corresponding to a molecular weight of 39,500 is obtained. A single region of enzymic activity corresponding to a molecular weight of 242,000 is obtained on sucrose gradients containing dithiothreitol (25).

With the crystalline enzyme, maleamic acid and formic acid have been identified as direct products of the oxidation of 2,5-dihydroxypyridine. The data show that 1 mole of oxygen is consumed and 1 mole of formate is produced per mole of substrate oxidized [Eq. (10)] which agree with those previously reported with a partially purified preparation (3). From studies

$$\text{(10)}$$

with $^{18}O_2$ and $H_2{}^{18}O$, the distribution of oxygen 18 in the products confirms the dioxygenase nature of the enzyme. Since N-formylmaleamic acid cannot be detected as an intermediate of the reaction and synthetic N-formylmaleamate is not a substrate for the enzyme, addition of molecular oxygen and hydrolytic cleavage of the N-formyl group appear to be catalyzed by a single enzyme (26).

D. Cysteamine Oxygenase

Cysteamine oxygenase catalyzes the conversion of cysteamine to hypotaurine with the incorporation of two atoms of molecular oxygen [Eq. (11)].

$$\text{(11)}$$

This enzyme has been extensively studied by Cavallini and his colleagues

(7–11). The enzyme is present in a number of different tissues of various animals. One of the tissues most suitable for extraction of the enzyme is horse kidney. The enzyme from this source has been purified to an almost homogeneous state based on criteria provided by starch-gel electrophoresis at different pH, sedimentation analysis, and chromatography on Sephadex G-200 (11, 12). The enzyme is stable for months when stored in the cold with 70% saturated ammonium sulfate. The absorption spectrum of the purified enzyme shows a single peak with a maximum at 278 nm. The enzyme possesses a molecular weight of approximately 100,000. When treated with 8 M urea or with 0.1% sodium dodecyl sulfate, the enzyme exhibits a molecular weight of 50,000 (12), suggesting that the enzyme consists of two subunits. C-Terminal amino acid analyses of cysteamine oxygenase reveal that alanine is the only amino acid present in the amount of 2 moles/mole of enzyme after 90 minutes of incubation with carboxypeptidase A. The sequence (Lys, Met, Ala)–Thr–Leu–Arg–Ala–COOH seems to be present at the C terminus of each of the subunits, suggesting that the enzyme is composed of two very similar, perhaps identical, subunits (19).

The enzymic activity is almost completely dependent on the presence of a catalytic amount of sulfide. Sulfide can be replaced by other compounds including elemental sulfur, elemental selenium, hydroxylamine, and redox dyes with $E^{0'}$ higher than $+0.011$ V. All the cofactor-like compounds are necessary in catalytic amounts and, with the exception of hydryxlamine, are inhibitory when added over a critical concentration. The following compounds are inactive as cofactors; sulfite, sulfate, dithionite, elemental tellurium, NAD, NADP, FMN, ascorbic acid, menadione, hematin, and redox dyes with $E^{0'}$ lower than $+0.011$ V (10, 11).

This enzyme contains 1 gm atom of iron per mole of enzyme based on a molecular weight of 83,000 (11). The ESR spectrum of the native enzyme shows a signal at $g = 4.3$, typical of high spin Fe(III) in a ligand field of rhombic symmetry. The intensity of this signal indicates that nearly all of the iron in the enzyme is in the ferric state. The $g = 4.3$ signal is the same in the presence or absence of oxygen but is modified by the addition of the substrate, cysteamine, in the absence of oxygen. The modified signal reverts to the original one by following the introduction of air. On adding the activator, sulfide ion, to the enzyme, the signal at $g = 4.3$ decreases and a new signal appears at $g = 7.25$. The latter signal is abolished by the addition of substrate in the presence of air. These results indicate that changes in the environment of the iron occur followed by the addition of either activator or substrate. These changes are specific and fully reversible. It is, therefore, proposed that the ferric form of nonheme iron is involved in the catalytic action of cysteamine oxygenase (73).

IV. REACTION MECHANISM

A. Reaction Sequence

As discussed above, catechol oxygenases can be classified into two major groups, the intradiol and extradiol types. The former contains a ferric form of iron, whereas the latter contains a ferrous form of iron (63). In these dioxygenase reactions, two substrates are involved: One is an organic substrate and the other is molecular oxygen. The relationship between the valence state of iron contained in these dioxygenases and the interaction of these enzymes with their substrates has been the subject of intensive investigation.

Since it could be argued on theoretical grounds that oxygen combines preferentially with ferrous ion to form $Fe^{2+}O_2$ which is in equilibrium with $Fe^{3+}O_2^-$, whereas catechol forms a complex with ferric ion easily, a sequential reaction mechanism for dioxygenases was postulated as follows (59): Ferric ion-containing enzyme combines with an organic substrate, resulting in the reduction of iron, which then reacts with oxygen to form a ternary complex that subsequently yields an oxygenated end product. On the other hand, it was proposed that ferrous ion-containing enzyme reacts with oxygen first and then with substrate to form a product. Binding studies with metapyrocatechase (66) and protocatechuate 3,4-dioxygenase (23), however, revealed that the free enzyme can combine with organic substrate to form an enzyme–substrate complex, but no evidence has so far been obtained to indicate an interaction of the enzyme with oxygen. These data suggest a sequential ordered mechanism in which the enzyme combines first with organic substrate followed by the addition of oxygen, rather than a random mechanism irrespective of the valence state of iron.

On the other hand, steady state kinetic studies of steroid dioxygenase containing the ferrous form of iron reveal that the ordered bi-uni mechanism of the reverse order is more likely, i.e., addition of molecular oxygen first, followed by organic substrate combination, and, finally, release of product (81). In order to clarify this discrepancy we recently carried out steady state kinetic analyses with metapyrocatechase and protocatechuate 3,4-dioxygenase containing ferrous and ferric forms of iron, respectively (40).

Double reciprocal plots of initial velocity with one substrate as the fixed substrate and the other as the variable substrate give rise to intersecting patterns of lines for both enzymes (Figs. 7a and 7b), indicating thereby that a ternary complex of the enzyme with oxygen and an organic substrate is involved in these dioxygenase-catalyzed reactions. o-Nitrophenol and

m-phenanthroline, dead end inhibitors for metapyrocatechase, each inhibits the enzyme competitively with respect to the organic substrate, catechol, and noncompetitively with respect to the other substrate, molecular oxygen (Figs. 7c and 7d). Likewise, protocatechualdehyde, a dead end inhibitor for protocatechuate 3,4-dioxygenase and a structural analog of the organic substrate, inhibits this enzyme competitively with respect to protocatechuic acid and noncompetitively with respect to oxygen. These results are consistent with an ordered bi-uni mechanism where the organic substrate (S) first combines with the enzyme (E), and then reacts with oxygen to form a ternary complex (ESO_2) [Eq. (12)], where I is a dead end inhibitor. The data on product inhibition are not sufficiently conclusive to

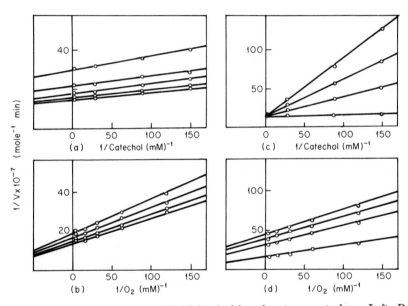

FIG. 7. Lineweaver-Burk plot of initial velocities of metapyrocatechase. Left: Reciprocal velocities are plotted as functions of reciprocal substrate concentrations. (a) Catechol concentrations were varied at fixed concentrations of oxygen. Oxygen concentrations in μM from top to bottom were 8.3, 15.7, 32.1, 63.3, and 250, respectively. (b) Oxygen concentrations were varied at fixed concentrations of catechol. Catechol concentrations in μM from top to bottom were 6.67, 11.1, 33.3, and 333, respectively. Right: Inhibition of metapyrocatechase by o-nitrophenol. (c) Catechol concentrations were varied at a fixed concentration of oxygen (250 μM). o-Nitrophenol concentrations in μM from top to bottom were 250, 167, 88.3, and 0, respectively. (d) Oxygen concentrations were varied at a fixed concentration of catechol (33 μM). o-Nitrophenol concentrations in μM from top to bottom were 333, 250, 167, and 0, respectively.

indicate whether the addition of substrates in the reaction is of an obligatory

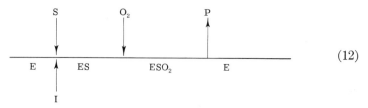

$$(12)$$

order or is random. Therefore, a rapid equilibrium random mechanism cannot be completely ruled out at this time.

B. Reaction Intermediates

Involvement of a ternary complex of oxygen, organic substrate, and enzyme has been proposed in various dioxygenase-catalyzed reactions (34, 42, 59, 66). However, such an oxygenated intermediate with nonheme iron-containing dioxygenases has not been detected as a discernible entity (66). In an attempt to identify reaction intermediates involved in the reaction of the nonheme iron-containing dioxygenases, spectral and kinetic studies were carried out with crystalline preparations of protocatechuate 3,4-dioxygenase, a trivalent nonheme iron-containing enzyme, by means of a rapid reaction method.

As described above, the enzyme has a red color and shows a broad absorption peak between 400 and 650 nm. When the substrate, proto-catechuic acid, is added under anaerobic conditions, the visible absorption shows an increase with a shift in the absorption peak from 465 to 480 nm, suggesting the formation of an enzyme–substrate complex (23) (Fig. 8a). On the other hand, the spectrum of the native enzyme is not altered by the addition of oxygen, nor does it change following exhaustive evacuation. It appears, therefore, that either the oxygen does not affect the absorption spectrum or the native enzyme is not oxygenated in the absence of organic substrate.

When the enzyme–substrate complex is mixed with a buffer solution containing oxygen, the absorbance at 470 nm decreases rapidly and reaches a minimum after 3 msec following which a gradual increase occurs so that the original enzyme–substrate complex is absorbed (Fig. 8b). Similar experiments conducted at different wavelengths results in a short-lived new spectral species which is reconstructed from the absorbance changes at each wavelength and the spectrum of the enzyme–substrate complex (Fig. 8a). Thus, a short-lived new spectral species of the enzyme is observed in the early stage of the reaction. This new species is characterized by a broad absorption band with a maximum between 500 and 520 nm, distinct

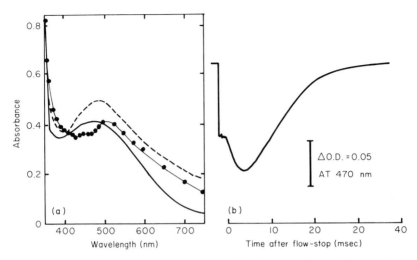

FIG. 8. The reconstructed spectrum appeared in the early stage of the reaction of (a) protocatechuate 3,4-dioxygenase and typical record of (b) a stopped flow experiment at 470 nm. Concentrations of protocatechuate 3,4-dioxygenase, protocatechuic acid, oxygen, and potassium phosphate buffer, pH 7.5, were 10.6 mg/ml, 1.25 mM, 140 μM, and 50 mM, respectively. The reaction was started by mixing the buffer solution containing oxygen with a mixture of enzyme and protocatechuic acid. (———) Absorption spectrum of the native enzyme; (- - -) after the addition of protocatechuic acid to the enzyme solution under anaerobic conditions; (-●-) the reconstructed spectrum of the intermediate which appeared 3 msec after the mixing of the enzyme–protocatechuate complex with a buffer solution containing oxygen. The spectrum was reconstructed from the absorbance change at different wavelengths and the spectrum of the enzyme–substrate complex. (Reproduced with permission from Fujisawa et al. (22).]

from those of the enzyme or the enzyme–protocatechuic acid complex. It can be demonstrated only in the presence of both protocatechuic acid and molecular oxygen (22).

The rate constant for the decomposition of the new spectral species is calculated to be 5,640 min⁻¹ per active site of the enzyme. Assuming that the enzyme contains eight active sites (23), the rate constant per mole of enzyme is calculated to be 45,120 min⁻¹, which agrees quite well with the turnover number of the enzyme (45,500 min⁻¹) determined by the overall reaction in a standard assay system (21).

A similar spectral species is also observed during the steady state of the reaction when enzymically active substrate analogs such as 3,4-dihydroxyphenylacetic acid or 3,4-dihydroxyphenylpropionic acid are used (Fig. 9). Further analysis of the detailed kinetics with these substrate analogs reveals that the rates of the overall reaction at different time intervals are proportional to the amount of intermediate present. The integrated sum of the intermediate present during the entire course of the reaction is also proportional to the initial concentration of oxygen (24).

These experimental results are consistent with the binding and steady state kinetic studies mentioned above. The new spectral species observed with protocatechuate 3,4-dioxygenase is really an obligatory intermediate and is a ternary complex of oxygen, substrate, and enzyme, i.e., an oxygenated intermediate. The decomposition of the oxygenated intermediate is the rate-limiting step for the overall reaction. In order to form the oxygenated intermediate, the presence of an organic substrate is necessary. Therefore, it seems reasonable to assume that the organic substrate combines with the enzyme first and then reacts with oxygen to form the oxygenated intermediate.

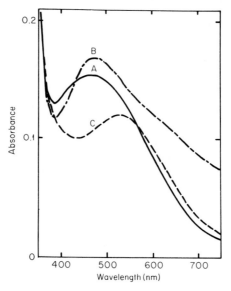

Fig. 9. Absorption spectra of the native, enzyme–substrate complex and oxygenated intermediate of protocatechuate 3,4-dioxygenase with 3,4-dihydroxyphenylpropionate as the substrate. The concentration of protocatechuate 3,4-dioxygenase, 3,4-dihydroxyphenylpropionic acid, oxygen, and potassium phosphate buffer, pH 7.5, were 3.87 mg/ml, 1, 0.26, and 50 mM, respectively. Curve A, spectrum of the native enzyme recorded without 3,4-dihydroxyphenylpropionic acid; curve B, spectrum of the enzyme–substrate complex recorded without oxygen; curve C, spectrum of the enzyme recorded during the steady state of the reaction. [Reproduced with permission from Fujisawa et al. (24).]

If the enzymic reaction proceeds via the mechanism given by Eq. (13),

$$E \underset{k_{-1}}{\overset{k_{+1}S}{\rightleftharpoons}} ES \underset{k_{-2}}{\overset{k_{+2}O_2}{\rightleftharpoons}} ESO_2 \xrightarrow{k_{+3}} E + P \qquad (13)$$

K_m values for oxygen and organic substrate should be expressed by $(k_{-2} + k_{+3})/k_{+2}$ and k_{+3}/k_{+1}, respectively, as the concentration of the other substrate in each case approaches infinity. In Table II, individual rate constants with various substrates are summarized together with Michaelis constants and enzyme turnover numbers determined from the overall reaction. Each value of k_{+3} is in good agreement with the turnover number based on the active site of the enzyme. With 3,4-dihydroxyphenylacetic acid as the substrate, the values of k_{+3}/k_{+1} and $(k_{-2} + k_{+3})/k_{+2}$ are almost equal to the K_m values for the organic substrate and oxygen, respectively. Similarly, the value of k_{+3}/k_{+1} with 3,4-dihydroxyphenylpropionic acid as the substrate coincides roughly with the K_m value (24).

Involvement of an oxygenated intermediate in oxygenase reactions has been reported for two heme-containing oxygenases: tryptophan 2,3-dioxygenase (41, 42) and cytochrome P-450 (18, 43), and for a flavoprotein, lysine monooxygenase (87). In the case of tryptophan 2,3-dioxygenase,

TABLE II

RATE CONSTANT AND MICHAELIS CONSTANTS[a]

Constants	Protocatechuic acid	3,4-Dihydroxy-phenylacetic acid	3,4-Dihydroxy-phenylpropionic acid
k_{+1}	—	$3.5 \times 10^4 \ M^{-1} \ S^{-1}$	$4.0 \times 10^3 \ M^{-1} \ S^{-1}$
k_{-1}	—	$7.7 \ S^{-1}$	$0.28 \ S^{-1}$
k_{+2}	—	$2.3 \times 10^5 \ M^{-1} \ S^{-1}$	—
k_{-2}	—	Almost zero	—
k_{+3}	$94 \ S^{-1}$	$0.18 \ S^{-1}$	$0.028 \ S^{-1}$
K_m for organic substrate	—	$6.7 \times 10^{-6} \ M$	$2.3 \times 10^{-5} \ M$
k_{+3}/k_{+1}	—	$5.1 \times 10^{-6} \ M$	$7.0 \times 10^{-5} \ M$
K_m for oxygen	—	$1.2 \times 10^{-6} \ M$	$4.5 \times 10^{-6} \ M$
$(k_{-2} + k_{+3})/k_{+2}$	—	$0.8 \times 10^{-6} \ M$	—
Turnover number[b]	$95 \ S^{-1}$	$0.15 \ S^{-1}$	$0.022 \ S^{-1}$

[a] Rate constants, k's, were obtained from the stopped flow experiments. K_m values and turnover numbers were determined from steady state kinetics of the overall reactions.

[b] These values were expressed per active site of the enzyme, based on the results that the enzyme had a molecular weight of 700,000 and was composed of eight identical subunits, each containing one substrate binding site.

a protoheme IX containing dioxygenase, an oxygenated intermediate has been spectrophotometrically established to be an obligatory intermediate (42). The results described in this chapter represent the demonstration of an enzymically active oxygenated form of a nonheme iron-containing enzyme. Hemerythrin, a nonheme iron-containing oxygen carrier in marine organisms, is considered to be a prototype for hemoglobin and can be oxygenated reversibly like hemoglobin (45). Protocatechuate 3,4-dioxygenase, however, cannot be oxygenated unless organic substrate is bound to the enzyme first. These findings are therefore analogous to the reaction mechanism of tryptophan 2,3-dioxygenase in which tryptophan combines with the enzyme first and then reacts with molecular oxygen to form an oxygenated intermediate (42). Moreover, they provide further support for our view that dioxygenase reactions in general involve a ternary complex of molecular oxygen, organic substrate, and enzyme (36, 66).

The nature of the oxygenated intermediate is still left unsolved. It could be a simple complex of oxygen, substrate, and enzyme; an activated state; or a state rather close to the enzyme–product complex. The elucidation of the basic characteristics of such an oxygenated intermediate may open the way to a better understanding of an active form of oxygen involved in these oxygenase-catalyzed reactions.

ACKNOWLEDGMENTS

The author would like to express his sincere appreciation to Professor O. Hayaishi for his many useful discussions and critical reading of this manuscript. Thanks are also due Professor A. Omachi for his aid in the preparation of the manuscript.

REFERENCES

1. Adachi, K., Takeda, Y., Senoh, S., and Kita, H. (1964). *Biochim. Biophys. Acta* **93,** 483.
2. Adachi, K., Iwayama, Y., Tanioka, H., and Takeda, Y. (1966). *Biochim. Biophys. Acta* **118,** 88.
3. Behrman, E. J., and Stanier, R. Y. (1957). *J. Biol. Chem.* **228,** 923.
4. Beychok, S. (1966). *Science* **154,** 1288.
5. Cain, R. B., and Cartwright, N. J. (1960). *Biochim. Biophys. Acta* **37,** 197.
6. Cain, R. B. (1962). *Nature (London)* **193,** 842.
7. Cavallini, D., De Marco, C., and Mondovi, B. (1961). *Nature (London)* **192,** 557.
8. Cavallini, D., De Marco, C., and Scandurra, R. (1962). *Ital. J. Biochem.* **11,** 196.
9. Cavallini, D., Scandurra, R., and De Marco, C. (1963). *J. Biol. Chem.* **238,** 2999.
10. Cavallini, D., Scandurra, R., and De Marco, C. (1965). *Biochem. J.* **96,** 781.
11. Cavallini, D., De Marco, C., Scandurra, R., Dupré, S., and Graziani, M. T. (1966). *J. Biol. Chem.* **241,** 3189.

12. Cavallini, D., Cannella, C., Federici, G., Dupré, S., Fiori, A., and Grosso, E. D. (1970). *Eur. J. Biochem.* **16,** 537.
13. Crandall, D. I. (1953). *Fed. Proc., Fed. Amer. Soc. Exp. Biol.* **12,** 192.
14. Crandall, D. I., Krueger, R. C., Anan, F., Yasunobu, K., and Mason, H. S. (1960). *J. Biol. Chem.* **235,** 3011.
15. Dagley, S., and Patel, M. D. (1957). *Biochem. J.* **66,** 227.
16. Dagley, S., and Stopher, D. A. (1959). *Biochem. J.* **73,** 16P.
17. Dagley, S., Geary, P. J., and Wood, J. M. (1968). *Biochem. J.* **109,** 559.
18. Estabrook, R. W., Hildebrandt, A. G., Baron, J., Netter, K. J., and Leibman, K. (1971). *Biochem. Biophys. Res. Commun.* **42,** 132.
19. Federici, G., Barra, D., Fiori, A., and Costa, M. (1971). *Physiol. Chem. Phys.* **3,** 448.
20. Flamm, W. G., and Crandall, D. I. (1963). *J. Biol. Chem.* **238,** 389.
21. Fujisawa, H., and Hayaishi, O. (1968). *J. Biol. Chem.* **243,** 2673.
22. Fujisawa, H., Hiromi, K., Uyeda, M., Nozaki, M., and Hayaishi, O. (1971). *J. Biol. Chem.* **246,** 2320.
23. Fujisawa, H., Uyeda, M., Kojima, Y., Nozaki, M., and Hayaishi, O. (1972). *J. Biol. Chem.* **247,** 4414.
24. Fujisawa, H., Hiromi, K., Uyeda, M., Okuno, S., Nozaki, M., and Hayaishi, O. (1972). *J. Biol. Chem.* **247,** 4422.
25. Gauthier, J. J., and Rittenberg, S. C. (1971). *J. Biol. Chem.* **246,** 3737.
26. Gauthier, J. J., and Rittenberg, S. C. (1971). *J. Biol. Chem.* **246,** 3743.
27. Gibson, D. T., Wang, K. C., Sih, C. J., and Whitlock, H. Jr. (1966). *J. Biol. Chem.* **241,** 551.
28. Gratzer, W. B., and Cowburn, D. A. (1969). *Nature (London)* **222,** 426.
29. Gross, S. R., Gafford, R. D., and Tatum, E. L. (1956). *J. Biol. Chem.* **219,** 781.
30. Guroff, G., and Ito, T. (1965). *J. Biol. Chem.* **240,** 1175.
31. Hayaishi, O., and Hashimoto, K. (1950). *J. Biochem. (Tokyo)* **38,** 371.
32. Hayaishi, O., Katagiri, M., and Rothberg, S. (1955). *J. Amer. Chem. Soc.* **77,** 5450.
33. Hayaishi, O., Katagiri, M., and Rothberg, S. (1957). *J. Biol. Chem.* **229,** 905.
34. Hayaishi, O. (1964). *Proc. Plenary Sessions, Int. Congr. Biochem., 6th, New York, 1964* **33,** 31.
35. Hayaishi, O. (1965). *In* "Oxidases and Related Redox Systems" (T. E. King, H. S. Mason, and M. Morrison, eds.), pp. 286–308. Wiley, New York.
36. Hayaishi, O. (1966). *Bacteriol. Rev.* **30,** 720.
37. Hayaishi, O. (1969). *Ann. N.Y. Acad. Sci.* **158,** 318.
38. Hayaishi, O., and Nozaki, M. (1969). *Science* **164,** 389.
39. Hirata, F., Nakazawa, A., Nozaki, M., and Hayaishi, O. (1971). *J. Biol. Chem.* **246,** 5882.
40. Hori, K., Hashimoto, T., and Nozaki, M. (1973). *J. Biochem. (Tokyo)* (in press).
41. Ishimura, Y., Nozaki, M., Hayaishi, O., Tamura, M., and Yamazaki, I. (1967). *J. Biol. Chem.* **242,** 2574.
42. Ishimura, Y., Nozaki, M., Hayaishi, O., Nakamura, T., Tamura, M., and Yamazaki, I. (1970). *J. Biol. Chem.* **245,** 3593.
43. Ishimura, Y., Ullrich, V., and Peterson, J. A. (1971). *Biochem. Biophys. Res. Commun.* **42,** 140.
44. Kawakami, K., Kita, H., and Senoh, S. (1967). *Int. Congr. Biochem., 7th Tokyo,* Abstr. IV, F-85.
45. Keilin, D. (1960). *Acta Biochim. Pol.* **7,** 415.

46. Kita, H. (1965). *J. Biochem. (Tokyo)* **58,** 116.
47. Kita, H., Kamimoto, M., Senoh, S., Adachi, T., and Takeda, Y. (1965). *Biochem. Biophys. Res. Commun.* **18,** 66.
48. Kita, H., Miyake, Y., Kamimoto, M., Senoh, S., and Yamano, T. (1969). *J. Biochem. (Tokyo)* **66,** 45.
49. Knox, W. E., and Edwards, S. W. (1955). *J. Biol. Chem.* **216,** 479.
50. Kojima, Y., Itada, N., and Hayaishi, O. (1961). *J. Biol. Chem.* **236,** 2223.
51. Kojima, Y., Fujisawa, H., Nakazawa, A., Nakazawa, T., Kanetsuna, F., Taniuchi, H., Nozaki, M., and Hayaishi, O. (1967). *J. Biol. Chem.* **242,** 3270.
52. Mehler, A. H. (1962). *In* "Oxygenases" (O. Hayaishi, ed.), pp. 87–99. Academic Press, New York.
53. Mitchell, R. A., Kang, H. H., and Henderson, L. M. (1963). *J. Biol. Chem.* **238,** 1151.
54. Nagami, K., and Miyake, Y. (1972). *Biochem. Biophys. Res. Commun.* **46,** 198.
55. Nagami, K. (1972). *Biochem. Biophys. Res. Commun.* **47,** 803.
56. Nakagawa, H., Inoue, H., and Takeda, Y. (1963). *J. Biochem. (Tokyo)* **54,** 65.
57. Nakazawa, A., Kojima, Y., and Taniuchi, H. (1967). *Biochim. Biophys. Acta* **147,** 189.
58. Nakazawa, A., Nakazawa, T., Kotani, S., Nozaki, M., and Hayaishi, O. (1969). *J. Biol. Chem.* **244,** 1527.
59. Nakazawa, T., Kojima, Y., Fujisawa, H., Nozaki, M., Hayaishi, O., and Yamano, T. (1965). *J. Biol. Chem.* **240,** PC3224.
60. Nakazawa, T., Nozaki, M., Hayaishi, O., and Yamano, T. (1969). *J. Biol. Chem.* **244,** 119.
61. Nozaki, M., Kagamiyama, H., and Hayaishi, O. (1963). *Biochem. Biophys. Res. Commun.* **11,** 65.
62. Nozaki, M., Kagamiyama, H., and Hayaishi, O. (1963). *Biochem. Z.* **338,** 582.
63. Nozaki, M., Kojima, Y., Nakazawa, T., Fujisawa, H., Ono, K., Kotani, S., Hayaishi, O., and Yamano, T. (1966). *In* "Biological and Chemical Aspects of Oxygenases" (K. Bloch and O. Hayaishi, eds.), pp. 347–368. Maruzen, Tokyo.
64. Nozaki, M., Fujisawa, H., and Kotani, S. (1967). *Int. Congr. Biochem., 7th Tokyo, 1964* p. 565.
65. Nozaki, M., Ono, K., Nakazawa, T., Kotani, S., and Hayaishi, O. (1968). *J. Biol. Chem.* **243,** 2682.
66. Nozaki, M., Nakazawa, T., Fujisawa, H., Kotani, S., Kojima, Y., and Hayaishi, O. (1968). *Advan. Chem. Ser.* (R. F. Gould, ed.), No. 77, pp. 242–251. Amer. Chem. Soc., Washington, D.C.
67. Nozaki, M., Kotani, S., Ono, K., and Senoh, S. (1970). *Biochim. Biophys. Acta* **220,** 213.
68. Ogasawara, N., Gander, J. E., and Henderson, L. M. (1966). *J. Biol. Chem.* **241,** 613.
69. Ono, K., Nozaki, M., and Hayaishi, O. (1970). *Biochim. Biophys. Acta* **220,** 224.
70. Ornston, L. N. (1966). *J. Biol. Chem.* **241,** 3787.
71. Peisach, J., Fujisawa, H., Blumberg, W. E., and Hayaishi, O. (1972). *Fed. Proc., Fed. Amer. Soc. Exp. Biol.* **31,** 448.
72. Priest, R. E., Bokman, A. H., and Schweigert, B. S. (1951). *Proc. Soc. Exp. Biol. Med.* **78,** 477.
73. Rotilio, G., Federici, G., Calabrese, L., Costa, M., and Cavallini, D. (1970). *J. Biol. Chem.* **245,** 6235.
74. Schepartz, B. (1953). *J. Biol. Chem.* **205,** 185.
75. Senoh, S., Kita, H., and Kamimoto, M. (1966). *In* "Biological and Chemical

Aspects of Oxygenases" (K. Bloch and O. Hayaishi, eds.), pp. 378–394. Maruzen, Tokyo.

76. Stanier, R. Y., and Ingraham, J. L. (1954). *J. Biol. Chem.* **210**, 799.
77. Stevens, C. O., and Henderson, L. M. (1959). *J. Biol. Chem.* **234**, 1188.
78. Suda, M., and Takeda, Y. (1950). *J. Biochem. (Tokyo)* **37**, 375.
79. Suda, M., and Takeda, Y. (1950). *J. Biochem. (Tokyo)* **37**, 381.
80. Tai, H. H., and Sih, C. J. (1970). *J. Biol. Chem.* **245**, 5062.
81. Tai, H. H., and Sih, C. J. (1970). *J. Biol. Chem.* **245**, 5072.
82. Takemori, S., Furuya, E., Suzuki, H., and Katagiri, M. (1967). *Nature (London)* **215**, 417.
83. Takemori, S., Furuya, E., Mihara, K., and Katagiri, M. (1968). *Eur. J. Biochem.* **6**, 411.
84. Takemori, S., Komiyama, T., and Katagiri, M. (1971). *Eur. J. Biochem.* **23**, 178.
85. Taniuchi, H., Kojima, Y., Kanetsuna, F., Ochiai, H., and Hayaishi, O. (1962). *Biochem. Biophys. Res. Commun.* **8**, 97.
86. Tokuyama, K. (1959). *J. Biochem. (Tokyo)* **46**, 1453.
87. Yamamoto, S., Hirata, F., Yamauchi, T., Nozaki, M., Hiromi, K., and Hayaishi, O. (1971). *J. Biol. Chem.* **246**, 5540.

5

α-KETOGLUTARATE-COUPLED DIOXYGENASES

MITCHEL T. ABBOTT and SIDNEY UDENFRIEND

I. INTRODUCTION

The α-ketoglutarate-dependent dioxygenases catalyze a variety of oxidations which range from the hydroxylation of a saturated carbon atom to the conversion of an aldehydic carbon to a carboxyl carbon. These dioxygenase reactions are described by the following general equation:

$$O_2 \;+\; S \;+\; \underset{\text{α-Ketoglutarate}}{\begin{array}{c} \text{COOH} \\ | \\ \text{CH}_2 \\ | \\ \text{CH}_2 \\ | \\ \text{COCOOH} \end{array}} \;\xrightarrow[\text{Fe}^{2+}]{\substack{\text{reducing} \\ \text{agent}}}\; SO \;+\; \underset{\text{Succinate}}{\begin{array}{c} \text{COOH} \\ | \\ \text{CH}_2 \\ | \\ \text{CH}_2 \\ | \\ \text{COOH} \end{array}} \;+\; CO_2 \qquad (1)$$

The requirement of α-ketoglutarate for enzyme-catalyzed hydroxylation was first described by Hutton et al. (83) for animal prolyl hydroxylase, the enzyme which hydroxylates specific prolyl residues of the collagen chain during its translation. Shortly thereafter, γ-butyrobetaine hydroxylase (114) and thymine 7-hydroxylase (4) were also shown to require α-ketoglutarate in addition to Fe^{2+} and ascorbate. Subsequent studies in a number of laboratories elucidated the role of α-ketoglutarate as a cosubstrate and demonstrated the stoichiometry shown above. Today we recognize the following α-ketoglutarate-dependent oxygenase reactions: prolyl hydroxylase, lysyl hydroxylase, γ-butyrobetaine hydroxylase, thymine 7-hydroxylase, 5-hydroxymethyluracil dioxygenase, 5-formyluracil dioxygenase, and pyrimidine deoxyribonucleoside 2'-hydroxylase. These enzymes, which occur in animals, plants, and microorganisms will be discussed individually. Although several articles have appeared which present the events leading up to our current knowledge of the prolyl and lysyl hydroxylases (8, 11, 28, 54, 60, 150, 176, 177), this chapter represents the first review in which the α-ketoglutarate-coupled dioxygenases which oxidize γ-butyrobetaine, pyrimidines, and nucleosides are treated in depth. The enzyme, p-hydroxyphenylpyruvate hydroxylase, is also discussed as a related α-ketoacid dioxygenase.

II. PROLYL HYDROXYLASE

$$\text{(2)}$$

A. Historical Background

Prolyl hydroxylase was first described in animal tissues associated with the formation of collagen; hence, the earlier names collagen proline hydroxylase (83), protocollagen hydroxylase (89), protocollagen proline hydroxylase (90), and peptidyl proline hydroxylase (127). As the specificity of the enzyme and the steps in collagen biosynthesis were elucidated, the name "prolyl hydroxylase" was proposed (178) as being less restrictive and more meanful. Prolyl hydroxylase, with properties comparable to those of the animal enzyme, is also found in plant tissues associated with the biosynthesis of hydroxyproline in cell wall proteins (160). A similar enzyme also appears to be present in certain microorganisms involved in the biosynthesis of hydroxyproline-containing actinomycins (87). The animal enzymes will be treated first.

B. Purification and Physical Properties

Prolyl hydroxylase has been purified most extensively from newborn rat skin (155) and chick embryo (14, 66, 140). The most purified preparations of enzyme show no absorption other than that in the far ultraviolet due to proteins (142, 143). Decarboxylation of α-ketoglutarate by prolyl hydroxylase is therefore not related to decarboxylation by an α-ketoglutarate dehydrogenase-type enzyme system which requires FAD, NAD, TPP, CoA, and lipoic acid.

The most purified rat skin enzyme formed 85–100 nmoles of hydroxyproline per minute per milligram of enzyme using reduced *Ascaris* cuticle collagen as substrate (155). The most purified chick embryo preparations

TABLE I

Amino Acid Compositions of Rat and Chick
Embryo Prolyl Hydroxylase

	Residues per 1000 residues	
	Rat skin[a]	Chick embryo[b]
Hydroxyproline	0	0
Aspartic acid	107	95
Threonine	57	50
Serine	66	100
Glutamic acid	149	180
Proline	50	35
Glycine	69	100
Alanine	83	75
Half-cystine	12	—
Valine	63	50
Methionine	11	15
Isoleucine	40	35
Leucine	95	65
Tyrosine	28	20
Phenylalanine	42	30
Lysine	77	85
Histidine	19	25
Arginine	44	35

[a] Data for the rat skin enzyme are from Rhoads
and Udenfriend (155).

[b] Chick embryo data are from Halme et al. (66).

TABLE II

Physical Properties of Prolyl Hydroxylase

Property	Rat skin enzyme[a]	Chick embryo enzyme[b]
Sedimentation coefficient	7.5	6.7
pH optimum	7.2	7.4
Isoelectric point	—	4.4

[a] From Rhoads and Udenfriend (155).

[b] From Halme et al. (66).

formed about 70–85 nmoles of hydroxyproline per minute per milligram of enzyme using the polytripeptide $(Pro–Gly–Pro)_n$, MW 6600–8000 (66).

The amino acid compositions and some other properties of the rat and chick embryo enzymes which had been purified almost to homogeneity are summarized in Tables I and II. Both enzymes contain relatively large amounts of aspartic and glutamic acid which account for their acidic character. The rat skin enzyme has a molecular weight of 130,000 and is apparently made up of two similar 65,000 subunits. Chick embryo enzyme preparations have been reported with widely divergent molecular weights ranging from 248,000 (142) to 350,000 (66). Subunits of about 113,000 were reported in the former study (142). More recently, Berg et al. (13) reported that chick embryo enzyme preparations yielded two dissimilar subunits of about 60,000 and 65,000. Olsen et al. (140) utilized electron microscopy to determine the shape and molecular dimensions of their chick embryo enzyme, but Pänkäläinen et al. (142) also carried out electron microscope studies and obtained quite different results. It should be kept in mind that chick embryos, though small, are made up of many tissues, some of which may contain different forms of prolyl hydroxylase; thus, embryo extracts may be made up of mixtures of various species of this enzyme. In fact, Cain and Fairbairn (26) have presented evidence for isozymes of prolyl hydroxylase in different tissues of Ascaris lumbricoides.

C. Factors Affecting Hydroxylase Activity

Ascorbate and Fe^{2+} were shown to be requirements of both the rat and chick enzymes at an early stage of purification. On further purification of the enzyme from rat skin, Hutton et al. (83) noted that partially purified preparations lost all activity on dialysis. Addition of ascorbate and Fe^{2+}, which had been implicated as requirements on less purified preparations, failed to restore activity. However, dialysates were capable of restoring full activity. Before embarking on the isolation of a possible new cofactor, attempts were made to see if any known chemical substances could replace dialysates. Many substances were tried, including all known cofactors. A mixture of all the Kreb's cycle acids was tried because of their known activation of another oxygenase, dopamine-β-hydroxylase (102). On further study it was found that only α-ketoglutarate was capable of restoring activity fully. Of the remaining acids only oxalacetate yielded even small amounts of activity (this disappeared on further purification). Activation by α-ketoglutarate was of great interest but did not establish it as the dialysate factor. Additional evidence for this was obtained when carbonyl reagents were shown to inhibit the reaction. The most direct evidence was obtained when dialysates were treated with α-ketoglutarate dehydrogenase in the presence of excess ammonia and NADPH. Under these conditions α-ketoglutarate is converted to glutamate. Such treatment destroyed the stimulatory activity which could then only be restored by adding back α-ketoglutarate.

The requirement for Fe^{2+} is very specific. This has been shown mainly by use of Fe^{2+} chelating agents which are inhibitory (16, 33) and through activation of purified enzyme by added Fe^{2+}. Other metal ions are inactive. However, even highly purified preparations of enzyme contain variable amounts of Fe^{2+} and are stimulated to varied extents by the metal ion. Kivirikko et al. (90) suggested that Fe^{2+} was bound loosely to the enzyme. From studies with chelators Hutton et al. (83) suggested that the ferrous ion was bound in a hydrophobic portion of the enzyme molecule. In contrast to the findings of Kivirikko et al. (90), Rencová et al. (152) have reported that Fe^{2+} is very firmly bound to the enzyme and cannot be removed by denaturation or treatment with chelating agents. On the other hand, the spectrophotometric studies of Pänkäläinen and Kivirikko (143) showed quite convincingly that highly purified chick embryo prolyl hydroxylase contains far less than one mole of iron per mole of enzyme (Table III). More recently, however, Hurych et al. (81) reported spectrophotometric evidence not only for the firm binding of ferrous iron to the enzyme but also for oxidation to the ferric form on interaction with molecular oxygen. It should be noted that the enzyme preparations of Kivirikko and his colleagues had been purified essentially to homogeneity. The purity of the preparations used by Hurych and his colleagues has not been indicated.

TABLE III

IRON CONTENT OF PREPARATIONS OF HIGHLY PURIFIED CHICK EMBRYO PROLYL HYDROXYLASE[a]

Preparation	Protein used (mg)	Iron found[b] (μg)	Moles of iron/mole of enzyme
1	0.15	<0.02	<0.06
2	1.00	0.07	0–0.3
3	1.70	0.05	0–0.1
Bovine serum albumin[c]	0.15	<0.02	—
Bovine serum albumin	1.00	0.07	—
Bovine serum albumin	1.70	0.09	—

[a] From Pänkäläinen and Kivirikko (143).

[b] Iron was determined on three different enzyme preparations by atomic absorption spectrophotometry.

[c] Highly purified bovine serum albumin was used as a control to indicate the limits of sensitivity of detection.

Purified enzyme also requires a reductant for activity. Of the reducing agents investigated, ascorbate is most effective. However, many other reductants can be used including ascorbic acid analogs, tetrahydropteridines, and reductones (83). Enzyme preparations can be activated to at least 75% of maximal activity obtained with ascorbate by adding high concentrations of dithiothreitol (155). Since many kinds of reductants can be used *in vitro*, one cannot determine which of the naturally occurring reducing agents participates in prolyl hydroxylation *in vivo*. Barnes and Kodicek (8) have reviewed the reasons to question a role for ascorbate in prolyl hydroxylase activity *in vivo*. Although the nature of the reductant is not certain, catalysis by prolyl hydroxylase does have an absolute requirement for a reductant. This may be to keep iron in the reduced form or perhaps to protect some sulfhydryl group in the enzyme (146). Apparent K_m values for the components of the prolyl hydroxylase system are shown in Table IV.

As with many other oxygenases, catalase was shown to be stimulatory. However, the degree of stimulation is variable and depends on the state of purification. The catalase effect is not exclusively the result of destruction of H_2O_2. Rhoads and Udenfriend (155) showed that the amounts of catalase which were stimulatory were far greater than were needed to destroy any conceivable amount of H_2O_2 which could be formed. Furthermore, even catalase, denatured by heat so that it would not destroy H_2O_2, was quite stimulatory. This latter type of stimulation is a protein effect which can best be shown with serum albumin or γ-globulins. As with most enzymes, the protein effect is most marked with highly purified preparations, par-

TABLE IV

APPARENT K_m VALUES FOR SUBSTRATES AND COFACTORS OF PROLYL HYDROXYLASE[a]

Reactant	K_m for rat skin enzyme (mM)	Ref.	K_m for chick embryo enzyme (mM)	Ref.
Fe^{2+}	0.001	126	0.005	88
Ascorbate	0.4[c]		0.3	90
α-Ketoglutarate	0.01	126	0.01	88
Prolyl substrate[c]	0.1	126	0.06	90
Oxygen	—		6 volume-%	90
			2.6 volume-%	83

[a] These values are apparent K_m values since they are not independent of one another.

[b] Unpublished observations, Cardinale, G. J., and Udenfriend, S.

[c] Utilizing (Pro–Gly–Pro)$_n$ substrates in the range 2000–8000 MW.

ticularly when the assay system is such that microgram quantities of hydroxylase are used. Large amounts of the enzyme are self-stimulatory and show less of a response to exogenous protein. Serum albumin, added at any time during an incubation, markedly stimulates microgram quantities of prolyl hydroxylase. The effect is not one of stabilizing the enzyme during storage but of enhancing the reaction during incubation. The mechanism of this stimulation by protein remains to be determined. Prolyl hydroxylase is also readily inhibited by sulfhydryl reagents such as p-chloromercuribenzoate indicating that it has a sulfhydryl group which may be near the reactive center (146). Purified chick embryo enzyme is stabilized during storage by large amounts of glycine (65). This effect is not nearly so marked with preparations of the rat skin enzyme.

Prolyl hydroxylase can hydroxylate proline residues in simple synthetic peptides. The only apparent requirement is that the proline precedes a glycine. The enzyme recognizes the sequence X–Pro–Gly–Y where X and Y can be any amino acid but glycine. Di-, tri-, and hexapolymers of the sequence $(Pro–Gly–Pro)_n$ (84) can serve as substrates. However, the enzyme has a higher affinity for substrates of high molecular weight with many susceptible prolyl groups (84, 88). Kinetic studies by Juva and Prockop (86) indicate that in polyvalent substrates each prolyl residue interacts with enzyme independently, the enzyme dissociating from the hydroxylated residue before attacking another prolyl group. Both K_m and V_{max} are also influenced by the nature of the amino acid residue in the vicinity of a susceptible prolyl group (126).

An interesting finding is that not all susceptible prolyl groups in collagen are hydroxylated *in vivo*. All collagens investigated thus far have been shown to be underhydroxylated to varying degrees and can be further hydroxylated when treated with purified prolyl hydroxylase (157). Since the susceptible prolyl groups of *Ascaris* cuticle collagen are almost totally unhydroxylated (53), it is an excellent substrate for prolyl hydroxylase and is actually utilized as the substrate in one of the standard assays for the enzyme (154).

D. Characterization of Reaction

Prolyl hydroxylase catalyzes the hydroxylation of specific prolyl residues in peptide linkage to *trans*-4-hydroxyprolyl residues. α-Ketoglutarate is an absolute requirement, serving as a cosubstrate, stoichiometric with the appearance of hydroxyprolyl groups (Fig. 1). Experiments were initially carried out with the chick embryo enzyme in the presence of $^{18}O_2$ to show that the reaction involves a direct oxygenation (52, 148, 149). Extension of these $^{18}O_2$ studies showed that molecular oxygen was incorporated into succinate as well as into hydroxyprolyl residues (27).

The overall reaction may be viewed as a sequence of reactions starting

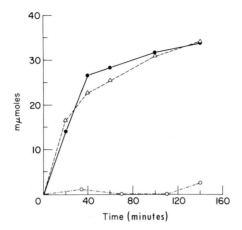

Fɪɢ. 1. Plot against time of hydroxyproline formed (△- - -△), α-ketoglutarate consumed (●—●), and α-ketoglutarate consumed in the absence of H(Pro–Gly–Pro)$_n$OH (○-··-○). The amount of α-ketoglutarate added was 37.8 mμmoles. From Rhoads and Udenfriend (154).

perhaps with the binding of iron to the enzyme, followed by interaction with oxygen, α-ketoglutarate, and prolyl substrate. Kivirikko et al. (90) pointed out the complexity involved in studying the kinetics of a reaction involving three or more substrates. They could not distinguish between a "sequential" or "ping-pong" type of mechanism. Whatever the kinetics of the reaction, the observed stoichiometry between utilization of α-ketoglutarate and prolyl substrate and the studies with $^{18}O_2$ are best explained by the formation of an intermediate peroxy compound bridging the 4 position of the prolyl residue to the 2 position of the α-ketoglutarate (Fig. 2). Cleavage of the peroxy compound would then yield the hydroxyprolyl group, succinate, and CO_2. This type of mechanism will be discussed in more general terms in Section VII.

Fɪɢ. 2. Hypothetical peroxy intermediate formed on interaction of α-ketoglutarate and a prolyl group in the presence of prolyl hydroxylase.

E. Enzymic Assays

Prolyl hydroxylase can be assayed in a number of ways, the method of choice depending on the state of purity of the enzyme and the information desired (156). The first and most widely used assay for the enzyme was based on the appearance of hydroxyproline, using proline as substrate. Labeled proline is incubated with slices or minces of collagen-forming tissues (chick embryo, rat skin, guinea pig granuloma, etc.) in the presence of an inhibitor of the hydroxylase, usually the Fe^{2+} chelator α,α'-dipyridyl. The labeled proline is incorporated into protein (protocollagen) which is extractable under the conditions used to extract collagen, but which contains little or no labeled hydroxyproline. This underhydroxylated collagen is incubated with enzyme and cofactors, and after hydrolysis the hydroxyproline is isolated and its radioactivity measured. This is a time-consuming method not well suited for enzyme purification. A variation of this method makes use of tritiated proline labeled in the 4-*trans* position (82). When such underhydroxylated collagen is isolated and incubated with prolyl hydroxylase, tritium is displaced by the entering hydroxyl group and is therefore released in proportion to the hydroxyproline formed. The tritium equilibrates with water which is distilled and assayed for radioactivity. This assay is quite sensitive and rapid once the substrate is prepared. One problem of all assays which utilize labeled protocollagen is that the specific activity of the substrate prolyl groups is not known. For this reason such assays yield enzymic activity only in arbitrary units which cannot be compared from laboratory to laboratory. An assay which utilizes carboxyl-labeled α-ketoglutarate (α-[1-^{14}C]ketoglutarate) is the simplest, most rapid, and yields data in molar units (154). The $^{14}CO_2$ evolved is equivalent to the hydroxyprolyl residues formed. Another advantage of this method is that any nonradioactive substrate can be used. Thus, the CO_2 assay is now used for all of the α-ketoacid dioxygenases. Modified *Ascaris* cuticle collagen* (123), an unhydroxylated form of collagen, is an excellent substrate for the assay of prolyl hydroxylase, but many synthetic substrates can also be used (see Section II,C). The CO_2 assay can only be used when the enzyme is purified sufficiently to remove all other α-ketoglutarate utilizing enzymes.

F. Regulatory Aspects of Prolyl Hydroxylase Activity

It was shown by Peterkofsky and Udenfriend (144) that hydroxylation begins after proline is incorporated into peptide linkage. Considerable evi-

---

* Reduced and carboxymethylated *Ascaris* cuticle collagen is commercially prepared by Dr. M. Chang, Korea Institute of Science and Technology, P. O. Box 131, Cheong Ryang 39-1, Hawolgok-Dong, Sungbuk-Ku, Seoul, Korea.

dence was then reported for hydroxylation occurring on prolyl groups in nascent chains of collagen (10, 55, 130). This was questioned by Prockop and coworkers (15) whose pulse label experiments with chick tibia preparations (85) suggested to them that hydroxylation did not begin until collagen chains were completed and separated from ribosomes. They coined the term "protocollagen" for unhydroxylated, completed collagen chains. Subsequent studies by Miller and Udenfriend (136), with minces of guinea pig granuloma, and by Lazarides and Lukens (101), with cultured fibroblasts, provided convincing evidence that hydroxylation of proline occurs largely during the process of translation. Nascent chains as small as 10,000 MW were shown to contain hydroxyproline. Conceivably some hydroxylation may occur after translation. However, protocollagen cannot be considered an obligatory intermediate of normal collagen biosynthesis. Recent studies by Al-Adnani and McGee (5a) with labeled antibodies to prolyl hydroxylase show that the enzyme is localized to the rough endoplasmic reticulum at the site of translation.

Studies in fibroblasts have revealed an inactive form of prolyl hydroxylase. This inactive form of the enzyme specifically cross reacts with antibody which was prepared by using pure prolyl hydroxylase as antigen (125). The cross reacting protein (CRP) is converted intracellularly to the active enzyme in the presence of lactate (34) or small amounts of ascorbate (169). Recently, CRP from mouse fibroblasts has been purified and shown to have a molecular weight of about 80,000–100,000 compared to active enzyme which has a molecular weight of 250,000–300,000 (128). Thus far it has not been possible to convert CRP to active enzyme except in intact cells. Immunological assay has revealed CRP in many animal tissues in amounts greater than enzyme. It may be that CRP is a precursor of the enzyme. However, the significance of this putative precursor and the mechanism for its conversion to enzyme are still to be determined. It may represent some regulatory mechanism, one which interestingly involves ascorbic acid.

G. Prolyl Hydroxylase in Plants

In 1951, Steward et al. (170) reported the presence of trans-4-hydroxyproline in hydrolysates of plant proteins. Subsequently, Lamport and Northcote (99) showed that the hydroxyproline-containing proteins (extensins) are synthesized at the growing tip of all higher plants and are then incorporated into the cell wall in glycosidic linkage with polysaccharides (98) through the hydroxyl group of hydroxyproline. Chrispeels (32) succeeded in showing unequivocally that the hydroxyproline was derived from proline already in peptide linkage, as is the case for collagen hydroxyproline, and Lamport (97) showed that the hydroxyl group arises from molecular oxygen. More recently, Sadava and Chrispeels (160) were able to purify prolyl hydroxylase from carrots about 24-fold. The purified enzyme required Fe^{2+} and ascorbate and also showed a requirement for α-ketoglutarate. The partially purified carrot enzyme did not utilize α-ketoglutarate exclusively but oxalacetate and pyruvate as well. It remains to be seen whether on further purification carrot prolyl hydroxylase will exhibit the absolute specificity for α-ketoglutarate that is shown by animal enzymes. Demonstration of stoichiometric conversion of α-ketoglutarate to succinate and CO_2 concomitant with peptidyl proline hydroxylation also requires further purification. Interestingly, the carrot enzyme was able to hydroxylate prolyl groups in underhydroxylated collagen, demonstrating that it recognizes the same substrate sequence as do the animal prolyl hydroxylases. Although the significance of extensin and other hydroxyproline-containing proteins in plants is not yet fully known, it is apparent that prolyl hydroxylase enzymes must have arisen very early in evolution.

III. LYSYL HYDROXYLASE

$$
O_2 + \begin{array}{c} CH_2NH_2 \\ | \\ CH_2 \\ | \\ (CH_2)_2 \\ | \\ R-NH-CH-COR' \end{array} + \begin{array}{c} \alpha\text{-Keto-} \\ \text{glutarate} \end{array} \xrightarrow[\text{Fe}^{2+}]{\begin{array}{c}\text{reducing}\\\text{agent}\end{array}} \begin{array}{c} CH_2NH_2 \\ | \\ CHOH \\ | \\ (CH_2)_2 \\ | \\ R-NH-CH-COR' \end{array} \begin{array}{c} + \text{ Succinate} \\ \\ + CO_2 \end{array} \quad (3)
$$

A. Purification

In most tissues lysyl hydroxylase activity is lower than that of prolyl hydroxylase. It has, therefore, been more difficult to purify. The similarity in cofactor requirements and the difficulty in separating the prolyl and lysyl hydroxylase activities from chick embryo extracts led Kivirikko and Prockop (89) to conclude that a single enzyme, protocollagen hydroxylase, was responsible for both activities. However, indirect evidence began to appear that the two activities were distinct (135, 155, 158). Complete separation of lysyl hydroxylase from prolyl hydroxylase was first reported by Miller (134) (Fig. 3). Newborn rat skin was used as the source. Shortly thereafter, Kivirikko and Prockop (91) and Popenoe and Aronson (147) reported separation of the two activities in chick embryo extracts.

The chick embryo enzyme has been purified over 250-fold by Kivirikko and Prockop (91) who reported fractions with molecular weights of approximately 550,000 and 200,000. Popenoe and Aronson (147) reported finding

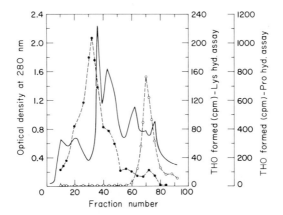

FIG. 3. Separation of lysyl hydroxylase from prolyl hydroxylase by DEAE-Sephadex chromatography. Ten milliliter fractions were collected from the DEAE-Sephadex column (2.5 by 40 cm). A 0.2-ml aliquot of each fraction was assayed for lysyl hydroxylase (●- - -●), and 0.005 ml aliquots were used in the prolyl hydroxylase assay (○- - -○). Also, the optical density at 280 nm was monitored (——). From Miller (134).

only one molecular species in their preparations of chick embryo enzyme but were not certain of the molecular weight. There is also a discrepancy between the pH optima reported for the two chick embryo preparations. Popenoe and Aronson reported a pH optimum between 8.0 and 8.4, whereas Kivirikko and Prockop gave a value of 7.4; both groups used Tris-HCl buffers. Conceivably, further purification of the enzyme will resolve this discrepancy.

The most purified preparations of Kivirikko and Prockop (91) were capable of hydroxylating 60 nmoles of lysyl residues per minute per milligram of protein at 37°, utilizing a synthetic substrate (92) and saturating concentrations of cofactors and cosubstrates.

B. Factors Affecting Hydroxylase Activity

Using crude extracts, Hausmann (70) was able to show requirements for α-ketoglutarate, Fe^{2+}, and ascorbate. Kivirikko and Prockop (91) obtained apparent K_m values of 0.05 mM for α-ketoglutarate, 0.001 mM for Fe^{2+}, and 0.05 mM for ascorbate. As with prolyl hydroxylase the activity of purified lysyl hydroxylase was stimulated by bovine serum albumin, catalase, and dithiothreitol. All of these are therefore included in the incubation mixtures used for assay. Popenoe and Aronson (147) reported that glycine afforded some stabilization of lysyl hydroxylase on storage. This was also reported previously for prolyl hydroxylase.

In a subsequent report Kivirikko *et al.* (92) utilized synthetic substrates to demonstrate that hydroxylation of lysyl groups was accompanied by a stoichiometric conversion of α-ketoglutarate to succinate and CO_2.

The substrate requirements for lysyl hydroxylation also appear to be comparable to those for prolyl hydroxylation. The synthetic substrates which have been prepared contain a Lys–Gly sequence (92). The length of the chain and the residues adjacent to this sequence influence the K_m and V_{max} of the substrates. Almost any peptide containing a Lys–Gly sequence can serve as substrate, and the vasoactive peptide vasopressin, which contains such a sequence, has been shown to be a good substrate. Additional factors may modify substrate utilization *in vivo*. Butler (25) has shown that cyanogen bromide peptides derived from the α chain of collagen contain mixtures of lysine and hydroxylysine in positions adjacent to glycine groups, indicating that these susceptible lysyl residues are not fully hydroxylated *in vivo*. Underhydroxylation of susceptible groups in completed collagen is considerable for both prolyl and lysyl groups, although the reason for this is not clear.

No studies have been reported as yet on the intracellular site for hydroxylation. It remains to be seen whether susceptible lysyl residues are hydroxylated on nascent chains as is the case for prolyl hydroxylation.

C. Enzymic Assays

Assays for lysyl hydroxylase are comparable to those used for prolyl hydroxylase. Incubation of tissue slices or minces with radioactive lysine in the presence of the Fe^{2+} chelator, α,α'-dipyridyl, yields collagen with susceptible lysyl groups which are not hydroxylated. This material is extracted with the collagen and used as substrate. If the lysine is labeled with tritium in the 5 position, then the entering hydroxyl groups displace the tritium resulting in the appearance of tritiated water which is a measure of the reaction rate. This method is used by Miller (135). If the lysine is labeled uniformly with ^{14}C then enzymic activity can be followed by isolation of the resulting [^{14}C]hydroxylysine by chromatographic methods as described by Popenoe and Aronson (147), or by treatment with periodate to yield [^{14}C]formaldehyde as described by Blumenkrantz and Prockop (17). With more purified preparations carboxyl labeled α-ketoglutarate has been used along with unlabeled substrate, and the assay followed by the appearance of $^{14}CO_2$ (92). As with prolyl hydroxylase, only the latter type of assay yields values for enzyme activity which can be compared from laboratory to laboratory and then only if the same, characterized synthetic peptide substrates are used.

D. Significance of Lysyl Hydroxylation

5-Hydroxylysine is found exclusively in the glycoprotein, collagen, which has been shown to contain residues of the disaccharide Gal–Glu (167). The monosaccharides are attached to the hydroxyl groups of hydroxylysine sequentially by specific enzyme systems (18, 168). Hydroxyproline groups are not substituted in this way. Cooper and Prockop (35) have presented evidence that hydroxylation is a prerequisite for secretion of collagen from the cell. Conceivably, collagen secretion is dependent on glycosylation of hydroxylysyl residues. As shown by Butler (25) normally synthesized and secreted collagen is far from being fully hydroxylated at susceptible lysyl residues. Obviously, if secretion of collagen requires hydroxylation and glycosylation, reaction of a small proportion of the susceptible lysyl residues is sufficient to satisfy the requirements of the secretory mechanisms. *Ascaris* cuticle collagen is secreted with little, if any, lysine hydroxylation.

IV. γ-BUTYROBETAINE HYDROXYLASE

$$O_2 + (CH_3)_3N^+CH_2CH_2CH_2COO^- + \alpha\text{-Ketoglutarate}$$
$$\gamma\text{-Butyrobetaine}$$

$$\xrightarrow[Fe^{2+}]{\text{reducing agent}} (CH_3)_3N^+CH_2CHOHCH_2COO^- + \text{Succinate} + CO_2 \quad (4)$$
$$\text{Carnitine}$$

A. Historical Background

Carnitine, which functions in both fatty acid oxidation and synthesis (51), has been found in several microorganisms as well as mammals (49, 172). In 1929, Linneweh (118) showed that the amount of carnitine in the urine increased when dogs were given γ-butyrobetaine and proposed that it is the precursor of carnitine. Bremer (21, 22) obtained further support for this proposal with experiments in which labeled potential precursors were injected into rats. An efficient conversion of γ-butyrobetaine to carnitine was demonstrated when γ-[1-^{14}C]butyrobetaine was injected into rats (22, 108) and mice (110). Lindstedt and co-workers prepared cell-free systems from rat liver (111) and from a strain of *Pseudomonas* obtained by enrichment culture on media containing γ-butyrobetaine (113, 117).

B. Purification

The most purified preparations of γ-butyrobetaine hydroxylase obtained from rat liver were reported to catalyze the hydroxylation of 3 nmoles of substrate per minute per milligram of protein. These preparations were purified by ammonium sulfate fractionation and hydroxyapatite chromatography procedures, but insufficient details of these procedures were presented to calculate the yield of enzyme obtained and the degree of purification (116). The enzyme from *Pseudomonas* has been purified 18-fold in 6% yield and has a specific activity of 220 nmoles/minute/mg (117). The enzymic assays were carried out either by measuring the rate of CO_2 formation from carboxy-labeled α-ketoglutarate, as described in Section II,E, or by resorting to chromatographic techniques after incubation with labeled γ-butyrobetaine (116, 117). Both the rat liver and *Pseudomonas* enzymes were found to have maximal activity at pH values near neutrality. Since the hydroxylases from the two sources are similar in many of their properties, the discussion which follows will pertain to both enzymes unless otherwise indicated.

C. Factors Affecting Hydroxylase Activity

Ferrous ion, ascorbate, and NADPH were found to stimulate the γ-butyrobetaine hydroxylase activity of the cell-free extract obtained from liver (109). NADPH could be replaced in the assay with an NADPH generating system, or in retrospect, an α-ketoglutarate generating system, since it contained isocitrate and isocitrate dehydrogenase. Some purification of the hydroxylase was effected with this type of supplementation (111, 113). Replacement of the generating system with α-ketoglutarate resulted in an equal or better stimulation of the hydroxylase activity (114).

With partially purified enzyme preparations (116, 117) the specificity of the requirement for α-ketoglutarate was found to be high. It could not be replaced by such compounds as glutarate, oxalacetate, 2-ketoadipate, glutamate, 2-hydroxyglutarate, and succinic semialdehyde. However, several of the analogs tested were inhibitory to the hydroxylase. α-Ketoglutarate is probably the cofactor *in vivo* since the addition of either glutamate-oxalacetate transaminase and asparate or of glutamate dehydrogenase, ammonium chloride, and NADPH inhibited the low level of hydroxylase activity observed in the absence of exogenous α-ketoglutarate. In those studies α-ketoglutarate had an apparent K_m value of about 0.5 mM.

The ferrous ion requirement was also found to be highly specific. 1,10-Phenanthroline was found to inhibit the hydroxylase activity of partially purified enzyme preparations in the absence of added iron. This inhibition could only be reversed by Fe^{2+}. Several of the metal ions tested inhibited the hydroxylase (112, 117). Apparent K_m values of 0.1 and 0.06 mM were obtained for Fe^{2+} with the rat liver and *Pseudomonas* hydroxylase systems.

Although ascorbate did have a stimulatory effect, appreciable hydroxylase activity was observed in the absence of added ascorbate, and it could be at least partially replaced by a number of reducing agents including tetrahydrofolate. Catalase stimulated the hydroxylase reaction, but the specificity of this effect was not investigated. Catalase was also shown to prevent the loss of activity which resulted from preincubation of the hydroxylase with a mixture of Fe^{2+} and ascorbate. Mercurials and other reagents which react with sulfhydryl groups were shown to inhibit the hydroxylase (116, 117).

Several analogs of γ-butyrobetaine were tested as substrates for the partially purified enzyme preparation from rat liver. Hydroxylation of 3-trimethylaminopropionic acid and of 4-dimethylaminobutyric acid did occur but at slower rates. However, 5-trimethylaminovaleric acid and quaternary amines without a carboxyl group were not hydroxylated to any extent (111). A number of substrate analogs were found to be inhibitory. Of these, 3-trimethylaminopropylsulfonate was the most effective (116). Bacterial γ-butyrobetaine hydroxylase preparations had a higher substrate specificity. A number of analogs of γ-butyrobetaine were found to be inhibitory, but none were hydroxylated themselves (117).

D. Characterization of Reaction

In the identification of carnitine as an enzymic product, nonradioactive carnitine was added as carrier to the hydroxylase system which had been incubated with radioactive γ-butyrobetaine. The mixture was purified and crystallized to constant specific activity as the gold salt of trimethylaminoacetone chloride (111). Lindstedt and co-workers (115) showed that α-ketoglutarate was degraded to succinate during the formation of carnitine. The succinate formed from α-[5-^{14}C]ketoglutarate in the γ-butyrobetaine hydroxylase reaction was identified by mixing the radioactive product with authentic succinate and achieving constant specific activity after subjecting the mixture to several chromatographic procedures as well as to recrystallizations (115, 116). It does not appear that the initial enzymic product of α-ketoglutarate is succinic semialdehyde since this compound was not able to substitute for α-ketoglutarate in the hydroxylase reaction and the formation of the aldehyde was not detected in incubation mixtures containing α-[5-^{14}C]ketoglutarate and nonradioactive succinic semialdehyde which had been added as a trapping agent (77).

In the studies of stoichiometry, the disappearance of α-ketoglutarate was measured enzymically with a spectrophotometric assay using glutamate dehydrogenase, ammonium chloride, and NADH. A 1:1 molar correlation between the utilization of α-ketoglutarate and the appearance of carnitine was demonstrated (115). A 1:1 molar correlation between the formation of CO_2 and carnitine was also established with the use of α-[1-^{14}C]keto-glutarate (116, 117).

The demonstration that the γ-butyrobetaine hydroxylase reaction was inhibited under anaerobic conditions suggested that the hydroxyl group of carnitine is derived from molecular oxygen (111, 113). That an intermediate peroxy compound is formed in the γ-butyrobetaine hydroxylase reaction gained strong support when Lindblad et al. (105) showed that one atom of molecular oxygen was incorporated into succinate. In that study the hydroxylase incubation mixture was incubated in the presence of oxygen gas enriched with [18]O. The succinate which was formed was converted to

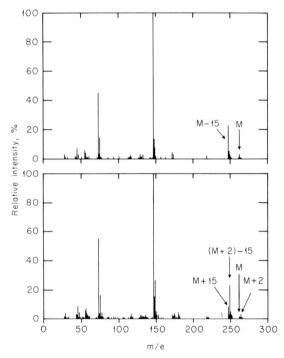

Fig. 4. Mass spectrum of the bis(trimethylsilyl)ester of succinic acid formed during the incubation of γ-butyrobetaine and α-ketoglutarate with γ-butyrobetaine hydroxylase in 90% [18]O_2 (lower graph) and of bis(trimethylsilyl)ester of authentic succinic acid (upper graph). M is the molecular ion and M-15 the positive ion after loss of one methyl radical from a trimethylsilyl group of the ester. The incorporation of one atom of [18]O in about 75% of the formed succinic acid is evident from the presence of the corresponding ions plus two mass units in the lower graph. From Lindblad et al. (105).

its bis(trimethylsilyl)ester and subjected to mass spectrometry. As shown in Fig. 4 about 75% of the succinate molecules were found to contain one atom of ^{18}O. The ^{18}O content of carnitine was not determined in that study.

V. DIOXYGENASES IN PYRIMIDINE AND NUCLEOSIDE METABOLISM

A. Historical Background

In their studies elucidating the reductive pathway of pyrimidine catabolism, Fink et al. (45) showed that a small amount of thymidine was converted to 5-hydroxymethyluracil and uracil-5-carboxylic acid in liver slices. Fink and Fink (46) showed that 5-hydroxymethyluracil and 5-formyluracil accumulated in the mycelia of Neurospora crassa during the conversion of thymidine to the uracil and cytosine of RNA in a process which does not involve fragmentation of the pyrimidine ring (47). On the basis of these findings they proposed an oxidative pathway of pyrimidine catabolism in which the methyl group of thymidine is removed in a series of oxidative steps (46). Abbott and co-workers have subsequently shown that enzymes preparations obtained from Neurospora catalyze the reactions shown in Fig. 5. Dioxygenases appear to catalyze each of the oxidative steps in the proposed pathway. In intact Neurospora cells, uracil and thymine may be formed via hydrolysis and phosphorylysis of deoxyribonucleosides and ribonucleosides although only the depicted hydrolysis of ribonucleosides was observed in the studies carried out with partially purified enzyme preparations (165).

Pyrimidineless mutants of Neurospora have been developed which are unable to utilize various intermediates in the pathway (186). The pyr-4, uc-1, uc-2 mutant of Neurospora can grow in media supplemented with thymine, 5-hydroxymethyluracil, 5-formyluracil or uracil but not when the supplement is thymidine, deoxyuridine, or deoxycytidine. The uc-2 mutation would appear to affect an enzyme which hydrolyzes, phosphorylyzes, or hydroxylates deoxyribonucleosides. Preliminary findings suggest that it is the hydroxylase which is defective (166). The pyr-4, uc-1, uc-3 mutant cannot utilize any compound preceding 5-formyluracil in the pathway. No mutant was obtained which either accumulated or utilized uracil-5-carboxylic acid (186). Additional organisms which may contain enzymes which oxidatively demethylate thymidine are Acetabularia (20), Spirogyra (132, 171), and Paramecium aurelia (12). The RNA of these organisms became labeled when they were grown in the presence of radioactive thymidine. Yeast has been shown to convert thymine to 5-hydroxymethyluracil and uracil-5-carboxylic acid (182). Little or no thymidine is in-

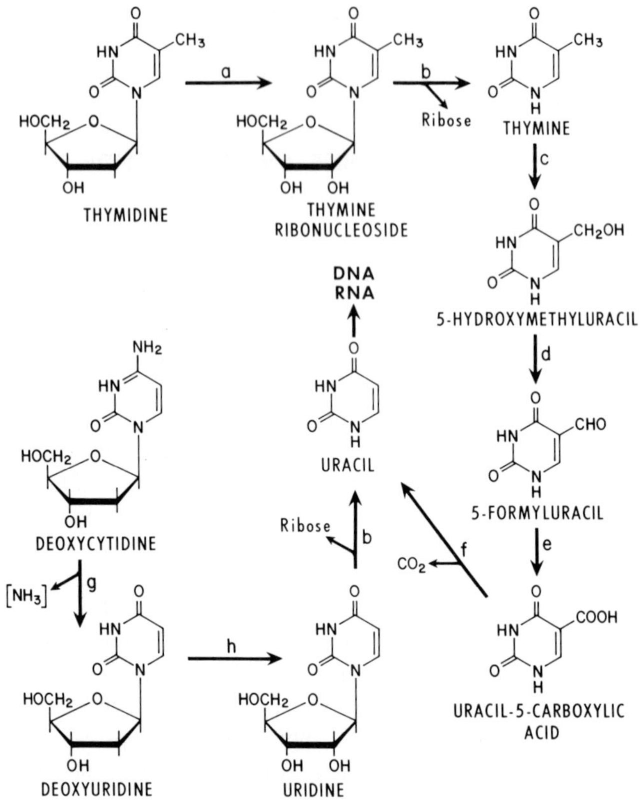

Fig. 5. Reactions catalyzed by enzyme preparations from *Neurospora crassa:* a, pyrimidine deoxyribonucleoside 2'-hydroxylase (163); b, hydrolase reaction (165); c, thymine 7-hydroxylase (3); d, 5-hydroxymethyluracil dioxygenase reaction (5); e, 5-formyluracil dioxygenase reaction (183); f, uracil-5-carboxylic acid decarboxylase (141); g, deaminase reaction (165); and h, pyrimidine deoxyribonucleoside 2'-hydroxylase (164).

corporated into RNA in mammals† (76) although, as mentioned above, 5-hydroxymethyluracil and uracil-5-carboxylic acid are formed. Uracil-5-carboxylic acid decarboxylase is apparently not present in mammals since following intraperitoneal injection of uracil-5-carboxylic acid into mice, the compound was quantitatively recovered in the urine (73). However, a pyrimidine deoxyribonucleoside 2'-hydroxylase reaction may occur in mammals since some direct transformation of deoxyuridine into the uridine

† Fink, R. M., personal communication.

of RNA without detachment of the deoxyribose moiety did appear to occur in experiments carried out with intact Ehrlich ascites cells (63).

B. Purification‡

Subjecting *Neurospora* extracts to the scheme shown in Table V resulted in the partial purification and separation of the enzymes involved in the oxidative demethylation of thymidine (119, 120). Uracil-5-carboxylic acid decarboxylase was removed from the rest of the enzymes in the pathway by the calcium phosphate gel fractionation procedure (steps 2 and 3). No significant separation of any of the other enzymes was effected until the DEAE-cellulose chromatography step had been carried out. Three peaks of enzymic activity were detected in the eluant from the DEAE-cellulose column as shown in Fig. 6. The peak, which was first eluted from the column, catalyzed the thymine 7-hydroxylase, 5-hydroxymethyluracil oxygenase, and 5-formyluracil oxygenase reactions, which had typical specific activities (units/mg) of 50, 40, and 25, respectively. This fraction, however, catalyzed neither the 2'-hydroxylase reaction nor the hydrolysis of either thymine ribonucleoside or uridine. The 2'-hydroxylation of thymi-

TABLE V

PURIFICATION OF PYRIMIDINE DEOXYRIBONUCLEOSIDE 2'-HYDROXYLASE
FROM *Neurospora*[a]

Step	Fraction	Protein (mg/ml)	Activity (units)	Specific activity (units/mg)	Yield(%)	Purification (-fold)
1	Extract	10	1480	0.8	100	1
2	Calcium phosphate I	2.5	1080	2.5	72	3
3	Calcium phosphate II	1.5	1700	8.7	115	11
4	Ammonium sulfate	11	1380	7.3	93	9
5	Sephadex G-150	0.95	1210	13	82	16
6	DEAE-cellulose	0.35	450	53	30	67

[a] From Liu *et al.* (119).

‡ A unit of activity with respect to the dioxygenases considered in this section is defined as the amount of enzyme which catalyzes the oxidation of 1 nmole of substrate per minute per milligram of protein under the conditions of the standard assay.

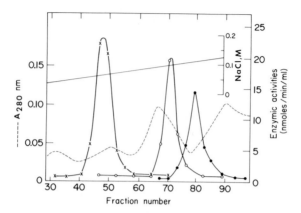

Fig. 6. Effect of DEAE-cellulose chromatography on thymine 7-hydroxylase and pyrimidine deoxyribonucleoside 2′-hydroxylase activities. The Sephadex G-150 enzyme preparation which was applied to the DEAE-cellulose column had a volume of 100 ml and a protein concentration of 0.85 mg/ml: (×) 7-hydroxylase activity, (○) 2′-hydroxylase activity, and (●) hydrolase activity. From Liu *et al.* (119).

dine and of deoxyuridine was catalyzed by the fraction constituting the second peak eluted from the column. In experiments in which the fractions containing the 2′-hydroxylase and 7-hydroxylase activities were recombined no inhibitory or stimulative effects were detected. The fraction constituting the third peak catalyzed the hydrolysis of both thymine ribonucleoside and uridine. This purification scheme has recently been refined and extended. Several crude extracts have been subjected to the new scheme, and a thymine 7-hydroxylase fraction has been obtained in satisfactory yield with a specific activity of 1300 units/mg (121). Lindstedt and coworkers (6, 7) have shown that DEAE-Sephadex chromatography can also separate the 2′-hydroxylase from the 7-hydroxylase. Their purification scheme involves subjecting a crude extract of *Neurospora* to chromatography through a hydroxylapatite column and subsequently through a DEAE-Sephadex column which yielded two enzymic fractions. One catalyzed the 2′-hydroxylase reaction and had a specific activity of 16 units/mg. No specific activities were reported for the other DEAE-Sephadex fraction which catalyzed the 7-hydroxylase, 5-hydroxymethyluracil oxygenase, and 5-formyluracil oxygenase reactions. With regard to the "hydroxylapatite fraction" these reactions had specific activities of 2.1 units/mg, 1.3 units/mg, and 0.2 unit/mg, respectively (79).

As described in Section II,E the activities of the oxygenases can be determined by measurement of the amount of CO_2 formed from carboxy-labeled α-ketoglutarate. This rapid and convenient assay was most instrumental in the development of the purification schemes. However, when crude enzyme preparations are assayed, the presence of other enzymes which decarboxylate α-ketoglutarate necessitate the use of assays which employ chromatographic techniques (7, 124) for determining other products of the reactions.

Gel filtration was used to determine that the molecular weight of the 2'-hydroxylase is 47,000 (7). The data of this determination also indicate that the molecular weight of the 7-hydroxylase is somewhat lower than that of the 2'-hydroxylase. Isoelectric focusing was used to determine that the isoelectric point of the 2'-hydroxylase is 4.6 and that of the 7-hydroxylase is 4.9 (7). The pH optima of the 2'-hydroxylase and 7-hydroxylase are about 6.5 (7) and 7.0 (124).

Whether the sequential reactions converting thymine to uracil-5-carboxylic acid are catalyzed by the same enzyme is not known, but no separation of the three activities was detected in any of the purification steps (6, 119, 124). In addition, a purification scheme carried out on the mutant, which utilized 5-formyluracil but not thymine or 5-hydroxymethyluracil, yielded a protein fraction in which only the 2'-hydroxylase activity was detected, whereas the protein fraction obtained by the same purification scheme carried out on the wild type strain contained the 2'-hydroxylase, 7-hydroxylase, 5-hydroxymethyluracil oxygenase and a small amount of the 5-formyluracil oxygenase activities (6). In preliminary experiments it was shown that intact cells of this mutant can convert 5-formyluracil to uracil-5-carboxylic acid (166). Perhaps a relatively non-specific enzyme is responsible for this conversion. The results of some competition-type experiments carried out with the 2'-hydroxylase suggest that both deoxyuridine and thymidine are hydroxylated by the same enzyme (166). That one enzyme is involved is also suggested by initial studies carried out with the mutant mentioned above which cannot utilize thymidine. Deoxyuridine and thymidine were not hydroxylated by either the intact cells of this mutant strain or by enzyme fractions prepared from them (166).

C. Thymine 7-Hydroxylase

$$O_2 + \text{Thymine} + \alpha\text{-Keto-glutarate} \xrightarrow[\text{Fe}^{2+}]{\text{reducing agent}} \text{5-Hydroxymethyluracil} + \text{Succinate} + CO_2 \quad (5)$$

1. Factors Affecting Hydroxylase Activity

Initially the cell-free conversion of thymine to 5-hydroxymethyluracil was demonstrable only when NADPH and GSH were added to the incubation mixture (2). Since this conversion was stimulated by molecular oxygen it appeared that thymine 7-hydroxylase belonged to the group of enzymes referred to as monooxygenases or mixed-function oxidases (1, 3). Ascorbate and Fe^{2+} were also found to be stimulatory, but recombination studies with fractions eluted from a Sephadex G-50 chromatography column indicated that an additional cofactor was required. The testing of α-ketoglutarate in this system (4) was prompted by reports of its requirement in the prolyl hydroxylase reaction (83).

The specificity of the α-ketoglutarate requirement appeared high (4) even when studied with the relatively impure calcium phosphate gel fraction (Table V). Little or no activity was observed when the following compounds were substituted for α-ketoglutarate and tested with this enzyme fraction: glutarate, levulinate, α-ketovalerate, pyruvate, glyoxylate, oxalacetate, isocitrate, glutamate, citrate, succinate, fumarate, NADPH, and NADH (133). α-Ketoglutarate had an apparent K_m value of 0.2 mM (78).

The Fe^{2+} requirement was also found to be highly specific when tested with the calcium phosphate gel fraction. Fe^{2+} could not be replaced by Mg^{2+}, Mn^{2+}, Cu^{2+}, Cd^{2+}, Ni^{2+}, Zn^{2+}, Ca^{2+}, Hg^{2+}, Al^{3+}, Co^{3+}, Na^+, K^+, Li^+, or Ag^+. Furthermore, many of these ions were inhibitory. No 7-hydroxylase activity was detected in the presence of 0.25 mM Fe^{2+} when Mg^{2+}, Cu^{2+}, Cd^{2+}, Ni^{2+}, Zn^{2+}, Hg^{2+}, Co^{3+}, or Ag^+ was included in the incubation mixture at a concentration of 0.5 mM. Appreciable hydroxylase activity was observed in the absence of added Fe^{2+}. However, this activity was eliminated by preincubation of the enzyme preparation with 1 mM EDTA at 0° for 15 minutes. The activity was fully restored to the EDTA-treated preparation by dialysis followed by incubation in the presence of Fe^{2+}. Both α,α'-dipyridyl (0.5 mM) and 1,10-phenanthroline (0.5 mM) also completely inhibited the activity observed in the absence of added Fe^{2+} (133).

The ascorbate requirement was of low specificity, but it appeared that ascorbate and isoascorbate were more stimulatory than the other reducing agents tested (Table VI). Catalase also had a stimulatory effect on the hydroxylase reaction (78, 121).

The substrate specificity of thymine 7-hydroxylase in the calcium phosphate gel fraction was examined using [2-^{14}C]thymidine, [2-^{14}C]thymine ribonucleoside (165), and 5-[^{3}H]methylcytosine (133). No hydroxylation of the methyl groups of these compounds was detected.

TABLE VI

SPECIFICITY OF THE REDUCTANT REQUIREMENT OF THE THYMINE
7-HYDROXYLASE REACTION[a]

Addition to 7-hydroxylase system[b]	Concn. (mM)	5-Hydroxy-methyluracil produced (nmoles)
None		0.3
Ascorbate	0.03	9.5
	1.0	18.7
	10	16.7
	100	4.5
Glutathione	1.0	1.6
	10	2.4
	100	2.0
Ascorbate (plus 1 mM glutathione)	0.03	10.0
	1.0	15.5
	10	10.4
Isoascorbate	1.0	17.6
Cysteine	1.0	1.2
2-Mercaptoethanol	10	3.1
Tetrahydrofolate[c] (plus 1 mM 2-mercaptoethanol)	1.0	2.8
NADPH	10	0.1
Catechol	0.5	0.7

[a] Midgett et al. (133).
[b] The 7-hydroxylase system contained 0.5 mM α-ketoglutarate, 0.5 mM Fe SO4, 0.25 mM [2-14C]thymine, and the calcium phosphate gel fraction (step 3). The incubations were carried out for 10 minutes at 30°.
[c] Tetrahydrofolate was prepared in 0.1 M 2-mercaptoethanol.

2. Characterization of Reaction

5-Hydroxymethyluracil was identified as an enzymic product of [2-14C] thymine in the following way: After incubation, unlabeled 5-hydroxymethyluracil was added to the deprotenized reaction mixture and the mixture was subjected to paper chromatography in seven diverse solvent systems. Coincidence of radioactive product and the authentic compound was demonstrated with radioautography (3). Succinate was identified as the product of α-[5-14C]ketoglutarate using silicic acid chromatography and paper chromatography in conjunction with radioautography (78, 124).

McCroskey *et al.* (124) showed that the conversion of thymine to 5-hydroxymethyluracil is coupled both to the conversion of α-[1-^{14}C]ketoglutarate to $^{14}CO_2$ and of α-[5-^{14}C]ketoglutarate to [^{14}C]succinate. The work of Holme *et al.* (78) also showed that the formation of radioactive CO_2 from α-[1-^{14}C]ketoglutarate is stoichiometric with the hydroxylation of thymine.

The hydroxyl group of 5-hydroxymethyluracil appeared to be derived from molecular oxygen since replacing the air in the incubation vessel with oxygen did not alter the enzymic activity, while an atmosphere of nitrogen reduced the activity essentially to zero (3). The crucial $^{18}O_2$ studies were carried out by Holme *et al.* (79). The ^{18}O content of the products was measured by means of mass spectrometry after gas chromatographic separation of the trimethylsilyl derivatives of 5-hydroxymethyluracil and succinate. One atom of molecular oxygen was shown to be incorporated into 5-hydroxymethyluracil and another into succinate.

D. 5-Hydroxymethyluracil Dioxygenase and 5-Formyluracil Dioxygenase Reactions

$$\tag{6}$$

$$\tag{7}$$

1. Factors Affecting Oxygenase Activities

When the calcium phosphate gel fraction (Table V) was used to study the highly active conversion of thymine to 5-hydroxymethyluracil, a trace amount of a new product was formed which appeared to be 5-formyluracil. Cofactors which are usually required by dehydrogenases did not stimulate the oxidation of 5-hydroxymethyluracil to 5-formyluracil. Unexpectedly, this conversion was only demonstrable when α-ketoglutarate, Fe^{2+}, and ascorbate were included in the incubation mixture (5). Similarly, the conversion of 5-formyluracil to uracil-5-carboxylic acid was stimulated by these three cofactors (183). Since the 5-hydroxymethyluracil dioxygenase and 5-formyluracil dioxygenase reactions are similar in many of their properties, the discussion which follows will pertain to both reactions unless otherwise indicated.

No activity was detected in the absence of all three cofactors, but some activity was obtained when one of the cofactors was omitted during incubation of the calcium phosphate gel fraction (step 3, Table V). When one of the cofactors was omitted from the reaction mixture, increasing the concentrations of the other two did not do away with the requirement for the omitted cofactor (5, 183). The specificity of the requirement for α-ketoglutarate appears high since it could not be replaced by such compounds as isocitrate, glutarate, NAD^+, and FAD. Although oxalacetate and glutamate did partially substitute for α-ketoglutarate in incubations with the calcium phosphate gel fraction (5, 183), when more highly purified enzyme fractions were tested, α-ketoglutarate was the only α-ketoacid which met this requirement (79). The appreciable dioxygenase activity observed in the absence of added Fe^{2+} was eliminated by pretreatment of enzyme preparations with EDTA. This metal ion requirement appears highly specific since no activity was observed when such metal ions as Cu^{2+}, Ni^{2+}, Co^{2+}, Zn^{2+}, or Cr^{3+} was added in place of Fe^{2+}. The ascorbate requirement was found to be the least specific. Such compounds as glutathione and dithiothreitol were at least partially effective as replacements for ascorbate (5, 183).

2. Characterization of Reactions

5-Formyluracil and uracil-5-carboxylic acid were identified as products of the 5-hydroxymethyluracil dioxygenase and 5-formyluracil dioxygenase reactions by paper chromatography in conjunction with radioautography and ion-exchange chromatography (5, 183). Succinate was identified as a product by paper chromatography (119) and mass spectrometry (79).

In studies of the stoichiometry of the 5-hydroxymethyluracil oxygenase reaction, the further oxidation of the 5-formyluracil produced by the 5-formyluracil oxygenase reaction did not have to be contended with since little or no uracil-5-carboxylic acid was formed when the incubation period was not prolonged. It is pertinent in this regard that the 5-formyluracil oxygenase reaction has a lower specific activity (see Section V,B) and seems to be inhibited by 5-hydroxymethyluracil. Liu et al. (119) showed that when 5-[7-^{14}C]hydroxymethyluracil and α-[1-^{14}C]ketoglutarate were used as substrates in experiments in which the time of incubation was varied, CO_2 and 5-formyluracil were produced in a 1:1 molar ratio. When unlabeled 5-hydroxymethyluracil and α-[1,5-^{14}C]ketoglutarate were used as substrates, CO_2 and succinate were produced in a 1:1 molar ratio. Similarly, in experiments in which 5-formyluracil was used as substrate, uracil-5-carboxylic acid, succinate, and CO_2 were produced in a 1:1:1 molar ratio (Fig. 7).

That molecular oxygen is incorporated into the formyl group of 5-formyluracil in the 5-hydroxymethyluracil dioxygenase reaction and into the carboxyl group of uracil-5-carboxylic acid in the 5-formyluracil dioxygenase reaction was suggested by studies which showed that these reactions were inhibited by exclusion of oxygen from the reaction mixtures (5, 183). In experiments with $^{18}O_2$, one atom of molecular oxygen was shown to be incorporated into uracil-5-carboxylic acid in its formation from 5-formyluracil. A similar attempt to show the incorporation of molecular oxygen into 5-formyluracil in its formation from 5-hydroxymethyluracil has not been made because of the anticipated rapid exchange of the oxygen of the formyl group with that of water. However, one atom of molecular oxygen has been shown to be incorporated into succinate in each of these reactions (79).

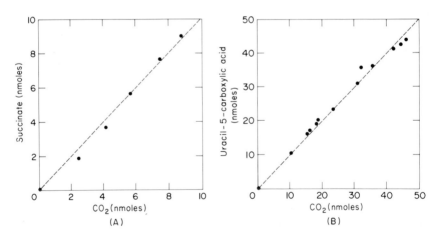

FIG. 7. The stoichiometry of the 5-formyluracil dioxygenase reaction. The standard incubation mixture contained the fraction eluted from the DEAE-cellulose column. The period of incubation was varied. (A) Incubation mixtures with nonradioactive 5-formyluracil and α-[1_5-^{14}C]ketoglutarate; (B) with 5-[2-^{14}C]formyluracil and α-[1-^{14}C]-ketoglutarate. From Liu et al. (119).

E. Pyrimidine Deoxyribonucleoside 2′-Hydroxylase Reaction

$$O_2 \ + \ \text{HOCH}_2 \text{(Thymidine)} \ + \ \begin{array}{c} \alpha\text{-Keto-} \\ \text{glutarate} \end{array} \xrightarrow[\text{Fe}^{2+}]{\text{reducing agent}} \text{HOCH}_2 \text{(Thymine ribonucleoside)} \ + \ \text{Succinate} \ + \ CO_2 \tag{8}$$

Thymidine → Thymine ribonucleoside

$$O_2 \ + \ \text{HOCH}_2 \text{(Deoxyuridine)} \ + \ \begin{array}{c} \alpha\text{-Keto-} \\ \text{glutarate} \end{array} \xrightarrow[\text{Fe}^{2+}]{\text{reducing agent}} \text{HOCH}_2 \text{(Uridine)} \ + \ \text{Succinate} \ + \ CO_2 \tag{9}$$

Deoxyuridine → Uridine

1. 2′-Hydroxylation without Detachment of Sugar Moiety

In the course of studies on the substrate specificity of the thymine 7-hydroxylase reaction, it was noted that although the methyl group of thymidine was not hydroxylated thymine ribonucleoside was formed. It did not appear that the formation of this compound occurred by the methylation of uracil at the RNA level (129) or by the ribosylation of thymine (131) since the capacity to form the ribonucleoside was not lost by enzyme preparations which had been subjected to some purification and passed through a Sephadex G-25 chromatography column. Further evidence that the conversion of thymidine to thymine ribonucleoside occurred without detachment of the deoxyribose was obtained with the use of thymine which was uniformly labeled with ^{14}C and enriched with ^3H in position 6 of the pyrimidine ring. Table VII shows that the ratio of the specific activities of ^3H to ^{14}C was the same for substrate and product. Furthermore, this ratio was not changed when nonradioactive, potential intermediates, e.g., ribose 1-phosphate, were included in the incubation

TABLE VII

RETENTION OF ^{14}C AND ^{3}H IN THE THYMINE
RIBONUCLEOSIDE PRODUCED[a]

	Specific radioactivity (Ci/mole)		Ratio
	^{14}C	^{3}H	^{3}H/^{14}C
Thymidine[b]	3.0	6.2	2.1
Thymine riboside[b]	3.1	5.9	1.9
Thymidine[c]	3.0	6.0	2.0
Thymine[b]	1.7	6.3	3.7

[a] From Shaffer *et al.* (163).

[b] Isolated, after 25 minutes of incubation, from standard incubation system containing the Sephadex-treated ammonium sulfate preparation (2.4 mg protein/ml). The amounts of thymine riboside and thymine produced were 5.0 and 2.0 nmoles, respectively.

[c] From stock solution of thymidine (uniformly labeled with respect to ^{14}C and only in the pyrimidine ring with ^{3}H) which was the source of the substrate for the standard incubation system.

mixture (163). The conversion of deoxyuridine to uridine has also been shown to occur at the nucleoside level (164, 165). Thus, pyrimidine deoxyribonucleoside 2′-hydroxylase catalyzes the hydroxylation of either thymidine or deoxyuridine in the only known direct transformation of deoxyribose to ribose. This reaction is in part the reverse of the ribonucleotide reductase reactions in which the 2′-hydroxyl group is reduced by dithiols. These reductases convert purine and pyrimidine ribonucleotide diphosphates and triphosphates to the corresponding deoxyribonucleotide phosphates in apparently irreversible reactions (75, 151).

2. Factors Affecting Hydroxylase Activity

The pyrimidine deoxyribonucleoside 2′-hydroxylase reaction was shown to be dependent on the presence of α-ketoglutarate, Fe^{2+}, and ascorbate in the incubation mixture (163). A number of analogs of α-ketoglutarate were tested but none was able to replace it (7). Some of the analogs, e.g., 3-ketoadipate and 2-ketoadipate, were moderately inhibitory to the reaction when they were included in the incubation vessel at a concentration ten times that of α-ketoglutarate. An apparent K_m value of about 0.2 mM was obtained for α-ketoglutarate. The metal ion requirement is also highly specific. Co^{2+}, Ni^{2+}, Cu^{2+}, Zn^{2+}, or Mn^{2+} could not replace Fe^{2+} in the 2'-hydroxylase reaction, and each of these ions was found to be inhibitory when included in reaction mixtures containing Fe^{2+}. An apparent K_m value of 0.45 mM was obtained for Fe^{2+} (7). The specificity of the ascorbate requirement has not been studied. Catalase was found to stimulate the 2'-hydroxylation of thymidine and deoxyuridine. The maximum stimulation, obtained by varying the concentration of catalase in the incubation mixture, was 3-fold (119). The specificity of the catalase effect has not been studied.

Shaffer et al. (164, 165) tested various radioactive compounds as substrates for the 2'-hydroxylase reaction. These were incubated in the presence and absence of the nonradioactive compounds which were the anticipated products and used as trapping agents. Since deoxycytidine is incorporated into RNA to about the same extent as is thymidine (47), it was suspected that deoxycytidine would also be hydroxylated. However, the calcium phosphate gel fraction did not catalyze the hydroxylation of deoxycytidine but instead its deamination to deoxyuridine which was found to undergo 2'-hydroxylation (Fig. 8). Deoxyadenosine and deoxyguanosine were also not hydroxylated by these preparations. Therefore, thymidine and deoxyuridine are the only commonly occurring deoxyribonucleosides which are substrates for the 2'-hydroxylase. Since neither deoxyuridylate nor deoxyribose was hydroxylated, the enzymic 2'-hydroxylation reaction appears to be specific for substrates at the nucleoside level (164, 165). Bankel et al. (7) confirmed these findings by testing the capacity of potential substrates to stimulate CO_2 production in the 2'-hy-

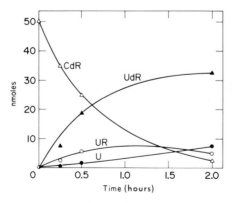

FIG. 8. Time course of deoxycytidine decomposition and product formation. The calcium phosphate gel enzyme preparation (1.8 mg of protein per milliliter of incubation mixture) was incubated with [2-^{14}C]deoxycytidine. CdR, deoxycytidine; UdR, deoxyuridine; UR, uridine; and U, uracil. From Shaffer et al. (165).

droxylation reaction carried out in the presence of α-[1-^{14}C]ketoglutarate. Thymidine and deoxyuridine had apparent K_m values of 0.09 and 0.19 mM and V_{max} values of 14 and 35 units/gm, respectively. Thymidylate appeared to be slightly active with a V_{max} one-thirtieth that of thymidine. In addition, the deoxyribose derivatives of 5-hydroxymethyluracil, 5-bromouracil, and 6-azathymine were found to be active substrates. However, substrate specificity data that are solely based on the assay for CO_2 should be interpreted with some caution since by this criterion a compound, which cannot serve as a substrate but which is contaminated with one, would appear to be active. Moreover, it may be possible to uncouple the decarboxylation of α-ketoglutarate from the hydroxylation step as has been done with the prolyl hydroxylase reaction (29) (see Section VII). The production of CO_2 was not detected when deoxycytidine, deoxyadenosine, deoxyguanosine, deoxyinosine, and deoxyribose were used as substrates (7).

3. Characterization of Reaction

The product of [2-^{14}C]thymidine with the hydroxylase system was combined with authentic thymine ribonucleoside and subjected to paper chromatography in ten diverse solvent systems. A borate-containing solvent system was especially useful since it separated thymine ribonucleoside from other compounds with sugar moieties which did not contain cis-hydroxyl groups. Radioautography was used to show that the radioactive product coincided with the ultraviolet absorbing marker compound. The experiments in which doubly labeled thymidine was used (Table VII) also helped identify the enzymically formed product. The ratio of the specific activities would not have been the same for substrate and product if the latter contained a lesser number of carbon atoms or if the substrate had been oxidized to a derivative of barbituric acid. In addition, the purified mixture of enzymically formed product and authentic thymine ribonucleoside was converted to a derivative which was shown to be homogeneous by chromatographic methods (163). More recent studies (7) have confirmed the conclusion that the product is thymine ribonucleoside. Paper chromatography (119) and silicic acid chromatography (7, 119) of deprotenized incubation mixtures containing α-[5-^{14}C]ketoglutarate showed that succinic acid is an enzymic product.

Liu et al. (119) showed in studies with α-[1-^{14}C]ketoglutarate, in which the time of incubation of the 2′-hydroxylase reaction mixture was varied, that CO_2 and thymine ribonucleoside are produced in a 1:1 molar ratio. In similar experiments with α-[1,5-^{14}C]ketoglutarate, CO_2 and succinate were shown to be produced in a 1:1 molar ratio. When deoxyuridine was used as substrate, uridine, succinate, and CO_2 were also shown to be produced in a 1:1:1 molar ratio (119). Moreover, in studies in which the concentration of either α-ketoglutarate or Fe^{2+} was made rate limiting, stoichiometric amounts of thymine ribonucleoside and CO_2 were produced (7).

The formation of thymine ribonucleoside in the 2′-hydroxylase reaction was markedly inhibited when the air of the incubation mixture was replaced with nitrogen. When the nitrogen gas was subsequently replaced with air or pure oxygen, full activity was restored (165). These findings have not been extended with studies using $^{18}O_2$.

F. Regulation

The uc-1 mutation developed by Williams and Mitchell (186) appears to affect a regulatory gene which controls the amount or activity of one or more enzymes involved in the oxidative demethylation of thymidine. This regulatory role is suggested by nutritional studies which show that the pyr-4, uc-1 strain of *Neurospora crassa* can grow on thymidine, thymine, 5-hydroxymethyluracil, or 5-formyluracil, whereas the pyr-4 strain, a pyrimidineless mutant, cannot use these compounds as sole pyrimidine sources unless growth is initiated with uridine or cytidine.

The atmosphere in which cultures of *N. crassa* are grown also appears to have a regulatory influence. The thymine 7-hydroxylase activity of cell-free extracts prepared from mycelia of *Neurospora*, grown under forced aeration in liquid media, is usually low or nondetectable. Studies *in vitro* and *in vivo* suggest that this lack of activity is a consequence of the cells containing less of the hydroxylase when grown under the aerated conditions (124). On the basis of these studies nonaerated conditions of growth have been developed which permit *Neurospora* to be grown in large batches which are a good source of thymine 7-hydroxylase (124). Whether there is a similar effect of aeration on the activities of the 5-hydroxymethyluracil and 5-formyluracil dioxygenase reactions has not yet been determined, but preliminary evidence (61) suggests there is no such effect on the 2′-hydroxylation reaction.

VI. p-HYDROXYPHENYLPYRUVATE HYDROXYLASE

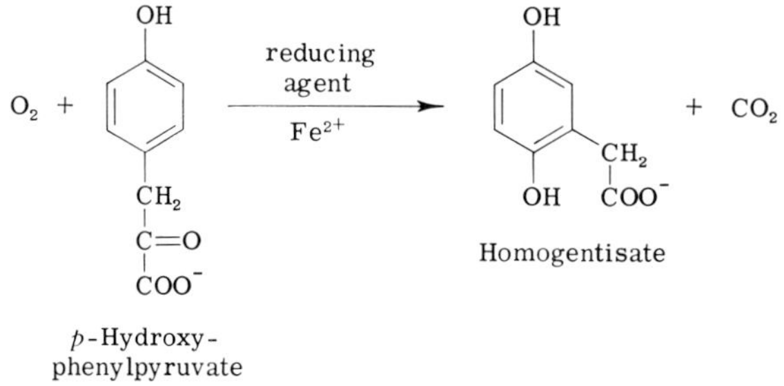

$$ (10) $$

p-Hydroxyphenylpyruvate

Homogentisate

A. Purification

Although evidence for the conversion of tyrosine to homogentisate was obtained at the end of the last century (187), it was not until the middle of this one that the enzymic conversion of p-hydroxyphenylpyruvate to homogentisate was shown conclusively. The migration of the side chain to the *ortho* position was proven by determining the isotopic labeling pattern of a metabolite, acetoacetate, formed from radioactive phenylalanine and tyrosine in experiments carried out *in vivo* and with liver slices (162, 184). The direct conversion of p-hydroxyphenylpyruvate to homogentisate was shown to be catalyzed by enzyme fractions from beef, pig, and dog liver (42, 64, 95). Other animals from which the oxygenase has been at least partially purified include the guinea pig (59), rat (44, 59, 173), rabbit (173), frog (100), man (107), chicken (43), monkey, opossum, and salmon (44). No enzymic activity has been observed in any tissue other than liver and kidney. The hydroxylase from pig liver (173) was purified 65-fold. Since throughout the purification procedure the ratio of the specific activities, with respect to the conversions of p-hydroxyphenylpyruvate to homogenisate and of phenylpyruvate to a o-hydroxyphenylacetate, was constant, it appeared that both reactions are catalyzed by the same enzyme. The conversion of fluorophenylpyruvate to 5-fluoro-2-hydroxyphenyl-acetate by the partially purified enzyme preparation indicated that the side chain migrated in this reaction too and thus was consistent with a single enzyme catalyzing both reactions. The p-hydroxyphenylpyruvate hydroxylase from frog liver appears to be present in an inactive form which can be activated by treatment with trypsin or by autolysis. After activation with trypsin the enzyme was purified 80-fold and its molecular weight was determined to be 85,000 by gel filtration (100). With the frog liver enzyme, p-hydroxyphenylpyruvate has an apparent K_m value of 0.5 mM which is higher than the values which were obtained using enzyme preparations from dog liver (0.04 mM) (190) and rat liver (0.02–0.05 mM) (44, 104). The hydroxylase from human liver has been purified extensively and has a molecular weight of 90,000–100,000 as determined by sedimentation equilibrium ultracentrifugation. The isoelectric point was found to be near pH 7 using isoelectric focusing in a sucrose density gradient (107). A preparation of the hydroxylase which has been obtained from chicken liver appears to be homogeneous on the basis of disc-gel electrophoresis carried out at several pH values and in the presence of sodium dodecyl sulfate. This enzyme preparation oxidized phenylpyruvate and 3,4-di-hydroxyphenylpyruvate as well as p-hydroxyphenylpyruvate (43).

In most of the earlier studies with the hydroxylase, the enzymic assays involved determination of the loss of substrates. As with the α-ketoglutarate-coupled oxygenases, a CO_2 assay has been developed for this hydroxylase, too, in which radioactive CO_2 formed from carboxy-labeled p-hydroxyphenylpyruvate is determined (44, 107). A spectrophotometric assay for homogentisate has also been developed. In this method homogentisate is condensed with cysteine to yield a 1,4-thiazine with an absorption maximum at 390 nm (44).

B. Factors Affecting Hydroxylase Activity

The p-hydroxyphenylpyruvate hydroxylase reaction was shown to be inhibited by high concentrations of p-hydroxyphenylpyruvate and O_2 (44, 64, 95). This inhibition by substrate can be reversed by a variety of reducing agents such as ascorbic acid, 2,6-dichlorophenolindophenol, tetrahydrofolic acid, and coenzyme Q_{10} (64, 93, 95, 193–195). The oxygenase was shown to be similarly inactivated and reactivated in vivo (58, 94, 191, 192, 194). Catalase also stimulated the hydroxylase in the presence of large amounts of substrate (44, 189). The kinetics of the inhibition were compatible with an inhibitory product being formed during the course of the reaction (190). p-Hydroxyphenylpyruvate and oxygen have been shown to generate hydrogen peroxide which may account for the inhibition (44). Moreover, the enzyme was inactivated by storage, purification, or treatment with oxidizing agents; this inactivation was also reversed by reducing agents (59). A metal ion was implicated in studies in which the hydroxylase was inhibited by 1,10-phenanthroline and other chelators but not by the relatively Cu^+-specific neocuproine. Of the large number of metal ions tested only Fe^{2+} restored activity to enzyme which had been inactivated with 1,10-phenanthroline. This restoration of activity with Fe^{2+} required a reducing agent. When reactivation was carried out with Fe^{2+} and ascorbate, reagents which react with sulfhydryl groups, such as N-ethymaleimide, did not inhibit the process. Thus, the oxidation and reduction of iron has been proposed as the cause of the reversible inactivation (59).

C. Characterization of Reaction

Neither 2,5-dihydroxyphenylpyruvic acid nor any other potential intermediate was detected in the p-hydroxyphenylpyruvate hydroxylase reaction (42, 96). The stoichiometry of the p-hydroxyphenylpyruvate reaction was demonstrated to be that indicated in Eq. (10) (42, 64, 95). Studies with $^{18}O_2$ were carried out which suggested that one oxygen atom of homogentisate was derived from molecular oxygen and two atoms from water, and thus the reaction seemed to be an intramolecular mixed-function oxidation (188). It appeared (36, 188) that the atom which was derived from molecular oxygen was incorporated into the new hydroxyl group, whereas the atoms derived from water were incorporated into the carboxyl group of homogentisate. Subsequently, however, Goodwin and Witkop (39, 56) proposed that both atoms of molecular oxygen are incorporated in a mechanism which involves a peroxide bridge being formed between the benzene ring and the side chain. Direct evidence for the formation of such a cyclic peroxide came from the laboratory of Lindstedt and co-workers (106) in studies in which the p-hydroxyphenylpyruvate system was incubated with $^{18}O_2$ or $H_2{}^{18}O$. Incubation of the hydroxylase system with $^{18}O_2$ was shown to result in the incorporation of one atom of ^{18}O into the carboxyl group of homogentisate, but only 30% of the homogentisate molecules were found to contain ^{18}O in their hydroxyl groups (Fig. 9). Thus it appeared that the molecular oxygen which was incorporated into the new hydroxyl group subsequently exchanged with water. In support of such an exchange was the demonstration that upon incubation of $H_2{}^{18}O$ with the hydroxylase system, ^{18}O was incorporated into the hydroxyl groups of 70% of the homogentisate molecules. This exchange may occur in a quinoid structure since an exchange in quinone oxygens with water has been reported (37, 38). Figure 9 also indicates that one atom of oxygen derived from water was incorporated into the carboxyl group of homogentisate. This incorporation can be explained by the exchange of the ketonic

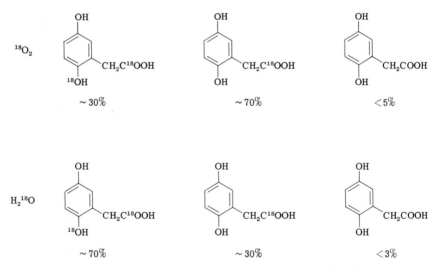

Fig. 9. Distribution of ^{18}O in the homogentisate formed during the p-hydroxyphenyl-pyruvate hydroxylase reaction in the presence of either $^{18}O_2$ or $H_2{}^{18}O$. From Lindblad et al. (106).

oxygen of p-hydroxyphenylpyruvate with water since after the hydroxylase system was incubated with $H_2^{18}O$, 82% of the residual p-hydroxyphenyl-pyruvate molecules were found to contain one atom of ^{18}O. Incubation of homogentisate with $H_2^{18}O$ in the complete system (minus p-hydroxyphenyl-pyruvate) was shown to result in ^{18}O being incorporated into less than 0.5% of the homogentisate molecules. As indicated by these data the p-hydroxy-phenylpyruvate hydroxylase reaction is quite complex. Their interpretation demanded special control experiments and much insight. Fortunately, unequivocal results were obtained with some of the α-ketoglutarate-coupled dioxygenases which have cosubstrates in which the oxygens are less exchangeable.

VII. GENERAL DISCUSSION

The α-ketoglutarate-coupled oxygenases were only recently discovered by investigators studying a variety of hydroxylase reactions. It is interesting to contemplate that these enzymes might have been discovered by investigators studying the metabolism of α-ketoglutarate. In that case a number of α-ketoglutarate decarboxylases would have appeared with such odd cofactors as thymidine, γ-butyrobetaine, and unhydroxylated collagen. It is highly probable that additional α-ketoglutarate-coupled oxygenases will be discovered.

Although the enzymes are found in animals, plants, and microorganisms and catalyze many different types of reactions, they have many features in common. In addition to α-ketoglutarate all of the enzymes require Fe^{2+} and a reducing agent. In all instances the Fe^{2+} requirement is highly specific, whereas the requirement for a reducing agent can be fulfilled by several substances. Ascorbate is frequently more effective than other re-ducing agents, but in no case is there direct evidence that ascorbate acts as the reducing agent *in vivo*. In fact, the hydroxylation of prolyl residues has been demonstrated in cultured fibroblasts which have been reported not to contain ascorbate (23, 145). The function of the reductant, as indicated in Section II, may be to maintain iron in the reduced form or to prevent oxidation of sulfhydryl groups of the enzyme. The reported (138) nonenzymic reaction of ascorbate and iron to produce superoxide ion (O_2^-) is perhaps germane. This ion has been shown to be involved in several enzymic reactions (50), but the studies of Hirata and Hayaishi (72, 74) with tryptophan 2,3-dioxygenase are especially intriguing. They have not only shown that this reaction is inhibited by superoxide dismutase but also that the ascorbate requirement can be met by superoxide gener-ating systems. Of course, these findings do not show that O_2^- reacts directly with the substrate. With respect to the chemical model systems which have been used to study hydroxylation, the Udenfriend system (175) is of special interest since it also has requirements for ascorbate and Fe^{2+}. This system not only hydroxylates aromatic compounds and epoxidizes olefins but also hydroxylates saturated hydrocarbons. (For reviews of this work and other model systems see refs. 67, 69, 139, 179, and 180.) Another common feature of the α-ketoglutarate-coupled dioxygenase reactions is that they are stimulated by catalase. A "catalase effect" was observed by Kaufman and co-workers (103) with phenylalanine hydroxylase and dopamine-β-hydroxylase, and the effect was shown to be a protective one, presumably against hydrogen peroxide. However, Rhoads et al. (153) have shown that many proteins markedly stimulate prolyl hydroxylase activity by some, as yet, unknown mechanism. Catalase, even when heated to destroy all enzymic activity, still shows such a protein effect (155).

The α-ketoglutarate aspect of the dioxygenase reaction mechanism has proven more vulnerable to attack. On the basis that the decarboxylation of α-ketoglutarate was stoichiometric with the hydroxylation of γ-butyrobetaine and that succinic semialdehyde did not appear to be a product, Lindstedt and co-workers proposed a mechanism (77) for the γ-butyrobetaine hydroxylase reaction. This coupling of decarboxylation to hydroxylation was also independently shown for the prolyl hydroxylase reaction by Rhoads and Udenfriend (154). In the proposed mechanism (Fig. 10) the peroxide anion of γ-butyrobetaine makes a nucleophilic attack on the carbonyl carbon of α-ketoglutarate. The subsequent demonstrations with γ-butyrobetaine hydroxylase by Lindblad et al. (105) and

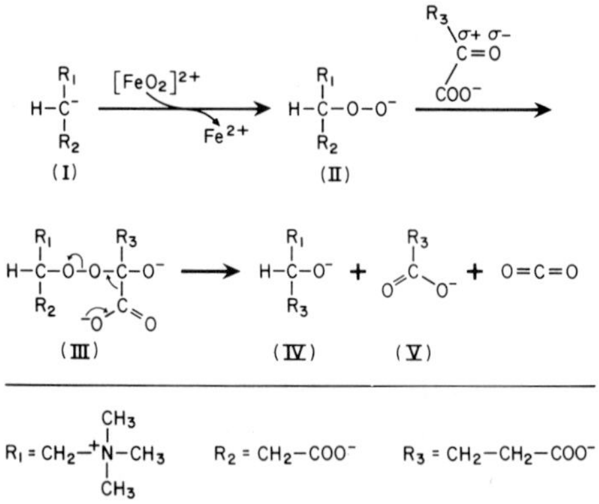

Fig. 10. Mechanism proposed for the α-ketoglutarate-coupled hydroxylation of γ-butyrobetaine. From Holme et al. (77).

with prolyl hydroxylase by Cardinale et al. (27) that $^{18}O_2$ was incorporated into the succinate formed in the reaction are in accord with this mechanism. At the present time the stoichiometry of all of the α-ketoacid dioxygenase reactions has been shown. The incorporation of molecular oxygen into succinate has been shown in all of the reactions except for those catalyzed by pyrimidine deoxyribonucleoside 2′-hydroxylase and lysyl hydroxylase. As further support for the mechanism depicted in Fig. 10, Lindstedt and co-workers claim (77) to have developed a model system in which t-butyl hydroperoxide and α-ketoglutarate react to produce t-butyl alcohol and succinic acid. Nonenzymic reactions of hydroperoxides with carbonyl groups to form hydroxy peroxides (24, 71, 161) and analogous peroxide fragmentation reactions (158, 159) are known. The nonenzymic oxidative decarboxylation of α-ketoacids by peroxides occurs even under the mild conditions tolerable to enzymes (68). However, Hamilton (68, 69) pointed out that it is difficult to explain how the substrates yield peroxides in the first place. The carbon atoms which are hydroxylated on many of the substrates such as proline and γ-butyrobetaine are completely inactivated. He suggested that it is far more likely that the initial oxidative attack is on the α-ketoglutarate. Interaction of O_2 with α-ketoglutarate could yield persuccinic acid which could then attack the other substrate through an oxenoid mechanism. Although there appears to be no example of a chemical model system in which a peracid is formed in this manner, peracids can attack saturated hydrocarbons. Persuccinic acid was prepared[§] and tested in the prolyl hydroxylase reaction (29) and subsequently in the pyrimidine deoxyribonucleoside 2′-hydroxylase reaction (121). When the peracid was substituted for α-ketoglutarate in both enzymic reactions, no hydroxylation was detected. Of course, enzyme-bound peroxysuccinate is not ruled out by such experiments. If O_2 initially attacks α-ketoglutarate, substrate amounts of the dioxygenase may effect a detectable decarboxylation of α-ketoglutarate in the absence of the other substrate. Although Lindstedt and Lindstedt (116) reported that in the γ-butyrobetaine hydroxylase reaction such an uncoupling of α-ketoglutarate decarboxylation from hydroxylation did not occur, uncoupling has been observed by Carbinale and Udenfriend (29) in experiments with substrate amounts of rat skin prolyl hydroxylase. The uncoupled reaction, like the coupled one, was found to be dependent on the active enzyme, Fe^{2+}, and ascorbate. This partial reaction yielded approximately 9 moles of CO_2 for each mole of prolyl hydroxylase during incubation periods of 20 minutes. Further studies on such partial reactions may help elucidate the

§ Monoperoxysuccinic acid was synthesized by Dr. B. A. Pawson of Hoffmann-La Roche, Inc., Nutley, New Jersey.

mechanisms of the α-ketoglutarate-coupled oxygenases. Large amounts of highly purified enzymes will be required to carry out such studies properly.

The 5-hydroxymethyluracil and 5-formyluracil dioxygenase reactions are unique in that no other dioxygenase is known that catalyzes either the conversion of an alcohol to an aldehyde or of an aldehyde to an acid. Although dehydrogenases which catalyze these types of oxidations come more readily to mind, it appears that similar oxidations are catalyzed by microsomal monooxygenases in NADPH-dependent reactions (9, 137, 174). When the cofactor requirements of the 5-hydroxymethyluracil and 5-formyluracil dioxygenase reactions were recognized, Watanabe *et al.* (183) proposed that these reactions proceed by a mechanism similar to that involved in the hydroxylation of thymine. The implication was that the hydroxymethyl group of 5-hydroxymethyluracil was hydroxylated to a dihydroxymethyl group which lost water to form a formyl group and that the conversion of 5-formyluracil to uracil-5-carboxylic acid occurred in a similar fashion. The stoichiometry of these reactions, the studies with $^{18}O_2$, and the possibility that the sequential reactions in the conversion of thymine to uracil-5-carboxylic acid are catalyzed by the same enzyme are consonant with this speculation.

As indicated earlier Goodwin and Witkop (56) had proposed, long before the discovery of α-ketoglutarate-dependent oxygenases, that a cyclic peroxide intermediate is formed in the *p*-hydroxyphenylpyruvate hydroxylase reaction (Fig. 11). However, the demonstration that the same enzyme converts phenylpyruvate to *o*-hydroxyphenylacetate argues against a mechanism which requires the *para*-hydroxyl group. Subsequently, Guroff *et al.* (62) suggested that the migration of the side chain proceeded by the NIH shift. This type of mechanism is shown in Fig. 12, as recently depicted by Daly *et al.* (41). Nevertheless, the studies of Daly *et al.* (40) indicate

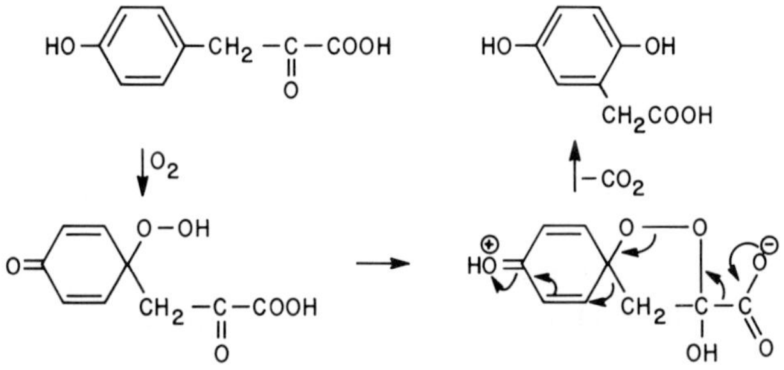

Fig. 11. Mechanism proposed for *p*-hydroxylphenypyruvate hydroxylase. From Daly and Witkop (39).

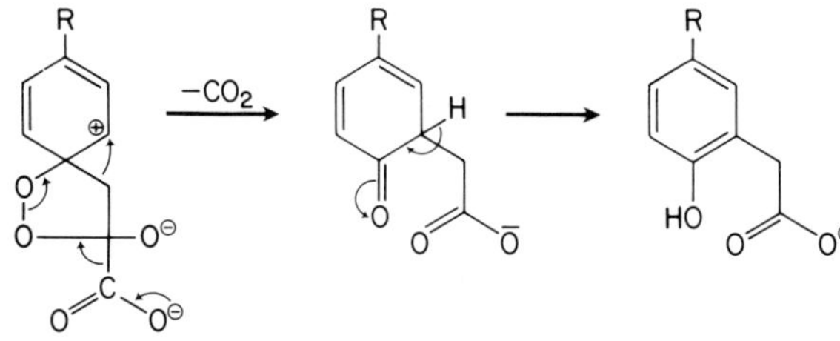

FIG. 12. Mechanism with NIH shift proposed for the p-hydroxyphenylpyruvate hydroxylase reaction. From Daly et al. (41).

that there may be a problem with such a mechanism in that a formal NIH shift should not occur with p-hydroxyphenylpyruvate since the *para*-hydroxyl group would be expected to stabilize the cationoid intermediate and bring about elimination of the side chain. Other mechanisms have been proposed for this reaction by Goodwin (57) and Hamilton (68). The oxenoid mechanism (Fig. 13) suggested by Hamilton is particularly attractive since it appears equally applicable to the other α-ketoacid-coupled dioxygenase reactions.

The mechanism proposed for the α-ketoacid dioxygenase reactions may be a more general phenomenon of oxygenase reactions; for example, one

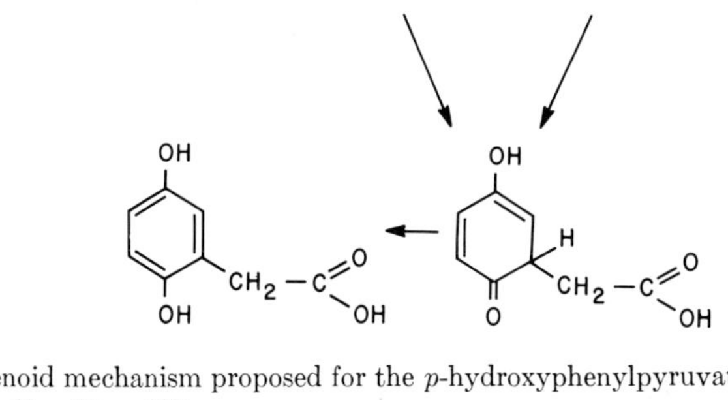

FIG. 13. Oxenoid mechanism proposed for the p-hydroxyphenylpyruvate hydroxylase reaction. From Hamilton (68).

might regard monooxygenase reactions as those requiring a reducing agent, such as NADPH, rather than the decarboxylation of an α-ketoacid, for fragmentation of a peroxide intermediate. Indeed, Chen and Lin (30) have reported that a hydroperoxide is formed in the conversion of tetralin to tetral-1-ol by rat liver extracts. Tetralol was shown to be formed from either tetralin hydroperoxide or tetralin in NADPH-dependent reactions. Moreover, when tetralin was used as substrate in reaction mixtures to which no NADPH was added, the hydroperoxide was shown to accumulate. In addition, Chen and Lin (31) have reported that fluorene-9-hydroperoxide is an intermediate in the hydroxylation of fluorene to fluorenol, although they were not able to detect an increase in the hydroperoxide in the absence of NADPH. Hydroperoxides have also been implicated in hydroxylation by reports that cytochrome P-450 acts as a microsomal peroxidase, converting steroid and other lipid peroxides to the corresponding alcohols (80, 181) and that alkyl hydroperoxides are reduced to alcohols in the rubredoxin-DPN reductase system (19). Another reaction in which decarboxylation appears to be involved in fragmentation of an intermediate hydroperoxide is that catalyzed by lactate oxygenase, i.e.,

$$O_2 + \text{lactate} \rightarrow \text{acetate} + CO_2 + H_2O$$

This is a monooxygenase reaction in which molecular oxygen is incorporated into the products, acetate and water. The oxygenase is a flavoprotein with FMN as the prosthetic group. In studies on the mechanism of this reaction, Massey and co-workers (48, 122) have obtained evidence that an enzyme $\cdot$ $FMNH_2 \cdot$ pyruvate complex reacts with oxygen to yield acetate, CO_2, H_2O, and enzyme-bound FMN. The reaction of this complex with oxygen appears to be analogous to the partial reaction demonstrated by Cardinale and Udenfriend (29) in which prolyl hydroxylase catalyzes the decarboxylation of α-ketoglutarate in the absence of the peptidyl substrate.

The α-ketoacid-coupled enzymes may represent a unique type of oxygenase mechanism. However, it is also possible that the fixation of oxygen into a peracid type compound by an activated molecule such as an α-ketoacid is a common feature of oxygenases. If so, then elucidation of the mechanisms of these α-ketoglutarate-coupled dioxygenases will have provided a clue basic to all oxygenase enzymes.

ACKNOWLEDGMENTS

The authors thank Dr. B. Witkop, Dr. George J. Cardinale, and Mr. Chen-Kao Liu for very helpful discussions pertaining to this manuscript. M. T. Abbott also wishes to thank National Institutes of Health, National Institute of Arthritis and Metabolic Diseases for financial support of his work (AM09314).

REFERENCES

1. Abbott, M. T. (1967). *Methods Enzymol.* **12A,** 47.
2. Abbott, M. T., and Fink, R. M. (1962). *Fed. Proc. Fed. Amer. Soc. Exp. Biol.* **21,** 377.
3. Abbott, M. T., Kadner, R. J., and Fink, R. M. (1964). *J. Biol. Chem.* **239,** 156.
4. Abbott, M. T., Schandl, E. K., Lee, R. F., Parker, T. S., and Midgett, R. J. (1967). *Biochim. Biophys. Acta* **132,** 525.
5. Abbott, M. T., Draglia, T. A., and McCroskey, R. P. (1968). *Biochim. Biophys. Acta* **169,** 1.
5a. Al-Adnani, M. J., Patrick, R. S., and McGee, J. O'D. (1973). *J. Cell Biol.* (submitted).
6. Bankel, L., Holme, E., Lindstedt, G., and Lindstedt, S. (1972). *Fed. Eur. Biochem. Soc. Lett.* **21,** 135.
7. Bankel, L., Lindstedt, G., Lindstedt, S. (1972). *J. Biol. Chem.* **247,** 6128.
8. Barnes, M. J., and Kodicek, E. (1972). *Vitam. Horm. (New York)* **30,** 1.
9. Bechtold, M. M., Delwiche, C. V., Comai, K., and Gaylor, J. L. (1972). *J. Biol. Chem.* **247,** 7650.
10. Bekhor, I., and Bavetta, L. A. (1967). *Proc. Nat. Acad. Sci. U.S.* **58,** 235.
11. Bennett, L. E. (1973). *Progr. Inorg. Chem.* **18,** 1.
12. Berech, J., Jr., and van Wagledonk, W. J. (1962). *Exp. Cell Res.* **26,** 360.
13. Berg, R. A., Olsen, B. R., Kivirikko, K. I. (1972). *Fed. Proc. Fed. Amer. Soc. Exp. Biol.* **31,** 479.
14. Berg, R. A., and Prockop, D. J. (1973). *J. Biol. Chem.* **248,** 1175.
15. Bhatnagar, R. S., Rosenbloom, J., Kivirikko, K. I., and Prockop, D. J. (1967). *Biochim. Biophys. Acta* **149,** 273.
16. Bhatnagar, R. S., Rapaka, S. S. R., Liu, T. Z., and Wolfe, S. M. (1972). *Biochim. Biophys. Acta* **271,** 125.
17. Blumenkrantz, N., and Prockop, D. J. (1969). *Anal. Biochem.* **30,** 377.
18. Bosman, H. B., and Eylar, E. H. (1968). *Biochem. Biophys. Res. Commun.* **30,** 89.
19. Boyer, R. F., Lode, E. T., and Coon, M. J. (1971). *Biochem. Biophys. Res. Commun.* **44,** 925.
20. Brachet, J. (1958). *Exp. Cell Res.* **14,** 650.
21. Bremer, J. (1961). *Biochim. Biophys. Acta* **48,** 622.
22. Bremer, J. (1962). *Biochim. Biophys. Acta* **57,** 327.
23. Bublitz, C., and Priest, R. E. (1967). *Lab. Inv.* **17,** 371.
24. Bunton, C. A. (1962). *In* "Peroxide Reaction Mechanisms" (J. O. Edwards, ed.), p. 16. Wiley (Interscience) New York.
25. Butler, W. T. (1968). *Science* **161,** 796.
26. Cain, G. D., and Fairbairn, D. (1971). *Comp. Biochem. Physiol.* **40B,** 165.
27. Cardinale, G. J., Rhoads, R. E., and Udenfriend, S. (1971). *Biochem. Biophys. Res. Commun.* **43,** 537.
28. Cardinale, G. J., and Udenfriend, S. (In press). *In* "Oxidases and Related Redox Systems" (T. E. King, H. S. Mason, and M. Morrison, eds.), *Proc. Int. Symp., 2nd.* Univ. Park Press, Baltimore, Maryland.
29. Cardinale, G. J., and Udenfriend, S. (In preparation).
30. Chen, C., and Lin, C. C. (1968). *Biochim. Biophys. Acta* **170,** 366.
31. Chen, C., and Lin, C. C. (1969). *Biochim. Biophys. Acta* **184,** 634.
32. Chrispeels, M. J. (1970). *Plant Physiol.* **45,** 223.
33. Chvapil, M., Ehrlichova, M., and Hurych, J. (1966). *Experientia* **22,** 584.

34. Comstock, J. P., and Udenfriend, S. (1970). *Proc. Nat. Acad. Sci. U.S.* **66**, 552.
35. Cooper, G. W., and Prockop, D. J. (1968). *J. Cell Biol.* **35**, 523.
36. Crandall, D. I. (1964). *In* "Oxidases and Related Redox Systems" (T. E. King, H. S. Mason, and M. Morrison, eds.), Vol. I, p. 275. Wiley, New York.
37. Dahm, H., and Aubort, J-D. (1968). *Helv. Chim. Acta* **51**, 1348.
38. Dahm, H., and Aubort, J-D. (1968). *Helv. Chim. Acta* **51**, 1537.
39. Daly, J. W., and Witkop, B. (1963). *Angew. Chem. Int. Ed.* **2**, 421.
40. Daly, J., Jerina, D., and Witkop, B. (1968). *Arch. Biochem. Biophys.* **128**, 517.
41. Daly, J. W., Jerina, D. M., and Witkop, B. (1972). *Experienta* **28**, 1129.
42. Edwards, S. W., Hsia, D. Y. Y., Knox, W. E. (1955). *Fed. Proc. Fed. Amer. Soc. Exp. Biol.* **14**, 206.
43. Fellman, J. H., Fujita, T. S., and Roth, E. S. (1972). *Biochim. Biophys. Acta* **268**, 601.
44. Fellman, J. H., Fujita, T. S., and Roth, E. S. (1972). *Biochim. Biophys. Acta* **284**, 90.
45. Fink, K., Cline, R. E., Henderson, R. B., and Fink, R. M. (1956). *J. Biol. Chem.* **221**, 425.
46. Fink, R. M., and Fink, K. (1962). *Fed. Proc. Fed. Amer. Soc. Exp. Biol.* **21**, 377.
47. Fink, R. M., and Fink, K. (1962). *J. Biol. Chem.* **237**, 2889.
48. Flashner, M. S., and Massey, V. (1973). Chapter 8, this volume.
49. Frankel, G., and Friedman, S. (1957). *Vitamins and Hormones* **15**, 74.
50. Fridovich, I. (1973). Chapter 12, this volume.
51. Fritz, I. B. (1963). *Advan. Lipid Res.* **1**, 285.
52. Fujimoto, D., and Tamiya, M. (1962). *Biochem. J.* **84**, 333.
53. Fujimoto, D., and Adams, E. (1964). *Biochem. Biophys. Res. Commun.* **17**, 437.
54. Gallop, P. M., Blumenfeld, O. O., and Seifter, S. (1972). *Annu. Rev. Biochem.* **41**, 617.
55. Goldberg, B., and Green, H. (1967). *J. Mol. Biol.* **26**, 1.
56. Goodwin, S., and Witkop, B. (1956). *J. Amer. Chem. Soc.* **79**, 179.
57. Goodwin, B. L. (1971). *Biochem J.* **125**, 17P.
58. Goswami, M. N. D., and Knox, W. E. (1961). *Biochim. Biophys. Acta* **50**, 35.
59. Goswami, M. N. D. (1964). *Biochim. Biophys. Acta* **85**, 390.
60. Grant, M. E., and Prockop, D. J. (1972). *New England J. Med.* **286**, 194; **286**, 242; **286**, 291.
61. Griswold, W. R., and Abbott, M. T. (1973). In preparation.
62. Guroff, G., Daly, J. W., Jerina, D. M., Renson, J., Witkop, B., and Udenfriend, S. (1967). *Science* **157**, 1524.
63. Habers, E., and Heidelberger, C. (1959). *J. Biol. Chem.* **234**, 1249.
64. Hager, S. E., Gregerman, R. I., and Knox, W. E. (1957). *J. Biol. Chem.* **225**, 935.
65. Halme, J., and Kivirikko, K. I. (1968). *Fed. Eur. Biochem. Soc. Lett.* **1**, 223.
66. Halme, J., Kivirikko, K. I., and Simons, K. (1970). *Biochim. Biophys. Acta* **198**, 460.
67. Hamilton, G. A. (1969). *Advan. Enzymol.* **32**, 55.
68. Hamilton, G. A. (1971). *Progr. Bioorganic Chem.* **I**, 83.
69. Hamilton, G. A. (1973). Chapter 11, this volume.
70. Hausmann, E. (1967). *Biochim. Biophys. Acta* **133**, 591.
71. Hawkins, E. G. E. (1961). *In* "Organic Peroxides," p. 274. van Nostrand Reinhold, Princeton, New Jersey.
72. Hayaishi, O. (1973). Chapter 10, this volume.
73. Heidelberger, C., Boohar, J., Kampschroer, N. (1965). *Cancer Res.* **25**, 377.
74. Hirata, F., and Hayaishi, O. (1971). *J. Biol. Chem.* **246**, 7825.

75. Hogenkamp, H. P. C. (1968). *Annu. Rev. Biochem.* **37**, 233.
76. Holmes, W. L., Prusoff, W. H., and Welch, A. D. (1954). *J. Biol. Chem.* **209**, 503.
77. Holme, E., Lindstedt, G., Lindstedt, S., and Tofft, M. (1968). *Fed. Eur. Biochem. Soc. Lett.* **2**, 29.
78. Holme, E., Lindstedt, G., Lindstedt, S., and Tofft, M. (1970). *Biochim. Biophys. Acta* **212**, 50.
79. Holme, E., Lindstedt, G., Lindstedt, S., and Tofft, M. (1971). *J. Biol. Chem.* **246**, 3314.
80. Hrycay, E. G., and O'Brien, P. J. (1972). *Arch. Biochem. Biophys.* **153**, 480.
81. Hurych, J., Hobza, P., Rencová, J., and Zahradnik, R. (1973). *In* "The Biology of Fibroblasts" (E. Kulmen and J. Pikkarainen, eds.), Academic Press, New York (in press).
82. Hutton, J., Tappel, A. L., and Udenfriend, S. (1966). *Anal. Biochem.* **16**, 384.
83. Hutton, J. J., Tappel, A. L., and Udenfriend, S. (1967). *Arch. Biochem. Biophys.* **118**, 231.
84. Hutton, J. J., Marglin, A., Witkop, B., Kurtz, J., Berger, A., and Udenfriend, S. (1968). *Arch. Biochem. Biophys.* **125**, 779.
85. Juva, K., and Prockop, D. J. (1966). *J. Biol. Chem.* **241**, 4419.
86. Juva, K., and Prockop, D. J. (1969). *J. Biol. Chem.* **244**, 6486.
87. Katz, E., Prockop, D. J., and Udenfriend, S. (1962). *J. Biol. Chem.* **237**, 1585.
88. Kivirikko, K. I., and Prockop, D. J. (1967). *J. Biol. Chem.* **242**, 4007.
89. Kivirikko, K. I., and Prockop, D. J. (1967). *Proc. Nat. Acad. Sci. U.S.* **57**, 782.
90. Kivirikko, K. I., Bright, H. J., and Prockop, D. J. (1968). *Biochim. Biophys. Acta* **151**, 558.
91. Kivirikko, K. I., and Prockop, D. J. (1972). *Biochim. Biophys. Acta* **258**, 366.
92. Kivirikko, K. I., Shudo, K., Sakakibara, S., and Prockop, D. J. (1972). *Biochemistry* **11**, 122.
93. Knox, W. E., and Le May-Knox, M. (1951). *Biochem. J.* **49**, 686.
94. Knox, W. E., and Goswami, M. N. D. (1960). *J. Biol. Chem.* **235**, 2662.
95. LaDu, B. N., and Zannoni, V. G. (1955). *J. Biol. Chem.* **217**, 777.
96. LaDu, B. N., and Zannoni, V. G. (1956). *J. Biol. Chem.* **219**, 273.
97. Lamport, D. T. A. (1963). *J. Biol. Chem.* **238**, 1438.
98. Lamport, D. T. A. (1967). *Nature (London)* **216**, 1327.
99. Lamport, D. T. A., and Northcote, D. H. (1960). *Nature (London)* **188**, 665.
100. Laskowska-Klita, T. (1969). *Acta Biochim. Pol.* **16**, 35.
101. Lazarides, E. L., Lukens, L. N., Infante, A. A. (1971). *J. Mol. Biol.* **58**, 831.
102. Levin, E. Y., Levenberg, B., and Kaufman, S. (1960). *J. Biol. Chem.* **235**, 2080.
103. Levin, E. Y., and Kaufman, S. (1961). *J. Biol. Chem.* **236**, 2043.
104. Lin, E. C. C., Pitt, B. M., Civen, M., and Knox, W. E. (1958). *J. Biol. Chem.* **233**, 668.
105. Lindblad, B., Lindstedt, G., Tofft, M., and Lindstedt, S. (1969). *J. Amer. Chem. Soc.* **91**, 4604.
106. Lindblad, B., Lindstedt, G., and Lindstedt, S. (1970). *J. Amer. Chem. Soc.* **92**, 7446.
107. Lindblad, B., Lindstedt, S., Olander, B., and Omfeldt, M. (1971). *Acta Chem. Scand.* **25**, 329.
108. Lindstedt, G., and Lindstedt, S. (1961). *Biochem. Biophys. Res. Commun.* **6**, 319.
109. Lindstedt, G., and Lindstedt, S. (1962). *Biochem. Biophys. Res. Commun.* **7**, 394.
110. Lindstedt, G., and Lindstedt, S. (1965). *J. Biol. Chem.* **240**, 316.
111. Lindstedt, G. (1967). *Biochemistry* **6**, 1271.
112. Lindstedt, G. (1967). *Biochim. Biophys. Acta* **141**, 492.

113. Lindstedt, G., Lindstedt, S., Midvedt, T., and Tofft, M. (1967). *Biochemistry* **6,** 1262.
114. Lindstedt, G., Lindstedt, S., Midvedt, T., and Tofft, M. (1967). *Biochem. J.* **103,** 19P.
115. Lindstedt, G., Lindstedt, S., Olander, B., and Tofft, M. (1968). *Biochim. Biophys. Acta* **158,** 503.
116. Lindstedt, G., and Lindstedt, S. (1970). *J. Biol. Chem.* **245,** 4187.
117. Lindstedt, G., Lindstedt, S., and Tofft, M. (1970). *Biochemistry* **9,** 4336.
118. Linneweh, W. (1929). *Hoppe-Seyler's Z. Physiol. Chem.* **181,** 42.
119. Liu, C. K., Shaffer, P. M., Slaughter, R. S., McCroskey, R. P., and Abbott, M. T. (1972). *Biochemistry* **11,** 2172.
120. Liu, C. K., Shaffer, P. M., Slaughter, R. S., McCroskey, R. P., and Abbott, M. T. (1972). *Fed. Proc. Fed. Amer. Soc. Exp. Biol.* **31,** 882.
121. Liu, C. K., and Abbott, M. T. (1973). In preparation.
121a. Liu, C. K., Hsu, C. A., and Abbott, M. T. (1973). *Arch. Biochem. Biophys.* (in press).
122. Lockridge, O., Massey, V., and Sullivan, P. A. (1972). *J. Biol. Chem.* **247,** 8097.
123. McBride, O. W., and Harrington, W. F. (1967). *Biochemistry* **6,** 1484.
124. McCroskey, R. P., Griswold, W. R., Sokoloff, R. L., Sevier, E., Lin, S., Liu, C. K. Shaffer, P. M., Palmatier, R. D., Parker, T. S., and Abbott, M. T. (1971). *Biochim. Biophys. Acta* **227,** 264.
125. McGee, J. O'D., Langness, U., and Udenfriend, S. (1971). *Proc. Nat. Acad. Sci. U.S.* **68,** 1585.
126. McGee, J. O'D., Rhoads, R. E., and Udenfriend, S. (1971). *Arch. Biochem. Biophys.* **144,** 343.
127. McGee, J. O'D., and Udenfriend, S. (1972). *Arch. Biochem. Biophys.* **152,** 216.
128. McGee, J. O'D., and Udenfriend, S. (1972). *Arch. Biochem. Biophys.* **152,** 216.
129. Mandel, L. R., and Borek, E. (1963). *Biochemistry* **2,** 555.
130. Manner, G., Kretsinger, R. H., Gould, B. S., and Rich, A. (1967). *Biochim. Biophys. Acta* **134,** 411.
131. Mantsavinos, R., and Zamenhoff, S. (1961). *J. Biol. Chem.* **236,** 876.
132. Meyer, R. R. (1966). *Biochem. Biophys. Res. Commun.* **25,** 549.
133. Midgett, R. J., Lin, S., Parker, T. S., McCroskey, R. P., Shaffer, P. M., and Abbott, M. T. Unpublished.
134. Miller, R. (1971). *Arch. Biochem. Biophys.* **147,** 339.
135. Miller, R. (1972). *Anal. Biochem.* **45,** 202.
136. Miller, R. L., and Udenfriend, S. (1970). *Arch. Biochem. Biophys.* **139,** 104.
137. Murphy, P. J., and West, C. A. (1969). *Arch. Biochem. Biophys.* **133,** 395.
138. Nakamura, S. (1970). *Biochem. Biophys. Res. Commun.* **41,** 177.
139. Norman, R. O. C., and Smith, J. R. L. (1965). *In* "Oxidases and Related Redox Systems" (T. E. King, H. S. Mason, and M. Morrison, eds.), p. 131. Wiley, New York.
140. Olsen, B. R., Jiminez, S. A., Kivirikko, K. I., and Prockop, D. J. (1970). *J. Biol. Chem.* **245,** 2649.
141. Palmatier, R. D., McCroskey, R. P., and Abbott, M. T. (1970). *J. Biol. Chem.* **245,** 6706.
142. Pänkäläinen, M., Aro, H., Simons, K., and Kivirikko, K. I. (1970). *Biochim. Biophys. Acta* **221,** 559.
143. Pänkäläinen, M., and Kivirikko, K. I. (1971). *Biochim. Biophys. Acta* **229,** 504.
144. Peterkofsky, B., and Udenfriend, S. (1963). *Biochem. Biophys. Res. Commun.* **12,** 257.
145. Peterkofsky, B. (1972). *Arch. Biochem. Biophys.* **152,** 318.

146. Popenoe, E. A., Aronson, R. B., and van Slyke, D. D. (1969). *Arch. Biochem. Biophys.* **133,** 286.
147. Popenoe, E. A., and Aronson, R. B. (1972). *Biochim. Biophys. Acta* **258,** 380.
148. Prockop, D., Kaplan, A., and Udenfriend, S. (1962). *Biochem. Biophys. Res. Commun.* **9,** 162.
149. Prockop, D., Kaplan, A., and Udenfriend, S. (1963). *Arch. Biochem. Biophys.* **101,** 499.
150. Prockop, D. J. (1970). *In* "The Chemistry and Molecular Biology of the Intercellular Matrix" (A. E. Balazs, ed.), Vol. I, p. 335, Academic Press, New York.
151. Reichard, P. (1967). *In* "The Biosynthesis of Deoxyribose," Ciba Lectures. Wiley, New York.
152. Rencová, J., Hurych, J., Rosmus, J., and Chvapil, M. (1968). *Fed. Eur. Biochem. Soc., Abstr. 5th Meet., Prague* p. 82.
153. Rhoades, R. E., Hutton, J. J., and Udenfriend, S. (1967). *Arch. Biochem. Biophys.* **122,** 805.
154. Rhoads, R. E., and Udenfriend, S. (1968). *Proc. Nat. Acad. Sci. U.S.* **60,** 1473.
155. Rhoads, R. E., and Udenfriend, S. (1970). *Arch. Biochem. Biophys.* **139,** 329.
156. Rhoads, R. E., Roberts, N. E., and Udenfriend, S. (1971). *Methods Enzymol.* **17,** 306.
157. Rhoads, R. E., Udenfriend, S., and Bornstein, P. (1971). *J. Biol. Chem.* **246,** 4138.
158. Richardson, W. H., and Smith, R. S. (1967). *J. Amer. Chem. Soc.* **89,** 2230.
159. Richardson, W. H., and Heesen, T. C. (1972). *J. Org. Chem.* **37,** 3416.
160. Sadava, D., and Chrispeels, M. J. (1971). *Biochim. Biophys. Acta* **227,** 278.
161. Sauer, M. C. V., and Edwards, J. O. (1971). *J. Phys. Chem.* **75,** 3377.
162. Schepartz, B., and Gurin, S. (1949). *J. Biol. Chem.* **180,** 663.
163. Shaffer, P. M., McCroskey, R. P., Palmatier, R. D., Midgett, R. J., and Abbott, M. T. (1968). *Biochem. Biophys. Res. Commun.* **33,** 806.
164. Shaffer, P. M., and Abbott, M. T. (1971). *Fed. Proc. Fed. Amer. Soc. Exp. Biol.* **30,** 1222.
165. Shaffer, P. M., McCroskey, R. P., and Abbott, M. T. (1972). *Biochim. Biophys. Acta* **258,** 387.
166. Shaffer, P. M., and Abbott, M. T., unpublished.
167. Spiro, R. G. (1969). *J. Biol. Chem.* **244,** 602.
168. Spiro, R. G., and Spiro, M. J. (1971). *J. Biol. Chem.* **246,** 4899.
169. Stassen, F. L. H., Cardinale, G. J., and Udenfriend, S. (1973). *Proc. Nat. Acad. Sci. U.S.* **70,** 1090.
170. Steward, F. C., Thompson, J. F., Miller, F. K., Thomas, M. D., and Hendricks, R. H. (1951). *Plant Physiol.* **26,** 123.
171. Stocking, C. R., and Gifford, E. M. (1959). *Biochem. Biophys. Res. Commun.* **1,** 159.
172. Strack, E., Aurich, H., Lorenz, I., and Rotzsch, W. (1960). *Protides Biol. Fluids.* **7,** 428.
173. Taniguchi, K., Kappe, T., and Armstrong, M. D. (1964). *J. Biol. Chem.* **239,** 3389.
174. Teschke, R., Hasumura, Y., Joly, J.-G., Ishii, H., and Lieber, C. S. (1972). *Biochem. Biophys. Res. Commun.* **49,** 1187.
175. Udenfriend, S., Clark, C. T., Axelrod, J., and Brodie, B. B. (1954). *J. Biol. Chem.* **208,** 731.
176. Udenfriend, S. (1966). *Science* **152,** 1335.
177. Udenfriend, S. (1970). *In* "The Chemistry and Molecular Biology of the Intercellular Matrix" (A. E. Balazs, ed.), Vol. 1, p. 371. Academic Press, New York.

178. Udenfriend, S., Cardinale, G. J., and Stassen, F. L. H. (1973). *Abstr. Int. Congr. Biochem., 9th.*

179. Ullrich, V., and Staudinger, Hj. (1966). *In* "Biological and Chemical Aspects of Oxygenases" (K. Block and O. Hayaishi, eds.), p. 235. Maruzen, Tokyo.

180. Ullrich, V., Ruff, H.-H., and Mimoun, H. (1972). *In* "Biological Hydroxylation Mechanisms" (G. S. Boyd and R. M. S. Smellie, eds.), p. 11. Academic Press, New York.

181. van Lier, J., Kan, G., Langloris, R., and Smith, L. L. (1972). *In* "Biological Hydroxylation Mechanisms" (G. S. Boyd, and R. M. S. Smellie, eds.), p. 21. Academic Press, New York.

182. Vitol, M. J., Vilks, S. R., Zabarovska, I. M., and Maurinia, K. A. (1970). *Dokl. Akad. Nauk SSSR* **192,** 908.

183. Watanabe, M. S., McCroskey, R. P., and Abbott, M. T. (1970). *J. Biol. Chem.* **245,** 2023.

184. Weinhouse, S., and Millington, R. H. (1948). *J. Biol. Chem.* **175,** 995.

185. Weinstein, E., Blumenkrantz, N., and Prockop, D. J. (1969). *Biochim. Biophys. Acta* **191,** 747.

186. Williams, L. G., and Mitchell, H. K. (1969). *J. Bacteriol.* **100,** 383.

187. Wolkow, M., and Baumann, E. (1891). *Z. Physiol. Chem.* **15,** 228.

188. Yasunobu, K., Tanaka, T., Knox, W. E., and Mason, H. S. (1958). *Fed. Proc. Fed. Amer. Soc. Exp. Biol.* **17,** 340.

189. Zannoni, V. G., and LaDu, B. N. (1956). *Fed. Proc. Fed. Amer. Soc. Exp. Biol.* **15,** 391.

190. Zannoni, V. G., and LaDu, B. N. (1959). *J. Biol. Chem.* **234,** 2925.

191. Zannoni, V. G., and LaDu, B. N. (1960). *J. Biol. Chem.* **235,** 165.

192. Zannoni, V. G., and LaDu, B. N. (1960). *J. Biol. Chem.* **235,** 2667.

193. Zannoni, V. G. (1962). *J. Biol. Chem.* **237,** 1172.

194. Zannoni, V. G., Jacoby, G. A., and LaDu, B. N. (1962). *Biochem. Biophys. Res. Commun.* **7,** 220.

195. Zannoni, V. G., Brown, N. C., and LaDu, B. N. (1963). *Fed. Proc. Fed. Amer. Soc. Exp. Biol.* **22,** 232.

NOTE ADDED IN PROOF

Liu *et al.* (121a), have obtained evidence which indicates that thymine 7-hydroxylase catalyzes the following three sequential reactions: thymine → 5-hydroxymethyluracil → 5-formyluracil → uracil-5-carboxylic acid. The enzyme was purified 1300-fold and had specific activities of 1200, 600, and 250 units/mg for the respective reactions.

6

MICROSOMAL CYTOCHROME P-450-LINKED
MONOOXYGENASE SYSTEMS IN MAMMALIAN TISSUES

STEN ORRENIUS and LARS ERNSTER

I. INTRODUCTION

The study of microsomal cytochrome P-450-linked monooxygenase systems is of relatively recent origin. The volume on "Oxygenases" of 1962 (1), the predecessor of this treatise, contains little mention of this topic, just as do most contemporary textbooks of biochemistry. Grown out of the early interest in the metabolism of certain carcinogenic dyes (2, 3) and various drugs (4–6), this field has expanded exponentially over the past decade, as witnessed by a large number of publications including numerous review articles (7–12) and symposia (13–19). Moreover, it has become a most important and fruitful meeting ground for scientists from an increasingly wide area of biology and medicine, ranging from basic biochemistry and molecular biology to clinical pharmacology and environmental health science.

In the sections that follow we shall concern ourselves primarily with various aspects of the cytochrome P-450-linked monooxygenase system of liver microsomes, including its catalytic components, interaction with substrates, and reaction mechanism. Subsequently, we shall consider information relating to the substrate specificity of the cytochrome P-450-linked monooxygenase system as present in the liver and in other mammalian tissues. Finally, we shall review some aspects of the relationship of the hepatic cytochrome P-450-linked monooxygenase system to the endoplasmic reticulum membranes as well as the substrate-induced synthesis of this system *in vivo*.

II. CATALYTIC COMPONENTS

A. Cytochrome P-450

In 1958, reports by Klingenberg (20) and Garfinkel (21) described the presence of a carbon monoxide binding pigment in mammalian liver microsomes. In the reduced form, this pigment was found to bind CO, giving rise to a prominent absorption band at 450 nm in the difference spectrum of the microsomes. Since the CO complexes of known hemoproteins had Soret absorption maxima at considerably shorter wavelengths (about 420 nm) and since the CO difference spectrum of the reduced liver microsomes showed no peaks other than that at 450 nm, there was no immediate clue as to the chemical nature of this newly discovered pigment.

Evidence that the CO-binding pigment is in fact a hemoprotein was, however, presented a few years later by Omura and Sato (22, 23), who found that, although the CO difference spectrum of the reduced pigment is atypical for a hemoportein, the isocyanide spectrum is characteristic of a hemoprotein compound. Introducing a stepwise solubilization procedure, Omura and Sato (24) were further able to prepare "CO-binding particles," free of cytochrome b_5, from liver microsomes. In these particles, the CO-binding hemoprotein was found to be partly converted into a spectrally different P-420 form. Further solubilization resulted in a complete conversion of cytochrome P-450 into P-420, which could be substantially purified (24). Spectral studies of the purified P-420 form revealed characteristics of a b type cytochrome, resembling bacterial cytochrome b_1, except that the reduced P-420 retained the property of reduced P-450 of being able to combine with CO. Reduced P-420 was further found to be strongly autoxidable, and its oxidation–reduction potential was determined to be -20 mV at pH 7.0. The molar extinction coefficient of the P-420–CO complex was calculated on a protoheme basis to be 111 m$M^{-1} \cdot$cm^{-1} at 420–490 nm, and it was now also possible to determine the extinction coefficient of the CO complex of microsomal, bound cytochrome P-450 (91 m$M^{-1} \cdot$cm^{-1} at 450–490 nm; reduced + CO *minus* reduced).

The elegant studies of Omura and Sato cited above thus provided evidence for the hemoprotein nature of the CO-binding pigment which received its unusual name—cytochrome P-450—from the peak position of its CO complex. They also provided a procedure for a reasonable purification of cytochrome P-420. Unfortunately, however, cytochrome P-420 was found to be a nonfunctional hemochrome form, which is readily formed from cytochrome P-450 upon treatment of microsomes not only with lipolytic and surface-active agents (23, 24) but also with chelating agents (25), sulfhydryl reagents (26), neutral salts (27), and various organic solvents (28). Although the detailed mechanism of conversion of cytochrome P-450 to P-420 still remains unknown, it has been suggested that the conversion of P-450 to P-420 caused by organic compounds may result from an interaction with its hydrophobic groups rather than from a change in the conformation of the P-450 molecule (28). Furthermore, studies on the interconversion of the two forms revealed a stabilization of P-450 by polyols as well as reconversion of P-420, produced by various mild treatments, to P-450 by polyols and reduced glutathione (29, 30), a finding which led to the suggestion that cytochrome P-450 may be of lipoprotein nature (28). Considering the possible common mechanism by which the diverse agents mentioned above act to convert cytochrome P-450 to the P-420 form, Imai and Sato (27) concluded that a disruption of the association between the hemoprotein and microsomal lipid may be the ultimate effect of each of the agents. This effect could be exerted either by an action on the lipid or on protein associated with the lipid.

The ready conversion of microsomal cytochrome P-450 into the P-420 form upon solubilization has long prevented the isolation and reconstitution of cytochrome P-450-linked microsomal functions. As discussed below, it is only recently that the solubilization of hepatic cytochrome P-450 as such and the reconstitution of an active liver microsomal cytochrome P-450-linked monooxygenase system have been reported, and up to now it has not been possible to obtain a microsomal cytochrome P-450 preparation of reasonable purity. This is one reason why even today so little is known of the chemistry of microsomal cytochrome P-450. While this lack of access to a purified preparation has been frustrating to workers in this field, valuable information on the mechanism of electron transport associated with the cytochrome P-450-linked monooxygenase system has been accumulated by applying spectrophotometric methods developed by Chance (31) to the study of this system in its membrane-bound form in microsomes *in situ*.

Except for traces of catalase in contaminating peroxisomes, cytochrome b_5 is the only other known hemoprotein present in the liver microsomal fraction. Thus, the spectral properties of cytochrome P-450 in its membrane-bound form can be studied once the spectral contribution of cytochrome b_5 is canceled. This situation has been achieved in either of two ways: (1) by the preparation of "cytochrome P-450 particles" which are freed of cytochrome b_5 and contain minimal amounts of cytochrome P-420 (32–35); or (2) by comparing microsomal preparations from phenobarbital-treated animals, containing a high concentration of cytochrome P-450 (see discussion below), with those from untreated animals, using the concentration of cytochrome b_5 for balancing the cuvettes (32). Some spectral properties of various cytochrome P-450 preparations are shown in Table I.

As briefly discussed above, the atypical CO difference spectrum of cytochrome P-450 suggested that it was a hemoprotein of unusual properties. This was further established in studies of cytochrome P-450 interaction with other ligands. Thus, the reduced form of the hemoprotein was found to bind ethyl isocyanide (23, 24, 37), aniline, and pyridine (38) to yield difference spectra with two distinct peaks in the Soret region, the relative heights of which are profoundly influenced by pH. Furthermore, KCN is able to interact with both the oxidized and the reduced forms of the hemoprotein to produce spectral changes (39). In summarizing these

TABLE I

ABSORPTION PEAKS OF "ABSOLUTE SPECTRA" OF VARIOUS CYTOCHROME P-450
PREPARATIONS FROM LIVER MICROSOMES

| Preparation | Conditions | | | |
	Oxidized	Reduced	Reduced + CO	Ref.
P-450 particles	650, 570, 535, 415	555, 420	555, 449, 424	34
P-450 particles	645, 568, 532, 414, 360	555, 416	555, 450, 423	35
Microsomes[a]	570, 535, 420	556, 423	555, 450, 425	32
Cytochrome P-450[b]	568, 537, 418, 360	545, 418	548, 450, 423	36

[a] Phenobarbital-treated minus control.
[b] Solubilized and partially purified.

and other anomalous properties of cytochrome P-450, Imai and Sato (40) have pointed out that these features are present only when the hemoprotein exists in a special state and is bound to the microsomal membrane and disappear when it is converted to the P-420 form. They may thus be dependent on the hydrophobic environment of the native cytochrome P-450 in the membrane.

The function of cytochrome P-450 as the terminal oxidase in a monooxygenation process was first shown with adrenal cortex microsomes. This fraction, which was known to catalyze the C21-hydroxylation of steroids, was found to be inhibited by carbon monoxide (41). Taking advantage of the observed light reversibility of hemoprotein (Fe^{2+})–CO complex in experiments designed according to Warburg (42), Estabrook et al. were able to show that cytochrome P-450 present in adrenal cortex microsomes is the terminal oxidase functioning in the C21-hydroxylation of 17α-hydroxyprogesterone (43). In 1964, the CO sensitivity of a liver microsomal monooxygenase reaction, i.e., aminopyrine N-demethylation, was reported (44), followed by evidence from induction studies (45) (see discussion below) which also strongly suggested the involvement of cytochrome P-450 in liver microsomal drug oxidation reactions. These reports were followed by the publication of photochemical action spectra of the CO-inhibited monooxygenation of codeine, 4-methylaminoantipyrine, acetanilide, and testosterone in liver microsomes, conclusively demonstrating the role of cytochrome P-450 as the terminal oxidase in these reactions (26, 46). Today, cytochrome P-450 is established to be the terminal oxidase involved in the monooxygenation of a great variety of foreign, as well as endogenous, lipid-soluble compounds (cf. ref. 8 and discussion below).

B. NADPH–Cytochrome P-450 Reductase

In the course of the monooxygenation process, reducing equivalents are transferred from NADPH to cytochrome P-450 via a NADPH-cytochrome P-450 reductase which most probably is closely related to the flavoprotein known since 1950 as NADPH–cytochrome c reductase (47). Evidence that the latter enzyme is in fact involved in the transfer of electrons from NADPH to cytochrome P-450 during the monooxygenation process comes from the early observations that drug oxidation in liver microsomes is inhibited by oxidized cytochrome c (48); that NADPH–cytochrome c reductase activity is enhanced parallel to the increase in drug oxidation activity during phenobarbital induction (45, 49); and, more conclusively, that specific antibodies prepared to purified NADPH–cytochrome c reductase inhibit cytochrome P-450 reduction by NADPH, as well as the overall monooxygenation process, catalyzed by liver microsomes (50–52).

The NADPH–cytochrome c reductase of liver microsomes has been solubilized, purified, and studied in great detail by several groups of investigators (53–55). The purified enzyme, which has an estimated molecular weight of 81,000 and contains 2 moles of FAD per mole (56), can reduce both one electron and two electron acceptors, including cytochrome c, ferricyanide, 2,6-dichlorophenolindophenol, and menadione. During electron transfer the purified enzyme shows the interesting property of shuttling between the half-reduced (free radical) and fully reduced states (57).

C. Additional Components

The involvement of additional catalytic components in the NADPH-linked monooxygenase system of microsomes had been considered on several grounds. Various reactions such as the NADPH-dependent reduction of neotetrazolium (58) or the iron pyrophosphate-catalyzed peroxidation of microsomal lipids (59, 60) which do not involve cytochrome P-450 (61, 62), but do involve the flavoenzyme NADPH-cytochrome P-450 reductase, have been found to exhibit a high sensitivity to the sulfhydryl reagent p-chloromercuribenzoate under conditions where the same flavoprotein measured as NADPH–cytochrome c or NADPH–2,6-dichlorophenolindophenol reductase was only partly sensitive (61, 62). The monooxygenase system as a whole is also highly sensitive to sulfhydryl reagents (61–64). This was taken as an indication that a component with essential SH groups may be involved as a catalytic link between the flavoprotein and cytochrome P-450 (61, 62). The possibility, however, was also considered that the essential SH groups are parts of the flavoprotein itself, required for neotetrazolium reduction and lipid peroxidation as well as for the interaction with cytochrome P-450, but not for the interaction with cytochrome c or 2,6-dichlorophenolindophenol, at least under certain conditions (61, 62). It has been shown, for example, that the sensitivity of the latter reaction to p-chloromercuribenzoate may vary depending on the order in which NADPH and the sulfhydryl reagent are added (54, 56). In fact, Franklin and Estabrook (65) recently obtained assay conditions under which NADPH-linked cytochrome c reduction and ethylmorphine N-demethylation activities were equally sensitive to the sulfhydryl reagent mersalyl, and assigned the locus of inhibition to the reduction, rather than the oxidation, of the flavoprotein. Thus, it now seems probable that the essential SH groups are parts of the flavoprotein itself. There remain, however, several observations, such as the decrease in neotetrazolium reductase activity upon solubilization of the flavoprotein (54) and the differences in response of the two activities to deoxycholate treatment of the microsomes (66), which indicate differences between the NADPH–cytochrome c and NADPH–neotetrazolium reductases. Such differences, when considered together with the failure of the extensively purified NADPH–cytochrome c reductase to function in the reconstructed, microsomal monooxygenase system (see discussion below), make it appear possible that at least the trypsin-solubilized cytochrome c reductase activity may reflect a modification of the enzyme similar to that found with mitochondrial NADH dehydrogenase (67).

Indications for the occurrence of additional component(s) in the liver microsomal monooxygenase have also been reported in studies of the postnatal development of this enzyme system. It has been shown that both the NADPH oxidizing flavoprotein (measured as cytochrome c reductase) and cytochrome P-450 increase rather rapidly after birth, whereas the overall monooxygenase activity (measured as oxidative demethylation of aminopyrine) increases relatively slowly (68). Thus, it has been suggested that a third component, intermediate between the flavoprotein and the cytochrome, may be rate limiting during the early development of the monooxygenase system (68).

For many years all attempts to solubilize and purify an enzymically functional cytochrome P-450 system from microsomes have been unsuccessful, and it is only recently that Lu and Coon and associates have been able to resolve the monooxygenase system from liver microsomes into three fractions (69, 70). In reconstruction studies with partially purified cytochrome P-450 and NADPH–cytochrome c reductase, they were able to show that a heat-stable lipophilic component is necessary for the interaction of the two enzymes in order to exhibit NADPH-linked cytochrome P-450 reductase and monooxygenase activities, the latter being measured with hydrocarbons, fatty acids, or various drugs as substrates (71–73). Significantly, extensively purified NADPH–cytochrome c reductase failed to function in the reconstituted system (71). Phospholipids, and in particular phosphatidyl choline, were found to replace the heat-stable lipophilic component (74). The degree of saturation of the constituent fatty acids of the phospholipid seems to be of some significance since the dioleoyl derivative was found to be much more active than the palmitoyl and stearoyl derivatives. On the other hand, substitution of stearate by saturated fatty acids of shorter chain length, such as laurate, restored the activity (74). In view of the pioneering studies of van Deenen (75), it would appear that the conformation of the phospholipid is the decisive factor in determining its reconstructive ability in the microsomal monooxygenase system by facilitating the interaction of the flavoprotein with the cytochrome. A possible hydrogen carrier function of phospholipids (76) and of ubiquinone (77) in microsomal electron transfer has been considered but is not supported by available evidence.

It is well established that the monooxygenase system of adrenal cortex mitochondria involves a nonheme iron sulfur component as a catalytic link between the NADPH oxidizing flavoprotein and cytochrome P-450 (78–81). Likewise, a nonheme iron sulfur protein called putida redoxin (82) is involved as electron carrier between the NADH-linked putida reductase flavoprotein and cytochrome $P-450_{cam}$ of the camphor oxidizing system of *Pseudomonas putida* (Chapter 7). Recently, EPR evidence has been obtained for the existence of a NADPH-reducible nonheme iron sulfur protein in rat kidney cortex microsomes (83).Although the possible occurrence of a nonheme iron sulfur protein in liver microsomes has been considered for some time, strict evidence for this conclusion, based on EPR data and excluding mitochondrial contamination, is still lacking.

Besides the cytochrome P-450-linked, NADPH-specific monooxygenase, liver microsomes (as well as microsomes from other mammalian tissues) contain another electron transport system consisting of cytochrome b_5 (84–86) and the flavoprotein NADH–cytochrome b_5 reductase (87–89). A similar system is also found in the outer membrane of mitochondria (cf. ref. 90). The metabolic function of the cytochrome b_5 — b_5 reductase system is not yet fully understood, although recent evidence indicates that in microsomes it is involved in fatty acid desaturation (91, 92). Despite the specificity of the purified cytochrome b_5 reductase for NADH, in microsomes cytochrome b_5 is also reduced by NADPH, although at a much lower rate than by NADH. Also, conversely, NADH is capable of reducing microsomal cytochrome P-450, whereas the purified NADPH oxidizing enzyme is unreactive with NADH; again, the reduction of microsomal cytochrome P-450 is much slower with NADH than with NADPH. A possible role of cytochrome b_5 in the cytochrome P-450-linked monooxygenase system has been suggested (see below).

III. SUBSTRATE INTERACTION WITH CYTOCHROME P-450

Following the observation by Narasimhulu and co-workers (93) that 17-hydroxyprogesterone produces spectral changes when added to a suspension of adrenal cortex microsomes, various xenobiotics were found to elicit similar characteristic absorbance changes in the difference spectrum of liver microsomes, and it was proposed that the spectral changes reflect the interaction of the added compound with cytochrome P-450 (94, 95). The majority of these xenobiotics were known substrates of the microsomal monooxygenase system, and it was subsequently found that endogenous substrates of this enzyme system produce analogous spectral changes when added to suspensions of liver microsomes (96, 97). The spectral changes which are obtained upon addition of this great variety of exogenous and endogenous compounds to liver microsomes in buffer systems, usually tris-Cl, pH 7.5, have been classified into three groups (96) (Fig. 1): a Type I spectral change, characterized by a peak at about 390 nm and a trough at about 420 nm (as exemplified by the hexobarbital-in-

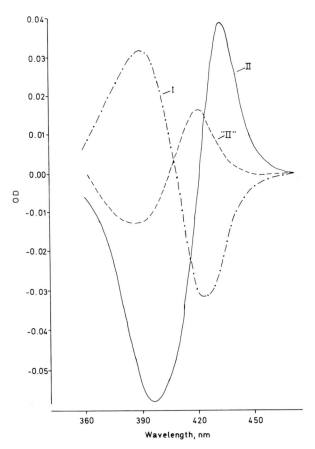

FIG. 1. Types of difference spectra produced in rat liver microsomes. Microsomes were suspended to a concentration of 1.5 mg protein per milliliter in a 25 mM sucrose-50mM tris-HCl, pH 7.5, medium. I, Type I spectral change (hexobarbital, 8.4 mM); II, Type II spectral change (aniline, 2.5 mM); "II," modified Type II spectral change (agroclavine, 1.6 mM).

duced spectrum); a Type II spectral change, exhibiting a trough near 390 nm and a peak between 425 and 435 nm (for example, the aniline-induced spectral change); and, finally, a modified Type II spectral change, characterized by a trough in the 390-nm region and a peak in the 420-nm region of the difference spectrum of the microsomes (represented in Fig. 1 by the agroclavine-induced spectral change). The modified Type II spectral change is the mirror image of the Type I spectrum and so has also been termed the "inverse" (98) or "reverse" (99) Type I spectral change.

Several lines of evidence support the assumption that these spectral manifestations do, in fact, reflect the interaction of the added substance with cytochrome P-450. It has been possible, for example, to show a relationship between the magnitude of the spectral change and the concentration of cytochrome P-450 in the microsomal membrane (96). Furthermore, "off balance absolute spectra" of cytochrome P-450 (32) also reveal that the drug-induced spectral changes relate to effects on cytochrome P-450 (100). Thus, in these preparations, the addition of compounds eliciting the Type I spectral change results in an increase in absorption at 384 nm and a reduction of the peak at 419 (or 416) nm. In addition, solubilized and partially purified cytochrome P-450 preparations from liver microsomes show the same drug-induced spectral changes as do the native microsomes (70). The most convincing evidence, however, that these spectral changes do reflect the interaction with cytochrome P-450 is perhaps the fact that they are obtained with purified cytochrome P-450$_{cam}$ from *Pseudomonas putida* (101).

The Type I spectral change was early associated with metabolism. An obvious reason for this was the finding that compounds which elicit this type of spectral change were known substrates of the monooxygenase system. Furthermore, observed similarities between the spectral dissociation and Michaelis constants for a number of substrates led to the suggestion that the Type I spectral change may in fact be the optical manifestation of the enzyme–substrate complex between cytochrome P-450 and the added compound (96). Further support for this assumption derives from EPR studies of the bacterial system, where it could be shown that the addition of the Type I substrate camphor to a purified preparation of cytochrome P-450$_{cam}$, which is in the low spin form when isolated, causes a conversion of the low spin to the high spin form (102). Also, in liver microsomes cytochrome P-450 exists as a low spin hemoprotein (103, 104) whose EPR characteristics have been shown to be affected by substrates (105–107).

Compounds which elicit the Type II spectral change are basic amines and generally do not serve as substrates for the cytochrome P-450 containing monooxygenase system. The Type II spectral manifestation has been shown to result from ferrihemochrome formation caused by the direct interaction of the heme iron with a basic nitrogen of the added compound (96). This interaction is competitive with carbon monoxide and is thus, in all probability, also competitive with oxygen. Such an interaction is hardly suggestive of metabolism since it implies that the substrate would compete with oxygen for its own oxidation. However, some compounds fitting into this class are metabolized by the monooxygenase system, aniline being the best known example. The finding that the binding constant calculated for aniline with respect to the Type II spectral change is approximately ten times higher than the K_m value obtained for aniline hydroxylation in the same microsomal system (96) raised further doubt about the significance of the Type II spectral change for metabolism. Moreover, a Type I component has recently been shown to be hidden in the aniline-induced Type II spectral change (108), and it may well be this component which is related to the metabolically active complex. Suggestive of such an interpretation is more recent work with didesmethylimipramine (109) and D-amphetamine (110), where a clear Type I spectral change was converted through indistinct spectral stages into a distinct Type II spectrum simply by increasing the concentration of the drug. Thus, there are examples of basic amines which are substrates of the monooxygenase system and which can interact to produce either the Type I or II spectral change. At least in the cases of didesmethylimipramine and D-amphetamine, Type I interaction reveals the higher affinity.

The group of compounds that elicits the modified Type II spectral change is chemically as heterogeneous as that producing the Type I spectral change. It includes various drugs, alkaloids, alcohols, steroids, and even the amino acid, tryptophan (96, 98, 111). Some of these compounds serve as substrates for the cytochrome P-450 containing monooxygenase system, while others have not, as yet, been shown to undergo oxidation in the microsomes.

Since the modified Type II spectral change is the mirror image of the Type I spectrum, the possibility that it arises through displacement of endogenously bound Type I substrates has been considered. Such a mechanism has been suggested to explain the modified Type II spectral change produced by n-butanol in liver microsomes (98). Studies with other compounds that produce the modified Type II spectral change and which are known substrates of the monooxygenase system, have, however, indicated that the modified Type II spectral change may sometimes be a composite spectrum which contains a hidden Type I component and that—analogous to the findings with the basic amines—it is this component that seems to be related to the metabolically active complex (111).

At the present stage, it thus seems that Type I interaction is of importance for the formation of the metabolically active cytochrome P-450–substrate complex. However, virtually nothing is, known about the underlying chemical mechanism, and many puzzling observations remain to be explained before the significance of the various types of spectral change can be fully understood. Examples of such observations are the findings that the same substrate may produce different types of spectral change in microsomes from different tissues (112), in liver microsomes from different species (113), or in liver microsomes from phenobarbital-treated rats as compared to controls (114). Furthermore, the Type I spectral change can also be obtained with compounds that do not serve as substrates such as fluorinated hydrocarbons (115) or, in the case of adrenal cortex microsomes, with androst-4-ene-3,17-dione (116).

IV. ON THE MECHANISM OF THE CYTOCHROME P-450-LINKED MONOOXYGENASE REACTION

In recent years the mechanism of the liver microsomal monooxygenase system has been revealed with respect to some details of the process, whereas other remain to be elucidated. Much of the progress made comes primarily from studies with the similar monooxygenase systems of non-microsomal origin, notably the steroid hydroxylating system of adrenal cortex mitrochondria and the camphor oxidizing system of *Pseudomonas putida*, in which the isolation and purification of the catalysts involved has proved easier than from microsomes. These systems are treated in separate chapters of this volume, and hence the present description will be confined to the microsomal cytochrome P-450-containing monooxygenase.

In all instances, including microsomes, it now appears well established that the initial reaction of the process consists of the rapid binding of substrate to the oxidized form of cytochrome P-450, with the transition of the latter from the low spin to the high spin form (102). This is often observable as a spectral shift of the cytochrome in the Soret region, yielding the characteristic difference spectrum with a maximum at 390 nm and a minimum at 420 nm ("Type I spectral change") (96). Most probably formation of the complex involves a stoichiometric binding of one molecule of substrate to a site near the iron atom of the heme protein (117) and is accompanied by a conformational change (96, 102, 105, 118) and an increase of the redox potential of the cytochrome from about -400 mV (119) to -170 mV (120).

Once the oxidized form of cytochrome P-450 has been bound to substrate giving the Type I spectral change, it can undergo rapid reduction by way of cytochrome P-450 reductase. The rate of this reduction, although suggested to be rate limiting for the overall monooxygenation of at least certain substrates (121–123), is considerably higher than the rate at which the free cytochrome is reduced by the reductase (122, 124), thus providing a regulatory device for the rate of oxygen consumption through this system according to the availability of substrate. The reduction of the cytochrome–substrate complex by the reductase presumably proceeds by way of one electron transfer, as also indicated by studies with the adrenal cortex mitochondrial and bacterial systems (125–127).

The next step is considered to consist of an interaction of the reduced cytochrome–substrate complex with molecular oxygen, giving rise to a ternary reduced cytochrome–substrate–oxygen complex. Evidence for the formation of this complex has been obtained first with the cytochrome P-450$_{cam}$ system (127, 128) and more recently also with liver microsomes (129). The ternary complex subsequently accepts a second electron with the formation of a yet unidentified active oxygen–cytochrome–substrate complex. After transfer within this complex of one oxygen atom to the substrate and uptake of two protons, the complex dissociates into oxidized cytochrome, H_2O, and product. A hypothetic reaction sequence, quoted from Ullrich (120), is shown in Fig. 2.

There is recent evidence for the involvement of the superoxide radical, O_2^-, in the above reaction sequence. The occurrence of this intermediate, which was first proposed by Hayaishi (130), is indicated by the findings of Strobel and Coon (131) that generation of superoxide by xanthine and xanthine oxidase can replace NADPH and NADPH–cytochrome c reductase in supporting monooxygenation catalyzed by solubilized liver

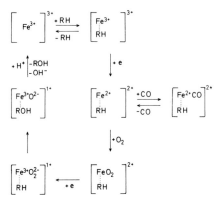

FIG. 2. A hypothetic reaction sequence of the cytochrome P-450-linked monooxygenation process, quoted from Ullrich (120).

microsomal cytochrome P-450 in the presence of phosphatidylcholine. Both the latter reaction and the one supported by the NADPH–cyto-chrome c reductase system were inhibited by superoxide dismutase (ery-throcupreine). Very recently, Aust and associates (132, 133) reported that purified NADPH–cytochrome c reductase from rat liver microsomes catalyzes the oxidation of epinephrine to adrenochrome (132), and, when supplemented with Fe^{3+}, ADP, and a critical concentration of EDTA, also the peroxidation of phospholipids (133). Both the epinephrine oxida-tion and the lipid peroxidation were inhibited by superoxide dismutase. These results point to the interesting possibility that the flavoenzyme NADPH–cytochrome c reductase, in addition to acting as a cytochrome P-450 reductase, may function as a generator of superoxide radical.

A puzzling problem concerns the sequential transfer of the two electrons, one before and one after the entry of oxygen. In the case of the cytochrome P-450$_{cam}$ system, it has been shown that the two electrons are transferred through the same nonheme iron sulfur protein (cf. ref. 82 and Chapter 7); this protein contains two atoms of iron and the possibility exists that one of these is involved specifically in the transfer of the first electron and the other of the second electron. In the case of liver microsomes, Estabrook and collaborators (134, 135) have proposed that cytochrome b$_5$ might be involved in the transfer of the second electron. This proposal was based on an observed enhancement of the rate and extent of the monooxygenase reaction by NADH in the presence of limiting NADPH on steady state shifts in the redox level of cytochrome b$_5$ and on an enhancement of the rate of NADH oxidation in the presence of NADPH. A critical evaluation of some of these findings has recently been made by Gillette et al. (11). In view of the latter and of the reported lack of inhibition of the mono-oxygenase system by an antibody against cytochrome b$_5$ (50), the sug-gested role of this cytochrome as an electron donor in the cytochrome P-450-linked monooxygenase system remains to be further substantiated.

It has for some time been rather generally assumed that the transfer of the first electron to the cytochrome P-450–substrate complex is rate determining for the overall monooxygenation process (121, 123). More recently, however, increasing evidence has accumulated indicating that this may not be the case, at least not with all substrates. Thus, in recent experiments, Baron and Peterson (136) have concluded that the rate of NADPH-linked reduction of the cytochrome P-450–substrate complex in liver microsomes from phenobarbital-treated rats proceeds much faster than the monooxygenation of the substrate studied. Moreover, the mere fact that an oxygenated intermediate can be seen to accumulate during the monooxygenation process (129) would indicate that the formation of this intermediate precedes the rate-limiting step. In fact, it has recently been reported that breaking of the C–H bond, which most probably is one of the last steps in the sequence of events in the monooxygenation process, is rate determining in certain microsomal fatty acid and steroid hydroxylations (137). From these and other studies it seems as if the rate-determining step may well vary with different substrates and different hydroxylations.

V. ON THE SUBSTRATE SPECIFICITY OF MICROSOMAL CYTOCHROME P-450-LINKED MONOOXYGENASE SYSTEMS

A. The Liver System

Characteristic of the liver microsomal cytochrome P-450-linked monooxygenase system, distinguishing it from those of steroidogenic animal tissues and of microorganisms, is its virtual lack of substrate specificity. Steroids, fatty acids (cf. Chapter 2), and heme (138–140), as well as a large number of foreign compounds including aliphatic and aromatic hydrocarbons, amines, alcohols, phenols, and thiophenes, can serve as substrates for the liver monooxygenase system. The reaction may result in hydroxylation, oxidative dealkylation or deamination, nitroxide or sulfoxide formation, all according to the nature of the substrate (cf. refs. 8 and 11 for reviews). Moreover, one and the same substrate may undergo several types of reaction; for example, testosterone may be hydroxylated in several different positions (1β, 2β, 6α, 6β, 7α, 15β, 16α, and 18) and fatty acids in two positions (ω1 and ω2); (cf. Chapter 2), whereas a drug such as chlorpromazine may undergo sulfoxidation and N-demethylation as well as hydroxylation in several positions of the phenothiazine nucleus (141–143). The proportion between the different alternative reactions seems to be dependent both on the nature of the substrate and on the species and physiological state of the animal.

Even though there still occur in the literature notations such as "aminopyrine demethylase" or "pentobarbital oxidase," it is now quite generally accepted that most of the substrates of the monooxygenase system interact directly with cytochrome P-450 without the intervention of any additional enzyme, formally thought to be needed for the activation of each specific substrate or group of substrates. On the other hand, the problem as to whether all substrates of the monooxygenase system interact with the same cytochrome P-450 does not yet appear to be conclusively resolved. Three major lines of evidence would seem to support the concept that indeed most substrates are oxygenated via a common cytochrome:

1. The demonstration of a competition among various substrates for binding to the cytochrome (144–146).

2. The demonstration of a competition among various substrates for metabolism via the monooxygenase system (97, 147, 148).

3. The demonstration that induction *in vivo* of the monooxygenase system by administration of one substrate results in an enhanced monooxygenase activity toward other substrates as well (cf. ref. 10 for review).

Several lines of evidence, however, present difficulties in envisaging the monooxygenase system of liver microsomes as a homogeneous functional entity with respect to various substrates:

1. The different effects on hydroxylation in various positions of the same substrate molecule by carbon monoxide and other inhibitors (149–151), by storage of the microsomes (152), by age, sex, and the physiological state of the animal (cf. Chapter 2), and by drug pretreatment of the animal (150, 151, 153).

2. The demonstration of two phenobarbital-inducible species of cytochrome P-450 with different stereoselectivity for the binding and monooxygenation of (+)- and (−)-methylphenobarbital (154).

3. The varying amounts of cytochrome P-450 bound at infinite concentrations of various substrates.

4. The behavior of one particular category of substrates, namely, polycyclic hydrocarbons.

It now seems clear that the extent of binding of various substrates to liver microsomal cytochrome P-450, measured as the magnitude of the Type I spectral change, may vary according to the substrate used and that not all of the cytochrome P-450 present participates in the binding of a particular substrate. For example, Ullrich (155) has estimated that only 12% of the cytochrome P-450 in liver microsomes from untreated rats is capable of binding cyclohexane. Furthermore, an investigation of 14 substituted barbiturates has revealed that the maximal spectral changes obtained by extrapolation to infinite substrate concentration showed a severalfold variation from one substrate to another (156). When two substrates with low maximal binding were combined, however, they showed no additivity, and there was also no clear correlation between the maximal extent of binding and the lipid solubility of the various barbiturates. These data thus seem to exclude the possibility that the unequal maximal binding found with various substrates results either from different populations of cytochrome P-450's, in which cases there should occur an additivity when two substrates are combined, or from different extinction coefficients of the various substrate–cytochrome complexes, or from different accessabilities of various portions of the microsomal cytochrome P-450 pool to different substrates, in which case one would expect a strict correlation between maximal extent of binding and lipid solubility. A final clarification of this problem seems to require a better knowledge of the chemical nature of the interaction between substrate and cytochrome.

Polycyclic hydrocarbons constitute an interesting exception with regard to the various parameters of substrate binding and metabolism via the liver microsomal monooxygenase system. In normal untreated animals, polycyclic hydrocarbons such as 3,4-benzpyrene serve as poor substrates for the isolated liver microsomes and give rise to little or none of the spectral shift characteristic of the formation of a complex with cytochrome P-450 (157). Administration of polycyclic hydrocarbons *in vivo* results in a drastic increase in the ability of liver microsomes to metabolize polycyclic hydrocarbons and a minor group of other drugs, whereas the activity toward most other substrates remains unchanged or even decreases (cf. ref. 9 for review). These changes concern both maximal velocities and affinities for the respective substrates (158–160) and are associated with an increase in absorbance of the cytochrome and a concomitant shift of the absorption maximum of the reduced, CO-bound complex from 450 to 448 nm (161). Thus, it has been suggested that induction of an enhanced 3,4-benzpyrene hydroxylation activity by polycyclic hydrocarbons is associated with the synthesis of a modified hemoprotein, designated P-448 (161), P-446 (162), or P_1-450 (163), having an increased affinity for polycyclic hydrocarbon hydroxylation and exhibiting differences in optical (36) as well as EPR (107) characteristics and in the response to different inhibitors (163, 164) when compared to cytochrome P-450. The mechanism by which polycyclic hydrocarbons bring about this effect is not known. From recent work on the isolation of the cytochrome from polycyclic hydrocarbon-treated animals (165), it appears, however, that the treatment leads to the formation of a modified cytochrome molecule rather than an altered environment of the cytochrome within the membrane (166) and that the mere binding of the inducer, or one of its metabolites, to preexisting cytochrome P-450 (157) could hardly explain all the effects of polycyclic hydrocarbon induction.

In the absence of added substrate, liver microsomes catalyze a slow oxidation of NADPH with the formation of H_2O_2 (48). The latter is thought to preceed by autoxidation, partly of reduced flavoprotein and partly of reduced cytochrome P-450 (167). The hydrogen peroxide so formed can be split by liver catalase present in peroxisomes into oxygen and water. If an aliphatic alcohol such as methanol or ethanol is present, this will interact with hydrogen peroxide under the influence of catalase with the formation of the corresponding aldehyde (168, 169). The kinetics of this reaction have recently been investigated by Chance and associates (167, 170) and may account, according to these authors, for an earlier postulated involvement of the liver microsomal monooxygenase system in the oxidation of methanol and ethanol (171–173). The kinetic studies of Chance and collaborators (167, 170) further reveal the difficulties in using catalase inhibitors to exclude the role of contaminating catalase in microsomal ethanol oxidation. On the other hand, Lieber (174) has recently reported ethanol oxidation activity in a solubilized microsomal system, free of contaminating catalase, which would suggest that the mechanism of microsomal ethanol oxidation still remains to be clarified.

B. Extrahepatic Systems

Microsomal cytochrome P-450-linked monooxygenase systems are found in a number of tissues besides liver, including kidney, lung, small intestine, and spleen and also in steroidogenic organs such as the adrenal cortex, testis, ovary, and placenta (175–177). Insofar as data are available, the catalytic components of these monooxygenase systems are similar to those of liver microsomes. The cytochrome P-450 contents and substrate specificities of the microsomal monooxygenases of various tissues are listed in Table II. The low concentration of the cytochrome in most extrahepatic tissues is obvious, and its measurement in these tissues is often made difficult by the presence of hemoglobin or mitochondria. It should be noted, however, that considerable species differences exist with respect to the concentration of cytochrome P-450 in microsomes from extrahepatic tissues. Rabbit lung microsomes, for example, have been reported to contain a much higher concentration of cytochrome P-450 than rat lung microsomes (179). In contrast to the liver monooxygenase, the monooxygenases present in the steroidogenic tissues and in spleen show a much more restricted array of substrate specificity, although the monooxygenase system of guinea pig adrenal cortex microsomes is an interesting exception in that it also catalyzes the N-demethylation of certain drugs (94). Also, the cytochrome P-450-linked monooxygenase system present in rat kidney cortex microsomes appears to have a higher substrate specificity than the liver system and catalyzes predominantly fatty acid hydroxylation in the $\omega 1$ and $\omega 2$ positions (178), whereas that of lung microsomes has been reported to catalyze the hydroxylation of various drugs at rates which are, however, lower than those obtained with liver microsomes. The mechanism underlying these apparent differences

TABLE II

SUBSTRATE SPECIFICITY OF CYTOCHROME P-450-LINKED, MICROSOMAL MONOOXYGENASE
SYSTEMS IN DIFFERENT MALE RAT TISSUES

Tissue	Cytochrome P-450[a]	Substrate	Hydroxylations	Ref.
Liver	0.60	Xenobiotics	Various	8, 11
		Steroids	1β, 2, 6, 7, 15, 16α, 18	b
		Fatty acids	ω1 and ω2	150
		Heme	Oxidation at α-methene bridge	138–140
Adrenal cortex	0.50	Steroids	17α, 21	b
Testis	0.10	Steroids	17α (21)	b
Kidney cortex	0.10	Fatty acids (Drugs)	ω1 and ω2 Various	178
Lung	0.035	Drugs	Various	179
Spleen	0.025	Heme	Oxidation at α-methene bridge	138–140

[a] In nmoles/mg microsomal protein (approximate concentration).
[b] See Chapter 2.

in substrate specificity between the cytochrome P-450-linked monooxygenase systems of various tissues is not known.

VI. RELATIONSHIP OF THE CYTOCHROME P-450-LINKED MONOOXYGENASE SYSTEM TO THE MICROSOMAL MEMBRANE

The association of various enzymes with the endoplasmic reticulum membranes raises the problem as to their topology and interaction with other membrane components. The electron transport systems are of special interest in this context since they involve the interaction of several catalytic components within the membrane. Studies based on the use of trypsin and other proteolytic enzymes to digest various protein components of liver microsomes have revealed that the flavoprotein component of the monooxygenase system, NADPH–cytochrome c reductase, is readily solubilized (54, 55, 180–182) with a parallel loss of the overall monooxygenase activity as well as other activities involving the NADPH-linked electron transport system such as cytochrome P-450 reductase, neotetrazolium reductase, and lipid peroxidation (182). Cytochrome

P-450 as such showed a low trypsin sensitivity and was converted into the P-420 form only after extensive trypsin treatment, without solubilization of the hemochromogen. Interestingly, the converse picture was found with the NADH–cytochrome c reductase system (86, 180–182) in that the cytochrome moiety of the system, cytochrome b_5, was readily solubilized upon trypsin treatment (although not as readily as NADPH–cytochrome c reductase), whereas its flavoprotein component, NADH–cytochrome b_5 reductase, proved to be relatively resistant to solubilization by trypsin. Again, the solubilization of cytochrome b_5 was accompanied by a loss of the overall NADH–cytochrome c reductase activity (183), and there are indications that this may result from a modification of cytochrome b_5 (181). More recently, it has also been shown that specific antibodies added to microsomes readily inactivate NADPH–cytochrome c reductase and cytochrome b_5 (50–52). The available information thus suggests that these two catalysts are located on the outer surface of the endoplasmic reticulum. It has been proposed (160) that the two microsomal electron transport enzyme systems may be localized in the membrane with an opposite transversal orientation, the NADH-linked system having its cytochrome component located near the outside of the membrane.

Little is known about the lateral mobility of the components of the electron transport systems within the ER membranes. Both cytochrome b_5 (25) and cytochrome P-450 (184) seem to be present in liver microsomes in a severalfold molar excess as compared to the flavoproteins involved in their reduction. This situation is similar to that found in mitochondria, where the interaction between flavoproteins and cytochromes is thought to be mediated by a mobile pool of ubiquinone (185). However, there is no evidence for the existence of a similar mobile mediator in the case of microsomes. Thus, unless one assumes a clusterlike arrangement of several molecules of cytochrome around one flavin (cf. ref. 65) it is necessary to invoke a sufficient degree of mobility of the individual flavin and/or cytochrome molecules within the membrane in order to account for their observed rates and extents of interaction. One way of explaining this interaction would be, in other words, in terms of the "fluid mosaic membrane model" proposed by Singer and Nicholson (186). The recent findings of Strobel et al. (74) that phospholipids containing unsaturated or short-chain fatty acids are most efficient in facilitating the interaction between isolated cytochrome P-450 and NADPH–cytochrome P-450 reductase suggest that the fluidity of the phospholipid is essential for the interaction.

Much work has recently been devoted to the question as to whether the enzymes associated with the microsomal membrane, including the two electron transport systems, are distributed at random along the endoplasmic reticulum or whether different enzymes are concentrated in different regions of the reticulum with a consequent functional specialization of the different regions. As far as the ribosome-carrying and ribosome-free ("rough" and "smooth") segments of the reticulum are concerned, it now appears that these are closely similar in terms of relative enzyme composition (66). Yet an important exception is the case when an enzyme or an enzyme system is in the stage of active induction, in which case there may be a disproportion in the enzyme contents between the two segments. Thus it is found, for example, that during the drug-induced synthesis of the hydroxylating enzyme system the concentration of its enzyme components is higher in the rough than in the smooth microsomes in an early stage of the induction (187), whereas the converse relationship holds in a later stage (187, 188).

Attempts have been made to demonstrate an uneven distribution of enzymes in subfractions of microsomes obtained by various procedures (189–192). Such studies have revealed a certain heterogeneity among the various fractions with respect to the relative contents of, e.g., the NADPH- and NADH-linked electron transport systems. This would indicate a functional heterogeneity of different parts of the endoplasmic reticulum. It has also been found that different microsomal enzymes such as NADPH–cytochrome c reductase and cytochrome b_5 (180, 182), and probably also cytochrome P-450 (193), have different turnover rates *in vivo*. It appears, therefore, that in the case of the endoplasmic reticulum, just as in those of other cellular structures, enzyme components may have their individual locations and lifetimes, and thus that a membrane-bound enzyme system represents a strictly organized but still highly dynamic entity with its components being renewed at different rates (194).

VII. SUBSTRATE-INDUCED SYNTHESIS OF THE LIVER MICROSOMAL MONOOXYGENASE SYSTEM

A large number of different drugs, insecticides, carcinogens, and other environmental chemicals, as well as steroid hormones, are known to induce an increased capacity of the liver microsomal monooxygenase system, when administered to animals *in vivo* (cf. refs. 9 and 10 for reviews). The inducers seem to share the property of being substrates for this enzyme system, and thus the phenomenon provides a model for studying the regulation by substrates of a mammalian multienzyme system. Since it was early realized that barbiturates, and in particular phenobarbital, are very potent inducers of the liver microsomal monooxygenase system (195, 196), phenobarbital has been widely used as a model compound in subsequent induction studies. The effects of phenobarbital also appear to be representative of those of most other inducers, except for the polycyclic hydrocarbons, discussed earlier.

With phenobarbital and similarly acting inducers, it is now well established that the induction process leads to a selective increase in the levels of both the flavoprotein and cytochrome components of the monooxygenase system (31, 45, 197) and involves increased rates of both protein and RNA syntheses, as indicated by its sensitivity to puromycin (198) and actinomycin D (31) as well as by increased RNA polymerase activity (199). Microsomes isolated from phenobarbital-treated animals were subsequently shown to be more active in endogenous L-[^{14}C]phenylalanine incorporation, and, after the removal of endogenous mRNA, to be more sensitive than control microsomes to polyuridylic acid-directed L-[^{14}C] phenylalanine incorporation, suggesting a phenobarbital-induced increase in both the endogenous microsomal mRNA content and the total number of microsomal binding sites for mRNA (200, 201). Turnover studies have further revealed an enhanced rate of synthesis of both the flavoprotein (182, 202) and cytochrome (193, 203) components of the liver monooxygenase system under the influence of phenobarbital. Simultaneously, there appears to occur a decrease in the rate of breakdown of the flavoprotein (182, 202) and, under certain conditions, also of the hemoprotein (203) which may further contribute to the increased enzyme level.

Although some other microsomal enzymes such as the glucuronyl transferase may increase as well under phenobarbital stimulation (204), the induction is selective and largely restricted to the components of the monooxygenase system. It thus leads to the formation of microsomal membranes, where the enzyme proportions clearly differ from those of the preexisting membranes. Using an elegant double-labeling technique, the selectivity of the process has recently been studied by Schimke and collaborators (205, 206).

Preceding the increase in the levels of the components of the microsomal monooxygenase following phenobarbital administration, there is a rapid increase in mitochondrial δ-aminolevulinic acid synthetase (207), the rate-limiting step in heme biosynthesis (208, 209). This observation and the finding that repressors of heme biosynthesis prevent the induction of the microsomal monooxygenase system by phenobarbital (207, 210) have led to the suggestion that heme biosynthesis plays an important role in the induction process. Whether the induction proceeds by way of a direct interaction of the inducing drug with the genetic system or by way of some mediator of cellular origin is, however, not known. There are indications of a regulatory effect of steroid hormones on the liver microsomal monooxygenase system as well as of a need for steroid hormones in phenobarbital induction (199). It has been speculated that, since drugs probably alter steroid metabolism by way of the monooxygenase system, steroid hormones might in turn act as cellular mediators in the drug-induced enzyme synthesis (199). It has, however, not been possible to simulate phenobarbital induction by *in vivo* administration of steroid hormones, and although various steroids such as spironolactone (211) and pregnenolone-16α-carbonitrile (212) are known to induce certain liver microsomal monooxygenase activities, the effects are usually less pronounced and more restricted than those of phenobarbital (213).

A striking phenomenon accompanying drug-induced enzyme synthesis is the proliferation of endoplasmic membranes (214). From the available evidence it appears that the induced enzyme synthesis begins in the rough-surfaced areas of the endoplasmic reticulum (187), and leads eventually to the accumulation of smooth-surfaced profiles with a high concentration of the enzyme components of the monooxygenase system (187, 188). The situation is similar to that found during the neonatal development of the liver endoplasmic reticulum which is characterized by a dramatic rise in membrane content and in certain enzymic activities, e.g., glucose-6-phosphatase (215). In both cases, the synthesis of new membranes seems to be geared to the formation of new enzymes by a mechanism that may involve an outgrowth and budding off of smooth-surfaced membrane profiles from the rough-surfaced endoplasmic reticulum at the sites where the enzyme synthesis takes place. The drug-induced membrane proliferation is accompanied by an increased rate of phospholipid synthesis (31, 216), although here again, a simultaneous slowing down of lipid catabolism appears to be a contributory factor (216, 217). The drug-induced synthesis of the microsomal hydroxylating system is a beautiful example of the now widely studied phenomenon of the induction of a membrane-bound multienzyme complex, and one of the few instances presently available where this phenomenon can be studied in a multicellular organism.

REFERENCES

1. Hayaishi, O., ed. (1962). "Oxygenases." Academic Press, New York.
2. Mueller, G. C., and Miller, J. A. (1948). *J. Biol. Chem.* **176,** 535.
3. Mueller, G. C., and Miller, J. A. (1953). *J. Biol. Chem.* **202,** 579.
4. Brodie, B. B., Axelrod, J., Cooper, J. R., Gaudette, L., La Du, B. N., Mitoma, C., and Udenfriend, C. (1955). *Science* **121,** 603.
5. La Du, B. N., Gaudette, L., Trousof, N., and Brodie, B. B. (1955). *J. Biol. Chem.* **214,** 741.
6. Brodie, B. B., Gillette, J. R., and La Du, B. N. (1958). *Annu. Rev. Biochem.* **27,** 427.

7. Gillette, J. R. (1963). *Progr. Drug Res.* **6**, 13.
8. Gillette, J. R. (1966). *Advan. Pharmacol.* **4**, 219.
9. Conney, A. H. (1967). *Pharmacol. Rev.* **19**, 317.
10. Kuntzman, R. (1969). *Annu. Rev. Pharmacol.* **9**, 21.
11. Gillette, J. R., Davies, D. S., and Sasame, H. A. (1972). *Annu. Rev. Pharmacol.* **12**, 57.
12. La Du, B. N., Mandel, H. G., and Way, E. L. (eds.) (1972). "Fundamentals of Drug Metabolism and Drug Disposition." Williams and Wilkins, Baltimore, Maryland.
13. King, T. E., Mason, H. S., and Morrison, M. (eds.) (1965). "Oxidases and Related Redox Systems," Vol. 2. Wiley, New York.
14. Symposium on Electron Transport Systems in Microsomes (1965). *Fed. Proc. Fed. Amer. Soc. Exp. Biol.* **24**.
15. Bloch, K., and Hayaishi, O. (eds.) (1966). "Biological and Chemical Aspects of Oxygenases." Maruzen, Tokyo.
16. Gillette, J. R., Conney, A. H., Cosmides, G. J., Estabrook, R. W., Fouts, J. R., and Mannering, G. J. (eds.) (1969). "Microsomes and Drug Oxidations." Academic Press, New York.
17. Shugar, D. (ed.) (1969). *Biochem. Aspects Antimetabolites Drug Hydroxylat. Fed. Eur. Biochem. Soc. Symp. Ser.* Vol. 16. Academic Press, New York.
18. Boyd, G. S., and Smellie, R. M. S. (eds.) (1972). *Biolog. Hydroxylat. Mech.* Biochem. Soc. Symp. Number 34. Academic Press, New York.
19. *Proc. 2nd Int. Symp. Microsomes Drug Oxidations, Palo Alto* (1973). *J. Drug Met. Disp.* Vol. 1, No. 1.
20. Klingenberg, M. (1958). *Arch. Biochem. Biophys.* **75**, 376.
21. Garfinkel, D. (1958). *Arch. Biochem. Biophys.* **77**, 493.
22. Omura, T., and Sato, R. (1963). *Biochim. Biophys. Acta* **71**, 224.
23. Omura, T., and Sato, R. (1964). *J. Biol. Chem.* **239**, 2370.
24. Omura, T., and Sato, R. (1964). *J. Biol. Chem.* **239**, 2379.
25. Mason, H. S., Yamano, T., North, J. C., Hashimoto, Y., and Sakagishi, P. (1965). *In* "Oxidases and Related Redox Systems" (T. King, H. S. Mason, and M. Morrison, eds.), p. 879. Wiley, New York.
26. Cooper, D. Y., Narasimhulu, S., Rosenthal, O., and Estabrook, R. W. (1965). *In* "Oxidases and Related Redox Systems" (T. King, H. S. Mason, and M. Morrison, eds.), p. 838. Wiley, New York.
27. Imai, Y., and Sato, R. (1967). *Eur. J. Biochem.* **1**, 419.
28. Ichikawa, Y., and Yamano, T. (1967). *Biochim. Biophys. Acta* **147**, 518.
29. Ichikawa, Y., Hagihara, B., Mori, K., and Yamano, T. (1966). *In* "Biological and Chemical Aspects of Oxygenases" (K. Bloch and O. Hayaishi, eds.), p. 211. Maruzen, Tokyo.
30. Ichikawa, Y., and Yamano, T. (1967). *Biochim. Biophys. Acta* **131**, 490.
31. Chance, B. (1957). *Methods Enzymol.* **4**, 273.
32. Kinoshita, T., and Horie, S. (1967). *J. Biochem. (Tokyo)* **61**, 26.
33. Nishibayashi, H., and Sato, R. (1968). *J. Biochem. (Tokyo)* **63**, 766.
34. Nishibayashi, H., Omura, T., Sato, R., and Estabrook, R. W. (1968). *In* "Structure and Function of Cytochromes" (K. Okunuki, M. D. Kamen, and I. Sekuzu, eds.), p. 658. Univ. Park Press, Baltimore, Maryland.
35. Ichikawa, Y., and Yamano, T. (1970). *Biochim. Biophys. Acta* **200**, 220.
36. Mannering, G. J. (1971). *In* "Fundamentals of Drug Metabolism and Drug

Disposition" (B. N. La Du, H. G. Mandel, and E. L. Way, eds.), p. 206. Williams and Wilkins, Baltimore, Maryland.
37. Nishibayashi, H., Omura, T., and Sato, R. (1966). *Biochim. Biophys. Acta* **118,** 651.
38. Imai, Y., and Sato, R. (1966). *Biochem. Biophys. Res. Commun.* **22,** 620.
39. Nishibayashi, H., and Sato, R. (1965). *J. Jap. Biochem. Soc.* **37,** 630.
40. Imai, Y., and Sato, R. (1967). *J. Biochem. (Tokyo)* **62,** 464.
41. Ryan, K. J., and Engel, L. L. (1957). *J. Biol. Chem.* **225,** 103.
42. Warburg, O. (1949). *In* "Heavy Metal Prosthetic Groups and Enzyme Action," Chapters XII and XIII. Oxford Univ. Press (Clarendon), London and New York.
43. Estabrook, R. W., Cooper, D. Y., and Rosenthal, O. (1963). *Biochem. Z.* **338,** 741.
44. Orrenius, S., Dallner, G., and Ernster, L. (1964). *Biochem. Biophys. Res. Commun.* **14,** 329.
45. Orrenius, S., and Ernster, L. (1964). *Biochem. Biophys. Res. Commun.* **16,** 60.
46. Cooper, D. Y., Levin, S., Narasimhulu, S., Rosenthal, O., and Estabrook, R. W. (1965). *Science* **147,** 400.
47. Horecker, B. L. (1950). *J. Biol. Chem.* **183,** 593.
48. Gillette, J. R., Brodie, B. B., and La Du, B. N. (1957). *J. Pharmacol. Exptl. Therap.* **119,** 532.
49. Orrenius, S., Ericsson, J. L. E., and Ernster, L. (1965). *J. Cell Biol.* **25,** 627.
50. Omura, T. (1969). Reported at the *Int. Symp. Microsomes Drug Oxidations, Tubingen.*
51. Masters, B. S. S., Baron, J., Taylor, W. E., Isaacson, E. L., and LoSpalluto, J. (1971). *J. Biol. Chem.* **246,** 4143.
52. Glazer, R. I., Schenkman, J. B., and Sartorelli, A. C. (1971). *Mol. Pharmacol.* **7,** 638.
53. Lang, C. A., and Nason, A. (1959). *J. Biol. Chem.* **234,** 1874.
54. Williams, C. H., and Kamin, H. (1962). *J. Biol. Chem.* **237,** 587.
55. Phillips, A. H., and Langdon, R. G. (1962). *J. Biol. Chem.* **237,** 2652.
56. Masters, B. S. S., Kamin, H., Gibson, Q. H., and Williams, C. H. (1965). *J. Biol. Chem.* **240,** 921.
57. Kamin, H., Masters, B. S. S., Gibson, Q. H., and Williams, C. H. (1965). *Fed. Proc. Fed. Amer. Soc. Exp. Biol.* **24,** 1164.
58. Williams, C. H., Gibbs, R. H., and Kamin, H. (1959). *Biochim. Biophys. Acta* **32,** 568.
59. Hochstein, P., and Ernster, L. (1964). *In* "Cellular Injury" (A. V. S. de Reuck and J. Knight, eds.), p. 123. Churchill, London.
60. Ernster, L., and Nordenbrand, K. (1967). *Methods Enzymol.* **10,** 574.
61. Orrenius, S. (1965). *J. Cell Biol.* **26,** 713.
62. Ernster, L., and Orrenius, S. (1965). *Fed. Proc. Fed. Amer. Soc. Exp. Biol.* **24,** 1190.
63. Conney, A. H., Miller, E. C., and Miller, J. A. (1957). *J. Biol. Chem.* **228,** 753.
64. Netter, K. J., and Jenner, S. (1966). *Naunyn-Schmiedebergs Arch. Pharmakol. Exp. Pathol.* **255,** 120.
65. Franklin, M. R., and Estabrook, R. W. (1971). *Arch. Biochem. Biophys.* **143,** 318.
66. Dallner, G. (1963). *Acta Pathol. Microbiol. Scand. Suppl. 166.*
67. Cremona, T., Kearney, E. B., Villavicencio, M., and Singer, T. P. (1963). *Biochem. Z.* **338,** 407.
68. Dallner, G., Siekevitz, P., and Palade, G. E. (1965). *Biochem. Biophys. Res. Commun.* **20,** 135.

69. Lu, A. Y. H., and Coon, M. J. (1968). *J. Biol. Chem.* **243**, 1331.
70. Coon, M. J., and Lu, A. Y. H. (1969). *In* "Microsomes and Drug Oxidations" (J. R. Gillette, A. H. Conney, G. J. Cosmides, R. W. Estabrook, J. R. Fouts, and G. J. Mannering, eds.), p. 151. Academic Press, New York.
71. Lu, A. Y. H., Junk, K. W., and Coon, M. J. (1969). *J. Biol. Chem.* **244**, 3714.
72. Lu, A. Y. H., Strobel, H. W., and Coon, M. J. (1969). *Biochem. Biophys. Res. Commun.* **36**, 545.
73. Lu, A. Y. H., Strobel, H. W., and Coon, M. J. (1970). *Mol. Pharmacol.* **6**, 213.
74. Strobel, H. W., Lu, A. Y. H., Heidema, J., and Coon, M. J. (1970). *J. Biol. Chem.* **245**, 4851.
75. van, Deenen, L. L. M. (1968). "Regulatory Functions of Biological Membranes" (J. Järnefelt, ed.) Vol. XI, p. 72. B.B.A. Library, Elsevier, Amsterdam.
76. Ernster, L., Nordenbrand, K., Orrenius, S., and Das, M. L. (1968). *Hoppe-Seyler's Z. Physiol. Chem.* **349**, 1604.
77. Leonhäuser, S., Leybold, K., Krisch, K., Staudinger, Hj., Gale, P. H., Atwood, C. P., and Folkers, K. (1962). *Arch. Biochem. Biophys.* **96**, 580.
78. Suzuki, K., and Kimura, T. (1965). *Biochem. Biophys. Res. Commun.* **19**, 340.
79. Omura, T., Sato, R., Cooper, D. Y., Rosenthal, O., and Estabrook, R. W. (1965). *Fed. Proc. Fed. Amer. Soc. Exp. Biol.* **24**, 1181.
80. Omura, T., Sanders, E., Estabrook, R. W., Cooper, D. Y., and Rosenthal, O. (1966). *Arch. Biochem. Biophys.* **117**, 660.
81. Kimura, T., and Suzuki, K. (1967). *J. Biol. Chem.* **242**, 485.
82. Cushman, D. W., Tsai, R. L., and Gunsalus, I. C. (1967). *Biochem. Biophys. Res. Commun.* **26**, 577.
83. Hoffström, I., Ellin, Å., Orrenius, S., Bäckström, D., and Ehrenberg, A. (1972). *Biochem. Biophys. Res. Commun.* **48**, 977.
84. Strittmatter, C. F., and Ball, E. G. (1952). *Proc. Nat. Acad. Sci. U.S.* **38**, 19.
85. Chance, B., and Williams, G. R. (1954). *J. Biol. Chem.* **209**, 945.
86. Strittmatter, P., and Velick, S. F. (1956). *J. Biol. Chem.* **221**, 253.
87. Hogeboom, G. H. (1949). *J. Biol. Chem.* **177**, 847.
88. Strittmatter, P., and Velick, S. F. (1957). *J. Biol. Chem.* **228**, 785.
89. Strittmatter, P. (1965). *Fed. Proc. Fed. Amer. Soc. Exp. Biol.* **24**, 1156.
90. Ernster, L., and Kuylenstierna, B. (1970). *In* "Membranes of Mitochondria and Chloroplasts" (E. Racker, ed.), p. 172. Van Nostrand Reinhold, Princeton, New Jersey.
91. Oshino, N., Imai, Y., and Sato, R. (1971). *J. Biochem. (Tokyo)* **69**, 155.
92. Oshino, N., and Sato, R. (1971). *J. Biochem. (Tokyo)* **69**, 169.
93. Narasimhulu, S., Cooper, D. Y., and Rosenthal, O. (1965). *Life Sci.* **4**, 2102.
94. Remmer, H., Schenkman, J. B., Estabrook, R. W., Sasame, H., Gillette, J. R., Narasimhulu, S., Cooper, D. Y., and Rosenthal, O. (1966). *Mol. Pharmacol.* **2**, 187.
95. Imai, Y., and Sato, R. (1966). *Biochem. Biophys. Res. Commun.* **22**, 620.
96. Schenkman, J. B., Remmer, H., and Estabrook, R. W. (1967). *Mol. Pharmacol.* **3**, 113.
97. Orrenius, S., and Thor, H. (1969). *Eur. J. Biochem.* **9**, 415.
98. Diehl, H., Schädelin, J., and Ullrich, V. (1970). *Hopper-Seyler's Z. Physiol. Chem.* **351**, 1359.
99. Orrenius, S., von Bahr, C., Jakobsson, S. V., and Ernster, L. (1972). *In* "Oxidation Reduction Enzymes" (A. Akeson and A. Ehrenberg, ed.), p. 309. Pergamon, Oxford .
100. Remmer, H., Schenkman, J. B., and Greim, H. (1969). *In* "Microsomes and Drug

Oxidations" (J. R. Gillette, A. H. Conney, G. J. Cosmides, R. W. Estabrook, J. R. Fouts, and G. J. Mannering, eds.), p. 371. Academic Press, New York.

101. Gunsalus, I. C. (1968). *Hoppe-Seyler's Z. Physiol. Chem.* **349**, 1610.
102. Tsai, R., Yu, C. A., Gunsalus, I. C., Peisach, J., Blumberg, W. E., Orme-Johnson, W. H., and Beinert, H. (1970). *Proc. Nat. Acad. Sci. U.S.* **66**, 1157.
103. Hashimoto, Y., Yamano, T., and Mason, H. S. (1962). *J. Biol. Chem.* **237**, PC 3843.
104. Mason, H. S., North, J. C., and Vanneste, M. (1965). *Fed. Proc. Fed. Amer. Soc. Exp. Biol.* **24**, 1172.
105. Cammer, W., Schenkman, J. B., and Estabrook, R. W. (1966). *Biochem. Biophys. Res. Commun.* **23**, 264.
106. Jefcoate, C. R. E., Gaylor, J. L., and Calabrese, R. L. (1969). *Biochemistry* **8**, 3455.
107. Peisach, J., and Blumberg, W. E. (1970). *Proc. Nat. Acad. Sci. U.S.* **67**, 172.
108. Schenkman, J. B. (1970). *Biochemistry* **9**, 2081.
109. Bahr, von, C., and Orrenius, S. (1970). *Xenobiotica* **1**, 69.
110. Hoffström, I., and Orrenius, S. (1973). *FEBS Lett.* **31**, 205.
111. Wilson, B. J., and Orrenius, S. (1972). *Biochim. Biophys. Acta* **261**, 94.
112. Kupfer, D., and Orrenius, S. (1970). *Mol. Pharmacol.* **6**, 221.
113. Mitani, F., and Horie, S. (1969). *J. Biochem. (Tokyo)* **65**, 269.
114. Bahr, von, C., Schenkman, J. B., and Orrenius, S. (1972). *Xenobiotica* **2**, 89.
115. Ullrich, V., and Diehl, H. (1971). *Eur. J. Biochem.* **20**, 509.
116. Narasimhulu, S. (1971). *Arch. Biochem. Biophys.* **147**, 391.
117. Peterson, J. A., Ullrich, V., and Hildebrandt, A. (1971). *Arch. Biochem. Biophys.* **145**, 531.
118. Yong, F. C., King, T. E., Oldham, S., Waterman, M. R., and Mason, H. S. (1970). *Arch. Biochem. Biophys.* **138**, 96.
119. Waterman, M. R., and Mason, H. S. (1970). *Biochem. Biophys. Res. Commun.* **39**, 450.
120. Ullrich, V. (1972). *Angew. Chem. Internat. Ed. Engl.* **11**, 701.
121. Davies, D. S., Gigon, P. L., and Gillette, J. R. (1969). *Life Sci.* **8**, 85.
122. Gigon, P. L., Gram, T. E., and Gillette, J. R. (1969). *Mol. Pharmacol.* **5**, 109.
123. Schenkman, J. B. (1972). *Mol. Pharmacol.* **8**, 171.
124. Schenkman, J. B. (1968). *Hoppe-Seyler's Z. Physiol. Chem.* **349**, 1624.
125. Ullrich, V., Cohen, B., Cooper, D. Y., and Estabrook, R. W. (1968). *In* "Structure and Function of Cytochromes" (K. Okunuki, M. D. Kamen, and I. Sekuzu, eds.), p. 649. Univ. Park Press, Baltimore, Maryland.
126. Cooper, D. Y. (1969). Reported at the *Int. Symp. Microsomes Drug Oxidat., Tübingen.*
127. Ishimura, Y., Ullrich, V., and Peterson, J. A. (1971). *Biochem. Biophys. Res. Commun.* **42**, 140.
128. Gunsalus, I. C. (1970). Reported at the *Wenner-Gren Symp. Structure Function Oxidat. Reduction Enzymes, Stockholm.*
129. Estabrook, R. W., Hildebrandt, A. G., Baron, J., Netter, K. J., and Leibman, K. (1971). *Biochem. Biophys. Res. Commun.* **42**, 132.
130. Hayaishi, O. (1964). Plenary Lecture, *Int. Congr. Biochem., 6th, New York.*
131. Strobel, H. W., and Coon, M. J. (1971). *J. Biol. Chem.* **246**, 7826.
132. Aust, S. D., Pederson, T. C., and Roerig, D. L. (1972). *Biochem. Biophys. Res. Commun.* **47**, 1133.
133. Pederson, T. C., and Aust, S. D. (1972). *Biochem. Biophys. Res. Commun.* **48**, 789.
134. Cohen, B., and Estabrook, R. W. (1971). *Arch. Biochem. Biophys.* **143**, 46.
135. Hildebrandt, A. G., and Estabrook, R. W. (1971). *Arch. Biochem. Biophys.* **143**, 66.

136. Baron, J., and Peterson, J. A., Personal communication.
137. Björkhem, I. (1972). *Eur. J. Biochem.* **27**, 354.
138. Tenhunen, R., Marver, H. S., and Schmid, R. (1969). *J. Biol. Chem.* **244**, 6388.
139. Tenhunen, R., Marver, H. S., and Schmid, R. (1969). *Trans. Ass. Amer. Phys.* **82**, 363.
140. Schachter, B. A., Nelson, E. B., Marver, H. S., and Masters, B. S. S. (1972). *J. Biol. Chem.* **247**, 3601.
141. Fishman, V., and Goldenberg, H. (1963). *Proc. Soc. Exp. Biol. Med.* **112**, 501.
142. Goldenberg, H., and Fishman, V. (1964). *Biochem. Biophys. Res. Commun.* **14**, 404.
143. Robinson, A. E. (1966). *J. Pharm. Pharmacol.* **18**, 19.
144. Leibman, K. C., Hildebrandt, A. G., and Estabrook, R. W. (1969). *Biochem. Biophys. Res. Commun.* **36**, 789.
145. Kupfer, D., and Orrenius, S. (1970). *Eur. J. Biochem.* **14**, 317.
146. Orrenius, S., Kupfer, D., and Ernster, L. (1970). *FEBS Lett.* **6**, 249.
147. Rubin, A., Tephly, T. R., and Mannering, G. J. (1964). *Biochem. Pharmacol.* **13**, 1007.
148. Tephly, T. R., and Mannering, G. J. (1968). *Mol. Pharmacol.* **4**, 10.
149. Conney, A. H., Levin, W., Ikeda, M., Kuntzman, R., Cooper, D. Y., and Rosenthal, O. (1968). *J. Biol. Chem.* **243**, 3912.
150. Björkhem, I., and Danielsson, H. (1970). *Eur. J. Biochem.* **17**, 450.
151. Frommer, U., Ullrich, V., Staudinger, Hj., and Orrenius, S. (1972). *Biochim. Biophys. Acta* **280**, 487.
152. Levin, W., Alvares, A., Jacobson, M., and Kuntzman, R. (1969). *Biochem. Pharmacol.* **18**, 883.
153. Kuntzman, R., Levin, W., Jacobson, M., and Conney, A. H. (1968). *Life Sci.* **7**, 215.
154. Bohn, W., Ullrich, V., and Staudinger, Hj. (1971). *Naunyn-Schmiedebergs Arch. Pharmak.* **270**, 41.
155. Ullrich, V. (1969). *Hoppe-Seyler's Z. Physiol. Chem.* **350**, 357.
156. Jansson, I., Orrenius, S., Ernster, L., and Schenkman, J. B. (1972). *Arch. Biochem Biophys.* **151**, 391.
157. Schenkman, J. B., Greim, H., Zange, M., and Remmer, H. (1969). *Biochim. Biophys. Acta* **171**, 23.
158. Gurtoo, H. L., Campbell, T. C., Webb, R. E., and Plowman, K. M. (1968). *Biochem. Biophys. Res. Commun.* **31**, 588.
159. Gnosspelius, Y., Thor, H., and Orrenius, S. (1969/70). *Chem.-Biol. Interactions* **1**, 125.
160. Orrenius, S., and Ernster, L. (1971). *In* "Cell Membranes: Biological and Pathological Aspects" (G. W. Richter and D. G. Scarpelli, eds.), p. 38. Williams and Wilkins, Baltimore, Maryland.
161. Alvares, A. P., Schilling, G., Levin, W., and Kuntzman, R. (1967). *Biochem. Biophys. Res. Commun.* **29**, 521.
162. Hildebrandt, A. G., Remmer, H., and Estabrook, R. W. (1968). *Biochem. Biophys. Res. Commun.* **30**, 607.
163. Sladek, N. E., and Mannering, G. J. (1969). *Mol. Pharmacol.* **5**, 186.
164. Wiebel, F. J., Leutz, J. C., Diamond, L., and Gelboin, H. V. (1971). *Arch. Biochem. Biophys.* **144**, 78.
165. Lu, A. Y. H., Kuntzman, R., West, S., Jacobson, M., and Conney, A. H. (1972). *J. Biol. Chem.* **247**, 1727.
166. Imai, Y., and Siekevitz, P. (1971). *Arch. Biochem. Biophys.* **144**, 143.

167. Boveris, A., Oshino, N., and Chance, B. (1972). *Biochem. J.* **128,** 617.
168. Keilin, D., and Hartree, E. F. (1945). *Biochem. J.* **39,** 293.
169. Chance, B. (1949). *Acta Chem. Scand.* **1,** 236.
170. Oshino, N., Oshino, R., and Chance, B. Personal communication.
171. Lieber, C. S., and De Carli, L. M. (1968). *Science* **162,** 917.
172. Lieber, C. S., and De Carli, L. M. (1970). *J. Biol. Chem.* **245,** 2505.
173. Lieber, C. S., Rubin, E., and De Carli, L. M. (1970). *Biochem. Biophys. Res. Commun.* **40,** 858.
174. Lieber, C. S. (1972). Reported at the *Int. Symp. Microsomes Drug Oxidat.,* Palo Alto.
175. Sato, R., Omura, T., and Nishibayashi, H. (1965). *In* "Oxidases and Related Redox Systems" (T. E. King, H. S. Mason, and M. Morrison, eds.) p. 861. Wiley, New York.
176. Ichikawa, Y., and Yamano, T. (1967). *Arch. Biochem. Biophys.* **121,** 742.
177. Garfinkel, D. (1963). *Comp. Biochem. Physiol.* **8,** 367.
178. Ellin, Å., Jakobsson, S. V., Schenkman, J. B., and Orrenius, S. (1972). *Arch. Biochem. Biophys.* **150,** 64.
179. Oppelt, W. W., Zange, M., Ross, W. E., and Remmer, H. (1970). *Res. Commun. Chem. Pathol. Pharmacol.* **1,** 43.
180. Omura, T., Siekevitz, P., and Palade, G. E. (1967). *J. Biol. Chem.* **242,** 2389.
181. Ito, A., and Sato, R. (1969). *J. Cell Biol.* **40,** 179.
182. Kuriyama, Y., Omura, T., Siekevitz, P., and Palade, G. E. (1969). *J. Biol. Chem.* **244,** 2017.
183. Orrenius, S., Berg, A., and Ernster, L. (1969). *Eur. J. Biochem.* **11,** 193.
184. Estabrook, R. W., Franklin, M. R., Cohen, B., Shigamatzu, A., and Hildebrandt, A. G. (1971). *Metabolism* **20,** 187.
185. Klingenberg, M., and Kröger, A. (1967). *In* "Biochemistry of Mitochondria" (E. C. Slater, Z. Kaniuga, and L. Wojtczak, eds.), p. 11. Academic Press, New York and Polish Scientific Publ., Warsaw.
186. Singer, S. J., and Nicholson, G. L. (1972). *Science* **175,** 720.
187. Orrenius, S. (1965). *J. Cell Biol.* **26,** 725.
188. Remmer, H., and Merker, H. J. (1963). *Science* **142,** 1657.
189. Dallner, G., Bergstrand, A., and Nilsson, R. (1968). *J. Cell Biol.* **38,** 257.
190. Dallman, P. R., Dallner, G., Bergstrand, A., and Ernster, L. (1969). *J. Cell Biol.* **41,** 357.
191. Glaumann, H., and Dallner, G. (1970). *J. Cell Biol.* **47,** 34.
192. Svensson, H., Dallner, G., and Ernster, L. (1972). *Biochim. Biophys. Acta* **274,** 447.
193. Greim, H., Schenkman, J. B., Klotzbucher, M., and Remmer, H. (1970). *Biochim. Biophys. Acta* **201,** 20.
194. Siekevitz, P. (1972). *Annu. Rev. Physiol.* **34,** 117.
195. Remmer, H. (1959). *Naunyn-Schmiedebergs Arch. Pharmak.* **235,** 279.
196. Conney, A. H., and Burns, J. J. (1959). *Nature (London)* **184,** 363.
197. Remmer, H., and Merker, H. J. (1965). *Ann. N.Y. Acad. Sci.* **123,** 79.
198. Conney, A. H., and Gilman, A. G. (1963). *J. Biol. Chem.* **238,** 3682.
199. Orrenius, S., Gnosspelius, Y., Das, M. L., and Ernster, L. (1968). *In* "Structure and Function of the Endoplasmic Reticulum in Animal Cells" (F. C. Gran, ed.), p. 81. Universitetsforlaget, Oslo.
200. Kato, R., Loeb, L. A., and Gelboin, H. V. (1965). *Biochem. Pharmacol.* **14,** 1164.
201. Kato, R., Jondorf, W. R., Loeb, L. A., Ben, T., and Gelboin, H. V. (1966). *Mol. Pharmacol.* **2,** 171.

202. Jick, H., and Shuster, L. (1966). *J. Biol. Chem.* **241,** 5366.
203. Greim, H. (1970). *Naunyn-Schmiedebergs Arch. Pharmakol. Exp. Pathol.* **266,** 261.
204. Zeidenberg, P., Orrenius, S., and Ernster, L. (1967). *J. Cell Biol.* **32,** 528.
205. Schimke, R. T., Ganschow, E., Doyle, D., and Arias, I. (1968). *Fed. Proc. Fed. Amer. Soc. Exp. Biol.* **27,** 1223.
206. Dehlinger, P. J., and Schimke, R. T. (1972). *J. Biol. Chem.* **247,** 1257.
207. Marver, H. S. (1969). *In* "Microsomes and Drug Oxidations" (J. R. Gillette, A. H. Conney, G. J. Cosmides, R. W. Estabrook, J. R. Fouts, and G. J. Mannering, eds.), p. 495. Academic Press, New York.
208. Granick, S., and Urata, G. (1963). *J. Biol. Chem.* **238,** 821.
209. Granick, S. (1966). *J. Biol. Chem.* **241,** 1359.
210. Marver, H. S., Schmid, R., and Stützel, H. (1968). *Biochem. Biophys. Res. Commun.* **33,** 969.
211. Solymoss, B., Classen, H. G., and Varga, S. (1969). *Proc. Soc. Exp. Biol. Med.* **132,** 940.
212. Solymoss, B., Werringloer, J., and Toth, S. (1971). *Steroids* **17,** 427.
213. Stripp, B., Hamrick, M., and Zampaglione, N. (1970). *Fed. Proc. Fed. Amer. Soc. Exp. Biol.* **29,** 346.
214. Remmer, H., and Merker, H. J. (1963). *Klin. Wochschr.* **41,** 276.
215. Dallner, G., Siekevitz, P., and Palade, G. E. (1966). *J. Cell Biol.* **30,** 97.
216. Holtzman, J. L., and Gillette, J. R. (1968). *J. Biol. Chem.* **243,** 3020.
217. Orrenius, S. (1968). *In* "The Interaction of Drugs and Subcellular Components in Animal Cells" (P. N. Campbell, ed.), p. 97. Churchill, London.

7

FLAVOPROTEIN OXYGENASES

MARCIA S. FLASHNER and VINCENT MASSEY

I. INTRODUCTION

The flavoprotein oxygenases form a fascinating and diverse group of enzymes. In recent years the availability of highly purified crystalline preparations has engendered many interesting studies. Although the substrates represent a diversity of structures ranging from acidic and basic aliphatic compounds to substituted aromatic rings, most of the flavo-protein oxygenases thus far studied catalyze monooxygenations. That is,

TABLE I

FLAVOPROTEIN OXYGENASES[a]

Electron donor	Enzyme	Source	Flavin prosthetic group	Reaction catalyzed
		Monooxygenases		
Internal	Lactate monooxygenase	M. phlei, M. smegmatis	FMN	Lactate + O_2 → acetate + CO_2 + H_2O
	Arginine monooxygenase	S. griseus	FAD	Arginine + O_2 → γ-guanidobutyramide + CO_2 + H_2O
	Lysine monooxygenase	P. fluorescens	FAD	Lysine + O_2 → δ-aminovaleramide + CO_2 + H_2O
	Tryptophan monooxygenase	P. savastanoi	b	Tryptophan + O_2 → indole-3-acetamide + CO_2 + H_2O
External NADH-preferring	Salicylate hydroxylase	P. putida, unidentified soil microorganism	FAD	Salicylate + NADH + H^+ + O_2 → catechol + NAD^+ + H_2O + CO_2
	m-Hydroxybenzoate-6-hydroxylase	P. aeruginosa	c	m-Hydroxybenzoate + NADH + H^+ + O_2 → gentisate + NAD^+ + H_2O
	Melilotate hydroxylase	Arthrobacter sp., Pseudomonas sp.	FAD	Melilotate + NADH + H^+ + O_2 → 2,3-dihydroxyphenylpropionate + NAD^+ + H_2O
	Orcinol hydroxylase	P. putida	FAD	Orcinol + NADH + H^+ + O_2 → 2,3,5-trihydroxytoluene + NAD^+ + H_2O
	Imidazoleacetate monooxygenase	Pseudomonas sp.	FAD	Imidazoleacetate + NADH + H^+ + O_2 → imidazaloneacetate + NAD^+ + H_2O
NADPH-preferring	p-Hydroxybenzoate hydroxylase	P. putida, P. desmolytica, P. fluorescens	FAD	p-Hydroxybenzoate + NADPH + H^+ + O_2 → protocatechuate + $NADP^+$ + H_2O

Enzyme	Source	Flavin	Reaction
m-Hydroxybenzoate-4-hydroxylase	P. testosteroni, A. niger	c FAD	m-Hydroxybenzoate + NADPH + H$^+$ + O$_2$ → protocatechuate + NADP$^+$ + H$_2$O
Kynurenine hydroxylase	Rat liver	FAD	Kynurenine + NADPH + H$^+$ + O$_2$ → 3-hydroxykynurenine + NADP$^+$ + H$_2$O
Mixed function amine oxidase	Pork liver	FAD	$$X-N-R_1{}^d + NADPH + H^+ + O_2 \rightarrow$$ $$X-N-R_1 + NADP^+ + H_2O$$ (with intermediate forms bearing R_2, O, and OH groups)

Dioxygenase

Enzyme	Source	Flavin	Reaction
External NADH-preferring	2-Methyl-3-hydroxy-pyridine-5-carboxy-late oxygenase	Pseudomonas sp.	FAD	2-Methyl-3-hydroxypyridine-5-carboxylate + NADH + H$^+$ + O$_2$ → α-(N-acetyl-aminomethylene)-succinate + NAD$^+$ + H$_2$O

^{a} References are given in the sections describing individual enzymes.

^{b} Unknown for certainty whether enzyme is a flavoprotein.

^{c} Flavin not yet identified.

^{d} R$_1$, R$_2$, methyl or ethyl groups usually; X, lipophilic alkyl or aryl group free from polar groups on the α-carbon atom.

of the two oxygen atoms of the oxygen molecule, only one is incorporated into the organic substrate, the other being reduced to water. There is one known exception, 2-methyl-3-hydroxypyridine-5-carboxylate oxygenase, which is a dioxygenase. In the reaction catalyzed by this enzyme, both atoms of the O_2 molecule are incorporated into the substrate in the formation of the product.

A convenient classification of oxygenases is on the basis of the electron donor. An external reductant such as NADH or NADPH is required in the first group of enzymes, which are hydroxylases or "mixed function oxidases" (1). In the second category (internal monooxygenases) the substrate itself serves as reductant and there is no requirement for reduced pyridine nucleotides. Table I lists the known flavoprotein oxygenases according to the above classification. At the present time many more external monooxygenases are known than internal.

Most of the enzymes discussed in this review are of bacterial origin, and of these many are induced by growing the microorganism in cultures which contain the substrate as the sole carbon source. This leads to the relatively facile isolation of large amounts of enzyme, often in highly purified, crystalline form. This fortunate circumstance has allowed the use of spectral studies, as well as rapid reaction techniques, to aid more conventional methods in elucidating the reaction mechanisms of the enzymes.

One of the most interesting aspects of the flavoprotein oxygenases is the question of how the reduced flavin of the enzyme "activates" molecular oxygen for incorporation into the product during the catalytic reaction. The mechanism of oxygen "activation" is of course central to hydroxylation reactions in general, whether they are catalyzed by enzymes containing flavin, heme, pteridines, or metals. A common form of activated oxygen may be operational in all these systems or structurally similar cofactor-oxygen compounds may be involved. The ready availability of pure flavoprotein oxygenases thus offers an opportunity to study this problem. Although no clear-cut answers are available, results with two enzymes (p-hydroxybenzoate hydroxylase and melilotate hydroxylase) indicate the transient existence of highly reactive peroxydihydroflavins in the course of catalysis.

This review will be limited to simple flavoprotein oxygenases which contain no other cofactors. Also omitted are the multienzyme hydroxylation systems, which contain flavin in addition to cytochrome P-450 and nonheme iron. These enzyme systems are reviewed elsewhere in this volume (2, 3). Several reviews concerned with flavoprotein monooxygenases have appeared previously (4–7).

II. INTERNAL FLAVOPROTEIN MONOOXYGENASES

The internal flavin monooxygenases catalyze oxidative decarboxylations in which an oxygen atom is inserted into the substrate and CO_2 released. The substrate itself serves as reductant to the flavin; thus, these reaction may be thought of as double oxidations in which the substrate is oxidized first at the expense of flavin reduction, then oxidized again by the insertion of oxygen. These reactions may be summarized as follows:

$$S\underset{COOH}{\overset{H_2}{\diagup}} + EF \longrightarrow \left[S\underset{COOH}{\diagdown} \cdot EFH_2 \right]$$

$$\left[S\underset{COOH}{\diagdown} \cdot EFH_2 \right] + O_2 \longrightarrow S\underset{O}{\overset{H}{\diagup}} + CO_2 + H_2O + EF$$

It is interesting to note that the mechanism by which the oxygen atom is inserted into the product may be quite different with this group of enzymes than with the external flavoprotein oxygenases. Reaction of reduced flavins and flavoproteins with molecular oxygen leads in most cases to the formation of H_2O_2. The nonenzymic decarboxylation of keto acids by H_2O_2 is a well-documented reaction, and similar reactivity of H_2O_2 and an imino acid would be expected. As will be discussed in detail in Section II, A, the possibility exists that the decarboxylation reactions catalyzed by this group of enzymes are due to the locally high concentrations of keto acid (or imino acid) and H_2O_2 at the active site undergoing further reaction to yield CO_2, H_2O, and decarboxylated product. If this suggestion is correct, the internal flavoprotein monooxygenases are in fact typical oxidases; the monooxygenase function would thus be an adventitious one because of the nature of the primary oxidase products and their known mutual reactivity.

A. Lactate Monooxygenase (Lactate Oxidase, Lactate Oxidative Decarboxylase)

Lactate monooxygenase catalyzes the incorporation of molecular oxygen into lactate, forming acetate. Historically, lactate monooxygenase was the first internal flavoprotein oxygenase studied. In 1954, Sutton found a lactate oxidase activity in crude extracts of *Mycobacterium phlei* and established the stoichiometry of lactate: O_2 consumed: CO_2 evolved as being 1:1:1(8). As reported in 1955, the enzyme was further purified, revealing a typical flavoprotein spectrum. Removal of flavin by acid ammonium sulfate treatment resulted in the loss of enzymic activity, which was regained on the addition of FMN (9). The oxygenase nature of the enzyme was established by Hayaishi and Sutton with ^{18}O studies (10). They found that the acetate formed enzymically was enriched 82% in ^{18}O when $^{18}O_2$ gas and $H_2^{16}O$ were used, while in the companion experiment using $^{16}O_2$ and $H_2^{18}O$, acetate did not contain appreciable ^{18}O. In the same year, using a crystalline preparation, Sutton demonstrated that pyruvate is the product when the enzyme is anaerobically reduced by lactate by identifying the 2,4-dinitrophenylhydrazone of pyruvate. Also, by the use of carbonyl fixatives, he was able to find no free carbonyl group formed during aerobic catalysis and concluded that any pyruvate formed during catalysis is enzyme-bound (11). A minimum molecular weight per flavin was calculated to be 125,700, using the ultraviolet absorption method of Kalckar (12). Later workers have calculated instead a minimum molecular weight of 55,000–56,000 on the basis of dry weight and biuret determinations (13, 14).

Sullivan (13) has purified and crystallized a lactate oxidase from *Mycobacterium smegmatis* with properties almost identical to the enzyme from *M. phlei*. The enzyme has a minimum molecular weight of 49,500 per FMN, while sucrose gradient and sedimentation velocity data give molecular weights of 300,000 and 341,000, respectively, indicating that the enzyme is composed of multiple subunits.

By use of a rapid scanning technique, Beinert and Sands (15), with enzyme from *M. phlei*, demonstrated the presence of a transient long-wavelength intermediate with absorption at 540 nm which appeared under both aerobic and anaerobic conditions. In 0.07 M phosphate buffer, pH 7.5, the intermediate was formed in a slow, biphasic manner and disappeared very slowly (cf. refs. 16 and 17). Stopped-flow measurements performed in 0.05 M phosphate buffer, pH 7.0, showed a similar pattern (18, 19); anaerobically a distinctly biphasic buildup of 550 and 530 nm absorbance followed by a slow decay was observed. As discussed below, these complexities have been found to result from the use of phosphate as a buffer.

Recently, Sullivan (16) and Lockridge *et al.* (17) have shown that lactate oxidase from *M. smegmatis* binds a variety of different anions which inhibit enzymic activity. This inhibition is competitive with L-lactate, and many of these anions cause perturbations in the visible spectrum of the enzyme. Phosphate ion is a competitive inhibitor of lactate oxidase and causes marked spectral changes as shown in Fig. 1 (17). The K_d calculated from this titration is 1.3 × 10^{-2} M at 25°. The effect of added phosphate on the anaerobic reduction of the enzyme by L-lactate is shown in

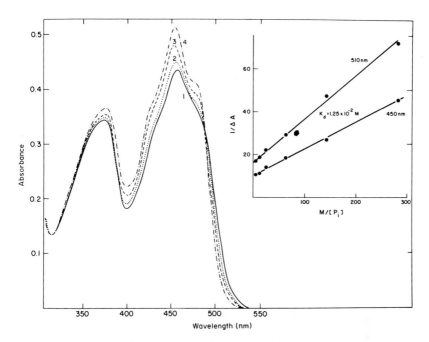

FIG. 1. Effect of phosphate on the absorption spectrum of lactate oxidase. Curve 1, 4×10^{-5} M enzyme (with respect to FMN) in 0.01 M imidazole, pH 7.0, 25°. Curve 2, in the presence of 7 mM phosphate. Curve 3, in the presence of 41 mM phosphate. Curve 4, in the presence of 197 mM phosphate. Inset: Dependence of the ΔA_{450} and $-\Delta A_{510}$ values on phosphate concentration (from ref. 17).

Fig. 2 (17). In the absence of phosphate, a rapid monophasic formation of intermediate is followed by a slow decay to fully reduced enzyme. Addition of phosphate produces a rapid increase in absorbance at 540 nm followed by a slower increase, then finally a slow decay of the intermediate. The simplest interpretation of these results is that lactate can reduce only enzyme uncomplexed with phosphate; thus, the slow formation of the intermediate results from the rate at which phosphate is released from the enzyme—phosphate complex (k_{off}).

$$E + phosphate \underset{k_{off}}{\overset{k_{on}}{\rightleftharpoons}} E \cdot phosphate$$

Lockridge et al. have also given compelling evidence that the intermediate seen transiently is an enzyme–FMNH$_2$–pyruvate complex and is the

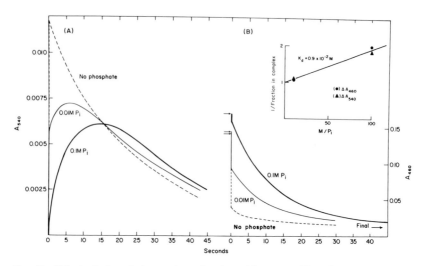

FIG. 2. Effect of phosphate on the reduction of lactate oxidase by L-lactate under anaerobic conditions. Lactate oxidase, 1.43×10^{-5} M with respect to bound FMN, was reacted with 0.01 M L-lactate at 25° under anaerobic conditions, in 0.01 M imidazole, pH 7.0, alone, or in the presence also of 0.01 M K-phosphate or 0.1 M K-phosphate, pH 7.0. (A) Changes monitored at 540 nm. In all cases the A_{540} before mixing with L-lactate was zero. In the absence of phosphate, the appearance of 540 nm absorbance is monophasic and complete within 200 msec. (B) Changes monitored at 460 nm. The arrows on the left ordinate of 3B show the initial A_{460} values in the presence and absence of phosphate. The inset shows a plot of the fraction complexed with phosphate versus the phosphate concentration (from ref. 17).

species reacting with O_2 in the catalytic cycle. The evidence presented is as follows:

1. Anaerobically, with lactate, the long-wavelength absorbing species is formed rapidly in a monophasic manner in the absence of competing anions.

2. The pseudo first-order rate constant of formation is dependent on lactate concentration with an extrapolated rate constant of 14,000 min^{-1}.

3. In equilibrium experiments, pyruvate reacts with reduced enzyme to produce a stable long-wavelength species spectrally identical to that seen transiently during stopped-flow experiments.

4. The reduced enzyme–pyruvate complex reacts with oxygen 200 times faster than uncomplexed reduced enzyme.

5. In stopped-flow turnover experiments, the enzyme is mostly in the oxidized form since the reaction of the intermediate with O_2 is much

faster than the rate of formation of the intermediate from oxidized enzyme and lactate.

6. When reduced enzyme is complexed with 2-^{14}C-labeled pyruvate and then reacted with O_2, [14]CH_3COOH is found in amounts stoichiometric with the reduced enzyme–pyruvate complex.

A minimal reaction mechanism for lactate oxidase on the basis of these and other results is shown in the following scheme (17):

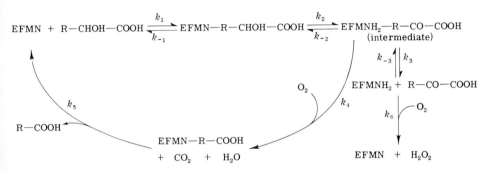

In this scheme, product dissociation (k_5) is considered to be partially rate limiting.

The finding that the catalytic intermediate which reacts with O_2 is a complex of $EFMNH_2$ and pyruvate suggests that the mechanism of decarboxylation may simply result from the enzyme providing locally high concentrations of reactants. It has been known for many years that H_2O_2 will decarboxylate pyruvate and other α-keto acids in free solution; however, high concentrations of the reactants are required for rapid reaction. The intriguing possibility exists with lactate oxidase that an $EFMNH_2$–pyruvate complex, in its rapid reaction with O_2, yields an $EFMN$–pyruvate complex, and that the H_2O_2 produced at the active site reacts with pyruvate before either product has a chance to be released from the enzyme. If this suggestion is correct, then lactate oxidase is really only an "adventitious monooxygenase." In this view, the primary products of the catalytic reaction, H_2O_2 and pyruvate, are typical oxidase products; the overall monooxygenase reaction is exhibited adventitiously because of the slow rate of release of the products, which at the locally high concentrations at the active center can react rapidly in the secondary decarboxylation reaction. The possibility exists that a similar mechanism may apply for all the internal flavoprotein monooxygenases and could readily explain why these enzymes can behave with some substrates as monooxygenases and with others as oxidases (see Section II,C).

By analogy with experiments performed with D- and L-amino acid oxidases (20), proton abstraction from substrate by the enzyme is proposed to be an early step in the reaction mechanism of lactate oxidase. This has been substantiated by the finding (21) that β-chlorolactate is also a substrate for lactate oxidase under anaerobic conditions. Like the analogous reaction of D-amino acid oxidase with β-chloroalanine (20, 22), this reaction leads anaerobically to the catalytic elimination of Cl^- with pyruvate as the oxidation product. Under aerobic conditions chlorolactate also behaves as a normal substrate, yielding chloroacetate, CO_2, and H_2O as reaction products.

Recent studies with an acetylenic substrate of lactate oxidase promise to offer further elucidation of the reaction mechanism of the enzyme. Walsh et al. (23), have shown that DL-2-hydroxy-3-butynoic acid is a substrate for lactate oxidase as well as an inactivator. Initial velocity measurements of oxygen uptake indicate an apparent K_m of 2 mM for this substrate. The rate of inactivation varies inversely with oxygen concentration, suggesting a partitioning of an enzyme–substrate intermediate between two reaction pathways—one leading to inactivation, the other leading to turnover. The inactivated enzyme complex exhibits a reduced-type flavin spectrum. Also, the spectrum of the flavin released from inactivated enzyme by methanol treatment is of the reduced type, indicating that a nonautoxidizable adduct of the reduced flavin is formed. Experiments with acetylenic acid labeled with tritium at the α-carbon showed no label in the inactivated protein, indicating that the substrate must lose its α-hydrogen before inactivation occurs. When the label is at carbon 4, inactivated enzyme contains 1 mole of ^3H per mole of active site, and this label was shown to be associated with the flavin. These results demonstrate that the inactivation is associated with a covalent modification of the flavin following the initial abstraction of the α-hydrogen of the acetylenic substrate.

B. Arginine Monooxygenase (Arginine Decarboxylase)

Arginine monooxygenase and lysine monooxygenase are closely related in both properties and function. Both enzymes catalyze similar reactions, i.e., the oxidative decarboxylation of basic amino acids. However, the substrate specificities for both are fairly rigid, so that arginine decarboxylase shows no activity toward L-lysine (24) and L-arginine is oxidized at only one-tenth the rate of L-lysine by lysine monooxygenase (25). Both enzymes have pH optima in the region of pH 9 and both exhibit sigmoidal substrate saturation curves (24, 26). Thoai and Olomucki (24) have isolated arginine monooxygenase from the washed mycelium of *Streptomyces griseus*, and detected similar activities in other *Acetomyces* species.

Studies with partially purified enzyme preparations (24) indicate that the enzyme is extrememly unstable in the presence of oxygen, completely losing activity after storage in the cold for 2–3 days. Addition of substrate (arginine) protects against inactivation as does storage in a nitrogen atmosphere. Reducing agents such as dithionite and thioglycolate also protect against inactivation by oxygen. Two other compounds similar in structure to arginine (canavanine and homoarginine) are substrates, yielding the corresponding amides (27).

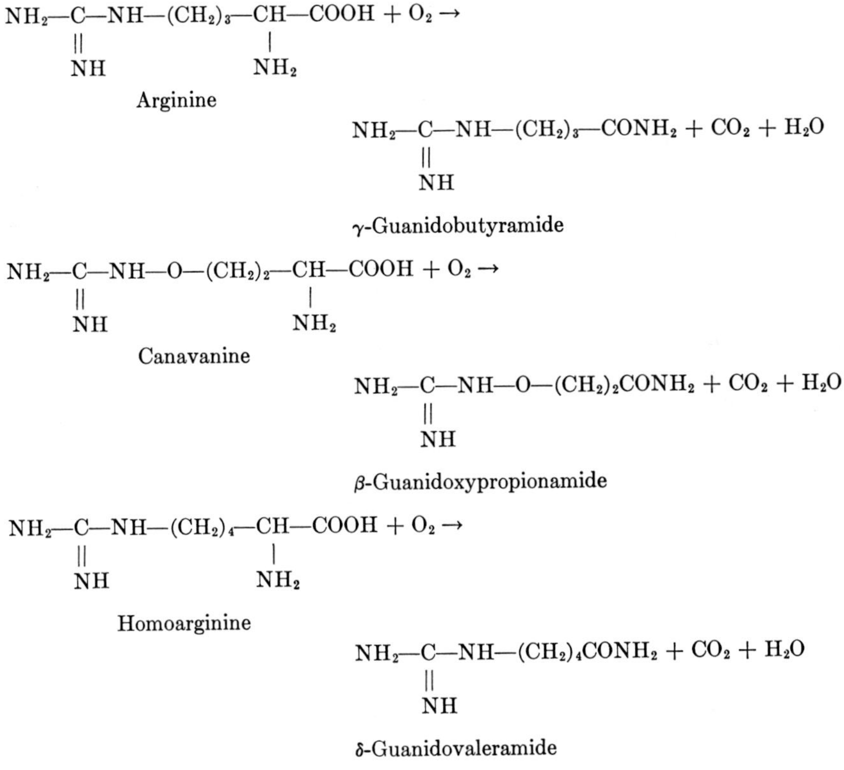

$$NH_2-C-NH-(CH_2)_3-CH-COOH + O_2 \rightarrow$$
$$\overset{\|}{NH} \qquad\qquad \overset{|}{NH_2}$$

Arginine

$$NH_2-C-NH-(CH_2)_3-CONH_2 + CO_2 + H_2O$$
$$\overset{\|}{NH}$$

γ-Guanidobutyramide

$$NH_2-C-NH-O-(CH_2)_2-CH-COOH + O_2 \rightarrow$$
$$\overset{\|}{NH} \qquad\qquad \overset{|}{NH_2}$$

Canavanine

$$NH_2-C-NH-O-(CH_2)_2CONH_2 + CO_2 + H_2O$$
$$\overset{\|}{NH}$$

β-Guanidoxypropionamide

$$NH_2-C-NH-(CH_2)_4-CH-COOH + O_2 \rightarrow$$
$$\overset{\|}{NH} \qquad\qquad \overset{|}{NH_2}$$

Homoarginine

$$NH_2-C-NH-(CH_2)_4CONH_2 + CO_2 + H_2O$$
$$\overset{\|}{NH}$$

δ-Guanidovaleramide

These compounds are poorer substrates than arginine and interestingly show different pH optima, 6.1 for canavanine and 8.0 for homoarginine. Homoarginine is a competitive inhibitor toward arginine, while canavanine is not.

The oxygenase nature of the enzyme was shown by $^{18}O_2$ experiments in which incorporation of molecular oxygen into substrate was demonstrated (28). Further studies on a more purified preparation revealed arginine monooxygenase to be a flavoprotein utilizing FAD as cofactor

(29). More recently, Thomé-Beau *et al.* (30) have shown that diethylpyro-carbonate inhibits arginine monooxygenase at pH 6.0 by modifying one histidyl residue per mole of enzyme-bound flavin. Addition of arginine completely protects the enzyme against inactivation at all concentrations of diethylpyrocarbonate tested. These data, together with a study of the pH dependence of the K_m for L-arginine, indicate that one histidyl residue per flavin is essential for activity, and that this histidyl residue is a part of the active center of the enzyme. In view of the finding that proton abstraction from substrate by some basic group on the protein is the primary step in catalysis by several flavoproteins (20–23), it is tempting to speculate that a similar mechanism may operate with arginine monooxygenase, with perhaps the histidine being the proton abstracting group.

C. Lysine Monooxygenase

Early work with partially purified preparations of lysine monooxygenase led to the conclusion that the enzyme catalyzed a rather complex series of reactions (i.e., oxidation, decarboxylation, and deamination) to yield δ-aminovaleric acid from L-lysine (31). Later studies (32), with a crystalline preparation from *Pseudomonas fluorescens*, indicated that crude preparations contained two enzymes, the first being lysine monooxygenase which forms δ-aminovaleramide and CO_2 via an oxidative decarboxylation:

$$
\begin{array}{l}
\text{CH}_2\text{—NH}_2 \\
| \\
(\text{CH}_2)_3 + O_2 \\
| \\
\text{CH—NH}_2 \\
| \\
\text{COOH}
\end{array}
\longrightarrow
\begin{array}{l}
\text{CH}_2\text{—NH}_2 \\
| \\
(\text{CH}_2)_3 + CO_2 + H_2O \\
| \\
\text{C—NH}_2 \\
|| \\
\text{O}
\end{array}
$$

$$\text{L-Lysine} \qquad\qquad \text{δ-Aminovaleramide}$$

The second enzyme present was δ-aminovaleramide deaminase, which catalyzes the following reaction:

$$
\begin{array}{l}
\text{CH}_2\text{NH}_2 \\
| \\
(\text{CH}_2)_3 + H_2O \\
| \\
\text{C—NH}_2 \\
|| \\
\text{O}
\end{array}
\longrightarrow
\begin{array}{l}
\text{CH}_2\text{NH}_2 \\
| \\
(\text{CH}_2)_3 + NH_3 \\
| \\
\text{COOH}
\end{array}
$$

$$\text{δ-Aminovaleramide} \qquad\qquad \text{δ-Aminovaleric acid}$$

[18]O studies were performed on partially purified lysine monooxygenase (33). They strongly suggest that molecular oxygen is incorporated into product.

The crystalline enzyme from *P. fluorescens* exhibits a typical flavoprotein absorption spectrum with maxima at 274, 385, and 460 nm (32). FAD was shown to be necessary for enzymic activity and cannot be replaced by FMN or riboflavin (32). Careful studies with metal chelating agents indicate that metals play no role in lysine monooxygenase (34), thereby confirming that the enzyme is a simple flavoprotein with FAD as the sole cofactor. Takeda *et al.* (35) calculated a molecular weight of 191,000 from sedimentation velocity experiments. However, their reported amino acid analysis fails to corroborate this value since summation of the amino acid residues as presented results in a molecular weight of 168,000.

Lysine monooxygenase has several unusual properties as compared with other flavoprotein oxygenases; for example, the anaerobic reduction of the enzyme-bound FAD by equimolar lysine is an extremely slow process, requiring from 7 to 9 hours for full reduction (35, 36). Such behavior is very puzzling and is further complicated by the recent report (37) that the reduction occurs in two to three consecutive first-order reactions. Under anaerobic conditions, the reduction of the enzyme by lysine results in the stoichiometric formation of α-keto-ϵ-aminocaproate (which is in equilibrium with the cyclic form, Δ^1-piperideine-2-carboxylate) (36–39).

$$
\begin{array}{ccc}
\underset{\text{L-Lysine}}{\begin{array}{l} CH_2NH_2 \\ | \\ (CH_2)_3 \\ | \\ CHNH_2 \\ | \\ COOH \end{array}} & \xrightarrow[\underset{H_2O \quad NH_3}{}]{FAD \quad FADH_2} & \underset{\substack{\alpha\text{-Keto-}\epsilon- \\ \text{aminocaproic} \\ \text{acid}}}{\begin{array}{l} CH_2NH_2 \\ | \\ (CH_2)_3 \\ | \\ C{=}O \\ | \\ COOH \end{array}} & \xrightleftharpoons[+H_2O]{-H_2O} & \underset{\substack{\Delta^1\text{-Piperideine-2-} \\ \text{carboxylic acid}}}{}
\end{array}
$$

The anaerobic formation of α-keto-ϵ-aminocaproate is analogous to the anaerobic formation of pyruvate with lactate monooxygenase (11, 17). A recent report by Nakazawa *et al.* (40) has revealed another interesting property of lysine monooxygenase. Aerobically, with L-ornithine or DL-2,8-diaminooctanoate as substrate, the enzyme catalyzes an oxidase reaction (with products being the corresponding α-keto acids and H_2O_2 without formation of CO_2, analogous to the anaerobic reaction with lysine described above). Interestingly, both oxidase- and oxygenase-type substrates exhibit sigmoidal saturation curves. The association of both oxygenase and oxidase activities with a single active site is indicated by several lines of evidence. A constant ratio of oxygenase to oxidase activities during purification was found, as were parallel losses of both activities caused by sulfhydryl inhibitors and sodium dodecyl sulfate. Similar pH activity profiles were found and ornithine abolishes the sigmoidicity in the lysine substrate saturation curve, activating at lower ornithine concentrations while inhibiting at higher concentrations (40). The K_m values for ornithine and 2,8-diaminooctanoate are higher than those for lysine and other oxygenase-type substrates by an order of magnitude. These results might indicate that substrates that are too large (8 carbon atoms) or too small (5 carbon atoms) to fit into the active center of the enzyme form a different type of enzyme–substrate complex in which the dehydrogenation of the substrate does not couple with "oxygen activation" and thus oxygen serves merely to reoxidize reduced flavin (40). On the other hand, in terms of the "adventitious oxygenase" hypothesis, the exhibition of both oxygenase and oxidase activities by a single enzyme may be merely a reflection of the strengths of binding of the primary oxidation products (imino acid and H_2O_2). Thus, oxygenase activity would be expected if *both* products were released slowly from the enzyme (relative to the rate of their chemical interaction at the active site), whereas oxidase activity would result if *either* product were released rapidly.

Yamamoto *et al.* (41) have utilized the stopped-flow technique to investigate the transient intermediates involved in both the aerobic and anaerobic reactions of lysine monooxygenase with lysine as substrate. They observed, during the anaerobic reduction of the enzyme, a transient long-wavelength absorbing species that appears in a biphasic manner and disappears very slowly to yield reduced enzyme. This species has a broad absorption band centered around 550 nm, and the initial rapid phase of its formation has a pseudo first-order rate constant that is great enough to account for catalysis. However, the significance of the biphasic nature of the formation of the intermediate is not clear at present. In the aerobic reaction of the enzyme with lysine, a slightly different intermediate spectrum was observed. The absorption peak was shifted by 5 nm from 460 to 455 nm and the broad absorption band was centered around 575 nm instead of 550 nm. This intermediate also decays to fully reduced enzyme. Further studies are required to determine whether the long-wavelength absorbing species formed anaerobically also participates in the aerobic reaction. The possibility exists that the species seen aerobically could be related to the formation of a ternary complex between enzyme, lysine, and oxygen (41).

D. Tryptophan Monooxygenase (Tryptophan Oxidative Decarboxylase)

Kosuge *et al.* (42) have purified a tryptophan oxidative decarboxylase from *P. savastanoi*. The reaction catalyzed is very similar to those of

Tryptophan   Indole-3-acetamide

the internal flavin monooxygenases already described. However, it is not clear at present whether this enzyme is indeed a flavoprotein. ^{18}O studies have shown that the reaction is a true oxygenation with ^{18}O enrichment in the product, indole-3-acetamide, only when gaseous $^{18}O_2$ is present (43).

III. EXTERNAL FLAVOPROTEIN MONOOXYGENASES

Most of the external flavoprotein oxygenases are monooxygenases, the only known exception being 2-methyl-3-hydroxypyridine-5-carboxylate oxygenase (44), which is a dioxygenase, and which will be discussed briefly at the end of this review. In the case of the monooxygenases, some remarkable similarities between the individual enzymes appear to be emerging. The chief of these is the dual role played by the substrate, which in all cases investigated appears to be an effector as well as a hydroxylatable substrate. The effector role is shown most dramatically by the increase in rate of reduction of the enzyme flavin by the external electron source (NADH or NADPH) when the enzyme is complexed with the substrate. The effector role has been corroborated by the finding of many examples in which substrate analogs increase the rate of reduction but do not serve as hydroxylatable substrates. In nearly all cases the combination of the oxidized enzyme with the effector (be it a substrate or a nonsubstrate effector) has been shown to result in marked changes in the physical properties of the enzymes (visible absorption spectrum, fluorescence, and circular dichroism spectrum). Such changes have permitted the determination of the dissociation constants for the complexes, which in all cases have been shown to consist of one equivalent of effector per equivalent of enzyme flavin. The effector role of the substrate is required for the *rapid* hydroxylation reaction. In the case of the nonsubstrate effectors, the effector role is displayed by an increased rate of the reduced pyridine nucleotide-oxygen reductase activity. The reduction product of oxygen in this case is H_2O_2 rather than H_2O, which is the product of the normal hydroxylase activity. Thus, an increased rate of O_2 consumption (or reduced pyridine nucleotide oxidation), coupled with the formation of H_2O_2, is a diagnostic test for nonsubstrate effectors. Some substrate analogs have been found to combine both substrate and nonsubstrate effector roles.

The effector role of the substrate has also been found in some cases to be exerted by enhancing the rate of reaction of the reduced enzyme with oxygen.

The effector role of the substrate may by expected to have a profound importance in the metabolic control of the cell. In the absence of such a phenomenon, NADH and NADPH oxidation would be continuously fast and uncontrolled; the fact that reduced pyridine nucleotide is permitted to react rapidly with the enzyme flavin only when the desired substrate is present thus offers a rather elegant control mechanism for NADH and NADPH oxidation (45).

A. Salicylate Hydroxylase

Historically, salicylate hydroxylase was the second flavoprotein mono-oxygenase to be discovered, and the first representative of the group of "external" flavoprotein hydroxylases (46). It was first purified by Katagiri and co-workers from *Pseudomonas putida* (47, 48) and shown to contain one molecule of FAD per molecule of protein of molecular weight 57,000. By the use of $^{18}O_2$ and product analysis, it was shown (49) that the stoichiometry of the reaction catalyzed by the enzyme is

$$\text{(salicylate)} + NADH + H^+ + {}^{18}O_2 \longrightarrow \text{(catechol)} + NAD^+ + CO_2 + H_2{}^{18}O$$

Recently, White-Stevens and Kamin (50–52) have reported studies on a salicylate hydroxylase isolated from a soil microorganism obtained by enrichment culture. This enzyme appears to differ significantly in many of its properties from the *Pseudomonas putida* enzyme; chief among these are its molecular weight (91,000) and the fact that it contains two molecules of FAD and is composed of two subunits of molecular weight 46,000. It also differs in the very fundamental property of exhibiting a benzoate-stimulated NADH–O_2 reductase activity, the benzoate not acting as a hydroxylatable substrate but as a substrate-like effector (50). Such a phenomenon has since been demonstrated with practically every external flavoprotein hydroxylase (see individual enzymes) but is apparently poorly shown with the *P. putida* enzyme, where benzoate was initially reported to be neither a substrate nor an inhibitor (47). More recent studies with the *P. putida* enzyme have revealed that benzoate does indeed stimulate NADH oxidation but only at high concentrations (53).

The *Pseudomonas putida* enzyme is reported to be quite specific in its requirement for NADH as electron donor, NADPH being ineffective (47). In addition to salicylate, the following hydroxybenzoates were reported as substrates: 2,3-dihydroxybenzoate, 2,4-dihydroxybenzoate, 2,5-dihydroxybenzoate, 2,6-dihydroxybenzoate, *p*-aminosalicylate, 1-hydroxy-2-napthoate (47), and 3-methylsalicylate (48). With the exception of the latter compound, which was shown to be hydroxylated to yield 3-methylcatechol (48), the remaining compounds were assayed for activity merely by following the rate of NADH oxidation in the presence of O_2. This test clearly does not distinguish whether the compounds were acting as true hydroxylatable substrates or whether they were acting as "effectors," stimulating the NADH-O_2 reductase activity in a manner similar to that found by White-Stevens and Kamin in the case of benzoate. The latter workers have reported on the specificity of their enzyme (51). Unlike the *P. putida* enzyme, their enzyme will utilize NADPH as electron donor with the same V_{max} value as NADH; the K_m value for NADPH, however, is higher by an order of magnitude than that for NADH.

This enzyme was also found to have its NADH–oxygen reductase activity stimulated by *o*-nitrobenzoate, *m*-hydroxybenzoate, *p*-hydroxybenzoate, and salicylamide; as in the case of stimulation by benzoate, these compounds were not hydroxylated. In addition, a number of substituted salicylates were found to act both as hydroxylatable substrates and nonsubstrate effectors (*p*-aminosalicylate, 3-methylsalicylate, 2,3-, 2,4-, 2,5-, and 2,6-dihydroxybenzoates) (51).

On mixing with salicylate, very marked changes in the visible absorption spectrum of salicylate hydroxylase are observed (48). Titration experiments revealed the formation of a 1:1 complex of enzyme and salicylate, with a dissociation constant at pH 7.0 and room temperature of 3.5×10^{-6} M. Extensive spectral changes were also observed in the complexes formed between enzyme and 2,5-dihydroxybenzoate or 2,3-dihydroxybenzoate; less marked changes were observed with 2,4-dihydroxybenzoate and 3-methylsalicylate. Similarly, complex formation of of holoenzyme with salicylate was shown in elegant fluorescence studies (54), which yielded a value of 3.2×10^{-6} M for the dissociation constant, in good agreement with the spectrophotometric titration results. By dramatic fluorescence changes, it was also shown that apoprotein can bind salicylate, various dihydroxybenzoates, and substituted salicylates.

The dissociation constant for the apoprotein–salicylate complex ($1.8 \times 10^{-6} M$) was even lower than that for the holoenzyme–salicylate complex (54). It was also shown that NADH and NADPH form complexes with apoprotein, with dissociation constants of $1.1 \times 10^{-5} M$ and $1.5 \times 10^{-5} M$, respectively. The fluorescence technique was also used to determine the dissociation constant for FAD binding to the apoprotein; a value of $4.5 \times 10^{-8} M$ was found (54).

A series of papers (48, 55) reported product analyses from reactions involving stoichiometric amounts of enzyme and demonstrated in a very clear fashion that the hydroxylation reaction is a result of the reaction of O_2 with the reduced enzyme—salicylate complex. The enzyme flavin was found to be reduced anaerobically by 1 mole NADH per mole EFAD, either in the presence or absence of salicylate. When enzyme saturated with salicylate was reduced with small quantities of NADH, amounts of catechol stoichiometric with the amounts of reduced enzyme were found on admission of O_2. These results were independent of the order of addition of enzyme, salicylate, and NADH, provided that O_2 was added last. The role of NADH was shown to be merely that of a source of reducing equivalents to produce $E \cdot FADH_2$; the latter produced by anaerobic dithionite titration or by the EDTA–light irradiation technique (56) also leads to catechol production when mixed with salicylate and O_2 (55). While no stable intermediates were detected in the anaerobic titration of the enzyme with NADH, long-wavelength intermediates were produced with substoichiometric amounts of dithionite and during EDTA-light irradiation. In the absence of salicylate, the red or anionic semiquinone (56) was produced in good yields, as confirmed by EPR spectrometry (55). On addition of salicylate, the spectrum changed to one indicative of the presence of a small amount of the blue or neutral semiquinone, but with no observed EPR signal. This puzzling result may be readily explained if the dissociation constant for the semiquinone–salicylate complex is much higher than that for the oxidized enzyme–salicylate complex, thereby favoring disproportionation of the semiquinone to oxidized and reduced enzyme. Thus, in the absence of salicylate, the equilibrium

$$2 \text{ EFADH}^{\cdot} \rightleftharpoons \text{EFAD} + \text{EFADH}_2$$

may lie much in favor of EFADH$^{\cdot}$, whereas in the presence of salicylate the equilibrium may be changed in disfavor of semiquinone if salicylate binds more strongly to EFAD (or EFADH$_2$) than to EFADH$^{\cdot}$. The dissociation constant for the EFAD–salicylate complex is known to be low, $3.5 \times 10^{-6} M$ (48, 54), making this a likely possibility. This explanation is also supported by the small amount of long-wavelength absorption found in the presence of salicylate. For most flavoprotein neutral semiquinones the extinction coefficient, ϵ_{600}, is of the order of 3000–5000 M^{-1} cm^{-1} (56); the observed ϵ_{600} on the addition of salicylate to the anion radical of salicylate hydroxylase may be estimated from the published data to be of the order of 600–700 M^{-1} cm^{-1}. From these considerations it would be clearly desirable to repeat the EPR experiments at higher concentrations of enzyme.

From the results described the following reaction pathway has been postulated (48, 57, 58):

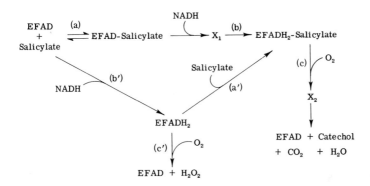

The evidence for individual steps in this reaction scheme has been obtained by the stoichiometric studies previously described or deduced from stopped-flow data (58). The K_d for reaction (a) has been determined from static titration experiments as 3.5×10^{-6} M; from stopped-flow data whose validity is difficult to assess (58), the "on" velocity for this reaction has been estimated as 1.8×10^7 M^{-1} sec^{-1} and the "off" velocity as 62 sec^{-1}. The reduction of EFAD by NADH in the absence of salicylate [step (b')] is reported to be a comparatively slow second-order reaction with a rate constant of 2.5×10^3 M^{-1} sec^{-1}; consequently, the pathway (b')-(a') is considered to be catalytically unimportant. Similarly, the reaction of O_2 with EFADH$_2$ is reported to be a comparatively slow second-order reaction with a rate constant of 2.9×10^4 M^{-1} sec^{-1}. Using a method which is not described in any detail and which is unknown to the present authors, step (b), the reaction of NADH with the EFAD–salicylate complex, is claimed to proceed via an intermediate X_1 with a limiting first-order rate of reduction of 200 sec^{-1}. It is claimed from analysis of stopped-flow turnover experiments that the reaction of EFADH$_2\cdot$ salicylate with O_2 proceeds via a second intermediate X_2, to yield reoxidized enzyme and products with a limiting first-order rate constant of 22 sec^{-1}. This latter value (referred to as "k_{ox}") was obtained by misapplication of a formula derived by Chance (59) to determine k_3 in the classic Michaelis-Menten and Briggs-Haldane theories

$$(E + S \underset{k_2}{\overset{k_1}{\rightleftharpoons}} ES \overset{k_3}{\rightarrow} E + P)$$

in cases where the concentration of ES can be measured. Equating this with the rate of oxidation of a reduced flavoprotein is not valid despite the fact that this is often done. The value of "k_{ox}" obtained from this formula merely gives an approximate estimate of the catalytic turnover number. It is not surprising, therefore, that the value of "k_{ox}" so obtained (22 sec^{-1}) is close to the turnover number (17 sec^{-1}) (58) obtained in conventional catalytic studies. It would seem clear that the elucidation of the reaction pathway and the nature of the proposed intermediates X_1 and X_2 must await more detailed rapid reaction studies.

A paper reporting rapid reaction and mechanistic studies with the White-Stevens and Kamin enzyme has recently appeared (52). Like the *P. putida* enzyme, marked spectral changes were observed on complex formation with substrates or nonsubstrate effectors such as benzoate, permitting the evaluation of dissociation constants. In general, the same basic scheme was proposed as that of Katagiri and co-workers. The forward and reverse rate constants for the complex formation between oxidized enzyme and salicylate were independently determined as 9×10^7 M^{-1} sec^{-1} and 130 sec^{-1}, respectively, yielding a K_d value of 1.5×10^{-6} M. The reduction by NADH of the enzyme–salicylate complex was also found to be a multistep process: the formation of a ternary complex ($K_d = 10^{-5}$ M) followed by two first-order reactions, the first with a rate constant in the range 150–600 sec^{-1} and the second 42 sec^{-1}. No long-wavelength absorbance was found associated with the reduction process; hence, the two first-order steps were ascribed to reduction of the enzyme flavin followed by a protein conformational change. The reaction of molecular oxygen with reduced enzyme was also studied. Surprisingly, no differences in reaction rate were found whether the reduced enzyme was in its free, uncomplexed state or whether it was complexed with benzoate, salicylate, 2,4-dihydroxybenzoate, or *p*-aminosalicylate. The reaction with O_2 was found to be second-order with a rate constant of approximately 1.7×10^4 M^{-1} sec^{-1} (i.e., with a pseudo first-order rate constant of 23 sec^{-1} at atmospheric O_2). While this rate constant is sufficient to account for the catalytic V_{max} with salicylate as substrate, it appears inadequate to account for the 3.6-fold greater rate found catalytically with 2,4-dihydroxybenzoate. The finding of a constant reoxidation rate led White-Stevens *et al.* (52) to propose the existence of a "nascent H_2O_2" bound to enzyme, which could either react further with bound substrate (to yield H_2O and hydroxylated product) or be dissociated (to yield free

H_2O_2 and unmodified "effector"). In view of the recent findings of oxygenated intermediates with p-hydroxybenzoate hydroxylase (60, 61) and melilotate hydroxylase (62), we would like to propose an alternate explanation. The concentrations of substrate and effector were only 10 times their apparent K_m values. The interpretation of White-Stevens et al. of a common oxidation rate relies on the assumption that the concentrations of effectors used were sufficient to saturate the enzyme in its fully reduced form. Since K_m values give no real clue to the actual dissociation constants of the reduced enzyme– effector complexes, this assumption is open to question. The possibility exists that they may have measured largely the reaction rate of free reduced enzyme with molecular oxygen. In similar experiments with p-hydroxybenzoate hydroxylase, Spector and Massey (61) found it necessary to equilibrate reduced enzyme with relatively high concentrations of substrate before mixing with O_2, in order to observe good yields of the transient oxygenated intermediates.

B. p-Hydroxybenzoate Hydroxylase

p-Hydroxybenzoate hydroxylase is one of the most thoroughly studied flavin hydroxylases. The enzyme has been obtained in crystalline form from four different species of pseudomonads, P. desmolytica (63), P. putida A 3.12 (64), P. putida M-6 (65), and P. fluorescens (45). In all cases the enzyme is obtained from p-hydroxybenzoate-adapted cells. The p-hydroxybenzoate hydroxylases from these various sources show striking similarities to each other in terms of molecular size, strict requirement for NADPH as an external reductant, the content of 1 mole of tightly bound FAD per mole or protein, Michaelis constant for p-hydroxybenzoate, and excess substrate inhibition, among others.

An important exception to the similarities is in terms of the stability of the enzymes. The crystalline enzyme preparations from both P. putida species extremely unstable requiring addition of stabilizing mixtures containing p-hydroxybenzoate, EDTA, and a small molecular weight thiol to protect against inactivation (64, 65). The enzyme from P. fluorescens, however, is stable for months, even in the absence of stabilizing agents, thus greatly facilitating spectral and anaerobic work (45).

Working with P. desmolytica enzyme, Yano et al. (63) succeeded in preparing two crystalline forms, namely, the holoenzyme and the holoenzyme–p-hydroxybenzoate complex. The enzyme–substrate (E–S) complex crystals displayed the same resolved visible spectrum as seen on adding p-hydroxybenzoate to holoenzyme, whereas the spectrum of the holoenzyme is unresolved (63). The sedimentation coefficient of the holoenzyme was found to be 5.1 S and that of the complex 5.0 S. That this small difference is real was confirmed by a rather sophisticated experiment in which an analytical ultracentrifuge was equipped with two cells, containing holoenzyme and E–S complex, respectively. When the volume of the holoenzyme was slightly larger than that of the E–S complex, two schlieren boundaries appeared which eventually fused. In the reverse case, the distance between the two boundaries increased with time (63). Therefore, it was concluded that a small but distinct conformational change is produced on binding of substrate.

Yano *et al.* (66) demonstrated the following stoichiometry for the catalytic reaction:

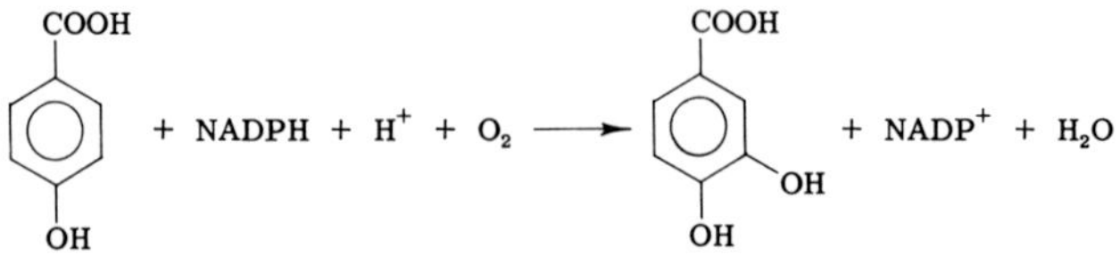

Stoichiometric binding of *p*-hydroxybenzoate to the holoenzyme was also found. Convincing evidence was also produced that the direct electron donor for the oxygenation reaction was enzyme-bound $FADH_2$; in experiments using high concentrations of enzyme and *p*-hydroxybenzoate, amounts of 3,4-dihydroxybenzoate were produced on admission of O_2 stoichiometric with the amount of $EFADH_2$, independent of whether the latter was produced by reduction with NADPH or dithionite.

In anaerobic stopped-flow experiments performed with enzyme containing less than a stoichiometric amount of *p*-hydroxybenzoate, reduction occurred in a distinctly biphasic reaction, which suggests that both activated and nonactivated enzyme species were present under the given conditions (67, 68). The values for k_{red} for the fast part of the reaction were of the order of 96 sec^{-1}, a value about 10^4-fold larger than for the slow reaction whose k_{red} was 0.010 min^{-1}. This supports the view that the "activator" is *p*-hydroxybenzoate itself and, together with the stoichiometry data, indicates that the active enzyme species is the oxidized enzyme–*p*-hydroxybenzoate complex.

Howell and Massey, with enzyme isolated from *P. fluorescens*, have also demonstrated a perturbation of the flavin absorption spectrum by addition of *p*-hydroxybenzoate (45, 69). Difference spectroscopy yielded a value for the dissociation constant of the complex of 2.9×10^{-5} M (pH 7.6 and 3.5°) with a titration curve fitting the theoretical curve predicted for 1:1 binding. Addition of *p*-hydroxybenzoate resulted in a 75% decrease of the flavin fluorescence of the enzyme. The K_d and stoichiometry from such fluorescence titrations were similar to those found spectrally.

As with the other p-hydroxybenzoate hydroxylases, the rate of anaerobic reduction of enzyme-flavin by NADPH is greatly enhanced in the presence of p-hydroxybenzoate, the extrapolated first-order rate constant being 0.41 min^{-1} in the absence and 1.52 $\times$ 10^4 min^{-1} in the presence of p-hydroxybenzoate, a rate stimulation of approximately 40,000-fold (45). Another effect of p-hydroxybenzoate is to decrease the dissociation constant for the enzyme–NADPH complex by approximately 13-fold.

A large number of analogs of p-hydroxybenzoate have been screened as possible substrates or inhibitors. Hesp *et al.*, using enzyme from *P. putida* M-6, found that salicylate and benzoate, rather mild inhibitors of enzymic activity, both markedly perturb the circular dichroism spectrum of the enzyme-bound FAD (65). In contrast, phenol, p-methoxybenzoate, and p-hydroxybenzoate methyl ester were all without significant effect on the CD spectrum, even though they are inhibitors of the same order as salicylate and benzoate. These studies suggest that all of the substrate analogs interact to some degree with the enzyme, even though the interaction might not lead to a modification of the circular dichroism properties. Further studies with this enzyme indicate that no prediction can be made about the magnitude of the circular dichroism perturbation from the inhibitory strength of a compound; for example, p-aminobenzoate strongly inhibits the catalytic reaction, yet is almost without effect on the CD spectrum (70). Clearly then, the inhibitors studied fall into two classes, those which affect the CD spectrum and those which do not. No clear-cut explanation appears to be available for this phenomenon. Teng *et al.* drew a positive correlation between the strength of inhibition and the Hammett substituent constant (σ) of the substituent on the inhibitory benzoate compound.[70] Recently, however, Spector and Massey examined a larger group of inhibitors of the *P. fluorescens* enzyme and found that many inhibitors deviate greatly from this type of correlation (71). It appears as if an aromatic ring with a carbonyl group is essential and sufficient for binding with a second substituent apparently not necessary. However, the finding by several groups that p-hydroxybenzoate is an inhibitor at high concentrations (64, 68, 71) would suggest that two binding sites are available on the enzyme, one for the carboxyl and the other for the hydroxyl group of the substrate. This is the usual explanation for excess substrate inhibition: i.e., at high concentration of substrate, "antiparallel" binging of two substrate molecules occurs, leading to a catalytically inactive species. This explanation appears to be too naive, however, in light of the recent finding that 2,4-dihydroxybenzoate, also a substrate of the enzyme, exhibits no inhibition at high concentrations (71).

Several compounds aside from p-hydroxybenzoate act as effectors by activating the NADPH oxidase activity as well as the rate of reduction under anaerobic conditions (67, 71). Howell and Massey have shown that 6-hydroxynicotinate is an activator of NADPH oxidase activity without itself being a substrate (45, 72). Steady state analysis together with the recovery of 6-hydroxynicotinate unchanged after the reaction have confirmed its effector role. A similar "uncoupling" of flavin reduction from substrate hydroxylation was first described with salicylate hydroxylase (50) (see Section III,A). More recently 3,4-dihydroxybenzoate, 2,4-dihydroxybenzoate, and benzoate have been found to activate NADPH oxidation by p-hydroxybenzoate hydroxylase (71). Product analysis showed that 2,4-dihydroxybenzoate is a substrate effector, itself being hydroxylated to 2,3,4-trihydroxybenzoate, while 3,4-dihydroxybenzoate is a nonsubstrate effector similar to 6-hydroxynicotinate. Benzoate is mainly a nonsubstrate effector, being hydroxylated to m-hydroxybenzoate with less than 5% efficiency (71). All the effectors mentioned perturb the visible fluorescence of the enzyme, and fluorimetric titrations allowed the calculation of the various dissociation constants. All effectors showed 1:1 binding.

Rapid reaction studies have allowed the identification of several intermediates in the anaerobic reduction of the enzyme by NADPH in the

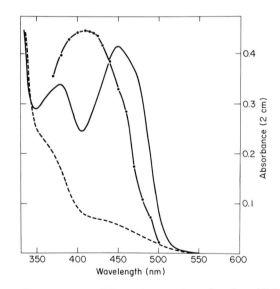

FIG. 3. Intermediate spectrum (15 msec) appearing after the addition of O_2 to the reduced p-hydroxybenzoate hydroxylase–2,4-dihydroxybenzoate complex. Time after mixing: (– – –) 0 (fully reduced), (. . .) 15 msec, and (—) 4.0 seconds (fully reoxidized). (from ref. 60).

presence of p-hydroxybenzoate or 2,4-dihydroxybenzoate (45, 60, 69). These intermediates appear to be of the charge-transfer type, exhibiting long-wavelength absorption bands. Most interesting, however, are the studies involving reoxidation of the reduced enzyme–substrate complex by oxygen. Immediately after the introduction of oxygen a novel spectral species is rapidly formed which is distinct from the oxidized E–S or ESOH complex. With 2,4-dihydroxybenzoate as substrate (60), this intermediate has an absorption maximum at 410 nm with a high extinction coefficient (Fig. 3). The intermediate is formed with a rate constant of 330 sec^{-1} under the conditions employed and decays to fully oxidized enzyme via another intermediate, presumably the oxidized enzyme–2,3,4-trihydroxybenzoate complex. The spectral characteristics of the 410 nm intermediate suggest that it is a reduced flavin–oxygen compound of the type previously postulated by Massey et al. (73). Furthermore, the similarity of its spectral characteristics to those of a model flavin substituted in the 10a position with a methoxy group (74) suggests that the intermediate may be substituted in the 10a position. However, recent molecular orbital calculations by Song (75) predict the following order of reactivity: N(5) > C(4a) > C(10a). The identification of the intermediate must await knowledge of the spectra of suitable model compounds. The three possible peroxydihydroflavin structures are:

N(5)-   C(4a)-   C(10a)-

Stopped-flow reoxidation studies with the reduced enzyme–p-hydroxybenzoate complex were performed at 2° in order to slow the reaction to measurable rates (61). Monitoring the reaction at various wavelengths revealed that an intermediate species was being formed during the reoxidation process. The calculated spectrum of the intermediate is somewhat different from that seen with 2,4-dihydroxybenzoate, being blue-shifted and lower in extinction. This difference might be a reflection of the profoundly different effects the two substrates have on the spectrum of the oxidized enzyme, or possibly the two different intermediate spectra seen result from O_2 being substituted in different positions of the isoalloxazine ring.

Enzyme in the presence of its nonsubstrate effectors, or uncomplexed enzyme, shows no evidence of spectral intermediates in the reoxidation process. However, the reoxidation rate in the presence of effectors is in all cases faster than that of uncomplexed enzyme.

C. Melilotate Hydroxylase

In 1965, while studying the metabolism of coumarin by an *Arthrobacter* species, Levy and Frost discovered an enzyme which uses molecular O_2 to hydroxylate melilotate to form 2,3-dihydroxyphenylpropionate (76).

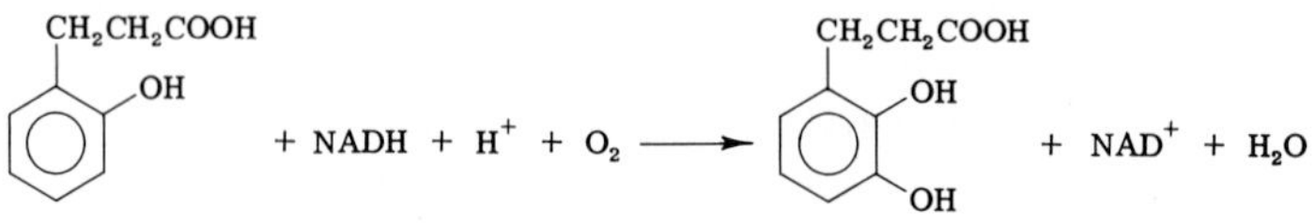

Levy subsequently purified the enzyme and found that it is a flavoprotein. It contains FAD as prosthetic group and has a molecular weight of 65,000 (77).

More recently, Strickland and Massey (78) have reported the isolation of melilotate hydroxylase from a *Pseudomonas* species. Like the *Arthrobacter* enzyme, this was also found to contain FAD as prosthetic group (1 molar equivalent per 65,000 gm protein), but unlike the *Arthrobacter* enzyme, it has a considerably higher molecular weight (238,000–250,000) and four subunits. This enzyme has been the subject of the first comprehensive kinetic study of a flavoprotein hydroxylase (62). In conventional steady state kinetic studies, the concentrations of each of the three substrates, melilotate, NADH, and O_2, were varied systematically. The results were found to be compatable only with the following reaction pathway:

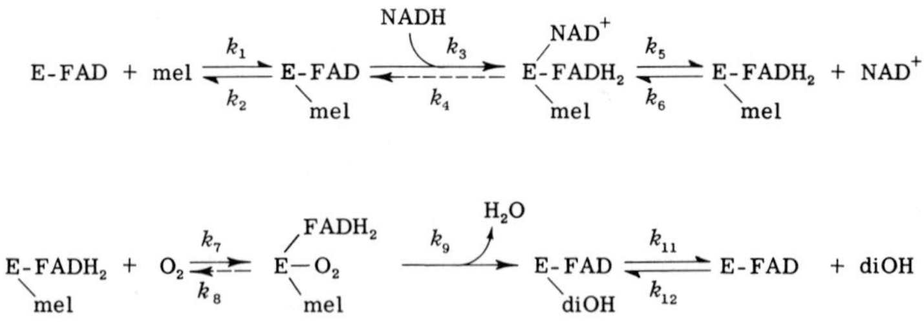

The rates of most of the individual steps in this pathway have been determined independently using relatively high concentrations of enzyme and employing rapid reaction spectrophotometry and fluorimetry to monitor changes in the state of the enzyme (62). These independently measured rate constants were then used to predict values for Michaelis constants of the substrates, the maximum catalytic velocity, and the dissociation constant of the oxidized enzyme–melilotate complex. Table II lists the observed kinetic constants and the predicted values. The predicted values are in remarkable agreement with those observed in the steady state kinetic analysis and provide very strong evidence for the validity of the proposed mechanism.

TABLE II

RATE CONSTANTS FOR MELILOTATE HYDROXYLASE (pH 7.3, 1°)

Step	Rate constant (from stopped-flow)
k_1	$5.7 \times 10^8 \; M^{-1} \, min^{-1}$
k_2	$2.2 \times 10^4 \; min^{-1}$
k_3	$1.4 \times 10^8 \; M^{-1} \, min^{-1}$
k_4	0
k_5	$1.3 \times 10^3 \; min^{-1}$
k_6	$9.0 \times 10^5 \; M^{-1} \, min^{-1}$
k_7	$1.6 \times 10^7 \; M^{-1} \, min^{-1}$
k_8	0
k_9	$1.4 \times 10^3 \; min^{-1}$
k_{10}, k_{11}, k_{12}	Not determined

	From steady state	Kinetic equivalent	Predicted from stopped-flow rate constants (above)
V_{max}	$735 \; min^{-1}$	$1/\phi_0 = \dfrac{1}{1/k_5 + 1/k_9 + 1/k_{11}}$	$680 \; min^{-1}$
$K_m(mel)$	$1.3 \times 10^{-6} \; M$	$\dfrac{1}{k_1 \cdot \phi_0}$	—
$K_m \; (NADH)$	$4.7 \times 10^{-6} \; M$	$\dfrac{(k_4 + k_5)}{k_3 k_5 \cdot \phi_0}$	$4.8 \times 10^{-6} \; M$
$K_m(O_2)$	$5.0 \times 10^{-5} \; M$	$\dfrac{(k_8 + k_9)}{k_7 k_9 \cdot \phi_0}$	$4.3 \times 10^{-5} \; M$

Several important mechanistic conclusions can be drawn from these results. First, melilotate combines with the oxidized enzyme before reaction with NADH. This is in keeping with the observations that a complex of melilotate and E–FAD can be observed spectrophotometrically and by changes in the circular dichroism (78). In this complex, each FAD-containing site was found to bind one equivalent of melilotate, with a K_d of 3.8×10^{-5} M. This complex is formed very rapidly ($k_1 = 5.7 \times 10^8$ M^{-1} min^{-1} at 1°, cf. Table II); the FAD of the resultant complex can be reduced very much more rapidly by NADH than that of the uncomplexed enzyme. Under anaerobic conditions in the absence of melilotate, E–FAD is reduced by NADH in a two-step reaction, an initial

complex being formed ($K_d = 3.8 \times 10^{-4}$ M) followed by a first-order reduction step with a rate constant of 1.4 min^{-1} (78). In contrast, when the enzyme is saturated with melilotate, NADH reacts to reduce the enzyme flavin without any detectable Michaelis complex, with a second-order rate constant of 1.4×10^8 M^{-1} min^{-1} (62). While no

complex has been detected kinetically, the enzyme species initially produced is clearly different from the binary complex

This latter species is produced at a rate independent of NADH concentration with a rate constant, k_5, of 1.3×10^3 min^{-1}. It is preceded by an initial species, which from the absorption spectrum clearly contains reduced flavin but possesses a distinctive long-wavelength absorption band centered at 750 nm. This species, seen transiently on reduction of the enzyme under anaerobic conditions, can be produced in static experiments by adding a high concentration of NAD$^+$ ($K_d = 1.45 \times 10^{-3}$ M) to the reduced enzyme–melilotate complex. Evidence is presented that it is a charge-transfer complex between

$$E\text{–}FADH_2$$
$$\diagdown$$
$$mel$$

and NAD^+.

The stopped-flow data also indicate that NAD^+ dissociates from the complex prior to the reaction with molecular oxygen (62). In addition to the dramatic enhancement of the reduction rate, complexing of the enzyme with melilotate also enhances significantly the reaction rate of the reduced flavin with O_2. In the absence of melilotate the rate is $9.7 \times 10^5\ M^{-1}\ min^{-1}$; in the presence of melilotate this is increased to $1.6 \times 10^7\ M^{-1}\ min^{-1}$ (62). Of particular interest is the intermediate detected on reaction of the reduced enzyme–melilotate complex with O_2. As in the case of p-hydroxybenzoate hydroxylase [complexed either with p-hydroxybenzoate (61) or 2,4-dihydroxybenzoate (60)] an intermediate with a distinctive absorption spectrum is formed rapidly, which decays more slowly to the spectrum of oxidized enzyme. The rate of appearance of the intermediate is very dependent on the oxygen concentration; its subsequent decay to oxidized enzyme (and presumably, products) is independent of oxygen concentration. These results suggest that the intermediate is an unstable (i.e., reactive) derivative composed of reduced flavin and oxygen, and is the species responsible for the insertion of the hydroxyl group into melilotate to yield 2,3-dihydroxyphenylpropionate.

The absorption spectrum of the intermediate, shown in Fig. 4, has many similarities to the analogous intermediates previously detected with p-hydroxybenzoate hydroxylase. As discussed in the section on p-hydroxybenzoate hydroxylase, the structure of this intermediate is still uncertain but probably represents an adduct of O_2 to the reduced flavin. The mechanism by which such a form of "activated" oxygen proceeds to carry out the actual hydroxylation of the substrate still remains to be elucidated. It has been suggested previously (73) that such an intermediate may carry out hydroxylation reactions by producing *in situ* (at the active site) reactive moities such as O_2^-, $OH^{\cdot}$, or, more likely, OH^+. An alternative proposal, based on a ring opening reaction has been put forward by Hamilton (79) and further detailed by him in Chapter 10 of this volume. Other displacement mechanisms may be postulated, also involving covalently linked flavin intermediates. This question will obviously provide a challenging area of research for some time to come.

D. Orcinol Hydroxylase

Ribbons and Chapman, in 1968 (80), isolated a strain of *Pseudomonas fluorescens* which uses orcinol as a sole carbon source. These cells con-

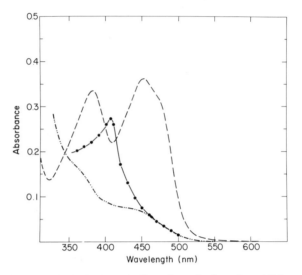

FIG. 4. The spectrum of the intermediate formed after the addition of O_2 to the reduced melilotate hydroxylase–melilotate complex. The spectrum of the intermediate was calculated by analog computer simulation using the determined rate constants k_7, k_8, and k_9 listed in Table II. (– ·· –) Reduced enzyme plus melilotate, (·—·) intermediate spectrum, and (– – –) reoxidized enzyme (from ref. 62).

sume oxygen in the presence of both orcinol and resorcinol, a close structural analog. Washed cell extracts require NADH or NADPH for oxygen consumption, NADPH being less efficient. From these observations, the authors concluded that the enzyme responsible for oxidizing orcinol is nonspecific, since m-cresol and resorcinol are oxidized as well (80, 81). Subsequent work (82, 83) has shown that m-cresol is a nonsubstrate effector, increasing NADH oxidase activity without being hydroxylated. Resorcinol behaves both as a substrate and a nonsubstrate effector. The reaction catalyzed by orcinol hydroxylase is as follows:

$$\text{Orcinol} + O_2 + NADH + H^+ \longrightarrow 2,3,5\text{-Trihydroxytoluene} + NAD^+ + H_2O$$

Orcinol hydroxylase isolated from *P. putida* 01 has been crystallized (84) and has been shown to be a flavoprotein consisting of a single polypeptide chain of molecular weight 60,000–70,000 containing 1 mole of FAD per mole of protein. The absorption spectrum is that of a simple flavoprotein with peaks at 273, 373, and 454 nm and a shoulder at 480 nm.

When resorcinol is used as a substrate for purified orcinol hydroxylase, some hydroxylated product (hydroxyquinol) is formed but not in sufficient quantities to account for the NADH and O_2 consumption (82). Furthermore, addition of catalase to the assay mixture results in the return of 40% of the oxygen consumed. With m-cresol as effector, no hydroxylation occurs, and catalase returns 50% of the oxygen. Addition of catalase had no effect on the NADH consumption as followed spectrophotometrically with either substrate analog. These results suggest that orcinol is the only true substrate with its hydroyxlation being tightly coupled to NADH oxidation, while resorcinol is hydroxylated to only a limited extent and m-cresol not at all.

Orcinol hydroxylase can utilize both NADH and NADPH as well as reduced 3-acetyl pyridine nucleotide as electron donors, although NADH is the best donor (85). Ribbons has studied the stereospecificity of hydride transfer from NADH to the enzyme-bound FAD using 4R and 4S-[³H]-NADH (85). The pro-R protium (A side) of NADH is stereospecifically transferred when orcinol is the aromatic substrate. With other effectors the results were less clear-cut. With m-cresol as effector, transfer of hydride rather than tritide appears to be preferred from either side of the reduced pyridine nucleotide, while with resorcinol, a "partial substrate," the specificity for the pro-R tritium is less than with orcinol yet more than with m-cresol. These results are quite unusual and await exact interpretation.

E. *m*-Hydroxybenzoate-6-hydroxylase

m-Hydroxybenzoate-6-hydroxylase is an inducible enzyme of *P. aeruginosa* which catalyzes the initial hydroxylation reaction in the gentisate pathway for m-hydroxybenzoate metabolism (86).

COOH ... + NADH + H⁺ + O₂ (NADPH) ⟶ HO COOH OH + NAD⁺ + H₂O (NADP⁺)

m-Hydroxy-
benzoic acid   Gentisic acid

It is a flavoprotein requiring NADH or NADPH as electron donor. As with the other aromatic hydroxylases previously discussed, reduced nicotinamide nucleotide oxidation is greatly facilitated by the presence of the substrate (effector) m-hydroxybenzoate (87).

F. m-Hydroxybenzoate-4-hydroxylase

Premkumar *et al.* (88) have reported the partial purification from *Aspergillus niger* of a *m*-hydroxybenzoate hydroxylase which converts *m*-hydroxybenzoate to protocatechuate, utilizing NADPH as electron donor:

The prosthetic group was found to be FAD.

A similar enzyme has also been obtained recently from *P. testosteroni*. This enzyme also functions with NADH, but NADPH is the preferred electron donor (87). Other aromatic substrates such as 2-fluoro-*m*-hydroxybenzoate, 4-fluoro-*m*-hydroxybenzoate, 3,5-dihydroxybenzoate, 2,3-dihydroxybenzoate, 3-hydroxyanthranilate and gentisate are effectors, forming varying amounts of hydroxylated products, and allowing H_2O_2 formation. The degree to which these compounds act as true substrates or nonsubstrate effectors is not clear at present.

G. Imidazoleacetate Monooxygenase

Imidazoleacetate monooxygenase is the only external flavin monooxygenase studied to date which acts on a nonaromatic substrate. The reaction catalyzed appears to be as follows (89).

Imidazoleacetate   Imidazaloneacetate

The enzyme is isolated from a *Pseudomonas* species and is part of the histidine catabolic pathway in which imidazaloneacetate is converted to aspartic acid by way of formiminoaspartic acid (90). In fact, early work on the enzyme was hindered by the presence of contaminating enzymes which yielded *N*-formiminoaspartic acid or formylaspartic acid as product (91, 92). Rothberg and Hayaishi established the monooxygenase nature of the enzyme with ^{18}O experiments (93).

Crystallization of imidazoleacetate monooxygenase (89, 90) revealed the flavoprotein nature of the enzyme with FAD being the only prosthetic group. The molecular weight was estimated to be between 87,000 and

90,000. NADH is required for activity, although NADPH can also act as electron donor in a less efficient manner. Okamoto *et al.* have studied the role of sulfhydryl groups in imidazoleacetate monooxygenase (94). They found two sulfhydryl groups titratable in the native protein by silver nitrate or *p*-chloromercuribenzoate. In the presence of imidazoleacetate only one SH group could be titrated; also, imidazoleacetate protected against inactivation by the mercurial. Interestingly, several sulfhydryl reagents increased the NADH oxidase activity two to threefold and caused production of hydrogen peroxide. Apparently, then, blocking of the imidazoleacetate binding site causes an "uncoupling" of the oxygenation reaction. As with the aromatic hydroxylases, imidazoleacetate monooxygenase exhibits a very weak NADH oxidase activity which is stimulated by the presence of substrate. What is unusual, however, is the observation that imidazoleacetate has no effect on the absorption spectrum of the enzyme nor on the CD or ORD spectra (89). This is in sharp contrast to the results found with other external hydroxylases.

H. Microsomal Amine Oxidase

All of the flavoprotein monooxygenases so far mentioned are bacterial in origin. In recent years Ziegler and his associates have reported on a monooxygenase present in liver microsomes of many vertebrate species including man (95–98). This enzyme appears to be a simple flavoprotein, with FAD as the prosthetic group, containing no significant quantities of metal ions or of heme proteins. The enzyme has been isolated from pig liver microsomes and is reported to have a molecular weight of approximately 500,000 (97). The FAD content was determined as 14 nmoles/mg protein, corresponding to a minimum molecular weight per molecule of flavin of 71,500. Thus, the enzyme would appear to be a polymeric, aggregating species.

The enzyme catalyzes the NADPH- and oxygen-dependent N-oxidation of a variety of secondary and tertiary amines. The secondary amines are oxidized to the corresponding hydroxylamines and the tertiary amines to amine oxides, e.g.,

$$\text{NH} - \text{CH}_3 + \text{NADPH} + \text{H}^+ + \text{O}_2 \longrightarrow \text{N(OH)} - \text{CH}_3 + \text{NADP}^+ + \text{H}_2\text{O}$$

N-Methylaniline   *N*-Methyl, phenyl hydroxylamine

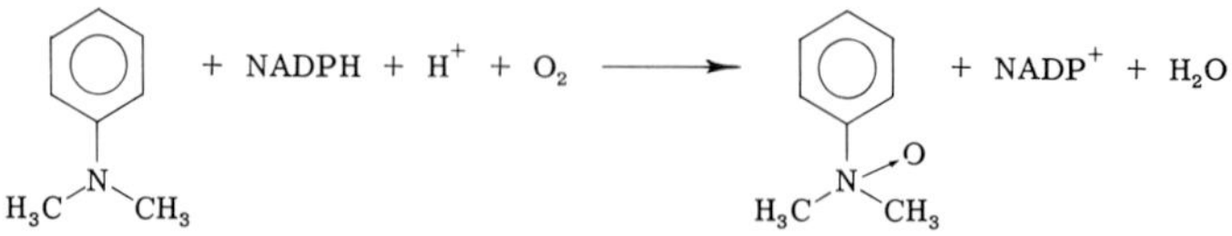

N, N-Dimethyl-aniline

N, N-Dimethyl-aniline-*N*-oxide

The primary amines, 1-naphthylamine and 2-naphthylamine, are also oxidized, but at much lower rates than secondary and tertiary amines (98). With one substrate, dimethyloctylamine, sigmoidal kinetics are found (97), suggesting the existence of a modifier site as well as the catalytic site. This idea is substantiated by the finding that a variety of compounds, which are not themselves substrates, can increase the rate of oxidation of true substrates (97).

I. Kynurenine-3-hydroxylase

Kynurenine-3-hydroxylase has been partially purified from rat liver mitochondria (99) with a 12-fold concentration of activity. Okamoto *et al.* showed that the enzyme is localized in the outer membrane of the mitochondria (100). Several workers have, in fact, utilized kynurenine hydroxylase as a mitochondrial marker (e.g., see refs. 101 and 102).

^{18}O experiments have established the reaction to be as follows (103, 104):

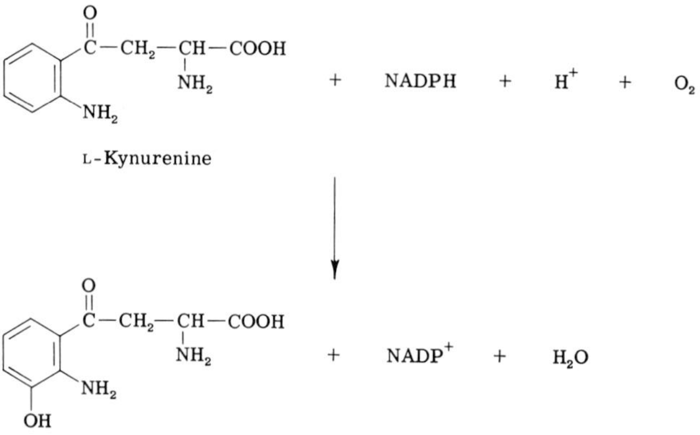

L-Kynurenine

3-Hydroxykynurenine

The enzyme can utilize NADH as well as NADPH (105). Acid ammonium sulfate treatment of outer membrane preparations resulted in a decrease in activity to 60% of the original level. Addition of FAD returned almost 100% of the activity, while FMN had a negligible effect (100, 104, 106). The enzyme thus has been shown to require FAD. However, the possibility of participation of other cofactors in the reaction of kynurenine hydroxylase cannot be eliminated since the enzyme is not pure. On subcutaneous administration of L-thyroxine to rats, Okamoto et al. have observed a decrease in the level of kynurenine hydroxylase (107, 108). This enzyme is thus under direct hormonal control and apparently plays a role in regulating the synthesis of NAD^+ and $NADP^+$.

IV. EXTERNAL FLAVOPROTEIN DIOXYGENASE

A. 2-Methyl-3-hydroxypyridine-5-carboxylate Oxygenase

Sparrow et al. (44) have purified, from a *Pseudomonas* species, an inducible oxygenase which functions in the degradative pathway of vitamin B_6. The enzyme catalyzes the opening of the pyridine ring of 2-methyl-3-hydroxypyridine-5-carboxylic acid as follows:

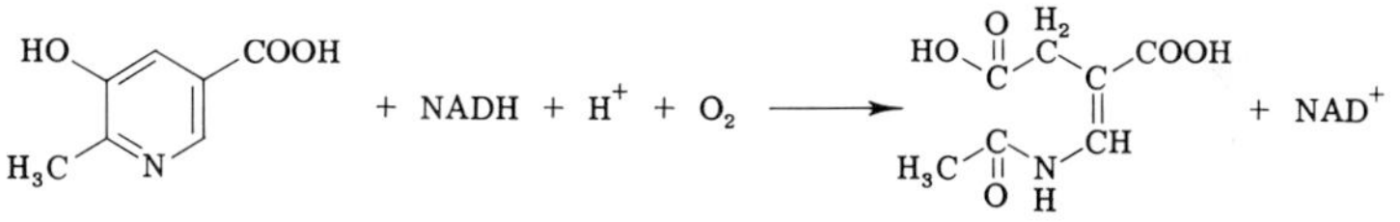

The crystalline oxygenase has been shown to be homogeneous by disc-gel electrophoresis and ultracentrifugation, and it has a molecular weight of 166,000 with 2 moles of FAD per mole of protein. Participation of metal ion in catalysis, however, has not been excluded. NADPH is almost as effective as NADH in the reaction, while only one other compound, 5-pyridoxic acid, showed any activity with the enzyme. Whether 5-pyridoxic acid functions as a true substrate, or as a nonsubstrate effector of the low NADH oxidase activity, remains to be seen.

While 2-methyl-3-hydroxypyridine-5-carboxylic acid has no effect on the visible absorption spectrum of the enzyme, it causes a marked stimulation in both the rate of reduction of enzyme-bound flavin by NADH and the rate of reoxidation by oxygen (44). Tracer experiments, carried out with $^{18}O_2$ and $H_2^{18}O$, show that molecular oxygen is incorporated mainly into the acetyl group of the product [α-(N-acetylaminomethylene) succinic acid]. Some $^{18}O_2$ is also incorporated into the newly formed carboxyl group of the product, indicating that the enzyme is best classified as a dioxygenase, although some ^{18}O from $H_2^{18}O$ is also incorporated into product. These data might be explained by a mechanism involving a cyclic peroxide compound such as proposed by Hayaishi (109) as an intermediate.

Such an intermediate could break down either by direct rearrangement to yield product containing both atoms of molecular oxygen or by a less direct route involving attack by a water molecule to yield product containing oxygens derived from both water and O_2 (44).

REFERENCES

1. Mason, H. S. (1957). *Science* **125**, 1185.
2. Orrenius, S., and Ernster, L., Chapter 6, this volume.
3. Gunsalus, I. C., Chapter 14, this volume.
4. Hayaishi, O. (1962). *In* "Oxygenases" (O. Hayaishi, ed.), p. 1. Academic Press, New York.
5. Hayaishi, O. (1969). *Ann. Rev. Biochem.* **38**, 21.
6. Hayaishi, O., and Nozaki, M. (1969). *Science* **164**, 389.
7. Hayaishi, O. (1966). *Bact. Rev.* **30**, 720.
8. Sutton, W. B. (1954). *J. Biol. Chem.* **210**, 309.
9. Sutton, W. B. (1955). *J. Biol. Chem.* **216**, 749.
10. Hayaishi, O., and Sutton, W. B. (1957). *J. Amer. Chem. Soc.* **79**, 4809.
11. Sutton, W. B. (1957). *J. Biol. Chem.* **226**, 395.
12. Kalckar, H. M. (1947). *J. Biol. Chem.* **167**, 461.
13. Sullivan, P. A. (1968). *Biochem. J.* **110**, 363.
14. Takemori, S., Nakazowa, K., Nakai, Y., Suzuki, K., and Katagiri, M. (1968). *J. Biol. Chem.* **243**, 313.
15. Beinert, H., and Sands, R. H. (1961). *In* "Free Radicals in Biological Systems" (M. S. Blois *et al.*, eds.), p. 35. Academic Press, New York.
16. Sullivan, P. A. (1971). *In* "Flavins and Flavoproteins" (H. Kamin, ed.), Vol. 3, p. 470. Univ. Park Press, Baltimore, Maryland.
17. Lockridge, O., Massey, V., and Sullivan, P. A., (1972). *J. Biol. Chem.* **247**, 8097.
18. Takemori, S., Nakai, Y., Katagiri, M., and Nakamura, T. (1969). *FEBS Lett.* **3**, 214.
19. Katagiri, M., and Takemori, S. (1971). *In* "Flavins and Flavoproteins" (H. Kamin, ed.), Vol. 3, p. 447. Univ. Park Press, Baltimore, Maryland.
20. Walsh, C. T., Schonbrunn, A., and Abeles, R. H. (1971). *J. Biol. Chem.* **246**, 6855.
21. Walsh, C. T., Lockridge, O., Massey, V., and Abeles, R. H., *J. Biol. Chem.* (in press).
22. Walsh, C. T., Krodel, E., Massey, V., and Abeles, R. H., (1973). *J. Biol. Chem.* **248**, 1946.
23. Walsh, C. T., Schonbrunn, A., Lockridge, O., Massey, V., and Abeles, R. H. (1972). *J. Biol. Chem.* **247**, 6004.
24. Thoai, N. V., and Olomucki, A. (1962). *Biochim. Biophys. Acta* **59**, 533.
25. Takeda, H., Yamamoto, S., Kojima, Y., and Hayaishi, O. (1969). *J. Biol. Chem.* **244**, 2935.
26. Nakazawa, T., Hori, K., and Hayaishi, O. (1972). *J. Biol. Chem.* **247**, 3439.
27. Thoai, N. V., and Olomucki, A. (1962). *Biochim. Biophys. Acta* **59**, 545.
28. Pho, D. B., Olomucki, A., and Thoai, N. V. (1966). *Biochim. Biophys. Acta* **118**, 299.
29. Olomucki, A., Pho, D. B., Lebar, R., Delcambe, L., and Thoai, N. V. (1968). *Biochim. Biophys. Acta* **151**, 353.
30. Thomé-Beau, F., Le-Thi-Lan, Olomucki, A., and Thoai, N. V. (1971). *Eur. J. Biochem.* **19**, 270.
31. Hagihara, H., Hayashi, H., Ichihara, A., and Suda, M. (1960). *J. Biochem.* **48**, 267.

32. Takeda, H., and Hayaishi, O. (1966). *J. Biol. Chem.* **241**, 2733.
33. Itada, N., Ichihara, A., Makita, T., Hayaishi, O., Suda, M., and Sasaki, N. (1961). *J. Biochem.* **50**, 118.
34. Yamamoto, S., Takeda, H., Maki, Y., and Hayaishi, O. (1969). *J. Biol. Chem.* **244**, 2951.
35. Takeda, H., Yamamoto, S., Kojima, Y., and Hayaishi, O. (1969). *J. Biol. Chem.* **244**, 2935.
36. Nakazawa, T., Yamamoto, S., Maki, Y., Takeda, H., Kajita, Y., Nozaki, M., and Hayaishi, O. (1968). *In* "Flavins and Flavoproteins" (K. Yagi, ed.), Vol. 2, p. 214. Univ. of Tokyo Press, Tokyo.
37. Yamamoto, S., Nakazawa, T., and Hayaishi, O. (1972). *J. Biol. Chem.* **247**, 3434.
38. Yamamoto, S., Maki, Y., Nakazawa, T., Kajita, Y., Takeda, H., Nozaki, M., and Hayaishi, O. (1968). *Advan. Chem. Ser.* **77**, 177.
39. Hayaishi, O., Yamamoto, S., Nakazawa, T., and Maki, Y. (1967). *Int. Congr. Biochem., 7th, Tokyo* Abstr. I, p. 163.
40. Nakazawa, T., Hori, K., and Hayaishi, O. (1972). *J. Biol. Chem.* **247**, 3439.
41. Yamamoto, S., Hirata, F., Yamauchi, T., Nozaki, M., Hiromi, K., and Hayaishi, O. (1971). *J. Biol. Chem.* **246**, 5540.
42. Kosuge, T., Heskett, M. G., and Wilson, E. E. (1966). *J. Biol. Chem.* **241**, 3738.
43. Hutzinger, O., and Kosuge, T. (1967). *Biochim. Biophys. Acta* **136**, 389.
44. Sparrow, L. G., Ho, P. P. K., Sundaram, T. K., Zach, D., Nyns, E. J., and Snell, E. E. (1969). *J. Biol. Chem.* **244**, 2590.
45. Howell, L. G., Spector, T., and Massey, V. (1972). *J. Biol. Chem.* **247**, 4340.
46. Katagiri, M., Yamamoto, S., and Hayaishi, O. (1962). *J. Biol. Chem.* **237**, P.C. 2413.
47. Yamamoto, S., Katagiri, M., Maeno, H., and Hayaishi, O. (1965). *J. Biol. Chem.* **240**, 3408.
48. Takemori, S., Yasuda, H., Mihara, K., Suzuki, K., and Katagiri, M. (1969). *Biochim. Biophys. Acta* **191**, 58.
49. Katagiri, M., Maeno, H., Yamamoto, S., Hayaishi, O., Kitao, T., and Oae, S. (1965). *J. Biol. Chem.* **240**, 3414.
50. White-Stevens, R. H., and Kamin, H. (1970). *Biochem. Biophys. Res. Commun.* **38**, 882.
51. White-Stevens, R. H., and Kamin, H. (1972). *J. Biol. Chem.* **247**, 2358.
52. White-Stevens, R. H., Kamin, H., and Gibson, Q. H. (1972). *J. Biol. Chem.* **247**, 2371.
53. Katagiri, M., Takemori, S., Nakamura, M., and Nakamura, T. (1971). *Int. Symp. Oxidases Related Redox Syst., 2nd Memphis* (T. E. King, H. S. Mason, and M. Morrison, eds.) (in press).
54. Suzuki, K., Takemori, S., and Katagiri, M. (1969). *Biochim. Biophys. Acta* **191**, 77.
55. Takemori, S., Yasuda, H., Mihara, K., Suzuki, K., and Katagiri, M. (1969). *Biochim. Biophys. Acta* **191**, 69.
56. Massey, V., and Palmer, G. (1966). *Biochemistry* **5**, 3181.
57. Katagiri, M., and Takemori, S. (1971). *In* "Flavins and Flavoproteins" (H. Kamin, ed.), Vol. 3, p. 447. Univ. Park Press, Baltimore, Maryland.
58. Takemori, S., Nakamura, M., Katagiri, M., and Nakamura, T. (1971). *In* "Flavins and Flavoproteins" (H. Kamin, ed.), Vol. 3, p. 464. Univ. Park Press, Baltimore, Maryland.
59. Chance, B. (1943). *J. Biol. Chem.* **151**, 553.
60. Spector, T., and Massey, V. (1972). *J. Biol. Chem.* **247**, 5632.

61. Spector, T., and Massey, V., (1972). *J. Biol. Chem.* **247**, 7123.
62. Strickland, S., and Massey, V., (1973). *J. Biol. Chem.* **248**, 2953.
63. Yano, K., Higashi, N., and Arima, K. (1969). *Biochim. Biophys. Res. Commun* **34**, 1.
64. Hosokawa, K., and Stanier, R. Y. (1966). *J. Biol. Chem.* **241**, 2453.
65. Hesp, B., Calvin, M., and Hosokawa, K. (1969). *J. Biol. Chem.* **244**, 5644.
66. Yano, K., Higashi, N., Nakamura, S., and Arima, K. (1969). *Biochim. Biophys. Res. Commun.* **34**, 277.
67. Nakamura, S., Ogura, Y., Yano, K., Higashi, N., and Arima, K. (1971). *In* "Flavins and Flavoproteins" (H. Kamin, ed.), Vol. 3, p. 475. Univ. Park Press, Baltimore, Maryland.
68. Nakamura, S., Ogura, Y., Yano, K., Higashi, N., and Arima, K. (1970). *Biochemistry* **9**, 3235.
69. Howell, L. G., and Massey, V. (1971). *In* "Flavins and Flavoproteins" (H. Kamin, ed.), Vol. 3, p. 499. Univ. Park Press, Baltimore, Maryland.
70. Teng, N., Kotowycz, G., Calvin, M., and Hosokawa, K. (1971). *J. Biol. Chem.* **246**, 5448.
71. Spector, T., and Massey, V. (1972). *J. Biol. Chem.* **247**, 4679.
72. Howell, L. G., and Massey, V. (1970). *Biochim. Biophys. Res. Commun.* **40**, 887.
73. Massey, V., Müller, F., Feldberg, R., Schuman, M., Sullivan, P. A., Howell, L. G., Mayhew, S. G., Matthews, R. G., and Foust, G. P. (1969). *J. Biol. Chem.* **244**, 3999.
74. Müller, F. (1971). *In* "Flavins and Flavoproteins" (H. Kamin, ed.), Vol. 3, p. 363. Univ. Park Press, Baltimore, Maryland.
75. Song, P. S., personal communication.
76. Levy, C. C., and Frost, P. (1966). *J. Biol. Chem.* **241**, 997.
77. Levy, C. C. (1967). *J. Biol. Chem.* **242**, 747.
78. Strickland, S., and Massey, V. (1973). *J. Biol. Chem.* **248**, 2944.
79. Hamilton, G. A. (1971). *Progr. Bioorg. Chem.* **1**, 83.
80. Ribbons, D. W., and Chapman, P. J. (1968). *Biochem. J.* **106**, 44 P.
81. Ribbons, D. W., and Ohta, Y. (1970). *Bacteriol. Proc.* 7.
82. Ribbons, D. W., and Ohta, Y. (1970). *FEBS Lett.* **12**, 105.
83. Ribbons, D. W., Ohta, Y., and Higgins, I. J. (1971). *J. Bact.* **106**, 702.
84. Ohta, Y., and Ribbons, D. W. (1970). *FEBS Lett.* **11**, 189.
85. Ribbons, D. W., Ohta, Y., and Higgins, I. J. (1972). *In Miami Winter Symp.* "The Molecular Basis of Electron Transport" (J. Schultz and B. F. Cameron, ed.), Vol. IV (in press).
86. Groseclose, E. E., and Ribbons, D. W. (1972). *Bacteriol. Proc.* **273**.
87. Ribbons, D. W., personal communication.
88. Premkumar, R., Subba Rao, P. V., Sreeleela, N. S., and Vaidyanathan, C. S. (1969). *Can. J. Biochem.* **47**, 825.
89. Maki, Y., Yamamoto, S., Nozaki, M., and Hayaishi, O. (1969). *J. Biol. Chem.* **244**, 2942.
90. Maki, Y., Yamamoto, S., Nozaki, M., and Hayaishi, O. (1966). *Biochem. Biophys. Res. Commun.* **25**, 609.
91. Hayaishi, O., Tabor, H., and Hayaishi, T. (1954). *J. Amer. Chem. Soc.* **76**, 5570.
92. Hayaishi, O., Tabor, H., and Hayaishi, T. (1957). *J. Biol. Chem.* **227**, 161.
93. Rothberg, S., and Hayaishi, O. (1957). *J. Biol. Chem.* **229**, 897.
94. Okamoto, H., Nozaki, M., and Hayaishi, O. (1968). *Biochem. Biophys. Res. Commun.* **32**, 30.
95. Machinist, J. M., Dehner, E. W., and Ziegler, D. M. (1968). *Arch. Biochem. Biophys.* **125**, 858.

96. Ziegler, D. M., Jollow, D., and Cook, D. E. (1971). *In* "Flavins and Flavoproteins" (H. Kamin, ed.), Vol. 3, p. 507. Univ. Park Press, Baltimore, Maryland.
97. Ziegler, D. M., and Mitchell, C. H. (1972). *Arch. Biochem. Biophys.* **150**, 116.
98. Ziegler, D. M., Poulsen, L. L., and McKee, E. M. (1971). *Xenobiotica* **1**, 523.
99. Okamoto, H. (1970). *Methods Enzymol.* **17**, 460.
100. Okamoto, H., Yamamoto, S., Nozaki, M., and Hayaishi, O. (1967). *Biochem. Biophys. Res. Commun.* **26**, 309.
101. Beattie, D. S. (1968). *Biochem. Biophys. Res. Commun.* **31**, 901.
102. Schnactman, C. A., and Greenwalt, J. W. (1968). *J. Cell Biol.* **38**, 158.
103. Saito, Y., Hayaishi, O., and Rothberg, S. (1957). *J. Biol. Chem.* **229**, 921.
104. Okamoto, H. (1968). *In* "Flavins and Flavoproteins" (K. Yagi, ed.), Vol. 2, p. 223. Univ. of Tokyo Press, Tokyo.
105. Hayaishi, O., and Okamoto, H. (1971). *Amer. J. Clin. Nutr.* **24**, 805.
106. Okamoto, H., and Hayaishi, O. (1967). *Biochem. Biophys. Res. Commun.* **29**, 394.
107. Okamoto, H. (1971). *Biochem. Biophys. Res. Commun.* **43**, 827.
108. Okamoto, H., Okada, F., and Hayaishi, O. (1971). *J. Biol. Chem.* **246**, 7759.
109. Hayaishi, O. (1964). *Int. Congr. Biochem., 6th, Proc. Plenary Sessions, IUB* **33**, 31.

8

PTERIN-REQUIRING AROMATIC AMINO ACID HYDROXYLASES

SEYMOUR KAUFMAN and DANIEL B. FISHER

I. INTRODUCTION

Since the original discovery that tetrahydrobiopterin is the natural cofactor for phenylalanine hydroxylase (89), two other enzymes, tyrosine (6, 91, 121) and tryptophan hydroxylase (109), have been demonstrated to require a reduced unconjugated pterin cofactor as an electron source.

285

These three aromatic amino acid hydroxylases all appear to be mixed function oxidases since they require atmospheric oxygen and an electron donor. These enzymes show many similarities and differences as will be demonstrated in this review. The most extensive studies of mechanism and physical properties have been carried out on phenylalanine hydroxylase because it is available in far greater quantities than the other two hydroxylases and is more readily purified. There is great interest in phenylalanine hydroxylase because it is the enzyme missing in a form of mental retardation, phenylketonuria. Tyrosine and tryptophan hydroxylase are of interest to neurochemistry because they catalyze the initial reactions in the formation of the neurotransmitters, norepinephrine, and serotonin, respectively. It is hoped that a deeper understanding of these enzymes will help illuminate the regulation of neural pathways.

II. PHENYLALANINE HYDROXYLASE

A. Introduction

The conversion of phenylalanine to tyrosine by phenylalanine hydroxylase is believed to play an obligatory role in the catabolism of phenylalanine for energy production. One of the products of tyrosine breakdown, fumarate, can be oxidized to CO_2 and water, or it can lead to the formation of glucose. Since gluconeogenesis takes place mainly in the liver, it seems appropriate that phenylalanine hydroxylase is present in highest concentrations in that tissue. It is also provacative that the only other organ capable of significant gluconeogenesis, the kidney, has also been shown to contain significant phenylalanine hydroxylase activity (158). It would be of interest to see if conditions which stimulate gluconeogenesis, such as hydrocortisone administration, also elevate phenylalanine hydroxylase activity.

Phenylalanine hydroxylase can also provide the organism with tyrosine under conditions of tyrosine deficiency. Tyrosine is not required in the diet of a rat if sufficient phenylalanine is present, and the dietary requirement for phenylalanine is nearly halved when tryosine is included in the diet (1, 170). By providing an endogenous source of tyrosine, phenylalanine hydroxylase also produces the precursor for the synthesis of melanin, thyroxin, norepinephrine, and epinephrine.

B. Components of the Phenylalanine Hydroxylating System

Since a detailed description of the isolation and identification of the components of the hepatic phenylalanine hydroxylating system has recently appeared (97), the following will be a condensed survey of these studies.

In 1952 the conversion of phenylalanine to tyrosine was demonstrated by an extract from rat liver which was supplemented with DPN^+ and an alcohol or aldehyde (162). In 1956 it was demonstrated that DPNH rather than DPN^+ was actually involved and that two protein fractions were necessary (116). Early studies from our laboratory indicated that the stoichiometry of the phenylalanine hydroxylation reaction was as follows (79):

$$TPNH + H^+ + phenylalanine + O_2 \rightarrow TPN^+ + tyrosine + H_2O \qquad (1)$$

The demonstration that an electron donor (i.e., TPNH) and O_2 were both required for and stoichiometrically consumed during the hydroxylation reaction led to the conclusion that phenylalanine hydroxylase is a mixed function oxidase and that the phenolic oxygen in tyrosine is derived from molecular oxygen rather than water. This conclusion was confirmed when the hydroxylation reaction was carried out in the presence of $^{18}O_2$ or $H_2{}^{18}O$; the phenolic group of tyrosine was labeled with ^{18}O in the former but not the latter case (98).

The phenylalanine hydroxylation reaction involves two essential enzymes; one of these enzymes was purified from rat liver (86) and the other was purified from sheep liver extracts (86). In this coupled enzyme system, TPNH was shown to be twice as effective as DPNH (79). This result contrasted with Mitoma's finding (mentioned above) that DPNH was more effective. Subsequently, it was found that an unconjugated pterin, tetrahydrobiopterin (Fig. 1) (87, 89), is an essential cofactor in the hydroxylation reaction and that Eq. (1) is actually the sum of two reactions: reaction (2), catalyzed by phenylalanine hydroxylase ("rat liver enzyme") and reaction (3), catalyzed by dihydropteridine reductase ("sheep liver enzyme") (90). The nucleotide specificity of reaction (3) was not determined although it was assumed to be the same as for the overall reaction (1).

$$Phenylalanine + tetrahydrobiopterin + O_2 \rightarrow$$
$$tyrosine + quinonoid\ dihydrobiopterin + H_2O \qquad (2)$$
$$TPNH + H^+ + quinonoid\ dihydrobiopterin \rightarrow TPN^+ + tetrahydrobiopterin \qquad (3)$$

The primary pterin oxidation product of the phenylalanine hydroxylase reaction was identified as a quinonoid isomer of dihydrobiopterin (90). Though spectral evidence favored the *para*-quinonoid structure, the primary oxidation product could be the *ortho*-quinonoid dihydrobiopterin (Fig. 1) (90). The cofactor isolated from rat liver, however, was 7,8-

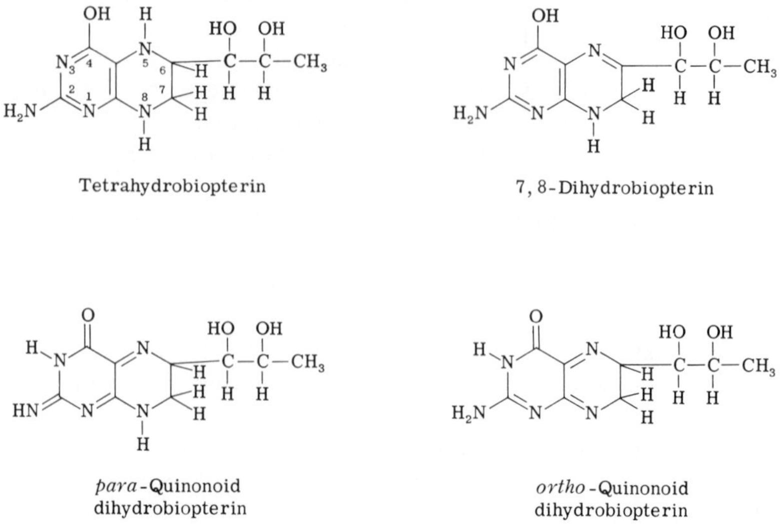

FIG. 1. Structure of 5,6,7,8-tetrahydrobiopterin and isomers of dihydrobiopterin.

dihydrobiopterin (Fig. 1) (89). In the presence of the 7,8-dihydro isomer, a third enzyme was shown to be an essential component of the hydroxylation system (94). This enzyme is dihydrofolate reductase, which catalyzes the TPNH-mediated reduction of 7,8-dihydrobiopterin to the tetrahydro form according to Eq. (4) (93):

$$\text{TPNH} + \text{H}^+ + 7,8\text{-dihydrobiopterin} \rightarrow \text{TPN}^+ + \text{tetrahydrobiopterin} \qquad (4)$$

This reaction initiates the hydroxylation cycle [i.e., Eqs. (2) and (3)] by converting the cofactor to the active tetrahydro form. The dihydrofolate reductase-catalyzed reaction may also serve to salvage any 7,8-dihydrobiopterin that might be formed from the unstable quinonoid dihydro isomer by nonenzymic isomerization (84).

Recent studies have demonstrated that with crude (127) and homogeneous (17) sheep liver dihydropteridine reductase, DPNH is far more active than TPNH. These results provide an explanation for why Mitoma found DPNH to be more active while Kaufman found that TPNH was more active in the phenylalanine hydroxylase system. In the crude fractions used by Mitoma, most of the endogenous cofactor was probably still in the tetrahydro form (92); thus, the TPNH-specific dihydrofolate reductase step played no role in the reaction. Therefore, the hydroxylation reaction was DPNH specific. On the other hand, as already mentioned, in partially purified enzyme fractions, the endogenous cofactor is in the 7,8-dihydro form and the TPNH-specific dihydrofolate reductase becomes essential. Therefore, the hydroxylation was TPNH specific with the partially purified enzymes used by Kaufman.

These results serve to emphasize the fact that the pyridine nucleotide specificity of the overall phenylalanine hydroxylating system will depend on the reduction level of the pterin cofactor as well as on the relative activities of the various enzymes; in the presence of 7,8-dihydrobiopterin (and limiting amounts of dihydrofolate reductase), the system will be more active with TPNH; in the presence of tetrahydrobiopterin, it will be more active with DPNH.

In vitro studies have shown that ascorbate, mercaptoethanol, and high concentrations of TPNH (83) can reduce the quinonoid dihydropterin to the tetrahydro level without dihydropteridine reductase. Studies with aminopterin indicate that *in vivo* the reduction step probably involves the dihydropteridine reductase–catalyzed reaction. Patients receiving levels of aminopterin sufficient to give a final concentration of 10^{-5} M had an impaired phenylalanine tolerance test (44). This concentration of aminopterin inhibits dihydropteridine reductase 50%, but has no effect on phenylalanine hydroxylase activity (17).

The presence of another component in the phenylalanine hydroxylase system was indicated by the early observations on the role played by glucose dehydrogenase. This enzyme (plus glucose) had been used to regenerate TPNH in the routine assay of phenylalanine hydroxylase. It was found, however, that the partially purified fractions that were used were contributing something in addition to the dehydrogenase: These fractions of dehydrogenase stimulated the hydroxylation reaction even in the absence of glucose or in the presence of an excess of TPNH (83). A clue to the identity of one of the active components in the glucose dehydrogenase preparation was provided by the observation that the dehydrogenase inhibited the rate of oxygen uptake associated with the aerobic nonenzymic oxidation of the 6,7-dimethyltetrahydropterin. Furthermore, when incubated under these conditions, the tetrahydropterin lost its cofactor activity, and the dehydrogenase preparation protected against this loss to a considerable extent (88). These results suggested the possibility that catalase was the active component in the glucose dehydrogenase preparation. Catalase was as effective as glucose dehydrogenase in stimulating the phenylalanine hydroxylation system and in protecting against aerobic inactivation of the cofactor. In experiments where the nonenzymic oxidation of tetrahydropterin was followed directly by measurement of the increase in absorbance as a result of dihydropterin formation, catalase was found to retard the rate of oxidation (Kaufman, unpublished).

A plausible mechanism for these effects of catalase is outlined in Eqs. (5)–(8).

$$XH_4 + O_2 \rightarrow \text{quinonoid-}XH_2 + H_2O_2 \tag{5}$$

$$XH_4 + H_2O_2 \rightarrow \text{quinonoid-}XH_2 + 2 H_2O \tag{6}$$

$$\text{Quinonoid-}XH_2 \rightarrow 7,8\text{-}XH_2 \tag{7}$$

$$2 H_2O_2 \rightarrow 2 H_2O + O_2 \tag{8}$$

The scheme is based on the fact that the aerobic oxidation of 6,7-di-methyltetrahydropterin is accelerated by low concentrations (0.001–0.002 M) of H_2O_2 (84),* which indicates that reaction (6) is faster than reaction (5), and on the observation that H_2O_2 is a product of the aerobic oxidation of tetrahydropterins. That the quinonoid dihydropterin is also a product of the H_2O_2-stimulated oxidation of tetrahydropterins was shown by the observation that H_2O_2 stimulated the dihydropteridine reductase–catalyzed oxidation of TPNH (84).

According to the scheme, catalase, by catalyzing the decomposition of H_2O_2 [Eq. (8)], would be expected to retard the rate of oxidation of tetrahydropterin. The inhibition by glucose dehydrogenase of oxygen consumption in the presence of the tetrahydropterin probably resulted from the release of oxygen as shown in Eq. (8). We are unable to reconcile our results with the claim that catalase does not retard the rate of aerobic oxidation of tetrahydropterins (126).

It should be pointed out that Eqs. (5)–(8) do not fully account for the protective effect of catalase. If the tetrahydropterin were oxidized by H_2O_2 exclusively via the quinonoid dihydropterin, one would predict that an excess of dihydropteridine reductase (plus TPNH) would obviate the need for catalase. This would be expected because a sufficient excess of the reductase should be capable of catalyzing the reduction of the quinonoid compound back to the tetrahydro level at such a rate that very little of it would be converted to the inactive 7,8-dihydro compound [Eq. (7)] and the level of tetrahydropterin would not be expected to decrease. Since catalase is required even in the presence of excess reductase, it seems likely that either (a) oxidized products other than the quinonoid dihydropterin are formed, or (b) H_2O_2 is capable of oxidizing some of the tetrahydro-

---

* It has been reported that the rapid oxidation of dihydroalloxazine by H_2O_2 does not proceed in the absence of oxygen (113). This observation has been interpreted as an indication that H_2O_2 does not react directly with the reduced flavin but rather with some oxidation product of it, for example, a semiquinone (113). If this observation is extended to tetrahydropterins, it would indicate that reaction (6) is an oversimplification and that the pterin species which reacts rapidly with H_2O_2 is not XH_4 but rather a semiquinone formed from XH_4 by interaction with oxygen.

pterins to the 7,8-dihydro compound by a pathway not involving the quinonoid dihydropterin as an intermediate.

Recently, another factor has been shown to participate in the hydroxylating system. This new factor, called the phenylalanine hydroxylase stimulator (PHS), has been purified to homogeneity from rat liver and has been shown to be a protein having a molecular weight of 51,000 (96; Huang, Max, and Kaufman, unpublished). Its properties will be discussed in detail in the following section.

C. Studies on the Mechanism of Action of Phenylalanine Hydroxylase

1. Kinetics and Stoichiometric Enzyme Studies

Having established the main components of the phenylalanine hydroxylase reaction, it was of interest to study the mechanism of hydroxylation in greater detail. A kinetic study was undertaken to determine the order of addition of the substrates and whether there were any intermediates in the reaction. Initial velocities were determined under conditions where the concentration of two of the substrates was varied at a constant concentration of the third substrate. As shown in Figs. 2a, b, and c, the three possible combinations of substrates all gave a pattern of intersecting lines when the reciprocal of the concentration of the variable substrate was plotted versus the reciprocal of the initial velocity at several concentrations of the fixed variable substrate (Fisher and Kaufman, unpublished). The results indicate that no product is released from the enzyme before all three substrates have combined with the enzyme; thus, the mechanism is of the sequential rather than ping-pong variety (13). Since raising the concentration of the third substrate did not increase the parallel character of the interesecting lines, it was also concluded that the substrates can add to the enzyme in at least a partially random fashion.

The results of our kinetic analysis of the reaction catalyzed by rat liver phenylalanine hydroxylase do not agree with those obtained by Zannoni et al. (105, 173, 174). With both crude rat liver and mouse liver hydroxylase preparations, they obtained parallel lines in double reciprocal plots of initial rates with phenylalanine and 6,7-dimethyltetrahydropterin (DMPH$_4$),† the variable substrates. They concluded that the mechanism was sequential with ordered addition of substrates. A possible explanation for the discrepancy between their results and ours is the fact that Zannoni et al. limited their analysis to such a narrow range of substrate concentra-

† The following abbreviations were used: DMPH$_4$, 6,7-dimethyltetrahydropterin; dopa, 3,4-dihydroxyphenylalanine; SDS, sodium dodecyl sulfate; PKU, phenylketonuria; 6-MPH$_4$, 6-methyltetrahydropterin; and DTT, dithiothreitol.

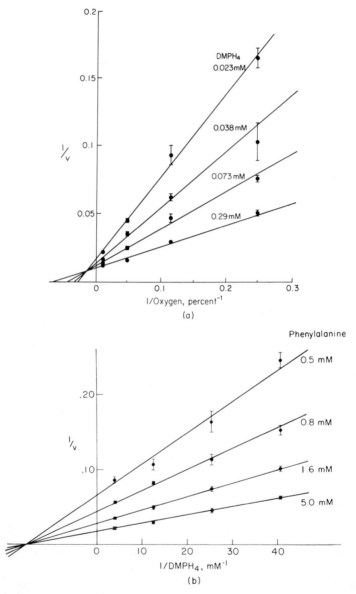

FIG. 2. (a) Double reciprocal plots of initial velocity versus oxygen concentration at several fixed concentrations of DMPH$_4$. The constant substrate was 4 mM phenylalanine. (b) Double reciprocal plots of initial velocity versus DMPH$_4$ concentration at several fixed concentrations of phenylalanine. The constant substrate was oxygen at 21%.

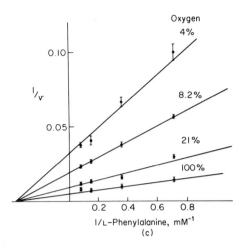

FIG. 2. (c) Double reciprocal plots of initial velocity versus phenylalanine concentrations at several fixed concentrations of oxygen. The constant substrate was DMPH₄ at 0.11 mM. Each value is the mean of three determinations ± standard error of the mean.

tions that it might have been difficult to distinguish between parallel and converging lines.

As will be discussed in greater detail below, tyrosine hydroxylase also appears to proceed by a sequential mechanism involving the formation of a ternary complex of substrates and enzyme. Therefore, in this respect, two pterin dependent hydroxylases, phenylalanine and tyrosine hydroxylase, appear to have a similar mechanism. The mechanism for these hydroxylases is different from the mechanism for dopamine β-hydroxylase, which is a ping-pong type (41) involving an initial reduction of the enzyme by the electron donor, ascorbate, to produce a reduced form of the enzyme and release of dehydroascorbate (37) before the other substrates (dopamine and O_2) add to the enzyme.

The results of experiments carried out with stoichiometric amounts of phenylalanine hydroxylase support the conclusion that a reduced form of the enzyme is not an intermediate in the hydroxylation reaction. The hypothetical reaction which was examined is shown in Eq. (9), where E and EH_2 are the hydroxylase and its reduced form, respectively, and XH_4 and XH_2 are the tetrahydropterin and dihydropterin, respectively.

$$XH_4 + E \rightleftharpoons EH_2 + XH_2 \tag{9}$$

If reaction (9) occurred, the amount of XH_4 oxidized to XH_2 would be a measure of the amount of reduced enzyme which had been formed. Since

XH_2 has an absorption maximum at 330–340 nm, whereas XH_4 does not (94), any conversion of XH_4 to XH_2 could be detected by an increase in the absorbance at 340 nm.

When substrate amounts of phenylalanine hydroxylase were incubated anaerobically with tetrahydropterin, no increase in absorbance ($\pm 10\%$ of expected increase) at 340 nm was observed (Fisher and Kaufman, unpublished). This result proves that phenylalanine hydroxylase cannot be reduced to a major extent by tetrahydropterin in the absence of the other substrates, phenylalanine and oxygen.

Since it has been reported that reduction of certain hydroxylases by their electron donating coenzymes does not occur rapidly in the absence of their substrates (153), attempts were made to detect the reduction of phenylalanine hydroxylase by tetrahydropterin in the presence of phenylalanine but the absence of oxygen. Under these conditions, slight increases in the absorbance at 340 nm occurred, but the increase was less than 10% of the theoretically calculated increase (assuming one active site per 50,000 MW).

The possibility that phenylalanine hydroxylase cannot be reduced by tetrahydropterin unless oxygen is present was ruled out by the finding that the enzyme did not accelerate the rate of oxidation of the reduced pterin. The enzyme also did not rapidly oxidize an equivalent of tetrahydropterin in the presence of oxygen. These results taken together indicate that if enzyme reduction occurs, both oxygen and phenylalanine must be combined with the enzyme. The possibility that the enzyme, as isolated, was already in the reduced form was eliminated by the demonstration that aerobic incubation of the enzyme with phenylalanine did not produce any tyrosine.

2. Use of Alternate Substrates

The mechanism has also been investigated with the use of alternative substrates. It was found that when p-fluorophenylalanine is used in place of phenylalanine, the ratio of TPNH oxidized to tyrosine produced is 3–4:1 instead of the predicted 2:1 (one equivalent of TPNH utilized in fluorine reduction) (85). Also, when tryptophan was used as the amino acid substrate and 6-methyltetrahydropterin as the cofactor, a ratio of 3:1 was found (150). Furthermore, when 7-methyltetrahydropterin was substituted for the tetrahydrobiopterin and phenylalanine was the amino acid substrate, the ratio of TPNH (or tetrahydropterin) oxidized to tyrosine formed was greater than the 1:1 found with the natural cofactor (Table I). We have used the term "uncoupled" or "loosely coupled" to describe the hydroxylation reaction when there is more tetrahydropterin

TABLE I

EFFECT OF PEROXIDASE ON THE STOICHIOMETRY OF
PHENYLALANINE HYDROXYLASE

		TPNH oxidized/tyrosine formed		
		Without peroxidase	With peroxidase	
Substrate	Tetrahydropterin		Theory	Found
Phenylalanine	6-Methyl	0.99	1.00	1.11
Phenylalanine	7-Methyl	2.96	4.92	5.31
4-F-phenylalanine	6-Methyl	3.44	4.88	5.11

oxidized than tyrosine formed, and "tightly coupled" when the ratio of tetrahydropterin oxidized and tyrosine formed is close to unity (102).

Phenylalanine must be present and has the same K_m for this excess oxidation as for tyrosine production, indicating that the uncoupled reaction is related to the hydroxylation reaction mechanism (150). This result also supports the findings with stoichiometric levels of enzyme that all three substrates must be on the enzyme before the flow of electrons from tetrahydropterin to oxygen can occur.

Molecular oxygen is the most likely acceptor for the extra electrons being consumed under conditions where this altered stoichiometry has been observed. The oxygen could be reduced to either H_2O or H_2O_2. The available evidence strongly supports the conclusion that O_2 is the acceptor for the extra electrons and that O_2 is reduced to H_2O_2.

If, under conditions where the ratio of TPNH oxidized to tyrosine formed is 3:1 and all the extra electrons consumed were used to reduce O_2 to H_2O_2, each 3 moles of XH_4 oxidized would lead to the formation of 1 mole of tyrosine, 1 mole of H_2O, and 2 moles of H_2O_2 in accordance with Eq. (10) where RH stands for phenylalanine and ROH for tyrosine.

$$3 XH_4 + RH + 3 O_2 \rightarrow 3 XH_2 + ROH + H_2O + 2 H_2O_2 \tag{10}$$

Preliminary experiments showed that peroxidase catalyzes the H_2O_2-mediated oxidation of XH_4 to quinonoid dihydropterin (XH_2) as shown in Eq. (11).

$$2 H_2O_2 + 2 XH_4 \rightarrow 2 XH_2 + 4 H_2O \tag{11}$$

In the presence of peroxidase, therefore, the expected TPNH:tyrosine ratio would be 5:1 as shown in Eq. (12) which is the sum of Eqs. (10) and (11).

$$5 XH_4 + RH + 3 O_2 \rightarrow 5 XH_2 + ROH + 5 H_2O \tag{12}$$

The results of such an experiment are shown in Table I, where it can be seen that in the presence of peroxidase ratios close to the expected ones were found (150). The finding that this action of peroxidase was blocked by excess catalase further supports the conclusion that H_2O_2 is produced under uncoupling conditions (Fisher and Kaufman, unpublished).

Additional evidence in favor of the idea that H_2O_2 is the product of oxygen reduction was obtained by determinations of the ratio of oxygen consumed to tetrahydropterin oxidized in the presence of catalase. One can calculate the expected ratio under these conditions by adding the reaction catalyzed by catalase [Eq. (13)] to Eq. (10). The sum of these two reactions is given in Eq. (14).

$$2 H_2O_2 \rightarrow O_2 + 2 H_2O_2 \tag{13}$$

$$3 XH_4 + RH + 2 O_2 \rightarrow 3 XH_2 + ROH + 3 H_2O \tag{14}$$

It can be seen that in the presence of catalase, the expected $O_2:XH_4$ ratio is 0.67. When the hydroxylation was carried out in the presence of 7-methyl-

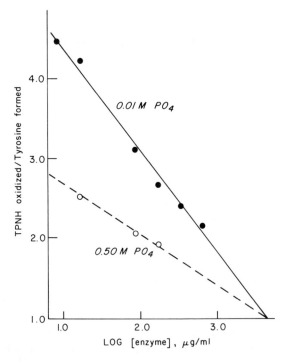

Fig. 3. Effect of enzyme concentration on the ratio of TPNH oxidized to tyrosine formed. The cofactor was 7-methyltetrahydropterin.

tetrahydropterin and catalase, the observed $O_2:XH_4$ ratio varied between 0.67 and 0.75, a range of values that is in fair agreement with the expected value of 0.67 (Hall and Kaufman, unpublished).

By varying the pterin structure it was possible to determine the structural requirements for tight coupling of pterin oxidation and tyrosine production. In addition to the natural cofactor, the 6-methyl- and 6,7-dimethyltetrahydropterin show tight coupling. In contrast, the 7-methyl-, 7-phenyl- or unsubstituted tetrahydropterin exhibit loose coupling. Therefore, the presence of an alkyl group at the 6 position is necessary for tight coupling to occur with phenylalanine hydroxylase. It is of interest to compare those results with a similar study of bovine adrenal tyrosine hydroxylase. As will be discussed later in more detail, with this enzyme the reaction is tightly coupled with tetrahydrobiopterin, 6-methyltetrahydropterin, and tetrahydropterin, whereas there is partial uncoupling with the 7-methyltetrahydropterin and 6,7-dimethyltetrahydropterin (143). Thus, the presence of an alkyl group in the 7 position causes a partial uncoupling of bovine adrenal tyrosine hydroxylose which cannot be prevented by a substitution at the 6 position. These results demonstrate a subtle difference between the active sites of tyrosine and phenylalanine hydroxylase.

The following scheme has been presented to explain these results (150):

$$E + XH_4 + RH + O_2 \rightarrow E(XH_4, O_2, RH) \tag{15}$$

$$E(XH_4, O_2, RH) \rightarrow E(XH_2, O_2^{2-}, RH) \tag{16}$$

$$E(XH_2, O_2^{2-}, RH) \begin{cases} \xrightarrow{a} E + XH_2 + RH + H_2O_2 \\ \xrightarrow{b} E + XH_2 + ROH + H_2O \end{cases} \tag{17}$$

According to this model, the relative rates of reactions a and b determine the degree of coupling of electron consumption to tyrosine (ROH) prohuction.

With 7-methyltetrahydropterin as the cofactor, the TPNH:tyrosine ratio may be lowered by increasing the enzyme concentration (Fig. 3), decreasing the reaction temperature, or increasing the pH (29). Presumably, these conditions favor reaction b in Eq. (17) over reaction a. At very low enzyme concentrations, where the ratio is nearly 5:1 (Fig. 3), phenylalanine hydroxylose is acting primarily as a phenylalanine-dependent, tetrahydropterin oxidase. Thus, the line between mixed function oxidases and classic oxidases is a fine one. Indeed, phenylalanine hydroxylase may have evolved from a pterin oxidase (97).

Subsequent to the discovery of the uncoupling with phenylalanine hydroxylase, it has been shown that severol other aromatic hydroxylases can also catalyze uncoupled reactions. A flavoprotein, salicylate hydroxylase, displays a 1:1 ratio of DPNH oxidized to salicylate hydroxylated. Substitution of benzene for salicylate results in complete uncoupling of oxygen reduction from benzene hydroxylation. As was found with phenylalanine hydroxylase, under uncoupled conditions the oxygen was reduced to hydrogen peroxide by salicylate hydroxylase (168, 169). Another flavoprotein, imidazoleacetate hydroxylase, contains 2 moles of titratable sulfhydryls per mole of enzyme. When these sulfhydryls are titrated with p-chloromercuribenzoate or silver nitrate, DPNH oxidation is uncoupled from imidazoleacetate hydroxylation. As was found with the other hydroxylases, the reaction produces hydrogen peroxide under conditions of uncoupling (129). Two other flavin hydroxylases, p-hydroxybenzoate hydroxylase (60) and orcinol hydroxylase (139), also show this uncoupling phenomenon. Therefore, the ability to reduce oxygen without hydroxylation of substrate appears to be a general property of pteridine and flavin hydroxylases. These findings also suggest that flavin and pteridine hydroxylases might activate oxygen by a similar mechanism. This possible similarity in mechanism may result from the structural resemblance between flavins and pterins. Indeed, the primary oxidation product of the pterin-dependent hydroxylation, p-quinonoid dihydropterin, has the same structure as the two heterocyclic rings of the oxidized form of flavin nucleotides. This parallel between flavin nucleotides and pteridines is further emphasized by the fact that the isoalloxazine form of flavins is one of the few compounds other than the quinonoid dihydropterin that will oxidize TPNH (145). Enzymically, flavins can act as mediators between a two-electron donor (pyridine nucleotide) and a one-electron acceptor system (cytochrome c) with the intermediate formation of a relatively stable free radical of the semiquinoid type. Perhaps the pterin plays a similar role in mediating the transfer of electrons from a reduced pyridine nucleotide to the enzyme-bound iron (phenylalanine hydroxylase is an iron enzyme as discussed below) with the formation of a pterin semiquinone as an intermediate.

3. Possible Role of a Pterin Hydroperoxide or Superoxide Radical

One of the most important aspects of mechanism of hydroxylation is the means by which oxygen is activated. Mager and Berends (112, 113) have isolated the autoxidation product of dihydroisoalloxazine. From the structure of this product, they have deduced that the autoxidation of dihydroisoalloxazine initially involves the formation of a hydroperoxide.

Based on the analogy between the heterocyclic rings of isoalloxazine and pteridines, Mager and Berends have suggested that a pterin hydroperoxide might participate in the enzymic hydroxylation reactions involving pterin cofactors. It is possible that this pterin hydroperoxide could break down to form hydrogen peroxide and dihydropterin under the conditions of loose coupling in the presence of an analog of the substrate or cofactor.

A new form of oxygen, a superoxide radical $O_2^- \cdot$, has been shown to be an intermediate in the xanthine oxidase–catalyzed reduction of cytochrome c (115). This conclusion was based, in part, on the finding that superoxide dismutases, which specifically break down superoxide radicals, inhibit the reduction of cytochrome c by xanthine oxidase. To test for the possible participation of the superoxide radical in phenylalanine hydroxylase reaction, the effect of the superoxide dismutase, erythrocuprein, on the hydroxylation reaction was examined. Erythrocuprein did not inhibit the phenylalanine hydroxylase reaction, suggesting that either the superoxide radical is not involved, or if it is, the dismutase is not accessible to it. By contrast, erythrocuprein inhibited the rate of autoxidation of tetrahydropterin by 50%. Therefore, part of the autoxidation reaction involves the participation of the superoxide radical (Holl, Fisher, and Kaufman, unpublished).

4. Hydroxylation-Induced Migration of Para Substituents of Phenylalanine

There is indirect evidence that an epoxide of phenylolanine might be an intermediate in phenylalanine hydroxylation. It has been demonstrated that phenylalanine hydroxylase causes a migration of tritium from the para to the meta position as a result of the hydroxylation reaction (51). This phenomenon has been termed the "NIH shift" and has been observed also when deuterium, chloride, bromide, or alkyl groups are originally present in the para position. This ability to cause intramolecular migration of a para substituent has been used as a criterion for evaluating the conformity of model chemical hydroxylating systems with enzymic hydroxylation. m-Chloroperoxybenzoic acid and peroxytrifluoracetic acid are model hydroxylating agents which cause the NIH shift (70). Studies with these reagents are consistent with a mechanism that involves electrophilic attack of the aromatic ring and, as shown in Fig. 4, the formation of an arene oxide. The arene oxide rearranges by a hydride shift forming a cationoid species (71). The deuterium or tritium is retained in preference to hydrogen because of their greater carbon bond strengths. In support of this postulated mechanism, deuterated oxides have been shown to rearrange to form phenols with migration and retention of the

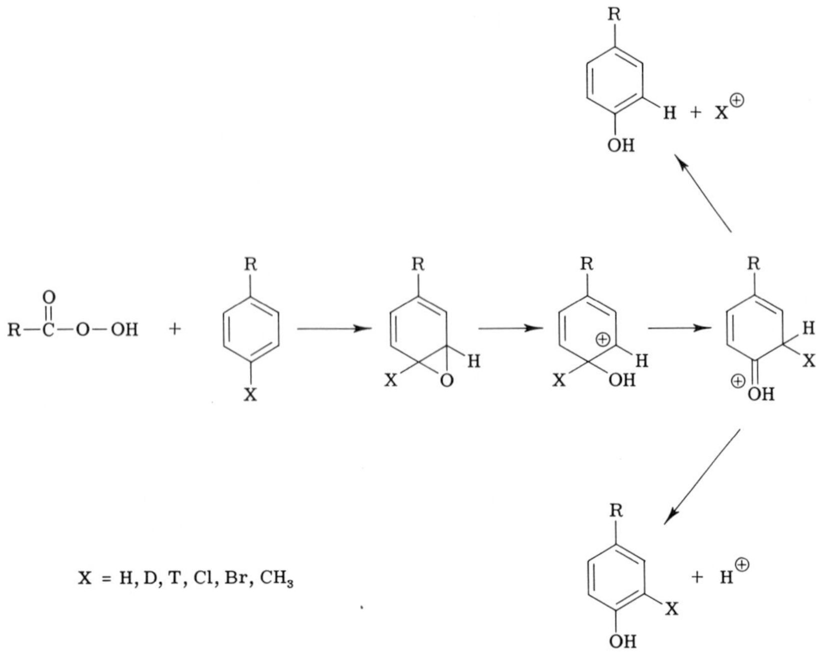

FIG. 4. Hypothetical mechanism for the NIH shift produced by model hydroxylating reagents such as *m*-chloroperoxybenzoic acid.

deuterium to the same extent as occurs enzymically (5). Furthermore, naphthalene oxide has been isolated as an intermediate in the hydroxylation of naphthalene (72). It should be emphasized that although this epoxide intermediate might be a good model for microsomal hydroxylations, the only evidence to favor this mechanism for phenylalanine hydroxylase is the fact that this enzyme induces the NIH shift in the same manner as do oxidizing reagents that lead to the formation of epoxides.

5. Kinetic Evidence for an Intermediate in the Hydroxylation of Phenylalanine

Recent kinetic studies in this laboratory lend support to the hypothesis that there is an intermediate formed during the hydroxylation catalyzed by phenylalanine hydroxylases. In the presence of less-than-saturating concentrations of tetrahydrobiopterin, and at high pH values (i.e., pH above 8), the specific activity of the hydroxylase has been found to decrease with increasing concentrations of hydroxylase (96). [This anomolous kinetic behavior is not seen when $DMPH_4$ is used in place of tetrahydrobiopterin (96).] Under these conditions, a protein isolated from rat liver extracts, PHS, markedly accelerates the rate of tyrosine formation; in the presence of PHS, the specific activity of the hydroxylase even at high pH values and low tetrahydrobiopterin concentrations is essentially constant. The phenylalanine hydroxylase stimulator has been purified to homogeneity. It is relatively heat stable and has a molecular weight of 51,000 ($\pm 2\%$) (Huang, Max, and Kaufman, unpublished).

Further investigation of this phenomenon led to two findings that shed some light on the mechanism of action of PHS as follows:

1. Under conditions of PHS stimulation (see above), but in the absence of PHS, the apparent K_m for tetrahydrobiopterin increases linearly with hydroxylase concentration, whereas in the presence of saturating amounts of PHS, the K_m for tetrahydrobiopterin is independent of hydroxylase concentration (see Fig. 5) (Huang and Kaufman, unpublished).

2. By measuring the amount of PHS (4 μg) necessary for full activation of purified phenylalanine hydroxylase (46 μg), as shown in Fig. 6, it was estimated that one PHS molecule can activate 10 hydroxylase molecules (Huang, Max and Kaufman, unpublished). This result suggests that PHS acts catalytically and may therefore be an enzyme.

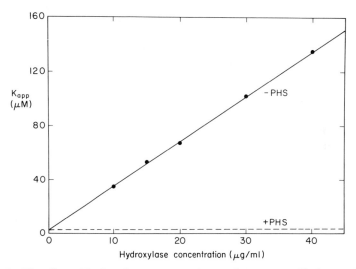

FIG. 5. The effect of hydroxylase concentration on the apparent K_m for tetrahydrobiopterin at pH 8.0. Note that in the presence of PHS the apparent K_m does not increase with hydroxylase concentration.

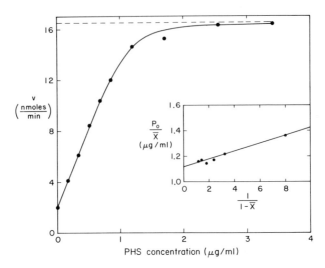

FIG. 6. Effect of increasing pure PHS concentration on the rate of tyrosine production by 46 μg of 90% phenylalanine hydroxylase.

To explain these results with PHS, the following scheme has been proposed for the mechanism of the hydroxylation reaction (Huang and Kaufman, unpublished), where S stands for tetrahydroblopterin:

$$E + S \rightarrow ES \underset{k_{-2}}{\overset{k_2}{\rightleftarrows}} ES' \overset{k_3}{\rightleftarrows} E + S' \rightarrow P \qquad (18)$$

This pathway can be described by the following kinetic equation:

$$\frac{v}{E_0} = \frac{k_2 S}{S + K_m[(1 + k_{-2}E_0)/k_3]} \qquad (19)$$

The essential features of this model can be described as follows: There is a free intermediate in the hydroxylation reaction, S', which can break down to product, P, either nonenzymically or enzymically; PHS is the enzyme that catalyzes the latter reaction. In the absence of PHS and at high enzyme and low S concentrations, S' competes with S; as a result of this competition, the rate of P formation is inhibited. According to this scheme, the anomolous protein concentration dependence of the specific activity of the hydroxylase is not seen at lower pH values either because k_3 is higher or k_{-2} is lower at pH 6.8 than at pH 8.0–8.2.

It can be seen that Eq. (19) predicts that when k_3 is much greater than k_{-2}, the apparent K_m for tetrahydroblopterin would be independent of

enzyme concentration, but at high E_0, the apparent K_m for tetrahydrobiopterin would increase linearly with enzyme concentration. The results
shown in Fig. 5 are in full accord with this prediction.

It is possible to draw some restrictive conclusions about the structure
of S′, the proposed intermediate in the hydroxylation readtion. Based on
the kinetic data obtained with tetrahydrobiopterin as the cofactor and
phenylalanine as the substrate, S′ could be a pterin–oxygen–phenylalanine complex, or a more immediate precursor of the quinonoid dihydrobiopterin or of tyrosine.

As already discussed, there is evidence suggestive of all epoxide intermediate in some aromatic hydroxylation reactions. We have therefore
considered the possibility that S′ is, in fact, the epoxide of phenylalanine.
Our kinetic results, however, are not compatible with this possibility. If
S′ were a simple epoxide of phenylalanine, the same S′ should be formed
with the same rate of nonenzymic breakdown to tyrosine regardless of the
type of tetrahydropterin used. To explain the fact that pronounced enzyme
concentration dependence of the specific activity of the hydroxylase is
only observed with tetrahydrobiopterin as the cofactor, one must assume
that the epoxide of phenylalanine behaves as a product inhibitor whose
binding to the enzyme is greatly enhanced by the presence of tetrahydrobiopterin but not by $DMPH_4$. But such an assumption is contradictory
to the observation that a saturating level of tetrahydrobiopterin eliminates
the enzyme concentration dependence. It may be concluded, therefore,
that whether or not an epoxide of phenylalanine is an obligatory precursor
of tyrosine, it is not the S′ that gives rise to the protein concentration
dependence of the specific activity of phenylalanine hydroxylase. Rather,
to account for the pterin specificity, S′ must be either a precursor of
quinonoid dihydrobiopterin or a pterin–oxygen–phenylalanine complex.

6. Involvement of Iron in the Hydroxylating Reaction

Recent studies have demonstrated that phenylalanine hydroxylase is
an iron enzyme (32). The first indication that a metal might be involved
in the enzymic reaction was the finding that metal chelators such as
o-phenanthroline (5×10^{-5} M) and 8-hydroxyquinoline (2×10^{-4} M)
produced 100% inhibition of the hydroxylase. Sucrose gradient centrifugation of 90% pure hydroxylase showed that a peak of iron coincided
with the activity and protein peak for the enzyme (Fig. 7). The 95% pure
phenylalanine hydroxylase obtained from a sucrose gradient was found to
contain nearly one atom of iron per 50,000 molecular weight subunit.
Treatment of the enzyme with o-phenanthroline and cysteine removed
50–80% of the iron and decreased the enzymic activity to nearly the

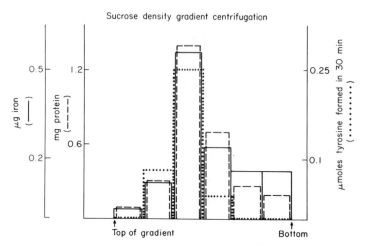

Fig. 7. An activity, protein, and iron profile of 90% pure phenylalanine hydroxylase which was subjected to sucrose gradient centrifugation: (—) iron, (– – –) protein, and (. . .) tyrosine.

same extent (Table II). Addition of $FeCl_2$ to this apoenzyme led to full restoration of enzymic activity. Mercurous, nickelous, cobalt, manganese, cupric, chromium, cadmium, and zinc ions were ineffective in restoring enzymic activity to the o-phenanthroline-treated enzyme (32). Therefore, the metal requirement for rat liver phenylalanine hydroxylase is highly specific for iron. This finding contrasts with the report that *Pseudomonas* phenylalanine hydroxylase is activated by mercuric, cadmium, cupric and cuprous, as well as ferrous ions (52). The finding that phenylalanine

TABLE II

IRON CONTENT AND ACTIVITY OF 95% PURE PHENYLALANINE HYDROXYLASE
BEFORE AND AFTER o-PHENANTHROLINE TREATMENT

Pretreatment	Addition to assay	Enzymic activity (% control)	Iron content (% control)
None	None	100	100
None	$FeCl_2$	120 ± 10[a]	
o-Phenanthroline	None	26 ± 5	35 ± 13
o-Phenanthroline	$FeCl_2$	105 ± 15	

[a] Mean of three experiments.

hydroxylase is a metalloenzyme contrasts with the report that several FAD-dependent hydroxylases contain neither iron nor copper (172).

In an attempt to determine whether iron is involved in the enzyme-catalyzed reaction or merely playing a structural role, ESR spectroscopy was employed. The highly purified phenylalanine hydroxylase (95% pure) gave a distinct signal at $g = 4.23$ in the presence of oxygen [Fig. 8, (32)]. A signal at this g value is characteristic of a high spin ferric ion. The high spin state of the d electrons of the ferric ion results from a small degree of splitting of the two groups of $3d$ orbitals. It is likely that this small degree of splitting is the result of a greater ionic than covalent binding between the iron and its ligands. When phenylalanine and DMPH$_4$ were added, 90% of the signal at $g = 4.23$ disappeared (Fig. 8), an indication that this iron is involved in the enzymic catalysis. The disappearance of the $g = 4.23$ ESR signal on addition of the substrates is consistent with the reduction of Fe^{3+} to Fe^{2+}. In light of the kinetic and stoichiometric enzyme studies cited above, however, this reduction of iron may not be a distinct step but rather a transient event occurring during a concerted mechanism. Such a mechanism would be in accord with Hamilton's pro-

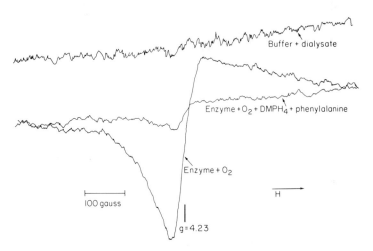

FIG. 8. ESR spectra of phenylalanine hydroxylase with and without its substrates. Three milligrams of sucrose gradient purified phenylalanine hydroxylase were quick frozen in liquid nitrogen and its ESR spectrum taken. The enzyme was thawed and phenylalanine and DMPH$_4$ were added. The enzyme was incubated 1 minute at 25°. Then the sample was refrozen in liquid nitrogen and the enzyme + O$_2$ + DMPH$_4$ + phenylalanine spectrum was taken. A spectrum of the same volume of dialysate and buffer was also taken as a blank.

posal that iron might form a link for the flow of electrons from the tetra-hydropterin to oxygen (56). Alternatively, the disappearance of the $g = 4.23$ signal upon addition of substrates may result from a change of the Fe^{3+} from a high to a low spin state caused by an alteration of the metal's ligand field.

Studies with model hydroxylation systems have suggested that iron may be more directly involved in the activation of oxygen. Oxotransition metals such as ferrate and manganate ions can catalyze the hydroxylation of olefins (142). Manganate ions, furthermore, promote the hydroxylation of aromatic compounds resulting in the NIH shift characteristic of the phenylalanine hydroxylase reaction. These findings have suggested that an oxoiron (Fe = O) species might be the active oxidant in hydroxylation reactions (142).

D. Substrates and Inhibitors of Phenylalanine Hydroxylase

1. Amino Acid Substrates

The enzyme requires an amino acid substrate with an unmodified side chain attached to an aromatic ring. The following compounds which have a phenyl group attached to an altered alanine side chain were inactive as substrates: D-phenylalanine, glycyl-DL-phenylalanine, β-phenyllactate, α-phenyl-α-alanine, benzylmalonate, phenylpyruvate, acetaminocinnamic acid, phenylglycine, β-phenylserine, phenylethylamine, and 1-phenyl-2-acetaminobutanone-3 (86). The aromatic ring, however, can be altered in a number of ways and, as long as the alanine side chain is intact, the molecule can serve as a substrate. The following compounds of this type have been shown to be substrates for phenylalanine hydroxylase: trypto-phan (33), β-2-thienylalanine (86), 4-chlorophenylalanine, 2-fluorophenyl-alanine, 3-fluorophenylalanine, 4-fluorophenylalanine (85), and p-methyl-phenylalanine (19). The products of the reaction with 4-fluorophenyl-alanine are tyrosine and F^- (85). The reaction rate with 4-fluorophenyl-alanine is 15% that observed with phenylalanine. As mentioned earlier, alternate substrates, such as 4-fluorophenylalanine and tryptophan, partially uncouple the hydroxylation reaction. 4-Chlorophenylalanine also leads to partial uncoupling. Indeed, nearly every amino acid substrate other than phenylalanine leads to a partially uncoupled reaction and has a reaction velocity much slower than the rate with phenylalanine. Tryp-tophan hydroxylation was originally reported for only an extract and a partially purified phenylalanine hydroxylase with the synthetic cofactor, $DMPH_4$ (33, 138). We have recently found that highly purified (90%

pure) rat liver phenylalanine hydroxylase with the natural cofactor, tetrahydrobiopterin, will hydroxylate tryptophan to 5-hydroxytryptophan at very high tryptophan concentrations (greater than 1 mM, Fig. 9). No reaction was detectable at serum levels of tryptophan (0.1 mM). When a substance which activates phenylalanine hydroxylase, lysolecithin, was added, significant tryptophan hydroxylation could be measured at serum levels of tryptophan. This rate, however, is only 0.1% of the rate of phenylalanine hydroxylation under the same conditions (Fisher and Kaufman, unpublished experiments). It is unlikely, therefore, that there is significant tryptophan hydroxylation by hepatic phenylalanine hydroxylase *in vivo*.

In contrast to an earlier finding (86), it has recently been reported that phenylalanine hydroxylase will catalyze the hydroxylation of *m*-tyrosine to form dopa at 20% the rate for phenylalanine (155) in the presence of DMPH$_4$. In confirmation of our earlier results, we have found (Fisher and Kaufman, unpublished) that *m*-tyrosine is hydroxylated at only 2% or less of the rate for phenylalanine when either DMPH$_4$ or tetrahydrobiopterin were used as cofactors. In the presence of lysolecithin, however, partially purified or 90% pure phenylalanine hydroxylase catalyzed the hydroxylation of *m*-tyrosine at 20% the rate of phenylalanine hydroxyla-

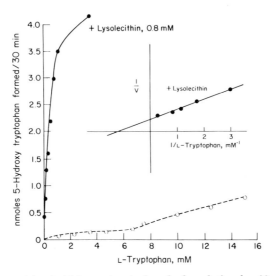

FIG. 9. Effect of lysolecithin on tryptophan hydroxylation by 60% pure phenylalanine hydroxylase. The cofactor was tetrahydrobiopterin at 0.025 mM.

tion (Table III). This reaction with m-tyrosine was completely coupled since 1 mole of dopa was formed for each mole of TPNH oxidized. The finding that lysolecithin preferentially increases m-tyrosine hydroxylation may explain the discrepancy between the results of Tong et al. and ourselves. Their enzyme preparation may have contained some lysolecithin or a similarly acting substance. We also tested the product of phenylalanine hydroxylase, p-tyrosine, as a substrate with and without lysolecithin. Without lysolecithin there was barely detectable TPNH oxidation and no dopa formation. When lysolecithin was added, however, p-tyrosine increased the rate of TPNH oxidation up to 10% of that observed with

TABLE III

COMPARISON OF PHENYLALANINE, m-TYROSINE, AND p-TYROSINE AS SUBSTRATES
FOR PHENYLALANINE HYDROXYLASE WITH AND WITHOUT LYSOLECITHIN

Substrate	Enzyme	Additions	Initial rate of TPNH oxidation (nmoles/min)	Dopa formed (nmoles)
L-Phenylalanine	First ammonium sulfate fraction	None	2.5	
	First ammonium sulfate fraction	Lysolecithin	13.0	
DL-m-Tyrosine	First ammonium sulfate fraction	None	0.0	
	First ammonium sulfate fraction	Lysolecithin	2.0	
L-Phenylalanine	Peak fraction of Sephadex G-200	None	5.0	
	Peak fraction of Sephadex G-200	Lysolecithin	27.0	
DL-m-Tyrosine	Peak fraction of Sephadex G-200	None	0.1	0.0[a]
	Peak fraction of Sephadex G-200	Lysolecithin	4.3	28
L-p-Tyrosine	Peak fraction of Sephadex G-200	None	0.1	0.0[b]
	Peak fraction of Sephadex G-200	Lysolecithin	2.4	0.0

[a] Dopa formed in 13 minutes.
[b] Dopa formed in 22 minutes.

phenylalanine. Even under these conditions there was still no dopa formation from tyrosine. Furthermore, there was no consumption of tyrosine.
These results indicate that in the presence of lysolecithin, p-tyrosine
completely uncouples electron consumption from hydroxylation.

The finding that in the presence of lysolecithin, tryptophan and p-tyro-
sine can serve as substrates for highly purified phenylalanine hydroxylase
suggests that there is a basic similarity between this hydroxylase and the
other two pterin-dependent hydroxylases, tryptophan and tyrosine
hydroxylases. As will be discussed below, it appears that lysolecithin
causes a subtle conformation change in the phenylalanine hydroxylase
structure. A similar conformational change (produced by either a different primary structure or some nonprotein conjugate such as a lipid or
carbohydrate) might differentiate tryptophan and tyrosine hydroxylases
from phenylalanine hydroyxlase.

2. Tetrahydropterin

Phenylalanine hydroxylase requires a reduced pterin as a source of
electrons in the hydroxylation reaction. No other class of reducing agents
has been found to be active. Ascorbate, dihydroxyfumarate, glutathione,
2-mercaptoethanol, and cysteine are all inactive (81).

The enzyme poorly tolerates any change in the structure of the pyrimidine ring of the pterin. The 2-amino-4-hydroxy configuration is essential;
the pterin, 2-hydroxy-4-amino-6-methyltetrahydropteridine, in which
these groups are reversed, is inactive (100).

Alkylation of the 2-amino or 8-nitrogen group leads to a severe or total
loss in cofactor activity. 2-Methylamino-4-hydroxy-6,7-dimethyltetra-
hydropteridine is one-third as active as the parent compound and both
2-dimethylamino-4-hydroxy-6,7-dimethyltetrahydropteridine and 2-ami-
no-4-hydroxy-6,7-dimethyl-8-hydroxyethyltetrahydropteridine   are   inactive (90).

Although the unsubstituted compound, tetrahydropterin,‡ has some
cofactor activity (80), alkylation of either the 6 or 7 carbon greatly enhances the cofactor activity. In descending order of activity the following
pterins in their tetrahydro form have been shown to have cofactor activity: 6-methylpterin, 6,7-dimethylpterin, 7-methylpterin, and   pterin.
Tetrahydrofolic acid was the first compound of known structure shown
to be active as a cofactor for phenylalanine hydroxylase (80). Recently,
however, it has been shown that at least 70% of tetrahydrofolate's activity results from a small amount of 6-methyltetrahydropterin present
in the tetrahydrofolate (107). Indeed, even the remaining 30% of the

‡ Pterin refers to the oxidized form of 2-amino-4-hydroxypteridine.

activity may result from breakdown of folate during its isolation. Thus, it remains to be seen whether conjugated pterins are active as cofactors with mammalian phenylalanine hydroxylase.

The natural cofactor isolated from rat liver is 7,8-dihydrobiopterin [7,8-d i h y d r o-2-a m i n o-4-h y d r o x y-6-[1,2-dihydroxypropyl-(L-*erythro*)] pteridine (89)]. There is some evidence, however, that in the liver it occurs predominantly in the tetrahydro form (137) and that it is oxidized to the dihydro compound during its isolation (89). Many of the earlier studies were carried out with 7,8-dihydrobiopterin prepared by chemical reduction (94) or with the 5,6,7,8-tetrahydrobiopterin prepared by catalytic hydrogenation (94). Carbon 6 of the pterin ring of tetrahydrobiopterin is asymmetric. The compound prepared by catalytic hydrogenation therefore, is a mixture of two diastereoisomers. To determine if the two isomers are active, the hydroxylation reaction was carried out in the absence of a tetrahydropterin-regenerating system. It was found that nearly 1.0 μmole of tyrosine was formed per μmole of DL-tetrahydrobiopterin (96). This result proves that both optical isomers are active with phenylalanine hydroxylase.

There are also two diastereoisomers of the dihydroxypropyl side chain. Recently, it has been demonstrated that the L-*erythro* form of tetrahydrobiopterin is four times more active than the D-*erythro* form (131).

Besides tetrahydrobiopterin, two other naturally occurring unconjugated pterins have high cofactor activity: tetrahydroneopterin (2-amino-4-hydroxy-6-[1,2,3-trihydroxypropyl (6-*erythro*)] tetrahydropteridine) and sepiapterin (2-amino-4-hydroxy-6-lactyl-7,8-dihydropteridine) (87). This latter compound, however, shows high activity only after its conversion to 7,8-dihydrobiopterin (114), a reaction catalyzed by sepiapterin reductase in the presence of TPNH.

3. Inhibitors

The most potent inhibitors of phenylalanine hydroxylase are catechols. Several catechol intermediates involved in the conversion of tyrosine to epinephrine, dopa, dopamine, and norepinephrine inhibit the enzyme with K_i values of about 10^{-5} M (10). Since hydrophobic metal chelating agents are also potent inhibitors, the inhibitory activity of catechols may result from their metal chelating properties. Halogenated phenylalanines such as the *para*-chloro (38) and *para*-fluoro derivatives (10) are moderate inhibitors of phenylalanine hydroxylation (K_i about 10^{-4} M). It is of interest to compare the inhibitors of phenylalanine, tyrosine, and tryptophan hydroxylase. All three enzymes are inhibited by catechols such as dopa and norepinephrine with similar K_i values of about

10^{-5} M (10, 69, 165). Halogenated amino acids, however, show selective inhibition. 3-Iodotyrosine, at 10^{-4} M, produces greater than 95% inhibition of bovine adrenal tyrosine hydroxylase (165), but it has no effect on rat liver phenylalanine hydroxylase (Fisher and Kaufman, unpublished). 6-Fluorotryptophan, at 10^{-3} M, causes 50% inhibition of brain tryptophan hydroxylase (Kappelman and Kaufman, unpublished) but inhibits rat liver phenylalanine hydroxylase no more than 5% (Fisher and Kaufman, unpublished). para-Chlorophenylalanine, however, is an equally good inhibitor of all three pterin-dependent hydroxylases.

E. Molecular Properties of Phenylalanine Hydroxylase from Rat Liver

Phenylalanine hydroxylase from rat liver has recently been obtained in a state of high purity. The procedure involves ethanol and ammonium sulfate precipitations, adsorption and elution from calcium phosphate gel, chromatography on DEAE-cellulose and gel filtration on Sephadex G-200 (99). By this procedure the enzyme was purified 400-fold with a 5% yield. This purified hydroxylase has a turnover number of 25 (μmoles tyrosine formed/minute/μmole hydroxylase) when assayed at 25° with K_m levels of 6,7-dimethyltetrahydropterin, phenylalanine, and oxygen. This turnover number is based on the assumption that there is one active site per 50,000 molecular weight subunit of the enzyme (see below).

By three criteria, the hydroxylase obtained by this procedure is 85–90% pure. In the standard polyacrylamide gel electrophoresis system the enzymic activity is associated with two major bands which comprise 85% of the total protein. Sucrose gradient centrifugation resolves the hydroxylase into two peaks of activity which represent 90% of the protein. Electrophoresis of the enzyme on SDS–polyacrylamide gel gives two major bands which contain 85% of the protein. An SDS-gel pattern of the sucrose gradient–purified enzyme shows the same two major bands which represent 95% of the protein on the gel.

1. Molecular Weight of the Enzyme

An analysis of the molecular weight of phenylalanine hydroxylase showed that the enzyme exists in several different polymeric forms. Gel filtration of the enzyme on Sephadex G-200 gives two peaks of activity. The major peak is more excluded and has a Stokes radius of 6.3 nm and the minor peak has a Stokes radius of 4.5 nm. Sucrose gradient centrifugation also gives two peaks of activity with the major form being the slower sedimenting form. The slower sedimenting form has a sedimentation constant of 5.9 S, and the more rapidly sedimenting form has a value of

8.15 S. The sedimentation constants, Stokes radii, and partial specific volume (determined from amino acid analysis to be 0.72) were used to calculate molecular weights of 110,000 and 210,000 for the two major forms of the enzyme (99).

It is not clear why the larger molecular weight form predominates on Sephadex and the smaller molecular weight form is most prevalent on the sucrose gradient. The difference in molecular weight distribution is probably not due to the sucrose because equilibrium centrifugation shows a majority of the enzyme in the lower molecular weight form (a molecular weight of 90,000 was obtained by this method, Fisher and Kaufman, unpublished). The difference in distribution among the larger and smaller forms may, however, be because we take the lower molecular weight form from Sephadex for further analysis since it is more highly purified than the higher molecular weight form.

Recently, it has been reported that incubation of rat liver phenylalanine hydroxylase with phenylalanine caused the majority of activity in sucrose gradient centrifugation to shift from the lower to the higher molecular weight form (157). It was suggested that this increase in molecular weight might accompany the increase in activity seen when the enzyme is incubated with phenylalanine. We have confirmed this finding and in addition found even higher molecular weight forms of the enzyme (greater than 210,000) in the presence of phenylalanine (Fisher and Kaufman, unpublished).

The molecular weights of the polypeptide chains of the hydroxylase were determined by carrying out nonstacking polyacrylamide gel electrophoresis in the presence of SDS. A broad, major band accounting for 95% of the total protein was observed to have a molecular weight of 51,000 (99). More recently, sucrose gradient–purified hydroxylase was electrophoresed by the stacking SDS method developed by Neville (124). Under these conditions, two closely migrating bands, accounting for at least 95% of the protein, were observed (Fig. 10, gel D). The molecular weights of these bands are 49,000 and 50,000 (Fisher and Kaufman, unpublished.

2. Isozymes of the Enzyme

Discontinuous electrophoresis of highly purified phenylalanine hydroxylase on 7% polyacrylamide gels with a tris-glycine pH 9.5 buffer at 0° showed that 85% of the protein (as measured by staining with Coomassie brilliant blue or amido black) was found in two bands of nearly equal intensity which migrated within 0.04 R_f unit (R_f's of major bands were about 0.36 and 0.40, Fig. 11). When the gel was sliced very finely, two

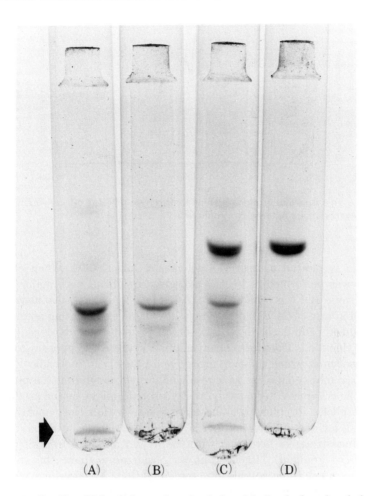

FIG. 10. Stacking SDS gel electrophoresis of phenylalanine hydroxylase before and after chymotrypsin treatment. (A) 10 μg of chymotrypsin-treated phenylalanine hydroxylase, (B) 5 μg of chymotrypsin-treated phenylalanine hydroxylase, (C) 5 μg of chymotrypsin-treated phenylalanine hydroxylase and 5 μg of untreated phenylalanine hydroxylase, and (D) 5 μg of untreated phenylalanine hydroxylase. Samples were resolved by sucrose gradient centrifugation prior to SDS electrophoresis. Marker dye migration is indicated by the arrow.

peaks of activity could be discerned corresponding to the two major protein bands. The molecular weights of these two major forms of the enzyme were determined by electrophoresis on gels of varying percent acrylamide. By this method the molecular weight of a protein is proportional to the slope of the line obtained when the logarithm of the protein's relative

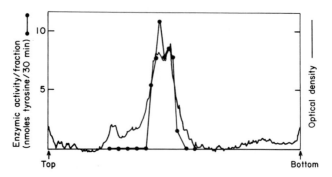

Fig. 11. Two isozymes of phenylalanine hydroxylase revealed by assaying fractions from a polyacrylamide gel after electrophoresis for 2 hours at 0°.

migration is plotted against the gel concentration (59). When electrophoresis was performed at 0°, the molecular weight of both major bands was the same and was found to be 110,000 ($\pm3\%$, N = 5). No new bands appeared at any of the different percent acrylamides indicating no proteins were hidden behind the hydroxylase bands under standard conditions (7% acrylamide). Therefore, phenylalanine hydroxylase appears to exist as two isozymes which differ in charge and possibly in molecular weight. These isozymes could arise as a result of proteolysis during the purification of the enzyme. Alternatively, these isozymes might be the product of different alleles for phenylalanine hydroxylase within the population of rats used for the purification. This molecular weight obtained by gel electrophoresis agrees favorably with the weight obtained from Sephadex and sucrose gradient, 110,000, but is considerably above the value of 90,000 obtained from equilibrium centrifugation. The average of these three values gives a molecular weight of 103,000.

When polyacrylamide electrophoresis was performed at 30°, the pattern observed depended on the amount of protein applied to the gel. Application of 40 μg of enzyme to the gels at 30° resulted in the same two major bands observed at 0° (R_f's 0.36 and 0.40). At lower concentrations, however, two new faster migrating bands appeared. The molecular weight was 55,000 ($\pm7\%$, N = 3) for both of these new faster migrating forms (99).

Sodium dodecyl sulfate gel electrophoresis separates polypeptides on the basis of molecular weight, rather than charge, since in the presence of SDS all proteins have a large negative charge. In order to determine how many subunits of differing charge were present in the two hydroxylase isozymes, a polyacrylamide electrophoresis was carried out at pH 9.5 in 8 M urea and 2 mM DTT at 40°. Under these conditions two major bands of equal staining intensity were present. Analysis with different percent acrylamide gels in 8 M urea showed that both of these bands had a molecular weight of 50,000 (Fisher and Kaufman, unpublished) which corresponds well with the values obtained in SDS. From all the results of studies on molecular weights of the enzyme forms we may tentatively suggest the following model for subunit composition. The smallest polypeptide chains are 49,000 and 50,000 and exist in equal amounts. These subunits can be called A and B. These subunits can aggregate with increasing enzyme concentration and decreasing temperature to form two electrophoretically separate dimers, both having molecular weights of about 103,000. These dimers are most likely AA and BB combinations of the subunits. The enzyme can also exist in a tetrameric form (when incubated with phenylalanine); it remains to be established how many charge isomers exist for this form.

Recently, the existence of three isozymes (pi, kappa, and upsilon) of phenylalanine hydroxylase has been reported (3). An extract of rat liver was fractionated on a calcium phosphate gel column, and three peaks of activity were observed. Each of these peaks had the same molecular weight on Sephadex G-200 (200,000) indicating that they differ in charge but not molecular weight. The authors claimed that the least prevalent isozyme, pi (10% of the total), is lost during the first step of our purification procedure. It was also found that this isozyme, pi, is the predominant form of phenylalanine hydroxylase in human fetal liver.

3. Amino Acid Analysis of the Enzyme

Amino acid analysis was performed on the 95% pure phenylalanine hydroxylase (final purification on sucrose gradient) as shown in Table IV. The enzyme is characterized by a small number of tryptophan (2) and methionine (4) residues per 50,000 molecular weight. Four of the free cysteines are buried inside the enzyme because the enzyme must be treated with a denaturing agent such as 8 M urea before Ellman's reagent (25) will react with them. One cysteine is partially buried and will slowly react with Ellman's reagent when the enzyme is in its native form (32). It should be pointed out that this is the amino composition for a mixture of the two nonidentical subunits, A and B. discussed above. From this amino acid composition and the partial specific volume contribution of each amino acid, it was calculated that phenylalanine hydroxylase has a partial specific volume of 0.72 (32).

TABLE IV

AMINO ACID ANALYSIS OF PHENYLALANINE HYDROXYLASE

Amino acid	Integral residues[a]	Amino acid	Integral residues[a]
Aspartic	71	Leucine	46
Threonine	20	Tyrosine	14
Serine	30	Phenylalanine	21
Glutamic	56	Lysine	23
Proline	28	Arginine	21
Glycine	26	Histidine	8
Alanine	31	Cysteic acid[b]	7
Valine	21	Cysteine	5
Methionine	4	Tryptophan	2
Isoleucine	19		

[a] Per 50,000 molecular weight subunit. Calculated from data obtained after hydrolysis for 24 and 72 hours.

[b] Obtained by peiformic acid oxidation.

F. Regulatory Properties of Phenylalanine Hydroxylase

1. Substrate Inhibition

The structure of the reduced pterin cofactor plays an important role in determining the affinity of the enzyme for its substrates, phenylalanine and oxygen. Changing from the synthetic cofactor, $DMPH_4$, to the natural cofactor, tetrahydrobiopterin, lowers the K_m for phenylalanine from 1.5 to 0.3 mM (96) and lowers the K_m for oxygen from 6 to 0.35% (30). In the presence of the natural cofactor, new regulatory properties of the enzyme emerge; e.g., under these conditions, phenylalanine and oxygen inhibit with much lower K_i values than those with the synthetic cofactor. With tetrahydrobiopterin, phenylalanine inhibits above 1.5 mM with a K_i of about 8 mM (96). When $DMPH_4$ is used, the K_i for phenylalanine is about 40 mM at 21% oxygen. It is of interest that with this cofactor, the K_i for phenylalanine is lowered to 15 mM when the oxygen concentration is raised to 100% (Fisher and Kaufman, unpublished). The fact that in the inhibitory range a plot of the reciprocal of the initial velocity versus the phenylalanine concentration gives a straight line (23) is consistent with the phenylalanine inhibition being due to the binding of a second substrate molecule at the active site with an affinity 30 times lower than the affinity of the enzyme for the first molecule of substrate. A similar observation has been made for oxygen. Oxygen inhibits above 4% with a K_i of about 50% when tetrahydrobiopterin is the cofactor (Fig. 12). In contrast, no inhibition at even 100% oxygen is observed when DMPH$_4$ is employed (30). As with phenylalanine, the nature of inhibition by oxygen is consistent with a second oxygen molecule combining at the active site.

2. Stimulation of Phenylalanine Hydroxylase by Lysolecithin and α-Chymotrypsin

It was recently reported that tryptophan displays a sigmoidal saturation curve for phenylalanine hydroxylase in the presence of DMPH$_4$ and that n-propanol converts this curve to the classic hyperbolic form (152). Since we had found that phenylalanine gave a sigmoidal saturation for the hydroxylase with tetrahydrobiopterin, we studied the effect of n-propanol on this reaction. n-Propanol (1 M) converted the curve for phenylalanine saturation from sigmoidal to hyperbolic and also stimulated the maximum velocity several fold (31). A survey of naturally occurring hydrophobic compounds showed that free fatty acids (C-18 and greater) and phospholipids greatly stimulated phenylalanine hydroxylation. Lysolecithin and lysophosphatidylserine were the most active. Like n-propanol, these

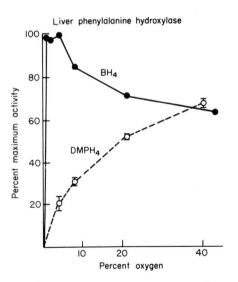

FIG. 12. Phenylalanine hydroxylase activity as a function of oxygen concentration in the presence of DMPH$_4$ or tetrahydrobiopterin (BH$_4$).

compounds converted the phenylalanine stauration curve to a hyperbolic form, but in contrast to *n*-propanol these lipids stimulated the maximum velocity 20–40-fold, and were effective in much lower concentrations (31). The effect of increasing lysolecithin concentration on enzymic activity is shown in Fig. 13. Maximum stimulation is achieved at 0.15 mM lysolecithin, which is the concentration at which micelles are formed (140). When DMPH$_4$ was substituted for tetrahydrobiopterin as cofactor, however, lysolecithin increased the maximum velocity only 1.15-fold and decreased the K_m for phenylalanine from 1.3 to 0.8 mM (31). Lysolecithin makes phenylalanine hydroxylase more active with tetrahydrobiopterin as cofactor than it is with DMPH$_4$. Without lysolecithin, the maximum velocity of the hydroxylase is 12 times higher with DMPH$_4$ than it is with tetrahydrobiopterin. In the presence of lysolecithin, however, the maxi-

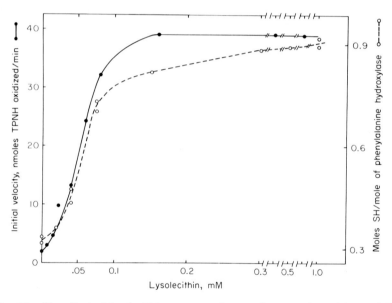

Fig. 13.   The effect of lysolecithin concentration on the enzymic activity and SH reactivity of phenylalanine hydroxylase. Half-maximum stimulation of the enzymic reaction was seen at 0.05 mM lysolecithin. For measuring the enzymic activity, the cofactor was tetrahydrobiopterin at 0.03 mM and the hydroxylase was present at 18 μg/ml and was 90% pure. Also shown is the exposure of a phenylalanine hydroxylase SH group by lysolecithin. The free sulfhydryl content of the hydroxylase was determined at various lysolecithin concentrations by incubating 80 μg/ml of 95% pure enzyme for 5 minutes with DTNB at pH 6.8. Half-maximum exposure of SH groups occurred at 0.055 mM lysolecithin. DTNB is 3,3′-dithiobis(6-nitrobenzoic acid).

mum velocity is four times higher with the natural cofactor than with the synthetic one.

At high concentrations of hydroxylase, lysolecithin produced a progressively smaller stimulation at pH 6.8. Since this result was reminiscent of the results obtained at higher pH where PHS was required, the lysolecithin stimulation was studied in the presence and absence of PHS at increasing hydroxylase concentrations. It is apparent from Fig. 14 that, as expected, at pH 6.8, there is nearly a linear relationship between enzyme concentration and initial velocity. Therefore, PHS has little effect on the enzymic activity even at high enzyme concentrations. In the presence of lysolecithin, however, the specific activity of the enzyme sharply decreases with increasing enzyme concentration. In the presence of PHS and lysolecithin the specific activity remained high at all enzyme concentrations and the full stimulation by lysolecithin was observed (Fisher and Kaufman, unpublished). Apparently, at pH 6.8, lysolecithin either increases the steady state concentration of S′ or increases the enzyme's affinity for S′ [Eq. (18)]. These results indicate that if lysolecithin stimulates the hydroxylase *in vivo*, PHS may also stimulate the enzyme *in vivo* at physiological pH values.

Another phospholipid, lecithin, is a potent inhibitor of phenylalanine hydroxylase with the natural, but not the synthetic, cofactor (31). Therefore, phospholipase A may play an important role in regulating phenyl-

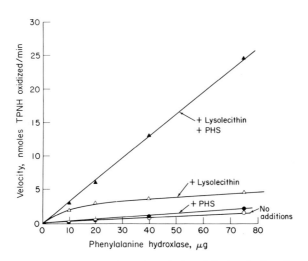

FIG. 14.   Enzymic activity as a function of phenylalanine hydroxylase (70% pure) concentration in the presence and absence of lysolecithin (1 mM) and PHS (36 μg/ml). The cofactor was tetrahydrobiopterin at 0.003 mM.

alanine hydroxylase by converting an inhibitor of the enzyme, lecithin, into a stimulator of the enzyme, lysolecithin. Phospholipase A is located mainly in lysosomes (21); thus, conditions which destabilize lysosomes, such as cell damage, could indirectly activate the hydroxylase.

A further search for naturally occurring compounds which might activate the hydroxylase led to the finding that the stimulation by free fatty acids (such as oleic acid) was synergistically enhanced by conjugated bile salts (taurochenodeoxycholic acid) as shown in Table V. These concentrations of bile salts are those found in the liver (128). Following a meal, the bile salt and free fatty acid concentration in the liver would be expected to increase and might stimulate phenylalanine hydroxylase at a time when the blood phenylalanine concentration would be elevated. It is also shown in Table V that 0.1 mM olelyl CoA stimulates the hydroxylase to the same extent as does lysolecithin. Fatty acid CoA levels in the liver vary greatly with changes in diet. High carbohydrate diet gives a low fatty acid CoA concentration (0.03 mM) in rat liver, and starvation elevates the fatty acid CoA level to 0.11 mM (159). This increase in fatty acid CoA level during starvation could stimulate phenylalanine hydroxylase at a time when catabolites of tyrosine breakdown are needed for gluconeogenesis.

TABLE V

Effect of Bile Salts, Free Fatty Acids, and Fatty Acid CoA on Phenylalanine Hydroxylase Activity

Additions	Activity (nmoles TPNH oxidized/min)
Experiment 1	
None	0.67
0.16 mM Taurocheno DOC[a]	0.67
0.4 mM Taurocheno DOC	0.75
0.04 mM Oleic acid	1.2
0.1 mM Oleic acid	3.2
0.16 mM Taurocheno DOC + 0.04 mM oleic acid	5.7
0.4 mM Taurocheno DOC + 0.1 mM oleic acid	12.0
Experiment 2	
None	0.3
0.1 mM Oleoyl CoA	7.5
0.1 mM Lysolecithin	7.5

[a] DOC stands for deoxycholic acid.

All the lipids which stimulate the hydroxylase are good detergents. In fact, a synthetic detergent, SDS, also stimulates the hydroxylase (31). Since detergents are known to dissociate protein subunits, we investigated the possibility that these detergents were dissociating the enzyme into monomers which were more active than the dimeric form. Sucrose gradient centrifugation and polyacrylamide gel electrophoresis studies in the presence and absence of lysolecithin, however, indicated that lysolecithin does not dissociate the enzyme into subunits (Fisher and Kaufman, unpublished). Lysolecithin does, however, increase the reactivity of an enzyme sulfhydryl with DTNB, 3,3'-dithiobis(6-nitrobenzoic acid). Lysolecithin increases sulfhydryl reactivity (Fig. 13) and enzymic activity with nearly identical concentration curves. This result indicates that lysolecithin produces a small conformational change in the enzyme structure that makes it a more efficient catalyst.

Our understanding of the mechanism of the lysolecithin activation has

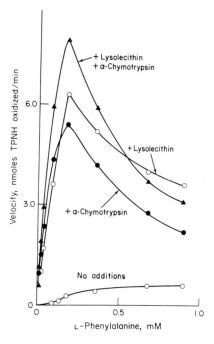

FIG. 15.   Influence of chymotrypsin digestion and lysolecithin on the phenylalanine saturation curves for phenylalanine hydroxylase. For chymotrypsin samples, 16 μg of 70% pure phenylalanine hydroxylase was preincubated with 8 μg of α-chymotrypsin for 10 minutes in a 20-μl volume and was then added to a complete reaction mixture. Lysolecithin concentration was 1.0 mM and tetrahydrobiopterin was 0.025 mM.

been aided by the finding that α-chymotrypsin increases the hydroxylase activity in a manner similar to lysolecithin. The chymotrypsin-activated form has a phenylalanine saturation curve similar to the curve for the lysolecithin-activated form (Fig. 15). Furthermore, after the enzyme has been activated by chymotrypsin treatment, lysolecithin produces little additional activation. These results indicate that lysolecithin and chymotrypsin are activating the enzyme in a similar fashion. The main difference between these two activators appears to be that lysolecithin acts reversibly and does not change the molecular weight of the hydroxylase, whereas chymotrypsin irreversibly stimulates the hydroxylases and reduces the molecular weight of this enzyme. When highly purified hydroxylase was treated with agarose-bound chymotrypsin and the protease-treated hydroxylase was sedimented by sucrose gradient centrifugation, the peak of hydroxylase activity had a sedimentation coefficient of 3.9 S (Fig. 16). This value is considerably lower than the coefficient of 5.9 S observed for the starting material (Fisher and Kaufman, unpublished). Combining this sedimentation coefficient with a Stokes radius (obtained by Sephadex G-200 gel filtration) showed that the chymotrypsin-activated hydroxylase has a molecular weight of 60,000. The stacking SDS

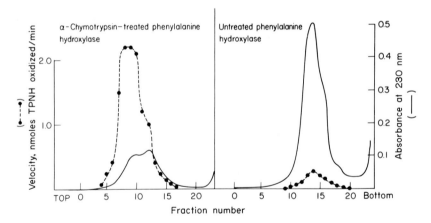

FIG. 16. Sucrose gradient centrifugation of α-chymotrypsin-treated and -untreated phenylalanine hydroxylase. Two milligrams of 90% pure hydroxylase were incubated 20 minutes with agarose-linked α-chymotrypsin. The hydroxylase was eluted from the agarose, concentrated with an Amicon membrane, and centrifuged through a sucrose gradient. One and a half milligrams of 90% pure hydroxylase were centrifuged for comparison. Protein content of the gradients at the end of the run was monitored with a Gilford flow cell at 280 nm (—). Fractions were assayed with 0.028 mM tetrahydrobiopterin (● – – – ●).

gel electrophoresis method was employed to analyze the subunit composition of the most active peak of the sucrose gradient–resolved chymotrypsin-treated hydroxylase. The chymotrypsin-treated hydroxylase has one major subunit with a molecular weight of 35,000 (Fig. 10, gels A and B). The minor, smaller molecular weight peptides might result from excessive digestion by chymotrypsin. The chymotrypsin-activated hydroxylase, therefore, appears to be a dimer whose subunits have identical molecular weights of 35,000. These results suggest that the phenylalanine hydroxylase isolated from rat liver is a proenzyme with a polypeptide on each subunit which can act as an internal inhibitor of the enzyme. This polypeptide chain can either be removed reversibly from its inhibitory site by the detergent action of a lipid, or it can be irreversibly removed by a protease. Apparently, this polypeptide is not inhibitory when the $DMPH_4$ is used as cofactor.

The above findings with chymotrypsin suggest another mechanism by which lysosomal enzymes might regulate the activity of phenylalanine hydroxylase. In addition to phospholipases, lysosomes contain a variety of proteases called "cathepsins" (21). It is possible that when lysosomes are disrupted, one of these cathepsins can activated phenylalanine hydroxylase in a manner similar to chymotrypsin.

G. The Nature of Phenylalanine Hydroxylase in Hyperphenylalaninemia and PKU

Phenylketonuria is a genetically determined disease which results in greatly elevated blood phenylalanine levels and mental retardation. The first clue that phenylalanine hydroxylase was missing in PKU individuals was provided by the finding of Jervis (73) that the administration of phenylalanine to normal animals or humans led to a prompt increase in tyrosine in the blood. When, on the other hand, phenylalanine was given to PKU patients, there was no increase in tyrosine. He concluded that the metabolic block in this disease was located between phenylalanine and tyrosine, i.e., in the hydroxylation of phenylalanine.

This conclusion was supported by the finding that the *in vivo* conversion of [14C]-phenylalanine to [14C]tyrosine was greatly reduced in this disease (160). Direct *in vitro* assay of the enzyme from the liver of PKU patients demonstrated that the phenylalanine hydroxylation system was lacking (74). After it was realized that at least two enzymes were involved in the conversion of phenylalanine to tyrosine, but before their roles in the hydroxylation system had been defined, the analysis of the defect in PKU was carried one step further when it was demonstrated that only the more labile enzyme was missing (117, 166).

In 1958, it was shown by direct assay that the cofactor was present in normal amounts in liver biopsy samples from PKU patients (82). It was also demonstrated that no significant phenylalanine hydroxylase activity was present in the PKU liver samples, whereas dihydropteridine reductase showed good activity. This work proved for the first time that the hydroxylase itself was the affected component of the hydroxylating system in PKU.

Recently, we have investigated the nature of the defective phenylalanine hydroxylase in PKU in more detail. An antibody to purified rat liver phenylalanine hydroxylase was shown to cross-react with human liver phenylalanine hydroxylase. In double immunodiffusion reactions with this antibody, no cross-reacting material was detected in a classic PKU liver (35). These studies could have detected a protein which is antigenically identical to human liver phenylalanine hydroxylase at 10% or more of the concentration of the hydroxylase in normal human liver. These results suggested the possibility that in PKU there is a complete deletion of the enzyme. The most sensitive assays for phenylalanine hydroxylase would have detected 1% of the activity found in normal liver. These assays were carried out with $DMPH_4$ (35).

Recently, we found that lysolecithin stimulates the normal human enzyme by about 350% and that in the presence of the natural cofactor, tetrahydrobiopterin, the K_m for L-phenylalanine of normal human liver phenylalanine hydroxylase is 20–40 times lower than in the presence of $DMPH_4$ (Friedman and Kaufman, unpublished). With the use of lysolecithin and tetrahydrobiopterin, and measuring [14C]phenylalanine conversion to [14C]tyrosine, the sensitivity of the assay for phenylalanine hydroxylase was increased 20-fold. Using this assay, we have been able to show that there is 0.25% of the normal level [level of hydroxylase in normal liver biopsy had been previously established (95)] of hydroxylase activity in a PKU liver (Friedman, Fisher, Kang, and Kaufman, unpublished). The 14C product was identified as p-tyrosine by its migration in two paper chromatographic systems, its elution from an ion exchange resin, and its conversion to [14C]dopa by tyrosine hydroxylase. The production of [14C] tyrosine by the PKU liver was completely dependent on tetrahydrobiopterin, an indication that a pterin-dependent hydroxylase was catalyzing the reaction. The enzyme was identified as phenylalanine hydroxylase by its sensitivity to inhibition by p-chlorophenylalanine and by its lack of sensitivity to an inhibitor of tyrosine hydroxylase, 3-iodotyrosine, or an inhibitor of tryptophan hydroxylase, 6-fluorotryptophan.

The detection of phenylalanine hydroxylase activity in a PKU liver suggests that the hydroxylase activity in PKU results from the presence of either a normal hydroxylase at an extremely low concentration or a structurally altered hydroxylase with low catalytic activity. Two facts support the latter interpretation. First, whereas the normal hydroxylase is stimulated 350% by lysolecithin, the PKU enzyme is stimulated only 100% by this phospholipid. Second, unlike the normal enzyme, the PKU enzyme does not show substrate inhibition at 0.15 mM phenylalanine (Fig. 17a). It can also be seen in Fig. 17a that the PKU enzyme has a K_m (0.037 mM) for phenylalanine which is similar to the K_m 0.028 mM found

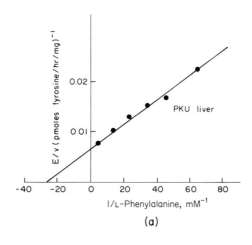

(a)

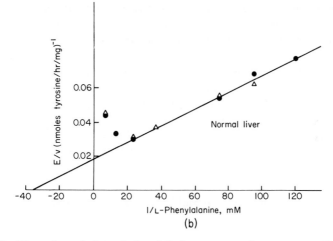

(b)

FIG. 17. The reciprocal plots of phenylalanine concentration versus specific activity for phenylalanine hydroxylase from a PKU (a) and normal (b) liver; (●) experiment I and (△) experiment II. Each reaction tube contained 0.03 mM tetrahydrobiopterin and 300,000 cpm [¹⁴C] phenylalanine. Tyrosine production was determined by paper chromatography.

for the normal hydroxylase under identical conditions (Fig. 17b) (Friedman, Fisher, Kang, and Kaufman, unpublished). The lack of substrate inhibition and decreased stimulation by lysolecithin with the PKU enzyme indicate that the very low hydroxylase activity is the result of the presence in PKU liver of an altered enzyme with low activity rather than a very low level of normal enzyme. Therefore, we can tentatively conclude that this case of PKU involves a mutation in the structural gene for phenylalanine hydroxylase. The possibility that the PKU enzyme has low activity because of a lack of iron was made unlikely by the finding that addition of $FeCl_2$ only increased the activity of the PKU enzyme by 15%. Furthermore, the PKU liver had a normal level of PHS. The low activity in PKU appears to result solely from a low level of phenylalanine hydroxylase.

Recently, there have been reports of patients with what might be called a milder variant of PKU, conceivably representing a third allele at the same genetic locus. The patients show either a transitory elevation of serum phenylalanine during the neonatal period or a moderate elevation persisting through several years of childhood. The clinical symptoms of PKU, however, are absent. These conditions have collectively been labeled hyperphenylalaninemia (76). Liver samples from several hyperphenylalaninemic individuals showed that they have 5% of the normal level of phenylalanine hydroxylase (77, 101). As with the case with the PKU hydroxylase, no material cross-reacting with the phenylalanine hydroxylase antibody was found. The phenylalanine hydroxylase present in hyperphenylalaninemia appears to be structurally altered because its stimulation by lysolecithin and its K_m for phenylalanine are less than that observed for the normal hydroxylase (35).

The fact that only 5% of the normal phenylalanine hydroxylase is needed to prevent the clinical symptoms of PKU is encouraging. It suggests that it might be possible to treat patients with PKU by administration of a pterin cofactor or by alteration of the hepatic phospholipid composition. Either, or both, of these treatments, instituted early enough, might elevate their residual hydroxylase to the level where part, if not all, of the mental retardation might be prevented.

H. The Developmental and Diurnal Variation of Phenylalanine Hydroxylase

All the foregoing studies deal with *in vitro* properties of phenylalanine hydroxylase. In the future, attention will probably turn more toward understanding the *in vivo* regulation of enzyme level and activity. Two such studies which have been undertaken are an analysis of the appearance of phenylalanine hydroxylase during development and its diurnal variation.

In 1958, it was reported that phenylalanine hydroxylase was a member of a group of liver enzymes that are essentially inactive at birth. The activity of these enzymes increases rapidly after birth, usually reaching adult levels within 1 or 2 days (103). In 1965, it was demonstrated that phenylalanine hydroxylase did not belong to this group of enzymes that are inactive at birth. It was found that when properly supplemented with dihydropteridine reductase and pterin cofactor, the hydroxylase in livers from newborn rats is equal to or higher than that of adults (7). The levels of the pterin cofactor and of dihydropteridine reductase, however, are relatively deficient in the newborn livers.

A study of the development of phenylalanine hydroxylase activity in fetal chicken liver showed that the activity appeared during the second half of the gestation period and approached adult levels at term (151). A more detailed study of the timing of appearance of phenylalanine hydroxylase in fetal rats revealed that no significant hydroxylase activity is detectable earlier than 24–36 hours prior to parturition. There follows a dramatic rise in activity until, later in labor, the hydroxylase activity reaches the adult value. The dihydropteridine reductase and pterin cofactor levels also increase dramatically during the 24 hours before birth (36). The guinea pig (4, 36) and human (36, 65) phenylalanine hydroxylase appeared earlier during gestation than the rat hydroxylase.

Several mammalian enzymes that metabolize amino acids follow regular changes in activity during each 24-hour period (135).Recently, it was found that the activity of rat liver phenylalanine hydroxylase shows circadian rhythmicity (12). Rats fed *ad libitum* had a minimum level of hepatic phenylalanine hydroxylase at 10 AM. After that time, the level rose slowly and reached a peak at 10 PM. The appearance of this peak of enzymic activity was independent of blood phenylalanine levels, although with high blood phenylalanine levels the peak enzymic activity was maintained for a longer period of time than in the controls.

III. TYROSINE HYDROXYLASE

The discovery of the essential role of unconjugated pterins in the enzymic hydroxylation of phenylalanine facilitated the study of certain aspects of the aerobic metabolism of the other aromatic amino acids.

Thus, although it had been known for some time that the first step in the sequence of reactions leading to norepinephrine synthesis in the adrenal medulla is the oxidation of tyrosine to dopa (43, 50, 163, 164) the enzyme responsible for this hydroxylation reaction remained elusive. In 1964, the enzymic conversion of L-tyrosine to L-dopa was demonstrated in particles isolated from adrenal medulla, brain, and other sympathetically innervated tissues (121). The properties of the enzyme, called "tyrosine hydroxylase," clearly distinguished it from tyrosinase, another enzyme that is capable of oxidizing tyrosine to dopa. With the latter enzyme, dopa is further oxidized to melanin.

With partially purified preparations of tyrosine hydroxylase from adrenal medulla, an absolute requirement for a tetrahydropterin was demonstrated (6, 121), thus establishing this enzyme as the second pterin-dependent aromatic amino acid hydroxylase. It appeared likely that the role of the tetrahydropterin in this hydroxylation system would prove to be the same as the one previously established in the phenylalanine hydroxylating system, an expectation that was fulfilled by subsequent work.

As already mentioned, tyrosine hydroxylase is present in brain, adrenal medulla, and sympathetically innervated tissues. Its intracellular location in these tissues, however, has been the subject of controversy. Udenfriend and his co-workers reported that in guinea pig brain all of the activity, and in beef adrenal medulla, most of the activity, was particle bound (121). They further concluded that the hydroxylase found in the soluble fraction of adrenal medulla was originally present in particles from which it had been released by the homogenization process. The conclusion that the bovine adrenal hydroxylase is particulate was strengthened by the finding of Petrack et al. (134) that about 90% of the activity was sedimentable after centrifugation for 1 hour at 144,000 g. These findings appeared to support the proposal that all of the enzymes involved in catecholamine biosynthesis are localized within a catecholamine-containing granule (161).

There is a growing body of evidence that contradicts the conclusion that tyrosine hydroxylase is present intracellularly within a granule (106, 118, 119, 149, 171). For the enzyme from both adrenal medulla and nerve tissue it has been found that a major part of the enzyme is present in the high-speed supernatant fraction and that the distribution of hydroxylase activity is a function of the composition of the homogenization medium; homogenization in isotonic KCl leads to more enzyme in the supernatant fraction than does homogenization in isotonic sucrose (171). More recently, evidence has been presented that indicates that rat brain tyrosine hydroxylase exists in two distinct forms, a soluble and a membrane-bound form (104).

At least part of the uncertainty about the intracellular localization of tyrosine hydroxylase is because of the tendency of the enzyme from bovine adrenal medulla, and probably from other tissues as well, to aggregate and adsorb to subcellular organelles (171). An unequivocal answer to this question of whether tyrosine hydroxylase is a cytoplasmic or particulate enzyme will probably have to await the development and application of a technique for intracellular localization of this enzyme that does not involve disruption of the tissue being examined.

Because soluble tyrosine hydroxylase has never been purified to any extent, almost all of our knowledge of the properties of the enzyme from bovine adrenal medulla has come from studies carried out with partially purified preparations that were solubilized by limited proteolysis with either trypsin (134) or chymotrypsin (143). Although there is some evidence that the molecular weight of the trypsin-treated enzyme may be less than that of the untreated one (120) (this point will be discussed in detail in a later section), there is little support for the expressed concern that studies of the kinetic properties of the solubilized enzyme would be "highly misleading" (120). Indeed, where the kinetic properties of the particulate and solubilized enzyme have been compared, they have been shown to be essentially the same (143).

A. Specificity of Tyrosine Hydroxylase

1. Tetrahydropterin

Just as with phenylalanine hydroxylase, tyrosine hydroxylase's requirement for a tetrahydropterin appears to be absolutely specific. The following compounds have been tested as possible substitutes for the reduced pterin and found to be completely inactive at a final concentration of 0.1 mM: ascorbate (6, 121), cysteine (6), glutathione, $FeCl_2$ (6, 121), flavin mononucleotide (6), FAD (6), and dopa (6, 121).

Within the pterin series, unconjugated tetrahydropterins are far more active than tetrahydrofolate (6, 121). Although the latter compound was reported to have good cofactor activity, especially in the presence of Fe^{2+} (121), subsequent work showed that most, if not all, of this activity is the result of contamination of the tetrahydrofolate with 6-methyltetrahydropterin (107).

As has been previously shown with phenylalanine hydroxylase (100), the 6-methylpterin is significantly more active than the 6,7-dimethyl compound (6). An unsubstituted N5 position is necessary for cofactor activity as is either a 2-amino or a 4-hydroxy substituent; a pteridine with neither of these substituents is completely inactive (24).

Tetrahydrobiopterin is the most active cofactor both with respect to its apparent K_m value and its relative activity in stimulating the rate of dopa formation. The apparent K_m values for tetrahydrobiopterin, 6,7-

dimethyltetrahydropterin, and 6-methyltetrathydropterin are shown in
Table VI for both the purified, chymotrypsin-solubilized, and the particulate adrenal enzyme (144). The K_m value for DMPH$_4$ is close to that
reported for the soluble adrenal enzyme (121). The relative rates of dopa
formation are shown in Table VII (143). It is apparent that none of these
kinetic parameters has been markedly altered by the protease treatment
used to solubilize the enzyme.

2. Amino Acid Substrate

The adrenal enzyme was originally thought to be absolutely specific
for L-tyrosine (121); none of the following compounds was active: D-tyro-
sine, tyramine, DL-m-tyrosine, or L-tryptophan (121). The first indication that the adrenal enzyme is not absolutely specific for L-tyrosine was
the finding that L-phenylalanine could serve as a substrate (64); parallel
inhibition of the rate of product formation showed that the two hydroxylation reactions were almost certainly catalyzed by the same enzyme. With
a relatively crude enzyme preparation and DMPH$_4$ as the cofactor, the
rate of hydroxylation of L-phenylalanine was about 5% of the rate of
hydroxylation of L-tyrosine under comparable conditions. It was also
shown that the initial product of phenylalanine hydroxylation was L-tyrosine.

The low rate of phenylalanine hydroxylation by tyrosine hydroxylase
in the presence of DMPH$_4$ indicated that this reaction was probably of
minor physiological significance. A more detailed study of the phenylalanine hydroxylating ability of tyrosine hydroxylase, however, led to
the surprising finding that in the presence of tetrahydrobiopterin the rate
of phenylalanine hydroxylation by both the highly purified, solubilized
enzyme and the particulate enzyme is equal to, or greater than, the rate
of tyrosine hydroxylation (143). Furthermore, as will be discussed in
greater detail later, whereas under these conditions the enzyme is sensitive to inhibition by excess tyrosine, there is no evidence for inhibition by

TABLE VI

PTERIN MICHAELIS CONSTANTS

Pterin used	Solubilized enzyme ($\times 10^{-5} M$)	Particulate enzyme ($\times 10^{-5} M$)
DMPH$_4$	30	30
6-MPH$_4$	30	30
Tetrahydrobiopterin	10	10

TABLE VII

RELATIVE RATE OF DOPA FORMATION WITH
DIFFERENT TETRAHYDROPTERINS

State of Enzyme	DMPH$_4$	6-MPH$_4$	Tetrahydrobiopterin[a]
Particulate enzyme	0.45	0.6	1.0
Soluble enzyme	0.53	0.9	1.0

[a] The rate with tetrahydrobiopterin has been set equal to 1.0.

excess phenylalanine. These results made it highly likely that *in vivo* L-phenylalanine could serve as an important precursor of norepinephrine, a prediction that has been recently fulfilled (78).

The original conclusion about the sharp amino acid specificity of tyrosine hydroxylase was modified still further by the recent studies of Tong *et al.* (155, 156). Using a crude preparation of the adrenal enzyme, they showed that it could catalyze the conversion of L-*m*-tyrosine to dopa at about 50% the rate at which L-tyrosine is converted to dopa. In addition, they reported that the preparation could catalyze the conversion of L-phenylalanine to *m*-tyrosine at about 15% of the rate at which L-phenylalanine is converted to tyrosine (155). Although it is not certain that tyrosine hydroxylase is the responsible enzyme, the sensitivity of the *m*-hydroxylation reaction to the tyrosine hydroxylase inhibitors, 3-iodotyrosine and α-methyltryosine, supports the conclusion that the same enzyme is catalyzing the hydroxylation of phenylalanine in the para and the meta positions. The conversion of phenylalanine to dopa via *m*-tyrosine is a potential alternate pathway for dopa formation although its *in vivo* significance relative to the phenylalanine–tyrosine–dopa pathway remains to be established.

A property of the pterin-dependent hydroxylases that is not fully appreciated is that the apparent K_m values for their substrates (both the amino acid substrate and oxygen) vary with the pterin used, a phenomenon that was first discovered with phenylalanine hydroxylase (96).

The summary of K_m values shown in Table VIII illustrates this point for tyrosine hydroxylase (144). For both the purified, solubilized, and the particulate enzyme, the K_m value for tyrosine is much lower in the presence of tetrahydrobiotperin than it is in the presence of DMPH$_4$.

The apparent K_m value for oxygen is also markedly lower in the presence of tetrahydrobiopterin than it is in the presence of DMPH$_4$. With the dimethylpterin, a value of about 6% has been reported (63), whereas in the presence of tetrahydrobiopterin the K_m for oxygen is about 1% (30).

TABLE VIII

MICHAELIS CONSTANTS FOR TYROSINE WITH TYROSINE
HYDROXYLASE FROM BOVINE ADRENAL

Pterin used	Solubilized enzyme ($\times 10^{-5} M$)	Particulate enzyme ($\times 10^{-5} M$)
DMPH$_4$	20	10
6-MPH$_4$	7	4
Tetrahydrobiopterin	1.5	0.4

B. Properties of Tyrosine Hydroxylase

1. Size

Since the enzyme has not yet been obtained in pure form, little is known about its physical properties. The molecular weight of the "native" adrenal enzyme has been estimated to be between 135,000 and 155 000, whereas a value of 34,000 has been reported for the trypsin-solubilized enzyme (119). The latter value is close to the molecular weight of about 40,000 reported for the 40% pure chymotrypsin-treated enzyme (143).

Based on the great difference in apparent molecular weight between the enzyme before and after trypsin treatment, Musacchio et al. (119) have concluded that trypsin treatment yields an enzyme that is "only a fragment of the native enzyme." The evidence in favor of this conclusion is not compelling. It is just as likely that treatment of the hydroxylase with either trypsin or chymotrypsin, under the carefully controlled conditions that have been used, leads to only a minor decrease in the size of the peptide chain(s) of the hydroxylase but that the protease-modified enzyme no longer aggregates. Indeed, attempts to induce aggregation of the trypsin-treated enzyme have been unsuccessful (119). The fact that all of the kinetic properties of the chymotrypsin-treated enzyme that have been examined are at least qualitatively the same as those of the "native enzyme" (see Tables VI, VII, and VIII) would appear to be more compatible with the latter interpretation than with the idea that the protease treatment leads to a fragment of the native enzyme.

2. Role of Iron

There have been repeated claims that tyrosine hydroxylase requires Fe^{2+} for activity. The first indication that Fe^{2+} might play a role in this hydroxylation reaction was the observation that in the presence of tetrahydrofolate, Fe^{2+} could stimulate the rate of dopa formation (121). The fact that Fe^{2+} stimulated only slightly when $DMPH_4$ was used in place of tetrahydrofolate was not consistent with a simple interpretation that the hydroxylase is a Fe^{2+} enzyme.

More recently, based on the observation that Fe^{2+} markedly and specifically stimulated the adrenal hydroxylase even in the presence of $DMPH_4$, Petrack et al. (134) concluded that Fe^{2+} participates in the hydroxylation reaction.

Studies on the mechanism of the stimulation by Fe^{2+} of the hydroxylation reaction have provided an alternative explanation for the role of Fe^{+2} in this system and have indicated that the conclusion that tyrosine hydroxylase is an iron enzyme is premature.

It has been known for some time that several mixed function oxidases, e.g., phenylalanine hydroxylase and dopamine β-hydroxylase, can be stimulated by catalase (88). The stimulation of phenylalanine hydroxylase by catalase has been amply confirmed (8, 66, 126). The protective effect of catalase in the two systems has been traced to the generation of peroxides under assay conditions by the nonenzymic oxidation of both ascorbate (the coenzyme for dopamine β-hydroxylase) and tetrahydropterins, and to the sensitivity to H_2O_2 of the two hydroxylases and of tetrahydropterins (88, 97).

Because of the similarities between phenylalanine and tyrosine hydroxylase, it was anticipated that the activity of tyrosine hydroxylase would also be stimulated by catalase. The data in Table IX show that not

TABLE IX

EFFECT OF IRON AND CATALASE ON TYROSINE
HYDROXYLASE FROM BOVINE ADRENAL MEDULLA[a]

	Hydroxylase	Additions	Δ counts/min
1	0	0	0
2	+	0	372
3	+	1 mM Fe^{2+}	1670
4	+	3 mM Fe^{2+}	1925
5	0	3 mM Fe^{2+}	0
6	0	Catalase	0
7	+	Catalase	1950
8	+	Catalase + 3 mM Fe^{2+}	1880

[a] Tyrosine hydroxylase activity was assayed by measurement of the release of tritium from 3,5-ditritiotyrosine by a modification of the method of Nagatsu et al. (121).

only was this expectation realized with purified, solubilized tyrosine hydroxylase but also that Fe^{2+} can replace catalase (143). The effect of Fe^{2+} under these conditions is probably the result of the well-known ability of Fe^{2+} to decompose H_2O_2 (125). The final point in the thesis that Fe^{2+} stimulates the hydroxylase by decomposing peroxide was the demonstration that the hydroxylase is sensitive to H_2O_2: A 10-minute preincubation of the enzyme with 0.5 mM H_2O_2 led to 40% inactivation. The ability of catalase to replace iron has been recently confirmed with brain tyrosine hydroxylase (104), although with that enzyme, catalase was reported to be only 80% as effective as 0.436 mM Fe^{2+}.

Although it is now clear that the observed stimulation of tyrosine hydroxylase by Fe^{2+} cannot sustain the conclusion that tyrosine hydroxylase is an iron enzyme, there are other reasons to suspect that the enzyme may, in fact, require Fe^{2+}. The adrenal enzyme is inhibited by iron chelators, such as α,α'-dipyridyl (121) and o-phenanthroline (24), but not by the nonchelating analog, m-phenanthroline (154).

Taylor $et\ al.$ (154) also reported that the inhibition by o-phenanthroline appeared to be noncompetitive with the tetrahydropterin. By contrast, there is evidence for some interaction between the pterin cofactor and o-phenanthroline. Thus, the extent of inhibition varies with the structure of the pterin cofactor used, the inhibition being less with tetrahydrobiopterin than with the dimethylpterin (143). In view of this relationship between the extent of inhibition and the structure of the pterin (which could be related to a structural resemblance between pterins and o-phenanthroline), and the reports that o-phenanthroline can inhibit enzymes that do not contain metals (172), it is apparent that even these inhibition data, suggestive though they are, do not prove that tryosine hydroxylase is an iron enzyme.

At the present state of our knowledge about tyrosine hydroxylase, perhaps the strongest hint about the ultimate answer to the question of whether this enzyme is an iron protein is provided by the recent demonstration that hepatic phenylalanine hydroxylase is an iron protein (32). The properties and kinetic characteristics of tyrosine hydroxylase are sufficiently similar to those of the latter enzyme that there is a high probability that tyrosine hydroxylase will also prove to be an iron protein.

C. Stoichiometry of Tyrosine Hydroxylase

Although a large number of studies of tyrosine hydroxylase have been published, the complete reaction stoichiometry has only recently been established (143). Nagatsu $et\ al.$ (121) have shown that with DMPH₄ as the cofactor the amount of dopa formed is equal to the amount of L-tyrosine consumed; neither the oxygen consumed nor the tetrahydropterin oxidized was measured in these studies. In general, it had been implicitly assumed that the stoichiometry would be identical with that found for liver phenylalanine hydroxylase (79) [see Eq. (1)]. When the complete stoichiometry was actually determined, it was found that this assumption was only partially correct.

In the presence of tetrahydrobiopterin, the partially purified, chymotrypsin-solubilized enzyme [this enzyme preparation (143) will be designated hereafter as the "solubilized" enzyme] was shown to catalyze the conversion of tyrosine to dopa in accordance with Eq. (20) where BH_4 stands for tetrahydrobiopterin (143):

$$\text{L-Tyrosine} + BH_4 + O_2 \rightarrow \text{dopa} + \text{dihydrobiopterin} + H_2O \qquad (20)$$

Under these conditions, therefore, the stoichiometry is, in fact, identical with that found for rat liver phenylalanine hydroxylase. Although the dihydrobiopterin product was not identified chemically, the finding that tyrosine hydroxylase can be coupled with dihydropteridine reductase (6) strongly supports the conclusion that the product is the quinonoid dihydro isomer (90). The stoichiometry of the reaction in the presence of pterins other than tetrahydrobiopterin will be discussed in the next section.

D. Mechanism of Tyrosine Hydroxylase

Since it is highly likely that all pterin-dependent hydroxylases share a common catalytic mechanism, and the phenylalanine hydroxylase mechanism has been discussed, only those aspects of the mechanism of tyrosine hydroxylase that appear to be peculiar to this enzyme will be considered here.

The demonstration that tyrosine hydroxylase required both an electron donor, i.e., a tetrahydropterin, and molecular oxygen (121) and that both of these components are stoichiometrically consumed during the hydroxylation reaction (143) provided indirect proof that this enzyme is a mixed-function oxidase or oxygenase. Direct proof was provided by Daley et al. (20) when they showed that ^{18}O from molecular oxygen is incorporated into the 3 position of the benzene ring during the enzymic conversion of tyrosine to dopa.

In view of the hydroxylation-induced migration of para substituents that accompanies various enzyme-catalyzed hydroxylations of aromatic compounds (53), a mechanism for dopa formation that involves introduction of a hydroxyl group on the 4 position of tyrosine followed by migration of one of these hydroxyl groups from the 4 to the 3 position has been considered. This possibility was eliminated with the demonstration that the dopa formed from [4-^{18}O]tyrosine (in the presence of $^{16}O_2$) contained only small amounts of ^{18}O in the 3 position, whereas the dopa formed from nonisotopic tyrosine and $^{18}O_2$ contained about 90% of the label in the 3 position (20). Although migration of the hydroxyl group does not occur, migration of tritium from the 4 to the 3 position does take place during the tyrosine hydroxylase–catalyzed hydroxylation of [4-^{3}H]phenylalanine. On the other hand, the hydroxylation of either [3,5-^{3}H]phenylalanine or [3,5-^{3}H]tyrosine by the enzyme leads to loss of one-half of the tritium (20). It is clear, therefore, that the first hydroxylation of [4-^{3}H]phenylalanine in the 4 position catalyzed by tyrosine hydroxylase leads to migration of the tritium to the 3 and 5 positions (just as it does during phenylalanine hydroxylase catalysis of this same reaction) and that the second hydroxylation of the resulting [3,5-^{3}H]tyrosine on either the 3 or 5 position is accompanied by the loss of the ^{3}H that originally occupied that position.

A detailed kinetic analysis of this hydroxylation reaction, carried out with the partially purified, soluble adrenal enzyme in the presence of DMPH$_4$, led to the conclusion that the reaction is of the ping-pong type, involving a reduced form of the enzyme as an intermediate (63). This conclusion was based mainly on the observation that double reciprocal plots of initial rates versus DMPH$_4$ concentrations at different concentrations of tyrosine, and of initial rates versus tyrosine concentrations at different concentrations of DMPH$_4$, gave a series of parallel lines. As the authors pointed out, their analysis indicated that the mechanism of this hydroxylation reaction is analogous to that of another hydroxylase, dopamine β-hydroxylase, where a reduced form of the enzyme was shown to be an intermediate (37). A sequence illustrating this kind of mechanism is shown in Eqs. (21) and (22), where E and EH$_2$ stand for the enzyme and its reduced form, respectively,

$$XH_4 + E \rightarrow EH_2 + XH_2 \qquad (21)$$

$$EH_2 + RH + O_2 \rightarrow E + ROH + H_2O \qquad (22)$$

In a subsequent analysis by Joh et al. (75), carried out with the solubilized adrenal enzyme, they concluded that the mechanism did not involve a reduced enzyme intermediate but rather a quaternary complex. Since their conclusions with the solubilized enzyme were at variance with those reported for the soluble enzyme (63), these authors postulated that the mechanism of action of the soluble and solubilized enzymes was different.

Shiman and Kaufman (unpublished results) have carried out a kinetic analysis of this hydroxylation reaction with both the particulate and the highly purified, solubilized enzyme. Their results do not support the idea that the mechanism is of the ping-pong type or that the solubilized and particulate enzymes differ in their mechanism of action.

The results of initial velocity measurements obtained by variation of the tyrosine concentration at several fixed concentrations of tetrahydropterin, and by variation of the tetrahydropterin concentration at several fixed concentrations of tyrosine, are shown in Fig. 18 as double reciprocal

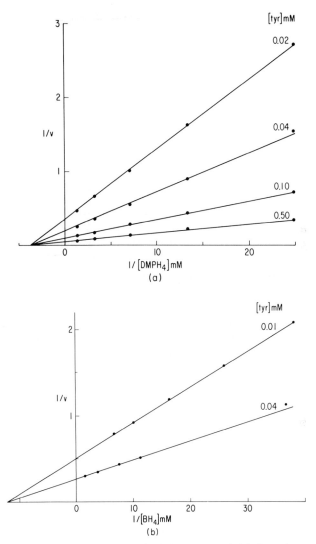

FIG. 18. Double reciprocal plots of the initial rates of (a) dopa formation versus $DMPH_4$ and (b) tetrahydrobiopterin (BH_4) concentrations. The highly purified solubilized enzyme from bovine adrenal medulla was used.

plots. The fact that the lines intersect indicates that the mechanism is sequential rather than ping pong and that the tyrosine must bind to the enzyme before the oxidized pterin is released. The results obtained with tetrahydrobiopterin are qualitatively the same as those with $DMPH_4$ except, as already noted (see Table VIII), that the K_m for tyrosine is smaller in the presence of tetrahydrobiopterin than it is in the presence of $DMPH_4$. Furthermore, the kinetic pattern observed with the particulate enzyme is the same (Fig. 19) as that obtained with the solubilized enzyme, i.e., converging, rather than parallel, lines in the double reciprocal plots.

These results strongly support the conclusion that the mechanism of action of the particulate and solubilized form of tyrosine hydroxylase from bovine adrenal glands is identical. There appears to be no evidence in favor of the postulate (119) that the two forms are characterized by differences in kinetic behavior. The results of the kinetic analysis of Shiman and Kaufman strongly indicate that the tyrosine hydroxylation reaction proceeds through a quaternary enzyme–substrate complex (hydroxylase–tyrosine–tetrahydropterin–oxygen) and that a reduced form of the enzyme, as depicted in Eq. (21), is not an intermediate in the reaction.

As already discussed in Section II,C. the phenomenon of partial uncoupling of hydroxylation from electron transfer, first discovered with phenylalanine hydroxylase (150), has provided important insights into the mechanism of pterin-dependent hydroxylation reactions.

At the time of its discovery, it was anticipated that the other pterin-dependent hydroxylases would exhibit the same phenomenon, an anticipation that has been realized with bovine adrenal tyrosine hydroxylase (143). Whereas uncoupling undoubtedly results from the analogous chemical reactions during phenylalanine and tyrosine hydroxylation, the structural requirements necessary to elicit uncoupling differ with the two hydroxylases.

The most striking difference is that with $DMPH_4$, the widely used cofactor analog, tight coupling is observed with phenylalanine hydroxylase (150), whereas there is loose coupling with tyrosine hydroxylase (143). Table X summarizes the results obtained with various tetrahydropterins and highly purified solubilized tyrosine hydroxylase. As already mentioned in a previous section, there is tight coupling with tetrahydrobiopterin as the cofactor (dopa/dihydrobiopterin = 1.0).

From this limited study of pterin analogs, it appears as if the presence of a methyl substituent in position 7 of the pterin ring leads to anomalous stoichiometry with tyrosine hydroxylase. By contrast, with phenylalanine hydroxylase it is the absence of a substituent in position 6 that appears to

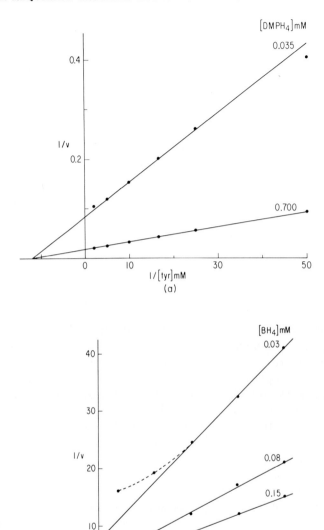

FIG. 19.   Double reciprocal plots of initial rates of dopa formation versus tyrosine concentration at several fixed concentrations of (a) DMPH$_4$ and (b) tetrahydrobiopterin. The particulate enzyme from bovine adrenal medulla was used.

TABLE X

STOICHIOMETRY OF TYROSINE HYDROXYLATION WITH MODEL COFACTORS

Tetrahydropterin	Dihydropterin/dopa	Range	No. of determinations
6MPH$_4$	1.10 ± 0.05	(1.05 − 1.12)	6
7MPH$_4$	1.60 ± 0.05	(1.56 − 1.62)	4
PH$_4$[a]	1.15 ± 0.05	(1.12 − 1.15)	4
DMPH$_4$	1.75 ± 0.15	(1.61 − 2.0)	10

[a] PH$_4$ stands for 2-amino-4-hydroxytetrahydropteridine.

be the structural requirement for the anomalous stoichiometry (150). This conclusion is based not only on the different results obtained for the two enzymes with DMPH$_4$ but also on those obtained with the unsubstituted pterin, PH$_4$; with phenylalanine hydroxylase, the hydroxylation reaction with this pterin gives an anomalous ratio (150), whereas the ratio is close to 1.0 with the tyrosine hydroxylase-catalyzed reaction. The mechanistic significance of these different structural requirements exhibited by the two hydroxylases is not clear.

With both hydroxylases, it has been shown that the excess oxidation of the tetrahydropterin that leads to ratios greater than unity is strictly dependent on the presence of the amino acid substrate (143, 150). These results indicate that the amino acid must combine with the enzyme in order for net electron transfer from tetrahydropterin to oxygen to occur. With phenylalanine hydroxylase, it has been shown that the K_m for phenylalanine is the same for the tightly coupled and the loosely coupled reactions (150), an indication that phenylalanine occupies the active site of the enzyme not only when it serves as the substrate to be hydroxylated but also when it functions to facilitate electron transfer between the tetrahydropterin and oxygen; i.e., the same phenylalanine molecule probably serves both functions. It is likely that this conclusion is also valid for tyrosine and tyrosine hydroxylase.

The conclusion that the amino acid substrate is required for the reaction between the tetrahydropterin and oxygen is consistent with the one already drawn from the kinetic analysis of the hydroxylation reaction, i.e., that it proceeds through a quaternary complex of the hydroxylase, amino acid substrate, oxygen, and tetrahydropterin.

An unusual feature of tyrosine hydroxylase is its ability to catalyze two successive hydroxylation reactions when phenylalanine serves as its substrate—phenylalanine is converted to tyrosine, which is then converted to dopa (64). [As has ready been mentioned, there is evidence that the enzyme can also catalyse the alternate sequence: phenylalanine → m-tyrosine → dopa (155, 156).] Substantial amounts of tyrosine can be formed under these conditions, proving that it must be a free intermediate in the reaction. A possibility that has not yet been eliminated is that at least part of the dopa formed from phenylalanine proceeds via the formation of the epoxide of phenylalanine, as an intermediate, followed by hydrolytic cleavage of the epoxide to form dopa. This mechanism would predict the incorporation of some $^{18}O_2$ from $H_2^{18}O$ into dopa.

E. Regulation of Tyrosine Hydroxylase

This discussion will be limited to those aspects of the regulation of the activity of tyrosine hydroxylase that do not involve alterations in the tissue levels of the enzyme; i.e., mechanisms of regulation that appear to involve synthesis or degradation of the enzyme will not be covered.

The first question that will be considered is whether the *in vivo* activity of tyrosine hydroxylase is normally limited by availability of its substrates and coenzyme. To attempt to answer this question, the K_m values of these compounds must be compared with their concentrations in the tissues under consideration. There are two obvious limitations to any firm conclusions that can be reached: (a) The K_m values have been determined with partially purified preparations of the enzyme and may not accurately reflect the properties of the enzyme *in vivo;* (b) in most cases, the concentration of the substrate in the intracellular compartment in which the enzyme might be located is not known. The only estimates that can be made are inadequate ones that are based on the weak assumption of uniform intracellular distribution of the substance in question. A final caveat that must be stated is that we have assumed in the following discussion that the naturally occurring pterin cofactor in adrenal medulla [see Lloyd and Weiner (108)] and brain tissue is either tetrahydrobiopterin or a closely related pterin.

One of the keys to a meaningful evaluation of the above question is the fact that with tyrosine hydroxylase, just as with phenylalanine hydroxylase, the K_m for tetrahydrobiopterin is not only lower than it is with $DMPH_4$ but also the K_m values of the other two substrates, tyrosine and oxygen, are lower in the presence of the naturally occurring pterin than in the presence of the synthetic cofactor analog.

The K_m of the adrenal enzyme for tetrahydrobiopterin is about 1×10^{-4} M (143) (see Table VI), whereas the concentration of biopterin in bovine adrenal medulla (assuming uniform intracellular distribution) can be estimated to be about 2×10^{-5} M (108). The same value has been reported for whole rat adrenal glands (136).

It would thus appear that the activity of the adrenal enzyme may be limited by availability of the cofactor but that the limitation is not as severe as might have been estimated from the K_m of DMPH₄.

On the other hand, the activity of brain enzyme would appear to be absurdly limited by low levels of the cofactor. The K_m value for tetrahydrobiopterin assayed with a partially purified preparation of the enzyme from brain is about 2.5×10^{-4} M (Lloyd and Kaufman, unpublished), whereas the level of biopterin in rat brain is about 1×10^{-6} M (137). This discrepancy suggests that in brain tissue one of the assumptions stated at the beginning of this section is probably invalid; i.e., (a) the K_m measured *in vitro* is not a true reflection of the value *in vivo*, or (b) there is a much higher concentration of tetrahydrobiopterin in some intracellular compartment in which the hydroxylase functions, or (c) the cofactor in brain is not tetrahydrobiopterin.

The K_m of the adrenal enzyme for tyrosine in the presence of tetrahydrobiopterin lies between 4×10^{-6} M (particulate enzyme) and 1.5×10^{-5} M (solubilized enzyme) (143) (see Table VIII), whereas the level of tyrosine in bovine adrenal medulla is about 0.5 to 1.0×10^{-4} M (55). It is unlikely, therefore, that the activity of the enzyme is limited by low concentrations of tyrosine. Indeed, as will be discussed later, a more likely possibility is that the tissue concentration of tyrosine is high enough to inhibit the enzyme.

The variation of the K_m of tyrosine with the pterin cofactor used provides a dramatic example of the importance of knowing the K_m in the presence of the naturally occurring pterin cofactor. The K_m for tyrosine in the presence of DMPH₄ is 13–25 times higher than it is in the presence of tetrahydrobiopterin (Table VIII). Based on these K_m values for tyrosine in the presence of DMPH₄, which are close to the tissue levels of the amino acid, one would have concluded that tyrosine hydroxylase is limited by low concentrations of tyrosine.

The K_m values of the other substrate of the adrenal enzyme, oxygen, vary in a similar way with pterin structure, being about 6% in the presence of DMPH₄ (63) and about 1% in the presence of tetrahydrobiopterin (30). With the brain enzyme, the K_m for oxygen is about 1% in the presence of DMPH₄ and less than 1% in the presence of tetrahydrobiopterin (30).

It has been reported that for rats breathing air at atmospheric pressure, the brain level of oxygen is 37 mM or 5% (67). Therefore, *in vivo*, the brain tyrosine hydroxylase is probably saturated with oxygen. The same conclusion is probably valid for the adrenal enzyme.

Tyrosine hydroxylase shares with phenylalanine hydroxylase the important property of being inhibited by excess concentrations of either of its substrates tyrosine (143) and oxygen (30). With both enzymes, marked inhibition is apparent only in the presence of tetrahydrobiopterin. The inhibition by tyrosine has been observed with both the particulate (see Figs. 19b and 20) and the solubilized adrenal enzyme (143). Inhibition by oxygen has been reported for both the solubilized adrenal enzyme and the brain enzyme (30).

The inhibition by tyrosine may be of physiological significance since the enzyme is about 30% inhibited (143) at only twice the normal tissue concentration of tyrosine (130). In contrast to tyrosine, phenylalanine (up to 0.5 mM) does not inhibit its own hydroxylation (143), although it will compete with tyrosine when both substrates are presented simultaneously to the enzyme (64). The latter finding has led to the suggestion (121, 122) that in PKU, inhibition of tyrosine hydroxylase by high levels of phenylalanine could account for the decreased amounts of epinephrine found in patients with the disease (167). Since it has been shown (143), however, that in the presence of tetrahydrobiopterin, dopa is formed at a high rate from phenylalanine, the proposed explanation is not plausible.

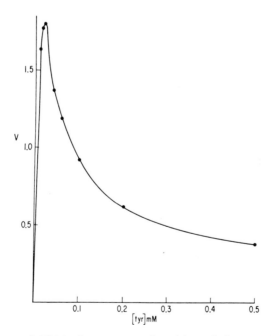

FIG. 20. Substrate inhibition by excess tyrosine with particulate tyrosine hydroxylase from bovine adrenal medulla. The tetrahydrobiopterin concentration was 0.1 mM.

It has been reported that hyperbaric oxygen (3–6 atm of pure oxygen) leads to a decrease in brain norepinephrine (26) and dopamine (54) as well as heart norepinephrine (9). At these levels of brain oxygen [for rats breathing 4 atm of pure oxygen, the brain level is 820 mm or 108% (67)], 50% inhibition of bovine tyrosine hydroxylase would be expected, an inhibition that could account for the decreased levels of norepinephrine and dopamine that have been observed following hyperbaric oxygen. Based on these findings, Fisher and Kaufman (30) have suggested that the inhibition of brain tyrosine hydroxylase by oxygen could account for some of the neurotoxic effects of the treatment of newborns with high concentrations of oxygen (58).

1. Product Inhibition

It has been reported that dopa, as well as a large variety of other catechols (11, 63, 165), inhibit tyrosine hydroxylase, the inhibition being

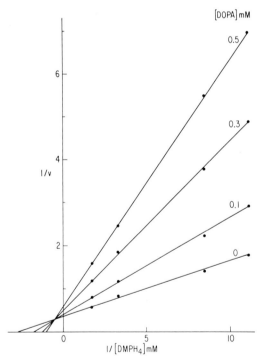

FIG. 21. Inhibition of particulate bovine adrenal tyrosine hydroxylase by dopa in the presence of DMPH$_4$.

competitive with the tetrahydropterin (63). Shiman and Kaufman (143) found that the inhibition by dopa that is observed with both the solubilized and particulate adrenal enzyme is noncompetitive with respect to both tyrosine and tetrahydropterin. The data for the soluble enzyme with tetrahydrobiopterin and the particulate enzyme with DMPH$_4$ are shown in Figs. 21 and 22. To forestall confusion, it should be noted that the extent of inhibition decreases as the concentration of tetrahydropterin increases, but that even at infinite concentrations of tetrahydropterin, inhibition is still manifest; in the accepted kinetic terminology, this pattern of inhibition is "noncompetitive." The same conclusion applies to the effect of tyrosine on the extent of inhibition.

Table XI summarizes the K_i (slope) values obtained from secondary plots of slope versus dopa concentration. The K_i values obtained when the tetrahydropterin was varied at fixed concentration of tyrosine are the same as those shown for tyrosine at fixed tetrahydropterin concentration. It is apparent from the data in Table XI that the K_i for dopa is very responsive to the structure of the tetrahydropterin used; the K_i with tetrahydrobiopterin is about one-fifth the K_i with DMPH$_4$. The change in K_i with pterin structure implies that at some time the pterin and dopa must be present simultaneously on the enzyme surface.

There is some evidence that tyrosine hydroxylase can be inhibited *in vivo* by catechols. When the monoamine oxidase inhibitor, pargyline, was

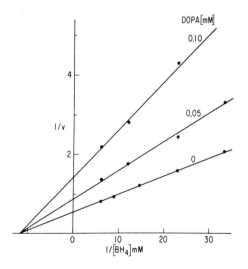

FIG. 22. Inhibition of solubilized bovine adrenal tyrosine hydroxylase by dopa in the presence of BH$_4$ (tetrahydrobiopterin).

TABLE XI

INHIBITION OF TYROSINE HYDROXYLASE BY L-DOPA

Tetrahydropterin	Particulate enzyme K_i (slope)[a]	Soluble enzyme K_i (slope)[a]
Tetrahydrobiopterin	40 μM	60 μM
DMPH$_4$	150 μM	500 μM

[a] Summary of K_i (slope) values obtained with two different pterins with dopa as the inhibitor. The K_i (slope) values were obtained from a replot of slope versus L-dopa. The K_i values obtained were identical within experimental error for inhibition against either pterin or L-tyrosine. The concentrations of the nonvaried substrates were at their K_m values.

administered to guinea pigs, the brain and heart levels of norepinephrine increased two- to threefold and this increase was accompanied by a four- to fivefold decrease in the rate of incorporation of [^{14}C]tyrosine into norepinephrine (148).

Although the elevated levels of norepinephrine might have inhibited one of the other enzymes in the biosynthetic pathway, the most likely explanation is that tyrosine hydroxylase, believed to catalyze the rate-limiting step in the pathway, is the site of inhibition.

An *in vivo* effect of dopa on tyrosine hydroxylase that cannot be fully evaluated at the present time has been reported by Dairman and Udenfriend (18). They found that the administration to rats of large amounts of L-dopa led to a decrease in the *in vivo* activity of tyrosine hydroxylase in the adrenals; after 4 and 7 days of dopa treatment, when the dopa-plus-dopamine levels were increased 126 and 136% (total catecholamines, i.e., epinephrine, norepinephrine, dopa, and dopamine, were increased about 10%), the hydroxylase levels were decreased 28 and 47%, respectively (18). Based on the small increase in total catecholamines, the authors concluded that inhibition of the activity of the enzyme in the *in vitro* assay could not explain their results.

The finding that the inhibition of the hydroxylase by dopa is much more severe in the presence of tetrahydrobiopterin than in the presence of DMPH$_4$ (Figs. 21 and 22) indicates that this conclusion may require re-examination. Although the amount of tissue used per assay was not stated, they have previously used the equivalent of 125 mg tissue per milliliter [see Table II, Nagatsu *et al.* (121)]. From their data on the dopa content of the adrenal glands, it can be calculated that this amount of tissue from the dopa-treated animals would have contained about 7 μg dopa- thus, the dopa concentration in the assay would have been about 0.035 mM. If one assumes that in the presence of tetrahydrobiopterin, dopa and dopamine are somewhat better inhibitors than the other catecholamines, an admittedly untested assumption, then it can be estimated that the concentration of dopa contributed by the adrenal tissue would have led to about 30% inhibition of the enzyme's activity. It can be seen that at least part, and perhaps all, of the observed inhibition could be accounted for by the dopa plus dopamine added with the tissue.

It is apparent from the studies of Shiman and Kaufman that two important regulatory properties of tyrosine hydroxylase are lost or attenuated when DMPH$_4$ is substituted for tetrabiopterin: inhibition by tyrosine and inhibition by dopa, the product of the reaction. It would be of interest to test the prediction that follows from these results, i.e., that DMPH$_4$ in vivo might lead to norepinephrine synthesis that is uncontrolled by these two restraints.

2. Dihydropteridine Reductase

The activity of tyrosine hydroxylase in any tissue could, in theory, be regulated by the activities of any one of the enzymes that have been shown to play a role in pterin-dependent hydroxylating systems: dihydropteridine reductase, dihydrofolate reductase, and catalase (97). One step further removed from these enzymes, but still potential regulators of the activity of the hydroxylases, are those enzymes that determine the tissue level of reduced pyridine nucleotides, presumably the secondary electron donors that are utilized *in vivo* to regenerate the tetrahydropterins from their dihydro derivatives.

It has been suggested that dihydropteridine reductase may play an important role in determining the rate of dopa formation from tyrosine in the adrenal medulla (119), The suggestion is based on the finding that the inhibition of adrenal tyroisne hydroxylase *in vitro* by a catechol, i.e., epinephrine, is decreased by increasing concentrations of the reductase. Since there is a decrease in the inhibition with catechols by increasing concentrations of tetrahydropterin (see Figs. 21 and 22), the partial relief by the reductase of the inhibition caused by epinephrine was thought to result from the higher levels of tetrahydropterin that could be maintained by the higher concentrations of the reductase.

The data in support of the conclusion that the inhibition by epinephrine is a function of dihydropteridine reductase levels are not convincing. The inhibition by 2×10^{-5} M epinephrine decreased from approximately 65 to 50% when the reductase concentration was increased by 400%, Since there is no basis for evaluation of the statistical significance of even this slight relief of the inhibition, the above conclusion would appear to be open to question. Furthermore, it should be noted that the inhibition of the hydroxylase by dopa (Figs. 21 and 22) observed by Shiman and Kaufman was measured in the presence of saturating concentrations of the reductase.

Finally, it should be mentioned that the activity of dihydropteridine reductase in adrenal medulla has been estimated to be 750 times greater than that of tyrosine hydroxylase (17), a finding that makes improbable the suggestion of Musacchio *et al.* (119) that the reductase limits norepinephrine synthesis in this tissue.

Since 7,8-dihydrobiopterin is a relatively poor substrate for dihydrofolate reductase (94), this enzyme would appear to be a more likely candidate than dihydropteridine reductase for playing an important role in the regulation of tyrosine hydroxylase activity.

3. Effect of Heparin on Brain Tyrosine Hydroxylase

The properties of tyrosine hydroxylase from rat brain have been reported to be dramatically altered by microgram quantities of the sulfated mucopolysaccharide, heparin (104). The enzyme is thought to occur in brain in both soluble and particulate forms. Although the two forms have identical K_m values for tyrosine (0.055 mM), the particulate enzyme has a higher V_{max} and a lower K_m (0.15 mM) for DMPH$_4$ than does the soluble enzyme ($K_m = 0.75$ mM). Another significant difference between the two forms of the enzyme is that the K_i for norepinephrine is much lower ($K_i = 0.1$ mM) for the particulate enzyme than for the soluble enzyme ($K_i = 0.7$ mM).

The addition of heparin to the soluble enzyme alters all of these kinetic properties so that it resembles the particulate enzyme. The authors have suggested that the site of binding of tyrosine hydroxylase on a membrane surface may contain a sulfated mucopolysaccharide and that intracellular binding of the enzyme at this site may be an important determinant of the regulatory properties of the enzyme.

In addition to this effect of heparin, these authors also observed that the soluble brain tyrosine hydroxylase is stimulated by high concentrations (>0.1 M ionic strength) of salt with sulfate ions being the most active. In contrast to the effect of heparin, sulfate ions did not alter the apparent K_m for DMPH$_4$, but only increased V_{max}.

In view of the critical dependence of many of the properties of the enzyme on the nature of the pterin cofactor used, it would be of great interest to see if these effects of heparin are exhibited when the enzyme is assayed in the presence of tetrahydrobiopterin.

IV. TRYPTOPHAN HYDROXYLASE

Considerably less is known about the third pterin-dependent aromatic amino acid hydroxylase, tryptophan hydroxylase, than is known about either phenylalanine or tyrosine hydroxylases. Although a preliminary publication reported that a tryptophan hydroxylase from rabbit brain stem had been purified about 100-fold (123), neither the details of the purification procedure nor results of studies of the properties of the highly purified enzyme has been forthcoming. There have been no other reports of studies carried out on extensively purified tryptophan hydroxylase.

The hydroxylase is believed to catalyze the rate-limiting step in the biosynthesis of the putative neurotransmitter, 5-hydroxytryptamine (serotonin) (45, 68). In view of the fact that the pathway involved in serotonin synthesis consists of only two consecutive steps, hydroxylation of tryptophan to 5-hydroxytryptophan, followed by decarboxylation of the hydroxylated amino acid to serotonin, and the fact that the decarboxylase has much higher *in vitro* activity than does the hydroxylase (61), the conclusion that the hydroxylation step is limiting is reasonable, if largely untested.

The first reports of the *in vitro* hydroxylation of tryptophan catalyzed by mammalian tissue preparations were published in 1961. In that year, Cooper and Melcer (15) reported the presence in intestinal mucosa of a particulate, ascorbate-dependent enzyme that could catalyze the anaerobic hydroxylation of tryptopahn. In the same year Freedland *et al.* reported the presence of a soluble tryptophan hydroxylase in a rat liver supernatant fraction that required oxygen and reduced pyridine nucleotide (33). The claim of Cooper and Melcer has never been confirmed, whereas the tryptophan hydroxylating activity in rat liver was traced to phenylalanine hydroxylase (138).

In 1964, Grahame-Smith (4S) reported the first evidence for a specific tryptophan hydroxylase in a normal mammalian tissue, The enzyme was detected in whole homogenates, but not in the high-speed supernatant fraction, of brain stem from dogs and rabbits. In dog brain, the activity was found in the hypothalamus-thalamus and the midbrain-medulla regions; no activity was detected in the cerebellum or cortex. The anatomical distribution of the enzyme closely paralleled the serotonin content of these areas, Tryptophan hydroxylation was not inhibited by amethopterin (5×10^{-5} M), 5-hydroxytryptamine (2×10^{-4} M), or 5-hydroxytryptophan (2×10^{-4} M). No other properties of the enzyme were reported.

After Grahame-Smith's demonstration that a specific tryptophan hydroxylase exists in the brain, the enzyme was partially purified from this tissue and from pineal glands. The following discussion will be restricted

TABLE XII

DISTRIBUTION OF TRYPTOPHAN HYDROXYLASE AND SEROTONIN IN VARIOUS
REGIONS OF THE BRAIN OF DIFFERENT ANIMALS

| Region | Rat brain[a] | | Guinea pig brain[b] | Cat brain[c] |
	Hydroxylase (nmoles/gm/hr)	Serotonin (μg/gm)	Hydroxylase (nmoles/gm/hr)	Hydroxylase (nmoles/gm/hr)
Cerebral cortex	0.89 ± 0.06	0.313 ± 0.01	0.122	1.0–1.9
Hippocampus	0.58 ± 0.07	0.278 ± 0.009	0.133	4.3
Striatum	1.14 ± 0.08	0.390 ± 0.017	—	22.8
Midbrain	3.35 ± 0.26	0.607 ± 0.027	—	10.8–17.3
Hypothalamus	4.81 ± 0.45	0.807 ± 0.023	0.146	14.7
Pons and medulla oblongata	2.83 ± 0.06	0.500 ± 0.015	0.206	4.0
Cerebellum	0.08 ± 0.01	0.056 ± 0.003	0.046	0.7

[a] Data for high-speed supernatant fractions from Deguchi and Barchas (22). Temperature of assay, 36°.

[b] Data for homogenates from Ichiyama et al. (61). Temperature of assay, 37°.

[c] Data for homogenates from Peters et al. (133).

to a consideration of the properties of the tryptophan hydroxylation system from these two tissues.

A. Distribution and Development of Tryptophan Hydroxylase in Brain

There have been several detailed studies of the anatomical distribution of tryptophan hydroxylase in the brains of different animals (see Table XII). The data obtained with rat brain confirm the earlier conclusion of Grahame-Smith (45), based on a less detailed analysis, that the distribution pattern of hydroxylase activity parallels that of serotonin in the various regions of the brain. There appear to be, however, marked differences in the pattern of distribution of the enzyme in the brains of the three different species. Deguchi and Barchas (22), e.g., found the highest activity in the hypothalamus of the rat brain, with very low activity in the hippocampus, whereas Ichiyama et al. (61) found the activity in these two regions of the guinea pig brain to be about the same. Peters et al. (133), on the other hand, reported that in the cat brain the striatum had the highest activity. There is general agreement that the cerebellum in all species examined is very low in tryptophan hydroxylase activity.

It is not known whether these apparent discrepancies are because of actual species differences in the regional distribution of the enzyme or of differences in the assay conditions employed in different laboratories. It should be noted that the activities in rat brain were determined on high-speed supernatant fractions, whereas those for the guinea pig and cat brains were determined on homogenates. Although it does not seem likely, this difference may account, in part, for the different distribution pattern.

The results in Table XII also appear to show that the specific activity of brain tryptophan hydroxylase varies markedly in the three different species that have been studied. Most of the variation, however, is more apparent than real, and can be traced to the use of different concentrations of tryptophan in the three studies, and to the fact that the activities reported for cat brain are V_{max} values. Thus, the tryptophan concentrations used by Ichiyama et al. (61) for guinea pig brain was 0.003 mM, whereas for rat brain it was 0.010 mM (22), and for cat brain it was 0.0165 mM (133). If all of the activities are expressed as V_{max} values [using a K_m for tryptophan of 0.05 mM (34)], the activities of the hippocampus region of the rat, guinea pig, and cat brains are 3.5, 2.2, and 4.3 nmoles/gm/hour, respectively.

In connection with the regional distribution of the enzyme in brain, the high tryptophan hydroxylase activity in the pineal gland deserves comment. This small organ, which is physically connected to the brain but anatomically distinct from it, contains 50 times as much serotonin per gram as the whole brain. It also contains the serotonin derivative melatonin (5-methoxy-N-acetyltryptamine), the factor that lightens the skin color, as well as the two enzymes that convert serotonin to melatonin. The specific activity of tryptophan hydroxylase in the rat pineal gland is about 10 times higher than that of rat brain stem (109). As will be discussed later, however, there are indications that the hydroxylase in the two different tissues are different enzymes.

It was also reported by Deguchi and Barchas (22) that the hydroxylase specific activity in rat brain stem at birth was about 30% of the adult value. For the next 10 days there was no increase in activity; between the tenth and twentieth days, the enzymic activity reached the adult levels. The enzymic activity in the cortex showed a similar pattern of development: 10% of the adult levels at birth, followed by a rapid increase between the tenth and thirtieth days with adult levels being reached at 30 days after birth. This pattern of increase in the specific activity of the enzyme is similar to that reported for the increase in serotonin content of the rat brain, where the level of birth is about 20–33% of the adult level and the adult levels are reached between 4 and 6 weeks after birth (see Deguchi and Barchas (22) and references therein).

B. Subcellular Distribution of Tryptophan Hydroxylase

The early studies on brain tryptophan hydroxylase indicated that the enzyme was particle-bound, most of the activity being located in the mitochondrial fraction (40, 49). More recent evidence points to its localization within the nerve ending granule (46).

Lovenberg and his colleagues confirmed the finding that most of the activity appeared to be particle-bound, but showed that after dialysis the major part of the enzyme was in the supernatant fraction (110). Dialysis appears to have at least two effects: It removes endogenous inhibitors (the total measurable activity increases), but it also leads to the ready solubilization of the activity that was originally particle-bound. Although these studies have identified some of the factors that can confound attempts to define the intracellular localization of the enzyme *in vitro*, they still leave unanswered the question of whether the enzyme is present exclusively in the cytoplasmic portion of the cell or in nerve ending particles from which it is released during homogenization, or both.

C. Properties of Tryptophan Hydroxylase

1. Tetrahydropterin Requirement

There has been considerable disagreement about the requirement for a tetrahydropterin for this hydroxylation reaction. The reason for the disagreement is that the particulate preparations of the enzyme used by the earlier workers catalyzed the reaction without any supplementation. Thus, Gal et al. (40), using a twice-washed rat brain mitochondrial fraction, were able to demonstrate a twofold stimulation of the reaction by TPNH, but no stimulation by DMPH$_4$; "reduced biopterin" added together with TPNH stimulated 20% over the rate with TPNH alone. Essentially the same results were obtained by Ichiyama et al. (62) with mitochondria from guinea pig brain. The addition of DMPH$_4$, or TPNH, or the combination of the two, to the crude mitochondrial fraction significantly inhibited the hydroxylation reaction. When the mitochondria were washed, the hydroxylation reaction was stimulated about threefold by the addition of TPNH plus DMPH$_4$; neither addition alone was effective (although DPNH alone stimulated almost as much as the above combination). Treatment of the mitochondrial fraction with 1-butanol, followed by dialysis, led to an enzyme preparation that was markedly stimulated by the addition of DMPH$_4$ plus reduced pyridine nucleotides. It is apparent from these results that the particle contains not only the hydroxylase but also an active cofactor and, assuming that the cofactor is a tetrahydropterin, a tetrahydropterin-regenerating system. It should be recognized that although it is reasonable to suspect that the hydroxylation cofactor contained within this particle is a pterin, this suspicion has not yet been verified experimentally.

In contrast to the particulate enzyme, the soluble enzyme from both brain and the pineal shows an absolute requirement for a pterin (109) and, indeed, shows the by-now-familiar characteristics of a pterin-dependent hydroxylase (97). In addition to the pterin requirement, Lovenberg et al. (109) showed that both enzymes required oxygen and 2-mercaptoethanol. The pineal, but not the brain enzyme, was also shown to be stimulated by twofold by Fe^{2+}. Both enzymes were reported to be inhibited by the iron chelators α,α-dipyridyl and o-phenanthroline (69). The activity of the rat brain stem enzyme was shown to be 1.7 times higher with tetrahydrobiopterin than with DMPH$_4$; the K_m values were reported to be 3×10^{-5} M and 5×10^{-6} M for DMPH$_4$ and tetrahydrobiopterin, respectively (69). Ichiyama et al (62) reported a K_m value of 6×10^{-5} M for DMPH$_4$ for the brain enzyme from guinea pigs.

A more detailed study of the pterin specificity of partially purified preparation of the soluble enzyme from rabbit hindbrain was recently reported by Friedman et al. (34) (see Table XIII). These workers confirmed the finding (69) that the K_m for tetrahydrobiopterin is much lower than it is for DMPH$_4$. The values that they reported for both pterins, however, are five times greater than those reported for the rat brain enzyme; the value for DMPH$_4$ is twice as high as that reported for the guinea pig

TABLE XIII

SUMMARY OF APPARENT K_m VALUES FOR VARIOUS SUBSTRATES FOR
RABBIT HINDBRAIN TRYPTOPHAN HYDROXYLASE[a]

Substrate	Cofactor	Tetrahydrobiopterin	DMPH₄	6-MPH₄
L-Tryptophan		50 μM	290 μM	78 μM
Pterin		31 μM	130 μM	67 μM
Oxygen		2.5%	20%	—

[a] All values were calculated from double reciprocal plots of initial
velocity versus variable substrate concentrations. The K_m for L-tryptophan
with tetrahydrobiopterin as the cofactor was obtained by extrapolation to
the abscissa of those points in the plot that fell on a straight line (tryptophan concentrations of 0.1 mM or less). When 6-MPH₄ concentration was
varied, the concentration of L-tryptophan was 0.7 mM. When L-tryptophan concentration was varied in the presence of 6-MPH₄, the concentration of 6-MPH₄ was 0.29 mM.

brain enzyme (62). It is not known whether these differences are characteristic of the enzyme in different species or whether they reflect differences in the assay conditions used.

The V_{max} for the rabbit brain enzyme in the presence of tetrahydrobiopterin is slightly (but perhaps not significantly) higher than it is with
DMPH₄ (34). This finding is in contrast to that of Jequier et al. (69) with
the rat brain enzyme. As already mentioned, they reported that the rate
of hydroxylation with tetrahydrobiopterin is almost 70% faster than it
is with DMPH₄, although it is not clear that their rates are V_{max} values.
The results obtained by Friedman et al. with tryptophan hydroxylase in
the presence of DMPH₄ and tetrahydrobiopterin are consistent with those
previously obtained with both phenylalanine and tyrosine hydroxylase;
they can be summarized by the following generalization: With all three
pterin-dependent hydroxylases, the major difference between the naturally
occurring cofactor and the cofactor analog, DMPH₄, is that the K_m is
much lower for the former compound, whereas the V_{max} with tetrahydrobiopterin is either about the same [tryptophan hydroxylase and tyrosine
hydroxylase (34, 143)[§] or it is somewhat lower [phenylalanine hydroxylase
(96)] than it is with DMPH₄.

[§] With solubilized bovine adrenal tyrosine hydroxylase, the rates of the oxidative
reaction catalyzed by the enzyme are the same with DMPH₄, 6MPH₄, and tetrahydrobiopterin (143). Since the reaction is partially uncoupled with the dimethylpterin,
however, the rate of dopa formation is less with DMPH₄ than with tetrahydrobiopterin.

As already discussed in connection with brain tyrosine hydroxylase, the hydroxylation cofactor in brain has not yet been identified so that the natural cofactor for brain tryptophan hydroxylase is not known with certainty. It seems probable that it will prove to be biopterin or a closely related compound.

Related to this question is an observation reported by Gal et al. (40). They reported that the pterin cofactor isolated from liver according to Kaufman's procedure (89) did not stimulate a particulate preparation of brain tryptophan hydroxylase, whereas chemically "reduced biopterin," presumably the tetrahydro compound, was very active, Although one can question the aptness of the phrase "very active" to describe a 23% stimulation, these observations left the question of biopterin's participation in this reaction needlessly beclouded. A likely explanation for these results is that the brain preparation used by Gal et al. is limiting in either TPNH, or dihydrofolate reductase, or both. It has been shown that the liver cofactor, in the form in which it is isolated, i.e., 7,8-dihydrobiopterin (89), must be converted to the tetrahydro derivatives before it is active with phenylalanine hydroxylase, and that this reductive step is catalyzed by dihydrofolate reductase in the presence of TPNH (94). The need for the dihydrofolate reductase with phenylalanine hydroxylase is obviated if tetrahydro, rather than 7,8-dihydrobiopterin, is used as the cofactor (94).

Thus, it seems likely that Gal et al. (40) were comparing the cofactor activity of 7,8-dihydrobiopterin from rat liver, which would require the dihydrofolate reductase–catalyzed reductive step before it was active, with chemically reduced tetrahydrobiopterin, which would not require this enzymic reaction. These considerations provide a plausible explanation for the observations reported by Gal et al. (40); hopefully, they will dispel some of the clouds of mystery that hang over the question of the role of pterins in tryptophan hydroxylation.

More recently, Gal (39) reported that a naturally occurring pterin-like compound could stimulate three- to sixfold the rate of tryptophan and tyrosine hydroxylation by a brain preparation, even in the presence of $DMPH_4$ or tetrahydrobiopterin.

Although it was reasonable to assume that tetrahydropterins function with tryptophan hydroxylase in the same way that they were first shown to function with phenylalanine hydroxylase (97), this assumption was not tested until recently. Indeed, it had never been shown that the pterin functions as a cofactor, i.e., that it functions catalytically during tryptophan hydroxylation. In 1972, Friedman et al. (34) proved this point for the first time when they showed that with 8.5 nmoles of tetrahydrobiopterin, 40 nmoles of 5-hydroxytryptophan were formed.

The demonstration of the catalytic role for the pterin implies that the tetrahydropterin must be capable of being regenerated during the hydroxylation reaction. Unfortunately, the use of 2-mercaptoethanol in the assay employed by some workers (110) has made it difficult to demonstrate the need for a tetrahydropterin-generating system. Sulfhydryl compounds can stimulate pterin-dependent hydroxylases not only by reducing quinonoid dihydropterins (83, 84) but perhaps also by stabilizing one of the enzymes in the system.

Thus, Lovenberg et al. (109), with a soluble enzyme preparation that showed a sharp requirement for $DMPH_4$, were unable to demonstrate a reduced pyridine nucleotide requirement, probably because 2-mercaptoethanol was functioning to keep the pterin reduced. On the other hand, Gal et al. (40), with a particulate preparation of the enzyme that showed only a minimal stimulation by a tetrahydropterin, were able to show a twofold stimulation of the rate of tryptophan hydroxylation with TPNH. Ichiyama et al. (62) were the first to show a clear-cut stimulation of the reaction by the combination of $DMPH_4$ and DPNH or TPNH. The first demonstration that tryptophan hydroxylation can be stimulated by highly purified dihydropteridine reductase, in addition to TPNH and tetrahydropterin, was reported recently by Friedman et al. (34). Since it had previously been established that the substrate for dihydropteridine reductase is a quinonoid dihydropterin (90), the stimulation of tryptophan hydroxylation by the reductase provides strong support for the conclusion that just as is the case with phenylalanine hydroxylase the quinonoid dihydropterin is the product of tetrahydropterin oxidation during the enzymic hydroxylation of tryptophan.

It is surprising that until recently the stoichiometry of the reaction catalyzed by tryptophan hydroxylase had not been determined. It was tacitly assumed that the stoichiometry was the same as that previously established for phenylalanine (79) and tyrosine (143) hydroxylases.

Friedman et al. (34) showed that with tetrahydrobiopterin this assumption is correct. They observed that 1 mole of 5-hydroxytryptophan is formed for each mole of tetrahydrobiopterin oxidized. This result, taken together with their demonstration that the hydroxylation reaction is stimulated by dihydropteridine reductase, led to the formulation of the reaction as shown in Eq. (23), where BH_4 stands for tetrahydrobiopterin and BH_2 stands for quinonoid dihydrobiopterin.

$$\text{L-Tryptophan} + BH_4 + O_2 \rightarrow \text{5-hydroxy-L-tryptophan} + BH_2 + H_2O \quad (23)$$

It is not yet known whether the stoichiometry with the widely used cofactor analog, $DMPH_4$, is the same as that shown in Eq. (23) or whether the analog leads to partial uncoupling as it does with tyrosine hydroxylase (143).

2. Amino Acid Specificity

The amino acid specificity of tryptophan hydroxylase has not been studied extensively. The pineal enzyme catalyzes the hydroxylation of phenylalanine at about the same rate as it does tryptophan, whereas the rat brain stem enzyme is essentially inactive with phenylalanine (110). Although it is not known whether a single enzyme is responsible for the hydroxylation of both amino acids with the pineal enzyme, the observation that phenylalanine inhibits tryptophan hydroxylation (110) suggests that this is the case. Results of recent *in vivo* experiments support the conclusion that pineal tryptophan hydroxylase can hydroxylate phenylalanine (2).

The K_m values for tryptophan with the beef pineal and the rat brain enzymes, both determined in the presence of $DMPH_4$, have been reported to be $5 \times 10^{-4} M$ and $3 \times 10^{-4} M$, respectively (110). The soluble guinea pig brain enzyme has a K_m for tryptophan of about $10^{-4} M$, whereas the particulate enzyme from the same tissue has a K_m of $2 \times 10^{-5} M$ (62), a finding that suggested that either two different tryptophan hydroxylases are present in this tissue or that the structure of the particle affects the properties of a single enzyme (62). A third alternative will be considered in a later section.

As can be seen in Table XIII, Friedman *et al.* (34) found with a partially purified enzyme from rabbit brain that the K_m for tryptophan varies with the tetrahydropterin used. A similar dependence of the amino acid substrate's K_m value on the structure of the pterin cofactor used had previously been reported for both phenylalanine and tyrosine hydroxylases (97, 143). The significance of this finding for tryptophan hydroxylase will be discussed in a later section.

3. Other Properties

The pH optimum for the enzyme from beef pineal, rat brain, and guinea pig brain has been reported to be 7.5, 7.5 (110), and 8.1 (62), respectively.

Most of the studies on tryptophan hydroxylase have included a mercaptan in the assay and have shown a marked stimulation of the hydroxylase activity by the sulfhydryl compound (62, 69). Since these assays did not include the dihydropteridine reductase–reduced pyridine nucleotide system for regeneration of the tetrahydropterin, and since it has been shown that mercaptans can effectively substitute for the enzymic regenerating system, it was not clear whether the observed stimulation by mercaptans resulted from their effects on the pterin cofactor or on the hydroxylase itself. In spite of this uncertainty, it was recently concluded that 2-mercaptoethanol functions not only to maintain the cofactor in a reduced form but also appears to be required by the enzyme for optimal activity (146).

Friedman *et al.* (34) have shown that in the presence of the enzymic tetrahydropterin-regenerating system, 2-mercaptoethanol is not only not required for optimal tryptophan hydroxylase activity but also that it actually inhibits the enzyme (46% inhibition at 0.05 M mercaptoethanol with the partially purified enzyme from rabbit brain).

Lovenberg *et al.* have reported that the activity of the hydroxylase from pineal glands is stimulated twofold by the addition of Fe^{2+} (109, 141), and have concluded that this result constitutes a direct demonstration of the role of iron in tryptophan hydroxylase (110). By contrast, the activity of the brain enzyme has been reported to be either not stimulated by Fe^{2+} (109) (rabbit brain enzyme) or only stimulated to the extent of 50% (62) (guinea pig brain enzyme). The enzymes from brain and pineal have been reported to be inhibited by iron chelators (62, 69).

Some recent observations made with the partially purified rabbit brain hydroxylase indicate that the conclusion that iron plays a role in tryptophan hydroxylase (110) may be premature. It was found that tryptophan hydroxylase resembles the other pterin-dependent hydroxylases in that its activity can be markedly stimulated by catalase (34), a finding that indicates that a component of the tryptophan hydroxylation system is sensitive to inactivation by H_2O_2 (H_2O_2 is generated by the nonenzymic oxidation of tetrahydropterins). It was also found that 1 mM Fe^{2+} can partially substitute for the catalase requirement (34), just as it can with tyrosine hydroxylase (143). With the latter enzyme, the requirement for Fe^{2+} has been shown to result from the ability of Fe^{2+} to protect the enzyme from inactivation by H_2O_2 (143).

As the above considerations show, the direct evidence in support of the conclusion that tryptophan hydroxylase is an iron enzyme is essentially nonexistent. In spite of this lack of evidence, the recent finding that rat liver phenylalanine hydroxylase is an iron enzyme (32), together with the many properties shared by the three pterin-dependent hydroxylases, leads to the strong expectation that tryptophan hydroxylase will also prove to be an iron enzyme.

D. Regulatory Properties of Tryptophan Hydroxylase

1. K_m Values and Tissue Levels of Substrates and Cofactor

The same restrictions that were used to limit the discussion of the regulatory properties of tyrosine hydroxylase will be used with this enzyme; i.e., we will not consider those phenomena that appear to involve alterations in the rates of synthesis or degradation of the hydroxylase. As with tyrosine hydroxylase, we will first attempt to correlate *in vivo* concentrations of substrates for the enzyme with the K_m values for these substrates, using the same assumptions as those used previously.

The early values of 300 μM reported for the K_m of this enzyme for tryptophan, all determined in the presence of $DMPH_4$, generated a considerable amount of speculative interest because this value is far above the plasma or the brain levels of tryptophan, both of which are about 30 μM for rats (28, 47). (The brain concentration was calculated from the reported value of 5 $\mu g/gm$ by assuming uniform distribution of the amino acid in the tissue.) These considerations led to the conclusion that tryptophan hydroxylase activity in brain is severely limited by availability of its amino acid substrate [Fernstrom and Wurtman (28), and references therein]. As has already been pointed out, however, the K_m of the enzyme for tryptophan varies significantly with the pterin cofactor used (see Table XIII). The value in the presence of tetrahydrobiopterin is 50 μM, a figure that is close to the estimated brain concentrations of this amino acid.

There is little doubt that a K_m value of 50 μM is more consistent with *in vivo* observations than is the previously accepted value of 300 μM. It has been shown (28) that the concentration of serotonin in the brain increases when the concentration of tryptophan in the brain is increased from 5 $\mu g/gm$ (about 30 μM) to 15 $\mu g/gm$ (about 90 μM), but further increases of tryptophan (up to 45 $\mu g/gm$) do not lead to further increases in serotonin content. The hydroxylating system in brain, therefore, appears to be saturated with tryptophan at approximately 90 μM. These *in vivo* observations agree very well with the apparent K_m value of 50 μM (34) but would be difficult to explain if one accepted the higher value of 300 μM.

Thus, it does not appear that this hydroxylase operates *in vivo* under any unique disadvantage of substrate limitation, but rather it appears to function, as probably most other enzymes do (14), with tissue concentrations of its substrate in the region of the K_m value of the substrate.

If the tissue concentration of tryptophan is close to the K_m value, conditions that either raise or lower its concentration should influence the rate of the hydroxylation reaction. There is considerable evidence that the rate of serotonin synthesis in brain can be altered by manipulations designed to change the tryptophan concentration.

It has already been pointed out that the serotonin content of the brain can be increased if the brain's tryptophan concentration is increased by administration of tryptophan to the animals.

The converse of this result also may have been observed. The injection of hydrocortisone into rats leads to an increase in tryptophan pyrrolase activity in the liver, which is followed by a decrease in the concentration of serotonin in the brain (48). The authors have interpreted these findings in terms of the following scheme: hydrocortisone → increased hepatic tryptophan pyrrolase activity → decreased tryptophan concentrations → decreased serotonin synthesis in the brain. The scheme is supported by the finding that inhibitors of the pyrrolase, allopurinol and yohimbine, prevented the hydrocortisone-induced decrease in serotonin concentration. It is regrettable that the actual decrease in tryptophan concentration (the second step in the above scheme) was not determined; had it been, one could have attempted to correlate observed changes with the K_m value for tryptophan.

Since there is convincing support for the idea that the activity of brain tryptophan hydroxylase can be regulated by blood levels of tryptophan, the well-known ability of serum proteins to bind tryptophan may be of potential significance in determining the rate of serotonin synthesis. It can be anticipated, therefore, that drugs such as aspirin, that can displace tryptophan from its binding sites on serum proteins (147), could have indirect effects on the rate of serotonin synthesis.

The lower K_m value for tryptophan in the presence of tetrahydrobiopterin is also relevant to the conclusion that the hydroxylation step is rate limiting in the biosynthesis of serotonin. One of the arguments used to support this conclusion is that the K_m of the decarboxylase for 5-hydroxytryptophan is much lower (20 μM) than that of the hydroxylase for tryptophan (16). Since Friedman et al. have shown that the apparent K_m of the hydroxylase for tryptophan may not be significantly different from 20 μM, this part of the argument is invalid, and the conclusion that the hydroxylation step is rate limiting, which is probably correct, must rest on other grounds.

Another puzzle that the lower K_m value for tryptophan in the presence of tetrahydrobiopterin may clarify is that reported by Ichiyama et al. (62). These workers found that the K_m value of soluble tryptophan hydroxylase for tryptophan was about 100 μM, whereas the K_m value of the particulate enzyme was approximately 20 μM, and suggested, as one possibility, that two different tryptophan hydroxylases, characterized by different K_m values for tryptophan, might be present in brain. It should be noted that these two K_m values are close to those that have been observed for tryptophan in the presence of DMPH$_4$ and tetrahydrobiopterin. It should also be mentioned that their soluble enzyme was assayed in the presence of DMPH$_4$, but the particulate enzyme, which showed no stimulation by added pterin, was actually assayed in the presence of endogenous cofactor. Since the brain pterin cofactor is probably identical with, or closely related to, tetrahydrobiopterin, a plausible explanation of their results is that the differences in K_m values that they measured are not characteristic of a soluble and particulate enzyme but are characteristic of the pterin cofactor with which the hydroxylase is functioning.

Tryptophan hydroxylase resembles tyrosine and phenylalanine hydroxylase in that the K_m for tetrahydrobiopterin is significantly lower than the K_m for DMPH$_4$ (see Table XIII). Even with the naturally occurring pterin, however, the K_m (31 μM) is considerably higher than the estimated tetrahydrobiopterin concentration in brain of 1×10^{-6} M (see Section III). This discrepancy suggests that either brain tryptophan hydroxylase is severely limited by its cofactor concentration or that the estimated brain concentration, based on the assumption of uniform distribution, is incorrect.

Just as with tryptophan, the K_m of the enzyme for oxygen is also strongly influenced by the pterin cofactor used (see Table XIII). The K_m for oxygen in the presence of tetrahydrobiopterin is 2.5%, a value somewhat less than the oxygen levels of 5% measured in the brain of animals inspiring air (67). It seems likely, therefore, that the enzyme is not normally limited by oxygen supply. It should be noted that this conclusion is the opposite of the one that would have been reached if one relied on the oxygen K_m value of 20% determined in the presence of the artificial cofactor, DMPH$_4$.

Another property that brain tryptophan hydroxylase shares with phenylalanine and tyrosine hydroxylases is its sensitivity to inhibition by high concentrations of its amino acid substrate (34). Just as with the other two hydroxylases, this inhibition (see Figs. 23 and 24) is apparent with tetrahydrobiopterin as the cofactor, but not with DMPH$_4$ (up to 2 mM tryptophan).

There is suggestive evidence that inhibition of brain tryptophan hydroxylase by high tryptophan levels can occur *in vivo*. It has been reported that the administration of high levels of tryptophan to rats increases the brain content of tryptophan (28, 47). In one of these studies (47), the tryptophan concentration achieved in the brain (about 0.5 mM) was high enough to cause inhibition of the hydroxylase if the hydroxylating system *in vivo* showed the same sensitivity to excess tryptophan as it does *in vitro* in the presence of tetrahydrobiopterin (see Fig. 24). Although the variation of serotonin accumulation as a function of tryptophan concentration was described by a hyperbolic curve, the data show inhibition of serotonin accumulation by high tryptophan concentrations (47).

Unlike the other two pterin-dependent hydroxylases, partially purified tryptophan hydroxylase from brain does not appear to be inhibited by excess oxygen in the presence of either tetrahydrobiopterin or DMPH$_4$ (34). Therefore, based on the properties of the isolated enzyme, it is not possible to account for the observed lowering of brain levels of serotonin

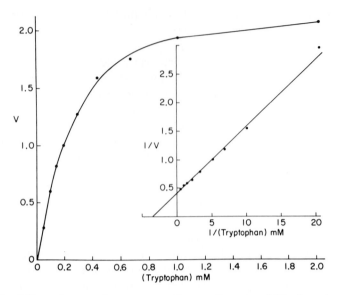

Fig. 23. Effect of tryptophan concentration on the rate of 5-hydroxytryptophan formation. Velocity (*V*) is expressed as nanomoles of 5-hydroxytryptophan formed in 25 minutes at 37°. The inset shows a double reciprocal plot of the data. DMPH₄ concentration was 0.33 m*M*.

by hyperbaric oxygen (27). In view of the observation that some preparations of brain tryptophan hydroxylase that have lost part of their activity do show inhibition by excess oxygen (34), however, it would be wise to regard this point as unsettled. It is possible that the sensitivity of the enzyme to inhibition by high oxygen concentrations depends on factors that have not yet been delineated.

2. End Product Inhibition

The possibility has been considered that serotonin (or its precursor, 5-hydroxytryptophan) inhibits tryptophan hydroxylase, thereby inhibiting its own synthesis. Jequier *et al.* (69) have shown, however, that concentrations of 5-L-hydroxytryptophan and serotonin as high as 1×10^{-4} *M* do not inhibit the enzyme from rat brain stem. Significant inhibition was observed with 1×10^{-3} *M* 5-hydroxytryptophan; but based on the high concentrations of the product required, the authors discounted the physiological importance of end product inhibition of the hydroxylase.

This conclusion appears to have been contradicted by evidence from both whole animals and rat brain (striatum) slices (57, 111). These workers have increased the brain concentration of serotonin either by the use of

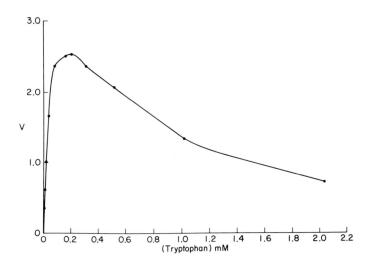

FIG. 24. Effect of tryptophan concentration on the rate of 5-hydroxytryptophan formation with tetrahydrobiopterin (83 μM) as the cofactor. Velocity (V) is expressed as nanomoles of 5-hydroxytryptophan formed in 25 minutes at 37°.

monoamine oxidase inhibitors or by exposing brain slices to high concentrations of serotonin. When the serotonin concentration was increased 2.5-fold, the accumulation of tritium-labeled serotonin (and its oxidation product, 5-hydroxyindole acetic acid) from tritium-labeled L-tryptophan was reduced by 37%. Since the specific activity of the tryptophan pool was not significantly reduced during the experiment, the authors concluded that end product inhibition of tryptophan hydroxylase must have occurred.

It is difficult to reconcile this conclusion with the relative insensitivity to inhibition by serotonin of partially purified tryptophan hydroxylase. The concentration of serotonin in the slices at which the apparent inhibition was observed was only 0.7–2 × 10⁻⁵ M (assuming uniform tissue distribution), values far below the 1 × 10⁻³ M required to demonstrate inhibition with the isolated enzyme. If the conclusion of Hamon et al. (57) and Macon et al. (111) is valid, one must assume that either (a) the concentration of serotonin at the site of the hydroxylase is almost 100 times higher than that calculated on the assumption of uniform distribution, or (b) there is something about the intracellular environment of the hydroxylase that confers on the enzyme high sensitivity to inhibition by serotonin. An obvious difference between the intracellular and "test tube" environment is that within the cell the enzyme probably functions with a cofactor that is closer in structure to tetrahydrobiopterin than to DMPH₄. Since the sensitivity toward product inhibition has been examined only with $DMPH_4$ (69), and since it has been shown that substrate inhibition is seen with tetrahydrobiopterin and not with $DMPH_4$ (34), we have looked at the inhibition by serotonin of rabbit brain tryptophan hydroxylase in the presence of tetrahydrobiopterin, No inhibition was seen at $5 \times 10^{-4} M$ serotonin in the presence of $3 \times 10^{-5} M$ tetrahydrobiopterin (Friedman, Kappelman, and Kaufman, unpublished observations). The alternative interpretation of the results obtained by the French group is that the apparent inhibition results from an indirect effect rather than from direct inhibition of the enzyme; e.g., serotonin is converted to, or causes the accumulation of, a more potent inhibitor of the enzyme. Since neither the intraneuronal concentration of tryptophan nor its specific radioactivity was determined (only "intraslice" values were reported), it is also possible that their results could still be explained by either a lower specific radioactivity or a higher concentration (see Fig. 24) of tryptophan at the site of the hydroxylase in the presence of the elevated concentrations of serotonin.

3. Inhibition by Catechols

Tryptophan hydroxylase from beef pineal glands is inhibited by the intermediates and end products of the norepinephrine biosynthetic pathway (69). At $1 \times 10^{-4} M$, dopa, dopamine, and norepinephrine all inhibit 60–75%. The rat brain stem enzyme is also sensitive to inhibition by catechols. With the pineal enzyme, it has been shown that norepinephrine is a noncompetitive inhibitor with respect to either substrate or cofactor (110). Since it has been found that Fe^{2+} can reverse the inhibition, it has been suggested that the catechols may inhibit by chelating an essential metal (69). In an environment that is presumably closer to its physiological one, i.e., in brain slices, the enzyme appears to be even more sensitive to inhibition by catechols. Dopamine at a concentration of $1 \times 10^{-7} M$ resulted in detectable inhibition of the conversion of tritiated tryptophan to tritiated serotonin in slices of brain stem and straitum; at $1 \times 10^{-5} M$, dopamine inhibited the conversion by 60–70% (42). Since dopamine did not inhibit the conversion of 5-hydroxytryptophan to serotonin, the authors concluded that the hydroxylase was the site of inhibition. The reason for the apparently greater sensitivity of the enzyme to inhibition by catechols in brain slices than in the partially purified state is unknown.

4. Inhibition by Phenylalanine

Both the rat brain and beef pineal enzymes are inhibited competitively (with respect to tryptophan) by phenylalanine with K_i values of $3 \times 10^{-4} M$

and 1×10^{-3} M, respectively (110). It has been suggested that this inhibition can account for the abnormally low blood serotonin levels and 5-hydroxyindoleacetic acid excretion seen in PKU children (132).

REFERENCES

1. Armstrong, M. D. (1955). *J. Biol. Chem.* **213,** 409.
2. Bagchi, S. P., and Zarycki, E. P. (1971). *Res. Commun. Chem. Pathol. Pharmacol.* **2,** 370.
3. Barranger, J., Geiger, P., Juzino, A., and Bessman, S. (1972). *Science* **175,** 903.
4. Berry, H. K., Cripps, R., Nicholls, K., McCandless, D., and Harper, C. (1972). *Biochim. Biophys. Acta* **261,** 315.
5. Boyd, D., Daly, J., and Jerina, D. (1972). *Biochemistry* **11,** 1961.
6. Brenneman, A. R., and Kaufman, S. (1964). *Biochem. Biophys. Res. Commun.* **17,** 177.
7. Brenneman, A. R., and Kaufman, S. (1965). *J. Biol. Chem.* **240,** 3617.
8. Bublitz, C. (1969). *Biochim. Biophys. Acta* **191,** 249.
9. Buckingham, S., Somer, S. C., and McNary, W. F. (1966). *Fed. Proc. Fed. Amer. Soc. Exp. Biol.* **25,** 566
10. Burkhard, W., Gey, K., and Pletscher, A. (1965). *Med. Pharmacol. Exp.* **13,** 388.
11. Carlson, A., Corrodi, H., and Waldeck, B. (1963). *Helv. Chim. Acta* **46,** 2270.
12. Castells, S., and Shirali, S. (1971). *Life Sci.* **10,** 233.
13. Cleland, W. W. (1963). *Biochim. Biophys. Acta* **67,** 188.
14. Cleland, W. W. (1967). *Ann. Rev. Biochem.* **36,** 77.
15. Cooper, J. R., and Melcer, I. (1961). *J. Pharm. Exp. Therap.* **132,** 265.
16. Cooper, J. R., Bloom, F. E., and Roth, R. H. (1970). *In* "The Biochemical Basis of Neuropharmacology," p. 147. Oxford Univ. Press, London and New York.
17. Craine, J. E., Hall, E. S., and Kaufman, S. (1972). *J. Biol. Chem.* **247,** 6082.
18. Dairman, W., and Udenfriend, S. (1971). *Science* **171,** 1022.
19. Daly, J. W., and Guroff, G. (1968). *Arch. Biochem. Biophys.* **125,** 136.
20. Daly, J. W., Levitt, M., Guroff, G., and Udenfriend, S. (1968). *Arch. Biochem. Biophys.* **126,** 593.
21. deDuve, C. (1963). *Ciba Found. Symp. Lysosomes,* 446.
22. Deguchi, T., and Barchas, J. (1972). *J. Neurochem.* **19,** 927.
23. Dixon, M., and Webb, E. C. (1964). *In* "Enzymes," p. 75. Academic Press, New York.
24. Ellenbogen, L., Taylor, R. J., Jr., and Brundage, G. B. (1965). *Biochem. Biophys. Res. Commun.* **19,** 708.
25. Ellman, G. (1959). *Arch. Biochem. Biophys.* **82,** 70.
26. Faiman, M. D., and Heble, A. (1966). *Life Sci.* **5,** 2225.
27. Faiman, M. D., Heble, A., and Mehl, R. G. (1969). *Life Sci.* **8,** 1163.
28. Fernstrom, J. D., and Wurtman, R. J. (1971). *Science* **173,** 149.
29. Fisher, D. B., and Kaufman, S. (1970). *Biochem. Biophys. Res. Commun.* **38,** 663.
30. Fisher, D. B., and Kaufman, S. (1972a). *J. Neurochem.* **19,** 1359.
31. Fisher, D. B., and Kaufman, S. (1972b). *J. Biol. Chem.* **247,** 2250.
32. Fisher, D. B., Kirkwood, R., and Kaufman, S. (1972). *J. Biol. Chem.* **247,** 5161.
33. Freedland, R. A., Wadzinski, I. M., and Waisman, H. A. (1961). *Biochem. Biophys. Res. Commun.* **5,** 94.

34. Friedman, P. A., Kappelman, A. H., and Kaufman, S. (1972). *J. Biol. Chem.* **247**, 4165.

35. Friedman, P. A., Kang, E. S., and Kaufman, S. (1972). *Nature (London)* **240**, 157 (1972).

36. Friedman, P. A., and Kaufman, S. (1971). *Arch. Biochem. Biophys.* **146**, 321.

37. Friedman, S., and Kaufman, S. (1965). *J. Biol. Chem.* **240**, 4763.

38. Gal, E. M., Roggeveen, A., and Millard, S. (1970). *J. Neurochem.* **17**, 1221.

39. Gal, E. M. (1971). Amer. Soc. for Neurochem., 2nd National Meeting, Hershey, Pennsylvania.

40. Gal, E. M., Armstrong, J. C., and Ginsberg, B. (1966). *J. Neurochem.* **13**, 643.

41. Goldstein, M., Joh, T., and Garvey, T. (1969). *Biochemistry* **7**, 2724.

42. Goldstein, M., and Frenkel, R. (1971). *Nature (London)* **233**, 179.

43. Goodall, McC., and Kirshner, N. (1957). *J. Biol. Chem.* **226**, 213.

44. Goodfriend, T. L., and Kaufman, S. (1961). *J. Clin. Invest.* **40**, 1743.

45. Grahame-Smith, D. G. (1964). *Biochem. Biophys. Res. Commun.* **16**, 586.

46. Grahame-Smith, D. G. (1967). *Biochem. J.* **105**, 351.

47. Grahame-Smith, D. G. (1971). *J. Neurochem.* **18**, 1053.

48. Green, A. R., and Curzon, G. (1968). *Nature (London)* **220**, 1095.

49. Green, H., and Sawyer, J. L. (1966). *Anal. Biochem.* **15**, 53.

50. Gurin, S., and Delluva, A. M. (1947). *J. Biol. Chem.* **170**, 545.

51. Guroff, G., Reifsnyder, C. A., and Daly, J. W. (1966). *Biochem. Biophys. Res. Commun.* **24**, 720.

52. Guroff, G., and Rhoads, C. A. (1967). *J. Biol. Chem.* **242**, 3641.

53. Guroff, G., Daly, J. W., Jerina, D. M., Renson, J., Witkop, B., and Udenfriend, S. (1967). *Science* **157**, 1524.

54. Haggendahl, J. (1965). *Acta Physiol. Scand.* **64**, 244.

55. Hall, G., Hillarp, N. A., and Thieme, G. (1961). *Acta Physiol. Scand.* **52**, 49.

56. Hamilton, G. A. (1964). *J. Amer. Chem. Soc.* **86**, 3391.

57. Hamon, M., Bourgoin, S., Morot-Gaudry, Y., and Glowinski, J. (1972). *Nature (London) New Biol.* **237**, 184.

58. Haugaard, N., (1968). *Physiol. Rev.* **48**, 311.

59. Hedrick, J. L., and Smith, A. J. (1968). *Arch. Biochem. Biophys.* **126**, 155.

60. Howell, L. G., and Massey, V. (1970). *Biochem. Biophys. Res. Commun.* **40**, 887.

61. Ichiyama, A., Nakamura, S., Nishizuka, Y., and Hayaishi, O. (1968). *Advan. Pharmacol.* **64**, 5.

62. Ichiyama, A., Nakamura, S., Nishizuka, Y., and Hayaishi, O. (1970). *J. Biol. Chem.* **245**, 1699.

63. Ikeda, M., Fahien, L. A., and Udenfriend, S. (1966). *J. Biol. Chem.* **241**, 4452.

64. Ikeda, M., Levitt, M., and Udenfriend, S. (1965). *Biochem. Biophys. Res. Commun.* **18**, 482.

65. Jakubovic, A. (1971). *Biochim. Biophys. Acta* **237**, 469.

66. Jakubovic, A., Woolf, L. I., and Chan-Henry, E. (1971). *Biochem. J.* **125**, 563.

67. Jamieson, D., and Van den Brenk, H. A. J. (1965). *J. Appl. Physiol.* **20**, 514.

68. Jequier, E., Lovenberg, W., Sjoerdsma, A. (1967). *Mol. Pharmacol.* **3**, 274.

69. Jequier, E., Robinson, D. S., Lovenberg, W., and Sjoerdsma, A. (1969). *Biochem. Pharmacol.* **18**, 1071.

70. Jerina, D., Daly, J. W., Landis, W., Witkop, B., and Udenfriend, S. (1967). *J. Amer. Chem. Soc.* **89**, 3347.

71. Jerina, D., Daly, J. W., and Witkop, B. (1971). *Biochemistry* **10**, 366.

72. Jerina, D., Daly, J. W., Witkop, B., Zaltzman-Nirenberg, P., and Udenfriend, S. (1970). *Biochemistry* **9**, 147.
73. Jervis, G. A. (1947). *J. Biol. Chem.* **169**, 651.
74. Jervis, G. A. (1953). *Proc. Soc. Exp. Biol.* **82**, 514.
75. Joh, T. H., Kapit, R., and Goldstein, M. (1969). *Biochim. Biophys. Acta* **171**, 378.
76. Justice, P., O'Flynn, M. E., and Hsia, D. Y. Y. (1967). *Lancet* **1**, 928.
77. Kang, E. S., Kaufman, S., and Gerald, P. S. (1970). *Pediatrics* **45**, 83.
78. Karobath, M., and Baldessarini, R. J. (1972). *Nature (London) New Biol.* **236**, 206.
79. Kaufman, S. (1957). *J. Biol. Chem.* **226**, 511.
80. Kaufman, S. (1958a). *Biochim. Biophys. Acta* **27**, 428.
81. Kaufman, S. (1958b). *J. Biol. Chem.* **230**, 931.
82. Kaufman, S. (1958c). *Science* **128**, 1506.
83. Kaufman, S. (1959). *J. Biol. Chem.* **234**, 2677.
84. Kaufman, S. (1961a). *J. Biol. Chem.* **236**, 804.
85. Kaufman, S. (1961b). *Biochim. Biophys. Acta* **51**, 619.
86. Kaufman, S. (1962a). *Methods Enzymol.* **5**, 802.
87. Kaufman, S. (1962b). *J. Biol. Chem.* **237**, PC2712.
88. Kaufman, S. (1962c). *In* "Oxygenases" (O. Hayaishi, ed.), p. 129. Academic Press, New York.
89. Kaufman, S. (1963). *Proc. Nat. Acad. Sci. U.S.* **50**, 1085.
90. Kaufman, S. (1964a). *J. Biol. Chem.* **239**, 332.
91. Kaufman, S. (1964b). *Trans. N.Y. Acad. Sci. Ser. II* **26**, 977.
92. Kaufman, S. (1964c). *In* "Pteridine Chemistry" (W. Pfleiderer and E. C. Taylor, eds.), p. 307. Pergamon, Oxford.
93. Kaufman, S. (1967a). *In Proc. Conf. Treatment Phenylketonuria Allied Dis.* (J. A. Anderson and K. F. Swaiman, eds.), p. 205. U.S. Govt. Printing Office, Washington, D.C.
94. Kaufman, S. (1967b). *J. Biol. Chem.* **242**, 3934.
95. Kaufman, S. (1969). *Arch. Biochem. Biophys.* **134**, 249.
96. Kaufman, S. (1970). *J. Biol. Chem.* **245**, 4751.
97. Kaufman, S. (1971). *Advan. Enzymol.* **35**, 245.
98. Kaufman, S., Bridgers, W. F., Eisenberg, F., and Friedman, S. (1962). *Biochem. Biophys. Res. Commun.* **9**, 497.
99. Kaufman, S., and Fisher, D. B. (1970). *J. Biol. Chem.* **245**, 4745.
100. Kaufman, S., and Levenberg, B. (1959). *J. Biol. Chem.* **234**, 2683.
101. Kaufman, S., and Max, E. E. (1971). *In* "Phenylketonuria" (H. Bickel, F. P. Hudson, and L. I. Woolf, eds.), p. 13. Verlag, Stuttgart.
102. Kaufman, S., Storm, C. B., and Fisher, D. B. (1970). *In* "Chemistry and Biology of Pteridines" (K. Iwai, M. Akino, M. Goto, and Y. Iwanami, eds.), p. 209. Int. Academic Printing, Tokyo.
103. Kenney, F. T., Reem, G. H., and Kretchmer, N. (1958). *Science* **127**, 86.
104. Kuczenski, R. T., and Mandell, A. J. (1972). *J. Biol. Chem.* **247**, 3114.
105. LaDu, B. N., and Zannoni, V. G. (1967). *In Proc. Conf. Treatment Phenylketonuria Allied Dis.* (J. A. Anderson and K. F. Swaiman, eds.), p. 193. U.S. Gov. Printing Office, Washington, D.C.
106. Laduron, P., and Belpaire, F. (1968). *Nature (London)* **217**, 1155.
107. Lloyd, T., Mori, T., and Kaufman, S. (1971). *Biochemistry* **10**, 2330.
108. Lloyd, T., and Weiner, N. (1971). *Mol. Pharmacol.* **7**, 569.
109. Lovenberg, W., Jequier, E., and Sjoerdsma, A. (1967). *Science* **155**, 217.

110. Lovenberg, W., Jequier, E., and Sjoerdsma, A. (1968). *Advan. Pharmacol.* **6A**, 21.
111. Macon, J., Sokoloff, L., and Glowinski, J. (1971). *J. Neurochem.* **18**, 323.
112. Mager, H. I. X., and Berends, W. (1965). *Rev. Trav. Chim. Pays-Bas* **84**, 1329.
113. Mager, H. I. X., Addink, R., and Berends, W. (1967). *Rev. Trav. Chim. Pays-Bas* **86**, 833.
114. Matsubara, M., Katoh, S., Akino, M., and Kaufman, S. (1966). *Biochim. Biophys. Acta* **122**, 202.
115. McCord, J., and Fridovich, I. (1969). *J. Biol. Chem.* **244**, 6049.
116. Mitoma, C. R. (1956). *Arch. Biochem. Biophys.* **60**, 476.
117. Mitoma, C. R., Auld, M., and Udenfriend, S. (1957). *Proc. Soc. Exp. Biol. N.Y.* **94**, 634.
118. Musacchio, J. M. (1968). *Biochem. Pharmacol.* **17**, 1470.
119. Musacchio, J. M., D'Angelo, G. L., and McQueen, C. A. (1971). *Proc. Nat. Acad. Sci. U.S.* **68**, 2087.
120. Musacchio, J. M., Wurtzburger, R. J., and D'Angelo, G. L. (1971). *Mol. Pharmacol.* **7**, 136.
121. Nagatsu, T., Levitt, M., and Udenfriend, S. (1964). *J. Biol. Chem.* **239**, 2910.
122. Nagatsu, T., and Takeuchi, K. (1967). *Experientia* **23**, 1.
123. Nakamura, S., Ichiyama, A., and Hayaishi, O. (1965). *Fed. Proc. Fed. Amer. Soc. Exp. Biol.* **24**, 604.
124. Neville, D. M., Jr. (1971). *J. Biol. Chem.* **246**, 6328.
125. Nicholls, P., and Schonbaum, G. R. (1963). *In* "The Enzymes" (P. D. Boyer, H. Lardy and K. Myrback, eds.), 2nd ed., Vol. 8, p. 147. Academic Press, New York.
126. Nielsen, K. H. (1969). *Eur. J. Biochem.* **7**, 360.
127. Nielson, K. H., Simonsen, V., and Lind, K. E. (1969). *Eur. J. Biochem.* **9**, 497.
128. Norman, S., and Sjovale, J. (1958). *J. Biol. Chem.* **233**, 872.
129. Okamoto, H., Nozaki, M., and Hayaishi, O. (1968). *Biochem. Biophys. Research Commun.* **32**, 30.
130. Olten, R. R., and Putnam, P. A. (1966). *J. Nutr.* **89**, 385.
131. Osani, M., and Rembold, H. (1971). *Hoppe-Seyler's Z. Physiol. Chem.* **352**, 1359.
132. Pare, C. M. B., Sandler, M., and Stacey, R. C. (1959). *Arch. Dis. Childh.* **34**, 422.
133. Peters, D. A. V., McGeer, P. L., and McGeer, E. G. (1968). *J. Neurochem.* **15**, 1431.
134. Petrack, B., Sheppy, F., and Fetzer, V. (1968). *J. Biol. Chem.* **243**, 743.
135. Potter, V. R., Gebert, R. A., and Pitot, H. C. (1966). *Advan. Enzyme Regul.* **4**, 247.
136. Rembold, H. (1964). *In* "Pteridine Chemistry" (W. Pfleiderer and E. C. Taylor, eds.), p. 465. Pergamon, Oxford.
137. Rembold, H., and Metzger, H. (1967). *Z. Naturforsch.* **22B**, 827.
138. Renson, J., Weissbach, H., and Udenfriend, S. (1962). *J. Biol. Chem.* **237**, 2261.
139. Ribbons, D. W., and Otha, Y. (1970). *Fed. Eur. Biochem. Soc. Lett.* **12**, 105.
140. Robinson, N., and Saunders, L. (1958). *J. Pharm. Pharmacol.* **10**, 755.
141. Robinson, D., Lovenberg, W., and Sjoerdsma, A. (1968). *Arch. Biochem. Biophys.* **123**, 419.
142. Sharpless, K. B., and Flood, T. C. (1971). *J. Amer. Chem. Soc.* **93**, 2316.
143. Shiman, R., Akino, M., and Kaufman, S. (1971). *J. Biol. Chem.* **246**, 1330.
144. Shiman, R., and Kaufman, S. (1970). *Methods Enzymol.* **17A**, 609.
145. Singer, T., and Kearny, E. (1950). *J. Biol. Chem.* **183**, 409.
146. Sjoerdsma, A., Lovenberg, W., Engelman, K., Carpenter, W. T., Wyatt, R. J., and Gessa, G. L. (1970). *Ann. Intern. Med.* **73**, 607.

147. Smith, H. G., and Lakatos, C. (1971). *J. Pharm. Pharmacol.* **23**, 180.
148. Spector, S., Gordon, R., Sjoerdsma, A., and Udenfriend, S. (1967). *Mol. Pharmacol.* **3**, 549.
149. Stjarne, L., and Lishajko, F. (1967). *Biochem. Pharmacol.* **16**, 1719.
150. Storm, C. B., and Kaufman, S. (1968). *Biochem. Biophys. Res. Commun.* **32**, 788.
151. Strittmatter, C. R., and Oakley, G. (1966). *Proc. Soc. Exp. Biol. Med.* **123**, 427.
152. Sullivan, P., Kester, M., and Norton, S. (1971). *Fed. Proc. Fed. Amer. Soc. Exp. Biol* **30**, 1067.
153. Takemori, S., Yasuda, H., Mihara, K., Suzuki, K., and Katagiri, N. (1969). *Biochim. Biophys. Acta* **191**, 56.
154. Taylor, R. J., Stubbs, C. S., and Ellenbogen, L. (1966). *Biochem. Pharmacol.* **18**, 587.
155. Tong, J. H., D'Iorio, A., and Benoiton, N. L. (1971a). *Biochem. Biophys. Research Commun.* **44**, 229.
156. Tong, J. H., D'Iorio, A., and Benoiton, N. L. (1971b). *Biochem. Biophys. Research Commun.* **43**, 819.
157. Tourian, A. (1971). *Biochim. Biophys. Acta* **242**, 345.
158. Tourian, A., Goddard, J., and Puck, T. T. (1969). *J. Cell. Physiol.* **73**, 159.
159. Tubbs, P., and Garland, P. (1964). *Biochem. J.* **93**, 550.
160. Udenfriend, S., and Bessman, S. P. (1953). *J. Biol. Chem.* **203**, 961.
161. Udenfriend, S. (1964). *Harvey Lect. Ser.* **60**, 57.
162. Udenfriend, S., and Cooper, J. R. (1952). *J. Biol. Chem.* **194**, 503.
163. Udenfriend, S., Cooper, J. R., Clark, C. T., and Baer, J. E. (1953). *Science* **117**, 663.
164. Udenfriend, S., and Wyngaarden, J. B. (1956). *Biochim. Biophys. Acta* **20**, 48.
165. Udenfriend, S., Zaltzman-Nirenberg, P., and Nagatsu, J. (1965). *Biochem. Pharmacol.* **14**, 837.
166. Wallace, H. W., Moldave, K., and Meister, A. (1957). *Proc. Soc. Exp. Biol. N.Y.* **94**, 632.
167. Weil-Malherbe, H. (1955). *J. Ment. Sci.* **101**, 733.
168. White-Stevens, R. H., and Kamin, H. (1970). *Biochem. Biophys. Res. Commun.* **38**, 882.
169. White-Stevens, R. H., and Kamin, H. (1972). *J. Biol. Chem.* **247**, 2358.
170. Womack, M., and Rose, W. C. (1946). *J. Biol. Chem.* **166**, 429.
171. Wurzburger, R. J., and Musacchio, J. M. (1971). *J. Pharm. Exp. Therap.* **177**, 155.
172. Yamamoto, S., Takeda, H., Maki, Y., and Hayaishi, O. (1966). *In* "Biological and Chemical Aspects of Oxygenases" (K. Bloch and O. Hayaishi eds.), p. 303. Maruzen, Tokyo.
173. Zannoni, V. G., and Moraru, E. (1970). *In Proc. Sixth Meeting Fed. Eur. Biochem. Soc.* (A. Sols and S. Grisolia eds.), p. 347. Academic Press, New York.
174. Zannoni, V. G., Rivkin, I., and LaDu, B. N. (1967). *Fed. Proc. Fed. Amer. Soc. Exp. Biol.* **26**, 840.

9

COPPER-CONTAINING OXYGENASES

WALTER H. VANNESTE and ANDREAS ZUBERBÜHLER

I. INTRODUCTION

Copper enzymes owe their catalytic role in oxygen metabolism principally to two assets: oxidation–reduction of the metal and binding with oxygen. In some instances coordination of the substrate to the copper may also be significant. Our view of the range of involvement of copper in biological oxidation was widened recently by the discovery of a copper-containing fungal dioxygenase (oxygen transferase), quercetinase, which catalyzes the oxidative cleavage of the heterocyclic ring of quercetin (I) to give 2-protocatechuoylphloroglucinolcarboxylic acid (II) and carbon monoxide (79, 112):

$$\text{(I)} + {}^{18}\text{O}_2 \longrightarrow \text{(II)} + \text{CO} \quad (1)$$

(I)   (II)

The monooxygenase nature (mixed-function oxidase requiring external electron donor) of the mammalian copper enzyme dopamine β-hydroxylase (3,4-dihydroxyphenylethylamine, ascorbate:oxygen oxidoreductase (hydroxylating), EC 1.14.2.1) was firmly established by Kaufman and co-workers (61, 63) a decade ago. The reaction, involving the hydroxylation of dopamine* (III) in the β position of the side chain and the coupled oxidation of ascorbate (85), constitutes the final step in the biosynthetic pathway from tyrosine to the neurotransmitter norepinephrine (IV):

$$+ O_2 + \text{Ascorbate} \longrightarrow \qquad + H_2O + \begin{array}{l}\text{Dehydro-}\\\text{ascorbate}\end{array} \qquad (2)$$

(III)   (IV)

Tyrosinase, longest known of all oxygenases, is a copper enzyme that participates in two very dissimilar reactions involving molecular oxygen, viz., the insertion of oxygen in the ortho position of a variety of monophenols (V) and the dehydrogenation of o-diphenols (VI) to o-quinones (VII):

$$+ O_2 \xrightarrow{\text{monophenolase}} \qquad + H_2O \qquad (3)$$

(V)   (VII)

$$2 \qquad + O_2 \xrightarrow{\text{diphenolase}} 2 \qquad + 2 H_2O \qquad (4)$$

(VI)   (VII)

Tyrosinases occur widespread in nature usually functioning in melanin biosynthesis, for example, from L-tyrosine in vertebrates, but also in

* The abbreviations and trivial names used are dopamine, 1-(3′,4′-dihydroxyphenyl)-2-aminoethane; dopa, 3,4-dihydroxyphenylalanine; dopachrome, 2,3-dihydroindole-2-carboxylic acid-5,6-quinone; norepinephrine (synonymous with noradrenaline), 1-(3′,4′-dihydroxyphenyl)-2-amino-1-ethanol; p-coumaric acid, 4-hydroxyphenylacrylic acid; caffeic acid, 3,4-dihydroxyphenylacrylic acid; ESR, electron spin resonance; CD, circular dichroism; and ORD, optical rotation dispersion.

sclerotization of arthropod cuticula (cf. refs. 91, 102, and 110). Various roles in plants, including biosynthesis of polyphenolic compounds, respiration, and antibiotic action of the quinone products, have been proposed (cf. refs. 22 and 91).

The monophenolase reaction as written in Eq. (3) satisfies the observed overall stoichiometry when properly substituted phenols (e.g., 4-*tert*-butylphenol) are used that yield stable reaction products (26). Furthermore, it expresses the monooxygenase character of monophenolase activity which will be discussed later (see Section II,C,1). The peculiar properties of tyrosinase have led to a sharp division of opinions about this enzyme throughout its history. Some workers have regarded the monophenolase function as truly enzymic (83, 92, 95, 109, 127), while others have favored a nonenzymic hypothesis (67, 72, 74, 116), according to which hydroxylation results from a secondary reaction between monophenol and the *o*-quinone product of diphenolase activity. Not knowing whether the rate with which *o*-quinone might oxidize monophenols in a nonenzymic system (23) is at all compatible with the enzymic rate, we note that the nonenzymic hypothesis meets considerable difficulties in explaining a vast number of experimental observations: (a) optical specificity, position specificity, and competitive inhibition by substrate analogs of the hydroxylation reaction (e.g., ref. 121); (b) laccase, which like tyrosinase catalyzes the formation of *o*-quinones (68), does not hydroxylate monophenols (7); (c) reducing agents reacting very quickly with quinones do not abolish hydroxylation (e.g., ref. 66); and (d) the source of oxygen incorporated in the substrate is molecular oxygen rather than water (92). Therefore, we will adopt the view that both the mono- and the diphenolase functions of tyrosinase depend on specific structural features of the protein.

The overall stoichiometry of tyrosinase-catalyzed *o*-diphenol oxidation [reaction (4)] was established by the work of Nelson and co-workers (cf. ref. 109). The role of tyrosinase in this reaction has served as a basis for its classification as catechol oxidase (*o*-diphenol:oxygen oxidoreductase, EC 1.10.3.1). Since, however, this activity is shared with the group of laccases, we will prefer to use here the trivial names "tyrosinase"† and "phenolase" in order to refer to that group of copper enzymes which in addition to *o*-diphenol oxidation display the hydroxylation function.

The object of this chapter will be to discuss our present knowledge and raise various questions concerning the structure and the mechanism of action of the above-mentioned copper enzymes, which have in common

† Tyrosinase should not be mistaken either for mammalian L-tyrosine hydroxylase, which also catalyzes the conversion of L-tyrosine to L-dopa but is not a copper enzyme and requires a tetrahydropteridine as cosubstrate (105), or for bacterial β-tyrosinase [L-tyrosine phenol lyase (deaminating)] (142).

the ability to catalyze the fixation of oxygen, derived from molecular oxygen, into their respective substrates.

II. TYROSINASES

A. Protein Structure and Multiple Forms

Progress in elucidating tyrosinase structure has been intimately connected with the study of molecular heterogeneity of this enzyme for which an explanation was sought in structural terms. Multiple forms of tyrosinase from a wide variety of sources at various phylogenetic levels have been reported. This heterogeneity has been related to protein association–dissociation behavior, conformational isomerism, or differences in amino acid sequence.

1. *Microbial and Plant Tyrosinases*

Extending the work of Smith and Krueger (139) on the separation of mushroom tyrosinase by chromatography on hydroxylapatite, Bouchilloux et al. (10) purified four forms of the enzyme having different monophenol: o-diphenol activity ratios. Three of the fractions (β, γ, and δ) had identical amino acid composition, sedimentation constant ($s_{20,w} = 7.25$ S), and molecular weight (123,800). The copper content was almost exactly four atoms per molecule in the β- and γ-enzyme, but was somewhat more variable in the δ form. Jolley and co-workers (59) found that the α-enzyme was significantly different from the other three with regard to sedimentation coefficient (6.66 S) and amino acid composition, which established the existence of at least two isozymes of different primary structure in the common mushroom, *Agaricus bispora*. Electrophoresis of the various mushroom isozymes in molecular sieving media revealed a pattern of components identifiable with polymeric states which to a certain extent were interconvertible (56–58). The leading component in gel electrophoresis had a molecular weight of 30,000 by ultracentrifugation, whereas the predominant component weighed 116,000 (58, 129). Heat, high ionic strength, dodecyl sulfate, and EDTA, all at neutral or slightly alkaline pH were effective in increasing the amount of low molecular weight species. The reverse process was induced by concentrating the protein, most successfully at lower pH. Calcium ions, although not bound firmly by tyrosinase, were shown to mediate the reassociation of the EDTA-dissociated sample. These and other experiments provided conclusive evidence for the occurrence of an association–dissociation equilibrium in the mushroom isozymes as prepared by Bouchilloux et al. (10); under conditions of neutral pH, low temperature, and low ionic strength, a tetrameric state strongly prevailed. A single NH_2-terminal amino acid, isoleucine, and a single COOH-terminal amino acid, valine, were analyzed in each isozyme. This falls in line with the presence in mushroom tyrosinase of only one type of subunit of molecular weight about 31,000, a fact which is strongly supported by the number of peptides in tryptic hydrolysates (59). It was pointed out that equilibration among polymeric forms must be slow under the experimental conditions for electrophoretic separation would not have been possible otherwise (57).

Using a more classic protein fractionation procedure, Kertész and Zito (75) purified tyrosinase from mushroom to a homogeneous state by sedimentation and electrophoretic criteria. The enzyme had $s_{20,w} = 6.3$ S and molecular weight 128,000. Dissociation could be brought about partially by 1% dodecyl sulfate and completely by 8 M urea ($s_{20,w} = 2.2$ S), but not by dilution, at least to 0.45 mg/ml (154). Although the amino acid compositions are very similar, it is not entirely clear how Kertész and Zito's enzyme correlates with the isozymes of Bouchilloux et al. A number of other tyrosinase preparations from mushroom have been reported to be in a monomeric (2, 73), dimeric (107), tetrameric (31, 80, 153), or a mixed oligomeric state (27).

Horowitz and Fling (50), followed by Fox and Burnett (35), discovered several forms of tyrosinase in *Neurospora crassa*. These forms were functionally indistinguishable (140) but responded differently to heat treatment. Horowitz and co-workers (51, 52) further observed that when *Neurospora* cultures are starved in phosphate buffer an unstable repressor protein, which controls tyrosinase synthesis at the level of DNA, disappears. This in turn leads to *de novo* synthesis of high quantities of the enzymes. This made possible the isolation of two isozymes in the crystalline state (29). Their properties were studied in great detail (29). One of the forms, called the S form, was rather stable to heat, the other, the L form, was more labile. The S and L isozymes also exhibited slight differences with regard to electrophoretic mobility but sedimented with the same speed in the ultracentrifuge and had identical diffusion constant. Sedimentation equilibrium analysis revealed a reversibly associating system with a range of molecular weights from 35,000 to 120,000. In agreement with this an effect of dilution on the sedimentation constant was observed (4.3 S by schlieren analysis at about 3 mg/ml; 3.6 S by sucrose gradient technique at about 30 μg/ml). On the basis of sedimentation and diffusion data, and considering the copper content, a monomer weight of 33,000 $\pm$ 2,000 was proposed, a value confirmed by the number of tryptic peptides in both S and L tyrosinase. Only one possibly significant difference in amino acid composition between S and L forms, notably the serine content, was detected, indicating that the divergent thermal properties are mainly the result of structural differences on the secondary and/or tertiary level.

Evidence for several types of tyrosinase in higher plants has been presented, but the basis for their existence is in general less clear. Patil and co-workers (118, 119) obtained from potato two fractions displaying different specific activities with both mono- and o-diphenols as substrates. Their behavior in gel filtration suggested a difference in molecular weight, but this difference was not quantitated. The larger species, which also had the highest specific activity, showed a sedimentation constant of 5.9 S. Balasingham and Ferdinand (6), employing the same source, recently have isolated at least three forms of the enzyme apparently differing in size. The most abundant and heaviest fraction was eluted at the void volume of Bio-Gel P-300 which characterizes its molecular weight as 300,000 or greater. However, half of the weight was found to consist of RNA so that this enzyme may well be a tetramer of subunits whose weight, as suggested by amino acid composition and copper content, lies around 36,000. Refiltration on Bio-Gel of this heavier component gave rise to small quantities of the lower molecular weight fractions, indicating that the original multiplicity may have resulted from a state of partial disaggregation.

Apple and broad bean (*Vicia faba* L.) are other plant sources in which heterogeneity of tyrosinases is thought to occur (46, 47, 130, 151). The various enzymes in both cases were characterized by their different electrophoretic mobility on starch gel. The enzymes from apple chloroplasts showed additional differences in substrate and inhibitor specificities and in K_m toward O_2 (47). Their behavior in gel filtration indicated molecular weights of about 35,000 ($s_{20,w} = 2.7$ S), 65,000, and 125,000, suggesting various oligomeric forms. Interconversion was shown to take place (46). Robb et al. (130) were unable to differentiate the broad bean enzymes by any other method except electrophoresis. Their $s_{20,w}$ values around 4 S correspond most likely with a predominantly monomeric state. Attempts to obtain interconversion of the different forms by unfolding and refolding were unsuccessful.

2. Insect Tyrosinases

Heyneman (48) isolated tyrosinase in its latent state from *Tenebrio molitor* hemolymph and found an $s_{20,w}$ value of 7.3 S and an equivalent weight per mole of copper of 29,000; its quaternary structure may be tetrameric. Mitchell and Weber (100) have separated three latent components from *Drosophila melanogaster* pupae by ammonium sulfate fractionation and gel electrophoresis, and treated them on the gel with a solution of natural activator. One component formed melanin from tyrosine and dopa, the other two from dopa only. Activation was considered to consist of a process of assembling the enzymes from protein subunits including the activator. Such a process, however, seems very unlikely because of the slow diffusion of protein in the gel as evidenced by the preserved sharpness of the bands (cf. Fig. 1 of ref. 100). An enzymic or low molecular weight activator better accounts for the result. Protyrosinase purified from silkworm, *Bombyx mori*, by Ashida appeared almost entirely dissociated in fresh preparations at moderately high ionic strength (3). Equilibrium centrifugation revealed a predominant molecular weight of 36,000 with evidence of some association. Exposure to low ionic strength or treatment with dodecyl sulfate gave rise to a single species of MW 72,000–80,000. Reversal of this aggregation could not be observed.

3. Mammalian Tyrosinases

Mammalian tyrosinase *in vivo* appears to be largely associated with subcellular constituents of the melanocyte, in particular melanosomes and smooth and rough endoplasmic reticulum (84, 133). Although the soluble or more readily solubilized tyrosinase makes up only a small percentage of the total, it has been studied in most detail. Early attempts to purify soluble mammalian tyrosinase from mouse melanoma disclosed multiple forms of the enzyme in an eluate from DEAE-cellulose (16, 17). Gel electrophoresis confirmed the presence of two active components in crude preparations from melanoma (121, 136). When separated, they displayed identical K_m values and molecular activities (121). Burnett and colleagues further demonstrated the occurrence of multiple soluble tyrosinases, named T^1 and T^2, in various mouse melanomas (20), in follicular melanocytes during hair growth (21), and in human metastatic melanoma (19). Final purification of T^1 and T^2 from Harding-Passey mouse melanoma was effected by a procedure employing preparative electrophoresis on polyacrylamide gels containing urea (18). The two forms were found to differ in amino acid composition. Molecular weights determined by gel electrophoresis were around 76,500 and 61,500 for T^1 and T^2, respectively. Equilibrium centrifugation yielded slightly lower but again significantly different values: 66,000 and 56,500, respectively. By a separate purification technique fractions of molecular weight 32,500 and 29,000, indistinguishable from T^1 and T^2 in amino acid composition, have been prepared. On the basis of this result Burnett has suggested that T^1 and T^2 are dimeric (18).

Particulate mammalian tyrosinase has been dispersed with varying degree of success by detergent treatment (20, 99), freezing and thawing, sonication, heat treatment (99), phospholipase digestion (99, 134), and protease action (135, 159). Exposure to 8 M urea released a fraction indistinguishable from T^1 (20). Protease digests contain two forms of tyrosinase resolvable by gel electrophoresis (135). It appears therefore not unlikely that soluble T^1 and T^2 represent cellular enzymes prior to incorporation in melanocytic membrane structures.

4. Heterogeneity: True or Artifact?

Mallette and Dawson (88), discussing the physical nature of mushroom tyrosinase, postulated a single native protein entity possessing well-defined activities toward mono- and diphenols; fragmentation during purification procedures led to differently active enzymes. Although this view has in part been refuted by the separation of isozymes of different primary structure (59), a more subtle effect (92) of preparative technique cannot be dismissed (75). The hydroxylative function in particular seems to depend on the maintenance of a fragile enzymic configuration (e.g., refs. 76 and 148). Numerous preparation procedures apply acetone fractionation (10, 17, 29, 31, 75) or detergent-assisted solubilization (20, 47). The deteriorative action of these operations has not been properly assessed by the customary activity measurements (e.g., ref. 75) because of the possible latency of tyrosinase prior to these steps. The phenomenon of auto-tanning, i.e., oxidation of exposed tyrosine residues or reaction of available NH_2 and SH groups with quinones formed from natural substrates in the tissue homogenate, may constitute an additional source of artifact. The δ fraction from mushroom presumably is of this type (59). Of course, it should also be kept in mind that distinguishing features may be destroyed instead of created during preparation of the enzyme.

The presence of tyrosinases of different primary structure appears to be firmly established in mushroom, *Neurospora*, and the mammalian melanotic melanocyte. Its genetic basis was studied in *Neurospora* (51). Tyrosinase multiplicity based either on occurrence of polymeric states or on conformational isomerism will very likely be affected by preparative steps. As shown by Jolley and Mason (57), the pattern of monomer-dimer-tetramer-octamer observed with fresh mushroom extract underwent changes in relative concentration during purification. The actual existence in plant and animal tissue of various multiple forms related to association–dissociation behavior, configurational variability, or charge distribution (130), although strongly indicated in the work of several authors, cannot be considered established at the moment. Also, the functional significance of tyrosinase multiplicity remains to be elucidated.

5. Amino Acid Composition, Secondary and Tertiary Structure, Latent Tyrosinase

Amino acid compositions of tyrosinases from sources as diverse phylogenetically as mushroom (27, 59, 154), *Neurospora* (29), potato (6), silkworm (3), and a mouse melanoma (18) show a remarkable resemblance. It seems justified to expect a close homology of primary structures. The content of hydrophobic residues is consistently about 30% (18, 58), a value too low to characterize tyrosinases as typical multichain proteins, yet too high for classification as single chain (145). From this feature originates the tendency to readily associate and dissociate.

As distinct from other tyrosinases, the *Neurospora* and broad bean enzymes do not contain cysteine (29, 130), and the latter does not contain methionine either (130). This finding eliminates sulfur as a possible ligand for copper in the broad bean enzyme, and in other tyrosinases too, inasmuch as all of them have identical copper coordination. The cysteine residues (two per monomer) in mushroom tyrosinase were shown to be linked in an intrachain disulfide bridge (59); in silkworm protyrosinase they appear in the sulfhydryl form rather inaccessible inside the protein (3). A disulfide bridge may be important in maintaining the catalytically active configuration of mammalian tyrosinase as evidenced by the effect of dithiothreitol on the activity, an effect which is nullified by reoxidation in air (18). However, we feel that this loss and reappearance of activity may as well be explained by the binding of dithiothreitol to enzyme copper and subsequent autoxidation of this ligand.

Little is known, except from CD and ORD measurements (18, 27) about tyrosinase at the level of secondary and tertiary structure. However, since the enzyme has been crystallized from at least one source, one might hope for an X-ray study in the next decennium, which would strongly enhance our understanding of oxidases in general and copper enzymes in particular.

Tyrosinase has been shown to occur in a latent form (protyrosinase) in various insects and plants from which the enzyme has been isolated in this state (for references see 3 and 141). A variety of conditions which one might term mildly denaturing (141) result in the appearance of activity. Since activation was not accompanied by a change in sedimentation behavior, and because the enzyme returned spontaneously to the inactive state upon interrupting the activating treatment—even after separating out any material of low molecular weight—Swain *et al.* (141) suggested that activation of *Vicia faba* protyrosinase is caused by a conformational change. A natural activating substance was discovered in several insects, but its nature and mode of action may vary with the species (5, 49, 53,

100, 132). Activation may be brought about also by limited proteolysis (69, 111, 132). However, the silkworm enzyme produced by α-chymotrypsin was different from that formed by a natural activator (111). Thus, more than one type of change appears to give rise to an active configuration.

6. Structure of Functional Unit

While much has been learned about the quaternary structure of tyrosinases in their isolated state, no conclusive evidence can be brought to bear on the nature of the minimal functional unit. The finding by sedimentation studies that *Neurospora* tyrosinases become purely monomeric with dilution (29) seemingly places beyond doubt the predominance of the monomer at the very low concentrations normally encountered in activity tests. Similarly, the fact that isolated monomeric tyrosinases exhibit activity on gel electrophoretograms (46, 58) apparently argues for an active unit containing only one copper atom. Unfortunately, these arguments fail to take into account an influence of substrate on the aggregation state of tyrosinase. Yet Jolley (56) demonstrated that 4-nitrocatechol, a strongly bound but slow substrate, effected the complete association of mushroom tyrosinase to the tetramer.‡ Arnaud (2), applying a novel technique of analytical ultracentrifugation, reached a similar conclusion. The sedimentation of a thin layer of enzyme from mushroom through a solution of catechol and NADH was followed by observing the coupled oixdation of NADH with 340 nm light. Proper conditions of substrate and pyridine nucleotide concentrations caused a return of NAD^+ to the reduced state behind the sedimenting enzyme thus clearing the background for watching catalytically active lighter units. As seen in Fig. 1, a preparation, which was shown by conventional analytical ultracentrifugation to contain essentially only monomer (3.0 S), when sedimenting in the presence of catechol, solely revealed activity accompanying a tetrameric weight

---

‡ In subsequent work, published after submission of our manuscript, Jolley and colleagues (158) observed no effect of 4-*tert*-butylcatechol on the state of aggregation of mushroom tyrosinase. Since diphenolase activity was found associated with tetrameric, dimeric, as well as monomeric fractions separated by gel filtration in the presence of substrate, the authors argue that all three forms represent active species. In band sedimentation experiments, under conditions comparable to those of Arnaud (2), activity was detected coincident with sedimenting monomer. It was further shown that no diphenolase activity loss resulted from the presence of 5 mM EDTA leading to dissociation into dimer and monomer. It appears tempting therefore to conclude that a single subunit of mushroom tyrosinase corresponds to the functional unit. However, it must still be shown that no change of quaternary structure undetected in band sedimentation or gel filtration because of its transient nature takes part in the tyrosinase reaction mechanism.

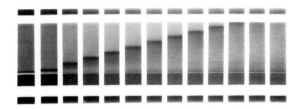

Fig. 1. Sedimentation of "monomeric" tyrosinase as tetramer in the presence of substrate. One microgram of enzyme sedimenting through 0.03 M phosphate buffer, pH 7.0, containing 2.5 mM catechol and 0.5 mM NADH. Sedimentation direction from bottom to top. Centrifugation speed, 60,000 rpm. Pictures ordered from left to right were taken every 8 minutes. Observation wavelength, 340 nm (from Arnaud, 2).

($s_{20,w}$ = 6.8 to 7.2 S). Enzyme incubated at 60°C for 1 hour and investigated with the same technique showed an active dimer (MW 62,000) and a very heavy but active polymer. It appeared from these experiments that both dimer and tetramer are functional. We, however, disagree with Arnaud as to his conclusion that the monomer is not functional. Since the observed s value indicates a complete shift of the aggregation equilibrium in favor of the tetramer, no appreciable monomer concentration was present in the experiment. Therefore, the functionality question remains unsettled. Another aspect that requires elaboration is whether the association is triggered by the phenolic substrate alone or whether O_2 is also involved.

Studying the binding of [^{14}C]benzoic acid to mushroom tyrosinase, Duckworth and Coleman (27) found at least—and unless their preparation was vastly degraded, not more than—two binding sites per tetramer. Occupation of both sites was required to completely inhibit catecholase activity. This result strengthens the idea that a dimeric unit has all the attributes to be functional.

Unequivocal evidence regarding the nature of the functional unit of both mono- and diphenolase is of prime importance to further our understanding of the catalytic mechanisms of tyrosinase. A dimer uniting two copper atoms, one from each subunit, into a single active site will be preferred on mechanistic grounds (92, cf. Section II,C,4). A structure of this hypothetical functional unit is schematically drawn in Fig. 2. The model has been given axial symmetry (101). The important features are (a) an active site located in the junction of two identical subunits and built from structural components on each side; (b) proximity of two copper ions promoting concerted function; (c) an appropriate cleft allowing access by substrates; and (d) complementarity between opposing subunit faces facilitating hydro-

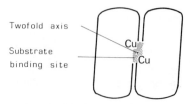

Twofold axis

Substrate
binding site

FIG. 2. Model showing the association of tyrosinase monomers to a functional dimer.

phobic, electrostatic, and/or nonbonded interactions that stabilize the assembly (isologous association). It follows from the twofold axis that if one phenol binding locus were found on the outward side of one copper atom a second such locus would occur symmetrically at the other end. Alternatively, of course, a single binding site may be localized in the center of the dimer.

The reconstitution of active tyrosinase from apoenzyme and copper was shown to require the addition of copper prior to the addition of substrate; when the order was reversed, activity was not restored (71). This result may be readily explained in terms of two features of tyrosinase referred to above, namely, the effect of substrate on enzyme aggregation and the masking of copper binding sites in the region of contact between two subunits. The long-known inactivation of tyrosinase in highly diluted solutions, which is offset by substrate (89), most likely represents another expression of the stabilizing effect of substrate on association.

7. Conclusion Concerning Tyrosinase Structure

Our present knowledge of protein structure of tyrosinases reveals that there is a striking likeness. Tyrosinases from widely different organisms resemble each other closely in amino acid composition, hydrophobic residue content, size of the single subunit, and association–dissociation behavior. This, together with the functional similarities among tyrosinases, warrants a common discussion of their structure–function relationships.

B. Active Center

Kubowitz's (80) classic experiment, later extended to tyrosinases from several sources (e.g., refs. 71 and 121), demonstrating the loss of activity upon removal of copper and reconstitution with added copper, implicated this metal in the function of tyrosinase. Furthermore, the inhibition of tyrosinase by carbon monoxide (67, 71, 107) indicated the participation of cuprous copper. Recent work in Mason's laboratory on an oxygenated form of tyrosinase reminiscent of oxyhemocyanin (60) has refocused attention on the likely function of copper in binding oxygen during catalysis (92). Functional redox behavior—another possible role of enzyme-bound copper—already suggested by Kubowitz (80) for diphenolase activity and an integral part of a mechanism of monophenolase by Mason (92), has not appeared in more recent discussions of mechanism (11, 74, 96) largely on the basis of chemical and spectral evidence obtained by Kertesz (71, 74) which seemed to indicate that tyrosinase copper was always monovalent no matter whether incubated aerobically or anaerobically, or in the presence of substrates. However, the $^{64}Cu^{2+}$ exchange experiments with functioning tyrosinase (26), to be discussed below (see Section II,B,3), strongly suggest a cupric state of the enzyme copper at one time or another during catalytic function.

The copper ion is tightly bound to tyrosinase. It follows that few coordination sites are available for substrate binding. Thus, any mechanism that requires the union of Cu, substrate, O_2, and activator in one complex (44, 45) may be possible in model systems using the naked ion but not in enzymic systems. The K_m for o-diphenol substrates of mushroom tyrosinase was found to decrease with increasing electron-withdrawing power of para substituents (27). This order is opposite to that expected if the phenol were to bind coordinatively with copper but would properly fit charge transfer complex formation with an electron-donating aromatic amino acid residue of the protein.

Above considerations argue for a site of phenol substrate attachment involving the protein rather than the copper. This does not mean, however, that the phenolic –OH is unimportant for binding. It clearly is, since o- and m-tyrosine and p-methoxyphenylalanine are neither substrates nor inhibitors (121). The participation of the protein in substrate binding is further supported by the binding of, and competitive inhibition by, such nonphenolic compounds as benzoic acid (27) and phenylalanine (121, 136) and the optical specificity particularly clear-cut with the mammalian enzyme (121). These arguments are of course thermodynamic and do not exclude a transient direct binding of the phenolic oxygen to copper in the course of reaction.

Current ideas about the state and function of copper in biological systems have been extensively reviewed (13, 87, 120). However, in contrast with iron proteins not a single active site of a copper enzyme has been characterized as to its structure and identity of the ligands. Thus, discussion of the properties of the active center in a copper enzyme is limited largely to the description of functional behavior and some phenomenological aspects if one does not want to be highly speculative. A review in that sense was published recently by Malkin and Malmström (87). In order to compare the state of copper in tyrosinase (Section II,B,1), dopamine β-hydroxylase (Section III,B), and quercetinase (Section IV) with that in other copper enzymes, the three different forms of this metal encountered in enzymes will be discussed briefly.

First, the "blue" cupric ion is characterized by its abnormally strong visible absorption of unknown origin with peak values ranging from 3,000 to 10,000 M^{-1} cm^{-1} near 600 nm, and by its ESR spectrum with unusual values of the $A_{||}$ parameter: 0.003–0.009 cm^{-1} as compared to 0.015–0.020 cm^{-1} in "normal" complexes. Asymmetry (8, 12) and/or a high degree of covalency (90, 103) of the coordination sphere has been suggested to explain these exceptional properties.

Second, the "nonblue" cupric ion seems to have no unusual properties. Its ESR spectra closely resemble those of many low molecular weight cupric complexes. Of course, nonblue cupric copper is not colorless; absorption bands from the d–d^* transitions in the d^9 system of Cu^{2+} will always be present. Characterization of enzymic copper as nonblue is normally based on ESR rather than optical spectrum. In many cases, no optical absorption data for nonblue copper enzymes have been reported, mainly because high protein concentrations are required to observe bands with extinction coefficients of 100 M^{-1} cm^{-1} or even less.

Finally, ESR nondetectable copper is characterized by the absence of an ESR signal. However convenient this classification may be, it is important to bear in mind that this term is purely operational and that absence of an ESR spectrum might have such diverse reasons as (a) dipolar interaction with a second Cu^{2+} or with another paramagnetic metal ion leading to extreme line broadening, (b) presence of diamagnetic Cu^{2+}–Cu^{2+} pairs, and (c) existence in the cuprous state. Clearly, it would be unjustified to assume that ESR nondetectable copper in different enzymes should have the same or even similar chemical properties. We might add that in discussing the valence state of copper in an enzyme, presence or absence of the d–d^* transition band and not absence of ESR spectrum is the crucial point.

1. Copper in Tyrosinase

It is easy to locate resting tyrosinase copper within the scheme discussed above as EST nondetectable. Only a small and variable amount of the total copper in tyrosinases can be seen by ESR at liquid nitrogen temperatures (10, 29, 75, 106, 107). Some or all of this copper is very likely adventitious or arises from denatured protein (75). Apoenzyme mixed with cupric copper shows a complex ESR spectrum that disappears while activity is recovered (160). The valence state of the active ESR nondetectable copper of tyrosinase remains open to discussion. Chemical determination of the valence state, inevitably ambiguous (10), has led to contradictory results. According to Kertész (74, 75), all copper in fresh mushroom tyrosinase occurs in the Cu^+ state, whereas Brooks (14) and Bouchilloux et al. (10) found 90% and variable amounts of Cu^{2+}, respectively. On the basis of the above ESR result and Kertész's chemical evidence, the description of tyrosinase as a cuprous enzyme has been preferred over the earlier conclusion of Kubowitz (80). However, recent findings of Mason's group on the reaction with H_2O_2 (60, cf. Section II,B,2) again point to the presence of cupric copper in the resting state of mushroom tyrosinase. Ashida (3) has recently observed that silkworm protyrosinase is weakly greenish blue ($\epsilon_{650} = 290$ M^{-1} cm^{-1}, including protein background) and shows the ESR spectrum of a nonblue cupric enzyme. Since the color disappears upon activation (4) and the active enzyme displays no ESR signal (106), it is tempting to conclude that copper exists in the Cu^{2+} state in latent tyrosinase and becomes reduced during the activation step. Unfortunately, the ESR spectrum of protyrosinase was not quantitated; thus, this spectrum and the blue color may perhaps represent only a minor part of the total copper in the preparation.

Pending a definitive answer to the valence question it seems most useful to discuss the possibilities of obtaining more information on that point. In the first place, establishing the presence or absence of an ESR signal resulting from copper in monomeric tyrosinase would furnish definite proof of the valence state in this form, yet conclusions regarding the dimeric and tetrameric forms would only be valid if oxidation of the metal ion during the aggregation were shown not to take place. Neurospora tyrosinase S has been reported to have no ESR signal (29). Since this enzyme was found to exist in the monomeric form (cf. Section II,A,1), it has been assumed that the absence of ESR spectrum proved its cuprous state (87). However, we consider the evidence inconclusive since association was apparent at lesser dilutions (29) and very likely was complete in the concentrated ESR sample.

In the second place, the Cu^{2+} $d-d^*$ absorption band could be used as valence diagnostic since any cupric complex will have some rather broad absorption between 500 and 800 nm. Absence of a visible band, as compared with apoenzyme, would be proof of the cuprous state. It must be realized that greater than millimolar concentrations of the enzyme (1 cm light path) and a sensitive spectrophotometer are needed for this experiment because of the expected low intensity and broad character of the absorption band. If absorption is found in the appropriate region, careful control of the experimental conditions will be necessary in order to correlate absorbance with the amount of ESR nondetectable copper in tyrosinase and to exclude artefact resulting from the presence of oxygenated form or of small amounts —1% may suffice—of some "blue" cupric protein. Reports of absorption spectra in the literature (10, 51, 71, 75, 107) fail to meet the concentration or light path requirement and are irrelevant to the valency question. Thus, this question remains open at the moment, but it might be settled if sufficient enzyme can be prepared to do the crucial experiments.

2. Oxytyrosinase

Jolley et al. (60) recently demonstrated that mushroom tyrosinase reacts with H_2O_2 and when under air develops an absorption spectrum similar to that of oxyhemocyanin with maxima at 345 nm ($\epsilon = 10,000\ M_{Cu}^{-1}\ cm^{-1}$) and at 600 nm ($\epsilon = 700\ M_{Cu}^{-1}\ cm^{-1}$) (see Fig. 3). The reaction required 1 H_2O_2 per 2 Cu and did not take place with apoenzyme. The newly formed absorption spectrum disappeared upon deaeration and reappeared when O_2 was readmitted, thus showing the existence of a reversible dioxygen–tyrosinase complex. The stoichiometry of H_2O_2 binding and the close similarity of the oxytyrosinase spectrum with that of oxyhemocyanin support the idea discussed above (Section II,A,6) of a dimeric active site containing two copper atoms. For kinetic reasons a spectrally nondetectable (i.e., in the 345-nm region) intermediate, T', had to be proposed. Altogether

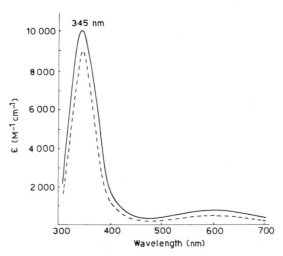

FIG. 3. Difference spectra in the near-ultraviolet and visible region of oxygenated versus resting tyrosinase [(—) from Jolley et al., 60] and of oxy- versus deoxyhemocyanin [(- - -) calculated from Cohen and van Holde, 156].

four forms of tyrosinase were required to explain the phenomena:

$$\text{Resting tyrosinase} \underset{}{\overset{H_2O_2}{\rightleftharpoons}} T' \underset{O_2}{\overset{}{\rightleftharpoons}} \text{oxytyrosinase} \rightleftharpoons \text{cuprotyrosinase}$$

Both mono- and o-diphenol substrates rapidly abolished the 345-nm band, indicating that oxytyrosinase is either itself a catalytic intermediate or can be fast transformed into one. This finding opens up entirely new avenues for the study of partial reactions of tyrosinase catalysis.

3. Different Centers of Monophenolase and Diphenolase Activity?

Every conceivable physical relationship of the monophenolase and diphenolase functions of tyrosinase has had its advocates: Both activities have been regarded at one time or another as expressions of a single catalytic center (80, 92) or of two distinct sites on the same (1, 26, 66, 94) or on different enzyme molecules (86). The lack of firm structural evidence renders a choice among these possibilities uncertain.

The isozymes from at least two plant sources (10, 47) in contrast with mammalian tyrosinases (17), display varying diphenolase:monophenolase activity ratios. However, three such isozymes from mushroom have revealed no significant differences in amino acid composition, end group identity, and tryptic peptide maps (59) that would have allowed a discrimination of structurally—and by inference functionally—different subunits. Differences at other than the primary level of structure have to be invoked to explain the observed activity patterns. Such differences may well be a consequence of preparative procedure. Thus, on the basis of our present structural knowledge it appears impossible to decide whether differently active multiple forms are caused by varying proportions of monophenolase and diphenolase specific subunits or by partial modification of an original bifunctional site.

Dressler and Dawson (26) found that mushroom tyrosinase exchanged enzymic copper against $^{64}Cu^{2+}$ only when actively engaged in oxidation of substrate. Moreover, catechols appeared to be more effective than monophenolic substrates, and preparations of relatively higher monophenolase activity exchanged less copper than more typically diphenolase enzymes. As interpreted by these authors the results suggested the existence of distinct active sites. Faster exchange with o-diphenols as opposed to monophenols may as well depend on the nature of the substrate as on the nature of the site. o-Diphenols, in contrast with monophenols, complex strongly with Cu^{2+}. It is conceivable that such complex formation facilitates exchange. Thus, the weight of the argument rests on observations with the same substrate but differently active enzymes. This leads back to our previous dilemma, i.e., whether different activity ratios are a result of different admixtures of inherently unequal sites or of a partial "opening" phenomenon (92). In addition, we note that in the work under discussion high monophenolase preparations are always more inactivated during exchange experiments than low monophenolase preparations. One wonders whether the results are not largely a reflection of the number of enzyme molecules actively participating in substrate oxidation—and therefore exchanging—in a particular preparation.

A number of kinetic investigations on tyrosinases have also been interpreted in terms of distinct sites on a single protein (1, 66, 117, 136) or on separate protein molecules (86). The assertions are usually based on the different response of mono- and diphenolase activities to the same inhibitor. However, these studies fail to take into account the multisubstrate nature of tyrosinase catalysis and the consequent complexity of the observed kinetics. Indeed, diphenolase function requires two phenolic substrate molecules to reduce one molecule of O_2, and monophenolase depends on both mono- and o-diphenol for its function. Therefore, K_m^{app} and V_{max} derived from Lineweaver-Burk plots acquire an intricate meaning in terms of reaction constants, which in addition differs for both activities. Divergent kinetic effects exerted by inhibitors should also be observed for a single site.

4. Reaction Inactivation

Tyrosinase acting on certain substrates becomes irreversibly inactivated (for complete coverage of earlier work cf. refs. 15, 54, and 109). Enzyme-bound copper is released during this process (15). Although the structure of the substrate, particularly with regard to blockade of the 4 and 5 ring positions which renders the quinone product resistent to nucleophilic substitution, was shown to be decisive, no direct inactivation by incubating tyrosinase with reactive o-quinone was observed. Likewise, when ascorbate was added to enzyme oxidizing pyrocatechol in order to reduce o-benzoquinone back to catechol as soon as it was formed, no change in the extent of inactivation was observed.

This enigmatic problem has received a reasonable explanation (15) on the basis of work by Wood and Ingraham (152) who found that [1-^{14}C]phenol, used as substrate, labeled tyrosinase during the course of reaction. It was suggested that newly formed o-quinone product may become covalently linked to a nucleophilic group before it leaves the active site. Ascorbate might not react quickly with the enzyme–quinone complex and therefore would be without effect. It was assumed that added o-quinone would largely react with aqueous solvent and be unable to build up an effective concentration at the active site. Still, one would expect at least some inactivation with added o-quinone. It might be considered therefore that the binding involves an electrophilic reaction intermediate rather than the final quinone product.

The labeling with substrate has been suggested for use in isolation of peptide fragments from the active site of tyrosinase (152). Also, from the foregoing the identification of the product of the labeling reaction seems of value from a point of view of mechanism. For our present purpose it seems important to note the presence of a nucleophilic group at the active center of tyrosinase. Interesting also, the reaction inactivation brought about by a catechol substrate equally affects the monophenolase activity (43), a fact in line with the single-site nature of the enzyme.

C. Mechanism

Not knowing the size of the functional enzymic unit, the structure of the active site(s), nor the valence state of copper, presents problems in discussing the mechanism of tyrosinase. From kinetic and other evidence we will attempt to define the most likely reaction course of monophenolase and diphenolase catalysis.

1. Monophenol Oxidation by Tyrosinase

In view of the diphenolase activity of tyrosinase the question whether o-diphenol or o-quinone is the direct product of monophenolase action is difficult to answer. It is related to whether tyrosinase could be a dioxygenase of the type 2 monophenol + $O_2 \rightarrow 2$ o-diphenol (94, 98). The identification of dopa during tyrosinase catalyzed oxidation of tyrosine [for the first time by Raper (128)] does not establish dopa as the first product since it may originate from dopa-quinone through secondary redox equilibration (28). Vaughan and Butt (146) have recently shown that spinach leaf tyrosinase catalyzing the conversion of p-coumaric acid to caffeic acid requires a two-electron donation per molecule of oxygen consumed. This observation together with the well-established need for o-diphenols to sustain mono-phenolase activity (see next section) strongly indicates the monooxygenase nature of tyrosinase. Thus, hydroxylation of monophenol appears to be coupled stoichiometrically to the oxidation of o-diphenol as suggested in the earlier reaction schemes of Mason (92, 93). Whether o-diphenol is released as a product or is immediately further oxidized to o-quinone may depend on kinetic factors. Dressler and Dawson (26) suggested that the smaller exchange of tyrosinase copper obtained with monophenols as opposed to o-diphenols may reflect the direct formation of o-quinone. However, their results could also be interpreted in terms of an o-diphenol product that participates as cosubstrate in monophenolase activity.

2. "Activation" of Monophenolase by Function

It has long been recognized that the monophenolase activity of tyrosinase requires a reducing agent for its initiation (109, 126). Work in this area has revealed a number of salient features of the induction phenomenon.

(a) The initial lag can be overcome spontaneously, in which case the needed reducing equivalents are generated by autoxidation of the monophenol substrate and/or by secondary reactions of initially formed quinone product (e.g., 2 dopaquinone → dopachrome + dopa). 3,4-Dimethylphenol yielding a relatively stable quinone accordingly shows an unusually long induction period (66). The view held sometimes (27) that the appearance of o-diphenol as the immediate product of monophenol oxidation results in overcoming the lag period is untenable since each molecule of o-diphenol forms at the expense of oxidizing another such molecule.

(b) When the added reducing agent is not an o-diphenol it most likely acts indirectly by reducing the accumulating o-quinone product. The reason for this will be given under (c). H_2O_2 in low concentrations initiates monophenolase activity (127). Since H_2O_2 would not reduce o-quinone, and in view of its reaction with tyrosinase leading to an oxygenated form (60), one may suggest that it, too, is acting in a direct way.

(c) The requirement for o-diphenols continues to exist when a steady rate of monophenol conversion has been reached. Naono and co-workers (108) found that pyrocatechase delays the onset of reaction in a mixture of phenol and tyrosinase. Because pyrocatechol is rapidly removed from this mixture the generation of appropriate reducing agent is blocked. Addition of ascorbate, known to initiate the reaction in the absence of pyrocatechase, was without effect in its presence unless a trace of dopa was added. o-Diphenols were therefore regarded as the obligatory direct "activators," in keeping with the observation that other reducing agents substantially shorten but not completely eliminate the lag phase (122, 147). Furthermore, the stimulation by L-dopa of the pyrocatechase inhibited system lasted only as long as L-dopa was present. Thus it appeared that monophenolase function in its linear phase remains dependent on o-diphenol.

(d) The "activating" effect of o-diphenols plateaus off at fairly low concentration levels, and the constant rate of monophenol conversion after the induction phase is independent of the initial o-diphenol concentration unless the point is reached where the o-diphenol shows competitive inhibition. For instance, mammalian tyrosinase is maximally stimulated for L-tyrosine oxidation by $10^{-4}\ M$ L-dopa whose K_m as substrate is $5 \times 10^{-4}\ M$ (121, 122). Thus, the effect is not truly autocatalytic and the initiation of monophenolase function is clearly not related to the "switching on" of diphenolase activity. Usually the diphenolase action will curtail the rise in o-diphenol concentration caused by nonenzymic secondary processes; a steady state concentration of L-dopa is established during L-tyrosine oxidation (70).

(e) Pomerantz and Warner (123) and Duckworth and Coleman (27) have described the "activation" as a saturation of an o-diphenol binding site on the enzyme. With L-dopa the K_m for this site was estimated to lie around 10^{-6} M, much lower than the K_m for L-dopa oxidation. It was suggested therefore that L-dopa binds on different sites when functioning as substrate or as activator. However, the data presented by these authors indicate that "activation" is more complex than simple binding of activator. If a simple binding were involved, then even at the lowest level of L-dopa employed (4×10^{-6} M) the enzyme should have been at least 75% saturated and the reaction should have started with the same fraction of the final rate. Yet the initial velocities of L–tyrosine conversion observed by Pomerantz and Warner (123) were all nearly zero except at the highest initial L-dopa concentrations. This suggests that the "saturation" has a kinetic rather than thermodynamic meaning.

(f) Increasing monophenol concentrations prolonged the lag period (27, 65, 84, 131); proportionately higher o-diphenol concentrations are required to equally shorten the lag time (27). This indicates a competition between the o-diphenol and monophenol for a reaction site on the enzyme. We suggest that the effect of hydroquinone on the lag phase, interpreted by Kertész as favoring the nonenzymic hypothesis (74), is of the same competitive type.

The many characteristics listed above, some of them only recently exposed, lend strong support to the view of Mason (92, 93) that reduction by o-diphenol is required in every turn of the catalytic cycle to produce the hydroxylating enzyme species. According to this view o-diphenol is not an activator but a cosubstrate of monophenol hydroxylation.

3. Kinetic Evidence

A sensitive and specific assay for the conversion of L-tyrosine into L-dopa developed by Pomerantz (122) measures ^{3}HOH formed from [3,5-^{3}H] L-tyrosine. In line with an earlier report (11), a slight tritium rate effect was observed. Comparison of ^{3}HOH and [^{3}H]L-dopa formation in the presence of ascorbate to prevent further oxidation of L-dopa showed that these products formed in nearly equal quantities, indicating that very little if any retention of ^{3}H took place.

L-Dopa was shown to be a competitive inhibitor of L-tyrosine oxidation catalyzed by mammalian tyrosinase. Substrate inhibition was revealed at high L-tyrosine concentrations and was found to be partially reversed by increasing the L-dopa concentration (122). Conversely, monophenols competitively inhibit o-diphenol oxidation (66, 117, 121). Metal complexing agents such as cyanide and diethyldithiocarbamate inhibit noncompetitively with respect to both mono- and o-diphenol substrates (27, 121, 122). Cyanide was shown to compete with O_2 at least in the diphenolase function (27). This behavior again suggests that the binding sites of O_2 and of phenol within the same active center are distinct, which confirms our previous conclusion that both of these substrates cannot bind to copper.

A great deal of kinetic information obtained with substrate analogs has been published (1, 27, 86, 121, 122, 136) but, as explained above (Section II,B,3), its bearing on reaction mechanism has been oversimplified.

4. Mechanistic Schemes

Although the problem is not definitely settled (cf. Section II,B,3), we will limit our discussion to the case of a single center capable of both monophenol and o-diphenol oxidation. We will also assume a single binding site for phenolic substrates in spite of assertions to the contrary in recent literature (27, 123) since we feel that the evidence for multiple sites can be readily explained by considering that a single site in different states (e.g., cupric, cuprous, or oxygenated) may display different affinities for substrate.

The most likely reaction scheme involving a dimeric functional unit with two copper atoms in the active center is depicted in Fig. 4. This scheme is based on an earlier proposal by Mason (92). The inner cycle represents diphenolase activity. It comprises two two-electron oxidations of o-diphenol in agreement with ESR results of Mason et al. (97) showing that semiquinone is not an intermediate. The first step (reactions 2 and 3) leads to cuprous enzyme, which temporarily stores two reducing equivalents.

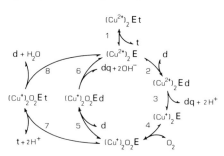

Fig. 4. Catalytic reaction cycles of monophenolase and diphenolase assuming a single active center on a dimeric functional unit. Abbreviations: d represents dopa or an o-diphenol substrate; t, tyrosine or a monophenol substrate; dq, dopaquinone or an o-quinone product; and E, tyrosinase. Single arrows indicate irreversible steps; double-headed arrows, reversible ones.

After reversible binding of dioxygen, the oxidation of a second substrate molecule results in direct reduction of dioxygen to water and regeneration of cupric enzyme. Intermediates such as $(Cu^+)_2O_2^{2-}E$ and $(Cu^+)_2OE$ might be conceived of during this process. Since such complexes would be extremely unstable and have not been observed so far, they are not included in the reaction scheme. The monophenolase function is described by the outer cycle in Fig. 4 which partly coincides with the diphenolase cycle. This overlap explains the need for o-diphenol to initiate and sustain monophenol hydroxylation. As shown, oxygen transfer takes place within a ternary complex of enzyme, dioxygen, and monophenol. Direct chemical activation of oxygen in this complex is not necessarily significant since many low molecular dioxygen adducts with transition metals show decreased rather than increased reactivity. Proximity and orientation effects of enzymic catalysis (78) may contribute more importantly to the rate of hydroxylation (reaction 8).

Reaction 1 provides for competition between monophenol and o-diphenol during "activation." It also explains substrate inhibition in monophenol oxidation and its reversal by increased concentration of o-diphenol. Competition between monophenols and o-diphenols for oxygenated tyrosinase (reactions 5 and 7) causes the reciprocal competitive inhibition behavior of these substrates.

The presence of two copper atoms per active center, and thus a dimeric active species, was assumed for the above mechanism. In the case of a monomeric functional unit a reaction cycle analogous to that proposed by Mason (93, 96) and discussed by Bright $et\ al.$ (11) should be preferred. The scheme involving an oxene complex is shown in Fig. 5. The active enzyme remains cuprous. The resting enzyme may still be cupric but would be converted to cuprous by o-diphenol. All of the kinetic characteristics of tyrosinase are expressed here and o-diphenol oxidation occurs by two-

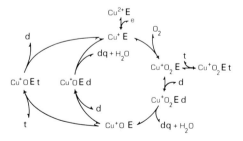

FIG. 5. Catalytic reaction cycles of monophenolase and diphenolase assuming a monomeric functional unit. Same symbols as in Fig. 4.

electron transfer. Contrary to other proponents of this mechanism (11, 93, 96) we do not believe it would operate with two atoms of copper per active center in view of the extreme instability of the system $(Cu^+)_2O$ that would instantaneously transform into $(Cu^{2+})_2$ and H_2O. Thus, n in the scheme of these authors should be 1. Because of the copper exchange and oxytyrosinase data (cf. Section II,B), which suggest the occurrence of Cu^{2+} during catalysis, we tend to favor the former mechanism. It is noteworthy that the mechanisms of tyrosinase-catalyzed o-diphenol oxidation considered here are fundamentally different from that proposed for laccase catalyzing the same reaction since the latter presumably involves one-electron transfer from the substrate and the participation of three structurally and functionally distinct types of enzyme copper (87).

o-Diphenols with less electron density at the –OH groups, although strongly bound, are less rapidly oxidized by tyrosinase (1, 27). Likewise, monophenols with electron withdrawing para substituents are very poor substrates (76). This suggests that the phenolic substrates are subject to electrophilic attack. Therefore, it appears that in analogy with other mixed-function oxidations (55) an arene oxide intermediate might be formed first which rearranges to the o-diphenol product. Since retention of tritium resulting from the NIH shift accompanying this rearrangement is expected to be very small for phenolic compounds, which indeed is the case in tyrosine hydroxylation catalyzed by tyrosinase (122), further evidence to support this step will be difficult to obtain.

III. DOPAMINE β-HYDROXYLASE

A. The Protein

Mammalian dopamine β-hydroxylase (DβH) is localized in catecholamine storage vesicles in both sympathetic nerve endings (124) and adrenal medula chromaffin cells (30, 77, 82). Within the vesicles it occurs in soluble and membrane-bound forms (30, 149). Purification of DβH from bovine adrenal medulla, the richest source of this enzyme, therefore requires a detergent for initial solubilization (30, 33, 40). However, DβH is not a lipoprotein unlike many other membraneous enzymes (30). The isolated enzyme has a molecular weight of about 250,000 by Sepharose chromatography (81) and 290,000 by sedimentation equilibrium analysis (33). No differences with respect to molecular weight, electric charge, and amino acid content were detected between membrane-bound and soluble DβH (30, 81), suggesting that the former is a membrane-adsorbed form of the latter. The unsharp and asymmetrical appearance of the DβH band in gel electrophoresis may be indicative of multiple enzyme forms (155). Helle (157) has presented evidence for the existence in DβH of subunits of molecular weight about 36,000 held together mainly by electrostatic forces.

B. Active Center

Dopamine β-hydroxylase is a copper protein (32, 33, 40), but the number of copper atoms and their state is incompletely known. "Nonblue" cupric copper is present in the resting enzyme as seen from the ESR spectrum with parameters: $g_x = 2.08$; $g_y = 2.05$; $g_z = 2.26$–2.28; $A_z = 0.017$ cm^{-1} (9, 34). Quantitative analysis showed that this spectrum represented all of the copper in a preparation passed through Chelex 100 resin [5.3 μg of copper per 12 mg of protein or two atoms per 290,000 daltons (9)]. However, the preparation of Goldstein et al. (40) used in that work was later judged to be of low purity (33) so that the actual stoichiometric ratio of the nonblue copper may exceed 2. A larger but varying copper content (2.9–6.3 atoms/molecule) was analyzed chemically in preparations of Friedman and Kaufman (33). On the basis of chemical valence determinations it was estimated that in the resting enzyme about two of the several atoms of copper are always in the cupric state. In view of the inherent ambiguity of the chemical method it appears unfortunate that this determination was not repeated with quantitative ESR (34). Solutions of DβH of up to 20 mg of protein per milliliter were found to be colorless (33). Thus, the extinction coefficient of the cupric copper cannot be especially high in keeping with its nonblue character.

The fact that the apoprotein is inactive (33) and that many complexing agents, among them carbon monoxide, inhibit the holoenzyme (33, 37, 38, 40, 42, 104), combined with the observed quantitative changes in ESR spectrum (9, 34) and proton relaxation rate (9) upon addition of ascorbate, and subsequently dopamine or phenylethylamine in the presence of O₂, point to an active role of the metal in the function of the enzyme. Nevertheless, present evidence concerning the state of copper in DβH yields no reason for believing that the metal is at the center of an atypical structure giving rise to a "state of entasis" which, as has been suggested (143), conditions metalloenzymes for catalytic activity. Unless evidence is found to the contrary, this may support the assumption (vide infra) that for DβH the protein plays another role in the catalytic action than the simple imposition of a distorted coordination geometry on the metal ion.

The catalytic action of DβH [reaction (2)] encompasses a variety of phenylethylamine derivatives (24, 36, 61). Substitution on the benzene ring by OH in para and/or meta position results in a better substrate, but OH in ortho position completely abolishes activity. Para substitution by OCH_3 is highly unfavorable, but meta substitution has a less drastic effect. A two-carbon side chain terminating with an amino group is greatly preferred to a three-carbon chain. Methylation of the nitrogen diminishes activity and tertiary amines are inactive. Since the above characteristics are also expressed in the same fashion in the mutual competitive inhibition of substrates (24, 36) and in the inhibitory effects of structurally related benzyloxyamine derivatives (25, 144), we conclude that the substitutions primarily affect the affinity of the substrates for $D\beta H$. The discrepancy between para and ortho substitution by OH with similar electronic effect and between para-OCH_3 and para-OH suggests that the attachment of the aromatic ring to the enzyme is largely governed by steric factors. The hydroxylation reaction catalyzed by $D\beta H$ is stereospecific since it yields natural levorotatory norepinephrine.§ Also, only one of the optical isomers of α-methylphenylethylamine and of α-methyldopamine—in either case the (+) enantiomorph—is hydroxylated (39,150). This stereospecificity together with above considerations regarding substrate and inhibitor effectiveness show that substrate binding involves several points of contact or bonds with the enzyme in which both the aromatic ring and the side chain of the substrate participate.

Binding appears to require a weakly basic group at an appropriate distance from the benzene ring. One might conceive of this group as a ligand to the copper. However, the finding that dopamine does not alter the proton relaxation rate enhancement by $D\beta H$ copper (9) suggests that if binding of this substrate takes place with the enzyme in the cupric state it does not likely involve the metal ion. The noncompetitive character of the inhibition by chelating agents further supports the nonparticipation of the metal in dopamine binding (42).

C. Reaction Mechanism

Kaufman and co-workers (63) identified molecular oxygen as the source of the hydroxyl oxygen in the β-hydroxylated product. This fact and the previous demonstration (85) of a stoichiometric oxidation of an electron donor, ascorbate, strictly coupled to the hydroxylation reaction, proved that $D\beta H$ is a monooxygenase.

---

§ The configuration of the product of $D\beta H$-catalyzed hydroxylation of dopamine has been incorrectly referred to as L-norepinephrine by the original workers (cf. ref. 61), the proper notation being D(−) or R(−) (sequence rule) as established for natural adrenaline (125).

In the absence of dopamine or an analog, but independently of the presence of dioxygen, the enzyme rapidly oxidizes an approximately equimolar quantity of ascorbate to dehydroascorbate (33). Although not unambiguously established it is generally believed that the electron accepting site on DβH is the cupric copper (9, 33, 34, 62). The chemical method, using 2,2′-biquinoline, by which the valence change was detected (33) cannot yield conclusive evidence since the reducing equivalent may be transferred to the copper concomitantly with the acid denaturation of the protein. Also, the diminution of the ESR spectrum of copper in the presence of ascorbate (9, 34), although strongly suggestive of a valence change, may result as well from an enhanced interaction between cupric copper and another paramagnetic center, e.g., a neighboring Cu^{2+} or the actual electron accepting site. On the other hand, the finding that no detectable SH groups are formed by reaction with ascorbate adds to the likelihood that the enzyme copper is the true acceptor.

The reduced enzyme under air, stable for several minutes without excess reducing cosubstrate present, was shown to be capable of hydroxylating an amount of phenylethylamine approximately equivalent to the amount of cosubstrate consumed in the preceding reduction (33). In the course of this process the ESR signal of copper is regenerated (34).

The results just described single out the reduction of the enzyme as the first step of the overall hydroxylation process (33). In support of this Goldstein et $al.$ (41) showed that the kinetics of β-hydroxylation are consistent with a ping-pong mechanism in which the reaction with ascorbate as first substrate must produce the reduced form of the enzyme and dehydroascorbate before the second substrate—O_2 or dopamine—may react. The kinetic data further indicated a sequential mechanism according to which both dioxygen and dopamine must combine with the reduced enzyme forming a ternary complex before the products are released. On the basis of these results a mechanistic scheme for DβH was proposed (33, 41), which is expressed in Eqs. (5)–(7).

$$(Cu^{2+})_2E + \text{ascorbate} \rightleftharpoons (Cu^{2+})_2E \text{ asc} \rightarrow (Cu^+)_2E + \text{dehydroascorbate} + 2\,H^+ \quad (5)$$

$$(Cu^+)_2E + O_2 \rightleftharpoons (Cu^+)_2O_2E \quad (6)$$

$$(Cu^+)_2O_2E + \text{dopamine} \rightleftharpoons (Cu^+)_2O_2E \text{ dopamine} \rightarrow$$

$$(Cu^{2+})_2E + \text{norepinephrine} + H_2O \quad (7)$$

We observe that this mechanism is analogous to that of Fig. 4 (outer cycle) describing tyrosine hydroxylation carried out by tyrosinase. In further analogy with tyrosinase a weak catechol oxidase activity associated with DβB has been detected (cf. ref. 62).

Nevertheless, several aspects of the mechanism remain unresolved, in particular the participation of two copper atoms per active center, the possible involvement of ascorbate radical in the reduction step (9), the formation of an oxygen adduct, the stimulation by carboxylates (41), and the details of the actual hydroxylation process. With regard to the last of these points, the possibility has been investigated that a hydroperoxide-substituted intermediate would be formed from the substrate (64). This intermediate would supposedly be reduced by the enzyme to yield the product and H_2O. However, the results showed that no ^3H is released from the substrate labeled in the β positions in contact with oxidized DβH. Therefore, if the proposed intermediate forms at all it requires the participation of reduced enzyme for its formation. As to an oxygenated enzyme intermediate, it seems noteworthy that the enzyme reduced under air, but thereafter equilibrated with N_2, was incapable of hydroxylating substrate unless air was readmitted. This shows that either O_2 does not complex with reduced enzyme alone or a reversible equilibrium is established (33). In the light of recent observations with tyrosinase (cf. Section II,B,2) the assigned role of copper in binding dioxygen should be checked by investigating the absorption spectrum of reduced DβH in the presence of dioxygen.

IV. QUERCETINASE (FLAVONOL 2,4-OXYGENASE)

Quercetinase is an inducible extracellular enzyme from *Aspergillus flavus* that catalyzes the oxidation of quercetin (I) to a depside (II) and CO [reaction (1)]. The enzyme as purified by Oka *et al.* (114) shows a molecular weight of approximately 110,000 and contains two atoms of copper per molecule (112). It is a glycoprotein containing 27.5% carbohydrate, most of which has been removed without affecting the enzymic properties (113). The copper of quercetinase probably belongs to the "nonblue" class since from ESR measurements it appears to be cupric (138) and only very concentrated solutions (30%) show a pale greenish color (112).

Krishnamurty and Simpson (79), using ^{18}O, unequivocally established the incorporation of two atoms of oxygen derived from dioxygen into the reaction product. Carbon-3 and its hydroxyl oxygen appeared in the carbon monoxide formed (79, 137). Studies of the substrate specificity disclosed a minimum requirement for a flavonoid substrate possessing a C-2–C-3 double bond, a hydroxyl group on C-3, and a keto function at C-4. Substitutions by hydroxyl on the flavonol nucleus affected both the K_m and the rate of turnover (115).

A bathochromic shift in the absorption spectrum of the substrates was revealed upon addition of stoichiometric quantities of the enzyme under nitrogen, indicating enzyme–substrate complex formation. Spectrophotometric titrations based upon this shift showed the presence of two substrate binding sites per molecule of quercetinase (115). The competitive nature of the inhibition by a metal chelating agent such as ethylxanthate and the fact that the bathochromic shift of substrate is prevented by this compound led Oka and co-workers (115) to the conclusion that the substrate forms a chelate with the enzyme copper. Moreover, the fact that a hydroxyl group at C-3 is essential to the bathochromic shift was evidence that the oxygen atoms on C-3 and C-4 participate in the binding.

Although from the nature of the reaction an intermediate peroxide bridging carbon atoms 2 and 4 appears mandatory (79, 115), more information is needed in order to understand how this adduct is formed and what role, if any, is played by the enzyme in the activation of dioxygen. Electron spin resonance measurements have shown that quercetinase copper remains in the cupric state after binding with substrate (138). Therefore, it appears unlikely that the metal is a preliminary site of attachment of dioxygen. It is noteworthy that the participation of free superoxide radical is excluded by the finding that superoxide dismutase does not inhibit the reaction (115).

V. CONCLUSION

For deeper penetration into the reaction mechanism of the copper enzymes discussed in this review enlightening information regarding their structures and the partial reactions of catalysis must be obtained. Some outstanding problems in this field include (a) the description of the active centers, particularly with regard to the number of copper atoms involved, the valence state and valence change of the metal ion, the nature of the ligands and geometry of the coordination sphere, and the substrate binding locus; and (b) the kinetic behavior of presumed catalytic intermediates, viz., the oxygenated forms and other enzyme–substrate complexes.

The similarity of the reaction schemes for tyrosinase and DβH-catalyzed hydroxylations reveals a unity of principle underlying the diversity of nature, which inspires hope for an economy of effort in resolving further mechanistic details of both enzymes.

REFERENCES

1. Aerts, F. E., and Vercauteren, R. E. (1964). *Enzymologia* **28,** 1.
2. Arnaud, Y. (1968). *J. Polym. Sci.* **16,** 4103.
3. Ashida, M. (1971). *Arch. Biochem. Biophys.* **144,** 749.

4. Ashida, M. (1971). Personal communication.
5. Ashida, M., and Ohnishi, E. (1967). *Arch. Biochem. Biophys.* **122**, 411.
6. Balasingam, K., and Ferdinand, W. (1970). *Biochem. J.* **118**, 15.
7. Bertrand, D. (1949). *Bull. Soc. Chim. Biol.* **31**, 1474.
8. Blumberg, W. E. (1966). *In* "The Biochemistry of Copper" (J. Peisach, P. Aisen, and W. E. Blumberg, eds.), p. 49. Academic Press, New York.
9. Blumberg, W. E., Goldstein, M., Lauber, E., and Peisach, J. (1965). *Biochim. Biophys. Acta* **99**, 187.
10. Bouchilloux, S., McMahill, P., and Mason, H. S. (1963). *J. Biol. Chem.* **238**, 1699.
11. Bright, H. J., Wood, B. J. B., and Ingraham, L. L. (1963). *Ann. N.Y. Acad. Sci.* **100**, 965.
12. Brill, A. S., and Bryce, G. F. (1968). *J. Chem. Phys.* **48**, 4398.
13. Brill, A. S., Martin, R. B., and Williams, R. J. P. (1964). *In* "Electronic Aspects of Biochemistry" (B. Pullman, ed.), p. 519. Academic Press, New York.
14. Brooks, D. W. (1966). *In* "The Biochemistry of Copper" (J. Peisach, P. Aisen, and W. E. Blumberg, eds.), p. 368. Academic Press, New York.
15. Brooks, D. W., and Dawson, C. R. (1966). *In* "The Biochemistry of Copper" (J. Peisach, P. Aisen, and W. E. Blumberg, eds.), p. 343. Academic Press, New York.
16. Brown, F. C., and Ward, D. N. (1957). *J. Amer. Chem. Soc.* **79**, 2647.
17. Brown, F. C., and Ward, D. N. (1958). *J. Biol. Chem.* **233**, 77.
18. Burnett, J. B. (1971). *J. Biol. Chem.* **246**, 3079.
19. Burnett, J. B., and Seiler, H. (1969). *J. Invest. Dermatol.* **52**, 199.
20. Burnett, J. B., Seiler, H., and Brown, I. V. (1967). *Cancer Res.* **27**, 880.
21. Burnett, J. B., Holstein, T. J., and Quevedo, W. C., Jr. (1969). *J. Exp. Zool.* **171**, 369.
22. Butt, V. S. (1972). *Hoppe-Seyler's Z. Physiol. Chem.* **353**, 131.
23. Califano, L., and Kertész, D. (1939). *Enzymologia* **6**, 233.
24. Creveling, C. R., Daly, J. W., Witkop, B., and Udenfriend, S. (1962). *Biochim. Biophys. Acta* **64**, 125.
25. Creveling, C. R., van der Schoot, J. B., and Udenfriend, S. (1962). *Biochem. Biophys. Res. Commun.* **8**, 215.
26. Dressler, H., and Dawson, C. R. (1960). *Biochim. Biophys. Acta* **45**, 515.
27. Duckworth, H. W., and Coleman, J. E. (1970). *J. Biol. Chem.* **245**, 1613.
28. Evans, W. C., and Raper, H. S. (1937). *Biochem. J.* **31**, 2162.
29. Fling, M., Horowitz, N. H., and Heinemann, S. F. (1963). *J. Biol. Chem.* **238**, 2045.
30. Foldes, A., Jeffrey, P. L., Preston, B. N., and Austin, L. (1972). *Biochem. J.* **126**, 1209.
31. Frieden, E., and Ottesen, M. (1959). *Biochim. Biophys. Acta* **34**, 248.
32. Friedman, S., and Kaufman, S. (1965). *J. Biol. Chem.* **240**, 552.
33. Friedman, S., and Kaufman, S. (1965). *J. Biol. Chem.* **240**, 4763.
34. Friedman, S., and Kaufman, S. (1966). *J. Biol. Chem.* **241**, 2256.
35. Fox, A. S., and Burnett, J. B. (1962). *Biochim. Biophys. Acta* **61**, 108.
36. Goldstein, M., and Contrera, J. F. (1962). *J. Biol. Chem.* **237**, 1898.
37. Goldstein, M., Anagnoste, B., Lauber, E., and McKereghan, M. R. (1964). *Life Sci.* **3**, 763.
38. Goldstein, M., Lauber, E., and McKereghan, M. R. (1964). *Biochem. Pharmacol.* **13**, 1103.

39. Goldstein, M., McKereghan, M. R., and Lauber, E. (1964). *Biochim. Biophys. Acta* **89**, 191.
40. Goldstein, M., Lauber, E., and McKereghan, M. R. (1965). *J. Biol. Chem.* **240**, 2066.
41. Goldstein, M., Joh, T. H., and Garvey, T. Q. (1968). *Biochemistry* **7**, 2724.
42. Green, A. L. (1964). *Biochim. Biophys. Acta* **81**, 394.
43. Gregg, D. C., and Nelson, J. M. (1940). *J. Amer. Chem. Soc.* **62**, 2500.
44. Hamilton, G. A. (1966). *In* "The Biochemistry of Copper" (J. Peisach, P. Aisen, and W. E. Blumberg, eds.), p. 357. Academic Press, New York.
45. Hamilton, G. A. (1969). *Advan. Enzymol.* **32**, 55.
46. Harel, E., and Mayer, A. M. (1968). *Phytochemistry* **7**, 199.
47. Harel, E., Mayer, A. M., and Shain, Y. (1965). *Phytochemistry* **4**, 783.
48. Heyneman, R. A. (1965). *Biochem. Biophys. Res. Commun.* **21**, 162.
49. Heyneman, R. A., and Vercauteren, R. E. (1968). *J. Insect Physiol.* **14**, 409.
50. Horowitz, N. H., and Fling, M. (1953). *Genetics* **38**, 360.
51. Horowitz, N. H., Fling, M., McLeod, H., and Watanabe, Y. (1961). *Cold Spring Harbor Symp. Quant. Biol.* **26**, 233.
52. Horowitz, N. H., Feldman, H. M., and Pall, M. L. (1970). *J. Biol. Chem.* **245**, 2784.
53. Inaba, T., and Funatsu, M. (1964). *Agr. Biol. Chem.* **28**, 206.
54. Ingraham, L. L. (1959). *In* "Pigment Cell Biology" (M. Gordon, ed.), p. 609. Academic Press, New York.
55. Jerina, D. M., and Daly, J. W. (1972). *Int. Symp. Oxidases Related Redox Syst., 2nd* (in press).
56. Jolley, R. L., Jr. (1967). *Advan. Biol. Skin* **8**, 269.
57. Jolley, R. L., Jr., and Mason, H. S. (1965). *J. Biol. Chem.* **240**, 1489.
58. Jolley, R. L., Jr., Robb, D. A., and Mason, H. S. (1969). *J. Biol. Chem.* **244**, 1593.
59. Jolley, R. L., Jr., Nelson, R. M., and Robb, D. A. (1969). *J. Biol. Chem.* **244**, 3251.
60. Jolley, R. L., Jr., Evans, L. H., and Mason, H. S. (1972). *Biochem. Biophys. Res. Commun.* **46**, 878.
61. Kaufman, S. (1962). *In* "Oxygenases" (O. Hayaishi, ed.), p. 129. Academic Press, New York.
62. Kaufman, S., and Friedman, S. (1965). *Pharmacol. Rev.* **17**, 71.
63. Kaufman, S., Bridgers, W. F., Eisenberg, F., and Friedman, S. (1962). *Biochem. Biophys. Res. Commun.* **9**, 497.
64. Kaufman, S., Bridgers, W. F., and Baron, J. (1968). *Advan. Chem. Ser.* **77**, 172.
65. Kean, E. A. (1964). *Biochim. Biophys. Acta* **92**, 602.
66. Kendal, L. P. (1949). *Biochem. J.* **44**, 442.
67. Keilin, D., and Mann, T. (1938). *Proc. Roy. Soc. Ser. B* **125**, 187.
68. Keilin, D., and Mann, T. (1940). *Nature (London)* **145**, 304.
69. Kenten, R. H. (1955). *Abstr. Int. Congr. Biochem. Brussels, 3rd* p. 102.
70. Kertész, D. (1953). *Bull. Soc. Chim. Biol.* **35**, 993.
71. Kertész, D. (1966). *In* "The Biochemistry of Copper" (J. Peisach, P. Aisen and W. E. Blumberg, eds.), p. 359. Academic Press, New York.
72. Kertész, D. (1969). *In* "Biological and Chemical Aspects of Oxygenases" (K. Bloch and O. Hayaishi, eds.), p. 300. Maruzen, Japan.
73. Kertész, D., and Zito, R. (1957). *Nature (London)* **179**, 1017.
74. Kertész, D., and Zito, R. (1962). *In* "Oxygenases" (O. Hayaishi, ed.), p. 307. Academic Press, New York.
75. Kertész, D., and Zito, R. (1965). *Biochim. Biophys. Acta* **96**, 447.

76. Keyes, M. H., and Semersky, F. E. (1972). *Arch. Biochem. Biophys.* **148**, 256.
77. Kirshner, N. (1957). *J. Biol. Chem.* **226**, 821.
78. Koshland, D. E. (1962). *J. Theor. Biol.* **2**, 75.
79. Krishnamurty, H. G., and Simpson, F. J. (1970). *J. Biol. Chem.* **245**, 1467.
80. Kubowitz, F. (1938). *Biochem. Z.* **299**, 32.
81. Kuzuya, H., and Nagatzu, T. (1972). *Biochem. Pharmacol.* **21**, 737.
82. Laduron, P., and Belpaire, F. (1968). *Biochem. Pharmacol.* **17**, 1127.
83. Lerner, A. B. (1953). *Advan. Enzymol.* **14**, 73.
84. Lerner, A. B., Fitzpatrick, T. B., Calkins, E., and Summerson, W. H. (1949). *J. Biol. Chem.* **178**, 185.
85. Levin, E. Y., Levenberg, B., and Kaufman, S. (1960). *J. Biol. Chem.* **235**, 2080.
86. Macrae, A. R., and Duggleby, R. G. (1968). *Phytochemistry* **7**, 855.
87. Malkin, R., and Malmström, B. G. (1970). *Advan. Enzymol.* **32**, 177.
88. Mallette, M. F., and Dawson, C. R. (1949). *Arch. Biochem. Biophys.* **23**, 29.
89. Mallette, M. F., and Dawson, C. R. (1947). *J. Amer. Chem. Soc.* **69**, 466.
90. Malmström, B. G., and Vänngård, T. (1960). *J. Mol. Biol.* **2**, 118.
91. Mason, H. S. (1955). *Advan. Enzymol.* **16**, 105.
92. Mason, H. S. (1956). *Nature (London)* **177**, 79.
93. Mason, H. S. (1957). *Advan. Enzymol.* **19**, 79.
94. Mason, H. S. (1965). *Ann. Rev. Biochem.* **34**, 595.
95. Mason, H. S. (1966). *In* "Biological and Chemical Aspects of Oxygenases" (K. Bloch and O. Hayaishi, eds.), p. 301. Maruzen, Japan.
96. Mason, H. S. (1966). *In* "The Biochemistry of Copper" (J. Peisach, P. Aisen, and W. E. Blumberg, eds.), p. 339. Academic Press, New York.
97. Mason, H. S., Spencer, E., and Yamazaki, I. (1961). *Biochem. Biophys. Res. Commun.* **4**, 236.
98. Massart, L., and Vercauteren, R. (1959). *Ann. Rev. Biochem.* **28**, 527.
99. Menon, I. A., and Haberman, H. F. (1970). *Arch. Biochem. Biophys.* **137**, 231.
100. Mitchell, H. K., and Weber, U. M. (1965). *Science* **148**, 964.
101. Monod, J., Wyman, J., and Changeux, J. P. (1965). *J. Mol. Biol.* **12**, 88.
102. Montagna, W., and Funan, H. (eds.) (1967). *Advan. Biol. Skin* **8.**
103. Morpurgo, G., and Williams, R. J. P. (1968). *In* "Physiology and Biochemistry of Haemocyanins" (F. Ghiretti, ed.), p. 113. Academic Press, New York.
104. Nagatsu, T. (1966). *In* "Biological and Chemical Aspects of Oxygenases" (K. Bloch and O. Hayaishi, eds.), p. 301. Maruzen, Japan.
105. Nagatsu, T., Levitt, M., and Udenfriend, S. (1964). *J. Biol. Chem.* **239**, 2910.
106. Nakamura, T., and Sho, S. (1964). *J. Biochem. (Tokyo)* **55**, 510.
107. Nakamura, T., Sho, S., and Ogura, Y. (1966). *J. Biochem. (Tokyo)* **59**, 481.
108. Naono, S., Kimoto, N., Katsuya, S., and Asanuma, K. (1955). *Med. J. Osaka Univ.* **6**, 161.
109. Nelson, J. M., and Dawson, C. R. (1944). *Advan. Enzymol.* **4**, 99.
110. Nicolaus, R. A. (1968). *In* "Melanins." Hermann, Paris.
111. Ohnishi, E., Dohke, K., and Ashida, M. (1970). *Arch. Biochem. Biophys.* **139**, 143.
112. Oka, T., and Simpson, F. J. (1971). *Biochem. Biophys. Res. Commun.* **43**, 1.
113. Oka, T., and Simpson, F. J. (1973). *Can. J. Microbiol.* (in press).
114. Oka, T., Simpson, F. J., Child, J. J., and Mills, C. (1971). *Can. J. Microbiol.* **17**, 111.
115. Oka, T., Simpson, F. J., and Krishnamurty, H. G. (1972). *Can. J. Microbiol.* **18**, 493.

116. Onslow, M. W., and Robinson, M. E. (1928). *Biochem. J.* **22**, 1327.
117. Osaki, S. (1963). *Arch. Biochem. Biophys.* **100**, 378.
118. Patil, S. S., and Zucker, M. J. (1965). *J. Biol. Chem.* **240**, 3938.
119. Patil, S. S., Evans, H. J., and McMahill, P. (1963). *Nature (London)* **200**, 1322.
120. Peisach, J., Aisen, P., and Blumberg, W. E. (eds.) (1966). "The Biochemistry of Copper." Academic Press, New York.
121. Pomerantz, S. H. (1963). *J. Biol. Chem.* **238**, 2351.
122. Pomerantz, S. H. (1966). *J. Biol. Chem.* **241**, 161.
123. Pomerantz, S. H., and Warner, M. C. (1967). *J. Biol. Chem.* **242**, 5308.
124. Potter, L. T., and Axelrod, J. (1963). *J. Pharm. Exp. Therap.* **142**, 299.
125. Pratesi, P., La Manna, A., Campiglio, A., and Ghislandi, V. (1958). *J. Chem. Soc.* 2069.
126. Pugh, C. E. M. (1929). *Biochem. J.* **23**, 456.
127. Pugh, C. E. M. (1930). *Biochem. J.* **24**, 1442.
128. Raper, H. S. (1926). *Biochem. J.* **20**, 735.
129. Robb, D. A. (1967). *Advan. Biol. Skin* **8**, 283.
130. Robb, D. A., Mapson, L. W., and Swain, T. (1965). *Phytochemistry* **4**, 731.
131. Sato, M. (1969). *Phytochemistry* **8**, 353.
132. Schweiger, A., and Karlson, P. (1962). *Hoppe-Seyler's Z. Physiol. Chem.* **329**, 210.
133. Seiji, M., and Iwashita, S. (1963). *J. Biochem. (Tokyo)* **54**, 103.
134. Seiji, M., and Yoshida, T. (1968). *J. Biochem. (Tokyo)* **63**, 670.
135. Seiji, M., Itakura, H., Miyazaki, K., and Irmajiri, T. (1970). *J. Invest. Dermatol.* **54**, 97.
136. Shimao, K. (1962). *Biochim. Biophys. Acta* **62**, 205.
137. Simpson, F. J., Talbot, G., and Westlake, D. W. S. (1960). *Biochem. Biophys. Res. Commun.* **2**, 15.
138. Simpson, F. J., and Oka, T. (1972). Personal communication.
139. Smith, J. L., and Krueger, R. C. (1962). *J. Biol. Chem.* **237**, 1121.
140. Sussman, A. S. (1961). *Arch. Biochem. Biophys.* **95**, 407.
141. Swain, T., Mapson, L. W., and Robb, D. A. (1966). *Phytochemistry* **5**, 469.
142. Uchida, M., Taketomo, Y., Kakihara, Y., and Ichihara, K. (1953). *Med. J. Osaka Univ.* **3**, 509.
143. Vallee, B. L., and Williams, R. J. P. (1968). *Proc. Nat. Acad. Sci. U.S.* **59**, 498.
144. van der Schoot, J. B., Creveling, C. R., Nagatsu, T., and Udenfriend, S. (1963). *J. Pharmacol. Exp. Therap.* **141**, 74.
145. van Holde, K. E. (1966). *In Symp. Mol. Architecture Cell Physiol.* (T. Hayashi and A. G. Szent-Györgi, eds.), p. 81. Prentice-Hall, Englewood Cliffs, New Jersey.
146. Vaughan, P. F. T., and Butt, V. S. (1967). *Biochem. J.* **104**, 65P; (1969). *ibid.* **113**, 109; (1972). *ibid.* **127**, 641.
147. Vaughan, P. F. T., and Butt, V. S. (1968). *Biochem. J.* **107**, 7P; (1970). *ibid.* **119**, 89.
148. Vaughan, P. F. T., and Butt, V. S. (1969). *Biochem. J.* **111**, 32P.
149. Viveros, O. H., Arqueros, L., Connett, R. J., and Kirshner, N. (1969). *Mol. Pharmacol.* **5**, 60.
150. Waldeck, B. (1968). *Eur. J. Pharmacol.* **5**, 114.
151. Walker, J. R. L., and Huhne, A. C. (1966). *Phytochemistry* **5**, 259.
152. Wood, B. J. B., and Ingraham, L. L. (1965). *Nature (London)* **205**, 291.
153. Yasunobu, K. T. (1959). *In* "Pigment Cell Biology" (M. Gordon, ed.), p. 583. Academic Press, New York.

154. Zito, R., and Kertész, D. (1969). *In* "Biological and Chemical Aspects of Oxygenases" (K. Bloch and O. Hayaishi, eds.), p. 290. Maruzen, Japan.
155. Ross, S. B , Weinshilboum, R., Molinoff, P. B., Vesell, E. S., and Axelrod, J. (1972). *Mol. Pharmacol.* **8,** 50.
156. Cohen, L. B., and van Holde, K. E. (1964). *Biochemistry* **3,** 1809.
157. Helle, K. B., and Brodtkorb, E. (1971). *Biochim. Biophys. Acta* **245,** 94.
158. Jolley, R. L., Nelson, R. M., Mason, H. S., and Ganapathy, K. (1972). *In* "Pigmentation: Its Genesis and Biologic Control" (V. Riley ed.), p. 607. Appleton, New York.
159. Miyazaki, K., and Seiji, M. (1971). *J. Invest Dermatol.* **57,** 81.
160. Kertész, D., Rotilio, G., Brunori, M., Zito, R., and Antonini, E. (1972). *Biochem. Biophys. Res. Commun.* **49,** 1208.

10

CHEMICAL MODELS AND MECHANISMS FOR OXYGENASES

GORDON A. HAMILTON

I. INTRODUCTION

It should be self-evident that mechanisms for enzymic reactions, including those for reactions catalyzed by oxygenases, must obey the laws of chemistry. Although a proposed enzymic mechanism may be different from any known chemical mechanism, it still must fall within the framework of known chemical laws, and if the mechanism is correct, then one should be able to devise and study a nonenzymic reaction which proceeds by a similar mechanism. Thus, *verification* of proposed enzymic mechanisms is one important and necessary reason for investigating chemical model systems; no biochemical mechanism can be considered proven until each of its component steps has a good chemical analogy.

An equally important reason for studying chemical model systems is that more educated *predictions* concerning the enzymic mechanisms can then be made. It may not be possible in this way to arrive at one unique mechanism for a given enzymic reaction, but the more likely ones can be pinpointed and the unreasonable ones eliminated. It is important to realize that not all "paper" mechanisms, consistent with the usually limited data on a particular enzyme, are equally likely. The vast amount of information available in the chemical literature, and that obtained from the detailed study of closely related model reactions, can be used to predict those that are preferred. Thus, the results from chemical studies can indicate the more likely characteristics to look for in the enzymic reactions and can aid in designing experiments to distinguish among the mechanistic alternatives.

Although a large number of model systems have been studied in attempts to *predict* and *verify* the mechanisms of oxygenases, the status of the field at present is that one has mainly suggestions rather than predictions and possibilities rather than verifications. However, considerable progress has been made since the publication of the first edition of this book in 1962. This article will focus mainly on describing the conclusions obtained from the study of various chemical model systems for oxygenases, and indicating how these conclusions, along with other chemical information, can be used to suggest rational chemical mechanisms for the different types of oxygenases. In order to limit the length of the article, only those aspects relating to oxygenases [i.e., enzymes which catalyze reactions in which one or more of the oxygen atoms of O_2 ends up in the organic product(s)] will be stressed; chemical models and mechanisms for oxidases (enzymes which catalyze reactions in which O_2 is reduced to H_2O_2 or $2H_2O$) have been considered in detail in earlier articles (1, 2) by the present author.

The usual classifications of oxygenases are based on the stoichiometry [monooxygenase and dioxygenase (3)] of the overall reaction and on the cofactors (flavin, heme, pteridine, metal ion, etc.) which are involved. Difficulties thus arise in organizing a chapter concerned mainly with mechanisms because it is now apparent that several enzymes in different classes have related mechanisms, and enzymes in the same class frequently have unrelated mechanisms. The compromise which has been reached in organizing this chapter is to begin with a general discussion of O_2 properties and reactions, then consider in detail one mechanism (the oxenoid mechanism) which appears to be involved in a large number of different oxygenase reactions, and, finally, discuss specific mechanisms and model systems for the various classes of oxygenases.

II. GENERAL CHARACTERISTICS OF O_2 AND O_2 REACTIONS

All oxygenase-catalyzed reactions of O_2, especially those in which the oxygen-oxygen bond is cleaved, are considerably exothermic. This is so because the sum of the bond energies (4, 5) of the carbon-carbon (*ca.* 80 kcal/mole), carbon-hydrogen (*ca.* 95 kcal/mole), and oxygen-oxygen (*ca.* 120 kcal/mole in O_2) bonds which are broken are less than the sum of those of carbon-oxygen (*ca.* 90 kcal/mole) and oxygen-hydrogen (*ca.* 110 kcal/mole) bonds which are formed. For these reasons oxygenase-catalyzed reactions are in general not reversible. Given the exothermicity of these reactions it is surprising that O_2 can exist in the atmosphere in the presence of organic compounds. Molecular chlorine, which is about as exothermic as O_2 in its reactions with organic compounds, is kinetically much more reactive and cannot coexist with most biological molecules at physiological temperatures. Thus, O_2 must have special properties which allow it to react only sluggishly with organic compounds at ordinary temperatures (of course, the reaction occurs very readily at elevated temperatures as, for example, in flames). In order to understand how oxygenases catalyze their specific reactions, it is obvious that the reasons for the low kinetic reactivity of O_2 with organic compounds should be examined very closely.

Probably the single most important reason (1, 2, 6, 7) for the low kinetic reactivity of O_2 is that O_2 has a triplet ground state—it is a diradical (i.e., has two unpaired electrons). On the other hand, the stable reduction products (H_2O_2 and H_2O) of O_2, and essentially all stable organic compounds, including the reactants and products of oxygenase reactions, are singlets (i.e., have all their electrons paired). *The direct reaction of a triplet molecule with a singlet to give singlet products is a spin-forbidden process,* i.e., it will not occur readily. This is so because angular momentum must be conserved in a chemical reaction and the coupling of electron spins with the environment is so weak that spin angular momentum is not readily exchanged with the environment. The time required for spin inversion to occur varies from about 1 to 10^{-9} second depending on the environment, but this is long relative to the time in which a chemical reaction occurs (approximately the time for one vibration or 10^{-13} second) when the reactants have sufficient energy to react. Thus, it is very unlikely that spin inversion will accompany the reaction of O_2 with the organic compound and the stability of the singlet product will not be obtained immediately; rather, a spin-allowed triplet product would be formed, and if this could exist for 1 to 10^{-9} second it could eventually go to the more stable singlet. However, the reaction of O_2 with an organic compound to give a triplet product is usually considerably endothermic and thus cannot occur with most biological molecules at physiological temperatures.

It has often been suggested that the reactive form of O_2 in biological reactions is an excited singlet state. It is known from chemical studies (8) that O_2 can be converted from its ground triplet state to a singlet state if energy is supplied, usually in the form of light in the presence of a photosensitizer. Such singlet state O_2 has a reactivity which is quite different from that of triplet O_2 (8); for example, singlet O_2 reacts very rapidly (in a fraction of a second) with alkenes at room temperature to give allylic peroxides [Eq. (1)] or with conjugated dienes to give cyclic peroxides [Eq. (2)]. The evidence is quite clear that the reactions of Eqs. (1) and (2)

$$
\begin{array}{c}
\underset{\substack{R \\ R}}{\overset{\substack{R \\ R}}{\bigg\diagup}}\!\!\!=\!\!\!\underset{CHR_2}{\diagup} + O_2 \text{ (singlet)} \longrightarrow R\!-\!\underset{HOO}{\overset{R}{\diagup}}\!\!\!\underset{CR_2}{\overset{R}{\diagup}}
\end{array} \qquad (1)
$$

$$
\bigcirc + O_2 \text{(singlet)} \longrightarrow \underset{O}{\overset{O}{\boxed{}}} \qquad (2)
$$

do not occur by a free radical mechanism. Ground state triplet O_2, on the other hand, will only react with simple alkenes at elevated temperatures and by a free radical chain mechanism.

Despite its greater reactivity, it is unlikely that singlet O_2 is involved in very many oxygenase reactions. For one thing, most of the reactions catalyzed by oxygenases bear little overall resemblance to known reactions of singlet O_2 (8) (see Section VI, B, 2 for one possible exception); for example, most reactions of singlet O_2 with organic compounds give peroxide products, whereas peroxides are the ultimate products of few oxygenase reactions. Also, singlet O_2 does not react with alkanes or unactivated aromatic compounds, both of which are frequently substrates for oxygenase reactions. However, the most persuasive argument against the involvement of singlet O_2 in biological reactions is that the lowest energy singlet state is 22 kcal/mole higher in energy than the ground state triplet, and it is not apparent how an enzyme could supply electronic energy of that magnitude. No chemical model system of which the author is aware indicates how this could be done.

How then do oxygenases cope with the fact that O_2 is a triplet and still get it to react with organic compounds? There appear to be two general methods which biological systems have evolved to circumvent this problem. One method involves complexing the triplet O_2 to a transition metal ion which itself has unpaired electrons. As has been discussed in detail previously (1, 2), because the electron orbitals on the O_2 overlap in such complexes with those on the metal ion, one cannot speak of the number of unpaired electrons on O_2 or on the metal ion but only of the number of unpaired electrons on the complex as a whole. It is a spin-allowed process for such a complex to react with a singlet organic compound to give singlet oxidized products if the number of unpaired electrons on the overall metal ion complex remains constant throughout the reaction. In such reactions, the O_2 does not react as singlet oxygen, but because it is complexed to a transition metal ion it can react by an ionic (nonradical) mechanism. Because it is not necessary to produce singlet O_2 per se such a mechanism does not have the high energy barrier (22 kcal/mole) associated with its production, and thus it would be expected to occur more readily. Several examples of this type of mechanism in chemical model systems, as well as its possible involvement in some enzymic reactions, have been discussed (1, 2, 7, 9), and some other examples will be considered later in this article.

The other apparent method by which biological systems circumvent the problem that O_2 is a triplet is to have the initial reaction of O_2 occur by a free radical mechanism. It is spin-allowed for a triplet to react with a singlet to give two doublets (i.e., two free radicals) and for the two doublets to recombine to give singlet products. The recombination is spin-allowed because the two unpaired spins originally on the triplet but now on two separate molecules do not remember their original relation to one another and thus can pair to give singlet products. In fact, the recombination is usually considerably exothermic because an extra bond and its associated bond energy are gained. The main reason that O_2 does not react with most organic compounds by this pathway at physiological temperatures is because the first step [Eq. (3) or (4)] is usually very endothermic. Con-

$$RH + O_2 \rightarrow R^{\cdot} + HO_2^{\cdot} \qquad (3)$$

$$RH + O_2 \rightarrow RH^{+} + O_2^{-} \qquad (4)$$

sider, for example, the reaction in Eq. (3). The hydrogen-oxygen bond energy in $HO_2 \cdot$ is only 47 kcal/mole (5) whereas the R–H bond energy is usually of the order of 90–100 kcal/mole (4) (if the hydrogen is bonded to carbon; N–H and O–H bonds have equal or higher bond energies). Thus, for most organic compounds the reaction of Eq. (3) [and also Eq. (4)] is endothermic by up to 50 kcal/mole and occurs to a negligible extent at ordinary temperatures.

However, there are several classes of organic molecules which do react rapidly at room temperature with triplet O_2 by one of the pathways of Eqs. (3) and (4) or a related pathway. Obviously some structural feature in these molecules allows step (3) or (4) to be considerably less endothermic. In all such cases the radical product ($R^{\cdot}$, RH^{+}, etc.) is part of a conjugated system and is highly stabilized by resonance delocalization. This resonance stabilization in the radical product decreases the R–H bond energy sufficiently that the reaction can and does occur at physiological temperatures. An example of particular relevance to biological chemistry is the reaction of fully reduced flavin (FH_2) with O_2. This reaction has recently

been shown (10, 11) to proceed by a radical mechanism; in the initial step
the flavin semiquinone radical (FH·) and superoxide are formed, and these
subsequently react to give fully oxidized flavin (F) and H_2O_2 [Eq. (5);
other ionized forms may be involved]. Both steps proceed very rapidly both

$$FH_2 + O_2 \rightarrow FH^{\cdot} + HO_2^{\cdot} \rightarrow F + H_2O_2 \tag{5}$$

enzymically and nonenzymically at room temperature. Thus, in the overall
reaction, a singlet (FH_2) reacts with the triplet O_2 to give singlet products,
but it only proceeds readily because the intermediate flavin semiquinone
radical is stabilized by extensive delocalization in the isoalloxazine ring
system.

In summary then, an ionic (nonradical) reaction of triplet O_2 with an
organic compound requires the presence of a transition metal ion catalyst,
and radical reactions of O_2 with organic compounds will only occur readily at
physiological temperatures if a highly resonance stabilized organic radical
can be formed as an intermediate. A search of the literature on the properties
of all enzymes which catalyze reactions of O_2 indicates that these two
general mechanisms are probably the only ones involved in the direct
reactions of O_2 with organic compounds. Thus, *every oxygenase* has one or
both of the following characteristics: (a) the enzyme requires a transition
metal ion (usually iron or copper) for activity; (b) the enzyme has a
cofactor or substrate, the reduced state of which gives a highly resonance
stabilized free radical by the loss of an electron or the equivalent of a
hydrogen atom.

For enzymes of the second type (which do not have metal ions) it now
becomes a little more clear why the cofactor is necessary. Molecular oxygen
is kinetically so poor an oxidant that it cannot react with most of the
substrates of these enzymes (the substrates cannot form stable intermediate
radicals) so the initial reaction is with the cofactor (by a radical mechanism),
and it is a *reactive product* of the cofactor-O_2 reaction which attacks the
substrate. Since such products would have the oxygen in a reduced or
partially reduced state, it thus seems appropriate to briefly summarize here
the known reactivity of the intermediates of oxygen reduction.

III. REACTIVITY OF REDUCED O_2 SPECIES

Oxygen is a four-electron oxidant, and each of the three intermediate
reduced states [Eq. (6); ionic forms of the intermediates are also possible]
has been chemically characterized. As might be expected the perhydroxyl

$$O_2 \underset{H^+}{\overset{1e}{\rightarrow}} HO_2^{\cdot} \underset{H^+}{\overset{1e}{\rightarrow}} H_2O_2 \underset{H^+ -H_2O}{\overset{1e}{\longrightarrow}} HO^{\cdot} \underset{H^+}{\overset{1e}{\rightarrow}} H_2O \tag{6}$$

radical ($HO_2\cdot$) shows radical characteristics in its reactions with organic compounds, but it is a relatively weak radical reagent. The reason for this is readily apparent when one considers the enthalpies of possible reactions of $HO_2\cdot$ with organic compounds; for example, the overall reaction of Eq. (7) is endothermic by 5–20 kcal/mole for most organic compounds because the hydrogen-oxygen bond energy of H_2O_2 [$ca.$ 90 kcal/mole (5)] is smaller than the R–H bond energies [95–110 kcal/mole (4)] usually encountered in

$$R\text{–}H + HO_2\cdot \rightarrow R\cdot + H_2O_2 \tag{7}$$

organic molecules. If the radical ($R\cdot$) is stabilized by resonance then reactions like that of Eq. (7) are considerably less endothermic, or even exothermic, so they will occur more readily. As a result $HO_2\cdot$ (and the similarly reactive peralkoxyl radical, $RO_2\cdot$) does not react readily at physiological temperatures with alkanes, unactivated aromatic compounds, etc., but the reaction does occur at elevated temperatures or if stabilized free radicals can be formed (12–14). Since most oxygenase substrates are ones which cannot stabilize intermediate free radicals, it is thus unlikely that the perhydroxyl or the related peralkoxyl radical is the species which directly attacks the substrate.

At physiological pH values the perhydroxyl radical [pK_a = 4.5 (15)] ionizes to the superoxide ion (O_2^-). Recent evidence indicates that O_2^- is ubiquitous to all aerobic organisms (16–18) and is involved in several enzymic reactions (9, 10, 16–22). Therefore, the question arises whether O_2^- might react with typical oxygenase substrates. However, the evidence again indicates that it is a weak reagent and only reacts readily with compounds that give stabilized free radical products. Thus, O_2^- will rapidly transfer an electron to tetranitromethane to give O_2, the stable NO_2 radical, and nitroform anion [$C(NO_3)_3^-$] (19), and it will readily accept an electron from ionized catechols to give peroxide and stable semiquinone radicals (23). But, O_2^- does not react with typical substrates of oxygenases such as alkanes, alkenes, alcohols, amines, and aromatic compounds. Therefore, although O_2^- or $HOO\cdot$ may be an intermediate in various oxygenase reactions, it is unlikely that either is involved in a direct reaction with an unactivated substrate. An exception to this would be if the substrate can be readily attacked by a nucleophile. One would expect O_2^- to be a good nucleophile and it, or a complexed form, could react with electrophilic sites on substrates (see, for example, Sections VI, A, 1 and VI, A, 2).

There is considerable confusion and many misconceptions stated in the literature concerning the reactivity of H_2O_2. The facts are that, under physiological conditions in the absence of transition metal ion catalysts, H_2O_2 is a very unreactive molecule except in situations where it can function as a nucleophile. The high nucleophilicity of H_2O_2 and especially the hydroperoxide ion (HOO^-) are well documented (24). Because peroxides are frequently used as initiators of free radical chain reactions (12, 13) the concensus seems to have arisen that peroxides cleave to radicals even under physiological conditions. However, this is not true; in order to get radicals from peroxides one of the following must apply: (a) the peroxide is irradiated with high energy radiation (e.g., ultraviolet light), (b) the temperature is elevated (to over 100°C), or (c) a transition metal ion in a lower valence state is present. The reason that peroxides do not dissociate spontaneously to radicals [Eq. (8)] at physiological temperatures is because the oxygen-oxygen bond energy is too high (51 and 43 kcal/mole) in H_2O_2 (5) and

$$ROOH \rightarrow RO^{\cdot} + OH^{\cdot} \tag{8}$$

ROOH (26), respectively); it is not apparent how an enzyme could supply energy of this magnitude for a homolytic cleavage reaction. If the roxy radical ($RO^{\cdot}$) formed by the cleavage of ROOH could be stabilized by resonance then the reaction of Eq. (8) would be less endothermic and occur at lower temperatures. No such stabilization is possible if R is alkyl, but a very stable radical would be formed if R is aryl. Perhaps for this reason aryl hydroperoxides have not been characterized chemically (25); they decompose too readily. However, one should be sensitive to their possible involvement as intermediates in some enzymic reactions.

Hydrogen peroxide is not a very effective electrophilic reagent either. Strong polarizable nucleophiles will attack H_2O_2 (26) and this can be considered an electrophilic reaction of H_2O_2, but simple nucleophiles do not react similarly; for example, even under strong acid catalysis H_2O_2 exchanges its oxygen with H_2O only very slowly (27), a reaction which would be simply a nucleophilic displacement by H_2O on one of the oxygens of H_2O_2. It is not surprising therefore that electrophilic attack of H_2O_2 on very weak nucleophiles such as alkenes, aromatic compounds, alcohols, and alkanes (i.e., typical oxygenase substrates) occurs only with extreme difficulty or not at all.

One encounters frequently the suggestion that HO^+ is the reactive species which attacks unactivated oxygenase substrates. At least on paper, one could envisage its being formed from H_2O_2 in the presence of acid as shown in Eq. (9). Such a species would be expected to be a very reactive elec-

$$\overset{H^+}{HOOH} \rightleftharpoons H_2O^+-OH \rightleftharpoons H_2O + HO^+ \tag{9}$$

trophilic reagent. However, in all of chemistry there is *no evidence* that HO^+ (or the related RO^+) is ever an intermediate in any chemical reaction which is not catalyzed by high energy radiation (26). The fact that H_2O_2 exchanges its oxygens only slowly with water in strong acid solution (see above) is one specific indication that HO^+ does not form readily. In acid-catalyzed rearrangements of alkyl hydroperoxides the evidence is consistent with a concerted alkyl migration and cleavage of the oxygen-oxygen bond [Eq. (10)], not with RO^+ being a discrete intermediate. Thus, the suggestion of free HO^+ or the related RO^+ species being involved in oxy-

$$R_2-\overset{\overset{\displaystyle R_1}{|}}{\underset{\underset{\displaystyle R_3}{|}}{C}}-O-OH \underset{}{\overset{H^+}{\rightleftharpoons}} R_2-\overset{\overset{\displaystyle R_1}{|}}{\underset{\underset{\displaystyle R_3}{|}}{C}}-\overset{+}{O}-\overset{+}{O}H_2 \longrightarrow R_2-\overset{\overset{\displaystyle }{|}}{\underset{\underset{\displaystyle R_3}{|}}{\overset{+}{C}}}-O-R_1 + H_2O$$

$$\Big\downarrow -H^+ \qquad\qquad (10)$$

$$\underset{R_3}{\overset{R_2}{>}}C=O \ + \ R_1OH \longleftarrow R_1-\overset{}{\underset{\underset{\displaystyle R_3}{|}}{C}}\overset{\displaystyle OR_1}{\underset{\displaystyle OH}{<}}$$

genase reactions must be considered a formalism; it cannot be considered a real mechanistic possibility which has a chemical analogy.

Although H_2O_2 itself, under physiological conditions, is reactive only as a nucleophile, it can be chemically altered to give compounds which are considerably more reactive as electrophiles or radical reagents. Several such altered peroxides [for example, peracids, peramides (imidoperacids), vinylogous ozone, and carbonyl oxides], which are believed to be involved in various oxygenase reactions, are discussed in detail in the following sections. An early chemical example of an altered peroxide, which is reactive enough to attack an unactivated alkane position, was reported by Corey and White (28). They observed that the peroxide (I) is converted to (III) [Eq. (11)] in low yield when reacted with *p*-nitrobenzenesulfonyl

(I)   (II)   (III)

chloride in cold pyridine-methylene chloride. Because the $ArSO_3^-$ is such a stable species the other peroxide oxygen of the perester, (II), presumably attains sufficient electrophilic character to attack the adjacent methyl group as indicated. Although this reaction is probably not directly analogous to any oxygenase reaction, it does indicate that alteration of peroxides to give reactive electrophilic species can be accomplished and that even unactivated alkane groups will react with a strongly electrophilic oxygen species.

There can be no doubt that the third of the intermediates of O_2 reduction, namely, the hydroxyl radical (HO·), is an extremely reactive reagent (12–14). It shows very little selectivity in its reactions; all known organic compounds including alkanes, alkenes, and aromatic compounds, react with HO·. The typical reactions it performs are hydrogen atom abstraction and addition to π systems; in each case the initial step is considerably exothermic; for example, hydrogen atom abstraction [Eq. (12)] by HO· is exothermic by at least 10–25 kcal/mole because the hydrogen-oxygen bond

$$R\text{–}H + HO· \rightarrow R· + H_2O \tag{12}$$

energy in H_2O [119 kcal/mole (5)] is greater than any known R–H bond energy. Therefore, the question concerning the possible involvement of HO· in oxygenase reactions is not whether it would be able to attack typical unactivated oxygenase substrates but whether it is likely to be formed at all.

One argument against the involvement of HO· in enzymic reactions is that such a reactive species would be very difficult for an enzyme to control; it would be expected to react with various enzymic groups at the active site and thus inactivate the enzyme. Also, it would be surprising for an enzyme to waste so much energy in producing such a high energy intermediate when lower energy compounds would suffice. However, the above arguments do not seem strong enough to eliminate the possibility of HO· involvement; consequently, the conditions required for HO· formation should be examined.

The most likely source of HO· would be H_2O_2, and as indicated earlier this would require one of high energy radiation, elevated temperatures, or a transition metal ion in a low valence state. The first two methods are not available to oxygenases under physiological conditions, but the third is certainly feasible for many metallo-oxygenases. The formation of HO· by the reaction of H_2O_2 with typical oxygenase metal ions such as Fe(II) [Eq. (13)] or Cu(I) is well documented (12–14). It has been reported (29)

$$Fe(II) + H_2O_2 \rightarrow Fe(III) + OH^- + HO· \tag{13}$$

that $O_2^{\bar{\cdot}}$ will react with H_2O_2 to generate HO·, but this mechanism for oxygenases is not likely because it would require two molecules of O_2 to generate the reactive species, and such characteristics are not shown by these enzymes. Therefore, if HO· is involved in any oxygenase reaction it can probably only be involved in those catalyzed by metallo-oxygenases.

In summary, from the known chemical reactivity of O_2, O_2 complexes, and reduced O_2 species, the following tentative rules concerning the most likely oxygen-containing species to react with oxygenase substrates can be stated:

1. For only the few oxygenase substrates which can form a highly resonance stabilized free radical by loss of an electron or hydrogen atom is the direct reaction of triplet O_2 with the substrate by a radical mechanism reasonable.

2. For oxygenase reactions which require a nucleophilic attack of an oxygen species on the substrate, the most likely nucleophiles are peroxide (ROOH, where R could be almost anything) and O_2^- (or complexed O_2^-).

3. For oxygenases which have no cofactor and whose substrates cannot stabilize an intermediate radical, a metal ion is required so that the substrate can react directly with O_2 by an ionic (nonradical) mechanism.

4. For oxygenases which have a cofactor and no metal ion and whose substrates cannot stabilize an intermediate radical, the O_2 must initially react with the cofactor by a free radical mechanism to give a reactive singlet product which then reacts with the substrate. If the only reasonable way the products from such substrates could be formed is by reaction with a highly electron deficient (electrophilic or radical) oxygen species then that species is most probably a chemically altered peroxide; there are no other chemically feasible alternatives.

5. For oxygenases which have a transition metal ion and whose products appear to be formed from their substrates by reaction with a highly electron deficient oxygen species, then the reactive oxygen species could be any of (a) a metal ion complexed O_2, (b) a chemically altered peroxide, or (c) possibly the hydroxyl radical.

IV. THE OXENOID MECHANISM

A. General Comments

In 1964, the present author (30) noted the remarkable similarity of many monooxygenase (mixed function oxidase) catalyzed reactions to carbene and nitrene reactions, and suggested that these enzymes catalyze their reactions by an oxygen atom transfer or oxenoid mechanism (1, 2). Since that time several enzymes, whose reactions have oxenoid characteristics, have been shown to be dioxygenases (see Section VI, A) rather than monooxygenases as originally thought. Also, it is now apparent that some monooxygenases (Section V, B) do not involve the oxenoid mechanism. These developments emphasize the point that the mechanisms of enzymic reactions frequently bear little relationship to their overall stoichiometries (on which the oxygenase classifications are based). In any event, it appears that a large fraction of oxygenase reactions involve the oxenoid mechanism at some stage, and thus various aspects of this mechanism will be considered here.

Carbenes (31) (CR_2) and nitrenes (32) (NR) are very reactive intermediates in many organic reactions. Some examples of the similarity between typical carbene (nitrenes give related reactions) and many oxygenase (1, 2, 33, 34) reactions are the following:

1. Carbenes will insert into completely unactivated alkane carbon-hydrogen bonds to give alkyl compounds [Eq. (14)] with overall retention of configuration; there are numerous examples of analogous oxygenase re-

$$-\overset{|}{\underset{|}{C}}-H \ + \ CR_2 \ \longrightarrow \ -\overset{|}{\underset{|}{C}}-CR_2H \tag{14}$$

actions where unactivated alkanes are converted to alcohols [Eq. (15)], and in those few cases where the stereochemistry has been studied the conversion occurs with retention of configuration (35).

$$-\overset{|}{\underset{|}{C}}-H \ + \ O_2 \ \xrightarrow{\text{oxygenase}} \ -\overset{|}{\underset{|}{C}}-OH \tag{15}$$

2. Carbenes add to alkenes to give cyclopropane compounds [Eq. (16)];

$$\tag{16}$$

the analogous oxygenases catalyze the epoxidation of alkenes [Eq. (17)].

$$\tag{17}$$

3. Carbenes react with aromatic compounds [Eq. (18)] to give alkyl benzenes (IV) and norcaradienes (V) which are in rapid equilibrium with

cycloheptatrienes (VI); various oxygenases catalyze the conversion of

$$(18)$$

aromatic compounds [Eq. (19)] to phenols (VII) and areneoxides (VIII) which are in rapid equilibrium with oxepines (IX).

$$(19)$$

Given this marked analogy of carbene and nitrene reactions to many oxygenase reactions one can consider the former as relatively distant chemical models for the enzymic oxidations. As a result these have been the starting point for the development and investigation of many of the more directly related chemical models discussed below.

Carbenes and nitrenes are species which have only six electrons in the outer valence shell of the atom. The close analogy of oxygenase reactions implies that the enzymic oxidant is an oxygen species which also has only six electrons in its outer shell. Such an oxygen species analogous to a *free* carbene or nitrene would be the oxygen atom, and oxygen atoms are certainly reactive enough (36) to attack typical oxygenase substrates. However, it is unlikely that a free oxygen atom is formed in the biochemical systems. The reaction [Eq. (20); AH_2 represents the reducing agent involved in these oxygenase reactions] of O_2 to give a ground state triplet oxygen atom is considerably endothermic [even the reaction of H_2 with O_2 to give

$$AH_2 + O_2 \rightarrow A + H_2O + O \text{ (triplet)} \tag{20}$$

H_2O and triplet O is somewhat endothermic (5)]. The formation of a singlet oxygen atom would be excessively endothermic (5, 36). Yet the oxygenase reactions resemble those of singlet carbenes and nitrenes; triplet carbenes and nitrenes have different characteristics (31, 32). Therefore, it is unreasonable to suggest that the *free* oxygen atom is responsible for the biological oxidations.

It is known that a free carbene is not involved in many reactions which show characteristics typical of carbene reactions. In such reactions, which are called "carbene transfer" or "carbenoid reactions," a carbon species with six outer shell electrons is merely transferred from the reagent to the substrate in the transition state (31). Therefore, it was suggested (30) that the oxidizing agent in many oxygenase reactions is a species which is capable of transferring an oxygen atom to the substrate. This mechanism is termed the "oxenoid mechanism."

Probably the strongest enzymic evidence which is in support of the oxenoid mechanism, and which has been obtained since the original suggestion of the mechanism, is the elegant work done by the NIH group on the "NIH shift" (37–41). These investigators have observed for a large number of enzymic hydroxylations of aromatic compounds that the hydrogen, on the position in the aromatic ring which eventually is hydroxylated, frequently shifts during the hydroxylation to a carbon adjacent to its original position. An impressive body of evidence has now been collected (37–41) which indicates that an areneoxide is formed as an intermediate in the enzymic hydroxylations. Hydrogen migration occurs when the areneoxide (X) rearranges to the keto tautomer (XI) of the phenol [Eq. (21)]. In the tautomerization of (XI) to the phenol a kinetic

(21)

isotope effect accounts for the retention of high percentages of the hydrogen isotope in the product.

Of particular interest to the present discussion is the initial step of Eq. (21), the formation of the areneoxide from the aromatic compound. The areneoxide is exactly the product expected if the oxidizing agent operates by an oxenoid mechanism. Since many enzymes which oxidize alkenes, alkanes, etc., are closely related to those which oxidize aromatic compounds, the work by the NIH group not only is strong evidence for the oxenoid mechanism for aromatic hydroxylations but also can be considered additional presumptive evidence that the mechanism holds for a whole group of oxygenases.

Although most of the presently available evidence indicates that the oxenoid mechanism is probably involved in a large number of oxygenase reactions, in no case has the structure of an actual enzymic oxenoid reagent been positively identified. One thing which is clear, however, is that the oxidizing agent is not exactly the same for all enzymes utilizing this mechanism; for example (1, 2, 33, 34), some such enzymes are metalloenzymes while others are not, various different prosthetic groups (e.g., heme, reduced flavin, and reduced pteridine) are required by different enzymes, and the cosubstrate or reducing agent, which makes it possible for the four-electron oxidant (O_2) to carry out a two-electron oxidation of the substrate, can vary. Until recently, few nonenzymic reactions with oxenoid properties had been examined, and thus it was difficult to make reasonable suggestions for possible structures and characteristics of the enzymic reagents. However, in the past few years several chemical systems have been studied in detail, and these will now be considered.

B. Chemical Model Studies

Udenfriend and co-workers in 1954 (42) reported that aromatic compounds are hydroxylated by O_2 in low yield at neutral pH and room temperature if Fe(II), ascorbic acid, and EDTA are also present. Because this nonenzymic system bears a marked resemblance to many monooxygenase reactions it has been the subject of numerous subsequent investigations (1, 14, 43–45). In the author's laboratory (43) it was shown that the Udenfriend system [ascorbic acid, Fe(II), and O_2] would not only hydroxylate aromatic compounds to phenols but would also hydroxylate saturated hydrocarbons to alcohols and epoxidize olefins. In addition, it was found (43) that the aromatic hydroxylation reaction would occur if the ascorbic acid is replaced by a specific pyrimidine derivative. This pyrimidine has a structure related to, and is at the same oxidation level as, the tetrahydropteridines which are involved in several monooxygenase reactions (46). Later, Bobst and Viscontini (47) showed that tetrahydropteridines themselves could replace ascorbic acid in the Udenfriend system. All of these results illustrate the apparent similarity of the Udenfriend system to various monooxygenase reactions.

Unfortunately, because of the complexity of the Udenfriend system, and because the system gives many other ill-defined reactions (yields of hydroxylated products are low), it has not been possible to establish definitely the detailed mechanism of the hydroxylations. It is now quite clear, however, that H_2O_2 is not an intermediate in this reaction and that the hydroxylating agent is not the hydroxyl radical (HO·) nor the perhydroxyl radical (HOO·). One peculiarity of the reaction is that substituted aromatic compounds, with substituents (e.g., the methoxy group of anisole) which usually direct ortho-para in both electrophilic and radical reactions, frequently give large amounts of the *meta*- phenol, and the relative amount of meta product is quite dependent on the reaction conditions (14, 43–45). Such results suggest that the aromatic compound initially reacts to give some intermediate and this intermediate subsequently leads to the phenol. Other evidence for an intermediate is that aromatic halides are frequently dehalogenated by the Udenfriend (48) or related (14) systems. In the author's laboratory (48) it was found that oxidation of chlorobenzene by the Udenfriend system gives small amounts of chlorophenols (ortho·meta·para approximately 1:1:1), but the main hydroxylated aromatic products are catechol and hydroquinone. It is difficult to rationalize the formation of these latter two products unless some intermediate is formed. If the Udenfriend system operates by an oxenoid mechanism (transfer of a singlet oxygen atom) then an areneoxide would be a reasonable intermediate. However, in the rearrangement of areneoxides to phenols, the NIH shift (37–41) is usually observed, and hydroxylations by the Udenfriend system show very little NIH shift in the products (38). Also, it is unlikely that *meta*-methoxyphenol would result from any areneoxide formed from anisole. Rather, if (XII) or (XIII) [Eq. (22)] formed one would expect them to open to the ortho and para products, respectively, because it is now known that, in the formation of phenols from areneoxides under both neutral and acid conditions, a carbonium ion is formed as an intermediate (49), and (XIV) and (XV) are expected to be the more stable carbonium ions. All these results indicate it is unlikely that areneoxides are intermediates in the hydroxylation of aromatic compounds by the Udenfriend system.

What then is the oxidizing agent in the Udenfriend system and what is the intermediate in aromatic hydroxylations by this system? As indicated earlier, it is not possible to state definitely what these are. However, the results seem consistent with the oxidizing agent being a complex of O_2,

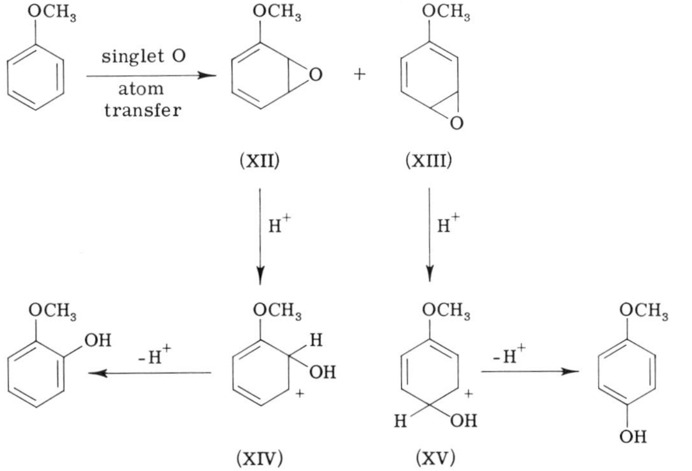

(22)

Fe(II), and ascorbic acid (1, 30) [(XVI, Eq. (23)] and that this transfers a *triplet* oxygen atom to the substrate. Therefore, the intermediate in aromatic hydroxylations would be a triplet precursor of the product. This precursor, following spin inversion and subsequent reactions, would

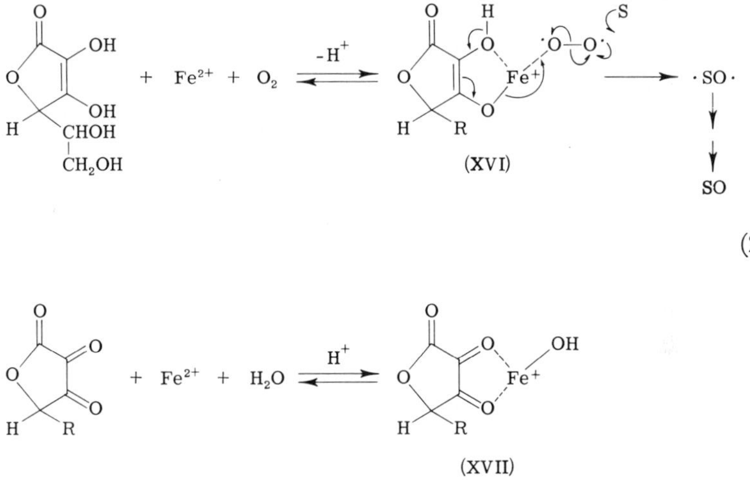

(23)

eventually lead to the observed singlet product, SO (phenols in the aromatic hydroxylations). It is difficult to predict what structure of the triplet ·SO· would be favored if anisole is the substrate, but attack at the meta position to give (XVIII) [(XVIIIa) and (XVIIIb), Eq. (24), are only two of the many possible resonance structures] would appear to be at least as favorable as attack at the ortho or para positions. Following spin inversion (XVIII)

(24)

would be expected to give *meta*-methoxyphenol. Several mechanistic alternatives could be suggested for the loss of halogen from a triplet intermediate formed from halobenzenes, but without more information to limit the number of possibilities they will not be considered further here. It might be expected that the triplet oxygenated intermediates would not give the NIH shift because the initial product following rearrangement would be a triplet dienone which would not have any special stability.

If the mechanism of Eq. (23) is the mechanism of the Udenfriend reaction one might inquire why (XVI) transfers a triplet oxygen atom rather than a singlet. According to the spin conservation rule the total number of unpaired electrons on (XVI) and on the substrate, S (which is a singlet and has none) must be the same as the total number on the products, (XVII) and ˙SO˙. Therefore, presumably the triplet ˙SO˙ is formed because the most stable form of (XVII) has two less unpaired electrons than (XVI). One could imagine, however, that other O_2 complexes similar to (XVI) could transfer a singlet oxygen atom if somehow the structure of the product (XVII) was altered (for example, by binding to other groups on an enzyme) so that it could be stabilized with the same number of unpaired electrons as (XVI). Thus, it is possible that the oxidizing agent in the Udenfriend system is analogous to that in some oxygenase reaction; the overall reaction characteristics may be somewhat different (triplet rather than singlet oxygen atom transfer) merely because of the relative stabilities of the different spin states of the metal ion complexes.

One of the basic requirements for oxygen atom transfer from a complex such as (XVI) is that O_2 is complexed to a transition metal ion, which is also connected to some reducing agent that can donate two electrons to the O_2 in the step where the oxygen atom is transferred to the substrate. One might expect that many different reducing agents would be able to replace the ascorbic acid in such a reagent. Presumably this is the reason that tetrahydropteridines (47) and pyrimidines (43) can function as reducing agents in the Udenfriend system; they have complexing and redox properties similar to ascorbic acid. It appears in many cases that reduced metal ions can perform the function of the reducing agent. It has been found in many model systems (14, 45, 50, 51) that no organic reducing agent is required to achieve hydroxylation of aromatic compounds by O_2 if a reduced metal ion is present in relatively large concentrations. Thus, under various conditions (pH, solvent, presence or absence of nonredox reactive complexing agents, etc.) it has been observed that O_2 will hydroxylate aromatic compounds if Fe(II), Cu(I), Ti(III), Sn(II), or V(II) is present. Some of those hydroxylations have characteristics very similar to those of the Udenfriend system; for example, aromatic compounds with substituents which usually direct orth-para give relatively high amounts of meta-substituted products. It is significant that the higher the concentration of the reduced metal ion the higher the ratio of meta product formed (50). As in the Udenfriend reaction, the yields of hydroxylated products in these metal ion systems are low, and thus it has not been possible to study them sufficiently quantitatively to be able to arrive at detailed mechanisms for the hydroxylations. However, for most of the systems [especially those involving Cu(I) or Fe(II)], the results are consistent with the hydroxylating agent being an O_2 complex of a binuclear (or polynuclear) metal ion compound [illustrated for Cu(I) in Eq. (25)].

$$\text{Cu(II)----Cu(II)}-\text{OH} \quad + \quad \cdot\text{SO}\cdot \qquad (25)$$

(XIX)

An oxygen atom transfer from (XIX) is possible if the two metal ions are electronically connected (2) by some bridging group which allows one molecular orbital to overlap both metal ions. Thus, each metal ion can be oxidized by one electron in the same step that one oxygen atom is transferred and the other reduced to water. Since the hydroxylation characteristics are similar to those of the Udenfriend system, presumably a triplet oxygen atom is transferred in these model systems but it is conceivable an enzyme could modify the binding of the metal ions to allow a singlet oxygen atom transfer.

Ullrich and his collaborators have studied in greater detail two of the metal ion-O_2 hydroxylation systems, one composed of Sn(II)–HPO_4^{2-}–O_2 (51, 52) and the other of Fe(II)–2-mercaptobenzoate–O_2 (53, 54). Both systems were found to oxidize alkanes to alcohols as well as aromatic compounds to phenols. The Sn(II) system is very unselective in its reaction with alkanes; primary, secondary, and tertiary carbon-hydrogen bonds react at the same rate. This is very similar to the selectivity shown by triplet oxygen atoms generated by irradiation of N_2O. Thus, presumably the oxidizing agent in the Sn(II) system is either a free triplet oxygen atom or a triplet oxygen atom transfer reagent. In its reaction with aromatic compounds the Sn(II) system gives very large relative amounts of meta-substituted phenols from compounds with substituents which usually give ortho-para products. This, therefore, can be taken as further evidence that the Udenfriend and related reactions proceed by a triplet oxygen atom transfer mechanism. The mechanism of the oxidations by the Sn(II) system is probably similar to that shown in Eqs. (24) and (25) except that one Sn(II) can act as a two-electron reductant to stabilize the oxygen atom not transferred. Thus, during the transfer of the triplet oxygen atom to the substrate the Sn(II) is converted to Sn(IV).

The Fe(II)–2-mercaptobenzoate–O_2 oxidation system shows many characteristics similar to oxidations by rat liver microsomes (53, 54). Thus, benzene compounds are hydroxylated, alkanes are oxidized (selectivity toward primary, secondary, and tertiary carbon-hydrogen bonds similar to microsomes), naphthalene is converted to naphthalene-1,2-dihydrodiol, and O-alkyl compounds are dealkylated. However, whereas liver microsomes give large amounts of the NIH shift, the Fe(II) model system gives very little (55). Therefore, the mechanism of the model system must be different from that of the enzymic one. Most probably the model system has a mechanism closely related to those in Eqs. (24) and (25).

All the chemical model systems discussed above have the advantage that like the biological oxidations O_2 is the reactant. However, from a mechanistic standpoint, they have the very serious disadvantage that they are not "clean" chemical systems, i.e., several different types of reactions are occurring. The particular reaction of interest, namely, the hydroxylation, proceeds in only very low yield. Thus, it has not been possible to apply to these systems the usual tools (e.g., kinetics and stoichiometry) of the mechanism chemist, and as a result any mechanistic conclusions can be considered as only tentative. As was discussed earlier (Sections II and III), it is now apparent that O_2 itself is frequently not the reagent which reacts with unactivated oxygenase substrates; rather, some reduced form of O_2 is the reactant. Since the reaction of interest to the present discussion is the attack of some oxygen-containing species on unactivated substrates by an oxenoid or other mechanism, it is evident that this step should be isolated as much as possible from those involving merely the formation of the oxenoid reagent from O_2. Then the actual substrate oxidation step can be studied separately, presumably the systems will be "cleaner," a more detailed picture of the characteristics of this step should be obtained, and hopefully more educated predictions concerning the biological oxenoid reagents can be made. This is the approach that the present author and his co-workers have followed in recent years.

Before discussing the recent work with simple oxenoid reagents, a relatively complex but surprisingly clean system in which aromatic compounds are hydroxylated by H_2O_2 in the presence of catalytic amounts of Fe(III) and catechol [hereinafter referred to as the Hamilton system (14)] should be considered briefly (56–60). This system has been of particular interest because it is believed to have clarified the nature of catalase intermediates (1, 59), but it seems possible that similar intermediates are involved in oxidations by P-450 enzymes (see Section V, A). It was concluded from a detailed kinetic and product study of the aromatic hydroxylation reaction that the hydroxylating agent in the Hamilton system is the resonance-stabilized intermediate (XXI) [Eq. (26); only a few

$$(26)$$

(XX)   (XXIa)   (XXIb)   (XXIc)   (XXId)

(XXIII)   (XXII)

of the possible resonance forms of (XXI) are illustrated]. This is believed to form by the loss of a water molecule as shown from a complex of catechol, Fe(III), and H_2O_2 (XX). Some of the evidence for these conclusions is the following: the oxidizing agent has properties different from those of any other known oxidizing agent (i.e., the $HO\cdot$ or $HOO\cdot$ radicals) which could be suggested for this system; the formation of the oxidizing agent is the rate-determining step in the reaction (i.e., aromatic compounds and alcohols are oxidized at the same rate, and the rate does not depend on their concentrations); and only diphenols capable of reversible oxidation and reduction are catalysts. Because (XXI) has a complexed organic ligand (the catechol) which can exist in oxidized and reduced states, one cannot say what the redox state of the ligand or what the valence state of the metal ion is; the complex is a resonance hybrid of many different valence bond structures involving oxidized and reduced forms of these species (1). Of particular interest is the state and reactivity of the other oxygen of (XXI). One could consider it as a bound oxygen atom as in structure (XXIb) but it would be stabilized considerably by resonance involving other valence bond structures, some of which are depicted in (XXIa), (XXIc), and (XXId). Probably the only reason such a species can form is because it is stabilized by resonance with a redox reactive ligand; if the catechol is replaced by nonredox reactive ligands the reaction does not proceed. In its reactions with aromatic compounds (XXI) acts as a very reactive radical reagent; it is less selective than the HO·. Thus, on reaction with aromatic compounds (XXII) is believed to form, and this by transfer of an electron and proton would give (XXIII) which is merely an Fe(III) complex of the product phenol and the catalyst catechol. According to this mechanism (58) one would not expect the reaction to give the NIH shift (an areneoxide is not an intermediate), and very little is observed for this model reaction (38, 61). The intermediate (XXI) is believed to be a model for catalase compound I (1, 59); porphyrins have redox properties similar to those of catechol. Since the model system can hydroxylate aromatic compounds, it is conceivable that the heme-containing monooxygenases, namely, the P-450 enzymes, could have a related oxidizing agent (see Section V,A).

As indicated earlier, in attempts to obtain more detailed information on the chemical characteristics of oxenoid reagents, some considerably simpler systems have recently been studied in the author's (62–70) and other laboratories (54, 71). From thermodynamic considerations, one would predict that there should be a large number of oxygen containing compounds (X=O) which potentially could react with organic compounds (S) to give overall oxygen transfer as illustrated in Eq. (27). For many cases where X is relatively stable in the oxidized and reduced states, the overall

$$X{=}O + S \rightarrow SO + X \qquad (27)$$

reaction is exothermic because the two carbon-oxygen and/or oxygen-hydrogen bonds in SO have greater energy than the double bond in X=O plus the original bond in S. However, it soon became clear from early experiments that there were other requirements for oxenoid reagents. In these preliminary experiments (69) various alkanes, alkenes, and aromatic compounds were heated with several potential oxenoid reagents such as sulfoxides, sulfones, selenoxides, arsine oxides, and amine oxides, and in no case was oxygen atom transfer to the organic compound observed. The reason for this lack of reactivity now seems fairly obvious. In all of these compounds the oxygen has an excess of electrons; since oxygen is more electronegative than the atom to which it is attached, the zwitterionic structure X^+-O^- will contribute considerably to the ground state structure of $X=O$. However, an electron-deficient species (electrophilic or radical) is probably required in order to obtain a reaction with alkanes, alkenes, or aromatic compounds. It is known that carbenes and nitrenes are electron deficient; in some reactions they act as electrophilic reagents and in others as radical reagents. Also, the best characterized oxenoid reaction, namely, the epoxidation of alkenes by peracids, is known to occur by an electrophilic attack of the reagent on the alkene (72). Thus, one important conclusion which was derived from this work is that any oxenoid reagent (enzymic or nonenzymic) which reacts with alkanes, alkenes, or aromatic compounds probably has to have an electron-deficient oxygen.

With this conclusion in mind it has been possible to develop and study in detail several oxidations, some of which show oxenoid character. Although amine oxides on heating do not transfer oxygen atoms to alkanes, alkenes, or aromatic compounds (see above), the transfer does occur with light catalysis (54, 66, 68, 71); presumably the oxygen is made more electron deficient by an n to π^* or related transition. Other reactions which have been studied extensively include ozonation (65, 67), oxidations by peroxytrifluoroacetic acid (54, 61, 70), oxidations by an adduct formed from carbenes and O_2 [Eq. (28)] (64, 71), and oxidations by chromate reagents (73–75). Several criteria have been applied in attempts to determine the char-

$$R_2C: + O_2 \rightarrow R_2\overset{.}{C}-O-\overset{.}{O} \text{ or } R_2C=\overset{+}{O}-\overset{-}{O} \tag{28}$$

acteristics of an oxenoid mechanism, but the three which have been the most useful are (1) the relative reactivity of primary, secondary, and tertiary carbon-hydrogen bonds in alkanes; (2) the stereochemistry of the alkane oxidation; and (3) the extent of the NIH shift observed. The results of these investigations plus some data taken from the literature (13, 31, 32, 62) on related characteristics of other alkane reactions are summarized in Table I.

The NIH shift data are important because they indicate that in reactions showing such a shift an areneoxide is an intermediate, presumably formed by an oxenoid mechanism. However, these data give little information concerning the detailed characteristics of the actual oxygen atom transfer step. For this the results from the alkane oxidations are more useful (62–70). One expected characteristic of an oxenoid reaction (singlet oxygen

TABLE I

CHARACTERISTICS OF SOME OXIDATIONS OF ALKANES AND AROMATIC COMPOUNDS

Reaction	Alkane relative reactivity per hydrogen			% Retention of configuration in alkane oxidation	Magnitude of the NIH shift[a] (%)
	Primary	Secondary	Tertiary		
Oxidations to alcohols and phenols					
Pyridine-N-oxide + $h\nu$	1	10	40	100	45
Pyridazine-N-oxide + $h\nu$	1	15	70	>95	34
Nitrobenzene + $h\nu$	1	20	300	0	—
Ozonation	1	13	110	60–70	—
Carbene-O_2	1	15	140	0	16
Peroxytrifluoroacetic acid	1	~30	~1000	100	9
Chromate	1	110	7000	70–100	30
Other alkane reactions					
Carbene insertions	1	1–8	1–21	100	—
Nitrene insertions	1	5–15	14–70	100	—
H· abstraction by RO·	1	12	44	—	—
H⁻ abstraction	1	~10^4	~10^8	—	—

[a] The magnitude varies with the isotope and aromatic compound used and with the conditions (61, 71); thus, these numbers should be considered only semiquantitative.

atom transfer) is that the conversion of an alkane to an alcohol will occur with retention of configuration. Equilibrated products (zero percent retention of configuration) are expected if a radical or ionic intermediate is formed by hydrogen abstraction from the alkane. Thus, the results given in Table I suggest that some of the alkane oxidations studied occur by an oxenoid mechanism, some do not, and at least one (the ozone reaction) by a mixed mechanism. Actually, however, there seems to be a gradual change from one stereochemical result to the other, and, at least for the first five reactions listed, a correlation between the selectivity of the reagent and the percent retention of configuration, the more selective the reagent the more equilibration obtained. Carbene and nitrene insertions (by singlet species) would fit into this same correlation; they show low selectivity (which varies with the structure of the carbene or nitrene) and 100% retention of configuration. The selectivity of the first five reactions listed in Table I is in the range of what is expected for radical reactions (similar to that for reaction of RO· with alkanes) but considerably lower than that expected for hydride ion abstraction. It is difficult to obtain an accurate estimate of the selectivity for hydride ion abstraction; the numbers given in the table are estimates based on the ease of carbonium ion formation from alkyl halides. They may not be very accurate but the important point is that the selectivity is expected to be very large. The fact that the chromate oxidation shows high selectivity and also high retention of configuration led Rocek to suggest that the transition state for this oxidation has considerable carbonium ion character (73). The results with peroxytrifluoroacetic acid would thus suggest that its transition state has less carbonium ion character than the chromate oxidation but more than the other oxenoid reactions.

The results of these investigations are all consistent with the general mechanism for alkane oxidation illustrated in Eq. (29) (X=O symbolizes a generalized oxidizing agent which in the light-catalyzed reactions would be the reagent after the absorption of light). It is suggested that in all cases the reactants give a transition state or intermediate, (XXIV), which is a

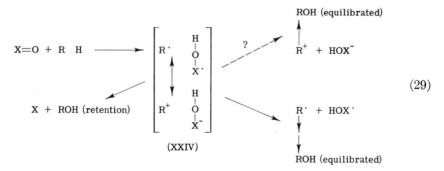

(29)

resonance hybrid of radical and ionic structures, and that the relative reactivity of the various carbon-hydrogen bonds and the sterochemistry observed for the reaction are determined by the characteristics of (XXIV). If HOX is not very stable as a radical (HOX·) or ion (HOX⁻) then (XXIV) would collapse rapidly to give alcohol with retention of configuration. If, on the other hand, HOX· is relatively stable the components of (XXIV) could diffuse apart to R· and HOX·. Reasonable mechanisms are available for the conversion of R· to ROH in the various systems, but since R· would equilibrate its stereochemistry rapidly, the ROH formed from it would be completely equilibrated. There is evidence that R· is formed and causes the partial equilibration in the ozone reaction (65, 67); presumably the reaction is proceeding completely by this pathway in the carbene-oxygen and nitrobenzene plus light systems. An examination of the structures for HOX· in these various systems indicates that the observed results are in the direction one might expect according to this mechanism; i.e., the cases where one predicts HOX· would be more stable give more equilibration.

The selectivity of the various oxidants is largely determined by the relative contribution of the radical and ionic structures to (XXIV); the more ionic structures contribute, the more selective is the reagent. Again those reagents (the peracid and chromate) which show greater selectivity are those where one might expect HOX⁻ to be more stable. In both these cases (XXIV) apparently collapses directly to alcohol to give retention, but it is possible that other cases will be found where (XXIV) separates to R⁺ and HOX⁻; if such were to happen, equilibrated alcohol would also be expected from the rapidly equilibrating R⁺.

In summary, then, the results of these investigations indicate that the oxidation of an alkane to an alcohol by an oxenoid mechanism proceeds through a transition state or intermediate which has some blend of radical and ionic character, which blend is determined by the characteristics of the compound (X) which is attached to the transferring oxygen atom. Overall insertion of the oxygen with retention of configuration will be observed if the transition state or intermediate collapses directly to products. One expects that the enzymic alkane oxidations occur in the same manner; presumably the enzyme, by binding both the oxidant and substrate, could prevent separation of the intermediate into radicals or ions, and thus it would ensure collapse to the alcohol. Also, it is now somewhat understandable why a variety of oxenoid reagents has evolved in biological systems; by having a variety of X groups the various enzymes have been able to alter the radical or ionic characteristics of the oxygen atom transfer so that the requirements for each individual reaction can be met.

Another product formed in the photolytic reaction of alkanes with pyridine-N-oxide deserves comment because it suggests that another group of enzymes (not oxygenases) may also be operating by an oxenoid mechanism. When pyridine-N-oxide is photolyzed in the presence of cyclohexane a major product is cyclohexene-oxide (68, 71) which has been shown (68) to form in two steps, the first step being a dehydrogenation of cyclohexane to cyclohexene, and the second, an epoxidation [Eq. (30)]. The first step is

$$\text{(30)}$$

particularly interesting because it is similar to reactions catalyzed by various desaturases (76). Furthermore, many of the characteristics of these enzymes are similar to those of oxygenases believed to involve the oxenoid mechanism. Consequently, the mechanism of these desaturases as well as that of the model reaction above is probably that shown in Eq. (31) (1). Thus, the oxenoid reagent, instead of inserting into the carbon-hydrogen

$$X{=}O \quad \xrightarrow{\quad} \quad X + H_2O + \text{(31)}$$

bond to give an alcohol, abstracts two hydrogens to give water and the alkene directly. Analogous carbene (77) and nitrene (78) reactions have been reported. The observation that at least in one enzymic desaturation the reaction occurs by a *cis* elimination of the two hydrogens (76) is consistent with this mechanism.

V. MONOOXYGENASES

A. Monooxygenases Apparently Operating by an Oxenoid Mechanism

1. Flavin-Requiring Hydroxylases

Several monooxygenases (2, 33, 34, 79), including salicylate decarboxylase, imidazoleacetate hydroxylase, p-hydroxybenzoate hydroxylase, and orcinol hydroxylase, require flavin as a cofactor, have no metal ion, and catalyze an oxidation which looks like an oxenoid reaction. In each case the enzyme, with the reduced form of flavin attached, interacts with O_2 and the substrate to give the oxidized form of flavin, the oxidized substrate, and water. The reduced form of the enzyme-bound flavin is regenerated for subsequent turnovers by reaction with a reducing agent (NADH or NADPH). Mager and Berends and co-workers (80) have studied a nonenzymic model system which may be related to the enzymic reaction. They observed that aromatic compounds are hydroxylated by O_2 when various reduced flavins are present. However, the yields are low and the mechanism of this model reaction has not been established.

Since these monooxygenases do not contain a transition metal ion, the only reasonable hydroxylating agent is a chemically altered peroxide (see Section III). Various investigators have suggested that the reduced flavin interacts with O_2 to give a flavin peroxide and that this is the reagent which attacks the substrate. However, as discussed earlier (Section III), alkyl peroxides are not good hydroxylating agents in the absence of metal ion catalysts. Therefore, it is unlikely that a flavin peroxide per se is the hydroxylating species. Recently, the present author proposed (2) that the oxygen species which attacks the substrate is the resonance-stabilized intermediate (XXVI) (only a few of the possible valence bond structures are illustrated) which is formed from the flavin peroxide (XXV) as shown in Eq. (32). The proposed oxenoid reagent (XXVI) can be considered a

$$(32)$$

carbonyl oxide but it is also a vinylogous ozone (62, 63, 81), i.e., a homolog of ozone having a carbon-carbon double bond between two of the oxygens. The homology can be seen by comparing valence structure (XXVIc) with the related one (XXVIIc) of ozone (XXVII). As discussed in Section

IV, ozone is a good oxenoid reagent and one would expect (XXVI) to be also, especially because the terminal oxygen atom can attain considerable electrophilic character resulting from valence bond structures such as

(XXVIb), (XXVIc), and (XXVId) which put negative charge on electronegative atoms.

The suggested overall mechanism (2) for these monooxygenases is illustrated in Eq. (33). As discussed in Section II, the reaction of the reduced flavin (FH$_2$) with O$_2$ to give (XXV) would have to proceed by

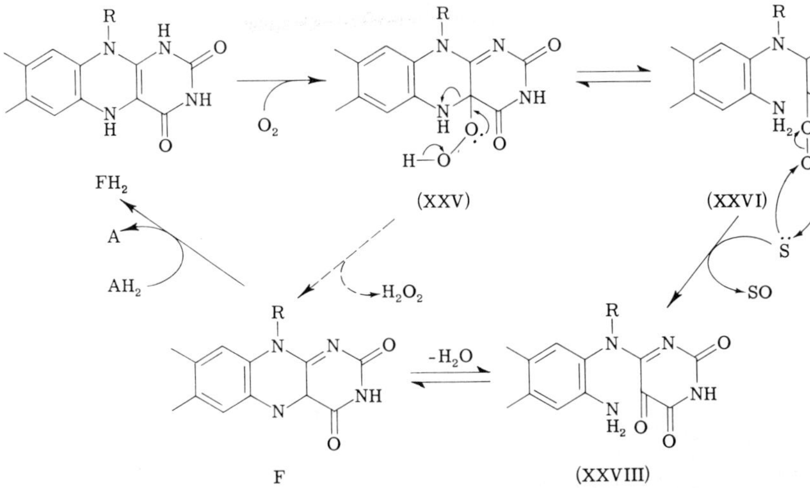

$$\text{(Eq. 33)}$$

FH$_2$

A

AH$_2$

H$_2$O$_2$

O$_2$

(XXV)

(XXVI)

S

SO

F

-H$_2$O

(XXVIII)

a free radical mechanism because of the spin conservation rule. The conversion of (XXV) to (XXVI) involves a proton transfer and cleavage of a nitrogen-carbon bond of the flavin. This should not be a difficult process, especially if appropriately placed acid and base groups on the enzyme assist the protonation and deprotonation. If (XXVI) transfers an oxygen atom to the substrate as shown, then (XXVIII) would be formed; (XXVIII) should regenerate the oxidized flavin (F) very readily, even nonenzymically, because it is just a Schiff base formation. To complete the cycle, F would be reduced back to FH$_2$ by NADH or NADPH (AH$_2$).

For many of these enzymes the substrate hydroxylation can be uncoupled from the oxidation of AH$_2$ if poor substrates are present. In such cases the enzyme catalyzes the overall reaction shown in Eq. (34). The mechanism of Eq. (33) can readily explain this result. If S is too unreactive to be

$$\text{AH}_2 + \text{O}_2 \rightarrow \text{A} + \text{H}_2\text{O}_2 \tag{34}$$

attacked by (XXVI), then the only reaction possible is for (XXV) to decompose to F and H$_2$O$_2$ [Eq. (33)]. One would expect (XXVI) and (XXV) to be in rapid equilibrium, and thus all of the reaction would eventually proceed according to Eq. (34).

From the known reactions of carbonyl oxides and vinylogous ozone (82, 83) one can suggest a very attractive detailed mechanism for the reaction of (XXVI) with the various substrates. It is noteworthy that *all* of the substrates for these flavin-requiring hydroxylases have a phenolic OH group (NH in the case of imidazoleacetate hydroxylase) adjacent to the site where they are attacked. This is a nucleophilic group, and it is known that carbonyl oxides and vinylogous ozones are very susceptible to attack by nucleophiles (82, 83). Thus, a phenol would be expected to react with (XXVI) to give (XXIX) [Eq. (35); for convenience only one ring of the

(35)

original flavin nucleus is illustrated in these structures]. The conversion of (XXIX) to (XXVIII) and (XXX) should occur readily; it is somewhat akin to the Claisen rearrangement (84) except that it is expected to proceed more rapidly because an oxygen-oxygen bond is broken and more stable carbon-oxygen bonds are formed. If X in (XXX) is hydrogen (as in most of these enzymic reactions), then enolization gives the product; if X is the carboxylate group (as in salicylate decarboxylase), then (XXX) is a β-ketoacid and should decarboxylate readily to the product diphenol. Thus, the sequence of reactions illustrated in Eq. (35) seems very plausible for the particular part of these enzymic reactions in which the substrate is hydroxylated. If such a mechanism is correct, then perhaps these reactions should not be referred to as oxenoid reactions because, although an overall oxygen atom transfer occurs in several steps, no one step in the sequence involves the direct transfer of an oxygen atom from a reagent to the substrate (oxenoid mechanism).

In the earlier article (2) the present author also discussed the possibility that the oxidizing agent for these flavin-requiring hydroxylases is a peramide (see Section V, A, 4) formed from a different flavin peroxide. This mechanism seems less likely for the reasons given (2) than that discussed above, but present evidence cannot exclude it, and it must still be considered a possibility.

2. Tetrahydropteridine-Requiring Hydroxylases

These monooxygenases (2, 33, 34, 46) have many characteristics which indicate that they have mechanisms similar to the flavin-requiring hydroxylases. For example, with poor substrates or an altered coenzyme the substrate hydroxylation can be uncoupled from the oxidation of the reducing agent so that H_2O_2 is the product. Also, tetrahydropteridines have chemical properties and structures closely related to reduced flavins; the main differences are that a nitrogen replaces a carbonyl oxygen (on carbon-2 of the flavin), and the aromatic ring is missing in the tetrahydropteridines. At one time it was thought that the tetrahydropteridine enzymes require a metal ion [Fe(II)], but recent evidence (85) indicates that the metal ion is not involved in the usual catalytic reaction but only protects the enzyme from inactivation. All these results indicate that the oxidizing agent in these enzymic hydroxylations is similar to that for the flavin-requiring hydroxylases. This has been recently suggested and discussed by the present author. (2) At least for some of these enzymes the detailed mechanism for the hydroxylation step shown in Eq. (35) is not possible, however, since some substrates (e.g., phenylalanine) do not have the requisite nucleophilic group. Therefore, a direct oxenoid transfer is suggested for the substrate oxidation step. This is consistent with the observation that these enzymic reactions give the NIH shift (37–41), and thus presumably the aromatic substrate is initially converted to the areneoxide. Apparently, therefore, the vinylogous ozone intermediate (2) in the tetrahydropteridine reaction has somewhat different properties from that formed from reduced flavin. This may result from the former's greater basisity (it has an extra nitrogen) which would allow it to be protonated, and thus the oxygen to be transferred would be more electrophilic.

3. Tyrosinase

Although tyrosinase (1, 86–88) (sometimes called polyphenol oxidase, phenolase complex, etc.) was one of the first enzymes catalyzing a reaction of O_2 to be studied, and the first shown to be an oxygenase (89), its mechanism is still obscure. Some of the reasons for this are the complexity of the enzyme and its reactions. The tyrosinases from various sources are multisubunit enzymes, and several atoms of copper [as Cu(I)] are required for activity. All these enzymes catalyze two apparently different oxidations by O_2, namely, the oxidation of catechols to *ortho*-quinones (catecholase activity) and the hydroxylation of phenols to *ortho*-diphenols (cresolase activity). This latter reaction is what categorizes tyrosinase as an oxygenase.

A nonenzymic system studied by Brackman and Havinga (90), in which phenols are converted to amine-substituted *ortho*-quinones by O_2 in the presence of Cu(II)–amine complexes, has been considered a model for tyrosinase since copper is a catalyst and the overall reaction bears some resemblance to the enzymic one. The proposed mechanism of the model reaction has been frequently discussed (86, 91). It is not known with certainty if the enzymic mechanism is similar, but it seems unlikely because H_2O_2 is required for the cresolase reaction in the model system, and no evidence for H_2O_2 as an intermediate or reactant in the enzymic reaction has ever been found. Based on conclusions derived from a detailed study of the Hamilton model hydroxylating system (see Section IV, B) the present author proposed (1) a mechanism for tyrosinase, but again enzyme-bound peroxide would have to be an intermediate. A different type of mechanism which avoids this problem was recently developed by the present author and is considered briefly below.

It now seems clear that tyrosinase must have a catechol tightly bound in order to catalyze the cresolase reaction (1, 88, 89), and it may also be necessary for the catecholase reaction (88). In addition, only phenols are substrates for the hydroxylation and the attack always occurs in a position ortho to the original phenol to give a catechol. In this respect the tyrosinase reaction is similar to that catalyzed by the flavin-requiring hydroxylases discussed above (Section V, A, 1); this suggests that related oxidizing agents are involved in the two different types of enzymes. Other pertinent evidence which any mechanism must be consistent with is the following: only *ortho*-diphenols are oxidized in the catecholase reaction, *para*-diphenols are unreactive; no evidence has been found for free radical intermediates during catalysis; and the copper is present as Cu(I) and does not change valence during the reaction. A mechanism consistent with all this evidence is illustrated in Eq. (36). Although not illustrated, the reactions are all considered to occur on the enzyme surface in the vicinity of at least one of the bound copper ions. The metal ion would especially be needed for the first step (see Section II). However, the other intermediates may also be complexed to the metal.

It is suggested that an enzyme-bound catechol and O_2 react to give (XXXI); a similar step is believed to occur in several dioxygenase reactions (1) (Section VI, B). The loss of water from (XXXI) gives the vinylogous ozone (XXXII) (only two of several possible resonance forms are illustrated). The reaction of (XXXII) with another phenol molecule (acting as

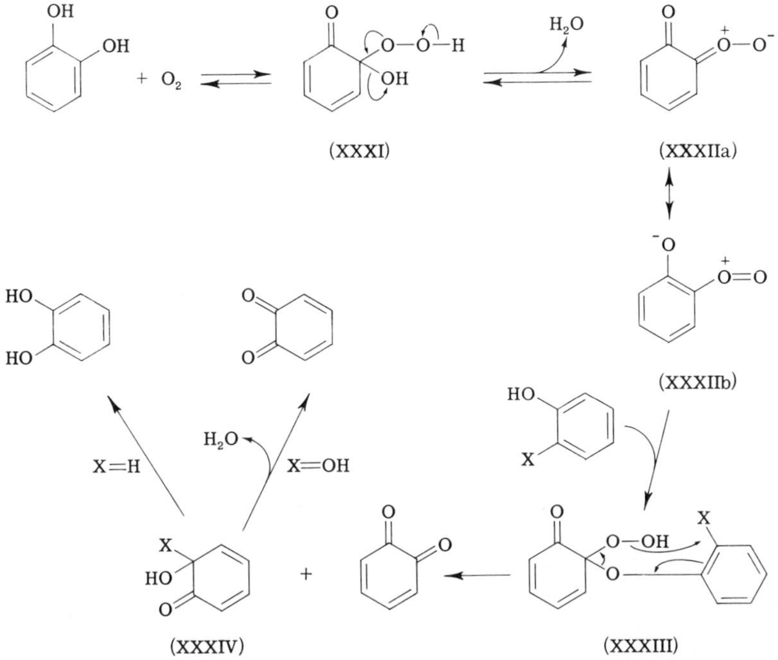

(36)

a nucleophile) to give (XXXIII), and its subsequent decomposition to (XXXIV) and an *ortho*-quinone, are similar to those previously discussed [Eq. (35)]. If X in (XXXIV) is H then tautomerization would give the catechol (cresolase activity), but if X is OH dehydration would give the *ortho*-quinone (catecholase activity). Both of these steps would be expected to be rapid. If X=OH a slight variation on the mechanism of Eq. (36) seems very probable, i.e., it is expected that (XXXIIII) would decompose very rapidly directly to two molecules of *ortho*-quinone as shown in Eq. (37).

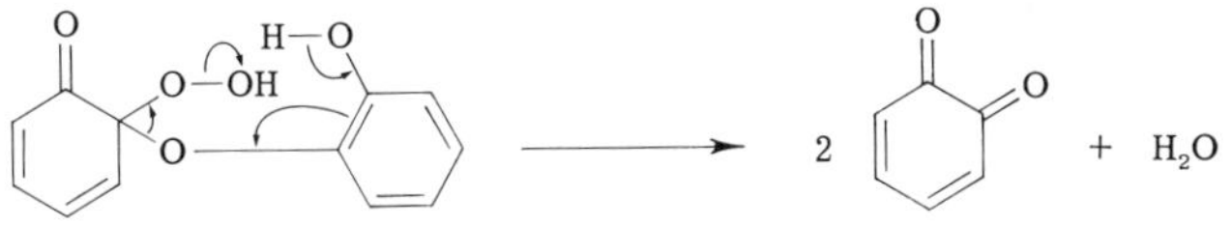

(37)

(XXXIII)

In any event, the mechanism of Eq. (36) gives a rationale for the two apparently different activities (catecholase and cresolase) of tyrosinase; their mechanisms are very similar in that a vinylogous ozone is involved in each.

If the vinylogous ozone mechanism for tyrosinase is correct, then it predicts that one of the oxygens of a catechol must be lost to water and replaced by an oxygen from O_2 during each turnover of the cresolase or catecholase reaction. Obviously, this experiment should be performed if feasible. However, problems which will probably be encountered in such an experiment include the following: the product *ortho*-quinone may exchange too rapidly with the solvent, and a tightly bound catechol may be the catalyst for many turnovers, its product *ortho*-quinone being merely reduced back to the catechol by other catechol molecules in solution.

4. P-450 Enzymes

The various hemoprotein monooxygenases which have been labeled P-450 enzymes (92, 93) catalyze the hydroxylation of a large number of aromatic and aliphatic compounds by what appears to be an oxenoid mechanism. One indication of this is that the aromatic hydroxylation reactions give the NIH shift (37–41). These frequently particulate monooxygenase systems have a complex overall mechanism, but the present discussion will focus on only one part, namely, the actual substrate oxidation step. There is now general agreement that this occurs following the transfer of an electron to an Fe(II) form of the enzyme which has substrate and oxygen bound to it, the latter apparently attached directly to the metal ion. Thus, at least formally, there would appear to be three possibilities for the oxenoid reagent: (1) an enzyme–Fe(I)–molecular oxygen complex, (2) an enzyme–Fe(III)–peroxide compound, and (3) an enzyme–Fe(V)–oxygen atom complex. It is important to stress that in the above the valence of the iron is given only in formal terms. The Fe(I) enzyme would presumably be a resonance hybrid of Fe(I)-porphyrin, Fe(II)-reduced porphyrin (by one electron), and Fe(III)-reduced porphyrin (by two electrons). Similarly, the formal Fe(V) intermediate would have resonance contributors involving Fe(IV) and Fe(III) with oxidized porphyrin structures (1, 58, 63).

Based on the present evidence, the second alternative of the three possibilities listed above for the oxenoid reagent in P-450 enzymes seems the most likely, but at present none can be eliminated. One reason why the first alternative seems less likely is that the bound oxygen would be expected to have an excess of electrons rather than being electron defficient. Regardless of whether the iron or porphyrin is reduced, some of the excess of electrons would probably be delocalized onto the complexed oxygen. However, it is thought that in some model systems [Eqs. (23) and (25)] the oxidizing agent is O_2 complexed to a metal ion which is either reduced or complexed to a reducing agent. Thus, this mechanism for P-450 enzymes must still be considered a possibility.

The third alternative, an enzyme–Fe(V)–oxygen atom complex, seems the least likely of the three. Such a reagent would presumably be similar to catalase compound I (1, 59) and to the intermediate in the Hamilton model hydroxylating system [Eq. (26)] so mechanisms would certainly be available for its formation. One argument against this as the oxenoid reagent, however, is that in aromatic hydroxylations by the P-450 enzymes the NIH shift is observed (37–41), whereas it is not observed in the Hamilton model system (38, 61). Perhaps a stronger argument against this reagent is that such a complex of an oxygen atom on an iron-porphyrin would not sterically be able to react with many of the known substrates by an oxenoid reaction; the steric requirements of the transition state appear to be too great.

Several structural possibilities can be envisaged for the most likely oxenoid reagent, an enzyme–Fe(III)–peroxide compound. However, a complex where the peroxide is bonded directly to the iron is not one of them, because Fe(III)–peroxide complexes do not readily react with alkanes or aromatic compounds. More likely structures would be a peracid (XXXIV) or peramide [imidoperacid (XXXV)] formed [Eqs. (38) and (39)] by nucleophilic attack of some oxygen species on an enzymic carboxylic acid or amide group. When another electron is fed into an $Fe(II) \cdot O_2$

$$(38)$$

$$(\text{XXXIV})$$

$$(39)$$

$$(\text{XXXV})$$

complex one would expect that a supernucleophile would be formed because even $Fe(II) \cdot O_2$ itself is expected to be somehat nucleophilic; $Fe(III) \cdot O_2^{\bar{\cdot}}$ is a significant resonance contributor to its structure. Thus, nucleophilic attack by such a species on an enzymic carboxylic acid or amide group to give (XXXIV) or (XXXV) seems very plausible. These are also very reasonable oxenoid reagents. As discussed in Section IV, peracids in chemical model systems are capable of performing essentially all the reactions that the P-450 enzymes do. Peramides are probably at least as reactive. They have never been isolated but are intermediates in the reaction of nitriles with H_2O_2, and such intermediates do carry out typical oxenoid reactions such as the epoxidation of alkenes (72).

In summary, the present author believes that a peracid (XXXIV) or peramide (XXXV) is the most likely substrate oxidizing agent in the reactions catalyzed by the P-450 enzymes, but other possibilities cannot be excluded. If a peracid or peramide is the oxenoid reagent then a group on the enzyme should pick up an oxygen atom from O_2 during the reaction. It is felt that, if feasible, such an isotope tracer experiment should be performed by investigators who have these enzymes in hand.

5. Dopamine β-Hydroxylase

The evidence seems quite clear that the reactions catalyzed by dopamine β-hydroxylase are those summarized in Eq. (40) (94, 95). Some reducing agent AH_2 (usually ascorbic acid) reduces two Cu(II)'s on the enzyme to

$$
\begin{array}{c}
\text{AH}_2 \qquad \text{A} \\
\text{E} \cdot \text{Cu (II)}\text{---}\text{Cu (II)} \quad \xrightarrow{\hspace{2cm}} \quad \text{E} \cdot \text{Cu (I)}\text{---}\text{Cu (I)} \\
\\
\text{H}_2\text{O} + \text{SO} \quad\quad\quad\quad\quad\quad\quad\quad\quad \text{S} + \text{O}_2 \qquad\qquad (40) \\
\\
\text{E} \cdot \text{Cu (II)}\text{---}\text{Cu (II)} \cdot \text{SO} \cdot \text{H}_2\text{O} \quad \longleftarrow \quad \text{E} \cdot \text{Cu (I)}\text{---}\text{Cu (I)} \cdot \text{O}_2 \cdot \text{S} \\
\text{(XXXVII)} \qquad\qquad\qquad\qquad\qquad \text{(XXXVI)}
\end{array}
$$

Cu(I)'s, and it is this reduced enzyme which reacts with O_2 and dopamine (S) to give norepinephrine and the enzyme with two Cu(II)'s. The step of interest to the present discussion is the substrate hydroxylation, (XXXVI) to (XXXVII). If the coppers are electronically connected (1), the mechanism suggested for various model reactions [Eq. (25)] seems reasonable except that the enzyme-bound copper–oxygen complex would presumably have a structure to allow a singlet oxygen atom transfer. Since a Cu(I)·O_2 complex would also have a Cu(II)·O_2^{-} resonance contributor, it could act as a nucleophile. Thus, it is possible that the oxenoid reagent is a peracid or peramide formed as illustrated earlier for the P-450 enzymes [Eqs. (38) and (39)]. As in the earlier case an oxygen isotope tracer experiment could distinguish between these two types of mechanisms.

B. Monooxygenases Apparently *Not* Operating by an Oxenoid Mechanism

1. Flavin-Requiring Oxidative Decarboxylases (of α-Hydroxy and α-Amino Acids)

It is quite clear that for several flavin-requiring monooxygenases (2, 79), which catalyze the overall reaction shown in Eq. (41) (X=O or NH), some mechanism other than the oxenoid mechanism is responsible for the incorporation of atmospheric oxygen into the product. The most probable

$$R-CH-COOH + O_2 \longrightarrow R-\overset{O}{\overset{\|}{C}}-XH + CO_2 + H_2O \qquad (41)$$
$$\underset{XH}{|}$$

mechanism [outlined in Eq. (42)] and some chemical model reactions have recently been discussed in detail by the present author (2). Some evidence

$$(42)$$

for this general mechanism is that under some conditions these enzymes can act as amino acid oxidases (2, 79, 96). Although it has not been established that peroxide is an intermediate in any of the enzymic decarboxylation reactions, the oxidative decarboxylation of α-ketoacids is known to proceed very readily nonenzymically by the mechanism (2) illustrated in Eq. (42). Thus, in these oxygenase reactions the oxygen from O_2 which ends up in the organic product almost certainly results from a nucleophilic attack of a reduced oxygen species (peroxide) on an intermediate of the overall reaction.

2. Inositol Oxygenase

The iron-containing monooxygenase, inositol oxygenase (1), catalyzes the oxidation by O_2 of myoinositol to D-glucuronate. This reaction shows no characteristics of an oxenoid mechanism. A reasonable mechanism and related chemical systems have been discussed by the present author (1). Enzymes which catalyze only glycol cleavage probably proceed by a related mechanism.

VI. DIOXYGENASES

A. Dioxygenases Apparently Operating by an Oxenoid Mechanism

1. Enzymes Utilizing α-Ketoglutarate as Cosubstrate

These oxygenases (2, 97), which catalyze the overall reaction shown in Eq. (43), are dioxygenases because one atom of oxygen ends up in the

$$
S + O_2 + \underset{\underset{\displaystyle CH_2-COOH}{|}}{CH_2-\overset{\overset{\displaystyle O}{\|}}{C}-COOH} \longrightarrow SO + \underset{\underset{\displaystyle CH_2-COOH}{|}}{CH_2-COOH} + CO_2 \tag{43}
$$

oxidized substrate (SO) and one in succinate. Typical substrates oxidized by such enzymes are alkane derivatives such as proline peptides, betaines, and the methyl group of thymine. The most likely mechanism for oxidation at unactivated alkane positions is the oxenoid mechanism. Recently, the present author suggested (2) that the oxenoid reagent in these enzymic reactions is persuccinic acid formed by oxidative decarboxylation of α-ketoglutarate [Eq. (44)]. As has been discussed (Section IV), peracids

$$
\underset{\underset{\displaystyle O}{\|}}{HOOCCH_2CH_2CCOOH} + O_2 \longrightarrow HOOCCH_2CH_2C\underset{O}{\overset{OOH}{<}} + CO_2 \tag{44}
$$

will convert alkanes to alcohols; thus, persuccinic acid is a plausible enzymic oxenoid reagent. Possible mechanisms for the reaction of Eq. (44) have been discussed (2); the first step in the conversion is probably a nucleophilic attack of a complexed O_2 on the keto group. Since these enzymes are Fe(II) enzymes, one would expect an O_2 complexed to the metal would have considerable nucleophilic character as a result of contributions by the resonance structure Fe(III)·O_2^-.

If persuccinic acid is formed in these dioxygenase reactions, then the initial reaction of O_2 should be with the α-ketoglutarate rather than the other substrate; thus, one might be able to observe uptake of O_2 and CO_2 formation in the absence of the other substrate. Preliminary experiments indicate this does occur with peptidyl proline hydroxylase (98). This is consistent with the proposed mechanism and not consistent with initial attack of O_2 on the other substrate (99).

2. p-Hydroxyphenylpyruvate Hydroxylase

This dioxygenase (2, 100, 101) appears in all respects to catalyze an intramolecular reaction similar to the α-ketoglutarate-requiring dioxygenases discussed in the previous section. An oxenoid mechanism, involving formation of a peracid as the oxenoid reagent, intramolecular formation of an areneoxide, and migration of an alkyl group by an NIH shift (37–41), has been suggested and discussed by the present author (2).

B. Dioxygenases Apparently *Not* Operating by an Oxenoid Mechanism

1. Enzymes Catalyzing the Cleavage of Aromatic Rings

A number of iron-containing dioxygenases (102, 103) (e.g., pyrocatechase and tryptophan pyrrolase) catalyze the conversion of aromatic compounds (usually phenols or indoles) to open chain compounds which contain both atoms of the O_2, one on each carbon of the bond cleaved. These reactions show no characteristics of an oxenoid mechanism. Since a common feature of these dioxygenase reactions is that an OH or NH is attached directly to the bond cleaved, possible mechanisms for the simplest molecule (phenol) with this minimum requirement are illustrated in Eq. (45). However, actual enzyme substrates such as catechol or tryptophan could react by related mechanisms.

$$\text{(45)}$$

One of the main problems in any mechanism for these reactions is to explain how the two atoms of O_2 end up on carbons that were adjacent to one another in the original substrate. Several investigators have suggested that a cyclic four-membered ring dioxetane (XL), possibly formed from the peroxide (XLI), is an intermediate. However, it can be calculated from simple bond energies that the conversion of (XXXVIII) and O_2 to (XL) is considerably endothermic. The exact amount of the endothermicity involved depends on the value taken for the strain energy in the dioxetane ring of (XL) (104). However, the formation of (XL) from (XXXVIII) and O_2 must be at least 10–20 kcal/mole endothermic. A further problem in the mechanism with (XL) as an intermediate is the extreme exothermicity (*ca.* 80 kcal/mole) associated with the conversion of (XL) to (XXXIX). In chemical systems where similar reactions occur the decomposition of dioxetanes to carbonyl compounds is accompanied by the emission of light (chemiluminescence) (83). Thus, some of the excess energy is dissipated by radiation. However, no light is observed in the enzymic reactions.

A different type of mechanism for these dioxygenase-catalyzed reactions has been proposed by the present author (1). It is suggested that the first step involves the conversion of (XXXVIII) and O_2 to a peroxy complex like (XLI). This step would be slightly exothermic (by *ca.* 10 kcal/mole) and thus is reasonable thermodynamically. It is then suggested that (XLI) undergoes a rearrangement to (XLII) as shown; there are proven chemical analogies to this type of reaction (1, 26, 105). The conversion of (XLI) to (XLII) is considerably exothermic (*ca.* 40 kcal/mole) but not excessively so. The subsequent opening of the ring of (XLII) to give (XXXIX) by another exothermic reaction would be expected to occur rapidly since it is just the reverse of the addition of a nucleophile to an aldehyde. Thus, a mechanism involving (XLI) and (XLII) as intermediates does not require excessive stabilization of intermediates by the enzyme, and for this reason seems more likely as a general mechanism for these dioxygenase-catalyzed cleavage reactions than a mechanism with (XL) as an intermediate. Both mechanisms are consistent with the oxygen-18 tracer results.

The enzymic metal ion [Fe(II) or Fe(III)] is probably mainly required for the first step of the proposed mechanism (1), the conversion of (XXXVIII) to (XLI). A complex of O_2 with the metal would be able to react with the phenol to give complexed (XLI) by an ionic mechanism [Eq. (46)] if the

$$\text{E} \cdot \text{Fe} \underset{\text{H}-\text{O}}{\overset{\text{O}=\text{O}}{\diagdown}} \longrightarrow \text{E} \cdot \text{Fe} \overset{\text{O}-\text{O}}{\diagdown} \underset{\text{O}}{\overset{\text{H}}{\diagup}} \tag{46}$$

product has the same number of unpaired electrons as the $\text{E} \cdot \text{Fe} \cdot O_2$ reactant (see Section II). One expects the enzyme binds the metal in such a way to allow this to happen during catalysis.

Several nonenzymic rearrangements of compounds like (XLI) (especially indole-3-hydroperoxides (106); such a compound would be the intermediate in the tryptophan pyrrolase reaction) to compounds like (XXXIX) have been characterized. However, they have not been studied sufficiently to be able to conclude whether they proceed through intermediates like (XL) or (XLII), or possibly by some other mechanism. Obviously, this is an area which needs further study.

2. Lipoxygenase

Since lipoxygenase (107, 108) has no metal ion, the only apparent mechanism available for the initial reaction of O_2 is by a free radical mechanism of some kind. The most direct such mechanism (108) is that shown in Eq. (47), namely, the reaction of O_2 with the 1,4-diene(RH) to

$$(47)$$

give HOO$\cdot$ and the resonance stabilized radical, R$\cdot$. These could then couple to give the product peroxide (ROOH). Presumably the binding of substrates and intermediates to the enzyme would give rise to the observed stereochemistry; in nonenzymic systems racemic and isomeric peroxides would be expected from such a radical coupling process. The second step of Eq. (47) is energetically very favorable and is expected to occur quite readily. However, the first step seems energetically unlikely for an enzymic reaction. There is no doubt that radicals like R$\cdot$ are intermediates in autoxidations of 1,4-dienes, but such autoxidations are free radical chain reactions, and the initiation step usually requires some other radical initiator reacting with the diene to give R$\cdot$ (12, 13). The direct reaction of O_2 with the diene is a very slow process at physiological temperatures. Therefore, the first step of Eq. (47) seems questionable.

Recently, good evidence has been obtained (109) that a bound product peroxide molecule is necessary for enzymic activity. With this observation in mind a mechanism [Eq. (48)] which eliminates the problem with the first step of Eq. (47) is proposed here. It is suggested that the required bound

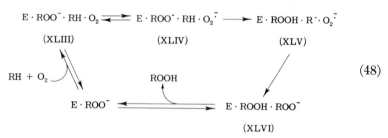

$$(48)$$

peroxide is present on the enzyme in its ionized form (ROO⁻), and the first step of the ternary complex (XLIII) is to transfer an electron from the ROO⁻ to O₂ to give an enzymic complex (XLIV) containing an alkylperoxy radical (ROO·) and O₂⁻. If the ROO· of (XLIV) abstracts a hydrogen atom from RH then (XLV) would result which by radical coupling gives (XLVI). Loss of peroxide from (XLVI) and binding of RH and O₂ regenerates (XLIII) for a further turnover.

The mechanism of Eq. (48) is preferable to that of Eq. (47) for several reasons, probably the most important one being that the hydrogen abstraction step (XLIV) to (XLV) is now expected to be rapid. This same step is a chain carrying step in nonenzymic autoxidations (12, 13) and thus must occur readily. However, it is presumably still the rate-determining step in the enzymic reaction because an isotope effect is observed. Because a new bond is formed, the radical coupling step (XLV) to (XLVI), like most radical couplings, is probably very rapid (diffusion controlled?). Although it is difficult to predict where the equilibrium of (XLIII) with (XLIV) will lie, it is expected that (XLIII) would be favored but not to the exclusion of (XLIV). It is informative to compare the overall exothermicity of (XLIII) to (XLV) with that of the first step of Eq. (47). Since HOO· is more acidic than ROOH by about seven orders of magnitude, then the conversion of (XLIII) to (XLV) is more exothermic by about 9 kcal/mole than the first step of Eq. (47). Thus, the important radical forming steps are much more favored in the mechanism of Eq. (48).

Several investigators have suggested that singlet oxygen (Section II) is the oxidant in lipoxygenase-catalyzed reactions, and the overall reaction is certainly similar to some singlet oxygen reactions (8). However, no direct evidence for the involvement of singlet oxygen has been forthcoming, and no satisfactory mechanism for its possible formation in an enzymic reaction has been presented. Thus, its involvement in the lipoxygenase reaction must be considered questionable. Consistent with the lipoxygenase mechanism in Eq. (48) is the evidence that radicals are formed during lipoxygenase reactions (108).

VII. CONCLUSIONS

In this and earlier (1, 2) articles the author has attempted to present a rational chemical basis for the many different enzymic reactions involving O_2. It is felt that with the ever-increasing amounts of information becoming available on the characteristics of oxygenases, oxidases, etc., one can consider chemically more realistic mechanisms than many of those such as "active oxygen" and HO^+ used heretofore. These have been useful terms, but it should be recognized that they are formalisms and not mechanisms in a molecular sense. The author is not presumptuous enough to feel that all, or even the majority, of the specific mechanisms proposed will turn out to be correct; rather, it is felt that they are representative of the level of sophistication one can now apply to reactions of this type. Information in the vast chemical literature on reactions of oxygen and oxygen derivatives puts limitations on the possible mechanisms of enzymic reactions of O_2, but it also can be used to predict the more likely mechanistic alternatives. In this article the author has attempted to indicate what some of these limitations and predictions are.

One conclusion to be derived from a survey of enzymic O_2 reactions is that there is no one species that can be labeled as "active oxygen." Perhaps this should not be unexpected because we do not speak, for example, of "active water"; rather, water reactivity depends on its environment, whether it is in a polar or nonpolar solvent, in the vicinity of an acid or a base, complexed to a metal ion, etc. So it is with oxygen; it can have differing reactivities depending on its environment. As the discussion in the preceding sections indicates, it is becoming increasingly clear that O_2 itself frequently does not react with the most visible substrate of the oxygenase reaction; rather, the O_2 is converted first to a considerably more reactive product (frequently at the peroxide oxidation level) by reaction with a cofactor or cosubstrate by mechanisms often encountered in nonenzymic O_2 reactions. Thus, in attempting to understand how usually unreactive substrates are oxidized by oxygenases, one should shift the focus from O_2 itself (which is probably reacting by a simple mechanism with a reactive cofactor or cosubstrate) to searching for the reactive reduced oxygen species.

ACKNOWLEDGMENTS

This research was supported by research grants from the National Science Foundation and The National Institute of Arthritis and Metabolic Diseases, Public Health Service.

REFERENCES

1. Hamilton, G. A. (1969). *Advan. Enzymol.* **32,** 55.
2. Hamilton, G. A. (1971). *In* "Progress in Bioorganic Chemistry" (E. T. Kaiser and F. J. Kezdy, eds.), Vol. 1, p. 83. Wiley, New York.
3. Hayaishi, O. (1964). *Proc. Int. Congr. Biochem., Plenary Sessions, 6th New York* **33,** 31.
4. March, J. (1968). "Advanced Organic Chemistry; Reactions, Mechanisms, and Structure," p. 26. McGraw-Hill, New York.
5. George, P. (1965). *In* "Oxidases and Related Redox Systems" (T. E. King, H. S. Mason, and M. Morrison, eds.), Vol. 1, p. 3. Wiley, New York; Benson, S. W., and Shaw, R. (1970). *In* "Organic Peroxides" (D. Swern, ed.), Vol. 1, p. 105. Wiley, New York.
6. Taube, H. (1965). *J. Gen. Physiol.* **49,** part 2, 29.
7. Collman, J. P. (1968). *Accounts Chem. Res.* **1,** 136.
8. Foote, C. S. (1968). *Accounts Chem. Res.* **1,** 104.
9. Hamilton, G. A., de Jersey, J., and Adolf, P. K. (1973). *In* "Oxidases and Related Redox Systems (*Proc. Int. Symp., 2nd*)" (T. E. King, H. S. Mason, and M. Morrison, eds.). Univ. Park Press, Baltimore, Maryland.
10. Massey, V., Strickland, S., Mayhew, S. G., Howell, L. G., Engel, P. C., Matthews, R. G., Schuman, M., and Sullivan, P. A. (1969). *Biochem. Biophys. Res. Commun.* **36,** 891.
11. Misra, H. P., and Fridovich, I. (1972). *J. Biol. Chem.* **247,** 188.
12. Walling, C. (1957). "Free Radicals in Solution." Wiley, New York.
13. Pryor, W. A. (1966). "Free Radicals." McGraw-Hill, New York.
14. Norman, R. O. C., and Lindsay Smith, J. R. (1965). *In* "Oxidases and Related Redox Systems" (T. E. King, H. S. Mason and M. Morrison, eds.), Vol. 1, p. 131. Wiley, New York.
15. Czapski, G. , and Dorfman, L. M. (1964). *J. Phys. Chem.* **68,** 1169.
16. Fridovich, I. (1973). Chapter 11, this volume.
17. McCord, J. M., Beauchamp, C. O., Goscin, S., Misra, H. P., and Fridovich, I. (1973). *In* "Oxidases and Related Redox Systems (*Proc. Int. Symp., 2nd*)" (T. E. King, H. S. Mason, and M. Morrison, eds.). Univ. Park Press, Baltimore, Maryland.
18. McCord, J. M., Keele, B. B., and Fridovich, I. (1971). *Proc. Nat. Acad. Sci. U.S.* **68,** 1024.
19. McCord, J. M., and Fridovich, I. (1969). *J. Biol. Chem.* **244,** 6049.
20. Keele, B. B., McCord, J. M., and Fridovich, I. (1970). *J. Biol. Chem.* **245,** 6176.
21. Greenlee, L., Fridovich, I., and Handler, P. (1962). *Biochemistry* **1,** 779.
22. (a) Hirata, F., and Hayaishi, O. (1971). *J. Biol. Chem.* **246,** 7825; (b) Strobel, H. W., and Coon, M. S., (1971). *Ibid.* **246,** 7826; (c) Rotilio, G., Calabrese, L., FinazziAgro, A., and Mondovi, B. (1970). *Biochim. Biophys. Acta* **198,** 618.
23. Misra, H. P., and Fridovich, I. (1972). *J. Biol. Chem.* **247,** 188.
24. Jencks, W. P. (1969). "Catalysis in Chemistry and Enzymology," p. 107, 499. McGraw-Hill, New York.
25. Hawkins, E. G. E. (1961). "Organic Peroxides," p. 72. Van Nostrand Reinhold, Princeton, New Jersey.
26. Edwards, J. O. (1962). *In* "Peroxide Reaction Mechanisms" (J. O. Edwards, ed.), p. 67. Wiley (Interscience), New York; Behrman, E. J., and Edwards, J. O. (1967).

Progr. Phys. Org. Chem. **4**, 93; Curci, R., and Edwards, J. O. (1970). *In* "Organic Peroxides" (D. Swern, ed.), Vol. 1, p. 199. Wiley, New York.

27. Anbar, M., and Guttmann, S. (1961). *J. Amer. Chem. Soc.* **83**, 2031, 2035.

28. Corey, E. J., and White, R. W. (1958). *J. Amer. Chem. Soc.* **80**, 6686.

29. Beauchamp, C., and Fridovich, I. (1970). *J. Biol. Chem.* **245**, 4641.

30. Hamilton, G. A. (1964). *J. Amer. Chem. Soc.* **86**, 3391.

31. Kirmse, W. (1971). "Carbene Chemistry." Academic Press, New York.

32. Lwowski, W. (1970). "Nitrenes." Wiley (Interscience), New York.

33. Mason, H. S. (1965). *Ann. Rev. Biochem.* **34**, 595.

34. Hayaishi, O. (1969). *Ann. Rev. Biochem.* **38**, 21.

35. Hayano, M. (1962). *In* "Oxygenases" (O. Hayaishi, ed.), p. 182. Academic Press, New York.

36. Cvetanovic, R. J. (1963). *Advan. Photochem.* **1**, 115.

37. Guroff, G., Daly, J. W., Jerina, D. M., Renson, J., Witkop, B., and Udenfriend, S. (1967). *Science* **157**, 1524.

38. (a) Daly, J., Guroff, G., Jerina, D., Udenfriend, S., and Witkop, B., (1968). *Advan. Chem. Ser.* **77**, 279; (b) Jerina, D. M., and Daly, J. W. (1973). *In* "Oxidases and Related Redox Systems (*Proc. Int. Symp., 2nd.*)" (T. E. King, H. S. Mason, and M. Morrison, eds.). Univ. Park Press, Baltimore, Maryland.

39. Jerina, D. M., Daly, J. W., Witkop, B., Zaltman-Nirenberg, P., and Udenfriend, S. (1970). *Biochemistry* **9**, 147.

40. Jerina, D. M., Kaubisch, N., and Daly, J. W. (1971). *Proc. Nat. Acad. Sci. U. S.* **68**, 2545.

41. Boyd, D. R., Daly, J. W., and Jerina, D. M. (1972). *Biochemistry* **11**, 1961.

42. Udenfriend, S., Clark, C. T., Axelrod, J., and Brodie, B. B. (1954). *J. Biol. Chem.* **208**, 731.

43. Hamilton, G. A., Workman, R. J., and Woo, L. (1964). *J. Amer. Chem. Soc.* **86**, 3390.

44. Staudinger, Hj., Kerékjárto, B., Ullrich, V., and Zubrzycki, Z. (1965). *In* "Oxidases and Related Redox Systems" (T. E. King, H. S. Mason, and M. Morrison, eds.), p. 815. Wiley, New York.

45. Ullrich, V., and Staudinger, Hj. (1966). *In* "Biological and Chemical Aspects of Oxygenases" (K. Bloch and O. Hayaishi, eds.), p. 235. Maruzen, Tokyo.

46. Kaufman, S. (1973). This volume Chapter 9.

47. Bobst, A., and Viscontini, M. (1966). *Helv. Chim. Acta* **49**, 884.

48. Hamilton, G. A., unpublished results.

49. Kasperek, G. J., Bruice, T. C., Yagi, H., and Jerina, D. M. (1972). *Chem. Commun.* 784.

50. Dearden, M. B., Jefcoate, C. R. E., and Lindsay Smith, J. R. (1968). *Advan. Chem. Ser.* **77**, 260.

51. Ullrich, V., and Staudinger, Hj. (1969). *In* "Microsomes and Drug Oxidations" (J. R. Gillette, A. H. Conney, G. J. Cosmides, R. W. Estabrook, J. R. Fouts, and G. J. Mannering, eds.), p. 199. Academic Press, New York.

52. Ullrich, V., and Staudinger, Hj. (1969). *Z. Naturforsch.* **24b**, 583.

53. Ullrich, V. (1969). *Z. Naturforsch.* **246**, 699.

54. Frommer, U., and Ullrich, V. (1971). *Z. Naturforsch.* **26b**, 322.

55. Jerina, D., private communication.

56. Hamilton, G. A., and Friedman, J. P. (1963). *J. Amer. Chem. Soc.* **85**, 1008.

57. Hamilton, G. A., Friedman, J. P., and Campbell, P. M. (1966). *J. Amer. Chem. Soc.* **88**, 5266.

58. Hamilton, G. A., Hanifin, J. W., Jr., and Friedman, J. P. (1966). *J. Amer. Chem. Soc.* **88**, 5269.

59. Hamilton, G. A. (1966). *In* "Hemes and Hemoproteins" (B. Chance, R. W. Estabrook, and T. Yonetani, eds.), p. 349. Academic Press, New York.

60. Snook, M. E., and Hamilton, G. A., unpublished results.

61. Jerina, D. M., Daly, J. W., Landis, W., Witkop, B., and Udenfriend, S. (1967). *J. Amer. Chem. Soc.* **89**, 3347.

62. Hamilton, G. A. (1974). *Accounts Chem. Res.* (in press).

63. Hamilton, G. A., Giacin, J. R., Hellman, T. M., Snook, M. E., and Weller, J. W. (1973). *Ann. N. Y. Acad. Sci.* **212**, 4.

64. Giacin, J. R., and Hamilton, G. A. (1966). *J. Amer. Chem. Soc.* **88**, 1584.

65. Hamilton, G. A., Ribner, B. S., and Hellman, T. M. (1968). *Advan. Chem. Ser.* **77**, 15.

66. Weller, J. W., and Hamilton, G. A. (1970). *Chem. Commun.* 1390.

67. Hellman, T. M. (1970). Ph.D. Thesis, Pennsylvania State Univ., University Park, Pennsylvania.

68. Weller, J. W. (1972). PhD. Thesis, Pennsylvania State Univ., University Park, Pennsylvania.

69. Giacin, J. R., and Hamilton, G. A., unpublished results.

70. Snook, M. E., and Hamilton, G. A., unpublished results.

71. Jerina, D. M., Boyd, D. R., and Daly, J. W. (1970). *Tetrahedron Lett.* 457.

72. Swern, D. (1971). *In* "Organic Peroxides" (D. Swern, ed.), Vol. II, p. 355. Wiley (Interscience), New York.

73. Rocek, J. (1962). *Tetrahedron Lett.* 135.

74. Wiberg, K. B. (1965). *In* "Oxidation in Organic Chemistry" (K. B. Wiberg, ed.), p. 69. Academic Press, New York.

75. Sharpless, K. B., and Flood, T. C. (1971). *J. Amer. Chem. Soc.* **93**, 2316.

76. Bloch, K. (1969). *Accounts Chem. Res.* **2**, 193.

77. Gutsche, C. D., Bachman, G. L., and Coffey, R. S. (1962). *Tetrahedron* **18**, 617.

78. Breslow, D. S., Prosser, T. J., Marcantonio, A. F., and Genge, C. A. (1967). *J. Amer. Chem. Soc.* **89**, 2384.

79. Massey, V., and Flashner, M. (1973). This volume, Chapter 8.

80. Mager, H. I. X., and Berends, W. (1965). *Rec. Trav. Chim. Pays.-Bas* **84**, 1329; (1967). *ibid.* **86**, 833; (1966). *Biochim. Biophys. Acta* **118**, 440.

81. Hamilton, G. A. (1971). Abstr. of Papers of the 162nd Nat. Meeting of the Amer. Chem. Soc., Washington, D. C., September, Biol. 2.

82. Bailey, P. S. (1958). *Chem. Rev.* **58**, 925.

83. Hiatt, R. (1971). *In* "Organic Peroxides" (D. Swern, ed.), Vol. II, p. 1. Wiley (Interscience), New York.

84. March, J. (1968). "Advanced Organic Chemistry; Reactions, Mechanisms, and Structure," p. 830. McGraw-Hill, New York.

85. Shiman, R., Akino, M., and Kaufman, S. (1971). *J. Biol. Chem.* **246**, 1330.

86. McDonald, P. M., and Hamilton, G. A. (1972). *In* "Oxidation in Organic Chemistry" (W. Trahanovsky, ed.), Part B, p. 97. Academic Press, New York.

87. Coleman, J. E. (1971). *Progr. Bioorg. Chem.* **1**, 159; Duckworth, H. W., and Coleman, J. E. (1970). *J. Biol. Chem.* **245**, 1613.

88. Vanneste, W., and Zuberbuhler, A. (1973). Chapter 9, this volume.

89. Mason, H. S., Fowlks, W. L., and Peterson, E. (1955). *J. Amer. Chem. Soc.* **77**, 2914.
90. Brackman, W., and Havinga, E. (1955). *Rec. Trav. Chim. Pays-Bas* **74**, 937, 1021, 1070, 1100, 1107.
91. Vercauteren, R., and Massart, L. (1962). *In* "Oxygenases" (O. Hayaishi, ed.), p. 355. Academic Press, New York.
92. Gunsalus, I. C. (1973). Chapter 14, this volume.
93. Various authors. (1973). *Ann. N. Y. Acad. Sci.* **212**, (1973). "Oxidases and Related Redox Systems (*Proc. 2nd. Int. Symp.*)" (T. E. King, H. S. Mason, and M. Morrison, eds.). Univ. Park Press, Baltimore, Maryland; (1969). "Microsomes and Drug Oxidations" (J. R. Gillette, A. H. Conney, G. J. Cosmides, R. W. Estabrook, J. R. Fouts, and G. J. Mannering, eds.). Academic Press, New York.
94. Friedman, S., and Kaufman, S. (1965). *J. Biol. Chem.* **240**, 4763.
95. Goldstein, M., Joh, T. H., and Garvey, T. Q. (1968). *Biochemistry* **7**, 2724.
96. Nakazawa, T., Hori, K., and Hayaishi, O. (1972). *J. Biol. Chem.* **247**, 3439.
97. Abbott, M., and Udenfriend, S. (1973). Chapter 5, this volume.
98. Cardinale, G. J., and Udenfriend, S. (1972). *In* "Oxidases and Related Redox Systems (*Proc. Int. Symp., 2nd*)" (T. E. King, H. S. Mason, and M. Morrison, eds.), p. 000. Univ. Park Press, Baltimore, Maryland.
99. Lindblad, B., Lindstedt, G., Tofft, M., and Linstedt, S. (1969). *J. Amer. Chem. Soc.* **91**, 4604.
100. Lindblad, B., Lindstedt, G., and Lindstedt, S. (1970). *J. Amer. Chem. Soc.* **92**, 7446.
101. Lindblad, B., Lindstedt, S., Olander, B., and Omfeldt, M. (1971). *Acta Chem. Scand.* **25**, 329.
102. Feigelson, P., and Brady, F. O. (1973). Chapter 3, this volume.
103. Nozaki, M. (1973). Chapter 4, this volume.
104. O'Neal, H. E., and Richardson, W. H. (1970). *J. Amer. Chem. Soc.* **92**, 6553.
105. Smith, P. A. S. (1963). *In* "Molecular Rearrangements" (P. deMayo, ed.), Part 1, p. 457. Wiley (Interscience), New York.
106. Sundberg, R. J. (1970). "The Chemistry of Indoles," p. 282. Academic Press, New York.
107. Hamberg, M., Samuelsson, B. I., Björkhem, I., and Danielsson, H. (1973). Chapter 2, this volume.
108. Tappel, A. L. (1963). *Enzymes* **8**, 275.
109. Smith, W. L., and Lands, W. E. M. (1972). *J. Biol. Chem.* **247**, 1038.

11

SUPEROXIDE DISMUTASE

IRWIN FRIDOVICH

I. INTRODUCTION

Oxygen is toxic. Obligate anaerobes exhibit their sensitivity at 0.2 atm of O_2 or less (1), whereas other organisms do not clearly show its deleterious effects until exposed to higher concentrations (2, 3). The basis of the toxicity of oxygen has long been pondered. Early proposals related this toxicity to the accumulation of the reduction product H_2O_2 (4, 5) and to the absence of catalase in obligate anaerobes (6). However, this theory did not prove entirely adequate because many organisms capable of aerobic growth do not contain catalase, e.g., the *Streptococci*, pneumococci, and the lactic acid bacteria (7). In addition, an aerobic soil organism, *Agromyces ramnosus*, has been found to produce H_2O_2 and yet to lack catalase (8, 9), whereas some obligate anaerobes have been found to contain catalase (10). An appreciation of the ease with which oxygen is reduced to the superoxide radical (O_2^-) by both enzymic and nonenzymic processes and the discovery of superoxide dismutase, which catalytically scavenges this radical (11–13), led to the proposal that the toxicity of oxygen relates to its conversion to O_2^-. A survey of the superoxide dismutase activity of aerotolerant and of obligately anaerobic microorganisms supported this proposal in that aerotolerant organisms contained superoxide dismutase whereas the obligate anaerobes did not (14).

It appears likely that the oxygen now so abundant in the earth's atmosphere is primarily a by-product of green plant photosynthesis and that essentially anaerobic conditions prevailed on this planet prior to the advent of green plants (15, 16). The many life forms, which must have evolved prior to the appearance of green plants, were therefore faced with a new opportunity and a serious challenge when oxygen began to accumulate in their environment. This opportunity was that of utilizing oxygen in energy-yielding and in biosynthetic reactions. The contents of this volume document the extent to which this opportunity has been exploited. The challenge which green plants threw down before all life forms, including themselves, was to evolve defenses against the reactivities of the radical species so readily generated from oxygen. The descendents of those organisms which failed to meet this challenge are now restricted to anaerobic niches and are known as obligate anaerobes. All the others, procaryote and eucaryote, which can live in contact with air and can utilize molecular oxygen have developed the necessary defenses.

We now take the position that oxygen radicals are generated by normal intracellular reactions and that the enzyme superoxide dismutase is one of the major elements in the defenses which have evolved to deal with these radicals. The following presentation will attempt to document the reasons for that proposition and to describe some of the work which has been done, to date, on the superoxide dismutase.

II. A HISTORICAL ACCOUNT OF THE DISCOVERY OF SUPEROXIDE DISMUTASE

Xanthine oxidase was observed to exhibit several rather unusual activities, all of which were dependent upon the simultaneous presence of substrate and of oxygen. It was the investigation of these oxygen-dependent activities which led to the realization that xanthine oxidase was causing the univalent reduction of oxygen to the superoxide radical and that this radical was actually the agent responsible for these peculiar activities. In the course of these studies, certain protein preparations were seen to powerfully and specifically inhibit these oxygen-dependent activities of xanthine oxidase. These protein preparations were then found to be acting catalytically, and the isolation of the inhibitory catalyst gave us superoxide dismutase. It may now be of interest to flesh out this story with a few details.

Xanthine oxidase, acting upon xanthine in the presence of oxygen, was able to initiate the aerobic oxidation of sulfite (17, 18), cause the reduction of cytochrome c (19), and induce the chemiluminescence of luminol or of lucigenin (20–22). Since the aerobic oxidation of sulfite was known to be a free radical chain reaction, which was readily initiated by reactive radicals, it was proposed (18) that O_2^-, generated by xanthine oxidase, was the initiating species. In accord with this proposal were the demonstrations that O_2^-, whether generated by the oxidation of Fe^{2+} (18) or by the electrolytic reduction of oxygen (23), was an effective initiator of sulfite oxidation. It has been proposed that the essentiality of oxygen for the reduction of cytochrome c by xanthine oxidase was the result of the production of H_2O_2, which was the actual reductant (24). This possibility was subsequently definitely excluded and O_2^- was proposed as the immediate reductant of cytochome c (25). O_2^-, generated by the aerobic action of xanthine oxidase, was also proposed as the agent responsible for the chemiluminescence induced by this enzyme (22).

It was noted that some preparations of cytochrome c were readily reduced by the aerobic xanthine oxidase system whereas others were not. Indeed, those preparations of cytochrome c which were not reduced inhibited the reduction of those that were, and this inhibitory action resulted from a nondialyzable, thermolabile component (26). Since this inhibitory component was eliminated by a purification procedure (27) which removed myoglobin from the cytochrome c, myoglobin was prepared from horse heart and found to competetively inhibit the reduction of cytochrome c by xanthine oxidase with an apparent K_i of 2×10^{-8} M, while having no effect on the rate of reduction of oxygen (26). The reductions of cytochrome c by the DPNH–cytochrome c reductase and by sulfite–cytochrome c reductase were seen to be unaffected by these preparations of myoglobin, whereas the reduction of cytochrome c by an enzyme similar to xanthine oxidase, i.e., the rabbit liver aldehyde oxidase (28), was susceptible to this competetive inhibition (26). Some measure of the specificity of this inhibition was provided by the observation that human hemoglobin, EDTA, RNA, and bovine serum albumin were without effect on the reduction of cytochrome c by xanthine oxidase (26).

Our conception of the problem at this time was that xanthine oxidase possessed two sites for the reduction of oxygen: a divalent site which produced H_2O_2 and a univalent site which made O_2^-. It was difficult at that time to conceive that so reactive a species as O_2^- would be liberated into free solution; thus, O_2^- was considered to remain on the enzyme and while bound to cause the reduction of cytochrome c, the initiation of sulfite

oxidation, and the chemiluminescence of luminol or of lucigenin (25). Inhibitors which specifically interfered with those activities which were dependent on O_2^- were thought to act by binding onto, and thus blocking
access to, those sites responsible for the univalent reduction of oxygen.

Interest in the action of protein inhibitors of the oxygen-dependent
cytochrome c reduction was revived after a hiatus of 5 years by the chance
observation that a preparation of bovine carbonic anhydrase competetively
inhibited the reduction of cytochrome c by xanthine oxidase and exhibited a K_i of 3×10^{-9} M (29). This inhibition was reminiscent of that
previously seen with myoglobin and it was pursued. Attempts to demonstrate the formation of xanthine oxidase–carbonic anhydrase complexes,
though carefully performed, yielded uniformly negative results. It was,
perhaps, a profound disaffection with the tedium of these binding studies
which led J. M. McCord, then a graduate student, to seek ways of conclusively demonstrating that there was no protein–protein interaction
involved in the inhibition of cytochrome c reduction by "carbonic anhydrase." This demonstration was accomplished simply and conclusively
by showing that K_m for cytochrome c and K_i for "carbonic anhydrase"
were dependent upon the concentration of xanthine oxidase (11). Thus,
if the reduced xanthine oxidase were reducing oxygen univalently:

$$EH_2 + O_2 \xrightarrow{k_1} EH\cdot + O_2^- + H^+ \tag{1}$$

and if the O_2^- so generated could either dismute or react with ferricytochrome c:

$$O_2^- + O_2^- + 2H^+ \xrightarrow{k_2} H_2O_2 + O_2 \tag{2}$$

$$O_2^- + Cytc^{3+} \xrightarrow{k_3} O_2 + Cytc^{2+} \tag{3}$$

Then, in the absence of cytochrome c, the steady state level of O_2^- would
depend upon a balance between reactions (1) and (2):

$$\frac{d(O_2^-)}{dt} = k_1(EH_2)(O_2) - k_2(O_2^-)^2 = 0 \tag{4}$$

The concentration of O_2 and of protons and reduced enzyme would be
essentially constant during the period of observation; thus, Eq. (4) reduces to

$$K(E_t) = k_2 (O_2^-)^2$$

and

$$(O_2^-) = [K/k_2(E_t)]^{1/2} \tag{5}$$

This predicted that the steady state level of O_2^- should vary with the
square root of the concentration of xanthine oxidase. The steady state

level of O_2^- was measured indirectly by assessing the rate of reduction of cytochrome c when the level of cytochrome c was too low to significantly peturb the level of O_2^-, and Eq. (5) was affirmed (11). In contrast, at saturating levels of cytochrome c, all of the O_2^- generated by reaction (1) would be consumed by reaction (3), and the rate of reduction of cytochrome c would then be a direct function of the concentration of xanthine oxidase and this, too, was affirmed (11). It was also shown that these preparations of carbonic anhydrase would inhibit the electrolytic initiation of sulfite oxidation and that the actual inhibitor could then be separated from the carbonic anhydrase activity by ion-exchange chromatography (11). It was apparent at this time that O_2^-, liberated into free solution by xanthine oxidase, was the reductant of cytochrome c and that the protein inhibitor exerted its effect by catalyzing the dismutation of these radicals.

Superoxide dismutase was subsequently isolated from bovine blood (12, 13). The myoglobin and carbonic anhydrase, which had been studied as inhibitors of cytochrome c reduction, could have contained no more than 0.2 and 0.8%, respectively, of the superoxide dismutase. The blue-green color of the isolated superoxide dismutase suggested copper, and its content of Cu^{2+} as well as its molecular weight and spectral properties suggested that it was identical to the hemocuprein previously isolated from this tissue without knowledge of its function (30). Blue-green copper proteins had been described from several sources (31–40), and these had been named on the basis of their copper content and source, i.e., hepatocuprein, erythrocuprien, cerebrocuprein, and cytocuprein. We obtained a sample of erythrocuprein through the generosity of H. F. Deutsch and found it to be identical to superoxide dismutase (12, 13). It is interesting but, in view of its esoteric action, not surprising that superoxide dismutase had been studied and well characterized as a protein over a period of 30 years before its catalytic function was known.

III. SUPEROXIDE DISMUTASES FROM EUCARYOTIC SOURCES

A. Physicochemical Properties

The molecular weight of superoxide dismutase has been estimated to be 32,500 by sedimentation velocity (32, 38, 40, 41), sedimentation equilibrium (13, 40–43), and gel exclusion chromatography (44). The apo- and holoproteins have the same molecular weight (42). Packed human erythrocytes contain close to 160 mg of superoxide dismutase per liter (35). This enzyme has been isolated by a variety of procedures (13, 30–33, 35–37, 40, 45, 46). The simplest procedure is that which exploited the unusual solubility properties of this enzyme. Thus, when potassium phosphate was added to the Tsuchihashi supernate of erythrocyte lysates, an organic phase rich in ethanol and chloroform was salted out and the enzymic activity was found, almost exclusively, in this organic phase (13). The enzyme contains two equivalents of Cu^{2+} (13, 32, 33, 36, 38, 40, 43, 44) and two of Zn^{2+} (42–44) and is composed of two subunits of equal size (43, 47) which appear to be joined by a disulfide bridge.

The metal prosthetic groups can be removed reversibly. The apoprotein prepared from horse liver was crystallized, and the Cu^{2+} and the color could then be restored by the addition of $CuSO_4$ (31). Dialysis of super-oxide dismutase against EDTA at pH 3.8 caused loss of color and of Cu^{2+}, which was accompanied by inactivation. Activity was restored by Cu^{2+} but not by Ni^{2+}, Co^{2+}, Hg^{2+}, Mg^{2+}, Fe^{2+}, or Zn^{2+} (13). Treatment with cyanide was shown to remove both Cu^{2+} and Zn^{2+} (42), and a method of preparing the apoprotein by gel filtration rather than dialysis was de-scribed (48). Dialysis against EDTA at pH 3.8 was shown to remove both Cu^{2+} and Zn^{2+} (44).

The amino acid compositions of human (32, 38, 40) and bovine (43, 44) superoxide dismutases have been reported. The bovine enzyme contains no tryptophan by colorimetric or by spectrophotometric analysis; how-ever, some component reminiscent of tryptophan has been detected fluorimetrically (44). This is probably the result of contamination with carbonic anhydrase. The human enzyme has been stated to be devoid of tryptophan (32, 38, 40), but this conclusion has been questioned (43). No amino termini have been detectable by the Edman or Sanger reagents (38, 44). It has been found (49) that the N-terminus of the bovine enzyme is blocked by acetylation. The amino acid sequence of this enzyme is cur-rently being determined (50).

The ultraviolet absorption spectrum of superoxide dismutase is dis-tinctly atypical. Thus, the bovine enzyme exhibits a maximum at 258 nm whose $E_m = 10,300$ (13, 43). This band resembles the absorption spectrum of phenylalanine and reflects the absence of tryptophan. It is certain, however, that the metal prosthetic groups contribute in some way to this ultraviolet absorption since the apoenzyme absorbs only half as much as does the holoenzyme and either Cu^{2+} or Zn^{2+} can augment the absorption in this region, but both Cu^{2+} and Zn^{2+} were required for full restoration of the spectrum (44). The spectrum of the bovine enzyme has been reported by several workers (47, 48). The human enzyme has been similarly investi-gated (38, 42). The color of superoxide dismutase is the result of a weak band at 680 nm (13, 38, 40, 47) which is eliminated when Cu^{2+} is removed. The weakness of this absorption band has been taken as an indication of a high degree of symmetry of the ligand field of the Cu^{2+} (44).

The Cu^{2+} of superoxide dismutase engenders an EPR signal which has been studied in numerous investigations (40, 43, 47, 51, 52). The parameters of this signal are $g_m = 2.080$, $g_{||} = 2.265$, and $A_{||} = 0.014$ cm^{-1}. The nine-banded super hyperfine structure of this signal has been interpreted as an indication of 3–4 nitrogenous ligands surrounding the Cu^{2+} (52). Reduction of the enzyme with dithionite, under anaerobic conditions, bleaches this Cu^{2+} signal and indicates that the enzyme can accommodate one electron per Cu^{2+} (47). Studies of the effects of azide and of cyanide of the spectral properties of the bovine enzyme have led to the suggestion that Cu^{2+} and Zn^{2+} are located in close proximity, perhaps as a ligand-bridged bimetal complex (53).

Superoxide dismutase has been subjected to electrophoresis, both moving boundary and in gels, and the effects of pH on mobility have been studied. The human enzyme has been reported to have isoelectric points in the range 4.6–5.2 (32, 38), while 4.95 was the value reported for the bovine enzyme (44). Multiple electrophoretic bands have been reported (38, 42, 44), and this heterogeneity has been found to increase with aging and treatment with chloroform-ethanol (38). By electrophoresis in acrylamide gels of graded porosity, two electrophoretically distinct bands of bovine superoxide dismutase were found to be size isomers (44). Preparations of the bovine enzyme have been reported, which were homogeneous by the criterion of disc-gel electrophoresis (13, 47).

B. Catalytic Properties

Superoxide dismutase catalyzes the following reaction (11–13)

$$O_2^- + O_2^- + 2H^+ \rightarrow H_2O_2 + O_2$$

The instability of its substrate dictates that routine assays be indirect. The strategy which has been used has coupled a reaction which generates O_2^- with a reaction which detects O_2^-. If superoxide dismutase is added to such a reaction system it will compete with the O_2^- detecting reaction and thus inhibit its rate. Several examples of this type of assay system will now be presented.

Xanthine oxidase, acting on xanthine in the presence of oxygen, generates O_2^- which, in turn, causes the reduction of cytochrome c (11–13). Superoxide dismutase inhibits this oxygen-dependent reduction of cytochrome c, and under specified conditions the degree of inhibition serves to quantitate the superoxide dismutase. The conditions of the assay must be specified carefully. Thus, the superoxide dismutase is competing not only with cytochrome c but with O_2^- itself. This is so because the spontaneous dismutation reaction is second order in O_2^-. One unit of superoxide dismutase was defined as that amount which caused 50% inhibition of the rate of cytochrome c reduction, and under the conditions originally specified this was 0.1 $\mu g/ml$ (13). Also, O_2^- causes the reduction of tetranitromethane (54–56) and the oxidation of epinephrine (13), and these reactions have also been used as a means of detecting O_2^- and thereby of assaying superoxide dismutase (13). It is indeed possible to prepare stable solutions of O_2^- as the tetrabutylammonium salt by the electrolytic reduction of O_2 in nonprotic solvents such as dimethyl sulfoxide or dimethyl formamide (57–59) and to use mechanical infusions of such solutions as the source of O_2^- in place of the xanthine oxidase reaction. This has been done, and superoxide dismutase has been assayed in terms of its ability to inhibit the reduction of cytochrome c by infusions of tetrabutylammonium superoxide (13).

Free radical chain reactions which can be initiated by O_2^-, or in which O_2^- can serve as a chain-propagating species, will be inhibited by superoxide dismutase and can thus be used to assay this enzyme. The oxidation of epinephrine to adrenochrome at elevated pH is such a reaction and is inhibited by superoxide dismutase. This fact has been useful in providing some information about the mechanism of this oxidation and has been exploited in providing a convenient and sensitive assay for superoxide dismutase (60). Although the aerobic oxidation of sulfite can be initiated by O_2^-, this radical is not one of chain-carrying species (11). In this situation, sulfite oxidation which is initiated by O_2^- will be inhibited by superoxide dismutase, whereas that which is initiated by other radicals will not be sensitive to this enzyme. Thus, the autoxidation of sulfite was strongly inhibited by superoxide dismutase in the presence of 5×10^{-6} M EDTA but was unaffected by this enzyme in the absence of EDTA. This effect of EDTA was related to the differences between outer sphere and inner sphere oxidation–reduction reactions, between metal cations and oxygen which generate O_2^- in the former case but not in the latter (61). The sensitivity of sulfite autoxidation to superoxide dismutase in the presence of low levels of EDTA provided an assay which could easily detect 0.001 $\mu g/ml$ of this enzyme (61).

It has been shown that O_2^- can be generated by photochemical processes (62) and that O_2^- can reduce tetrazolium dyes to the insoluble formazans (63, 64). These facts were utilized in devising assays for superoxide dismutase, which could be applied in free solution or on acrylamide gels. Illumination of solutions containing riboflavin and TEMED (tetramethylethylenediamine) leads to photooxidation of the TEMED and photoreduction of the riboflavin. The reduced flavin then interacts with oxygen to generate O_2^-. If nitro blue tetrazolium is also present, it will be reduced to the blue formazan, and superoxide dismutase, if present, will prevent this blueing by intercepting the O_2^- (65). When applied to acrylamide gels, zones containing superoxide dismutase remain achromatic, while the rest of the gel turns blue. As little as 0.016 μg of the enzyme could be detected on acrylamide gel electrophoretograms by this procedure, and the existence of isozymes of superoxide dismutase was discovered by the application of this method to crude extracts of *Escherichia coli* (65). More recently, extracts of *Streptococcus mutans* (66), wheat germ, and chicken liver have been found to contain isozymes of superoxide dismutase.

The ability of superoxide dismutases to prevent the blueing of tetrazolium by photochemically generated O_2^- was actually observed in 1967 (67) but was not correctly interpreted. Thus, the achromatic zones that developed on starch gel electrophoretograms, which had been treated with phenazinium methyl sulfate and exposed to light, were ascribed to a tetrazolium oxidase which was supposed to catalyze the oxidation of the reduced tetrazolium (67). A genetic variant of the human "tetrazolium oxidase" was discovered (67) and tetrazolium oxidase has since been studied in numerous organisms including salmon and trout (68), rockfish (69), dogs (70), *Drosophila* (71), and soybeans (72). The ease with which tetrazolium oxidase can be visualized on gel electrophoretograms makes it an attractive marker for genetic analysis, and it has been much used for this purpose. A line of human cells gave one band of tetrazolium oxidase activity on gel electrophoretograms, and a line of mouse cells gave a different band. Hybrids of these two cell lines gave three bands of activity which may be interpreted in terms of a hybrid molecule of the tetrazolium oxidase (73). A protein inhibitor of tetrazolium reduction has been independently reported (74). It appears almost certain that the tetrazolium oxidases studied in so many sources, as well as the protein inhibitor (74), are, in fact, superoxide dismutase.

Direct assays of superoxide dismutase are difficult but not impossible. The requirements are for a means of introducing O_2^- into aqueous solution and for a method of detection of O_2^- which is sensitive enough to measure the concentrations achieved. Thus, O_2^- was generated by the reoxidation of reduced flavins in a stopped-flow system and was quantitated by electron paramagnetic resonance (EPR), and in this system the activity of superoxide dismutase was measured in terms of its acceleration of the rate of disappearance of the EPR signal of O_2^- (75). The ability of indirect assays to detect nanogram per milliliter levels of superoxide dismutase had indicated a great catalytic efficiency. A turnover number of at least 3×10^6 was estimated from stopped-flow measurements (75). A second-order rate constant for the reaction of O_2^- with superoxide dismutase of 6×10^9 M^{-1} sec^{-1} has been calculated on theoretical grounds (76). More recently, high concentrations of O_2^- have been generated in aqueous solutions by pulse radiolysis, and the action of superoxide dismutase has been directly assayed by spectrophotometric methods (77). Under these conditions the catalytic action of superoxide dismutase was affirmed, and a rate constant of approximately 2×10^9 M^{-1} sec^{-1} was measured and found to be almost independent of pH in the range 4.8–9.7. Furthermore, the catalytic dismutation reaction was first order in O_2^- (77). An independent study (78) of superoxide dismutase, using pulse radiolysis has resulted in similar observations. The enzyme showed no sign of saturation with O_2^- in the 10^{-5} M range. On this basis the K_m for O_2^- was estim'ted to be greater than 5×10^{-4} M. Because increasing the viscosity of the solution with glycerol decreased the rate of the superoxide dismutase reaction, this reaction was assumed to be diffusion-controlled. These workers (78) also made the interesting observation that exposure of the enzyme to pulse of O_2^- caused a decrease in both its optical absorption at 650 nm and in its EPR signal. It appears possible that these changes may reflect a reduction of the Cu_2^+ on the enzyme and may thus support a mechanism in which Cu_2^+ is alternately reduced and reoxidized by successive interactions with O_2^-.

Superoxide dismutase is a remarkably stable enzyme. It is thus stable for several hours at room temperature in the organic phase salted out of the Tsuchihashi mixture (13) and will tolerate 95% ethanol for like periods of time. Indeed, it can actually be assayed in the presence of 9.5 M urea, and although 6.0 M guanidanium chloride inhibits its activity, most of this inactivation is reversed when the concentration of the guanidinium is reduced by dilution. The enzyme is stable to pH 11.5 but is irreversibly altered at pH 12.0 (52). Superoxide dismutase is irreversibly destroyed by brief boiling, and at 75° in 0.10 M potassium phosphate + 1×10^{-4} EDTA the enzyme suffered a first-order inactivation whose half-time was 11 minutes (61). Cu^{2+} is essential for catalytic action while Zn^{2+} is dispensable (13). Thus, the enzyme may be dialyzed against EDTA at pH 3.8 and essentially complete removal of Zn^{2+} and of Cu^{2+} achieved. The apoenzyme is completely inactive and at least 80% of the initial activity can be restored by the addition of Cu^{2+}, whereas addition of Zn^{2+} alone causes no reactivation. The ligand field of the Cu^{2+} appears to be composed of histidine residues. This deduction rests upon the comparative effects of group-specific reagents upon the apo- and holoenzymes. Thus, holo superoxide dismutase (bovine) was resistant to methylene blue–photosensitized oxidation and to diazotized sulfanilamide, whereas the apoenzyme was irreversibly inactivated by these treatments. Apoenzyme was, of course, assayed after restoration of Cu^{2+}. The inactivation of the apoenzyme by these reagents was paralleled by a loss of histidine residues, while all other amino acid residues were unchanged. Complete inactivation of the apoenzyme by the Pauly reagent occurred when 3.6 residues of histidine per subunit had been derivitized (79). Bovine superoxide dismutase contains a total of 8 histidine residues per subunit. It may be inferred that the Cu^{2+} at the active site of this enzyme is surrounded by three to four imidazole rings.

There are some puzzling aspects of the stability of the bovine superoxide dismutase which remain to be explained. Thus, the enzyme may be briefly exposed to $0.1N$ HCl without loss of activity, and dilution from $0.1 N$ HCl into neutral buffer gives a preparation which is fully active. If EDTA is present in the $0.1N$ HCl solution of the enzyme, then it will be inactive following dilution into neutral buffer until free Cu^{2+} is added. Consequently, it appears that $0.1N$ HCl reduces the affinity for Cu^{2+} but does not irreversibly alter the protein so that upon neutralization activity is restored. This protein, which is stable to $0.1N$ HCl or to $9.5 M$ urea or $6.0 M$ guanidinium chloride, is nevertheless irreversibly inactivated by brief boiling or by $0.1 M$ NH_2OH. It appears possible that bovine superoxide dismutase may contain a peculiar covalent bond which is attacked by heat or by NH_2OH.

The bovine superoxide dismutase does interact with cyanide and azide, and spectroscopic methods have been used in studies of these interactions (53). It was concluded that one azide can bind to each Zn^{2+} and one cyanide to each Cu^{2+} (53). Cyanide at $5 \times 10^{-4} M$ does inhibit superoxide dismutase, whereas azide is without effect. This result is fully in accord with the reconstitution experiments which demonstrated that Cu^{2+} restored activity to the apoenzyme whereas Zn^{2+} did not. It is very likely that in the holoenzyme Cu^{2+} constitutes the catalytically active center, whereas Zn^{2+} plays some ancillary role. The inhibition of superoxide dismutase by cyanide and the lack of inhibition by azide have also been demonstrated by the direct assay method using O_2^- generated by pulse radiolysis (78).

IV. SUPEROXIDE DISMUTASES FROM PROCARYOTIC SOURCES

The supposition that superoxide dismutase was evolved to protect organisms against the deleterious actions of O_2^- indicated that it should be present in all aerobic cells. Surveys of a variety of sources indicated this to be the case (14). The superoxide dismutase of *Escherichia coli* was isolated, and although its specific activity was virtually the same as that previously seen with the mammalian enzyme its visible color and ultraviolet absorption spectrum indicated that it was a totally different protein. Indeed, the *E. coli* superoxide dismutase was found (80) to be wine red, to contain two equivalents of manganese, to have a molecular weight of close to 40,000, and to be composed of two subunits of equal size whose association entirely resulted from noncovalent interactions. The manganese in the native *E. coli* superoxide dismutase was EPR-silent, and the six-banded EPR spectrum of Mn_2^+ was not evident until the enzyme was denatured. There was no indication of Cu^{2+} or Zn^{2+} in this enzyme. The valence of the manganese in the native enzyme is unknown. The amino acid composition of the *E. coli* enzyme indicated no close similarities between it and the mammalian enzyme. Unlike its eucaryotic counterpart, the *E. coli* superoxide dismutase is not inhibited by cyanide. It may be assayed by the same strategies which were successful with the mammalian enzyme.

Disc-gel electrophoresis of crude extracts of *E. coli* indicated two well-separated bands of superoxide dismutase of approximately equal intensity, whereas the purified enzyme gave only one band (65). Evidently one of these isozymes had been eliminated during the isolation. Superoxide dismutase has since been isolated from *Streptococcus mutans*. In this case there were again two isozymes, both of which were purified and found to contain manganese (66). The superoxide dismutase of *Streptococcus faecalis* is also under study, and based upon its resistance to cyanide inhibition and the fact that enriching the growth medium with Mn^{2+} augments the yield of this enzyme, we tentatively conclude that it, too, contains manganese in place of Cu^{2+} and Zn^{2+}. On the basis of this admittedly limited data, we conclude that all procaryotes contain the reddish manganese-containing superoxide dismutase, whereas all eucaryotes will be found to contain the blue-green Cu^{2+}- and Zn^{2+}-containing enzyme.

V. BIOLOGICAL SIGNIFICANCE OF SUPEROXIDE DISMUTASE

When superoxide dismutase was isolated from bovine erythrocytes, a variety of tissues were assayed and the activity was found at comparable levels in bovine heart, brain, and liver and in horse heart and porpoise muscle (13). Disc-gel electrophoresis indicated that the enzyme in bovine heart, lung, brain, and erythrocytes was identical (65). A number of other sources were sampled including plants, fungi, birds, and insects, and the enzyme was found to be present in all. Its catalytic activity, coupled with the reactivity of O_2^- and the liklihood that O_2^- would be generated by diverse processes in virtually all aerobic cells, suggested that superoxide dismutase served a protective function. Several approaches were tried in attempts to validate or to disprove this proposal.

Thus, obligate anaerobes should have no need for superoxide dismutase because there is no possibility that they will make O_2^-. Indeed, we propose that an important element in their inability to tolerate oxygen is their lack of this enzyme. Accordingly, a wide range of microorganisms, including those capable of utilizing oxygen as the major electron acceptor, as well as microaerotolerant and obligately anaerobic types were surveyed for their contents of superoxide dismutase and of catalase (14). Those organisms capable of growing in oxygen were found to contain superoxide dismutase, whereas the obligate anaerobes had none.

These encouraging results prompted a more actively experimental approach to the problem. Thus *Escherichia coli* K 12-C600 was mutagenized with *N*-methyl-*N'*-nitro-*N*-nitrosoguanidine, and after washing and brief growth in a rich medium the cells were plated and colonies were allowed to grow out at 30° in air. Two replicas were then prepared from each master plate, and one of these was incubated at 42° in air while the other was grown at 42° anaerobically. Visual comparison of these replicas allowed selection of colonies which could grow at 30° in air but not at 42° in air. The mutants selected were thus obligate anaerobes at 42° but not at 30°. Seventeen such mutants were selected and grown in liquid culture at 42° under strictly anaerobic conditions. These cells were harvested, sonicated, and assayed for superoxide dismutase. It is possible to imagine multiple defects which could engender a sensitivity toward oxygen; thus, it was not surprising that seven of these mutants showed normal levels of superoxide dismutase. The remaining mutants did, however, exhibit varying deficiencies in this enzyme, and one was selected which showed less than 20% of the normal activity. This mutant, referred to as TS-2, was compared to the wild type with respect to ability to grow aerobically and anaerobically, in the range 30°–42°. Whereas the wild type grew equally well at all temperatures whether oxygen was present or not, TS-2 showed a progressive inability to grow aerobically as the temperature was elevated. The mutant and the wild type, grown anaerobically in the range 30°–42°, were also compared with respect to their content of superoxide dismutase. The wild type cells contained equal concentrations of this enzyme at all temperatures, but TS-2 contained less superoxide dismutase, the higher the temperature of growth. Indeed, the decrease in superoxide dismutase nicely paralleled the decreased ability to grow aerobically as a function of elevated temperature. It thus appears that TS-2 had a temperature-sensitive defect in superoxide dismutase which corrllated with a temperature-sensitive intolerance toward oxygen (81). These dismutase-defective mutants are unstable and have been lost, but the correlation which was observed could hardly have been fortuitous.

An independent demonstration of the biological role of superoxide dismutase might be obtained if its induction were possible. *Streptococcus faecalis* has been grown at a wide range of oxygen pressures ranging from completely anoxic conditions up to 20 atm. This organism is able to grow under these conditions, and its content of superoxide dismutase has been found to increase in proportion to the partial pressure of oxygen in the growth medium. Disc-gel electrophoresis of crude extracts of this organism demonstrated a single band of superoxide dismutase whether the cells were grown anaerobically or in 20 atm of O_2. These cells were acatalasic under any conditions of growth. The superoxide dismutase of *Streptococcus faecalis* was not inhibited by CN^-, and we assume on this basis that it is the manganese-containing enzyme found in other procaryotes.

Although we know of many reactions, both enzymic and nonenzymic, which are capable of generating O_2^-, we have no knowledge of the quantitatively significant sources of O_2^- inside respiring cells, and it is difficult to devise ways of obtaining the information. One approach, which has been tested on human full term placental tissue, exploits the fact that the eucaryotic superoxide dismutase is cyanide sensitive, whereas the procaryotic enzyme is not and that O_2^- readily reduces nitro blue tetrazolium to the blue formazan. Thus, a crude soluble extract of placenta was treated with 1×10^{-4} M cyanide to inhibit endogenous superoxide dismutase, and the rate of aerobic reduction of tetrazolium by TPNH was then measured in the presence and absence of *E. coli* superoxide dismutase. The extent of tetrazolium reduction which was oxygen-dependent and inhibited by the manganese-superoxide dismutase was taken as a measure of a TPNH–O_2^- reaction. There was a great deal of this activity in placental tissue, thus demonstrating quantitatively significant O_2^- production in this source.

If superoxide dismutase has been essential to most organisms since the advent of green plant photosynthesis then this enzyme is obviously a prime candidate for comparative studies. Isolation from diverse sources was also necessary to delineate the biological boundary between the copper and zinc dismutases and the manganese dismutases. Superoxide dismutase has been isolated from wheat germ, garden peas (46), yeast, *Neurospora crassa* (60), and chicken liver, and in all of these the enzyme was b'ue-green and contained Cu^{2+} and Zn^{2+}.

VI. APPLICATIONS OF SUPEROXIDE DISMUTASE

There have been numerous occasions when the evidence at hand led investigators to consider the involvement of O_2^- in some process. Almost invariably, however, they have lacked a method for either clearly demonstrating or for eliminating the possibility of such involvement. Superoxide dismutase provides a tool for making these discriminations. Thus, if

O_2^- is an intermediate in a process, superoxide dismutase will inhibit that process. We will now relate a few illustrative cases in which superoxide dismutase has been successfully used in this way.

a. O_2^- in Sulfite Oxidation. The aerobic oxidation of sulfite is a free radical chain reaction (82) which can be initiated by O_2^- (17, 23). Whether or not O_2^- was a chain-propagating species in this chain reaction could be answered by testing the action of superoxide dismutase under various conditions. If O_2^- was the initiating species, then superoxide dismutase must inhibit, whether or not O_2^- is also a chain-propagating species. On the other hand, if initiation was accomplished by any means not involving O_2^-, then superoxide dismutase would inhibit only if O_2^- was a chain-propagating radical. Superoxide dismutase did inhibit sulfite oxidation when initiation was accomplished by O_2^-, which was generated by xanthine oxidase or by cathodic reduction, but it had no effect on sulfite oxidation initiated by methylene blue plus light (11). We conclude that O_2^- is not involved in propagating the sulfite oxidation chains. Dimethyl sulfoxide was found to initiate sulfite oxidation, and superoxide dismutase was useful in probing the mechanism of this very complex reaction (61). Thus, when the ratio of sulfite:DMSO was high, superoxide dismutase was a potent inhibitor; but when this ratio was low, superoxide dismutase was actually an activator of sulfite oxidation. Superoxide dismutase was thus able to uncover a balance between oxidizing radicals and reducing radicals which otherwise would have gone undetected.

Sulfite oxidation is readily initiated by traces of transition metal cations (83), and this catalysis is actually responsible for the autoxidation of sulfite. The autoxidation of sulfite is unaffected by speroxide dismutase, but when 5×10^{-6} M EDTA was present this autoxidation became exquisitely sensitive to superoxide dismutase, being appreciably inhibited by 0.001 $\mu g/ml$ of this enzyme (61). Thus, in the presence of EDTA, but not in its absence, initiation of sulfite oxidation by transition metals involved the generation of O_2^- as the chain-initiating species. This observation, whose mechanistic basis has been discussed (61), could not readily have been made without the availability of superoxide dismutase.

b. O_2^- in Ethylene Production. The role of ethylene in controlling the ripening of fruit has led to interest in the mechanism of production of ethylene by plants. It was proposed that the action of O_2^- or OH· on methional could give rise to ethylene (84, 85). If O_2^- could, by its action on methional, generate ethylene, then the reaction mixture xanthine oxidase, xanthine, methional, and O_2 should generate ethylene and superoxide dismutase should inhibit. This was found to be the case (86), but the time course of ethylene production showed a lag which suggested the involvement of some stable product of the xanthine oxidase reaction. This was shown to be H_2O_2 by the inhibitory action of catalase and by the elimination of the lag by added H_2O_2 (86). Thus, both H_2O_2 and O_2^- were required for the generation of ethylene from methional, and this in turn suggested (87–89) that OH· was really the species which reacted with methional. Ethanol or benzoate, both of which are known to very effectively scavenge OH· (90), did inhibit the production of ethylene in the xanthine oxidase plus methional system (86). These results may or may not have relevance to the actual mechanism of production of ethylene in plants; but they do, in any case, demonstrate the utility of superoxide dismutase in probing radical reactions, and they suggest that superoxide dismutase would be required in aerobic cells not only because of the direct reactivities of O_2^- but also because O_2^- can, by reaction with H_2O_2, generate the hydroxyl radical, which is the most powerful oxidant known.

c. O_2^- Generated by Autoxidation of Ferredoxins. The generation of O_2^- during the autoxidation of reduced ferredoxins has been explored (91). In this case O_2^- was detected by its ability to cause the oxidation of epinephrine, and superoxide dismutase was used to verify the role of O_2^- by its ability to inhibit the oxidation. Both spinach and clostridial ferredoxins were shown, in this manner, to produce O_2^-. The extent of O_2^- production increased with pO_2 and with pH in a way reminiscent of the behavior of xanthine oxidase (92).

d. O_2^- Generated by Autoxidation of Reduced Flavins, Quinones, and Dyes. A number of electron carriers have been noted to mediate the reduction of cytochrome c by enzymic systems (25, 93, 94). The ability of these electron carriers to augment the rate of reduction of cytochrome c anaerobically as well as aerobically had led to the view that oxygen played no role in these systems other than that of a relatively ineffective competing electron acceptor (93, 95). Superoxide dismutase was used to explore these mediated reductions of cytochrome c, and it was shown that in the presence of oxygen the favored reaction pathway involved O_2^- when menadione, FMN, or FAD was used as the electron carriers. But this was not the case when methylene blue, 2,6-dichlorobenzenone, or N-methyl phenazinium methosulfate was used (96). The reduced forms of the latter three compounds presumably reacted directly with cytochrome c more rapidly than they reacted with oxygen. It was possible to demonstrate that reduced methylene blue caused the univalent reductions of O_2 by using epinephrine as the detector of O_2^- (96). The effects of pH and pO_2 on the univalent reduction of O_2 by reduced flavins and quinones have been explored (97). More recently, the univalent reduction of O_2 by reduced N-methyl phenazinium methosulfate has been demonstrated (98). Superoxide dismutase was also used to demonstrate the involvement of O_2^- in the reoxidation of reduced tetraacetyl riboflavin (75, 99).

e. O_2^- and Hydroxylations. It appears possible that O_2^- may be involved in the mechanism of certain hydroxylation reactions. Thus, the air oxidation of dihydroxyfumarate is catalyzed by iron salts and causes the concomitant hydroxylation of p-cresol (81). Superoxide dismutase inhibits this hydroxylation. Further study of this system has demonstrated that H_2O_2 is also involved and that $OH\cdot$, generated by a Haber-Weiss reaction (87–89), is probably the hydroxylating species in this model system (100). O_2^-, generated photochemically, has been shown to cause the hydroxylation of p-hydroxybenzoate, and superoxide dismutase inhibited this hydroxylation (101). It was considered unlikely in these experiments that $OH\cdot$ was actually the hydroxylating species because catalase did not inhibit this hydroxylation and because mannitol and ethanol did not inhibit hydroxylation (101). Because infusion of electrolytically generated O_2^- did not cause hydroxylation in the absence of flavin, the possibility that O_2^- itself is not the hydroxylating species but an intermediate essential to its production was also considered (101).

Reaction mixtures containing cytochrome P-450, the TPNH–P-450 reductase, TPNH, oxygen, and dilauroyl glycerylphosphorylcholine were observed to cause the hydroxylation of benzphetamine. This hydroxylation was partially inhibited by superoxide dismutase, and in the presence of 0.6 M NaCl it was completely inhibited by superoxide dismutase (102). Furthermore, a known enzymic source of O_2^- was able to replace both TPNH and the TPNH–P-450 reductase in these reaction mixtures. A scheme was proposed in which O_2^- bound to P-450 is a normal intermediate in the activation of oxygen for hydroxylation reactions (102).

f. O_2^- Generated by Ultrasonication. Ultrasonication has been observed to cause a variety of chemical changes in compounds of biochemical importance, and the lysis of water into $H\cdot$ and $OH\cdot$ within the sonically generated cavities has been proposed in explanation of sonochemical transformations (103, 104). The sonochemical yield of H_2O_2 was enhanced by O_2, and in the presence of volatile scavengers of $OH\cdot$ was entirely dependent upon the presence of O_2 (105–108). A series of reactions was proposed involving the radicals $H\cdot$, $OH\cdot$, and $HO_2\cdot$. Among these radicals, the most long lived, and therefore the one most readily able to escape from the cavities and thus affect nonvolatile solutes, would be $HO_2\cdot$ or its conjugate base O_2^-. This notion has been directly tested by using cytochrome c to trap O_2^- and superoxide dismutase to intercept O_2^- (109). Thus, ultrasonication in buffered and aerated solutions leads to the reduction of ferricytochrome c which is inhibited by superoxide dismutase (109).

g. O_2^- Generated by Photolysis of Water. Short ultraviolet light is sufficiently energetic to cause the cleavage of water, and in the presence of oxygen the hydrogen atoms so generated should yield HO_2 and O_2^-. In-Indeed, one can write a series of reactions including the following:

$$HOH + h\nu \rightarrow H\cdot + OH\cdot$$

$$OH\cdot + OH\cdot \rightarrow H_2O_2$$

$$H\cdot + H\cdot \rightarrow H_2$$

$$H\cdot + O_2 \rightarrow HO_2 \rightarrow H^+ + O_2^-$$

$$OH\cdot + O_2^- \rightarrow OH^- + O_2$$

One would anticipate that irradiation of buffered solutions of ferricyto-chrome c might lead to their reduction because of the reactivities of $H\cdot$ and of O_2^- and that in the presence of oxygen this reduction of cytochrome c should be inhibited by superoxide dismutase. All of these consequences have been affirmed (110). As expected, scavengers of $OH\cdot$ were found to increase the rate of superoxide dismutase-inhibited reduction of cytochrome c.

h. O_2^- and Tryptophan Dioxygenase. Superoxide dismutase has been found to partially inhibit the activation of tryptophan dioxygenase by diverse reductants, thus implicating O_2^- in the reductive activation of this enzyme (111). Superoxide dismutase did not inhibit the pseudomonad tryptophan dioxygenase if added after the process of reductive activation had been completed, thus indicating that free O_2^- was not obligatorily involved in the catalytic cycle of this enzyme. In contrast, the tryptophan dioxygenase of rabbit intestine was inhibited by superoxide dismutase added at any stage of the reaction, thus implicating O_2^- in the mechanism of action of this enzyme (112). Superoxide dismutase has been shown to inhibit the oxidation of DPNH by myeloperoxidase plus 2,4-dichloro-phenol (113).

This listing of the applications of superoxide dismutase is not exhaustive, and considering the short time since the discovery of this enzyme it must be considered as a small indication of the possibilities which will certainly be explored in the years ahead.

VII. SOME CHEMICAL PROPERTIES OF OXYGEN AND ITS REDUCTION PRODUCTS

Problems of excessive reactivity are posed neither by molecular oxygen nor by its ultimate reduction product water, but rather by intermediates on this reduction pathway. The reduction of O_2 to H_2O requires four electrons and could involve the intermediates O_2^-, H_2O_2, and OH·. Molecular oxygen, in the ground state, does exhibit a propensity for univalent reduction which makes the production of radical intermediates a likely occurrence. The preference of O_2 for free radical reaction pathways depends upon the fact that it contains two parallel electronic spins. It is thus paramagnetic and a biradical, and its reaction with an exogenous electron pair would face a spin restriction. This spin restriction can pose a formidable barrier to reaction because the time required for a change of spin state is orders of magnitude greater than the lifetime of a collisional complex. Free radical pathways circumvent this spin restriction and are therefore often favored (114). This is not to say that the reduction of oxygen is always accomplished by successive univalent steps, but rather that the univalent reduction of oxygen will be favored, whenever energetically feasible, unless there are special means for circumventing the spin restriction faced by the vibalent reaction. For example, the reaction of SO_3^{2-} with O_2 proceeds exclusively by a free radical pathway, and the direct divalent interaction between SO_3^{2-} and O_2 cannot be observed even when the free radical chain pathway is inhibited (115). In contrast, the reduction of O_2 by Fe^{2+} can be divalent or univalent depending on the conditions. In perchlorate medium the rate equation $V = k\ (O_2)\ (Fe^{2+})^2$ (116) implies that two atoms of Fe^{2+} cooperate, in a rate-limiting step, to bring about the divalent reduction of O_2, whereas in phosphate buffered solutions the rate equation $V = k\ (O_2)\ (H_2PO_4^-)^2\ (Fe^{2+})$ (117) indicates that a univalent reduction of O_2 by a single Fe^{2+} was the rate-limiting step.

The univalent reduction of O_2 gives rise to the hydroperoxyl radical or its conjugate base, the superoxide anion. This species has been generated by a variety of means including the cathodic reduction of oxygen in non-protic solvents (57–59) and in water (118), the oxidation of H_2O_2 by ceric ions (119), the reduction of oxygen with hydrogen atoms or hydrated electrons generated by the photolysis of water (120, 121), or by the use of a beam of electrons (54). It has been prepared in the gas phase by passing a mixture of H_2O_2, H_2O, O_2, and H_2 through a radiofrequency discharge (122). Superoxide radicals have been detected by a number of physical methods including conductimetry (123), optical spectroscopy (124), electron spin resonance spectroscopy (125–128), and mass spectrometry (122). The chemical methods which have been used for detection depend upon the ability of this radical to reduce tetranitromethane (54–56) or cytochrome c (11–13), to oxidize epinephrine (13), or to initiate free radical chain reactions (13, 23).

The pK_a for the hydroperlxyl radical is 4.88 ± 0.1 (129). These radicals undergo a spontaneous dismutation to yield $H_2O_2 + O_2$, which is most rapid at pH 4.8. A complete explanation for the variation of dismutation rate with pH is provided by the rate constants for the following reactions, coupled with the known pK_a for HO_2 (129); thus:

$$HO_2\cdot + HO_2\cdot \rightarrow H_2O_2 + O_2 \qquad k_2 = 7.6 \times 10^5 \ M^{-1}\ sec^{-1} \qquad (6)$$

$$HO_2\cdot + O_2^- + H^+ \rightarrow H_2O_2 + O_2 \qquad k_2 = 8.5 \times 10^7 \ M^{-1}\ sec^{-1} \qquad (7)$$

$$O_2^- + O_2^- + 2H^+ \rightarrow H_2O_2 + O_2 \qquad k_2 < 10^2 \ M^{-1}\ sec^{-1} \qquad (8)$$

The lethargy of reaction (8) probably relates to electrostatic repulsion which hinders close approach of two superoxide anions. It will be recalled that the rate constant of the enzyme-catalyzed dismutation is virtually independent of pH in this range (77, 78). Univalently reduced oxygen can react very rapidly with other radicals, i.e.,

$$HO_2\cdot + H\cdot \rightarrow H_2O_2 \qquad k_2 = 2 \times 10^{10} \ M^{-1}\ sec^{-1}$$

$$HO_2\cdot + OH\cdot \rightarrow H_2O + O_2 \qquad k_2 = 1.5 \times 10^{10} \ M^{-1}\ sec^{-1} \ (130)$$

The properties of oxygen radicals have recently been thoroughly reviewed (131).

VIII. PROJECTED STUDIES

It is possible, from the vantage point of currently available information, to make several proposals concerning the role of O_2^- and of superoxide dismutase in such phenomena as radiation sensitivity, oxygen toxicity, and senescence. In each of these phenomena O_2^- radicals may play a role which could be interdicted by superoxide dismutase. It may be interesting to briefly survey a few of these projected studies.

Oxygen is known to enhance the radiosensitivity of a variety of cells (132–134). Since ionizing radiation passing through water would certainly generate hydrated electrons and hydrogen atoms, both of which react very rapidly with O_2 to give O_2^-, and since O_2^- is a relatively long-lived radical, it seems likely that in oxygenated solutions the only radicals able to diffuse significantly from the ionization spurs would be O_2^-. It appears likely that the oxygen enhancement of radiation damage relates to this production of O_2^-. One approach to this question would be to manipulate the concentration of superoxide dismutase within cells and then to compare the oxygen enhancement ratio of low dismutase cells with that of high dismutase cells. The level of superoxide dismutase within microorganisms can be manipulated in several ways so these experiments are certainly feasible. Growing yeast in a copper-rich medium has been reported to increase radio resistance (135), and we have observed that the superoxide dismutase of yeast is a copper- and zinc-containing enzyme (136) whose level in these organisms can be elevated by growing them in copper-rich medium. Oxygen induces superoxide dismutase synthesis in *Streptococcus faecalis*, and the level of superoxide dismutase in these cells can simply be manipulated by varying the partial pressure of O_2 under which they are grown. Alternately, mutants containing abnormally low or high superoxide dismutase could be compared with the parental strain for oxygen enhancement ratio.

The toxicity of hyperbaric oxygen could also relate to enhanced production of O_2^- at elevated concentrations of oxygen, and here, too, a comparison of low dismutase and high dismutase cells would allow a test of this proposal. Senescence may reflect the accumulation of irreparable chemical damage, and for cells living aerobically it now appears very reasonable that part of that chemical damage should be the result of the uncontrolled reactivities of oxygen radicals. Manipulation of the flux of O_2^- by means of varying the partical pressure of O_2, or manipulation of the endogenous level of superoxide dismutase by means already discussed, might allow an approach to this facet of senescence.

It would be extremely useful to have a small copper complex which displayed superoxide dismutase activity. Such an artificial catalyst might readily gain entry to cells and thus allow direct demonstrations of the role of O_2^- in the phenomena we have been discussing. This hypothetical synthetic superoxide dismutase might enable obligate anaerobes to grow in the presence of oxygen and might reduce the oxygen enhancement ratio of aerobes. Still another possibility would utilize transduction to introduce the superoxide dismutase gene into obligate anaerobes. How would this affect their oxygen tolerance? It is clear that much remains to be done, but the door has been opened and many investigators will now certainly attack the problems exposed to view.

ACKNOWLEDGMENTS

The author would like to acknowledge the great good fortune which brought him to mentors (Dr. A. Mazur at the undergraduate level and Dr. P. Handler at the graduate and postgraduate level) who taught him to follow the tortuous reserach trails with diligence and open eyes and which brought to him students and research associates from whom he learned at least as much as he could teach. Among the latter, those who have been directly involved in the work on superoxide dismutase include L. L. Greenlee, J. M. McCord, C. Beauchamp, B. B. Keele, Jr., H. P. Misra, S. Goscin, M. Gregory, H. Forman, and R. Weisiger. Special mention must be made of J. M. McCord since it was his special insight which first drew back the blinds.

REFERENCES

1. Stanier, R. Y., Doudoroff, M., and Adelberg, E. A. (1970). "The Microbial World," 3rd ed., p. 75. Prentice-Hall, Englewood Cliffs, New Jersey.
2. Haugaard, N. (1968). *Physiol. Rev.* **48,** 311.
3. Mizrahi, A., Vosseller, G. V., Yagi, K., and Moore, G. E. (1972). *Proc. Soc. Exp. Biol. Med.* **139,** 118.
4. McLeod, J. W., and Gordon, J. (1923). *J. Pathol. Bacteriol.* **26,** 332.
5. Callow, A. B. (1923). *J. Pathol. Bacteriol.* **26,** 320.
6. Rywosch, D., and Rywosch, M. (1907). *Zentralbl. Bakteriol. Parasitenk. Infectionskr. Hyg. Abt. Orig.* **44,** 295.
7. Stanier, R. Y., Doudoroff, M., and Adelberg, E. A. (1970). "The Microbial World," 3rd ed., p. 663–664. Prentiss-Hall, Englewood Cliffs, New Jersey.
8. Gledhill, W. E., and Casida, L. E. Jr. (1969). *Appl. Microbiol.* **18,** 340.
9. Jones, D., Watkins, J., and Meyer, D. J. (1970). *Nature (London)* **226,** 1249.
10. Prevot, A. R., and Thouvenot, H. (1952). *Ann. Inst. Pasteur Paris* **83,** 443.
11. McCord, J. M., and Fridovich, I. (1968). *J. Biol. Chem.,* **243,** 5753.
12. McCord, J. M., and Fridovich, I. (1969). *Fed. Proc. Fed. Amer. Soc. Exp. Biol.* **28,** 346.
13. McCord, J. M., and Fridovich, I. (1969). *J. Biol. Chem.* **244,** 6049.
14. McCord, J. M., Keele, B. B., Jr., and Fridovich, I., (1971). *Proc. Nat. Acad. Sci. U. S.* **68,** 1024.
15. Berkner, L. V., and Marshall, L. C. (1964). *Discuss. Faraday Soc.* **37,** 122.
16. Berkner, L. V., and Marshall, L. C. (1965). *J. Atmos. Sci.* **22,** 225.
17. Fridovich, I., and Handler, P. (1958). *J. Biol. Chem.* **233,** 1578.
18. Fridovich, I., and Handler, P. (1958). *J. Biol. Chem.* **233,** 1581.
19. Horecker, B. L., and Heppel, L. A. (1949). *J. Biol. Chem.* **178,** 683.
20. Totter, J. R., Medina, V. J., and Scoseria, J. L. (1960). *J. Biol. Chem* **235,** 238.
21. Totter, J. R., deDugros, E. C., and Riveiro, C., (1960). *J. Biol. Chem.* **235,** 1839.
22. Greenlee, L., Fridovich, I., and Handler, P. (1962). *Biochemistry* **1,** 779.
23. Fridovich, I., and Handler, P. (1961). *J. Biol. Chem.* **236,** 1836.
24. Weber, M. M., Lenhoff, H. M., and Kaplan, N. O. (1956). *J. Biol. Chem.* **220,** 93.
25. Fridovich, I., and Handler, P. (1962). *J. Biol. Chem.* **237,** 916.
26. Fridovich, I. (1962). *J. Biol. Chem.* **237,** 584.
27. Margoliash, E. (1954). *Biochem. J.* **56,** 529.
28. Rajagopalan, K. V., Fridovich, I., and Handler, P. (1962). *J. Biol. Chem.* **237,** 922.
29. Fridovich, I., (1967). *J. Biol. Chem.* **242,** 1445.
30. Mann, T. and Keilin, D. (1939). *Proc. Roy. Soc. Ser. B. Biol. Sci.* **126,** 303.
31. Mohamed, M. S., and Greenberg, D. M. (1953). *J. Gen. Physiol.* **37,** 433.
32. Kimmel, J. R., Markowitz, H., and Brown, D. M. (1959). *J. Biol. Chem.* **234,** 46.
33. Markowitz, H., Cartwright, G. E., and Wintrobe, M. M. (1959). *J. Biol. Chem.* **234,** 40.
34. Porter, H., and Ainsworth, S. (1959). *J. Neurochem.* **5,** 91.
35. Shields, G. S., Markowitz, H., Klassen, W. H., Cartwright, G. E., and Wintrobe, M. M. (1961). *J. Clin. Invest.* **40,** 2007.
36. Porter, H., Sweeney, M., and Porter, E. M. (1964). *Arch. Biochem. Biophys.* **105,** 319.
37. Stansell, M. J., and Deutsch, H. F. (1965). *J. Biol. Chem.* **240,** 4299.
38. Stansell, M. J., and Deutsch, H. F. (1965). *J. Biol. Chem.* **240,** 4306.

39. Bannister, W. H., Salisbury, C. M., and Wood, E. J. (1968). *Biochem. Biophys. Acta* **168**, 392.
40. Carrico, R. J., and Deutsch, H. F. (1969). *J. Biol. Chem.* **244**, 6087.
41. Wood, E., Dalgleish, D., and Bannister, W. (1971). *Eur. J. Biochem.* **18**, 187.
42. Carrico, R. J., and Deutsch, H. F. (1970). *J. Biol. Chem.* **245**, 723.
43. Keele, B. B., Jr., McCord, J. M., and Fridovich, I. (1971). *J. Biol. Chem.* **246**, 2875.
44. Bannister, J., Bannister, W., and Wood. E. (1971). *Eur. J. Biochem.* **18**, 178.
45. Porter, H., and Folch, J. (1957). *Arch. Neurol. Psychiat.* **77**, 8.
46. Sawada, K., Ohyama, T., and Yamazaki, I., (1971). *Seikagaku (Japan)* **43**, 532.
47. Weser, U., Bunnenberg, E., Commack, R., Djerassi, C., Flohé, L. Thomas G., and Voelter, W. (1971). *Biochem. Biophys. Acta* **243**, 203.
48. Weser, U., and Hartman, H. J. (1971). *FEBS Lett.* **17**, 78.
49. Evans, H. J., and Hill, R. L., personal communication.
50. Hill, R. L., personal communication.
51. Malmström, B. and Vanngard, T. (1960). *J. Mol. Biol.* **2**, 118.
52. Rotilio, G., Finazzi, A., Calabrese, L., Bossa, F., Guerrieri, P., and Mondovi, B. (1971). *Biochemistry* **10**, 616.
53. Fee, J. A., and Gaber, B. P. (1972). *J. Biol. Chem.* **247**, 60.
54. Czapski, G. H., and Bielski, B. H. J. (1963). *J. Phys. Chem.* **67**, 2180.
55. Rabani, J., Mulac, W. A., and Matheson, M. S. (1965). *J. Phys. Chem.* **69**, 53.
56. Bielski, B. H. J., and Allen, A. O. (1967). *J. Phys. Chem.* **71**, 4544.
57. Maricle, D. L., and Hodgson, W. G. (1965). *Anal. Chem.* **37**, 1562.
58. Sawyer, D. T., and Roberts, J. L., Jr. (1966). *J. Electroanal. Chem.* **12**, 90.
59. Poever, M. E., and White, B. S. (1961). *Electrochim. Acta* **11**, 1061.
60. Misra, H. P., and Fridovich, I. (1972). *J. Biol. Chem.* **247**, 3170.
61. McCord, J. M., and Fridovich, I. (1969). *J. Biol. Chem.* **244**, 6056.
62. Massey, V., Strickland, S., Mayhew, S. G., Howell, L. G., Engel, P. C., Matthews, R. G., Schuman, M., and Sullivan, P. A. (1969). *Biochem. Biophys. Res. Commun.* **36**, 891.
63. Rajagopalan, K. V., and Handler, P. (1964). *J. Biol. Chem.* **239**, 2022.
64. Miller, R. W. (1970). *Can. J. Biochem.* **48**, 935.
65. Beauchamp, C., and Fridovich, I. (1971). *Anal. Biochem.* **44**, 276.
66. Vance, P. G., Keele, B. B., Jr., and Rajagopalan, K. V. (1972). *J. Biol. Chem.* **247**, 4782.
67. Brewer, G. J. (1967). *Amer. J. Hum. Genet.* **19**, 674.
68. Utter, F. M. (1971). *Comp. Biochem. Physiol.* **39B**, 891.
69. Johnson, A. G., Utter, F. M., and Hodgins, H. O. (1970). *Comp. Biochem. Physiol.* **37**, 281.
70. Baur, E. W. (1969). *Science* **166**, 1524.
71. Richmond, R. C., and Powell, J. R. (1970). *Proc. Nat. Acad. Sci. U.S.* **67**, 1264.
72. Larsen, A. L., and Benson, W. C. (1970). *Crop Sci.* **10**, 493.
73. Khan, P. M. (1971). *Arch. Biochem. Biophys.* **145**, 470.
74. Fried, R., Fried, L. W., and Babin, D. (1970). *Eur. J. Biochem.* **16**, 399.
75. Ballou, D., Palmer, G., and Massey, V. (1969). *Biochem. Biophys. Res. Commun.* **36**, 898.
76. Fee, J. A. and Gaber, B. P. (1971). *Int. Symp. Oxidative Enzymes Related Redox Reactions, 2nd Memphis, Tenn.*, June.
77. Klug, D., Rabani, J., and Fridovich, I. (1972). *J. Biol. Chem.* **247**, 4839.

78. Rotilio, G., Bray, R. C., and Fielden, E. M. (1972). *Biochem. Biophys. Acta* **268,** 605.

79. Forman, H. J., Evans, H. J., Hill, R. L., and Fridovich, I. (1973). *Biochemistry,* **12,** 823.

80. Keele, B. B. Jr., McCord, J. M., and Fridovich, I. (1970). *J. Biol. Chem.* **245,** 6176.

81. McCord, J. M., Beauchamp, C. O., Goscin, S., Misra, H. P., and Fridovich I. (1971). *Int. Symp. Oxidative Enzymes Related Redox Reactions, 2nd, Memphis, Tenn.* June.

82. Fuller, E. C., and Crist, R. H. (1941). *J. Amer. Chem. Soc.* **63,** 1644.

83. Abel, E. (1951). *Monatsh. Chem.* **82,** 815.

84. Yang, S. F. (1967). *Arch. Biochem. Biophys.* **122,** 481.

85. Yang, S. F. (1969). *J. Biol. Chem.* **244,** 4360.

86. Beauchamp, C. and Fridovich, I. (1970). *J. Biol. Chem.* **245,** 4641.

87. Haber, F., and Willstatter, R. (1931). *Berichte* **64,** 2844.

88. Haber, F., and Weiss, J. (1932). *Naturwissenschaften* **20,** 298.

89. Haber, F., and Weiss, J. (1934). *Proc. Roy. Soc. (London)* **A-147,** 332.

90. Neta, P., and Dorfman, L. M. (1968). *Advan. Chem. Ser.* **81,** 222.

91. Misra, H. P., and Fridovich, I. (1971). *J. Biol. Chem.* **246,** 6886.

92. Fridovich, I. (1970). *J. Biol. Chem.* **245,** 4053.

93. Mahler, H. R., Fairhurst, A. S., and Mackler, B. (1955). *J. Amer. Chem. Soc.* **77,** 1514.

94. Muraoka, S., Enomato, H., Sugiyama, M., and Yamazaki, H. (1967). *Biochim. Biophys. Acta* **143,** 408.

95. Singer, T. P., and Kearney, E. B. (1950). *J. Biol. Chem.* **183,** 409.

96. McCord, J. M., and Fridovich, I. (1970). *J. Biol. Chem.* **245,** 1374.

97. Misra, H. P., and Fridovich, I. (1972). *J. Biol. Chem.* **247,** 188.

98. Nishikimi, M., Rao, N. A., and Yagi, K. (1972). *Biochem. Biophys. Res. Commun.* **46,** 849.

99. Massey, V., Palmer, G. and Ballou, D. (1971). *Int. Symp. Oxidative Enzymes Related Redox Reactions, 2nd, Memphis, Tenn.* June.

100. Goscin, S., and Fridovich, I. (1972). *Arch. Biochem. Biophys.* **153,** 778.

101. Strickland, S., and Massey, V. (1971). *Int. Symp. Oxidative Enzymes Related Redox Reactions, 2nd Memphis, Tenn.* June.

102. Strobel, H. W., and Coon, M. J. (1971). *J. Biol. Chem.* **246,** 7826.

103. Alexander, P., and Fox, M. (1954). *J. Polym. Sci.* **12,** 533.

104. Lindstrom, O. (1955). *J. Acoust. Soc. Amer.* **27,** 654.

105. Del Luca, M., Yeager, E., Davies, M. O., and Hovorka, F. (1958). *J. Acoust. Soc. Amer.* **30,** 301.

106. Weissler, A. (1959). *J. Amer. Chem. Soc.* **81,** 1077.

107. Anbar, M., and Pecht, I. (1964). *J. Phys. Chem.* **68,** 352.

108. Spurlock, L. A., and Reifsneider, S. B. (1970). *J. Amer. Chem. Soc.* **92,** 6112.

109. Lippitt, B., McCord, J. M., and Fridovich, I. (1972). *J. Biol. Chem.* **247,** 4688.

110. McCord, J. M., and Fridovich, I., (1973) *Photochem. Photobiol.* **117,** 115.

111. Brady, F. O., Forman, H. J., and Feigelson, P. (1971). *J. Biol. Chem.* **246,** 7119.

112. Hirata, F., and Hayaishi, O. (1971). *J. Biol. Chem.* **246,** 7825.

113. Odajima, T. (1971). *Biochem. Biophys. Acta* **235,** 52.

114. Taube, H. (1965). *In Oxygen-Proc. Symp. Sponsored N. Y. Heart Ass.* p. 29. Little, Brown, Boston, Massachusetts.

115. Alyea, H. N., and Backstrom, H. L. J. (1929). *J. Amer. Chem. Soc.* **51,** 90.

116. George, P. (1954). *J. Chem. Soc.* 4349.
117. Cher, M., and Davidson, N. (1955). *J. Amer. Chem. Soc.* **77,** 793.
118. Forman, H. J., and Fridovich, I. (1972). *Science* **175,** 339.
119. Saito, E., and Bielski, B. H. J. (1961). *J. Amer. Chem. Soc.* **83,** 4467.
120. Baxendale, J. H. (1962). *Radiat. Res.* **17,** 312.
121. Adams, G. E., Boag, J. W., and Michael, B. D. (1965). *Proc. Roy. Soc. London Ser. A,* **289,** 321.
122. Foner, S. N., and Hudson, R. L. (1962). *Advan. Chem. Ser.* **36,** 34.
123. Ander, S. (1967). *Strahlen Therap.* **132,** 135.
124. Czapski, G., and Dorfman, L. M. (1964). *J. Phys. Chem.* **68,** 1169.
125. Saito, E., and Bielski, B. H. J. (1961). *J. Amer. Chem. Soc.* **83,** 4467.
126. Knowles, P. F., Gibson, J. F., Pick, F. M., and Bray, R. C. (1969). *Biochem. J.* **111,** 53.
127. Bray, R. C., Pick, F. M., and Samuel, D. (1970). *Eur. J. Biochem.* **15,** 352.
128. Kroh, J., Green, B. C., and Spinks, J. W. T. (1961). *J. Amer. Chem. Soc.* **83,** 2201.
129. Behar, D., Czapski, G., Rabani, J., Dorfman, L. M., and Schwarz, H. A. (1970). *J. Phys. Chem.* **74,** 3209.
130. Schwarz, H. A. (1964). *Radiat. Res. Suppl.* **4,** 89.
131. Czapski, G. (1971). *Ann. Rev. Phys. Chem.* **22,** 171.
132. Gray, L. H., Conger, A. D., Ebert, M., Hornsey, S. and Scot, O. C. (1953). *Brit. J. Radiol.* **26,** 638.
133. Alderson, T., and Scott, B. R. (1971). *Nature (London) New Biol.* **230,** 45.
134. Cramp., W. A., Watkins, P. K., and Collins, J. (1972). *Nature (London) New Biol.* **235,** 76.
135. Gesswagner, D., Altmann, H., Suilvinyi, A. V., and Kaindl, K. (1968). *Int. J. Appl. Radiat. Isotopes* **19,** 152.
136. Goscin, S., and Fridovich, I. (1972). *Biochim. Biophys. Acta* **289,** 276.

12

Cytochrome c Oxidase

PETER NICHOLLS and BRITTON CHANCE

I. INTRODUCTION

✔ Photochemistry signaled the way toward basic discoveries on the nature of cell respiration and vision. The pioneering studies of Koenig (1) on the photochemical action spectra of visual receptors were logically followed by related work on the photochemical action spectra of yeast respiration by Warburg (2). These two studies afford the first "pure" spectra of chromoproteins in their state in nature and have provided paragons for subsequent attempts to duplicate the data from the photochemical action spectra by absorption spectra of the purified materials.

✔ The principle of the method as employed by Warburg is simple. The respiratory rate of yeast cells in the presence of carbon monoxide depends critically upon the concentration of ferrocytochrome oxidase (a_3^{2+}), which

479

in turn can be regulated by the light-induced photodissociation of $a_3^{2+}CO$. The effectiveness of a particular wavelength of light in photolyzing the a_3CO bond depends upon its specific absorption by the compound. In this way, the absorption spectrum of a_3CO and the respiratory rate are directly related for yeast cells as determined for selected bright lines in arc spectra (Fig. 1A). Some years later, Castor and Chance (3) developed a method

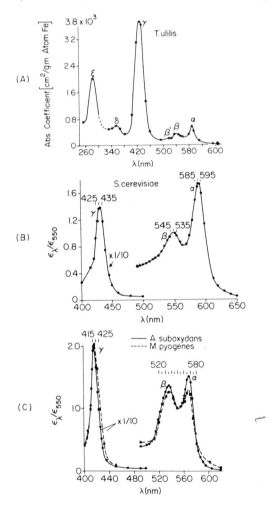

Fig. 1. Spectrum of the carbon monoxide compound of cytochrome oxidase. (A) Photochemical action spectrum (yeast enzyme). From Fig. 1 of Warburg (2). (B) Photochemical act on spectrum (cytochrome a_3 of yeast). From Castor and Chance (3). (C) Photochemical action spectrum (cytochrome o of two bacteria). From Castor and Chance (3).

for using all wavelengths and applied it to baker's yeast (Fig. 1B) and to the bacterial oxidase, cytochrome o, whose CO compound has a spectrum (Fig. 1C) resembling that of carboxymyoglobin.

As is often the case in experimental work, the timing was right for intensive studies of the photochemistry of respiration. Haldane and Smith had discovered that mixtures of carboxy and oxyhemoglobin showed different characteristics when exposed to summer and winter light, and from this observation they had identified the photodissociability of carboxyhemoglobin. Although according to their work, Warburg should have used very different mixtures of carbon monoxide and oxygen in his experiments on the inhibition of yeast respiration by carbon monoxide, the requirements of respirometry led him to the proper conditions for experimental success. However serendipic the experiment, it was this single elegant result that transferred Warburg's activities from charcoal models to effective studies of the catalysis of cellular respiration in living cells.

✓At the same time, the work of Keilin (4) on the cytochromes of cells and tissues opened up the consideration of the cytology and biochemistry of respiration. Keilin (5) provided the concept of a respiratory chain transporting electrons from succinate to oxygen via a chain of cytochromes (6) with a terminal cytochrome oxidase. Nevertheless, Keilin himself was unable to obtain any evidence for a carbon monoxide binding cytochrome which corresponded to the characteristics of the CO compound of Warburg's *atmungsferment*. Warburg and Negelein themselves could have bridged this gap when, using the hand spectroscope, they observed the Soret band of cytochrome a_3 at 445–447 nm in the thorax of the bee (7). As fate would have it, they failed to follow up their own work and to examine the spectrum in the presence of carbon monoxide. Had they done so, they would have been the first to observe the disappearance of the 445-nm band and the appearance of the 430-nm band of cytochrome a_3 upon its conversion to a_3CO. These incomplete observations caused Warburg not only to ignore the relationship between Keilin's work and his own but also to overlook the fact that cytochrome a_1 of bacteria, which responded directly to oxygen and carbon monoxide (8), provided a model for the mammalian or yeast oxidase. Instead, 10 years of polemic ensued during which Warburg asserted that cytochrome a_1 was the only functional entity and the others degraded *atmungsferment*. Keilin (9), on the other hand, recognized early the need for a separate oxidase which he termed "indophenol oxidase"; but he considered that this had to be something quite different from a cytochrome—a copper enzyme, perhaps (10). His discovery, together with Hartree 5 years later (11a, 11b), of cytochrome a_3, whose carbon monoxide spectrum (12) is close to that predicted by Warburg, did much to reconcile the two schemes, except for the salient fact that

Keilin, using carbon monoxide–oxygen mixtures predicted on the basis of Barcroft's studies of hemoglobin, failed to make the essential identification of cytochrome a_3CO as a photodissociable hemoprotein compound, a point that was later settled by the use of very low carbon monoxide concentrations and high light intensities.

The relationship of cytochrome a_3 to cytochrome a, already identified in 1925 by Keilin (13) and known to be close to the site of oxygen attack (14), remained uncertain. We now know that the two cytochromes, a and a_3, are closely associated but functionally distinguishable in all active preparations of cytochrome oxidase. Warburg, in 1933, had already postulated one or two iron-containing oxidase species intervening between cytochrome a and oxygen (15). Keilin's observations with inhibitors clearly showed that another catalytic compound was needed for the reaction of cytochrome a with oxygen. Shibata (16), who had been thinking in terms of oxygenation of the cytochromes, suggested that cytochrome a might be identical with the *atmungsferment*. Later, he realized (17) that the yeast oxidase must be more like a_1 of bacteria than like cytochrome a. By 1937, Tamiya and Ogura (18) had definitely relegated cytochrome a to the status of a subordinate electron carrier, placing no fewer than three components between it and oxygen—cytochrome c, the cyanide-sensitive "indophenolase," and the carbon monoxide–sensitive "oxygen-transporting enzyme". When Keilin and Hartree abandoned copper as the terminal oxidase and identified cytochrome a_3, Yakushiji and Okunuki (19) modified the previous theory (18), placing a single oxidase between c and oxygen, and left cytochrome a out of the main electron transfer sequence.

The studies of Okunuki and his group eventually led to the preparation (20) of a mitochondrial cytochrome oxidase free from other cytochromes (b, c, and c_1). This preparation contains cytochrome a as well as cytochrome a_3 and has been variously referred to as "cytochrome a", "cytochrome a + a_3," and "cytochrome aa_3." It was first suggested by these workers (21) that the properties attributed to cytochrome a_3 resulted from interaction between cytochromes a and c; more recently, schemes have been put forward that retain the critical function of cytochrome c but include roles for two a-type hemes in the oxidase (22). Conversely, those who originally believed in separable a and a_3 species have come to accept their close interrelationship even in the isolated system. Although it is thus now agreed that the oxidase (aa_3) system contains two heme a groups, their catalytic role and chemical behavior remain areas of disagreement.

The historical development of these ideas has been presented recently by Lemberg (23) and by Wainio (24), and the viewpoint of the Japanese school by Okunuki (21,25) and by Sekuzu and Takemori (26).

II. PHYSICOCHEMICAL PROPERTIES

Figure 2 illustrates the spectra of cytochrome aa_3 isolated from beef heart. Four derivatives of the oxidase are shown: the ferrous ($a^{2+}a_3^{2+}$) and ferric ($a^{3+}a_3^{3+}$) forms in absence of cyanide, and the ferrous ($a^{2+}a_3^{3+}$ HCN), and ferric ($a^{3+}a_3^{3+}$HCN) forms obtained in the presence of cyanide (27). In addition to these species, cyanide added at higher concentrations to anaerobic reduced enzyme combines to form cyanferrous a_3 ($a^{2+}a_3^{2+}$HCN) with an α peak shifted from 605 to 601 nm and a slight decrease of the Soret absorption. It was such spectra, seen initially in intact particles with the microspectroscope, that helped persuade Keilin and Hartree of the separate existence of cytochrome a_3.

Figure 3 shows the spectra of the derivatives observed in the presence of carbon monoxide (cf. Fig. 1). A CO complex is formed with ferrocytochrome a_3 whether cytochrome a is ferrous or ferric, with rather similar appearance in either case (28). The oxidation state of cytochrome a has no effect on the spectrum of a_3^{2+}CO. Similarly, the photodissociation of a_3^{2+}CO cannot be achieved by illuminating cytochrome a (Fig. 1) showing

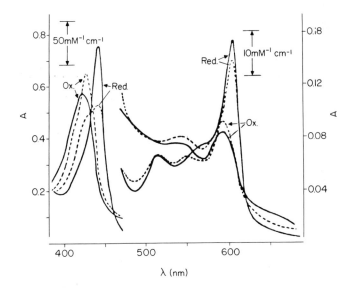

FIG. 2. Absorption spectra of cytochrome aa_3 and cyanocytochrome aa_3; 3.5 μM enzyme from beef heart in 100 mM tris-sulfate (pH 8.0), 0.5% cholate, 0.5% Tween-80. Reduced spectrum of cyanocytochrome aa_3 measured 15 seconds and of cytochrome aa_3 15 minutes after addition of $Na_2S_2O_4$. (—) Cytochrome aa_3 and (– – –) cyanocytochrome aa_3. From Fig. 9 of van Buuren et al. (27).

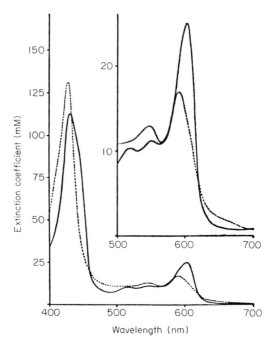

FIG. 3.   Spectra of cytochrome aa₃ containing reduced cytochrome a₃—carbon monoxide complex and cytochrome a in reduced and oxidized forms. Extinction values are based on the total heme a content. (- - -) Reduced cytochrome a₃—carbon monoxide complex and oxidized cytochrome a, and (—) reduced cytochrome a₃—carbon monoxide complex and reduced cytochrome a. From Horie and Morrison (28).

that the latter is not part of the cytochrome a_3 chromophore and that the energy transfer between the two separate hemes is not feasible. On the other hand, spectroscopic changes suggesting heme–heme interactions are seen in the presence of azide (cf. Fig. 4) (29). This ligand, like cyanide (Fig. 2) reacts with both $a^{2+}a_3^{3+}$ and $a^{3+}a_3^{3+}$. With azide as inhibitor, however, the spectrum attributed to ferrocytochrome a is modified, the α peak being shifted some 4 nm toward the blue (29, 30).

Electron paramagnetic resonance (EPR) spectra of cytochrome oxidase have also been obtained in different oxidation and ligand states. There are characteristic signals attributable both to the copper (31) and to iron atoms present (29). Figure 5 shows the EPR spectra of the enzyme in the fully oxidized state and in the partially reduced state (with added ascorbate) (29). This disappearance of the copper signal at $g = 1.99$ and of the iron signal at $g = 3.0$ is accompanied by the appearance of a new signal, characteristic of high spin ferric iron at $g = 6.0$. All the signals, resulting both from copper and from iron, disappear on full reduction of the enzyme as determined spectrophotometrically. Quantitatively, it has been found that only part of the iron and only part of the copper ($\sim 50\%$ in each case) are detectable by EPR in the fully oxidized enzyme. Hence, it has been proposed that two kinds of copper atom (EPR-detectable and EPR-undetectable) are present in the enzyme, as well as two kinds of heme group (31, 32).

Magnetic susceptibility studies (33) showed fully oxidized enzyme ($a^{3+}a_3^{3+}$) to be much more paramagnetic than $a^{2+}a_3^{2+}$ CO. The species $a^{2+}a_3^{3+}$ HCN was also almost completely diamagnetic, while the reduced $a^{2+}a_3^{2+}$ and oxidized cyanide derivatives had intermediate susceptibilities.

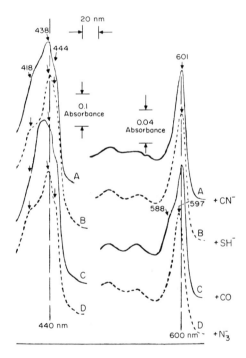

FIG. 4. Absolute spectra of cytochrome aa_3 in the aerobic steady state in the presence of inhibitors; 16 μM oxidase plus 30 mM ascorbate, 200 μM TMPD and inhibitor. Aerated for several minutes before being frozen at liquid N_2 temperature. Reference solution, 0.5% Tween-80, 0.1 M phosphate (pH 7.4) plus 30 mM ascorbate and 200 μM TMPD. (—) Curve A, plus 1 mM cyanide; (- - -) curve B, plus 1 mM sulfide; (—) curve C, bubbled with CO gas; (- - -) curve D, plus 1 mM azide. From Fig. 6 of Gilmour *et al.* (29).

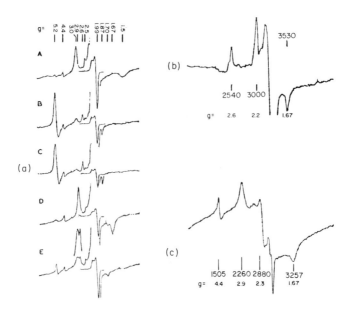

Fig. 5. (a) EPR spectra of 0.5 mM cytochrome aa$_3$. A, Isolated enzyme; B, 50% reduced with 1.5 mM ascorbate under anerobic conditions; C, 70% reduced anerobically with ascorbate; D, 50% reduced anerobically with ascorbate plus 0.1 M sodium azide; and E as D but only approximately 20% reduced. Temperature 81°K. Copper signal recorded at a 20-fold diminished amplifier gain. In E the signal at $g = 3$ was recorded at a 4-fold increased amplifier gain. (b) EPR spectrum of cytochrome aa$_3$ approximately 70% reduced anerobically with NADH and phenazine methosulfate showing the low spin ferric signal of an intermediate in the absence of inhibitors. Field markers in gauss; frequency 9250 MHz. (c) EPR spectrum of cytochrome c oxidase approximately 50% reduced anerobically with NADH and phenazine methosulfate in the presence of 0.1 M sodium azide, showing the low spin ferric signal of an intermediate in the presence of azide. Frequency 9242 MHz, temperature 85°K. From Figs. 1–3 of van Gelder and Beinert (32).

The results suggest that none of the paramagnetic species is fully coupled in the oxidized form, although interactions may occur that broaden the EPR signals, and that ferric and ferrous a$_3$ are both high spin, while ferric and ferrous a are low spin.

The prosthetic group, heme a, is the same for both cytochromes a and a$_3$ (Fig. 6) (23). The protein groups to which it is liganded, as well as the groups associated with the copper atoms, are not known. In addition to heme and copper, isolated cytochrome oxidase also contains phospholipid. Lipid depletion induces loss of activity which is restored when mitochondrial phospholipids are added back (34). Evidence has been obtained for

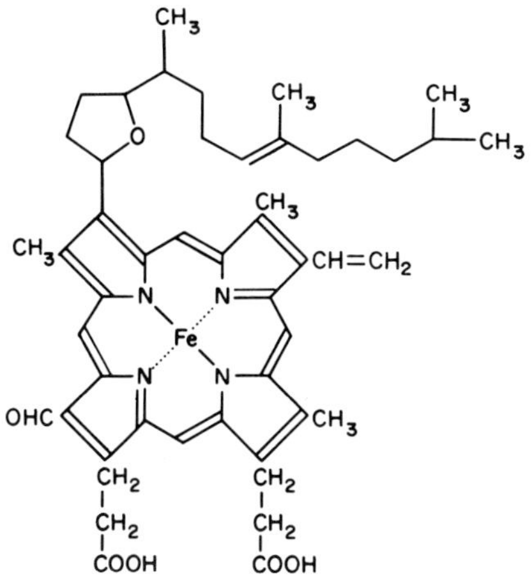

FIG. 6. Possible structure of native heme a. From Lemberg (23).

a role of negatively charged phospholipids (cardiolipin) in the binding of cytochrome c to the oxidase (35), although whether this accounts for the whole activating effect is uncertain. There is little doubt that 1–2 moles of cardiolipin per heme a equivalent are tightly bound to cytochrome oxidase (36, 37) (i.e., are not removable by a 2:1 chloroform–methanol mixture) and that further quantities of cardiolipin can be added to produce full activity in such preparations. No one has prepared a completely lipid-free oxidase, active or inactive, but the Yonetani preparation (38) appears to be the most reliable.

The minimum molecular weight of the enzyme appears to be about 250,000.* This minimum unit contains two heme a groups (a and a_3), two copper atoms, and between 2 and 100 moles of phospholipid (2–30% w/w) (39, 40a). In the electron microscope, the oxidase molecule appears to be approximately cylindrical (70 Å diameter by 95 Å long) and capable of aggregating along the long dimension to form sheets or partial lamellar structures (40a, 40b). Studies (41, 42) of oxidase subunits have indicated that one equivalent of enzyme (cytochrome aa_3) must contain at least ten polypeptide chains. The chains associated with the copper may be separable from those associated with the hemes (43), but the detailed relationship remains uncertain.

---

* The range of molecular weights determined by direct methods is about 250,000–600,000 for the mammalian enzyme. The range of equivalent weights on a two-heme or aa_3 basis is 160,000–280,000 in grams of protein per 2 moles heme a.

III. REACTION CHARACTERISTICS OF THE ISOLATED ENZYME

Although there have been claims for the existence of catalytically active "monomeric" oxidase (44—46), extensive studies have been carried out only with preparations of a molecular weight greater than 200,000. These contain both cytochromes a and a_3 as well as copper atoms. Surprisingly, there is still no complete study of the overall kinetics of reaction (1) (i.e., systematically varying the concentrations of both cytochrome c and oxygen). A considerable amount of piecemeal information is available.

$$4 \text{ cytochrome c } Fe^{2+} + 4 H^+ + O_2 \rightarrow 4 \text{ cytochrome c } Fe^{3+} + 2 H_2O \qquad (1)$$

A. Overall Reaction Kinetics

When ascorbate (with or without mediators such as TMPD)† is used to reduce cytochrome c, a plot of oxidase activity against cytochrome c concentration has the conventional Michaelis form (7, 47–49). Nevertheless, even at saturating levels of c in the spectrophotometric assay measuring c oxidation directly, the time course of disappearance of reduced c is first order (50, 51). Such a deviation from the expected zero-order kinetics is characteristic of product inhibition by ferric cytochrome c as the reaction proceeds. This product inhibition is shown by swollen mitochondria and submitochondrial particles as well as by the soluble enzyme (52). It also seems to be characteristic of cytochrome c peroxidation by yeast cytochrome c peroxidase under some conditions (53, 54), although in other circumstances the latter enzyme (55) as well as horse radish peroxidase (56) show zero-order oxidation kinetics. At pH values below 8.0, the binding constants for the oxidized and reduced forms of cytochrome c appear to be almost identical (57). Several kinetic models can be fitted to these observations, but the simplest one that accounts for the behavior of the oxidase both in isolation and as part of the respiratory chain is given by Eq. (2), where $K_m \simeq K_i = k_{-1}/k_1$.

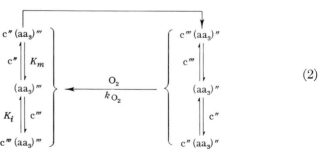

$$(2)$$

† The following abbreviations are employed: TMPD, N,N,N',N'-tetramethyl-p-phenylenediamine; PMS, phenazine methosulfate.

✓ Three features of this mechanism can be determined by overall kinetic methods: the maximum turnover of the system, the binding constant for the reaction with cytochrome c, and the rate of reaction with oxygen. The first preparations of oxidase (20) had somewhat lower turnovers than found in the mitochondrion *in situ*. Yonetani's enzyme had a maximum turnover of 60 sec^{-1} (electrons per heme a in 50 mM phosphate pH 7 at 25°) (58, 59), while more recent preparations have turnovers almost twice this value (60, 61). Table I lists the range of values now obtained for isolated enzyme preparations.

✓Maximal rates of oxidation are dependent on pH (about a twofold increase occurring per unit decrease of pH between pH 8 and pH 6) (57, 64), on temperature (an 80% increase per 10° rise, equivalent to an activation energy of 11 kcal) (62), and on ionic strength (optimal ionic strength is between 0.1 and 0.2 at pH 7.4) (56, 63). Highest turnover numbers are obtained by extrapolating values obtained in the presence of ascorbate, TMPD, and c to infinite concentrations of reductant and c; extrapolations to 100% reduction and infinite c concentration in the spectrophotometric assay may give slightly lower values. These values for isolated oxidase are comparable with those obtained for the oxidase *in situ* (below).

✓The apparent optimum pH for the enzyme declines with increasing ionic strength from about 6.6 in 10 mM phosphate to 6.0 in 100 mM phosphate and above (64, 65). The observations in each case are consistent with the existence of a group with a pK about 0.6 pH units higher than the

✓TABLE I

TURNOVER OF CYTOCHROME OXIDASE[a]

pH	% Activity[b]	T(°C)	% Activity[c]	mM Phosphate	% Activity[d]
8.0	45	20	56	5	18
7.4	80	25	75	10	27
7.0	(100)[e]	30	(100)[e]	20	50
6.3	160	35	135	30	70
6.0	185	40	180	100	(100)[e]

[a] Data from Smith and Newton (62), Mason and Ganapathy (60), Yonetani (59), Yonetani and Ray (57), and Nicholls (63).

[b] 30°C, 0.06 to 0.10 M phosphate.

[c] pH 7, 0.06 to 0.10 M phosphate.

[d] pH 7 to 7.4, 25° to 30°C.

[e] Turnover at pH 7.0, 0.1 M phosphate, 30°C [= 400 sec^{-1} (electrons/second/cytochrome aa$_3$) or 200 sec^{-1} in terms of heme a], taken as "100%" in each case.

"optimal" pH (on the alkaline side) and a rather less well defined group on the acid side. It is probably necessary to assume a true effect of ionic strength on the overall maximal velocity, although it is possible that the effect might be attributable to a greater influence of ionic strength on the acid than on the alkaline pK.

✓The apparent Michaelis constant for cytochrome c is almost directly proportional to buffer concentration (52, 56). Studies of the effects of various metal cations and of organic polycations (below) suggest that one reason for this is a competition between cytochrome c and cations, including the monovalent K^+ and Na^+, for a binding site on the oxidase (66, 67).

✓Table II lists some of the K_m values obtained for isolated enzyme preparations. Both ferrous and ferric cytochrome c interact with the oxidase, the binding constants (K_m and K_i) diverging only at pH 8, where the ferric form is more tightly bound (57). Between pH 8 and pH 6 the K_m for cytochrome c in potassium phosphate buffer is given by Eq. (3).

$$K_m(\text{cyt. c}) = K_m' + \alpha[K^+] \qquad (3)$$

The observed value of α varies from 0.12 to 0.18 $\times$ 10^{-3}; the value of K_m' (the binding constant at zero ionic strength) is uncertain, but less than 1 μM. Changes in pH are almost without effect throughout this range

✓TABLE II

Apparent K_m Values for Cytochrome c Oxidation[a]

pH	mM Phosphate	mM [K+]	Reported K_m (μM)[b]
6.0	78	89	8.5
6.3	66	82	8.2
7.0	45	72	8.3
7.4	38	68	7.1
8.0	35	68	6.0
	5	9	~1
	10	18	~3
7.4	18	32	6
	30	54	11
	100	180	30
	350	630	77

[a] Data from Yonetani and Ray (57) and Nicholls (56).

[b] For reaction of ferrous cytochrome c with enzyme under catalytic conditions: 25°C, Yonetani-type anzyme, Smith-Conrad assay procedure.

(57, 68). The slight *increase* of K_m with decreasing pH observed by Yonetani and Ray (57) may be attributed to their preserving constant ionic strength (and hence increasing the total phosphate concentration as pH was lowered); a slight *decrease* of K_m with decreasing pH observed by McGuinness and Wainio (68) may be attributed to their preserving constant phosphate concentration (and hence decreasing the cation concentration as pH was lowered). No studies of the effect of temperature on K_m or K_i seem to have been made.

✓Ionic and pH changes also affect cytochrome c itself. At alkaline pH values, ferric cytochrome c is transformed into a species inert to enzymic reduction and reduction by ascorbate (69, 70). This form of c (pK for transition ~8.7) may be responsible for the decrease of K_i with increasing pH (Table III), implying that the inactive form is bound more tightly than normal ferricytochrome c. Both ferric and ferrous cytochrome c are also capable of binding ions (71) at neutral pH. The dissociation constant for phosphate binding (obtained from the decline in reduction rate by ascorbate with increasing phosphate concentrations) is 2–3 mM. If the ferrocytochrome c phosphate complex cannot readily be oxidized by the oxidase, this would result in a very similar dependence of K_m on ionic strength to that given in Eq. (3), which assumes cation binding to the enzyme. At present we can be certain only that *both* cation binding to enzyme and anion binding to c cannot be kinetically important between 0.01 M and 0.2 M phosphate buffer concentration [if they were, Eq. (3) would have to be replaced by an expression including square terms in ion concentrations].

TABLE III

APPARENT K_d VALUES FOR CYTOCHROME c BINDING

Method	Conditions	Apparent K_d or K_i (μM)	References
A[a]	pH 6, $\mu = 0.1$, 25°	8.3	
	pH 7, $\mu = 0.1$, 25°	7.9	Yonetani and Ray (57)
	pH 8, $\mu = 0.1$, 25°	3.5	
A[a]	pH 7.4, 5 mM phosphate	0.5–1.0	Mochan and Nicholls (78)
B[b]	pH 7.4, 5 mM phosphate	0.3–0.5	[also refs. (59), (56), and
C[c]	pH 7.4, 5 mM phosphate	0.1–0.4	(78)]

[a] By measurement of inhibition of the cytochrome oxidase activity by ferric cytochrome c (K_i value).

[b] By following the decline in ascorbate reducibility on binding to the oxidase (K_d value).

[c] Estimates based on co-chromatography of cytochrome c and oxidase (K_d value).

Under conditions of saturation at pH 7.4 (25°) (TN $\sim$240 sec^{-1}),* the apparent K_m for molecular oxygen is between 0.5 and 1.0 μM (72, 73). The velocity constant for the oxygen reaction is given by Eq. (4).

$$k(O_2) = TN/4 \, K_m \, (M^{-1} \, sec^{-1}) \tag{4}$$

The value of k for the isolated enzyme is thus of the order of $10^8 \, M^{-1}$ sec^{-1}. This rate constant appears to be pH-independent, but its dependence on ionic strength is not known. Direct measurement of the oxidation of reduced enzyme by O_2 has given the same value ($1.0 \times 10^8 \, M^{-1} \, sec^{-1}$ at 28°C) with a very small temperature coefficient ($Q_{10} < 1.4$) (74).

The corresponding rate constant for the interaction of cytochrome c with the enzyme must also be large. From Eq. (2) we have a maximum rate (moles O_2/aa_3/second) given by Eq. (5) for the reaction in the presence of ascorbate and TMPD:

$$k_{max} = k_c k_o/(4k_o + k_c) \tag{5}$$

where k_c is the rate of oxidation of one equivalent of c and k_o the slowest step in the reduction of a molecule of oxygen. In the spectrophotometric assay, four extra reactions (dissociations of c, governed by k_{-1}) must occur, giving:

$$k'_{max} = k_{-1}k_c k_o/(4k_c k_o + 4k_{-1}k_o + k_c k_{-1}) \tag{6}$$

Since k'_{max} is not less than $0.8 \, k_{max}$ (Table I), k_{-1} must be at least $4 \times k_c$ or $\geq 10^3$ sec^{-1}. If $K_m = 8 \, \mu M$ (Table II), k_1 must be at least $10_4 \, M^{-1} \, sec^{-1}$, of the same order of magnitude as the rate of reaction between cytochrome c and cytochrome c peroxidase (55, 75).

The rates of reaction of oxygen and cytochrome c with the oxidase are thus of similar magnitude. In both cases diffusion limitations probably play a large part in controlling the rate. The mechanism of Eq. (2) fits the experimental observations at 25°, pH 6.3, in 66 mM phosphate, if we assume $K_m = K_i = 8 \, \mu M$ (with $k_1 > 10^8 \, M^{-1} \, sec^{-1}$ and $k_{-1} > 10^3$ sec^{-1}), k_c (turnover) $= 480$ sec^{-1}, and k_{O_2} (apparent) $= 4 \times 10^8 \, M^{-1}$ sec^{-1} (four times the rate constant for reaction of O_2 with ferrous cytochrome a_3).

B. Intermediate Reactions of the Enzyme

Studies of the partial reactions involved in cytochrome oxidase activity have included observations on:

(a) binding of cytochrome c,
(b) paths of reduction of fully oxidized enzyme ($a^{3+} \, a_3^{3+}$),
(c) the stoichiometric titration of oxidized enzyme with reductant,

* TN = turnover number in moles cytochrome c per mole aa_3 (two heme) per second.

(d) paths of oxidation of fully reduced enzyme (a^{2+} a_3^{2+}),

(e) apparent binding of oxygen (the so-called oxygenated form),

(f) steady states of a, a_3, and Cu (in the presence of O_2 and c), and

(g) the redox potentials of groups on the enzyme.

A chemical mechanism for the oxidase must extend the scheme of Eq. (2) to account for the reaction with four molecules of cytochrome c and to describe the fates of the four reducing equivalents on their way to oxygen, while retaining the experimentally observed simplicity of the overall kinetics.

Direct studies of cytochrome c binding have been made by chromatographic (76, 77) and ultracentrifuge (56) techniques as well as by following the decline in the ascorbate reducibility of bound cytochrome c. Nicholls (56) reported 0.9–1.5 moles cytochrome c bound per mole of cytochrome aa_3 (0.4–0.7 moles per equivalent of oxidase heme a). By sonicating their mixtures, Kuboyama et al. (77) obtained up to 2 moles c bound/mole aa_3. Similarly, Orii et al. (76, 77) believed that they had obtained a complex containing 1 mole c/oxidase heme a equivalent. More recently, Mochan and Nicholls (78) again reported a value of 1 mole c/mole aa_3 on the basis of ascorbate reducibility changes. The latter ratio seems to correspond with that usually found in mitochondria. As discussed by Lemberg, however, larger amounts of c are sometimes found (23). To account for such ratios, Nicholls proposed two kinds of binding site of different affinities for the "intact" respiratory chain (65). So far no evidence has been adduced for two binding sites on the oxidase itself. The complex formed by sonication (77), however, seems to be less readily reversible than that formed by simple mixing (56).

The affinities shown by the oxidase in such direct binding studies are similar to the K_m and K_i values obtained kinetically (Table III). The intact membrane (below) displays similar binding behavior (79) presumably resulting from its oxidase components; some evidence suggests somewhat tighter binding by the intact membrane (75).

Cytochrome c binding is largely electrostatic with some contribution from hydrophobic forces (66, 67, 77). Acetylated cytochrome c has lost its catalytic activity (80) as well as the ability to inhibit the activity toward normal ferrocytochrome c. Guanidylated c retains both catalytic and inhibitory capabilities (81). Polycations, both natural (salmine) (52, 82) and synthetic (high molecular weight polylysine) (66, 83, 84, 86) inhibit cytochrome oxidase competitively toward cytochrome c. Lysine copolymer studies have implicated other groups in the inhibitory activity (85, 86). The very high affinity shown by the isolated oxidase and by oxidase-containing mitochondrial fragments has permitted the titration of cytochrome c binding sites by high molecular weight polycations; the same number of anionic sites bind polylysine as bind cytochrome c in Keilin–Hartree particles (83, 87). Specificity in the cytochrome c reaction itself may be achieved either because the positively charged "patch" on the molecule is correctly positioned for activity or because a subsequent mobility of c on the protein or membrane [the idea of "reduction of dimensionality" put forward by Adam and Delbrück (88)] ensures the final "fit." The very long synthetic polycations (MW 150,000+) may also show "super-stoichiometric" behavior by blocking more than one c binding site in solution or on the mitochondrial membrane (87).

Cytochrome c in the bound state has a normal spectrum in both oxidized and reduced forms (56, 77, 89). Reactivity with cyanide is retained, as well as a normal rate of cyanide dissociation from the cyanferrocytochrome c derivative (78, 90). The rate of reduction by ascorbate (78), however, is diminished by at least 80 or 90%. No effect of cytochrome c binding on the spectrum of either heme component of the oxidase has been shown [although it is possible that the rate of electron transfer from cytochrome a to a_3 is affected (25), and Lemberg (91) has claimed that the rate of transition from the "oxygenated" (428 nm) form to the ferric form is accelerated by cytochrome c]. In the intact membrane systems, an apparent decrease in the redox potential of bound c has been reported (implying a tighter binding of the ferric form) (92); this cannot be the case for the isolated oxidase, which binds both oxidation states equally well (57). Table IV summarizes the apparent ("allotopic") changes of c on enzyme or membrane binding.

After cytochrome c has formed a complex with the oxidase, a rapid transfer of electrons takes place, in which cytochrome c is oxidized and

TABLE IV

ALLOTOPIC CHANGES IN CYTOCHROME c PROPERTIES ON
BINDING TO OXIDASE OR MEMBRANE

Property	Effect of binding	Reference
1. Oxidized spectrum	None	Keilin and Hartree (89)
2. Reduced spectrum	None	Keilin and Hartree (89)
3. 695 nm band (Fe^{3+})	No effect (?)	Chance et al. (93)
4. CN^- reaction	No effect	Nicholls et al. (90)
5. Ascorbate reducibility	Declines by 90%	Mochan and Nicholls (78)
6. Redox potential	30 mV more negative	Dutton et al. (92)
7. CO binding	Increases in complex (?)[a]	King et al. (94)

[a] Not confirmed for membrane-bound c.

cytochrome a reduced [governed by the constant k_c in Eq. (2)]. This is followed in the aerobic system by the slow attainment of a steady state (95, 96). The initial reduction of cytochrome a by cytochrome c has a bimolecular rate constant of at least 10^7 M^{-1} sec^{-1}. However, despite the high levels of cytochrome a reduction attained, subsequent rates of electron transfer to oxygen are quite slow. This was first pointed out by Yakushiji and Okunuki (19) and confirmed for the isolated oxidase by Yonetani (58). Presteady state kinetics (95, 96) have led to similar conclusions, namely:

(a) The reduction of a is sometimes faster than the overall turnover of the enzyme.

(b) Reduced a in the a^{2+} a_3^{3+} system does not transfer electrons to a_3 or to O_2 as rapidly as in the a^{2+} a_3^{3+} system generated by oxygen [Eq. (7)].

(c) The steady state flux to oxygen depends upon reduction of other enzyme components (a_3 or Cu) by cytochrome c and not by cytochrome a.

Gibson et al. (95) concluded that oxidized cytochrome c blocked electron transfer within the oxidase molecule itself. It is, however, likely that ferric c merely blocks the binding of ferrous c to the enzyme; under the reported conditions only a small part of the enzyme was in the form of a c enzyme complex. Antonini et al. (96) proposed that only the completely reduced form $a^{2+}a_3^{2+}$ could react with oxygen. But their system was incapable of generating appreciable quantities of the species $a^{3+}a_3^{2+}$, and therefore the simpler hypothesis that the reaction with oxygen is dependent only on the presence of a_3^{2+} has not been ruled out. Lemberg and Cutler (97) found that two electrons are rapidly transferred from ferrocytochrome c to the oxidase (a + Cu) while the reduction of ferric a_3 and its copper atom was much slower. They believed that the ferricytochrome a_3 formed by the reaction with oxygen (a_{3x}^{3+}, an "oxygenated" species) is, however, much more readily reduced by cytochrome c under turnover conditions. Malmström (98) has similarly proposed that the reduction of a^{3+} + Cu^{2+} is the initial event, the oxidation of a_3^{2+} + Cu^+ by oxygen the final event, and that the binding of oxygen to the enzyme promotes electron transfer from the aCu to the a_3Cu system, in addition to causing the concerted oxidation of the latter pair of electron carriers. However, Chance, in collaboration with Erecinska (99), has adapted the rapid flow technique for the study of the kinetics of cytochromes in intact mitochondria at temperatures down to $-25°$. Oxygen pulses added to the reduced cytochromes showed oxidations at considerably slower rates than those observed at room temperature and, in addition, hitherto unidentified spectroscopic species. A hypothetical reaction sequence involving a stepwise reduction of oxygen to water, in which cytochrome a_3 is the electron donor, is indicated in Eq. (7). The first two intermediates are tentatively identified by their absorption maxima; the remaining two have not been observed spectroscopically and are therefore bracketed.

$$
\begin{array}{ccccccc}
 & (1) & & & (2) & & \\[2pt]
 & k_1 \sim 5 \times 10^6\ M^{-1}\ \mathrm{sec}^{-1} & & & k_2 \sim 20\ \mathrm{sec}^{-1} & & \\
a_3{}^{2+} + O_2 & \longleftarrow\!\!\!\longrightarrow & a_3{}^{2+}\!\cdot O_2 & \longleftarrow\!\!\!\longrightarrow & a_3{}^{3+}\!\cdot O_2{}^{-} \\
(445\ \mathrm{nm}) & ? & (428\ \mathrm{nm}) & ? & (410\ \mathrm{nm})
\end{array}
$$

$$
\begin{array}{cccc}
 & k_3 \sim 5\ \mathrm{sec}^{-1} & ? & \\
a_3{}^{3+}\!\cdot O_2{}^{-} & \longrightarrow & [a_3{}^{2+}\!\cdot O_2{}^{-}] \longrightarrow [a_3{}^{3+}\!\cdot O_2{}^{2-}] \\
(410\ \mathrm{nm}) & \underset{Cu^{1+}}{a^{2+}}\!\diagdown\!\underset{Cu^{2+}}{a^{3+}} & (?) \\
 & (3) & (4)
\end{array}
$$

$$(7)$$

The observed absorption maxima are shown together with the suggested chemical species. The proposed reaction sequence consists of:

(1) A second-order reaction of $a_3{}^{2+}$ with oxygen (about a hundredfold slower than that at room temperature). The reaction product has an absorption maximum at 428 nm and other spectroscopic characteristics similar to those of $a_3{}^{2+}\!\cdot CO$, and is thus tentatively identified as $a_3{}^{2+}\!\cdot O_2$.

(2) This compound is observed to maintain a steady state level during its conversion to a reaction product absorbing at 410 nm, different from that of the $a_3{}^{3+}$ species and tentatively identified as $a_3{}^{3+}\!\cdot O_2{}^{-}$. However, in this case there is no spectroscopic analog on which to base the identification, and it is assumed that this is the one-electron oxidation product of ferricytochrome oxidase with oxygen, the superoxide anion remaining bound to the iron atom.

(3) The slower reduction of $a_3{}^{3+}\!\cdot O_2{}^{-}$ by either copper or cytochrome a (only preliminary kinetic data are available to determine which) leads to the formation of ferrocytochrome a_3 and, according to this mechanism, a species which is still bound to the oxidation reduction product.

(4) A second electron is then transferred to oxygen, reducing it to the level of peroxide and reoxidizing cytochrome a_3 to the ferric form. The velocity constant for this reaction is not known, and no evidence is available on the spectrum of cytochrome oxidase combined to oxygen at the "peroxide" level.

The mechanism rests upon the kinetic observations that the first product of cytochrome $a_3{}^{2+}$ oxidation is not a ferric species but rather a species absorbing at 428 nm, and that the reaction product has an absorption maximum that is displaced to a shorter wavelength than is recognized for the absorption band of $a_3{}^{3+}$. This mechanism identifies $a_3{}^{2+}$ as the principal electron donor to oxygen and favors a sequential rather than a concerted mechanism in intact mitochondria. Thus, the mechanism of Lemberg and Cutler (97) and Malmström (98) may apply only to the isolated oxidase.

Whatever the intermediate steps involved, the overall reaction of a molecule of oxygen [Eq. (1)] involves four reducing equivalents. Since the oxygen reaction can take place in the absence of cytochrome c, these four reducing equivalents must be stored in one or more molecules of oxidase (100). Earlier ideas (101, 102) that more than one a_3 heme might be implicated in the reaction with a single oxygen molecule now seem unlikely. Each a_3 moiety is therefore associated with at least four one-electron donors—conventionally assumed to be the a_3 heme itself, cytochrome a, and two copper atoms.

Van Gelder and Muijsers (103, 104) have demonstrated the existence of four such groups in titrations with stoichiometric amounts of NADH catalyzed by PMS. Under these conditions cytochromes a and a_3 are reduced concurrently, as almost linear increases in absorbance are seen at 605 and 445 nm. The corresponding EPR studies of van Gelder and Beinert (32) show that the intermediate form (Fig. 5) is not a mixture of fully reduced and fully oxidized enzyme but contains partially reduced species ($a^{2+}a_3^{3+}$ and $a^{3+}a_3^{2+}$). Reduction of EPR-detectable copper is not linear with added NADH, but sigmoidal (32). Most of the signal disappears over the range in which between one and three electrons are added per aa_3 "unit." If the calculations of van Gelder and Muijsers are correct, this implies that the EPR-silent copper is the first species to be reduced. Measurements at 830 nm are not conclusive. It is also possible that the reduction of the other paramagnetic species (a_3^{3+}?) responsible for the silence of the second copper atom may render that atom EPR-detectable (thus "buffering" the decline in signal with reduction).

The signals from the "half-reduced" species (retaining two electrons per aa_3 unit) may be expected to be complex (six half-reduced forms are possible, or four if the copper atoms are indistinguishable). At least one set of new EPR signals is seen (32, 105) corresponding to the appearance of a high spin ferric iron at $g = 6$ and $g = 2.6$, presumably reflecting a_3 hemes in a new interacting state with a or copper (or *released* from paramagnetic interaction in the fully oxidized form). Whether the various partially reduced forms are in partial thermodynamic equilibrium, or whether they represent a kinetic ensemble without interconvertibility, is uncertain. Redox potential measurements (below) do not rule out the latter possibility.

Transfer of electrons from the reduced enzyme to oxygen proceeds in several steps. In the initial reaction about 65% of the absorbance (reduced minus oxidized) disappears at 445 nm, 35% at 605 nm (74), but relatively little at 830 nm (106). At low O_2 concentrations this takes place with a rate constant of 10^8 M^{-1} sec^{-1} at 25°, showing little temperature effect At higher oxygen concentrations, a limiting rate of 30,000 sec^{-1} is seen ("K_m" for oxygen of 300 μM), but the spectrum remains that of fully reduced enzyme (107). The lifetime of any intervening oxyferrous intermediate must therefore be less than 10 μsec at 25°, although at $-25°$ (99) a possible species of this type has been detected [Eq. (7)].

Following the initial oxygen-dependent reaction (also in intact mitochondria), a complex pattern of oxidations occurs at 445, 605, and 830 nm. Gibson and his co-workers (106, 107) have dissected these into three processes—a fast step ($t_{1/2}$, $\sim$0.1 msec) involving the appearance of absorption at 830 nm and also at 605 nm (attributed to $Cu^+ \rightarrow Cu^{2+}$) (106), a slower step ($t_{1/2}$, $\sim$1 msec) at 605 nm and 445 nm (attributed to $a^{2+} \rightarrow a^{3+}$) (74, 106), and a very slow step ($t_{1/2}$, $\sim$30 sec) involving changes in the Soret region (attributed to the appearance of low spin ferric species like the "oxygenated" form) (108). Table V lists the rates at various wavelengths and the species believed to be responsible.

These findings are consistent with the model of Eq. (8)—an alternative to Eq. (7)—where a_3^{3+}, which may be a 428 nm species, retains one or three additional oxidizing equivalents:

$$
\begin{array}{ll}
\text{(a)} & a^{2+}Cu_2^+a_3^{2+} \xrightarrow{O_2} a^{2+}Cu_2^+a_3^{2+}(O_2) \\[4pt]
\text{(b)} & a^{2+}Cu_2^+a_3^{2+}(O_2) \rightarrow a^{2+}Cu_2^+\overline{a_3^{3+}} \\[4pt]
\text{(c)} & a^{2+}Cu_2^+\overline{a_3^{3+}} \rightarrow a^{2+}Cu_2^{2+}a_3^{3+} \\[4pt]
\text{(d)} & a^{2+}Cu_2^{2+}\overline{a_3^{3+}} \rightarrow a^{3+}Cu_2^{2+}a_3^{3+} \\[4pt]
\text{(e)} & a^{3+}Cu_2^{2+}a_3^{3+} \xrightarrow{\text{slow}} a^{3+}Cu_2^{2+}a_3^{3+} \text{ (low spin)}
\end{array}
\qquad (8)
$$

Whether both copper atoms are oxidized together (as indicated here) or not is unknown. Equation (8′) gives an alternative formulation of the last two reactions [Lemberg (23)]:

$$
\begin{array}{ll}
\text{(d′)} & a^{2+}Cu_2^{2+}\overline{a_3^{3+}} \rightarrow \overline{a^{3+}Cu_2^{2+}a_3^{3+}} \text{ (low spin)} \\[4pt]
\text{(e′)} & \overline{a^{3+}Cu_2^{2+}a_3^{3+}} \xrightarrow{\text{cyt. c}} a^{3+}Cu_2^{2+}a_3^{3+} \text{ (normal ferric)}
\end{array}
\qquad (8′)
$$

The final reaction product from the fully reduced soluble enzyme is a compound with an absorption maximum at 428 nm instead of the 418–423 nm characteristic of the ordinary ferric form (25, 111, 112). This 428 nm compound was originally thought to represent a complex between the enzyme and molecular oxygen, and was therefore called "oxygenated"

TABLE V

RATES OF OXIDATION FOR DIFFERENT COMPONENTS OF THE
OXIDASE BY MOLECULAR OXYGEN

Preparation	$k(O_2)^a$ $M^{-1} sec^{-1}$	$k(a_3)^b$ sec^{-1}	$k(a)^b$ sec^{-1}	$k(Cu)^b$ sec^{-1}	$k(c)^b$ sec^{-1}
$aa_3{}^c$	1.0×10^8	3×10^4	700	7×10^3	?
$RLM_w{}^d$	1.5×10^8	$>3 \times 10^3$	1200	?	300
$PHM_w{}^e$	0.33×10^8	$>10^3$	70	?	140

[a] The second-order rate constant observed for all components at low O_2 concentrations.
[b] First-order rate constants observed at 445 nm (initial phase), 605 nm (final phase), 830 nm, and 550 nm, respectively, at high O_2 concentrations.
[c] Gibson and Greenwood (74,106) (isolated oxidase; flow-flash).
[d] Chance et al. (109) (rat liver mitochondria; pulsed flow).
[e] Chance and Erecinska (110) (pigeon heart mitochondria; flow-flash).

oxidase (25). Lemberg later proposed that it might be a peroxide compound similar to metmyoglobin peroxide or complex II of peroxidase [cf. the products of reactions (8b) and (8c)] (113). Most recently the view has been put forward that in the 428-nm compound both hemes are ferric and both copper atoms cupric, but that one heme (probably that of cytochrome a_3) is a more low spin than usual [cf. the products of reactions (8d) and (8e)] (114).

Although the existence of such low spin ferric configurations is probably now generally accepted, their role in the overall oxidase reaction is quite uncertain. Lemberg and his co-workers find the 428-nm compound to be the first oxidation product at times less than 100 msec (115). Wharton and Gibson (108) reported "ferric" enzyme at such a reaction time, with the production of "oxygenated" form occurring much more slowly ($t_{1/2} \sim 30$ seconds) as in reaction (8e). Both Davison and Wainio (112) and Lemberg and Gilmour (91) consider the 428-nm compound thermodynamically unstable. Its relaxation to ferric enzyme is accelerated by oxidants and reductants, especially cytochrome c (either ferrous or ferric) (91). But according to Lemberg (23) the ferric form is not a necessary overall reaction intermediate since the 428-nm compound itself can be reduced to ferrous enzyme by dithionite (although perhaps not by c or by NADH + PMS). He therefore inverts the sequence of Eqs. (8d) and (8e) as indicated in Eq. (8′), where $\overline{a^{3+}\ Cu_2{}^{2+}\ a_3{}^{3+}}$ represents the 428 or "oxygenated" form.

Gilmour et al. (115) produced mixtures of ferric and 428 nm $a_3{}^{3+}$ by oxygenation under various conditions but still concluded that in the presence of excess oxygen most of the product has the 428-nm band. With

Vanneste (116) (who based his conclusions on static spectra) they assigned a 426-nm peak to ferric cytochrome a and a 414-nm peak to cytochrome a_3^{3+}. Accordingly, the 428-nm species must primarily involve the latter. Most recently, Erecinska and Chance (117) have identified the early appearance of a 428-nm band on oxygenation of intact mitochondria at low temperatures preceding the oxidation of cytochrome a; they concluded [cf. Eq. (7)] that, unlike the other 428-nm species (91, 108), this may be a kinetically competent $a_3^{2+}O_2$ compound. The spin state has not yet been determined by EPR because of the instability of the compound.

Some of the disagreement may in fact be resolved following the discovery of multiple "oxygenated" forms of the isolated enzyme by Muijsers *et al.* (118). The early (Lemberg and Chance) and late (Wharton and Gibson) appearance of 428 nm absorption may thus be assignable to different reactions. But another problem is presented by the fact that with both isolated and intact mitochondrial enzymes, the initial fast oxidation involves greater changes at both 445 and 605 nm than would be predicted if (a) cytochrome a_3 were the only species concerned, and (b) its spectrum is the same as that given by static difference methods (Figs. 2–4) (74, 117). Thus, 40% of the 605-nm peak in the difference spectrum disappears in the initial step, although the calculated contribution of a_3 at this wavelength is only 15%. Four kinds of explanation may be envisaged:

(a) that the product $(\overline{a_3^{3+}})$ has a much lower absorption than its static counterpart (a_3^{3+});

(b) that ligand (CN^-, N_3^-, and CO) binding to a_3 increases the absorbance of a^{2+} by about 25% [cf. Wilson *et al.* (119)],

(c) that the initial fast reaction nevertheless includes a partial oxidation of cytochrome a (cf. the maximal rate of oxidation of c^{2+} in its complex with yeast peroxidase of 14,000 sec^{-1}) (55), and

(d) that the static spectrum has a contribution from the blue Cu^{2+} species which remains reduced in the initial oxygen reaction (ΔE mM values at 605 nm of 22 for a, 9 for a_3, and 7 for Cu would be needed).

Explanations (a) and (d) are unlikely to be adequate (the 830-nm copper is *reduced* in the a^{2+} a_3^{3+} HCN complex) but a choice between (b) and (c) will require more thought and experimentation (see below).

In the steady state [Eq. (9)], in the presence of both oxygen and reduced cytochrome c,

$$AH_2 \rightarrow c \rightarrow aa_3 \rightarrow O_2 \qquad (9)$$

it is possible to correlate the flux with the percentage reduction of the several intermediates. The rate of reduction of c by ascorbate, for example, is proportional to the product of ascorbate and ferric cytochrome c concen-

trations [Eq. (10a)], while the rate of oxidation is given by the equation
for the mechanism of Eq. (2) above [Eq. (10b)]:

(a) $\quad v = k_r\,[\text{AH}_2]\,[\text{c}^{3+}]$

(b) $\quad v^1 = k_0\,[\text{c}^{2+}]\,[\text{aa}_3] / ([\text{c}^{2+}] + [\text{c}^{3+}] + K_m)$ $\qquad$ (10)

In the steady state the two velocities are equal and the ratio $[\text{c}^{2+}]/[\text{c}^{3+}]$
given by

$$[\text{c}^{2+}]/[\text{c}^{3+}] = k_r\,[\text{AH}_2]\,([\text{c}^{2+}] + [\text{c}^{3+}] + K_m)/k_0\,[\text{aa}_3] \qquad (11)$$

This kind of relationship seems to hold over a wide range of experimental
conditions (58, 61, 120, 121).

The steady state reduction of cytochrome a, measured at 605 nm, indi-
cates a more complex situation (58). When $[\text{c}^{2+}]/[\text{c}^{3+}]$ is small, the corresponding reduction of a is larger, indicating a redox potential [assuming
quasiequilibrium (121)] about 75 mV more positive than that of c (i.e.,
about $+330$ mV). When c is more reduced the reduction of the 605-nm
band increases to a maximum of about 60%. The remaining 40% (only
part of which can be assigned to a_3 by classic criteria) remains oxidized
even at quite high c^{2+} concentrations or, in absence of c, at high concentrations of TMPD and ascorbate (58). Only in the presence of inhibitors
(below) and possibly in the frozen steady state (122) and in the glycerol-
inhibited steady state (123) is full reduction seen.

In all such systems, the percentage reduction at 445 nm is about half
that at 605 nm. This implies that under these conditions all the 445-nm
absorption can be assigned to cytochrome a and that the reduction of
a_3 is negligible until the oxygen concentration has fallen to the micromolar
level (19). For a system obeying Eq. (4) above, at a turnover of 200 $\sec^{-1}$
with a $k(\text{O}_2)$ of $10^8\ M^{-1}\ \sec^{-1}$, only 0.2% reduction will be needed to carry
the flux in air-saturated buffer (250 μM). Claims to have measured changes
in steady state reduction of cytochrome a_3 in aerobic systems—involving
either isolated aa_3 or intact mitochondria—are thus implausible. No
measurements have been made of the steady state reduction of the third
oxidase component (copper) either at 830 nm or by EPR.

Redox potentials of the oxidase, in completely static systems, have
been obtained for the 605-, 445-, and 830-nm bands and for EPR-detectable copper. The original observation of Ball (124) that cytochrome a
at 605 nm had a midpoint potential of $+280$ mV was confirmed by Minnaert (125), who also found that the value of n at the midpoint was about
0.5, indicating that the reduction was not a simple one-electron exchange
between a and c. Horio and Ohkawa (126) showed that this could be accounted for if the oxidase contained heme species of differing redox po-

tential. Since the proportions of high and low potential species are similar at 605 and 445 nm, they cannot be simply identified with cytochromes a and a_3 (indeed Horio and Ohkawa believed three species to be present). Attempts to make this identification (127) involve postulating changes in extinction coefficients as well as ligand interactions with both a and a_3 (119).

Alternatively, it is possible that the potential of cytochrome a (and a_3) depends on the redox state of other oxidase components, especially the other heme group (128, 129); in fact, King observed such interactions in the copper heme signals (130). The copper measured by light absorption shows small or negligible interactions (43), titrating with an n value of 1.0 and a midpoint potential of $+280$ mV. Van Gelder et al. (131, 132) suggested that cytochromes a and a_3 initially have closely similar redox potentials, which diverge on partial reduction in a cytochrome c–dependent reaction. The results of Wilson et al. (105, 127) support the view that cytochrome a_3 has a midpoint potential which is always a little more positive than that of cytochrome a, but a contribution of both a and a_3 to both high and low potential species is still possible. Table VI summarized the observed redox equilibrium results in the absence of terminal inhibitors.

In the presence of inhibitors such as cyanide, azide, and carbon monoxide, the unliganded component titrates with a single midpoint potential and $n = 1.0$. First observed by Tzagoloff and Wharton (133) for CO, this was confirmed for azide and CO by Wilson et al. (119) The data of Caswell (134) indicate a similar situation for the cyanide inhibited state. Table VII lists redox potentials observed in these systems. According to classic

TABLE VI

Redox Potentials of Cytochrome Oxidase in the Absence of Added Ligands

Potential (mV)	Measured at (nm)	Attributed to	Reference
$+200$ to $+220$[a]	605 and 445	a (or a$\longleftrightarrow$$a_3$ equilibrium mixture)[b]	Wilson and Dutton (127)
$+350$ to $+400$[a]	445 and 605	a_3 (or $a_3\longleftrightarrow$a equilibrium mixture)[b]	Wilson and Dutton (127)
$+280$[c]	830	Cu	Tzagoloff and MacLennan (43)

[a] Intact mitochondria.
[b] Cf. Nicholls (129).
[c] Isolated aa_3.

TABLE VII

REDOX POTENTIALS OF CYTOCHROME OXIDASE IN THE PRESENCE
OF ADDED LIGANDS

Ligand	Potential (mV)	Measured at (nm)	Reference
HCN	+330[a]	445	Caswell (134)
HN_3	+350[a]	605	Wilson et al. (119)
CO	+255[a]	605	Wilson et al. (119)
CO	+250[b]	605	Tzagoloff and Wharton (133)

[a] Intact mitochondria.
[b] Isolated aa₃.

criteria (Keilin and Hartree) (11b) all these values should be assigned to "cytochrome a." As with uninhibited enzyme (Table VI), "cytochrome a" may have a more positive potential in the presence of ferric a_3^{3+} (a_3^{3+} HN_3, a_3^{3+} HCN) than ferrous a_3^{2+} (a_3^{2+} CO). Wilson et al. (119) have put forward an alternative explanation in which azide binds a^{3+} rather than a_3^{3+} (discussed below).

One odd consequence of the redox potential measurements may be noted. If, as suggested by Greenwood and Gibson (74) and by Antonini et al. (96), the reaction with oxygen is a quasi-concerted one with a fully reduced (a^{2+} Cu_2^+ a_3^{2+}) enzyme, then the overall kinetics show that there must be at least one component with a redox potential much more positive than any of those seen directly. Were this not so, the equilibration of enzyme and c at high levels of ferric c would decrease the amount of completely reduced oxidase (and hence the reaction rate) in proportion to the fourth power of $[c^{2+}]$. In practice, the time course remains exponential at potentials greater than +400 mV. The unknown high potential (>500 mV) species could be EPR-invisible copper, which according to van Gelder et al. (131) seems to titrate early with NADH + PMS. Such high potential copper is present, for example, in the copper enzyme laccase (135). Alternatively, the exponential kinetics obtaining with an enzyme possessing only low potential groups may indicate that the quasi-concerted mechanisms are wrong.

C. Inhibitors of Cytochrome Oxidase

There are four categories of oxidase inhibitors: (a) inhibitors competitive with oxygen (CO and possibly NO); (b) heme-binding inhibitors noncompetitive with both O_2 and cytochrome c (HCN, HN_3, and H_2S), (c) noncompetitive inhibitors not affecting the heme groups (phosphate ions, alkaline pH, and possibly SH compounds), and (d) inhibitors competitive with cytochrome c (cations and polycations).

The last two categories have been treated in Section III,A. The heme ligands in the first two categories have given more direct evidence bearing on the oxidase mechanism.

1. Carbon Monoxide

Figure 1 showed the spectrum of a_3^{2+} CO "dissected out" by the photochemical action spectrum method of Warburg (2). Comparison with the directly observed spectra of the intact oxidase (Fig. 3) showed that the action spectrum represented only one component of the system, the other, the classic cytochrome a, retaining its ordinary capability for oxidation and reduction. Vanneste (116) has derived separate spectra for cytochromes a and a_3 in various oxidation and ligand states by appropriate additions and subtractions of the photochemical and difference spectra. The results correlate reasonably well with estimates based on cyanide difference spectra (97, 136).

Carbon monoxide binding was originally expressed in terms of competition with oxygen, as a partition constant,

$$K = \frac{n}{(1 - n)} \frac{[CO]}{[O_2]} \tag{12}$$

where n is the ratio of respiration in the presence of CO to that in its absence (127). The value of K in the dark for both heart muscle and yeast was about 8. Because CO is about 75% as soluble as O_2 this is equivalent to a 50% inhibition concentration (at 250 μM O_2) of 1.5×10^{-3} M. Direct estimates of 1.2 mM CO and 0.3 mM CO have been made for succinate oxidation by normal and c-deficient heart muscle submitochondrial particles, respectively [(137); and P. Nicholls, unpublished experiments] and 0.6 mM for cytochrome c oxidase activity [Wainio (138)].

For the system of Eq. (13), the relationship between this parameter (K_i) and the dissociation constant (K_{CO}) for the reaction of a_3^{2+} with CO will be given by Eq. (14):

$$\text{AH}_2 \rightarrow \overset{k_1 \ c^{3+}}{\underset{c^{2+}}{\updownarrow}} \rightarrow \overset{k_2 \ a_3^{3+}}{\underset{a_3^{2+}\underset{K_{CO}}{\overset{CO}{\rightleftharpoons}} a_3^{2+}CO}{\updownarrow}} \overset{k_3}{\rightarrow} O_2 \tag{13}$$

where $k_2 \simeq 200$ sec^{-1} and $k_3 = 10^5$ sec^{-1} (at 250 μM O_2)

$$K_{CO} = \frac{10^{-3} k_1 (200 + 2 k_1)}{(200 + k_1)^2} K_i \tag{14}$$

If $k_1 = 50$ sec^{-1} (corresponding to a succinate oxidase system turning over at 40 sec^{-1}), then $K_{CO} = 2.4 \times 10^{-4}$ K_i, or about 0.3 μM. At infinite k_1, K_i would be $500 \times K_{CO}$, or 0.15 mM.

Measurements of CO binding to the fully reduced enzyme (a^{2+} a_3^{2+}) by Gibson and Greenwood (100) have given $k_{on} = 8 \times 10^4$ M^{-1} sec^{-1} and $k_{off} = 0.023$ sec^{-1} at pH 7.4, 20°C, equivalent to $K_{CO} = 0.2$–0.3 μM. Direct and catalytic measurements of CO binding are thus in good agreement. Nitric oxide binds more rapidly and more tightly to the oxidase than does CO, while HCN, which gives rise to an autoxidizable a_3^{2+} HCN species, binds less well (Table VIII)(139–141). Wohlrab and Ogunmola (141) have obtained a pH independent value of 0.4 μM for K_{CO} in uncoupled rat liver mitochondria. In coupled mitochondria plus ATP, this value increased to about 0.9 μM. They suggested that this may reflect an ATP-dependent conformational change in the oxidase.

Titrations of reduced enzyme with CO at first gave disconcertingly low CO: heme a ratios (100, 142). These seem to have resulted from the presence of inactive enzyme since later results gave CO:heme a ratios of 0.5 (i.e., CO:aa$_3$ = 1.0). (143, 144) This showed that a$_3$ was not a minor catalytic fraction of the total heme a but a component present in a concentration approximately equal to that of cytochrome a.

The a^{2+} a_3^{2+} CO complex is inert toward oxygen (11b). It can, however, be oxidized to a^{3+} a_3^{2+} CO by the addition of ferricyanide because a_3^{2+} CO has a very positive effective redox potential (and dissociates rather slowly to give the vulnerable a_3^{2+} species) (28, 145). The redox potential of cytochrome a, which titrates as a simple one-electron acceptor in this system, is $+260$ mV. NO added to a^{3+} a_3^{3+} gives rise to the a^{3+} a_3^{2+} NO complex (145). No evidence has been obtained for an a_3^{3+} NO (a_3^{2+} NO$^+$) species in this enzyme.

TABLE VIII

BINDING OF LIGANDS TO FERROUS CYTOCHROME a$_3$[a]

Ligand	k_{on} (M^{-1} sec^{-1})	k_{off} (sec^{-1})	K_d (μM)
CO	0.8–1.2 $\times$ 10^5	0.02	0.25
HCN	120–150	0.015–0.07[b]	100–500
NO	4 $\times$ 10^7	≤ 1	≤ 0.05
O$_2$	$\sim$10^8	< 0.01	< 10

[a] pH 7.0–7.4; temperature from 20° to 25°C. All data for isolated oxidase preparations.

[b] Larger values from Antonini et al. (139) and Gibson and Greenwood (100); smaller values from van Buuren et al. (140).

2. Cyanide

Cyanide is the only inhibitor that reacts with both ferrous and ferric cytochrome a_3. The reactions and their products are, however, quite different. As Table VIII shows, the ferrous enzyme binds HCN in a reversible reaction to give an a_3^{2+} HCN complex with maxima at 445 and 590 nm. Keilin and Hartree found that this complex, unlike that with carbon monoxide, remained autoxidizable (11b). Indeed, the reaction of Eq. (15) was the strongest initial evidence for the autoxidation of cytochrome a_3 itself.

$$a^{2+}a_3^{2+}HCN + O_2 \rightarrow a^{2+}a_3^{3+}HCN + (O_2^-)? \tag{15}$$

This reaction, which takes place in both the isolated oxidase and the intact mitochondrion (146, 147), has two unexpected features:

(a) It involves the oxidation of a_3^{2+} alone, cytochrome a remaining fully reduced in the product (146);

(b) the rate seems to be of the same order of magnitude as with the free ferrous enzyme (147).

As indicated in Eq. (15), the reaction stoichiometry is uncertain. In titrations of the oxidized cyanocytochrome aa_3 complex, however, to give rise to what should be the same partially reduced species, two reducing equivalents appear to be involved (104, 146). One mole of O_2 would accordingly be capable of oxidizing two moles of a^{2+} a_3^{2+} HCN [cf. the reaction with limiting amounts of oxygen reported by Gibson and Greenwood (100)].

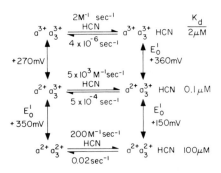

Fig. 7. Reaction scheme for cyanide and cytochrome oxidase. From Nicholls and Kimelberg (75).

The species a^{2+} a_3^{3+} HCN is inert toward molecular oxygen and is the actual inhibited form of the oxidase present in systems containing both substrate (c^{2+}) and O_2. Figure 7 shows possible reactions between oxidase and cyanide and the alternative routes giving rise to a^{2+} a_3^{3+} HCN. In addition to the reaction of Eq. (15), these include the reduction of a^{3+} a_3^{3+} HCN and cyanide binding by uninhibited a^{2+} a_3^{3+}.

The upper reaction of Fig. 7 is very slow in the isolated enzyme and in Keilin–Hartree particles (139, 140). Rate constants below 10 M^{-1} sec^{-1} have been reported (140). In pigeon heart mitochondria higher rates up to 2500 M^{-1} sec^{-1} were found (148). The rate of binding by isolated enzyme is increased by azide and possibly by cytochrome c (140, 149). In all these systems, the final high → low spin state transition (Fig. 8) occurs as a first-order process with a cyanide independent rate of between 0.02 and 0.08 sec^{-1} ($t_{1/2}$ from 30 to 10 seconds) (140). The cyanide binding may thus occur in the two steps of Eq. (16):

$$a^{3+}a_3^{3+} + \text{HCN} \underset{k_{-1}}{\overset{k_1}{\rightleftharpoons}} a^{3+}a_3^{3+}(\text{HCN}) \underset{k_{-2}}{\overset{k_2}{\rightleftharpoons}} a^{3+}a_3^{3+}\text{HCN} \tag{16}$$

Here the intermediate complex—a^{3+} a_3^{3+} (HCN)—must involve a trapping of the cyanide molecule that does not involve the d electrons of the a_3 iron.

The final complex, cyanocytochrome aa_3, is sufficiently stable to permit its chromatographic separation from excess cyanide and overnight dialysis (27, 150). Equilibrium dialysis and direct spectroscopy suggest a value of 1–2 μM for the overall dissociation constant ($K_d = k_{-1} k_{-2}/k_1 k_2$), implying an "off" constant ($\sim k_{-2}$) of about 2×10^{-6} sec^{-1} for the isolated enzyme (149). Cytochrome a and the EPR-detectable (830 nm) copper remain reducible in the complex, but cytochrome a_3 and (probably) the EPR-silent copper remain oxidized for a reasonable interval, even in the presence of dithionite (32, 104). Only 1 mole of cyanide is bound per mole aa_3 (i.e., 0.5 mole CN$^-$/heme a equivalent) in the isolated complex (27, 146), although various claims for more than one cyanide binding site have been put forward (139, 150).

The central equilibrium of Fig. 7 is probably responsible for cyanide inhibition of the catalytic reaction (146). Under such conditions, cyanide reacts with an "on" constant of up to 5000 M^{-1} sec^{-1} and a K_i of about 0.1 μM. When dithionite is added to a^{3+} a_3^{3+} HCN, the a_3^{3+} is reduced at a rate of 1 to 2×10^{-3} sec^{-1}. Similarly, when a^{3+} a_3^{3+} HCN is added to a catalytic system containing c^{2+}, activity is restored with a "lag" time of several minutes (27, 150). The product $K_i k_{on}$ has a value of 5×10^{-4}

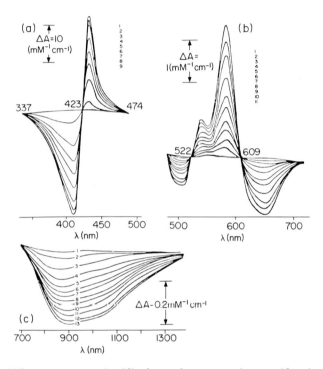

Fig. 8. Difference spectra of oxidized cytochrome aa₃ plus cyanide minus oxidized cytochrome aa₃. (A) Soret region, 7.5 μM cytochrome aa₃ and 2 mM cyanide. 9, Without cyanide; 8, immediately after mixing; 7, after 1.5 minutes; 6, after 3.0 minutes; 5, after 4.5 minutes; 4, after 7.5 minutes; 3, after 10.5 minutes; 2, after 30 minutes; 1, after 4 hours. (B) Visible region, 40 μM cytochrome aa₃ and 0.5 mM cyanide. 11, Without cyanide; 10, immediately after mixing; 9, after 2.5 minutes; 8, after 5.0 minutes; 7, after 7.5 minutes; 6, after 12.5 minutes; 5, after 20 minutes; 4, after 30 minutes; 3, after 50 minutes; 2, after 90 minutes; 1, after 4 hours. (C) Near infrared region, 140 μM cytochrome aa₃ and 1.0 mM cyanide. 1, Without cyanide; 2, immediately after mixing; 3, after 2.5 minutes; 4, after 5.0 minutes; 5, after 7.5 minutes; 6, after 10.0 minutes; 7, after 12.5 minutes; 8, after 15 minutes; 9, after 20 minutes; 10, after 30 minutes; 11, after 45 minutes; 12, after 60 minutes; 13, after 4 hours. From van Buuren et al. (140).

sec^{-1}, in reasonable agreement with direct estimates. Table IX summarizes the various kinetic parameters obtained.

The inhibition pattern given in Fig. 7 can also account for the "biphasic" approach to the inhibited steady state (146). At low levels of reduction the equilibrium will be dominated by the upper reaction, but the approach to equilibrium will be catalyzed by the central and possibly lower reactions.

TABLE IX

BINDING OF CYANIDE TO FERRIC CYTOCHROME a_3 [a]

Enzyme species reacting	k_{on} (M^{-1} sec^{-1})	k_{off} (sec^{-1})	K_d (μM)
a^{3+} $a_3{}^{3+}$	2 [b]	(4×10^{-6}) [c]	$1.0-2.0$ [d]
a^{2+} $a_3{}^{3+}$	5×10^3	5×10^{-4}	$0.05-0.1$ [e]

[a] pH 7.4, 0.1 M phosphate, 25°.
[b] In isolated oxidase. Faster rates are observed in the presence of azide or cytochrome c, or in intact mitochondria.
[c] Indirect estimate (order of magnitude arrived at by stability to dialysis, etc.).
[d] By prolonged equilibration or equilibration dialysis.
[e] All data obtained from kinetics of inhibition.

As inhibition develops, cytochromes c and a become progressively more reduced and the overall equilibrium will shift to that given by the central reaction with the smaller K_i. Such inhibition patterns have been seen with isolated oxidase, Keilin–Hartree particles, and intact mitochondria (128, 146).

3. Azide and Hydroxylamine

Azide was introduced as a ferric heme ligand and terminal respiratory inhibitor by Keilin in 1936 (151). In 1939, Keilin and Hartree (11b) classed azide with cyanide with respect to its effect on ferric a_3. More recently clear differences have been shown between the two inhibitors (128). Although azide, like cyanide, binds to the ferric oxidase and prevents the reduction of a_3 (and part of the copper?) the reaction is rapid ($t_{1/2} < 5$ seconds) and involves only a small change in the spectrum of the enzyme (152). Both the inhibitory K_i and the observed K_d decrease with decreasing pH (as in the case of cyanide, the inhibitory species is the free acid) (153) and they are of similar magnitude (Table X).

The effect of azide on the steady state, and its transition from the uninhibited to the inhibited condition, is therefore simpler than that of cyanide (128). But the inhibited species, a^{2+} $a_3{}^{3+}$ HN_3, is unusual in two respects:

(a) The spectrum of reduced cytochrome a is shifted toward the blue, both in the α region (605 → 602 nm) and in the Soret region (although the characteristic split γ peak is retained) (29, 122, 154).

TABLE X

Binding of Azide to Ferric Cytochrome a_3 [a]

Method[b]	k_{on} (M^{-1} sec^{-1})	k_{off} (sec^{-1})	K_d (μM)
1. Direct[c]	$\sim 4 \times 10^3$	0.2	50
2. Catalytic reaction[d]	$> 10^4$	>1	70–120
3. α-Peak shift[e]	4.5×10^3	0.5	100–200
4. Redox potential[f]	—	—	250

[a] pH 7.4; 20–25°C. $k_{on} \propto$ [H$^+$]; $K_d \propto 1/$[H$^+$].

[b] Methods 1 and 2, isolated enzyme; methods 3 and 4, mitochondria.

[c] From changes in Soret region and at 558 nm in fully oxidized enzyme (neglecting slow secondary changes with rates of 0.025 sec^{-1} independent of azide concentration) (130, 151).

[d] Inhibition of enzymic activity ($a^{3+}a_3^{3+}/a^{2+}a_3^{3+}$ mixtures) (151).

[e] Rate of 605 → 602 nm in $a^{2+}a_3^{3+}$ (153).

[f] From azide concentrations needed to modify redox titration curves ($a^{3+}a_3^{3+}/a^{2+}a_3^{3+}$ mixtures): attributed to a^{3+} by Wilson et al. (119).

(b) It shows a unique EPR signal at $g = 2.9$, not seen in either the $a^{3+} a_3^{3+}$ or $a^{+2} a_3^{3+}$ HCN species (32, 105).

The scheme for azide binding given in Fig. 9 may thus involve even more complicated processes. Indeed, Muijsers et al. (152) have indicated that the rates of change at 420 and 558 nm in the difference spectrum (Fig. 10) are different with the former independent of, the latter dependent on, HN$_3$ concentration [cf. Eq. (16)]. The inhibition of the intact mitochondrial system by azide also shows oligomycin-like behavior in that state 3 is considerably more sensitive than state 3u (uncouplers can therefore "release" azide inhibition) (30, 155, 156). The explanation for this is uncertain but may involve

(a) changes in steady state reduction, and hence inhibitor sensitivity, on transition from state 3 to state 3u (75, 157);

(b) inhibition of ATPase by azide (158);

(c) accumulation of azide by coupled mitochondria (159); and

(d) energy-dependent azide-sensitive oxidase species (155).

A similar increased sensitivity of state 3 respiration is seen with hydroxylamine inhibition (160, 161). The mode of hydroxylamine action is, however, even more uncertain than that of azide (161). It has at least three effects on the oxidase: (1) as a ligand for ferric a_3 (a_3^{3+} NH$_2$OH), (2) as a reductant (of c and a), and (3) by generating a ligand (NO) for ferrous

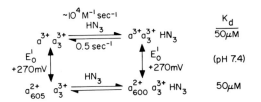

FIG. 9. Reaction scheme for azide and cytochrome oxidase. From Nicholls and Kimelberg (75).

a_3. Like cyanide its reaction with the isolated enzyme is slow and appears to depend on a preferential binding to the $a^{2+} a_3^{3+}$ species (162). Binding of nitric oxide as a consequence of incubation with NH_2OH has been shown by Blokzijl-Homan and van Gelder (163). The hyperfine structure provides evidence for a proximal nitrogen atom associated with a_3 heme iron.

4. Fluoride and Sulfide

Ferric a_3 ligands that seem to act reversibly without inducing unusual spectroscopic changes include fluoride and sulfide. Fluoride induces a low → high spin shift in the isolated ferric enzyme and acts as a weak inhibitor in the catalytic system (164). Sulfide acts rather like cyanide (29) but binds more rapidly with a K_i of about 10 μM.

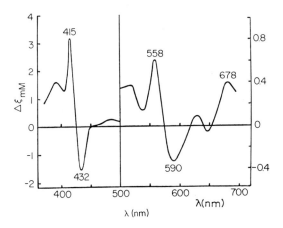

FIG. 10. Difference spectrum of oxidized cytochrome aa_3 plus azide minus oxidized cytochrome aa_3. Beef heart oxidase ±1 mM azide in 100 mM potassium phosphate (pH 7.3), 0.7% cholate, 25°C. Spectrum taken 3 minutes after azide addition. Extinction coefficients in terms of heme a. Courtesy A. O. Muijsers, B. C. P. Jansen Institute, University of Amsterdam (unpublished data).

Table XI summarizes the reactions of various inhibitors with ferric cytochrome a_3.

D. Heme–Heme Interactions in Cytochrome Oxidase

To account for all these observations it is necessaey to postulate some form of heme–heme or heme–copper interaction. There are presently three such theories suggested by Wilson and co-workers (119), by van Gelder's group (131), and by Nicholls (128, 129). At the moment is is not easy to suggest critical experiments to distinguish them because the prosthetic groups of cytochromes a and a_3 are identical and unique assignments of particular properties are therefore impossible.

Wilson et al. (119) considered that the hemes of cytochromes a and a_3 are titrimetrically distinguishable but spectroscopically indistinguishable in the reduced form. Cytochrome a_3 is then redefined as the species reduced with a midpoint potential of 340–380 mV, while cytochrome a is reduced with a midpoint potential of 190–220 mV. Carbon monoxide binds a_3^{2+}, leaving a with a potential of 260 mV. Azide, according to this view, can bind a^{3+}, leaving a_3 with a potential of 340 mV. The EPR signal of the oxidized enzyme ($g = 3$) is assigned to cytochrome a_3, and that of the partially reduced enzyme ($g = 6$) to cytochrome a. The absorption at

TABLE XI

REACTIONS OF LIGANDS WITH FERRIC CYTOCHROME a_3

Inhibitory ligand	Apparent affinity[a] (μM)	Spectroscopic effect	EPR signal	Effect on cytochrome a
HCN	0.1–1.0	High→low spin[b]	None	Increases E_0'
HN$_3$	0.5	Anomalous[c]	$g = 2.9$ (only in $a^{2+}\ a_3^{3+}$)	Shifts ferrous α peak
HF	1.0[d]	Low→high spin[e]	$g = 6.2$	?
H$_2$S	~10	?	None ?	No effect?

[a] Of undissociated acid [apparent reactive form in each case, cf. Stannard and Horecker (153)].

[b] In both visible and Soret region (cf. Fig. 8) (139).

[c] Low→high spin in Soret region but high→low spin characteristics in visible region (cf. Fig. 10) (151).

[d] 25 mM at pH 7.4.

[e] In both visible and Soret region [cf. Muijsers et al. (164)].

605 nm in the presence of cyanide or carbon monoxide is attributed to a heme–heme interaction increasing the extinction coefficient of ferrous cytochrome a.

van Gelder (131) believed that the hemes of cytochromes a and a_3 are both titrimetrically and spectroscopically indistinguishable in the absence of cytochrome c (with similar extinction coefficients and a redox potential of $+280$ mV). Cytochrome c, however, catalyzes a change in the relative potentials of a and a_3 whereby the previously indistinguishable hemes take on the characteristic properties of a and a_3. Like that of Wilson, this theory involves changes in extinction coefficients of cytochrome a when cytochrome a_3 is liganded. Unlike Wilson's theory, it permits interactions involving redox potentials, and hence does not postulate the existence of an azide complex with ferric cytochrome a.

Nicholls (128, 129) supposed that the hemes of cytochromes a and a_3 are spectroscopically distinguishable but titrimetrically indistinguishable. He therefore retained the classic assignments of extinction coefficients (reduced minus oxidized) at 605 and 445 nm ($a:a_3$ ratios of 85:15 at 605 and 47:53 at 445 nm) and the original definition of a_3 as the component reacting with azide, cyanide, and carbon monoxide. In order to explain the redox titration curve on this theory it is necessary to postulate that the oxidation state of cytochrome a_3 affects the redox potential of cytochrome a, and vice versa, according to Eq. (17):

$$\begin{aligned} \text{(a)} \quad & a^{3+}a_3^{3+} + c^{2+} \rightleftharpoons c^{3+} + a^{2+}a_3^{3+} \text{ (or } a^{3+}a_3^{2+}) \\ \text{(b)} \quad & a^{3+}a_3^{2+} \text{ (or } a^{2+}a_3^{3+}) + c^{2+} \rightleftharpoons c^{3+} + a^{2+}a_3^{2+} \end{aligned} \qquad (17)$$

where $E_o'(a) \simeq +340$ mV and $E_o'(b) \simeq +220$ mV.

This viewpoint receives support from the finding that cyanide binding to a_3 is a function of the redox state of a. For this implies that the redox potential of a associated with the a_3^{3+} cyanide complex may be about 60 mV more positive than that of a associated with ordinary ferric a_3. Shifts in potential when a_3 is reduced, bound to azide, or bound to CO (Table VII) are therefore quite plausible. However, this hypothesis proposes that the "half reduced" form of the enzyme is a mixture of $a^{2+} a_3^{3+}$ and $a^{3+} a_3^{2+}$. Spectroscopically, this species is reported to show the single Soret band of cytochrome a_3 (122), and kinetically it seems to react homogeneously with oxygen (156).

On the other hand, the theory of Wilson has difficulties in treating the action of azide (postulated as binding sometimes to a and sometimes to a_3) and the EPR data, indicating the appearance of high spin ferric heme (a_3?) in the half reduced state (32). The three interaction theories are not mutually exclusive, and eventually some combination of such ideas may be required. Meanwhile it is worth noting that, despite differences of detailed interpretation, everyone now agrees that a and a_3 must interact "allosterically" as well as by electron transfer. Only the quantitative contribution of such interactions, compared to the contribution due to initial chemical differences between cytochromes a and a_3, is in dispute.

The redox potential measurements also indicate that the copper atoms are titrated independently of the hemes (although n values greater than 1.0, seen at 830 nm in the absence of cytochrome c, suggest interactions between the copper species themselves) (131). This is inconsistent with the rather reasonable viewpoint of Beinert (165), which explains the absence of an a_3 heme signal and of 50% of the copper EPR signal, in the fully oxidized state, in terms of a copper–heme interaction. Again, a simple and complete model of the system is not easy to construct.

Table XII summarizes the interpretations of the various states of the enzyme made by the different interaction models.

E. Reaction Mechanisms

Two basic types of reaction mechanism have been proposed (23), the quasiconcerted [Eq. (18), cf. Eq. (8)] and the sequential [Eq. (19), also Eq. (7)]:

$$4\ c^{3+} \rightarrow \quad \rightarrow a^{2+}\ a_3^{2+}\ Cu_I^+\ Cu_{II}^+ \quad \rightarrow O_2$$
$$4\ c^{2+} \rightarrow \quad \rightarrow a^{3+}\ a_3^{3+}\ Cu_I^{2+}\ Cu_{II}^{2+} \quad \rightarrow 2\ H_2O \tag{18}$$

$$c^{3+} \rightarrow \quad a^{2+} \rightarrow \quad Cu^{2+} \rightarrow \quad a_3^{2+} \rightarrow \quad O_2$$
$$c^{2+} \rightarrow \quad a^{3+} \rightarrow \quad Cu^+ \rightarrow \quad a_3^{3+} \rightarrow \quad O_2^- \tag{19}$$

The former type was preferred by Gibson and Greenwood (74, 106), the latter type by Chance and Erecinska (110). The quasi-concerted mechanism is favored by the observations that

(a) ferrous cytochrome a does not reduce ferric cytochrome a_3 rapidly;
(b) in the reaction of fully reduced enzyme with oxygen, cytochrome a_3 does not seem to enter a Fe^{2+}/Fe^{3+} steady state; and
(c) no satisfactory evidence is available for redox intermediates other than ferric and ferrous a_3.

However, it fails to account for the facts that

(a) the first-order rate constants for oxidation are not the same for all three components;

TABLE XII

Postulated Patterns of Heme–Heme Interaction

Property	Theory A[a]	Theory B[b]	Theory C[c]
Ferric enzyme (state of a and a_3)	Distinguishable hemes	Indistinguishable hemes	Distinguishable hemes
Redox titrations	E_0' (a_3):340 mV E_0' (a):220 mV	E_0' (a_3):220 mV? E_0' (a):340 mV?	Both a and a_3 contribute to each E_0'
CN^- binding	To $a_3{}^{3+}$ (affects a spectrum)	To $a_3{}^{3+}$ (affects a spectrum)	To $a_3{}^{2+}$ (affects E_0' for a)
$N_3{}^-$ binding	To a^{3+} (affects a_3 spectrum) EPR signal = $a^{3+}N_3{}^-$	To $a_3{}^{3+}$ (affects a spectrum) EPR signal = $a_3{}^{3+}N_3{}^-$	To $a_3{}^{3+}$ (affects a spectrum) EPR signal = $a_3{}^{3+}N_3{}^-$
CO binding	To $a_3{}^{2+}$ (affects a spectrum)	To $a_3{}^{2+}$ (affects a spectrum)	To $a_3{}^{2+}$ (affects E_0' for a)
Species seen at partial reduction	High spin a^{3+}	High spin $a_3{}^{3+}$	High spin $a_3{}^{3+}$

[a] From ref. (119).
[b] From ref. (130).
[c] From ref. (129).

(b) the immediate product of autoxidation of $a^{2+}a_3{}^{2+}$ HCN is $a^{2+}a_3{}^{3+}$ HCN and not $a^{3+}a_3{}^{3+}$ HCN;

(c) the reaction does not "back up" at high ratios of $[c^{3+}]/[c^{2+}]$, as would be expected if all the components were in redox equilibrium with c but only the fully reduced form were active; and

(d) at low temperatures cytochrome a_3 is still oxidized rapidly (99) by O_2 but the subsequent oxidation of both c and a are markedly slowed (117).

The sequential mechanism is favored by classic analogy and by the observations showing marked qualitative and quantitative differences in the oxidation pathways of the several oxidase components. Its weak points are that it does not give a satisfactory explanation of the way in which the four-electron reduction of oxygen is coupled to the one-electron oxidation of cytochrome c and that it fails to provide convincing reasons for the number of redox components in the oxidase as isolated and for their close chemical relationships. Nor has it been possible to demonstrate the coupled oxido reduction of any of the three pairs of components linked in such schemes (a/Cu, a_3/Cu, and a/a_3).

Any satisfactory mechanism must in addition take into account the behavior of cytochrome oxidase as a membrane-bound energy-conserving system. The results obtained *in situ* inevitably further complicate the picture derived from studies with the enzyme as isolated.

IV. INTERACTIONS OF OXIDASE WITH THE MEMBRANE

The natural milieu of the enzyme is the inner mitochondrial membrane. As pointed out in Section III,B this membrane would therefore be the preferred material for oxidase studies were it not for the complications produced by the presence of other pigments and catalysts. Even in their presence, certain information is obtainable only from the intact system.

A. Allotopy and Topology

The phospholipid bilayer can affect the enzyme in two ways—allotopically or topologically; that is, the membrane as a matrix may modify the chemistry of the protein (allotopy), while as a physical structure with sidedness it may orient the molecule in a specific direction (topology).

Relatively few intrinsic allotopic properties have been proved for cytochrome oxidase. Zahler and Fleischer (36) have found effects of phospholipids on both V_{max} and K_m for cytochrome c oxidation, but turnover and cytochrome c binding for the best soluble preparations are very similar to the same properties in membraneous systems. Correspondingly, the quantitative effects of inhibitors such as cyanide and azide, both catalytic and spectroscopic, are the same in the two types of enzyme. Even the heme–heme interactions (Section III), which are likely to involve conformational change, occur in the same way *in situ* and in solution. On the other hand, some evidence of differences exists. The multiplicity of membrane binding sites—of both high and low affinity (63, 78)— does not seem to occur with the isolated enzyme (although the significance of these sites is uncertain). One molecule of polylysine (MW 10^5) is capable of inhibiting more equivalents of soluble than of membrane-bound oxidase, perhaps because the latter cannot be brought as close as the former (87).

Cyanide, which reacts very slowly with the isolated ferric enzyme (139, 140), binds more rapidly to the mitochondrial enzyme (148). But the presence of cytochrome c increases the reactivity of the isolated enzyme by about twentyfold (149) and this may account entirely for the increase *in situ*. Similarly, the differences between the redox potential obtained by comparative titrations of mitochondria, particles, and soluble enzyme can be mimicked by the addition of small amounts of cytochrome c to the soluble enzyme (132).

Wilson and Leigh (105) have shown considerable differences in the EPR spectra of mitochondrial as compared with isolated enzyme. Whether cytochrome c can modify the EPR characteristics of the latter is not known. Erecinska and Chance (117) reported differences in the kinetics of reaction with oxygen. Both the initial product $(a_3^{2+}O_2?)$ (117) and the rate of cytochrome a oxidation (110) were modified when compared with the isolated enzyme ($\pm$ cytochrome c).

The topology of the mitochondrial oxidase has been the subject of considerable comment but perhaps of less incisive experiment. The idea that the enzyme is "plugged through" the membrane has been the most popular (167–170), although the possibility of other configurations is not excluded. This concept received strong support from the experiments of Hinkle et al. (171) with reconstituted artificial vesicles containing "inlaid" [Jasaitis et al. (172)] cytochrome oxidase. Cytochrome c added to the outside of vesicles prepared by dialyzing mixtures of cholate, oxidase, and phospholipids can be oxidized in an uncoupler-dependent reaction. Similar vesicles can be prepared (173) which will oxidize internally trapped cytochrome c. If the oxidase is oriented transversely as visualized by Mitchell (167), Racker (168, 173), and others, half the molecules, when assembled randomly, should be accessible from the outside and the other half should be accessible from cytochrome c internally. Alternatively, as intimated by Muijsers et al. (132), cytochrome c may induce asymmetry of the a/a_3 type in a previously symmetrical molecule.

B. Energy-Linked Properties

Topology may link electron transport in the oxidase with energy conservation. Ascorbate oxidation in the vesicles prepared by Hinkle et al. (171) and Jasaitis et al. (172) is dependent on the addition of uncouplers or valinomycin plus K^+. Yong and King (130) have claimed a similar uncoupler dependence for a solubilized oxidase—cytochrome c complex (174), but this and other properties of their "soluble" complex may be attributable to its microvesicular character. Can all the energy-dependent behavior of the mitochondrial aa_3 system be linked to its probable "vectorial" activity? At present it is difficult to make this simple correlation, as will be seen from the following.

The first observation of an in situ energy-dependent reaction was that of the crossover point (175) between cytochromes c and a. When respiration is slowed in state 4, cytochrome c is usually more reduced and cytochrome a more oxidized, than in the steady state (state 3 or 3u). Ramirez (176), in intact heart, and Muraoka et al. (177), with isolated mitochondria, have demonstrated the possibility of another crossover point between cytochrome a (and perhaps a_3) and oxygen. At least two reaction steps must therefore be energy-dependent (directly or indirectly). Slater (179) has argued on thermodynamic grounds that the energy conservation step must lie between a and oxygen; the classic crossover point between a and c would then be the consequence of an indirect effect, for example, a decreased reactivity between ferous cytochrome c and a high energy form of a ($a^{3+} \sim I$) (175). Addition of ATP to state 5 mitochondria causes the oxidation of cytochromes c, a, and probably a_3, by reversed electron transfer (180). Knowledge of redox potentials ought to permit assignment of interaction sites in such a system. Unfortunately (see Section III,D), the measured redox potentials are not simply attributable to either a or a_3. Moreover, the redox titrations are themselves affected by the presence of ATP. The ATP effect has been variously described as being exerted on cytochrome a_3, on cytochrome a, and on both of them. And in each case, the observations may be accounted for *either* by assuming unique new "high energy" forms of the cytochrome involved *or* by assuming that the redox mediators do not interact directly but via an energy-dependent pathway.

Table XIII summarizes the observations on crossover sites in the cytochrome aa_3 system.

$$\text{Substrate} \xrightarrow{k_1} c \xrightarrow{k_2} a \xrightarrow{k_3} O_2 \tag{20}$$
$$a_3$$

At high flux (178) the crossover point lies at k_3 in Eq. (19), i.e., between cytochrome a and O_2. If the flux is slowed by adding azide, cyanide, or sulfide, the crossover point moves toward the substrate end of the chain, to k_2. In the former case, ADP addition causes the oxidation of both c and a; in the latter, c is oxidized but a reduced [as observed by Chance and Williams (175)].

Such observations can also be explained by a model in which cytochrome a is on a side path [Eq. (21)]:

$$\text{Substrate} \xrightarrow{k_1} c \xrightarrow{k_3} O_2 \tag{21}$$
$$K_{eq} \updownarrow \ a_3$$
$$a$$

This model, however, also requires two effects of ADP: on the velocity of c oxidation (governed by k_3) and on the equilibration between c and a (governed by K_{eq}) which must move in favor of c reduction and a oxidation in state 4.

ATP certainly affects the apparent redox potentials of the cytochrome aa_3 system (127). The uncoupled potentials of about 380 and 220 mV (Table VI) decrease to 250 and 160 mV. A decrease also occurs in the

TABLE XIII

SITES OF ADP AND UNCOUPLER ACTION ON CYTOCHROME OXIDASE *in Situ*

| | Crossover[a] observed at | | | |
System	$c \rightarrow a$	$a \rightarrow a_3$	$a_3 \rightarrow O_2$	Reference
Liver mitochondria (NADH-linked subtrate or succinate)	+	0	0	175
Intact toad heart	0	0	+?	176
Mitochondria (ascorbate + TMPD)	+	0	+?	178
Mitochondria (succinate)	+	+	?	177

[a] A decrease in reduction of the first component and an increase in reduction of the second as flux increases.

potential measured in the presence of carbon monoxide. Table XIV lists the observed potential shifts induced by ATP and the interpretation given by different authors. There are at least three unresolved questions:

1. Is the change in potential associated only with a_3 [as favored by Wilson and Dutton (127)], only with a [as suggested by Hinkle and Mitchell (181)], or with both [as proposed by Wikström (182)]?

2. Does the change in potential reflect the occurrence of new high energy species, or an effect of membrane potential, or an energized reduction pathway?

3. Does the $+380$ mV component shift in potential to $+160$ mV, or do both components shift their potential ($+380$ to $+250$ and $+220$ to $+160$ mV)?

TABLE XIV

EFFECT OF ATP ON REDOX POTENTIALS OF CYTOCHROME OXIDASE

System	Potential (uncoupled)	Potential (+ATP)	Reference
aa_3	$+380$ mV	$+250$ mV[a] ($+160$ mV)[b]	127, 182
aa_3	$+220$ mV	$+160$ mV[a] ($+250$ mV)[b]	127, 182
$aa_3 + CO$	$+250$ mV	$+210$ mV	181
Cu	$+280$ mV	No change (?)	105

[a] As assigned by Wikström (182).
[b] Alternative assignment of Wilson and Dutton (127).

The occurrence of at least two crossover sites (Table XIII) suggests that both a and a_3 may be affected. Moreover it should be remembered that in a single turnover [Eq. (1)], two molecules of ATP are synthesized. There is also chemical and spectroscopic evidence for energy-dependent changes in both the constituent cytochromes.

Both Wilson (154) and Wikström (160) considered that the inhibition of mitochondrial cytochrome oxidase by azide and by hydroxylamine provided evidence for a high energy form of cytochrome a (or a_3) reacting with the inhibitor or with uncouplers. Titrations of azide-inhibited system with uncouplers gave 1:1 stoichiometries with cytochrome a. Although the interpretation of these results may be uncertain (see Section III,C,3), other observations have reinforced the suggested energy-linked behavior of cytochrome a. Chance et al. (183) observed a blue shift (λ_{max}, 595 nm, λ_{min}, 602 nm at 77°K) on adding phosphate to ATP-energized submito-chondrial particles. Similarly, Wikström (180) has reported a blue shift (λ_{max}, 600 nm, λ_{min} 614 nm at room temperature) on adding oligomycin to ATP-energized mitochondria in state 5. He has most recently (182) identified the species involved as a low potential form whose α maximum varies from 607 nm (uncoupled) to 609 nm (in the presence of ATP). The high potential aa_3 species, whose redox potential shows a more marked ATP-dependence (Table XIV) retains an α peak at 603–604 nm in both high and low energy states. Similar observations have been made by Lindsay and Wilson (184).

Although the assignment of these effects to a or a_3 depends on the model chosen (cf. Table XII), Hinkle and Mitchell (181) have demonstrated un-ambiguously the effect of ATP on the redox potential of a in the CO-in-bited system. It would be interesting to know whether the α-peak position is affected by ATP in this case, where a_3 is "clamped" in the a_3^{2+} CO form. In the converse situation, where a_3 is "clamped" as a_3^{3+} HCN or as a_3^{3+}HN$_3$, we have information on neither the redox nor the spectroscopic effect of ATP upon cytochrome a.

Spectroscopic effects of ATP on cytochrome a_3 have also been seen (184–186). ATP addition to fully oxidized mitochondria (ferricyanide treated) induces the appearance of a 515-nm band and a red shift in the Soret region, suggesting a high to low spin state transition similar to that which distinguishes "ferric" from "428 nm" oxidase in vitro (Section III,B) (187). This 575 nm band is discharged by uncouplers and its appearance is blocked by oligomycin (185). Most interesting, however, is its "competi-tive" relationship with cyanide (186). This inhibitor, which also induces a high to low spin transition in a_3 (Fig. 8), blocks the action of ATP. Conversely, if the ATP species is formed first, the reaction with cyanide (which can be seen quite clearly in intact mitochondria) is greatly slowed.

There also seem to be interactions between ATP and cyanide in the reaction with fully reduced mitochondria ($a^{2+}a_3^{2+} \rightarrow a^{2+}a_3^{2+}$ HCN) (186) although the results are less clear cut than in the oxidized case.

The existence of ATP effects on both a and a_3, combined with the evidence for heme–heme interactions (Section III,D) allows theories of indirect energization effects (analogous to the indirect effects of HCN and HN$_3$ on cytochrome a) to be constructed. Such theories remain highly speculative.

Is there a direct relationship between the oxidase (c–a–a$_3$) system and the ATP synthetase complexes? According to the chemiosmotic hypothesis, no such relationship should exist. However, there remains some awkward evidence for interrelationships between c, oxidase, and ATPase, brought forward some time ago (188). Wilson (189) has drawn attention to the (accidental?) stoichiometry that often holds between ATP synthetase and oxidase units. Yong and King (130) have claimed ADP control of oxidase activity in a system constructed with solubilized oxidase. Some kinetic evidence also exists for the occurrence of limited "domains" of coupled electron transfer and energy conservation (190). The properties of doubly inlaid liposomes (with ATP synthetase and cytochrome aa$_3$) as prepared by Jasaitis *et al.* (172), should throw more light on the problem. Table XV lists the known spectroscopic effects of energization upon the aa$_3$ system.

Three reconstituted oxidase preparations which exhibit various features of the hemoprotein in the natural membrane have been studied, and a summary of the properties of these preparations is afforded by Table XVI. Considering the preparations according to the name of the investi-

TABLE XV

SPECTROSCOPIC AND CHEMICAL EFFECTS OF ATP ON CYTOCHROME OXIDASE

System	Spectroscopic change upon energization	Reactivity change upon energization	Reference
$a^{2+}a_3^{3+}$	Red shift of 605 nm peak	Decrease in redox potential (a$_3$?)	127, 182
$a^{2+}a_3^{2+}$	Same(?)	Decrease in redox potential (a$_3$?)	127, 182
$a^{2+}a_3^{2+}$CO	Not known	Decrease in redox potential (a) and CO affinity (a$_3$)	140, 181
$a^{3+}a_3^{3+}$	575 nm peak (high to low spin change)	Decrease in rate of cyanide binding	186
$a^{2+}a_3^{2+}$CN$^-$	Decrease in extinction at 590 nm (?)	Increase in affinity for cyanide	186

TABLE XVI

Energy Coupling and H^+ Binding in Cytochrome Oxidase Preparations[a]

Compositions	Principal investigator		
	Racker (173)	Hinkle (171)	King (130)
1. Lipid	Soybean lipid extract vesicles	Soybean lipid extract vesicles	No added lipid
2. Lipid/cytochrome oxidase (mole/mole)	~4000	~4000	~50[b]
3. Type of cytochrome oxidase (nmoles/mg protein)	Yonetani (37) 10	Yonetani (38) 10	King (174) 11
4. Cytochrome c/cytochrome a (mole/mole)	~25	0	0.5
5. Coupling factors		None	None
A. Hydrophobic proteins	8×10^5 gm/mole cytochrome oxidase		
B. OSCP	1 mole/mole cytochrome oxidase[c]		
C. F_1	0.5 mole/mole cytochrome oxidase[c]		
6. Vesicles	+	+	Not examined

Functions			
7. Respiratory activity			
A. TN for heme a	0.8/second	$\sim$80/second	2/second
B. TN for heme c	0.03/second	0.04/second	4/second
8. Efficiency			
A. P/O ratio	0.13	0	0.4–0.6[d]
B. H$^+$/O	0.5[e]	1.9	—
9. Figure of merit [7A × 8A]	0.1	0	1.2
10. Energy-linked ANS response	Fluorescence enhancement	—	—

[a] Compiled by Dr. C. P. Lee.
[b] Based on the lipid content of cytochrome oxidase preparation.
[c] F_1 (200,000 was used as the molecular weight of F_1 for calculation) was added during the assays, and an additional 2 nmoles of OSCP/0.42 nmole cytochrome oxidase was added during assays of oxidative phosphorylation.
[d] No detectable ATPase activity as measured by ATP hydrolysis.
[e] P. Hinkle, personal communication.

gator, Racker's (173) represents the present "state of the art" in reconstitution of the third phosphorylation site. Soybean lipid extract vesicles and highly purified cytochrome oxidase [Yonetani preparation (38)] are combined in a mole-to-mole ratio of lipid to cytochrome oxidase of about 4000:1. In addition, coupling factors of a weight many times that of the oxidase are added. Oxidation of cytochrome c is measured at cytochrome c levels considerably in excess of those of cytochrome a (25:1) resulting in a turnover number of 0.8 sec^{-1} and a phosphorylation efficiency of 0.13 [at an oxidase concentration of 0.4 μM this value is increased to 0.5 (E. Racker, personal communication)]. Racker's vesicles transport electrons in the "submitochondrial configuration" and thus hydrogen ions would be expected to be bound during respiration, but no published observations are available. The vesicles also exhibit a significant energy-dependent ANS fluorescence enhancement.

Hinkle's (171) preparation is in many ways similar to Racker's, except that electron transfer occurs in the "mitochondrial" configuration and therefore hydroxyl ions are bound, with an efficiency of 2 gm atoms OH^{-} per oxygen atom. External cytochrome c is used, and the turnover number under the usual experimental conditions is 80 sec^{-1} or 100-fold greater than that of the Racker preparation. So far, neither phosphorylation activity nor energy-dependent ANS fluorescence enhancement has been reported for these vesicles.

King (130) has approached the problem from quite a different aspect and sonicates the oxidase and cytochrome c in the ratio of 2:1 with respect to cytochrome a, without adding lipid. Thus, the lipid content is about 50 per cytochrome oxidase. In the assay procedure for oxidative phosphorylation, cytochrome c is not added. The turnover number in energy coupling is 2 sec^{-1}, about three times that of Racker's preparation. No coupling factors are added, and a high phosphorylation efficiency is observed, 0.4–0.6, with $\sim$1 μM heme a concentrations used in the assay. Neither H^{+} uptake nor ANS fluorescence enhancement is reported, nor has the preparation been examined to ensure that vesicles are present. No further confirmation of these observations has been published.

In order to compare the degree to which cytochrome turnover may be coupled to oxidative phosphorylation in these preparations, the last line of Table XVI uses, as a "Figure of merit," the product of the turnover number of heme a and the P:O ratio. A value of 0.1 is obtained for Racker's preparation, 0 for Hinkle's preparation, and 1.2 for King's preparation. If these data are fully confirmed, there is the possibility that energy coupling may be studied in a very simple system containing only purified electron transfer components.

C. The Oxidase and the Respiratory Chain

In mammalian mitochondria, such as those from rat liver or beef heart, there are about two cytochrome oxidase complexes (Complex IV) for every Complex III (bc_1) (191):

$$b_K \, nhFe \rightarrow c \rightarrow aCu_2a_3 \text{ (Complex IV)}$$
$$\text{(Complex III)} \hspace{6em} (22)$$
$$b_T \hspace{2em} c_1 \rightarrow c \rightarrow aCu_2a_3 \text{ (Complex IV)}$$

Whether this stoichiometric fact has some special mechanistic significance we do not know. Kinetically, the oxidase is also usually in "excess," having a potential turnover higher than that of any other part of the respiratory chain [except possibly the antimycin A site (192)].

The ability of a small part of the oxidase to carry the major portion of the electron flux to oxygen creates the possibilities (193) known as cushioning" and "branching." Cushioning [or 'temporal" buffering (63)] can occur in isolated respiratory complexes containing only one oxidase molecule; the addition of a terminal inhibitor increases the steady state reduction of earlier members of the chain (e.g., cytochrome c). Even though, say, half the complexes are blocked, the net activity will be more than half-maximal because the blocked chains (reduced) will be more active on dissociation of the inhibitor *provided* that the rate of dissociation is not much smaller than the turnover of the system. Branching [or "spatial" buffering (63)] can occur only with numerous complexes attached to the same membrane: electron flow may then occur in the presence of a terminal inhibitor from one chain to another, bypassing the blocked chains. In this case, even an irreversible inhibitor will have a decreased effectiveness in systems with a low turnover. Both phenomena have been observed with oxidase inhibitors in intact mitochondria (193) and in submitochondrial particles (63).

In addition to steady state effects, transient responses can distinguish inhibited from active chains in mitochondria. With particulate preparations, Holmes (194) surprisingly found that one mole of carbon monoxide could slow the oxidation of up to 3 moles of cytochrome c (suggesting concerted action of more than one chain). More recent results (141) with intact mitochondria or phosphorylating particles indicate complete independence of respiratory chains in this respect. Carbon monoxide (141) and cyanide binding (195) studies show no evidence for interchain heme–heme interaction such as has been postulated for cytochrome b in the presence of antimycin A (196). However, a slow ($k \leq 1$ sec^{-1}) branching reaction seems to take place between oxidase molecules themselves while a much faster ($k \geq 100$ sec^{-1}) branching reaction occurs at the c_1–c region (63, 193, 197).

Wohlrab (197) has shown that the earlier reaction is dependent on the presence of cytochrome c. At least three mechanisms can be envisaged:

1. direct electron transfer from one molecule of c to another,
2. indirect ("hetero") transfer via common molecules of c_1 or (less likely) a; thus, $c_\alpha \to c_1 \to c_\beta$, or $c_\alpha \to a \to c_\beta$, etc., and
3. "hopping" transfer by rapid dissociation and binding of cytochrome c molecules exchanging ferrous for ferric forms mechanically.

Although the bimolecular rate of electron exchange between molecules of cytochrome c in solution is slow ($\sim 10^4$ M^{-1} sec^{-1}) (198), the monomolecular rate is unknown; thus, the likelihood of mechanism 1 cannot be estimated. The observed stoichiometry [Eq. (22)] seems to demand the occurrence of mechanism 2 for pairs of oxidases, but whether more extensive branching (involving $c_{1\alpha} \to c \to c_{1\beta}$) can take place is less certain. The estimated dissociation and binding rates [Eqs. (5) and (6)] are large enough for mechanism 3. The degree of mobility of cytochrome c in the intermembrane space (especially the intracristal pockets) may, however, be rather low.

Wagner et al. (199, 200) have successfully accounted for the kinetics of CO-inhibited systems by assuming free interaction (quasi-solution) between cytochromes c_1, c, and a. And both Chance (201) and Green (202) have postulated a certain mobility of cytochrome c in the membrane, involving at least vibrational and rotational freedom, and possibly translational movement. The latter could involve transfer of a molecule from one binding site to another without dissociation into solution (by Delbrückian two-dimensional diffusion). Margoliash (203, 204) and others (205) have, on the other hand, emphasized the probable need for most of the cytochrome c to be held for a finite time in binding sites involving a fixed orientation of the molecule toward oxidase and reductase systems. Does the consecutive reduction and oxidation of a given molecule of c demand appreciable (more than ordinary thermal) movement of the molecule? Both yes (201, 206) and no (63, 204, 205) answers have been given to this question.

As the apparent K_i values for inhibitors are increased by the "buffering" effects of low turnover and low steady state reduction at the cytochrome c level, so the apparent K_m value for oxygen is *decreased* by the same phenomena. The value of 0.5–1.0 μM given above (Section III,A) should therefore, in the simplest systems, be the maximal value. At lower turnovers, the K_m declines proportionately (207). Thus, beef heart particles turning over at 18 sec^{-1} (electrons/heme a) showed an apparent

K_m of 0.05 μM and k (for the O_2–a_3 reaction) of 0.6×10^8 M^{-1} sec^{-1}. More recent measurements (208) in uncoupled mitochondria oxidizing glutamate and malate gave $K_m < 0.05$ μM. As the oxygen concentration falls through the sensitive range, the percentage reduction of a_3 rises from its initial low ($<1\%$) level. Most of the results with uncoupled or poorly controlled ADP activated systems conform to the relationship:

$$\nu = 4k_1 [O_2] [a_3Fe^{2+}] = k_2[a_3Fe^{3+}] [cFe^{2+}] \tag{23}$$

Depending on the initial (high $[O_2]$) steady state reduction of c, the reduction of a_3 can then rise to a relatively high value before respiration is reduced by 50% (the data of Chance et al. (209) but not of Holton (210) are consistent with this requirement).

Two features complicate the situation. First, as pointed out by Chance (207), the buffering characteristics of the respiratory chains ensure that the relationship between O_2 concentration and rate is not of the Michaelis–Menten type but a sharper curve resembling a "titration," obeying the equation:

$$K_m = \alpha(TN)/4k_1 \tag{24}$$

Experimentally, α has a value of 0.6–0.7 (compared to the expected values of 1.0 for Michaelis–Menten kinetics and 0.5 for a strictly linear "titration"), as predicted for a linear chain of, say, four cytochromes turning over at 15–20% of the maximal rate. Second, the simple kinetics of Eqs. (22) and (23) do not apply to controlled systems (in state 4 or state 3 when this is markedly slower than the uncoupled state). Such systems seem to have abnormally high K_m values. Thus, rat liver mitochondria in state 4 were reported as having a K_m of 0.5 μM with a turnover of 14 sec^{-1}, equivalent to k_1 of 7×10^6 M^{-1} sec^{-1} (for 4×10^6 M^{-1} sec^{-1} if $\alpha = 0.6$), about 5% of the uncoupled rate (up to 1.5×10^8 M^{-1} sec^{-1}) (208).

On the other hand, ATP does not affect the observed reaction of reduced a_3 with oxygen (117) as would be expected if the apparent slowing were due to the presence of unreactive (high energy?) ferrous a_3, suggested by Slater and others (177, 178) on the basis of steady state observations. An alternate explanation of the results may be found in the occurrence of high energy (unreactive) (186) forms of ferric a_3 or of reversed electron transfer from a_3. Anaerobiosis in state 4 leads to incomplete reduction of cytochromes a and a_3; complete reduction follows only upon the addition of uncoupler to the energized state 5 system.

Jöbsis (211) has recently reviewed some of the problems involved in experimentation at low oxygen tensions, and Oshino et al. (212) have made further precise determinations of the oxygen affinity of mitochondria according to the luminous bacteria technique employed earlier by Schindler

(73). It is clear that Eq. (24) holds for a variety of conditions for those electron carriers whose mid-potentials are not dependent upon the ATP level.

V. CONCLUSIONS

Cytochrome oxidase is the only membrane-bound enzyme

(a) whose catalytic activity can be studied both *in situ* and in solution,

(b) whose prosthetic groups have been clearly identified,

(c) whose allosteric behavior can be monitored spectroscopically in the membrane,

(d) whose chemical reactivity is modified by membrane energization,

As such it must soon be the subject of intensive study that will resolve the problems described here and rapidly pose others that we have not even considered.

ACKNOWLEDGMENT

One of the authors (B. C.) acknowledges the support of USPHS GM 12202, P.N. of the Stanley Elmore Fund, Sidney Sussex College, Cambridge, and the support and encouragement of Dr. A. D. Bangham of the Biophysics Unit at Babraham. The authors wish to thank Drs. C. Greenwood, A. O. Muijsers, D. F. Wilson, and B. F. van Gelder for manuscripts of unpublished papers and for discussions.

REFERENCES

1. Koenig, A. (1894). "Über den menschlichen Sehrpurpum und seune Bedeutung." Akademie der Wissenschaften, Berlin.
2. Warburg, O. (1932). *Angew. Chem.* **45,** 1.
3. Castor, L. and Chance, B. (1955). *J. Biol. Chem.* **217,** 453.
4. Keilin, D. (1929). *Proc. Roy. Soc., Ser. B* **104,** 206.
5. Keilin, D. (1966). "A History of Cell Respiration and Cytochrome." Cambridge Univ. Press, London and New York.
6. Chance, B. (1967). *Biochem. J.* **103,** 1 (Second Keilin Lecture).
7. Warburg, O., and Negelein, E. (1931). *Biochem. Z.* **233,** 486.
8. Warburg, O., and Negelein, E. (1933). *Biochem. Z.* **262,** 237.
9. Keilin, D. (1930). *Proc. Roy. Soc., Ser. B* **106,** 418.
10. Keilin, D. (1933). *Nature (London)* **132,** 783.
11a. Keilin, D., and Hartree, E. F. (1938). *Nature (London)* **141,** 870.
11b. Keilin, D., and Hartree, E. F. (1939). *Proc. Roy. Soc., Ser. B* **127,** 167.
12. Chance, B. (1953). *J. Biol. Chem.* **202,** 383.
13. Keilin, D. (1925). *Proc. Roy. Soc., Ser. B* **98,** 312.
14. Laki, K. (1938). *Z. Physiol. Chem.* **254,** 27.
15. Warburg, O., Negelein, E., and Haas, E. (1933). *Biochem. Z.* **266,** 1.

16. Shibata, K., and Tamiya, H. (1933). *Acta Phytochim.* **7**, 191.
17. Shibata, K. (1935). *Ergeb. Enzymforsch.* **4**, 348.
18. Tamiya, H., and Ogura, Y. (1937). *Acta Phytochim.* **9**, 123.
19. Yakushiji, E., and Okunuki, K. (1940). *Proc. Imp. Acad. Tokyo* **16**, 299.
20. Okunuki, K., Sekuzu, I., Yonetani, T., and Takemori S. (1958). *J. Biochem. (Japan)* **45**, 847.
21. Okunuki, K. (1962). *In* "Oxygenases" (O. Hayaishi, ed.), p. 409. Academic Press, New York.
22. Tsudzuki, T., and Okunuki, K. (1971). *J. Biochem. (Japan)* **69**, 909.
23. Lemberg, M. R. (1969). *Physiol. Rev.* **49**, 48.
24. Wainio, W. W. (1970). "The Mammalian Respiratory Chain." Academic Press, New York.
25. Okunuki, K. (1966). *In* "Comprehensive Biochemistry" (M. Florkin and E. H. Stotz, eds.), Vol. 14, Biological Oxidations, p. 232. Elsevier, Amsterdam.
26. Sekuzu, I., and Takemori, S. (1972). *In* "Treatise on Electron Transport" (T. E. King and M. Klingenberg, eds.). Dekker, New York.
27. van Buuren, K. J. H., Zuurendonk, P. F., van Gelder, B. F., and Muijsers, A. O. (1972). *Biochim. Biophys. Acta* **256**, 243.
28. Horie, S., and Morrison, M. (1963). *J. Biol. Chem.* **238**, 2859.
29. Gilmour, M. V., Wilson, D. F., and Lemberg, M. R. (1967). *Biochim. Biophys. Acta* **143**, 487.
30. Wilson, D. F. (1967). *Biochim. Biophys. Acta* **131**, 431.
31. Beinert, H., Griffiths, D. E., Wharton, D. C., and Sanadi, R. W. (1962). *J. Biol. Chem.* **237**, 2337.
32. van Gelder, B. F., and Beinert, H. (1964). *Biochim. Biophys. Acta* **189**, 1.
33. Ehrenberg, A., and Yonetani, T. (1961). *Acta Chem. Scand.* **15**, 1071.
34. Brierley, G. P., and Merola, A. (1962). *Biochim. Biophys. Acta* **64**, 205.
35. Tzagoloff, A., and MacLennan, D. H. (1965). *Biochim. Biophys. Acta* **99**, 476.
36. Zahler, W. L., and Fleischer, S. (1971). *J. Bioenerg.* **2**, 209.
37. Awasthi, Y. C., Chuang, T. F., Keenan, F. W., and Crane, F. L. (1971). *Biochim. Biophys. Acta* **226**, 42.
38. Yonetani, T. (1967). *Methods Enzymol.* **10**, 30.
39. Orii, Y., and Okunuki, K. (1967). *J. Biochem. (Japan)* **61**, 388.
40a. Seki, S., Hiyashi, H., and Oda, T. (1970). *Arch. Biochem. Biophys.* **138**, 110.
40b. Seki, S., and Oda, T. (1970). *Arch. Biochem. Biophys.* **138**, 122.
41. Chuang, T. F., and Crane, F. L. (1971). *Biochem. Biophys. Res. Commun.* **42**, 1076.
42. Shakespeare, P. G., and Mahler, H. R. (1971). *J. Biol. Chem.* **246**, 7649.
43. Tzagoloff, A., and MacLennan, D. H. (1966). *In* "The Biochemistry of Copper" (J. Peisach, P. Aisen, and W. E. Blumberg, eds.), p. 253. Academic Press, New York.
44. Criddle, R. S., and Bock, R. M. (1959). *Biochem. Biophys. Res. Commun.* **1**, 138.
45. Love, B., Chan, S. H. P., and Stotz, E. (1970). *J. Biol. Chem.* **245**, 6664.
46. Chan, S. H. P., Love, B., and Stotz, E. (1970). *J. Biol. Chem.* **245**, 6669.
47. Minnaert, K. (1961). *Biochim. Biophys. Acta* **54**, 26.
48. Stotz, E., Altschul, A. M., and Hogness, T. R. (1938). *J. Biol. Chem.* **124**, 745.
49. Slater, E. C. (1949). *Biochem. J.* **44**, 305.
50. Smith, L., and Conrad, H. (1956). *Arch. Biochem. Biophys.* **63**, 403.
51. Minnaert, K. (1961). *Biochim. Biophys. Acta* **50**, 23.
52. Davies, H. C., Smith, L., and Wassermann, A. R. (1964). *Biochim. Biophys. Acta* **85**, 238.

53. Beetlestone, J. (1960). *Arch. Biochem. Biophys.* **89**, 35.
54. Nicholls, P., and Mochan, E. (1971). *Biochem. J.* **121**, 55.
55. Yonetani, T., and Ray, G. S. (1966). *J. Biol. Chem.* **241**, 700.
56. Nicholls, P. (1964). *Arch. Biochem. Biophys.* **106**, 25.
57. Yonetani, T., and Ray, G. S. (1965). *J. Biol. Chem.* **240**, 3392.
58. Yonetani, T. (1960). *J. Biol. Chem.* **235**, 3138.
59. Yonetani, T. (1962). *J. Biol. Chem.* **237**, 550.
60. Mason, H. S., and Ganapathy, K. (1970). *J. Biol. Chem.* **245**, 230.
61. van Buuren, K. J. H., Eggelte, T. A., and van Gelder, B. F. (1971). *Biochim. Biophys. Acta* **234**, 468.
62. Smith, L., and Newton, N. (1968). *In* "Structure and Function of Cytochromes" (K. Okunuki, M. D. Kamen, and I. Sekuzu, eds.), p. 153. Tokyo and Univ. Park Press, Baltimore, Maryland.
63. Nicholls, P. (1965). *In* "Oxidases and Related Redox Systems" (T. E. King, H. S. Mason, and M. Morrison, eds.), p. 764. Wiley, New York.
64. Yonetani, T. (1961). *J. Biol. Chem.* **236**, 1680.
65. Wainio, W. W., Eichel, B., and Gould, A. (1960). *J. Biol. Chem.* **235**, 1521.
66. Smith, L., and Minnaert, K. (1965). *Biochim. Biophys. Acta* **105**, 1.
67. Estabrook, R. W. (1961). *In* "Haematin Enzymes" (J. E. Falk, M. R. Lemberg, and R. K. Morton, eds.), p. 276. Pergamon, Oxford.
68. McGuinness, E. T., and Wainio, W. W. (1962). *J. Biol. Chem.* **237**, 3273.
69. Greenwood, C., and Palmer, G. (1965). *J. Biol. Chem.* **240**, 3660.
70. Czerlinski, G., and Bracokova, V. (1971). *Arch. Biochem. Biophys.* **147**, 707.
71. Margoliash, E., Barlow, G. H., and Byers, V. (1970). *Nature (London)* **228**, 723.
72. Kiese, M., and Reinwein, D. (1953). *Biochem. Z.* **324**, 51.
73. Chance, B., and Schindler, F. J. (1965). *In* "Oxidases and Related Redox Systems" (T. E. King, H. S. Mason, and M. Morrison, eds.), p. 921. Wiley, New York.
74. Greenwood, C., and Gibson, Q. H. (1967). *J. Biol. Chem.* **242**, 1782.
75. Nicholls, P., and Kimelberg, H. K. (1972). *In* "Biochemistry and Biophysics of Mitochondrial Membranes" (G. F. Azzone, E. Carafoli, A. L. Lehninger, E. Quagliariella, and N. Siliprandi, eds.), p. 17. Academic Press, New York.
76. Orii, Y., Sekuzu, I., and Okunuki, K. (1962). *J. Biochem. (Japan)* **51**, 204.
77. Kuboyama, M., Takemori, S., and King, T. E. (1962). *Biochem. Biophys. Res. Commun.* **9**, 534.
78. Mochan, E., and Nicholls, P. (1972). *Biochim. Biophys. Acta* **267**, 309.
79. Kimelberg, H. K., and Nicholls, P. (1969). *Arch. Biochem. Biophys.* **133**, 327.
80. Takemori, S., Wada, K., Ando, K., Hosokawa, M., Sekuzu, I., and Okunuki, K. (1962). *J. Biochem. (Japan)* **52**, 28.
81. Hettinger, T. O., and Harbury, H. A. (1964). *Proc. Nat. Acad. Sci. U. S.* **52**, 1469.
82. Smith, L., and Conrad, H. (1961). *In* "Haematin Enzymes" (J. Falk, M. R. Lemberg, and R. K. Morton, eds.), p. 260. Pergamon, Oxford.
83. Nicholls, P., Kimelberg, H. K., Mochan, E., Mochan, B. S., and Elliott, W. B. (1971). *In* "Probes of Structure and Function of Macromolecules and Membranes" (B. Chance, C. P. Lee, and J. K. Blasie, eds.), Vol. I, p. 431. Academic Press, New York.
84. Person, P., and Fine, A. S. (1961). *Arch. Biochem. Biophys.* **94**, 392.
85. Person, P., Mora, P. T., and Fine, A. S. (1963). *J. Biol. Chem.* **238**, 4103.
86. Person, P., Zipper, H., Fine, A. S., and Mora, P. T. (1964). *J. Biol. Chem.* **239**, 4159.
87. Mochan, B. S., Elliott, W. B., and Nicholls, P. (1973). *J. Bioenergetics* **4**, 329.

88. Adam, G., and Delbrück, M. (1968). *In* "Structural Chemistry and Molecular Biology" (A. Rich and N. Davidson, eds.), p. 198. Freeman, San Francisco, California.

89. Keilin, D., and Hartree, E. F. (1955). *Nature (London)* **176,** 200.

90. Nicholls, P., Mochan, E., and Kimelberg, H. K. (1971). *In* "Energy Transduction in Respiration and Photosynthesis" (E. Quagliariello, S. Papa, and C. S. Rossi, eds.), p. 339. Adriatica Editrice, Bari.

91. Lemberg, M. R., and Gilmour, M. V. (1967). *Biochim. Biophys. Acta* **143,** 500.

92. Dutton, P. L., Wilson, D. F., and Lee, C. P. (1970). *Biochemistry* **9,** 5077.

93. Chance, B., Lee, C. P., Mela, L., and Wilson, D. F. (1968). *In* "Structure and Function of Cytochromes" (K. Okunuki, M. D. Kamen, and I. Sekuzu, eds.), p. 353. Tokyo and Univ. Park Press, Baltimore, Maryland.

94. King, T. E., Kuboyama, M., and Takemori, S. (1965). *In* "Oxidases and Related Redox Systems" (T. E. King, H. S. Mason, and M. Morrison, eds.), p. 707. Wiley, New York.

95. Gibson, Q. H., Greenwood, C., Wharton, D. C., and Palmer, G. (1965). *J. Biol. Chem.* **240,** 888.

96. Antonini, E., Brunori, M., Greenwood, C., and Malmström, B. G. (1970). *Nature (London)* **228,** 936.

97. Lemberg, M. R., and Cutler, M. E. (1970). *Biochim. Biophys. Acta* **197,** 1.

98. Malmström, B. G. (1973). *Trans. Biochem. Soc.* **1,** 47.

99. Chance, B. (1972). *In* "The Molecular Basis of Electron Transport" (J. Schultz and B. F. Cameron, eds.), p. 65. Academic Press, New York.

100. Gibson, Q. H., and Greenwood, C. (1963). *Biochem. J.* **86,** 541.

101. King, T. E., and Lee, C. P. (1960). *Biochim. Biophys. Acta* **37,** 342.

102. Chance, B. (1961). *In* "Haematin Enzymes" (J. Falk, M. R. Lemberg, and R. K. Morton, eds.), p. 316. Pergamon, Oxford.

103. van Gelder, B. F. (1966). *Biochim. Biophys. Acta* **118,** 36.

104. van Gelder, B. F., and Muijsers, A. O. (1966). *Biochim. Biophys. Acta* **118,** 47.

105. Wilson, D. F., and Leigh, J. S. (1972). *Arch. Biochem. Biophys.* **150,** 154.

106. Gibson, Q. H., and Greenwood, C. (1965). *J. Biol. Chem.* **240,** 2694.

107. Gibson, Q. H., and Greenwood, C. (1965). *J. Biol. Chem.* **240,** PC 957.

108. Wharton, D. C., and Gibson, Q. H. (1968). *J. Biol. Chem.* **243,** 702.

109. Chance, B., DeVault, D., Legallais, V., Mela, L., and Yonetani, T. (1967). *In* "Fast Reactions and Primary Processes in Chemical Kinetics" (S. Claessen, ed.), p. 437. Wiley (Interscience), New York.

110. Chance, B., and Erecinska, M. (1971). *Arch. Biochem. Biophys.* **143,** 675.

111. Sekuzu, I., Takemori, S., Yonetani, T., and Okunuki, K. (1959). *J. Biochem. (Japan)* **46,** 43.

112. Davison, A. J., and Wainio, W. W. (1968). *J. Biol. Chem.* **243,** 5023.

113. Lemberg, M. R., and Mansley, G. E. (1966). *Biochim. Biophys. Acta* **118,** 19.

114. Williams, G. R., Lemberg, M. R., and Cutler, M. E. (1968). *Can. J. Biochem.* **46,** 1371.

115. Gilmour, M. V., Lemberg, M. R., and Chance, B. (1967). *Biochim. Biophys. Acta* **172,** 37.

116. Vanneste, W. H. (1966). *Biochemistry* **5,** 838.

117. Erecinska, M., and Chance, B. (1972). *Arch. Biochem. Biophys.* **151,** 304.

118. Muijsers, A. O., Tiesjema, R. H., and van Gelder, B. F. (1971). *Biochim. Biophys. Acta* **234,** 481.

119. Wilson, D. F., Lindsay, J. G., and Brocklehurst, E. S. (1972). *Biochim. Biophys. Acta* **256**, 277.
120. Smith, L. (1955). *J. Biol. Chem.* **215**, 833.
121. Minnaert, K. (1961). *Biochim. Biophys. Acta* **54**, 26.
122. Nicholls, P., and Kimelberg, H. K. (1968). *Biochim. Biophys. Acta* **162**, 11.
123. Baum, H., and Rieske, J. S. (1966). *Biochem. Biophys. Res. Commun.* **24**, 1.
124. Ball, E. G. (1938). *Biochem. Z.* **295**, 262.
125. Minnaert, K. (1965). *Biochim. Biophys. Acta* **110**, 42.
126. Horio, T., and Ohkawa, J. (1968). *J. Biochem. (Japan)* **64**, 393.
127. Wilson, D. F., and Dutton, P. L. (1970). *Arch. Biochem. Biophys.* **136**, 583.
128. Nicholls, P. (1968). *In* "Structure and Function of Cytochromes" (K. Okunuki, M. D. Kamen, and I. Sekuzu, eds.), p. 76. Tokyo and Univ. Park Press, Baltimore, Maryland.
129. Nicholls, P. (1972). *Biochem. J.* **128**, 98P.
130. Yong, F. C., and King, T. E. (1972). *Biochem. Biophys. Res. Commun.* **47**, 380.
131. Tiesjema, R. H., Muijsers, A. O., and van Gelder, B. F. (1973). *Biochim. Biophys. Acta* **305**, 19.
132. Muijsers, A. O., Tiesjema, R. H., Henderson, R. W., and van Gelder, B. F. (1971). *Biochim. Biophys. Acta* **267**, 216.
133. Tzagoloff, A., and Wharton, D. C. (1965). *J. Biol. Chem.* **240**, 2628.
134. Caswell, A. H. (1968). *J. Biol. Chem.* **243**, 5827.
135. Fee, J. A., Malkin, R., Malmström, B. G., and Vanngård, T. (1969). *J. Biol. Chem.* **244**, 4200.
136. Horie, S. (1964). *J. Biochem. (Japan)* **56**, 57.
137. Keilin, D., and Hartree, E. F. (1938). *Proc. Roy. Soc.* **B125**, 171.
138. Wainio, W. W., and Greenlees, J. (1960). *Arch. Biochem. Biophys.* **90**, 18.
139. Antonini, E., Brunori, M., Greenwood, C., Malmström, B. G., and Rotilio, G. C. (1971). *Eur. J. Biochem.* **23**, 396.
140. van Buuren, K. J. H., Nicholls, P., and van Gelder, B. F. (1972). *Biochim. Biophys. Acta* **256**, 258.
141. Wohlrab, H., and Ogunmola, G. (1971). *Biochemistry* **10**, 1103.
142. Morrison, M., and Horie, S. (1965). *J. Biol. Chem.* **240**, 1354.
143. Gibson, Q. H., Palmer, G., and Wharton, D. C. (1965). *J. Biol. Chem.* **240**, 925.
144. Mansley, G. E., Stanbury, J. T., and Lemberg, M. R. (1966). *Biochim. Biophys. Acta* **113**, 33.
145. Nicholls, P. (1963). *Biochim. Biophys. Acta* **73**, 667.
146. Nicholls, P., van Buuren, K. J. H., and van Gelder, B. F. (1972). *Biochim. Biophys. Acta* **275**, 279.
147. Storey, B. T. (1970). *Plant Physiol.* **45**, 455.
148. Wilson, D. F., Erecinska, M., and Brocklehurst, E. S. (1972). *Arch. Biochem. Biophys.* **151**, 180.
149. van Buuren, K. J. H., and van Gelder, B. F. (1973). Personal communication.
150. Camerino, P. W., and King, T. E. (1966). *J. Biol. Chem.* **241**, 970.
151. Keilin, D. (1936). *Proc. Roy. Soc., Ser. B* **121**, 165.
152. Muijsers, A. O., Slater, E. C., and van Buuren, K. J. H. (1968). *In* "Structure and Function of Cytochromes" (K. Okunuki, M. D. Kamen, and I. Sekuzu, eds.), p. 129. University Park Press, Baltimore, Maryland.
153. Stannard, J. N., and Horecker, B. L. (1948). *J. Biol. Chem.* **172**, 599.
154. Wilson, D. F. (1971). *In* "Probes of Structure and Function of Macromolecules and

Membranes" (B. Chance, T. Yonetani, and A. S. Mildvan, eds.), Vol. II, p. 573. Academic Press, New York.
155. Wilson, D. F., and Chance, B. (1966). *Biochem. Biophys. Res. Commun.* **23**, 751.
156. Wilson, D. F., and Chance, B. (1967). *Biochim. Biophys. Acta* **131**, 421.
157. Nicholls, P., and Wenner, C. (1972). *Arch. Biochem. Biophys.* **151**, 206.
158. Mitchell, P., and Moyle, J. (1971). *J. Bioenergetics* **2**, 1.
159. Palmieri, F., and Klingenberg, M. (1967). *Eur. J. Biochem.* **1**, 439.
160. Wikström, M. K. F., and Saris, N. E. L. (1969). *Eur. J. Biochem.* **9**, 160.
161. Wilson, D. F., and Brooks, E. (1970). *Biochemistry* **9**, 1060.
162. Yoshikawa, S., and Orii, Y. (1970). *J. Biochem. (Japan)* **68**, 145.
163. Blokzijl-Homan, M. F. J., and van Gelder, B. F. (1971). *Biochim. Biophys. Acta* **234**, 493.
164. Muijsers, A. O., van Gelder, B. F., and Slater, E. C. (1966). *In* "Hemes and Hemoproteins" (B. Chance, R. W. Estabrook, and T. Yonetani, eds.), p. 467. Academic Press, New York.
165. Beinert, H. (1966). *In* "The Biochemistry of Copper" (J. Peisach, P. Aisen, and W. E. Blumberg, eds.), p. 213. Academic Press, New York.
166. Tsudzuki, T., and Wilson, D. F. (1971). *Arch. Biochem. Biophys.* **145**, 149.
167. Mitchell, P., and Moyle, J. (1970). *In* "Electron Transport and Energy Conservation" (S. Papa, J. M. Tager, E. Quagliariello, and E. C. Slater, eds.), p. 575. Adriatica Editrice, Bari.
168. Racker, E. (1972). *In* "The Molecular Basis of Electron Transport" (J. Schultz and B. F. Cameron, eds.), p. 45. Academic Press, New York.
169. Nicholls, P., Mochan, E., and Kimelberg, H. K. (1969). *FEBS Lett.* **3**, 242.
170. Greville, G. D. (1969). *In* "Current Topics in Bioenergetics" (D. R. Sanadi, ed.), Vol. 3, p. 1. Academic Press, New York.
171. Hinkle, P., Kim, J. J., and Racker, E. (1972). *J. Biol. Chem.* **247**, 1338.
172. Jasaitis, A. A., Nemeček, I. B., Severina, I. I., Skulachev, V. P., and Smirnova, S. M. (1972). *Biochim. Biophys. Acta* **275**, 485.
173. Racker, E., and Kandrach, A. (1971). *J. Biol. Chem.* **246**, 7069.
174. Kuboyama, M., Yong, F. C., and King, T. E. (1972). *J. Biol. Chem.* **247**, 6375.
175. Chance, B., and Williams, G. R. (1956). *Advan. Enzymol.* **17**, 65.
176. Ramirez, J. (1959). *J. Physiol.* **147**, 14.
177. Muraoka, S., and Slater, E. C. (1969). *Biochim. Biophys. Acta* **180**, 227.
178. Grimmelikhuijzen, C. J. P., and Slater, E. C. (1972). *Biochim. Biophys. Acta* **256**, 24.
179. Slater, E. C. (1969). *In* "The Energy Level and Metabolic Control in Mitochondria" (S. Papa, J. M. Tager, E. Quagliariello, and E. C. Slater, eds.), p. 255. Adriatica Editrice, Bari.
180. Wikström, M. K. F., and Saris, N. E. L. (1970). *In* "Electron Transport and Energy Conservation" (S. Papa, J. M. Tager, E. Quagliariello, and E. C. Slater, eds.), p. 77. Adriatica Editrice, Bari.
181. Hinkle, P., and Mitchell, P. (1970). *J. Bioenergetics* **1**, 45.
182. Wikström, M. K. F. (1972). *Biochim. Biophys. Acta* **283**, 385.
183. Chance, B., Lee, C. P., and Schoener, B. (1966). *J. Biol. Chem.* **241**, 4574.
184. Lindsay, J. G., and Wilson, D. F. (1973). *Biochemistry* **11**, 4613.
185. Wilson, D. F., Erecinska, M., and Nicholls, P. (1972). *FEBS Lett.* **20**, 61.
186. Erecinska, M., Wilson, D. F., Sato, N., and Nicholls, P. (1972). *Arch. Biochem. Biophys.* **151**, 188.

187. Nicholls, P., Erecinska, M., and Wilson, D. F. (1973). *In* "Mechanisms in Bio-energetics" (G. F. Azzone, E. Carafoli, A. L. Lehninger, E. Quagliariello, and N. Siliprandi, eds.), p. 561. Academic Press, New York.
188. Glaze, R. P., and Wadkins, C. L. (1964). *Biochem. Biophys. Res. Commun.* **15**, 194.
189. Wilson, D. F. (1969). *Biochemistry* **8**, 2475.
190. Baum, H., Hall, G. S., Nalder, J., and Beechey, R. B. (1971). *In* "Energy Transduction in Respiration and Photosynthesis" (E. Quagliariello, S. Papa, and C. S. Rossi, eds.), p. 747. Adriatica Editrice, Bari.
191. Klingenberg, M. (1968). *In* "Biological Oxidations" (T. B. Singer, ed.), p. 3. Wiley, New York.
192. Rieske, J. S., Baum, H., Stoner, C. D., and Lipton, S. H. (1967). *J. Biol. Chem.* **242**, 4854.
193. Chance, B. (1965). *In* "Oxidases and Related Redox Systems" (T. E. King, H. S. Mason, and M. Morrison, eds.), p. 929. Wiley, New York.
194. Holmes, W. F. (1960). Ph.D. Dissertation, Univ. of Pennsylvania, Philadelphia, Pennsylvania.
195. Wrigglesworth, J., Baum, H., and Nicholls, P. (1973). *FEBS Lett.* (in press).
196. Berden, J. A., and Slater, E. C. (1972). *Biochim. Biophys. Acta* **256**, 199.
197. Wohlrab, H. (1970). *Biochemistry* **9**, 474.
198. Kowalsky, A. (1965). *Biochemistry* **4**, 2382.
199. Wagner, M., and Erecinska, M. (1971). *Arch. Biochem. Biophys.* **147**, 666.
200. Wagner, M., Erecinska, M., and Pring, M. (1971). *Arch. Biochem. Biophys.* **147**, 675.
201. Chance, B., Erecinska, M., and Lee, C. P. (1970). *Proc. Nat. Acad. Sci. U. S.* **66**, 928.
202. Green, D. E., Wharton, D. C., Tzagoloff, A., Rieske, J. S., and Brierley, G. P. (1965). *In* "Oxidases and Related Redox Systems" (T. E. King, H. S. Mason, and M. Morrison, eds.), p. 1032. Wiley, New York.
203. Margoliash, E. (1972). *In* "Harvey Lectures," Ser. 66, p. 177. Academic Press, New York.
204. Dickerson, R. E., Takano, T., Eisenberg, D., Kallai, O. B., Samson, L., Cooper, A., and Margoliash, E. (1971). *J. Biol. Chem.* **246**, 1511.
205. Nicholls, P., and Mochan, E. (1971). *Nature (London) New Biol.* **230**, 276.
206. Smith, L., and Camerino, P. (1963). *Biochemistry* **2**, 1432.
207. Chance, B. (1965). *J. Gen. Physiol.* **49**, 163.
208. Degn, H., and Wohlrab, H. (1971). *Biochim. Biophys. Acta* **245**, 347.
209. Chance, B., Schoener, B., and Schindler, F. (1964). *In* "Oxygen in the Animal Organism" (F. Dickens and E. Neil, eds.), p. 367. Pergamon, Oxford.
210. Holton, F. A. (1964). *In* "Oxygen in the Animal Organism" (F. Dickens and E. Neil, eds.), p. 390. Pergamon, Oxford.
211. Jöbsis, F. F. (1972). *Fed. Proc. Fed. Amer. Soc. Exp. Biol.* **31**, 1404.
212. Oshino, N., Oshino, R., Tamura, M., Kobilinsky, L., and Chance, B. (1972). *Biochim. Biophys. Acta* **273**, 5.

13

Peroxidase

ISAO YAMAZAKI

I. INTRODUCTION*

Peroxidases are enzymes, whose primary function is to oxidize molecules at the expense of hydrogen peroxide. Probably because of their wide distribution, especially in plants, and their dramatic catalysis of colored product formation, these enzymes have been among the most extensively investigated since the beginnings of enzymology. A detailed historical review of these enzymes was given by Saunders et al. (1). The names "peroxidase" and "oxygenase" are analogous because of their strong specificities for peroxide and oxygen, respectively. From the point of view of function, however, peroxidase is rather similar to oxidase, the name of which is now used in a narrow sense for electron transfer oxidases. The name "oxygenase" is restricted to enzymes that catalyze the incorpora-

---

* Abbreviations; HRP, horseradish peroxidase; CCP, cytochrome c peroxidase; TP, tryptophan pyrrolase; IAA, indole-3-acetic acid; DHF, dihydroxyfumaric acid; NADH, reduced form of nicotinamide-adenine dinucleotide; and ESR, electron spin resonance.

tion of atmospheric oxygen into substrate molecules (2). Based on the acceptor specificity and by analysis of the final reaction products, the distinctions between electron transfer oxidase, oxygenase, and peroxidase appear to be established.

Reactions of mixed types such as monooxygenase (mixed function oxidase by Mason's terminology, refs. 3 and 4) and peroxidase–oxidase have been reported. If peroxide is an intermediate product of O_2 reduction it might be said that in many cases peroxide metabolism is involved as a part of the overall oxygen metabolism. Detailed analyses of the mechanism of O_2 metabolism have revealed that three types of reactions are correlated in complicated ways.

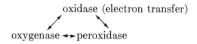

The recently established idea of oxyferroperoxidase structure for so-called Compound III would make it easy to relate the function of peroxidase with those of electron transfer oxidase and oxygenase.

II. FUNCTIONS OF PEROXIDASES

Besides peroxidatic oxidation of electron donor molecules various reactions have been found to be catalyzed by peroxidase. These are aerobic oxidations of DHF (5, 6), IAA (7–9), triose reductone (10), NADH (11, 12) and naphtohydroquinone (13, 14), hydroxylation of aromatic molecules (15, 16), formation of ethylene from methional (17), halogenation (18, 19), and antimicrobial activity (20, 21). The common feature of these reactions appears to be an involvement of H_2O_2. Thus, it is likely that their mechanisms are controlled by a unique feature of the peroxidase reaction.

A. Higher Oxidation States of HRP

1. HRP Compounds Formed in the Catalytic Reaction

Peroxidase catalysis is characterized by the one-electron oxidation of donor molecules. The following reactions have been confirmed (22–24):

$$\text{Peroxidase} + H_2O_2 \rightarrow \text{Compound I} \tag{1}$$

$$\text{Compound I} + AH_2 \rightarrow \text{Compound II} + AH\cdot \tag{2}$$

$$\text{Compound II} + AH_2 \rightarrow \text{peroxidase} + AH\cdot \tag{3}$$

$$2\,AH\cdot \rightarrow A + AH_2 \text{ (or AH–AH)} \tag{4}$$

Compounds I and II are thus found to be obligatory enzyme intermediates in the overall peroxidase reaction. Judged by the hyperfine structure of ESR spectra, the free radicals derived from several donor molecules in the above reactions were considered to be free in solution (24, 25). These free radicals are very reactive, and their reactions will result in a variety of peroxidase functions.

During aerobic oxidations of DHF and NADH catalyzed by HRP the enzyme was converted into an another compound, called Compound III (5, 6, 12). The compound had been identified as a product formed in the presence of excess H_2O_2 (26, 27).

2. Formation of HRP Compound III

An oxyferroperoxidase structure has been suggested for peroxidase Compound III (3, 28–31). This idea, however, was not generally accepted since reduced peroxidase was thought to be oxidized to the ferric enzyme without an intermediate similar to Compound III (32–34). Recently, the oxyferroperoxidase structure of Compound III has been confirmed by the following experiments. (a) Ferroperoxidase did react with oxygen to form Compound III when excess hydrosulfite that reduced Compound III was not present (35–37). (b) Compound III was formed by photolysis of an aerobic solution of CO-ferroperoxidase (38, 39). (c) Titrimetric experiments showed that Compound III was at a three-equivalent oxidized state above the ferric enzyme (37, 39, 40).

It was shown by Chance (41) and George (27) that Compound III was formed from the reaction of Compound II and H_2O_2. One more path of Compound III formation via reaction between the ferric enzyme and superoxide anion was suggested by Yamazaki and Piette (30). The latter reaction has been considered to be involved in the formation of Compound III during aerobic oxidations of DHF (30, 31) and NADH (12) catalyzed by peroxidase. Participation of superoxide anions in the above reactions was finally confirmed in the cases of HRP (42) and myeloperoxidase (43, 44) using superoxide dismutase (45) which accelerates the decomposition of superoxide anions.

3. Reactions of Higher Oxidation Forms of HRP

Figure 1 shows the relationship between five redox forms of HRP. All these forms could be obtained in fairly stable states under suitable experimental conditions and have been crystallized using a basic HRP preparation, named "isoenzyme F" (46). Higher oxidation forms of HRP could be reduced to the ferric enzyme by various electron donors. The most active form is Compound I. Chance and his colleague showed that nitrous acid

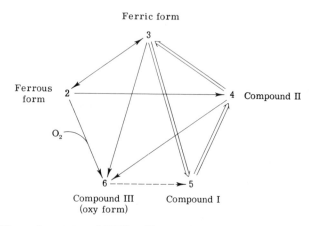

Fig. 1. Five redox states of HRP. The numbers formally indicate the effective oxidation number of the peroxidase heme: $2 + O_2 \rightharpoonup 6$, $2 + H_2O_2 \rightharpoonup 4$, $3 + HO_2 \rightharpoonup 6$, $3 + H_2O_2 \rightharpoonup 5$, and $4 + H_2O_2 \rightharpoonup 6$; others are one-electron steps. Reaction paths, $3 \rightharpoonup 5 \rightharpoonup 4 \rightharpoonup 3$ indicate a catalytic cycle of the enzyme.

(47) and p-aminobenzoic acid (48) reduced Compound I 100 and 25 times faster than Compound II. Similar results were obtained by Cormier and Prichard (49) with luminol, by Dunford and colleagues with ferrocyanide (50) and iodide (51, 52), and by Yamazaki et al. (53) with ascorbic acid and anthranilic acid. The conversion of HRP Compound I to the ferric enzyme without an appreciable formation of Compound II was reported by Björksten (54) and Roman and Dunford (52) using iodide as an electron donor.

Horseradish peroxidase Compound III also reacted with various electron donors which are not autoxidizable (31, 39, 55). Compound III was less reactive with electron donors than Compound II, while in the absence of such donors Compound III was less stable than Compound II. The stability of Compound II varied greatly with enzyme preparations, but that of Compound III was almost independent of the purity of enzyme preparations (39).

Horseradish peroxidase Compound III underwent a spontaneous decay to the ferric enzyme without detectable intermediates like Compounds I and II (36, 37, 39). This might be explained by assuming that the oxidative decomposition of Compound III occurs in the presence of one-electron oxidants such as Compounds I and II.

$$\text{Compound III} + e^- \rightarrow \text{peroxidase} + O_2 \qquad (5)$$

Rate constants of the reactions of Compound III with Compounds I and

II could be measured (39). However, the release of O_2 during the reaction has not been confirmed. For the decomposition of Compound III a mechanism in which dissociation of Compound III into Compound II and H_2O_2 is rate limiting was proposed (39), but this needs further proof.

The reaction of Compound III with electron donors would be of particular interest. Undoubtedly, O_2 is activated when it combines with ferroperoxidase. It might be reasonable to assume that Compound III is reduced by electron donors via the intermediates Compounds I and II. As mentioned above, these intermediates are much more active oxidants for the electron donors so far used. Therefore, an ingenious device would be needed to identify the intermediates in the reduction process of Compound III.

4. Oxidation–Reduction Potentials

Harbury (33) measured the oxidation–reduction potential of HRP for the reaction, ferric HRP $\rightleftharpoons$ ferrous HRP. The potential was -0.271 at pH 7.0. For the reaction, ferric HRP $\rightleftharpoons$ HRP Compound II, the reversible potential has not been obtained. George (56, 57) showed that HRP could be oxidized by a variety of one- and two-electron oxidants to compounds similar to Compounds I and II. Fergusson (58), however, suggested a possibility that the formation of such compounds occurred via the intermediate formation of H_2O_2. Observing that HRP Compound II was reduced by chloroiridite ions (1.02 V) but not by ferrous tris-2,2'-dipyridyl and tris-o-phenanthroline ions (1.06 and 1.14 V), George (22) proposed a tentative value of about 1.0 V for the oxidation–reduction potential of the Compound II and ferric HRP couple. He also suggested that Compound I could be a more powerful oxidizing agent than Compound II by about 0.3–0.6 V.

It might be generally said that the reactions shown in Fig. 1 tend to proceed almost to completion under physiological conditions, and it is difficult to measure these equilibrium constants with the usual methods except for the ferrous and ferric couple. The direct conversion from ferrous HRP to Compound II was demonstrated by Noble and Gibson (59).

B. IAA Oxidation

1. General Mechanism of Peroxidase–Oxidase Reaction

Among numerous substrates of peroxidase–oxidase reactions the IAA reaction has been most intensively studied. The principal feature of the peroxidase–oxidase reaction appeared to be the same irrespective of substrate molecules. The earlier studies have been reviewed by Nicholls (60).

The mechanism of Fig. 2 was proposed by Yamazaki and Piette (30). The common properties of the reaction are (a) a catalytic amount of H_2O_2 is necessary to initiate the reaction by forming the free radicals of donor molecules (catalase inhibits the reaction); (b) redogenic molecules (60, 61) such as ascorbate inhibit the reaction by consuming H_2O_2 or by inter-action with active intermediates having oxidizing equivalents (61); (c) oxidogenic molecules (60, 61) such as phenols promote the reaction by in-creasing the rate of free radical formation of substrate molecules (61); and (d) Mn^{2+} ions promote the reaction, probably through interaction with superoxide radical species or with free radicals of oxidogenic molecules.

The extent of activation or inhibition caused by a given molecule varies greatly with substrates and experimental conditions. Although this mecha-nism appeared to be essentially valid for the reaction with IAA, evidence has been accumulated which indicates that the reaction of HRP with this substrate differs from those with other substrates such as NADH and DHF.

2. Problems in IAA Oxidation

It is very interesting to note here a feature of IAA–HRP reactions that is rather similar to an oxygenase type. Except for IAA the oxidation prod-

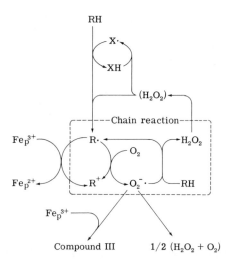

FIG. 2. General mechanism of peroxidase–oxidase reaction. XH is an oxidogenic molecule which promotes the reaction. The mechanism will vary with RH molecules depending upon the reactivity with Compound III, the stability of a product, ROO· radical, etc. Fe_p^{2+} and Fe_p^{3+} denote ferrous and ferric HRP, respectively.

ucts of peroxidase–oxidase reactions are principally two-electron oxidized forms. By the efforts of many workers (7, 62–66) the following stoichiometry has been confirmed:

$$(6)$$

The products were found to be indole-3-aldehyde and 3-methylene oxindole. Morita *et al.* (64) showed that the dominant product was indole-3-aldehyde when higher enzyme concentrations were used. Apparently the reaction is similar to the lactate oxidative decarboxylase reaction, which is a monooxygenase type (67) or an internal mixed function oxidase type (4). Indole-3-acetic acid was found to react with HRP Compound III at a relatively high rate (31, 39, 55). In this respect the reaction is similar to the tryptophan pyrrolase reaction, in which the oxygenated enzyme was considered to be an obligatory intermediate (68, 69). Many features of the mechanism of the IAA reaction, however, still remain to be elucidated. The key points of the mechanism would be summarized in the following way.

a. The IAA free radical formed by peroxidase catalysis is an important intermediate (9, 65, 66, 70–74). Hinman and Lang (65) proposed a mechanism in which the radical reacts with molecular oxygen to form 3-methylene oxindole as a main product through several nonenzymic steps. Though it is unknown whether the radical is free or attached to the enzyme, it may react with external oxidants such as ferric cytochrome c and ferric o-phenanthroline complexes (70). A question which arises is why the nature of the final products depends upon the concentration of enzymes in the reaction (64–66).

b. During the reaction with IAA a part of the enzyme is observed as an intermediate, spectrally similar to Compounds II and III (66, 70, 73–77). From the visible spectrum between 500 and 600 nm the intermediate is judged to be Compound III at around pH 4.0 (66) and Compound II at pH 5.0 (75, 78). It has been concluded that Compound III accumulated during the reactions of DHF and NADH is not a Michaelis type of intermediate (6, 12, 30). Since IAA reacts with HRP Compound III at a significantly high rate, the formation of Compound III during the IAA oxidation must have a positive meaning in the catalysis. However, it is still uncertain whether or not the main reaction proceeds via Compound II or III, mostly because of the difficulty in distinguishing between these two HRP compounds.

c. A small amount of H_2O_2 eliminates the lag phase and promotes the O_2-consuming oxidation of IAA under certain experimental conditions (7, 66, 70–72, 79). It is also true that the inhibition by catalase is negligible under certain experimental conditions (75, 78). Consequently, the role of H_2O_2 does not appear to be essential even as an initiator of the reaction. Using superoxide dismutase it has recently been found that superoxide anions are not involved in the reaction of IAA with HRP (77, 78). This fact seems to be a peculiar property of the IAA reaction since the oxidations of NADH and DHF by HRP are strongly inhibited by superoxide dismutase (78). This fact is compatible with the mechanism (65, 72) that the IAA peroxide radical instead of the superoxide anion radical is a product of the reaction between the IAA radical and O_2. Questions to be answered are how the IAA radical can be formed in the abesnce of H_2O_2 and how HRP Compound III is formed without an involvement of superoxide anions.

If the absolute necessity for H_2O_2 in the activation process does not apply to the IAA reaction, the mechanism would be different from the general peroxidase—oxidase reaction shown in Fig. 2. Fox et al. (75) have proposed a mechanism which involves a rapid equilibrium exchange of biradical molecular oxygen with water through a coordination position of the ferric iron to yield an oxygenated form of HRP. They suggested that oxygenated HRP attacks IAA to form an intermediate compound similar to Compound I. Though the mechanism involving an activation process without H_2O_2 participation is not necessarily supported by unambiguous evidence, it is suggestive. On the other hand, Ricard and Nari (73) showed that HRP was reduced directly by IAA under anaerobic conditions at acidic pH. This reduction seems to be too slow to explain the rapid oxidation of IAA (78) but is in accord with the known properties of HRP. Since interactions between HRP and donor molecules have been demonstrated (53, 80), it would be reasonable to assume an equilibrium between HRP and IAA as the first step of the reaction. The rapid oxidation of IAA accompanied by the formation of HRP peroxide compounds after the addition of IAA would be tentatively explained by a mixed mechanism

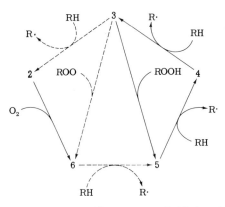

FIG. 3. Mechanism of IAA (RH) oxidation (78). Solid lines indicate maintenance reactions. A radical chain process may be involved, with propagation steps which can be generalized as

$$R\cdot + O_2 \rightharpoonup ROO\cdot$$

$$ROO\cdot + RH \rightharpoonup ROOH + R\cdot$$

as shown in Fig. 3. In this mechanism it is assumed that IAA peroxide (ROOH) and its free radical (ROO·) serve as H_2O_2 and $HO_2\cdot$, respectively, in the general peroxidase–oxidase reaction of Fig. 2. According to Yonetani et al. (81), CCP can react with relatively large peroxide such as 2,5-dimethylhexane 2,5-dihydroperoxide to form its peroxide compound. Schonbaum (82) has shown that reaction of HRP with m-nitroperoxybenzoate yields Compound I with the simultaneous release of m-nitrobenzoate.

The mechanism proposed by Ricard and Nari (73, 74) might be involved to some extent in the reaction at acidic pH.

3. Heme Degradation during IAA Oxidation

Yokota and Yamazaki (55, 83) reported that during the oxidation of IAA, HRP was converted into a compound similar to Compound IV which was found by Chance (84) in the presence of excess methyl peroxide. This fact appears to be in accord with the observation of Fox et al. (75) that a considerable amount of HRP is inactivated during the oxidation of IAA. Further studies (85) revealed that one mole of CO is released for one mole of HRP transformed and the product of HRP is similar to choleglobin (86), details of which have been reviewed by Lemberg (87) and O'hEocha

(88). Since the efficiency of the transformation was higher when IAA reacted with Compound I than Compound II or III (78), a tentative scheme may be proposed as shown in Fig. 4. This transformation was specific for the isoenzymes of HRP and was observed in the case of isoenzymes HRP B and C but not HRP A and F.

Bagger and Williams (89) observed a compound with an intense absorption band in the near infrared when excess H_2O_2 was added to HRP. They suggested that this compound, called P-940, is an intermediate in the conversion of HRP into P-670 caused by excess H_2O_2; P-940 was also observed as an intermediate in the reaction of HRP with IAA (78).

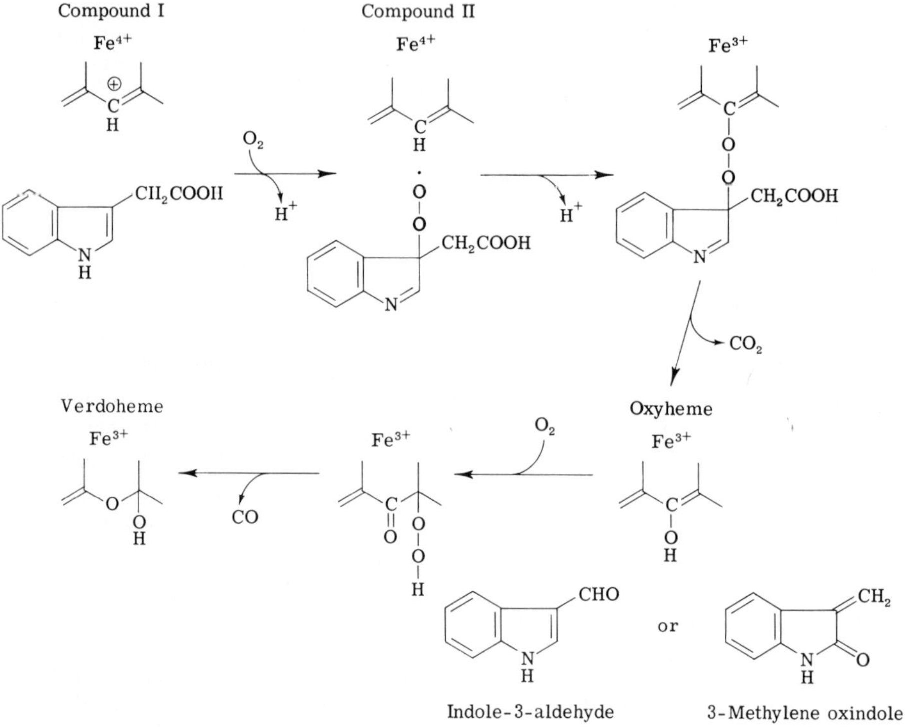

FIG. 4. Heme transformation in the reaction with IAA. The mechanism is based on the following observations (78): (1) During the catalytic oxidation of IAA the disappearance of Compound II and the appearance of verdoheme were observed at the same time; (2) in the stoichiometric reaction (not catalytic) the heme transformation was most effective when IAA was added to Compound I. The latter half mechanism is from Lemberg (87).

III. STRUCTURE OF PEROXIDASES

A. Chemical Structure

1. Prosthetic Group

The prosthetic groups of HRP, Japanese radish peroxidase, CCP, and chloroperoxidase are known to be ferriprotoporphyrin IX. The heme groups in animal peroxidases are much more tightly bound to the protein than are the hemes in plant peroxidases, and their chemical structures have not been thoroughly clarified. Thyroid peroxidase has a heme group which is not covalently linked to the apoenzyme and is closely related to protoheme (90–92). The prosthetic group of milk peroxidase is bound by covalent bonds to the protein and can be separated from the protein by reductive cleavage with HI. The isolated heme gave rise to mesoporphyrin IX, and it was suggested that the heme in milk peroxidase is directly related to protoheme (93). The spectral properties of myeloperoixdase are quite different from other peroxidases. It was reported that the pyridine hemochromogen spectrum was similar to that of heme a (94, 95).

2. Protein

Little is known about the chemical structure of peroxidases. Amino acid analyses were carried out for five isoenzyme preparations of HRP (96) and for two isoenzyme preparations of Japanese radish peroxidase (97, 98). The conformation of two HRP isoenzymes A_1 and C was investigated by means of circular dichroism, and it was suggested that the active sites were similar in both isoenzymes but that small differences did exist (99). The environment surrounding the heme of CCP was found to be hydrophobic (100). This might be the case with all plant peroxidases. The hydrophobic structure of the heme environment of myoglobin and hemoglobin has been disclosed by X-ray analyses.

A possibility that the fifth coordination position in HRP is occupied by an imidazole group of the protein was suggested by several workers (101–104). By elaborate experiments using [14]NO, Yonetani (104) concluded that the fifth ligand of the heme iron in CCP and HRP may be the imidazole group of a histidine residue and that the unpaired electron of NO sits in the sixth coordination position although it is considerably delocalized to the heme iron and the proximal nitrogen in these NO compounds of ferrous peroxidases.

B. Electronic Structure of the Heme

The electronic structure of the heme iron has been a subject of considerable interest. Most of the earlier studies were carried out by Theorell's group (32, 105) using the technique of magnetic susceptibility. Theorell and Ehrenberg (106) measured the number of odd electrons present in some peroxide compounds of myoglobin, peroxidase, and catalase. Electron spin resonance spectroscopy was also successfully used for peroxidases in frozen solutions (107, 108). The major advance, however, has recently come from the work of Yonetani and his colleagues. Through a series of elegant ESR studies, mostly on CCP, they were able to measure the electronic structure of the heme iron in various states of the enzyme (109, 110).

1. High Spin and Low Spin States of Ferric Peroxidases

The spin state of the ferric heme iron can be investigated by means of paramagnetic susceptibility and optical, ESR, and Mössbauer spectroscopies. The measurements of paramagnetic susceptibility of CCP (111, 112) and HRP (113, 114) gave precise information on a thermal equilibrium between high and low spin states. The information on the anisotropy of paramagnetic susceptibility in the heme plane can be provided by ESR spectroscopy (109, 115, 116).

2. Ferryl Structure in Compounds I and II

On the basis of the oxidation equivalent and the comparison with model systems, George (27) suggested a ferryl ion, FeO^{2+}, as a possible structure of the iron atom of Compound II. The magnetic susceptibility data (106) which indicated the existence of two unpaired electrons is consistent with this idea. The conclusive support for the idea was given by means of Mössbauer spectroscopy (117–120).

The electronic structure of Compound I had been a complete mystery until Mössbauer spectroscopic data showed that the electronic structure of the heme iron was the same in Compounds I and II (117, 119, 120). One of the two oxidizing equivalents in Compound I was thus accounted for by the loss of an electron from the iron atom. Using metal–porphyrin complexes, Dolphin et al. (121) showed that the optical absorption spectra of Compounds I of catalase and HRP were similar to those of π-cation radicals of porphyrin. They concluded, therefore, that the second oxidizing equivalent in Compound I originates in the π-cation radical of the porphyrin. The failure to observe an ESR signal arising from the radical of Compound I was explained by an exchange interaction of a single electron localized on the porphyrin ring with the spin of electrons localized on the iron.

Yonetani et al. (122) clearly observed an ESR signal of a free radical in the primary peroxide compound of CCP. This compound possesses two oxidizing equivalents above the ferric enzyme but gives an optical spectrum similar to those of Compounds II of catalase and HRP. In the primary peroxide compound of CCP the second oxidizing equivalent might be localized in an amino acid residue of the protein. The formation of a small amount of free radical was also observed in the reaction of HRP with H_2O_2 (108).

3. Compound III (*Oxyperoxidase*)

The electronic structure of the heme of oxyhemoproteins has been discussed for a long time, mostly with hemoglobin and myoglobin. Oxyhemoglobin is known to be diamagnetic (123). However, it has been suggested by several workers (124–126) that the migration of an electron occurs from the heme iron to the oxygen. As discussed by Peisach et al. (126), two unpaired electrons are not necessarily localized in the iron and oxygen atoms and the spins could combine to cancel their paramagnetism. It would seem difficult to confirm the migration of an electron from the iron to the oxygen by Mössbauer spectroscopy. The oxygen molecules in oxy-TP and HRP Compound III are considered to be more activated than those in oxyhemoglobin and oxymyoglobin, but at present no effective physical technique has been reported to distinguish the electronic structure of these hemoproteins.

IV. RELATIONSHIP BETWEEN STRUCTURE AND FUNCTION

Obviously, the information on the chemical structure of peroxidase is quite insufficient to discuss the relationship between the structure and function. There are a number of protein groups which contain common prosthetic atoms or molecules such as iron, copper, flavin, heme, and pyridoxal. New functions of these groups acquired by incorporation into specific environments of proteins have attracted considerable attention. Of these, the heme group seems to be one of the most extensively investigated molecules because of its biological importance and its characteristic properties which can be measured by various physical and chemical methods.

Most hemoproteins contain protoheme as the prosthetic group. Peroxidases of plants contain protoheme while those of animals contain heme more or less different from protoheme. Accordingly, it might be supposed that the function of hemoproteins is not strongly influenced by modifications in the heme moiety.

A. Peroxidases with Unnatural Heme Groups

Since Theorell (127) found that HRP was reversively split into apo-HRP and protoheme, a number of experiments have been carried out in order to clarify the effect of heme modification on the functions of hemoproteins. Artificial hemoproteins with unnatural hemes have been prepared for hemoglobin (128), myoglobin (129), HRP (132–136), CCP (137, 138), and cytochrome b_5 (139). These results are summarized in Fig. 5.

It may be concluded from the results that substituents at positions 2 and 4 of deuteroheme have no significant effect on the functions of hemoproteins. Phillips (140) showed that the electron-withdrawing capacities of 2,4 substituents of deuteroporphyrin are correlated with properties such as spectroscopic behavior of the porphyrin and basicity of the pyrrole

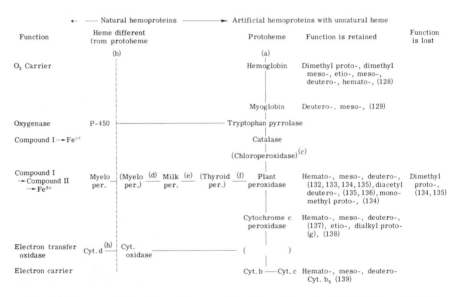

FIG. 5. Hemes and functions of hemoproteins. Parentheses imply that the position is still in question. (a) Left and right sides indicate red and blue shifts in absorption maxima, respectively. (b) Left side of the dotted line means a possible involvement of a structure of chlorine type or other unusual types. (c) Chloroperoxidase has strong catalase-type functions (130), and higher oxidation states other than Compound I have not been identified (131). (d) Cytochrome oxidase-like derivative of myeloperoxidase formed by mild acid treatment (see text). (e) Salivary peroxidase is found to be the same as milk peroxidase (141), and intestinal peroxidase is also very similar to milk peroxidase (144). (f) Higher oxidation states have not been observed. (g) The reaction with ferrocytochrome c is lost. (h) A chlorine type.

nitrogen. The basicity has been measured in terms of pK_3, defined as

$$pK_3 = pH - \log(PH_2)/(PH_3^+) \tag{7}$$

where PH_2 and PH_3^+ are neutral and monocationic species of the porphyrin, respectively. Makino and Yamazaki (135) found that the basicity of the pyrrole nitrogen is reflected directly in the basicity of the heme iron in HRP, which can be expressed in terms of the proton dissociation constant, K_a

$$pK_a = pH - \log(\text{alkaline HRP})/(\text{acid HRP}) \tag{8}$$

Although the value of pK_a varied with different isoenzyme preparations of HRP, the dependence of pK_a of each isoenzyme preparation upon 2,4 substituents was identical as can be seen in Fig. 6. For given 2,4 substituents the following equations were given to the relation between pK_3 of the porphyrin and pK_a of HRP containing the heme:

$$\text{For HRP-A,} \qquad\qquad pK_a(A) = pK_3 + 4.4 \tag{9}$$

$$\text{For HRP-(B + C),} \qquad pK_a(B + C) = pK_3 + 6.0 \tag{10}$$

The difference of 1.6 between two constants in the above equations may be attributable not only to the difference of the local charge distribution of the protein but also to the specific amino acid residue which might be responsible for the proton dissociation reaction. Thus, artificial HRP containing a 2,4-diacetyldeuterohemin group had pK_a values much different

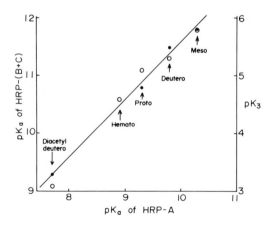

FIG. 6. Correlation of dependence of heme substitution upon pK_a (135). (○) Correlation between HRP-A and HRP-(B + C) and (●) pK_3 values (140) plotted against pK_a of HRP-A.

from that of natural HRP. The stability of Compounds II and III of 2,4-diacetyldeutero-HRP increased, while that of Compound I decreased as compared with natural HRP (135).

B. Natural Peroxidases with Heme Groups Different from Protoheme

It is of special interest to note that peroxidases so far isolated from animal tissues contain prosthetic groups different from protoheme. The common feature of animal peroxidases is that the heme is tightly bound to the protein. Except for thyroid peroxidose (92) the binding is considered to be a covalent type. By immunodiffusion analysis salivary peroxidase was found to be identical with milk peroxidase (141, 142). Eosinophil peroxidase (143) and intestinal peroxidase (144) had spectroscopic properties very similar to milk peroxidase. Morell and Clezy (145) concluded that the heme in milk peroxidase contains a strongly electrophilic substituent, conjugated to the porphyrin ring, which is labile to strong alkali. Judging from the spectral property the electronic structure of the heme in milk peroxidase appears to be similar to that of 2,4-diacetyldeutero-HRP. In this connection it should be noted that Compound III of milk peroxidase is as stable as Compound III of 2,4-diacetyldeutero-HRP and the decay time at 2° might be more than a day (40, 135).

Of the known peroxidases myeloperoxidase has unique spectral properties quite different from other peroxidases. The possibility that myelo-- peroxidase contains a heme group similar to heme a was suggested by Schultz and Shmukler (94) and Newton *et al.* (95). Odajima and Yamazaki (146) found that by acid treatment myeloperoxidase of normal pig leukocytes was converted to a derivative still possessing peroxidase activity (60% of the original) but having absorption spectra quite similar to cytochrome oxidase. The spectral comparison was applied to their respective forms such as ferric, ferrous, cyanide-ferric, and CO-ferrous. Like cytochrome oxidase myeloperoxidase probably contains two iron atoms per molecule (147–150) which have different reactivities with H_2O_2 (151) and cyanide (150). It was concluded (146) that the electronic structure of the heme in myeloperoxidase is distorted by interactions with the protein and the distortion can be removed by mild modification of the protein structure to form a spectrally usual type of hemoprotein containing heme a analogues. The mechanism of the conversion might be analogous to that of the conversion from P-450 to P-420 (152, 153); for instance, on modification the absorption maxima of reduced myeloperoxidase (475 and 637 nm) were shifted to shorter wavelengths by about 30 nm (146). From the appearance of these properties myeloperoxidase looks much more similar to cytochrome oxidase than HRP. The function of the myeloperoxidase heme is, however, found to be essentially the same as that of HRP. The existence of five redox states of myeloperoxidase has been confirmed (154).

C. Comparison of the Function of HRP with Other Hemoproteins

In plant tissues (96, 155, 156) and in leukocytes of normal blood (157, 158) multiple forms of peroxidase have been observed. Though the amino acid composition varies greatly for the isoenzyme preparations, the basic catalytic property of these preparations described in Fig. 1 is essentially the same. On the other hand, the pattern of physiological functions of the heme is characteristic of the group of hemoproteins, and the distinction between these groups is obviously ascribed to the protein moiety (Fig. 5). Based on the known three-dimensional structures of hemoglobin and myoglobin the dependence of the catalytic properties of the heme upon the protein structure will be extensively studied in the future. Since the catalytic properties of five redox states of the heme are clarified to a considerable extent in the case of HRP it might be worthwhile to compare them with those of other hemoproteins. The outline is shown in Fig. 7.

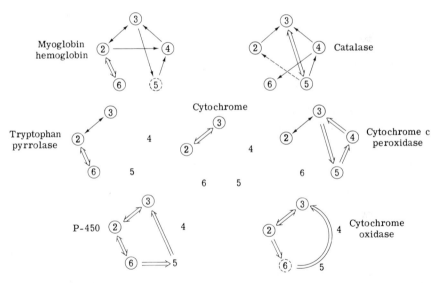

FIG. 7. Five redox states and functions of hemoproteins. Solid circles, identified; dotted circles, still in question; ⟹, main function. Catalase: The details are reviewed by Nicholls and Schonbaum (159). The reaction from 5 to 2 needs sodium azide or hydroxyl amine. Cytochrome c peroxidase: according to Coulson et al. (160). Cytochrome oxidase: according to Okunuki (161). P-450: according to Ishimura et al. (162). Hemoglobin, myoglobin: The addition of H_2O_2 to myoglobin forms ferrylmyoglobin (163). The state 5 is stable at pH 8 and the reaction from 2 to 4 occurs by H_2O_2 in the case of myoglobin (164). Tryptophan pyrrolase: according to Ishimura et al. (69).

1. Oxidation–Reduction Potentials of the Ferric and Ferrous Couple

It is known that unlike other hemoproteins native catalase cannot be reduced to the ferrous form by hydrosulfite. The oxidation–reduction potential of most hemoproteins varies from -0.27 V (HRP) to $+0.28$ V (cytochrome a) at pH 7. The potentials of hemoglobin, myoglobin, and cytochromes are between these limits. Williams (165) suggested that the potential depends on the ligand electronegativity. The electronegativity of substituents conjugated to the porphyrin ring may profoundly affect the potential.

2. Compound II

Compound II of myoglobin is fairly stable (163). Compound II of catalase has little catalytic activity (159). Compound II of CCP is an obligatory intermediate in the peroxidase reaction of CCP, but it is too reactive to isolate as a single species (160).

3. Compound I

The form and stability of this oxidation state is an important criterion which indicates the dependence of the catalytic function of the heme upon the protein structure. In the reaction of ferric hemoproteins with H_2O_2 three types are known:

$$\text{HRP, catalase} + H_2O_2 \rightarrow \text{(ferryl} + \pi\text{-cation radical in porphyrin)}$$

$$\text{CCP} + H_2O_2 \rightarrow \text{(ferryl} + \text{free radical in protein moiety)}$$

$$\text{Hb, Mb} + H_2O_2 \rightarrow \text{(ferryl)} + \text{nonspecific products derived from HO}$$

Since one oxidizing equivalent is retained in the iron atom of these hemoproteins, the difference between them can be ascribed to the state of a second oxidizing equivalent.

Unlike HRP, a main function of catalase is the reduction of Compound I to the ferric form directly by two-electron donors such as H_2O_2 and ethanol (159). According to the mechanism of Ishimura et al. (162) a similar reaction might be expected in the case of P-450. This is an example of external mixed function oxidase (4). Now, what kind of structure would be expected for an intermediate after the oxygenated P-450 (Compound III type) accepts an electron from the external source? Could it be something like Compound I of catalase, HRP, or CCP? The question remains to be solved whether an intermediate species formed immediately after HRP Compound III accepts an electron from donor molecules is Compound I or something else. It is likely that such an intermediate possesses a transferable oxygen atom as suggested by Ullrich and Staudinger (166).

4. *Compound III (Oxyhemoproteins)*

It is known that the oxy forms of hemoglobin, myoglobin, catalase, tryptophan pyrrolase, and HRP have similar absorption spectra. This fact would imply that the electronic structure of the oxygen bound to the heme iron of these hemoproteins is essentially the same. A Fe^{3+}-O_2^- type of binding might be the case with all oxyhemoproteins. If the activation of molecular oxygen cannot be explained by the electronic structure, the proximity effect will become the probable mechanism to explain the incorporation of molecular oxygen into tryptophan on the surface of the tryptophan pyrrolase molecule (see Chapter 3 by Feigelson).

Compound III of CCP has not been found. The oxygen in Compound III of HRP and milk peroxidase was found not to be dissociable in a measurable time (40, 135).

V. CONCLUSION

A characteristic feature of peroxidase catalysis is known to be high specificity for the electron acceptor and extremely low specificity for the electron donor. This fact might be related to the difference in the nature of the primary product of the reaction (167). In the ordinary peroxidase reaction the free radicals of two-electron donor molecules are freed from the enzyme as a primary product, but hydroxyl radicals cannot be a primary product. Peroxidase is a kind of iron complex which can hold every chemical species present in the reduction process from O_2 to H_2O.

A rough outline of the story has now been given. The most interesting step of the O_2 reduction on the heme iron of peroxidase seems to be the conversion from the oxy form (Compound III) to Compound I, and at present the reaction is far beyond understanding. It can be said that HRP is a useful model to analyze the mechanism of electron transfer reactions and O_2 metabolism including oxygenase. A systematic study of the effect of heme substitution upon the various functions of HRP appears to be a promising approach.

Among various donor molecules poorly specific for HRP, IAA behaves in a unique manner, like a true physiological substrate. However, the problem still remains to be solved whether the reaction arises from specific interaction between IAA molecule and the HRP protein or only from the reactivity of IAA molecule and its oxidized intermediates. Since the mechanism of IAA oxidation depends on the experimental conditions, a typical reaction of an oxygenase type would be expected under the specified conditions.

REFERENCES

1. Saunders, B. C., Holmes-Siedle, A. G., and Stark, B. P. (1964). "Peroxidase." Butterworths, London and Washington, D.C.
2. Hayaishi, O. (1964). *In* "Oxygenase" (O. Hayaishi, ed.), p. 1. Academic Press, New York.
3. Mason, H. S. (1957). *Advan. Enzymol.* **19,** 79.
4. Mason, H. S. (1965). *Ann. Rev. Biochem.* **34,** 595.
5. Swedin, B., and Theorell, H. (1940). *Nature (London)* **145,** 71.
6. Chance, B. (1952). *J. Biol. Chem.* **197,** 577.
7. Kenten, R. H. (1955). *Biochem. J.* **59,** 110.
8. Ray, P. M. (1958). *Ann. Rev. Plant Physiol.* **9,** 81.
9. Maclachlan, G. A., and Waygood, E. R. (1956). *Can. J. Biochem. Physiol.* **34,** 1233.
10. Yamazaki, I., Fujinaga, K., Takehara, I., and Takahashi, H. (1956). *J. Biochem.* **43,** 377.
11. Akazawa, T., and Conn, E. E. (1958). *J. Biol. Chem.* **232,** 403.
12. Yokota, K., and Yamazaki, I. (1965). *Biochim. Biophys. Acta* **105,** 301.
13. Klapper, M. H., and Hackett, D. P. (1963). *J. Biol. Chem.* **238,** 3736.
14. Klapper, M. H., and Hackett, D. P. (1963). *J. Biol. Chem.* **238,** 3743.
15. Mason, H. S., Onopryenko, I., and Buhler, D. R. (1957). *Biochim. Biophys. Acta* **24,** 225.
16. Buhler, D. R., and Mason, H. S. (1961). *Arch. Biochem. Biophys.* **92,** 424.
17. Yang, S. F. (1967). *Arch. Biochem. Biophys.* **122,** 481.
18. Morrison, M., Bayse, G., and Danner, D. J. (1970). *In* "Biochemistry of the Phagocytic Process" (J. Schultz, ed.), p. 51. North–Holland Publ., Amsterdam.
19. Hager, L. P., Thomas, J. A., and Morris, D. R. (1970). *In* "Biochemistry of the Phagocytic Process" (J. Schultz, ed.), p. 67. North–Holland Publ., Amsterdam.
20. Steele, W. F., and Morrison, M. (1969). *J. Bacteriol.* **97,** 635.
21. Klebanoff, S. J. (1970). *In* "Biochemistry of the Phagocytic Process" (J. Schultz, ed.), p. 89. North–Holland Publ., Amsterdam.
22. George, P. (1952). *Nature (London)* **169,** 612; (1953). *Biochem. J.* **54,** 267.
23. Chance, B. (1952). *Arch. Biochem. Biophys.* **41,** 416.
24. Yamazaki, I., Mason, H. S., and Piette, L. H. (1960). *J. Biol. Chem.* **235,** 2444.
25. Piette, L. H., Yamazaki, I., and Mason, H. S. (1961). *In* "Free Radicals in Biological Systems" (M. S. Blois *et al.*, eds.), p. 195. Academic Press, New York.
26. Keilin, D., and Hartree, E. F. (1951). *Biochem. J.* **49,** 88.
27. George, P. (1953). *J. Biol. Chem.* **201,** 427.
28. George, P. (1952). *Adv. Catal. Relat. Sub.* **4,** 367.
29. Mason, H. S. (1958). *Proc. Int. Symp. Enzyme Chem.* p. 220. Maruzen, Tokyo.
30. Yamazaki, I., and Piette, L. H. (1963). *Biochim. Biophys. Acta* **77,** 47.
31. Yamazaki, I., Yokota, K., and Nakajima, R. (1965). *In* "Oxidases and Related Redox Systems" (T. E. King, H. S. Mason, and M. Morrison, eds.), p. 485. Wiley, New York.
32. Theorell, H. (1947). *Advan. Enzymol.* **7,** 265.
33. Harbury, H. A. (1957). *J. Biol. Chem.* **225,** 1009.
34. Chance, B. (1965). *In* "Oxidases and Related Redox Systems" (T. E. King, H. S. Mason, and M. Morrison, eds.), p. 504. Wiley, New York.
35. Yamazaki, I., and Yokota, K. (1965). *Biochem. Biophys. Res. Commun.* **19,** 249.
36. Yamazaki, I., Yokota, K., and Tamura, M. (1966). *In* "Hemes and Hemoproteins"

(B. Chance, R. W. Estabrook, and T. Yonetani, eds.), p. 319. Academic Press, New York.

37. Wittenberg, J. B., Noble, R. W., Wittenberg, B. A., Antonini, E., Brunori, M., and Wyman, J. (1967). *J. Biol. Chem.* **242**, 626.
38. Blumberg, W. E., Peisach, J., Wittenberg, B. A., and Wittenberg, J. B. (1968). *J. Biol. Chem.* **243**, 1854.
39. Tamura, M., and Yamazaki, I. (1972). *J. Biochem.* **71**, 311.
40. Yamazaki, I., Yamazaki, H., Tamura, M., Ohnishi, T., Nakamura, S., and Iyanagi, T. (1968). *Advan. Chem. Ser.* **77**, 290.
41. Chance, B. (1952). *Arch. Biochem. Biophys.* **41**, 404.
42. Yokota, K., to be published.
43. Odajima, T. (1972). *Biochim. Biophys. Acta* **235**, 52.
44. Odajima, T., and Yamazaki, I. (1972). *Biochim. Biophys. Acta* **284**, 355.
45. McCord, J. M., and Fridovich, I. (1969). *J. Biol. Chem.* **244**, 6049.
46. Nakajima, R., and Yamazaki, I., to be published.
47. Chance, B. (1952). *Arch. Biochem.* **41**, 416.
48. Chance, B., and Fergusson, R. R. (1954). *In* "The Mechanism of Enzyme Action" (W. D. McElroy and B. Glass, eds.), p. 389. Johns Hopkins Press, Baltimore, Maryland.
49. Cormier, M. J., and Prichard, P. M. (1968). *J. Biol. Chem.* **243**, 4706.
50. Hasinoff, B. B., and Dunford, H. B. (1970). *Biochemistry* **9**, 4930.
51. Roman, R., Dunford, H. B., and Evett, M. (1971). *Can. J. Chem.* **49**, 3059.
52. Roman, R., and Dunford, H. B. (1972). *Biochemistry* **11**, 2076.
53. Yamazaki, I., Nakajima, R., Miyoshi, K., Makino, R., and Tamura, M., *In* "Oxidases and Related Redox Systems (*Proc. Int. Symp., 2nd*)" (T. E. King, H. S. Mason, and M. Morrison, eds.). Univ. Park Press, Baltimore, Maryland (in press).
54. Björksten, F. (1970). *Biochim. Biophys. Acta* **212**, 396.
55. Yokota, K., and Yamazaki, I. (1965). *Biochem. Biophys. Res. Commun.* **18**, 48.
56. George, P. (1953). *Arch. Biochem. Biophys.* **45**, 21.
57. George, P. (1953). *J. Biol. Chem.* **201**, 413.
58. Fergusson, R. R. (1956). *J. Amer. Chem. Soc.* **78**, 741.
59. Noble, R. W., and Gibson, Q. H. (1970). *J. Biol. Chem.* **245**, 2409.
60. Nicholls, P. (1962). *In* "Oxygenases" (O. Hayaishi, ed.), p. 273. Academic Press, New York.
61. Yamazaki, I. (1958). *Proc. Int. Symp. Enzyme Chem.* p. 224. Maruzen, Tokyo.
62. Ray, P. M., and Thimann, K. V., (1956). *Arch. Biochem. Biophys.* **64**, 175.
63. Ray, P. M. (1956). *Arch. Biochem. Biophys.* **64**, 193.
64. Morita, Y., Kameda, K., and Mizuno, M. (1962). *Agr. Biol. Chem. (Tokyo)* **26**, 442.
65. Hinman, R. L., and Lang, J. (1965). *Biochemistry* **4**, 144.
66. Morita, Y., Kominato, Y., and Shimizu, K. (1967). *Mem. Res. Inst. Food Sci. Kyoto Univ.* **28**, 1.
67. Bloch, K., and Hayaishi, O. (1966). *In* "Biological and Chemical Aspects of Oxygenases" (K. Bloch and O. Hayaishi, eds.), preface. Maruzen, Tokyo.
68. Ishimura, Y., Nozaki, M., Hayaishi, O., Tamura, M., Yamazaki, I. (1967). *J. Biol. Chem.* **242**, 2574.
69. Ishimura, Y., Nozaki, M., Hayaishi, O., Tamura, M., Nakamura, T., and Yamazaki, I. (1970). *J. Biol. Chem.* **245**, 3593.
70. Yamazaki, I., and Souzu, H. (1960). *Arch. Biochem. Biophys.* **86**, 294.

71. Ray, P. M. (1960). *Arch. Biochem. Biophys.* **87,** 19.
72. Ray, P. M. (1962). *Arch. Biochem. Biophys.* **96,** 199.
73. Ricard, J., and Nari, J. (1966). *Biochim. Biophys. Acta* **113,** 57.
74. Ricard, J., and Nari, J. (1967). *Biochim. Biophys. Acta* **132,** 321.
75. Fox, L. R., Purves, W. K., and Nakada, H. I. (1965). *Biochemistry* **4,** 2754.
76. Degn, H. (1969). *Biochim. Biophys. Acta* **180,** 271.
77. Miller, R. W., and Parups, E. V. (1971). *Arch. Biochem. Biophys.* **143,** 276.
78. Yamazaki, H., and Yamazaki, I. (1973). *Arch. Biochem. Biophys.* **154,** 147.
79. Shin, M., and Nakamura, W. (1962). *J. Biochem.* **52,** 444.
80. Critchlow, J. E., and Dunford, H. B., (1972) *J. Biol. Chem.* **247,** 3703.
81. Yonetani, T., Schleyer, H., Chance, B., and Ehrenberg, A. (1966). *In* "Hemes and Hemoproteins" (B. Chance, R. W. Estabrook, and T. Yonetani, eds.), p. 293. Academic Press, New York.
82. Schonbaum, G. R. (1970). *Fed. Proc. Fed. Amer. Soc. Exp. Biol.* **29,** 2748.
83. Yamazaki, I., Sano, H., Nakajima, R., and Yokota, K. (1968). *Biochem. Biophys. Res. Commun.* **31,** 932.
84. Chance, B. (1949). *Arch. Biochem.* **21,** 416.
85. Yamazaki, H., Ohishi, S., and Yamazaki, I. (1970). *Arch. Biochem. Biophys.* **136,** 41.
86. Lemberg, R., Legge, J. W., and Lockwood, W. H. (1941). *Biochem. J.* **35,** 328.
87. Lemberg, R. (1956). *Rev. Pure Appl. Chem. Roy. Aust. Chem. Inst.* **6,** 1.
88. O'hEocha, C. (1968). *In* "Porphyrins and Related Compounds" (T. W. Goodwin, ed.), p. 91. Academic Press, New York.
89. Bagger, S., and Williams, R. J. P. (1971). *Acta Chem. Scand.* **25,** 976.
90. Hosoya, T., and Morrison, M. (1967). *J. Biol. Chem.* **242,** 2828.
91. Taurog, A., Lothrop, M. L., and Estabrook, R. W. (1970). *Arch. Biochem. Biophys.* **139,** 221.
92. Krinsky, M. M., and Alexander, N. M. (1971). *J. Biol. Chem.* **246,** 4755.
93. Hultquist, D. E., and Morrison, M. (1963). *J. Biol. Chem.* **238,** 2843.
94. Schultz, J., and Shmukler, H. W. (1964). *Biochemistry* **3,** 1234.
95. Newton, N., Morell, D. B., and Clarke, L. (1965). *Biochim. Biophys. Acta* **96,** 476.
96. Shannon, L. M., Kay, E., and Lew, J. Y. (1966). *J. Biol. Chem.* **241,** 2166.
97. Morita, Y., and Kameda, K. (1959). *Bull. Agr. Chem. Soc. Japan* **23,** 28.
98. Shimizu, K., and Morita, Y. (1966). *Agr. Biol. Chem. (Tokyo)* **30,** 149.
99. Strickland, E. H., Kay, E., Shannon, L. M., and Horwitz, J. (1968). *J. Biol. Chem.* **243,** 3560.
100. Asakura, T., and Yonetani, T. (1969). *J. Biol. Chem.* **244,** 537.
101. Brill, A. S., and Williams, R. J. P. (1961). *Biochem. J.* **78,** 246.
102. Nakamura, Y., Tohjo, M., and Shibata, K. (1963). *Arch. Biochem. Biophys.* **102,** 144.
103. Shin, J. H. C., Shannon, L. M., Kay, E., and Lew, J. Y. (1971). *J. Biol. Chem.* **246,** 4546.
104. Yonetani, T., *In* "Oxidases and Related Redox Systems (*Proc. Int. Symp., 2nd*)" (T. E. King, H. S. Mason, and M. Morrison, eds.). Univ. Park Press, Baltimore, Maryland (in press).
105. Theorell, H. (1942). *Ark. Kemi Mineral. Geol.* **16A,** No. 3.
106. Theorell, H., and Ehrenberg, A. (1952). *Arch. Biochem. Biophys.* **41,** 442.
107. Ehrenberg, A. (1962). *Ark. Kemi* **19,** 119.
108. Morita, Y., and Mason, H. S. (1965). *J. Biol. Chem.* **240,** 2654.

109. Yonetani, T., and Schleyer, H. (1967). *In* "Structure and Function of Cytochromes" (K. Okunuki, M. D. Kamen, and I. Sekuzu, eds.), p. 535. Univ. of Tokyo Press, Tokyo.

110. Iizuka, T., and Yonetani, T. (1970). *Advan. Biophys.* **1**, 155.

111. Iizuka, T., Kotani, M., and Yonetani, T. (1968). *Biochim. Biophys. Acta* **167**, 257.

112. Yonetani, T., Iizuka, T., and Asakura, T. (1972). *J. Biol. Chem.* **247**, 863.

113. Tamura, M. (1971). *Biochim. Biophys. Acta* **243**, 239.

114. Tamura, M. (1971). *Biochim. Biophys. Acta* **243**, 249.

115. Kotani, M. (1969). *Ann. N. Y. Acad. Sci.* **158**, Art. 1, 20.

116. Berzofsky, J. A., Peisach, J., and Blumberg, W. E. (1971). *J. Biol. Chem.* **246**, 3367.

117. Maeda, Y., and Morita, Y. (1967). *In* "Structure and Function of Cytochromes" (K. Okunuki, M. D. Kamen, and I. Sekuzu, eds.), p. 523. Univ. of Tokyo Press, Tokyo.

118. Bearden, A. J., Ehrenberg, A., and Moss, T. H. (1967). "Structure and Function of Cytochromes" (K. Okunuki, M. D. Kamen, and I. Sekuzu, eds.), p. 528. Univ. of Tokyo Press, Tokyo.

119. Moss, T. H., Ehrenberg, A., and Bearden, A. J. (1969). *Biochemistry* **8**, 4159.

120. Maeda, Y., Morita, Y., and Yoshida, C. (1971). *J. Biochem.* **70**, 509.

121. Dolphin, D., Forman, A., Borg, D. C., Fajer, J., and Felton, R. H. (1971). *Proc. Nat. Acad. Sci.* **68**, 614.

122. Yonetani, T., Schleyer, H., and Ehrenberg, A. (1966). *J. Biol. Chem.* **241**, 3240.

123. Pauling, L., and Coryell, C. D. (1936). *Proc. Nat. Acad. Sci.* **22**, 210.

124. Weiss, J. J. (1964). *Nature (London)* **202**, 83.

125. Viale, R. O., Maggiora, G. M., and Ingraham, L. L. (1964). *Nature (London)* **203**, 183.

126. Peisach, J., Blumberg, W. E., Wittenberg, B. A., and Wittenberg, J. B. (1968). *J. Biol. Chem.* **243**, 1871.

127. Theorell, H. (1941). *Ark. Kemi Mineral. Geol.* **14B**, No. 20.

128. Sugita, Y., and Yoneyama, Y. (1971). *J. Biol. Chem.* **246**, 389.

129. Brunori, M., Antonini, E., Phelps, C., and Amiconi, G. (1969). *J. Mol. Biol.* **44**, 563.

130. Thomas, J. A., Morris, D. R., and Hager, L. P. (1970). *J. Biol. Chem.* **245**, 3129.

131. Thomas, J. A., Morris, D. R., and Hager, L. P. (1970). *J. Biol. Chem.* **245**, 3135.

132. Theorell, H., and Akeson, A. (1943). *Ark. Kemi Mineral. Geol.* **17B**, No. 7.

133. Paul, K. G. (1959). *Acta Chem. Scand.* **13**, 1239, 1240.

134. Tamura, M., Asakura, T., and Yonetani, T. (1972). *Biochim. Biophys. Acta* **268**, 292.

135. Makino, R., and Yamazaki, I. (1972). *J. Biochem.***72**, 655.

136. Chance, B., and Paul, K. G. (1960). *Acta Chem. Scand.* **14**, 1711.

137. Yonetani, T., and Asakura, T. (1968). *J. Biol. Chem.* **243**, 4715.

138. Asakura, T., and Yonetani, T. (1969). *J. Biol. Chem.* **244**, 4573.

139. Ozols, J., and Strittmatter, P. (1964). *J. Biol. Chem.* **239**, 1018.

140. Phillips, J. N. (1963). *In* "Comprehensive Biochemistry" (M. Florkin and E. H. Stotz, eds.), p. 34. Elsevier, Amsterdam.

141. Morrison, M., and Allen, P. Z. (1963). *Biochem. Biophys. Res. Commun.* **13**, 490.

142. Morrison, M., Allen, P. Z., Bright, J., and Jayasinghe, W. (1965). *Arch. Biochem. Biophys.* **111**, 126.

143. Archer, G. T., Air, G., Jackas, M., and Morell, D. B. (1965). *Biochim. Biophys. Acta* **99**, 96.

144. Stelmaszynska, T., and Zgliczynsky, J. M. (1971). *Eur. J. Biochem.* **19**, 56.
145. Morell, D. B., and Clezy, P. S. (1963). *Biochim. Biophys. Acta* **71**, 157.
146. Odajima, T., and Yamazaki, I. (1972). *Biochim. Biophys. Acta* **284**, 368.
147. Agner, K. (1958). *Acta Chem. Scand.* **12**, 89.
148. Ehrenberg, A., and Agner, K. (1958). *Acta Chem. Scand.* **12**, 95.
149. Felberg, N. T., and Schultz, J. (1972). *Arch. Biochem. Biophys.* **148**, 407.
150. Okajima, T., and Yamazaki, I. (1972). *Biochim. Biophys. Acta* **284**, 360.
151. Agner, K. (1963). *Acta Chem. Scand.* **17**, S 332.
152. Omura, T., and Sato, R. (1964). *J. Biol. Chem.* **239**, 2370.
153. Williams, R. J. P. (1967). *In* "Structure and Function of Cytochromes" (K. Okunuki, M. D. Kamen, and I. Sekuzu, eds.), p. 645. Univ. of Tokyo Press, Tokyo.
154. Odajima, T., and Yamazaki, I. (1970). *Biochim. Biophys. Acta* **206**, 71.
155. Paul, K. G. (1958). *Acta Chem. Scand.* **12**, 1312.
156. Morita, Y., Yoshida, C., Kitamura, I., and Ida, S. (1970). *Agr. Biol. Chem. (Tokyo)* **34**, 1191.
157. Schultz, J., Felberg, N., and John, S. (1967). *Biochem. Biophys. Res. Commun.* **28**, 543.
158. Himmelhoch, S. R., Evans, W. H., Mage, M. G., and Peterson, E. A. (1969). *Biochemistry* **8**, 914.
159. Nicholls, P., and Schonbaum, G. R. (1963). *In* "The Enzymes" (P. D. Boyer, H. Lardy, and K. Myrback, eds.) 2nd ed., Vol. 8, p. 147. Academic Press, New York.
160. Coulson, A. F. W., Erman, J. E., and Yonetani, T. (1971). *J. Biol. Chem.* **246**, 917.
161. Okunuki, K. (1963). *Compr. Biochem.* **14**, 232.
162. Ishimura, Y., Ullrich, V., and Peterson, J. A. (1971). *Biochem. Biophys. Res. Commun.* **42**, 140.
163. George, P., and Irvine, D. H. (1952). *Biochem. J.* **52**, 511.
164. Peisach, J., and Wittenberg, J. B., private communication.
165. Williams, R. J. P. (1959). *In* "The Enzymes" (P. D. Boyer, H. Lardy, and K. Myrback, eds.), 2nd ed., Vol. 1, p. 391. Academic Press, New York.
166. Ullrich, V., and Staudinger, Hj, *In* "Biological and Chemical Aspects of Oxygenases" (K. Bloch and O. Hayaishi, eds.), p. 235. Maruzen, Tokyo.
167. Yamazaki, I. (1971). *Advan. Biophys.* **2**, 33.

14

BACTERIAL MONOXYGENASES—THE P450 CYTOCHROME SYSTEM*

I. C. GUNSALUS, J. R. MEEKS, J. D. LIPSCOMB, P. DEBRUNNER, and E. MÜNCK

> For more than a decade my work was devoted to the elucidation of the mechanism of the apparently simple reaction 2 H + O = H$_2$O.
>
> *Szent-Györgyi*, 1937

* A memoriam to Randolph Tsai, scholar, colleague, talented investigator. His life and work brilliant; all too brief.

Abbreviations

Ad	adrenodoxin	Lip(SH)$_2$	dihydrolipoic acid
cam or CAM	camphor (D-isomer)	Mb	myoglobin
CPO	chloroperoxidase	MbO$_2$	oxygenated myoglobin
cyt b	cytochrome b	MSG	monosodium glutamate
cyt b$_5$	cytochrome b$_5$	MW	molecular weight
cyt b$_{562}$	cytochrome b from *E. coli*	mV	millivolt
cyt c	cytochrome c	MV	methyl viologen
δ	isomeric shift	NTA	nitrilotriacetate
DPN$^+$	diphosphopyridine nucleotide	P450	cytochrome, pigment, with λ_{max} for ferrous CO complex 450 nm, b type
DPNH	reduced diphosphopyridine nucleotide	P450$_{cam}$	cytochrome P450, specific for camphor hydroxylation
DTE	dithioerythritol		
ΔE_Q	quadrupole splitting		
E$_A$	flavoprotein reductase	P450$_{cam} \cdot$S	cytochrome P450$_{cam}$ substrate bound
E$_B$	putidaredoxin		
E$_C$	cytochrome P450$_{cam}$	P450$_{cam}^r \cdot$O$_2$	reduced cytochrome P450$_{cam}'$ oxygenated
E$_a$	energy of activation		
E_o'	electromotive potential, 25°C, pH 7.0 (redox potential)	P450$_{cam}^r \cdot$S$\cdot$O$_2$	reduced cytochrome P450$_{cam}$ substrate bound, oxygenated
eff	effector	P450$_{cam}^o$	oxidized cytochrome P450$_{cam}$
ϵ_{mM}	millimolar extinction coefficient	Pd	putidaredoxin (2 Fe–2 S protein)
e_1	first electron reduction (by putidaredoxin, etc.)	Pd^o	oxidized putidaredoxin
e_2	second electron reduction (by putidaredoxin, etc.)	Pd^r	reduced putidaredoxin
EDTA	ethylenediaminetetracetic acid	Pd$\cdot$des$\cdot$Trp	putidaredoxin treated with carboxypeptidase A
ENDOR	electron nuclear double resonance	PF	proflavin
		PpG1	*Pseudomonas putida*, Gunsalus collection, strain 1
FAD	flavin adenine dinucleotide	redox	reduction–oxidation
FADH$_2$	reduced flavin adenine dinucleotide	Rd	rubredoxin (subscript indicates species of origin)
Fd	ferredoxin	$\vec{S}$	spin vector
fp	flavoprotein reductase	S	substrate (camphor)
g	gyromagnetic ratio ($h\nu = g\beta H$)	S–OH	hydroxylated substrate
		TON	turnover number
$g\perp, g_{\|\|}$	g tensor components	TPN$+$	triphosphopyridine nucleotide
Hb	hemoglobin		
HbO$_2$	oxygenated hemoglobin	TPNH	triphosphopyridine nucleotide reduced
HiPIP	high potential iron protein		

I. INTRODUCTION

Prefatory phrases nearly four decades ago framed the critical roles of oxygen. There follows in Chapter 1 (of Ref. 105) consideration of the binding, activation, and the separation by cytochrome of the four reductive valences $O_2 \rightarrow 2 \ OH_2$ to avoid peroxide formation as Szent-Györgyi struggled to see and to separate the energy-reaction continuum established by biological systems.

Mixed function oxidation, before 1960 clearly defined by Mason (74) and by Hayaishi, presented a two valence reduction of O_2, soon related to protein P450, to cytochrome, and to a second separation of reductive valences to avoid peroxide formation, i.e., $=CH_2 + O_2 + 2 \ e \rightarrow =CHOH + OH_2$.

The aerobic bacteria, perhaps the precursors of mitochondria, are energy machines. They convert "inert" molecules to dehydrogenase susceptible ones with active oxygen. Their activity:weight ratio is high; the resultant supply of energy and of biosynthetic precursors supports a short generation time and the high enzymic activities and concentrations render the critical mechanisms visible and their separations simplified. The bacteria have contributed to the monoxygenase systems: a cytochrome P450 and reductive enzyme components in quantity, all pure, characterized and combinable to a fully active system. Critical to the rate of information retrieval, too, have been lysable mutants, free proteins, and a cytochrome P450 crystalline for structure–function definition.

This chapter is devoted to a statement of the current problems and the level of understanding of the chemistry, physics, biology, and genetics of oxygen reduction. A substrate-selective P450 cytochrome has been isolated and studied in depth. A two-electron reduction of dioxygen is accompanied by stereospecific transformation of a selected methylene-carbon to a secondary alcohol. The cytochrome has been characterized in four stable states of an oxygenase reaction cycle. The reductant and effector proteins that complement the cytochrome in the complete native system have been replaced by simple chemical reagents. The final reaction step(s) of product formation from the heme iron oxygen substrate complex have been penetrated with native, modified, and synthetic effectors.

The unusually large fraction of the total data and deductions supplied by a single biological system results from a sustained commitment and concerted effort to elucidate the reaction sequence and mechanism of oxygen reduction. The susceptibility of the monoxygenase mechanism and the unusually appropriate properties found in the system chosen have fostered rapid progress. The variety of investigative skills available and their impact in revealing the fine structure of the catalytic centers and the genetic organization and regulation of peripheral metabolism have been fortunate. An enhanced value of the chemical and physical properties and constants derived from this pure bacterial cytochrome accrues from their general applicability to the cytochrome P450's recovered from mammalian organelles or studied in their intact tissues (25, 33, 81).

The two most clearly identified of the mammalian cytochrome hydroxylase systems are those from the mitochondria of the adrenal cortex and of liver. The first, a substrate and stereo selective biosynthetic hydroxylase, contains three components similar to those of the microbial system and is regulated in a similar manner. The "less-specific" or "nonspecific" hepatic system is partially inducible and bears similar cytochrome and flavine reductive components, though not the iron sulfide reductant-effector proteins observed in the microbial and adrenal hydroxylases (26).

Whether the languages of present-day physics and biochemistry will merge to communicate new understanding in the near, or distant, future is not entirely clear. The simultaneous examination of biological catalysts, and interpretation along both lines, is an act of faith which has some basis of scientific substantiation and has offered definite encouragement. The time required for the language of biochemistry to be infiltrated with knowledge of useful physical probes, to discard her many empirical summations, to learn to deal in time and space dimensions of catalysis with binding forces between substrates and proteins and among proteins, and to define clearly the transitions which occur is uncertain. The facts as set down here, some with interpretation and known precision, are assembled from the present state of our continuing attempt to understand the mechanisms developed by organisms in their adaptation to variety and provision of life's essentials.

This chapter was constructed and edited in parts by several investigators of different skills. The authors also acknowledge the help and collaboration of their many colleagues both within and outside the University and the advances from criticisms by the many investigators and leaders attracted to this area through their contacts with biological materials and problems. It is hoped that the language barrier has been overcome sufficiently to let the message reach an array of biochemist-biologists and that the jumps and blanks in knowledge will not be too disturbing.

II. BACTERIAL MONOXYGENASES

A. Cytochrome Couplings in Bacteria

Coupling looks here in two directions. How are respiratory pigments organized in microbial cells? How do cytochromes participate in dioxygen cleavage and in cellular work?

An attractive hypothesis attributes the origin of eucaryotic mitochondria to the invasion and modification of microbial ancestors. If so, a dimension new to biology originated therein and the selection of suitable pro- or eucaryotic examples should lead to an understanding of the essentials of both questions.

Aerobic bacteria and mitochondria share a high respiratory and work function through funneling dehydrogenase energy to reduction of the pyridine nucleotides TPN and DPN. A small fraction of this energy is used by monoxygenases in the activation of dioxygen for reactions on functional groups not readily dehydrogenated. These processes are required to prepare unreactive functional groups for dehydrogenase action, for the biosynthesis of essential regulatory metabolites, and for the excretion, or further metabolism, of foreign chemicals. The energy is thought to be transmitted through a separate reductive, or electron transfer, system without oxidative energy coupling for cellular work.

Cytochrome in the monoxygenases binds the dioxygen, the carbon substrates, and the reductive protein components. The binding can be monitored by optical measurements of spectral changes in the visible region, by fluorescence, and by circular dichroism. Changes in size and mobility of the proteins during isolation or on reconstruction of the enzyme systems have also revealed protein coupling. In these instances, coupling concerns the organization of the protein components plus their interaction in catalysis.

The oxygen combining and redox properties of heme proteins enjoy an extensive literature. The X-ray crystal structures of oxygenated and deoxygenated hemoproteins and of ferric–ferrous cytochromes, e.g., c and c', furnish models to build hypotheses extendible to other proteins [see, for example, the summaries by Dickerson (30) and by Kraut and co-workers (94)].

The cell-organelle location of the monoxygenase proteins in microorganisms is unclear. In the mammalian systems the microsome and mitochondrial organelle location is known; the fine structure remains obscure. In both microbe and mammal, induction of monoxygenase components occurs, in some cases with large changes in the activity and concentration of the P450 cytochromes and their redox couples. In bacteria, the regulatory mechanisms permit greater changes and higher activities than is customarily seen in mitochondria. This is particularly true of enzymes in the principal energy and carbon-yielding pathways. If the monoxygenase proteins in bacteria are of membrane location, as occasionally assumed, on induction appreciable changes must occur in membrane size and structure including effects on the levels of "essential" respiratory couples.

The b-type cytochromes, other than P450, have not been associated with monoxygenases in bacteria; in the mammal, cytochrome b_5 is found in microsomes with cytochrome P450. A function may exist in the P450 reductant chain for the second electron or, alternatively, as recently suggested by Sato *et al.* (95), b_5 may play a role in alkane and fatty acid desaturation. The effector activity of rabbit liver cytochrome b_5 in the $P450_{cam}$ hydroxylase may be fortuitous or foretell a similar biological reaction and mechanism (see Section IV, C).

To the second question of how cytochromes cleave dioxygen, the four stable states of cytochrome P450 in the two-electron reduction-oxygenase reaction cycle furnish an opening. The energy levels of the first and second reductions, e_1 and e_2, are very different, presumably dictated by the structure of dioxygen and of the derivatives formed in the individual reduction steps. One can assume similar energy differences for the four-electron reduction of $O_2 \rightarrow 2\,H_2O$ and an elucidation by systems reconstructed from pure enzymes to be more illuminating than the studies with membranes which resist the separation and analysis required to identify the composition of reactants and of the reaction states.

B. The $P450_{cam}$ Model

A pure cytochrome, $P450_{cam}$, catalyzes the stepwise reduction of molecular oxygen to water in the presence of a selected substrate and two additional proteins, a DPNH-driven reductant and a reductant-effector protein. The DPNH flavoprotein reductase is readily replaced by chemical reductants and the reductant-effector, an iron–sulfide protein, recently has been replaced with simple chemical reductant-effectors to illuminate further the mechanism of the final reaction in oxygenation.

The enzymic reductants support *in vitro* hydroxylation rates comparable to those calculated for the whole cell. The separation of states in the reduction of dioxygen by pure proteins of known composition and interactions has resolved the mechanism of one of the most important of the fundamental reactions of aerobic, and thus of mammalian, life. This model is extendible to the cytochrome monoxygenase systems of other cells and organisms. The energy-coupled oxidative reactions that drive the chemical work of cells require similar definition.

The cytochrome $P450_{cam}$ reaction center is classed as a b-type from the spectrum and binding of one molecule of ferriprotoporphyrin IX to a single polypeptide, molecular weight, 45,000 daltons. The P450 cytochromes share some iron and heme properties with other cytochromes. The principal difference lies in their serving both the oxygen-binding role of heme proteins and the ferrous–ferric redox role of cytochromes. The oxygenated P450 resembles the oxyheme proteins MbO_2 and HbO_2 of mammalian tissues.

The CO adduct of ferrous P450 differs from all other known cytochromes and oxygen-carrying heme proteins in a shift of the Soret maximum to the 450-nm region (125, 126). The immediate biological electron donor to the $P450_{cam}$ cytochrome is the specific $Fe_2S_2^*Cys_4$ iron–sulfide protein, putidaredoxin, with affinity for the homologous cytochrome in the micromolar range. The ultimate reductant, DPNH, is activated by an FAD reductase flavoprotein; the reduction rate of the iron-sulfide protein is not rate limiting for the hydroxylation process.

Figure 1a illustrates the cytochrome reaction cycle in reductive oxygenation. Figure 1b shows a replacement of the DPNH-flavoprotein energy couple by a light-activated chemical reductant on a time scale of 2–10 μsec. Figure 1c illustrates an added bypass, of the specific iron–sulfide protein, for the first reduction equivalent, e_1, not the second, e_2. Thus, the ferric–ferrous reduction is rapid and nonselective. The second, e_2, reduction is part of a currently unresolved multiatomic process; two substrates release two products and require two iron active sites. The e_2 product-forming sequence, a primary focus of our current attention, is explored with the dynamic measurements (Section IV,C).

$P450_{cam}$ has supplied information on the following aspects of the mechanism of monoxygenase action:

1. The formation of active oxygen.
2. Separation of the reductant energy levels for e_1^- and e_2^- at ~ -170 and 0 mV.
3. Identification of four stable states in the reaction cycle of the heme iron prosthetic group.
4. Rich resonance spectra responsive to substrate and redox states.
5. A two protein *in vitro* model with *in vivo* activity.
6. A one protein dioxygen oxygenation-reduction with small molecule reductant-effectors.

These data have proved critical to the elucidation of the energy levels and single reactions of a complex in which only the rate limiting steps were visible to reveal primary events. The admonition of Szent-Györgyi (105) that cytochromes participate in stepwise single valence changes of dioxygen appears most pertinent. The systematic reminder by Boyd (8) of the increased bond distances in dioxygen on each electron addition with accompanying decreases in energy levels is observable in stable intermediate steps; the fine structure is now discernible.

These reactions and properties of the $P450_{cam}$ system are reviewed at

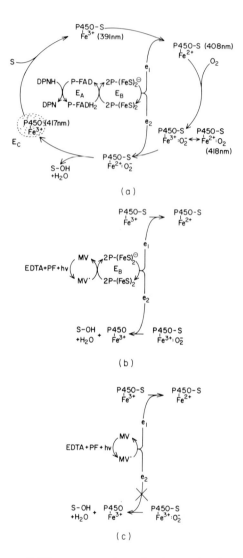

FIG. 1. Cytochrome P450$_{cam}$ oxygenation-reduction cycle: (a) native system; (b) photoreduction—without reductace; and (c) one equivalent, e_1, photoreduction without putidaredoxin.

least once a year in progress reports to repeated symposia in expression of the active interest of workers in this field, as, for example, at Konstanz (37), Oxford (112), Memphis, Edinburgh, Skokloster (43, 46, 47), Miami, New York, and Stanford (42, 45, 68).

C. Component States

The reaction cycle of the cytochrome P450$_{cam}$ is shown in Fig. 1; the details of the final step are extended in Section IV, C. The optical absorption spectra of the three protein components singly in their important reaction states are shown in Fig. 2, the principal absorption maxima and extinction coefficients in Table I.

The flavoprotein, E_0' − 285 mV (73) undergoes a two-electron reduction by DPNH. Following a chemical reduction, for example, by dithionite, the addition of DPN$^+$ causes an increase in absorption in the 650-nm region indicative of complex formation (Fig. 2a).

The redoxin in the oxidized form, Pd^o, contains two ferric atoms in a sulfide-mercaptide reactive center. The latter possesses a strong absorption maximum in the 280 nm-region and three lesser maxima in the 300–500-nm region (Fig. 1b) (26) with the extinction values shown in Table I. A one-electron reduction, E_0' − 240 mV, to Pd^r by the flavoprotein or by chemical reductants converts one of the iron atoms to the ferrous state with a bleaching of the absorption fairly uniformly above 300 nm to expose a new band at 550 nm; the other resonance spectra are considered in Section IV. Removal of the single tryptophan, plus a glutamine, from the C-terminus

TABLE I

ABSORBANCE DATA OF METHYLENE *exo*-5-HYDROXYLASE COMPONENTS[a]

| | P450$_{cam}$ | | | | Redoxin | | Reductase | |
| | E_C | | +S(50 μM) | | E_B | | E_A | |
	λ_{nm}	ϵ_{mM}	λ_{nm}	ϵ_{mM}	λ_{nm}	ϵ_{mM}	λ_{nm}	ϵ_{mM}
Oxidized	417	105	391	87	280	22.5	275	72
	535	10	540	10	325	15.0	378	9.7
	571	10.5	646	4.5	415	10.0	454	10.0
					455	9.6	480	8.5
Reduced[b]	411	71	408	73	545	5.0		
	540	14	540	14				
+O$_2$[c]			418	62				
			552	14				
+CO[c]			447	106				
			550	12				

[a] 50 mM potassium phosphate buffer (pH 7.4) at 25°; S = D-camphor. From Gunsalas *et al.* (47).

[b] Dithionite, 5 equivalents.

[c] At 5°, O$_2$ or CO bubbled.

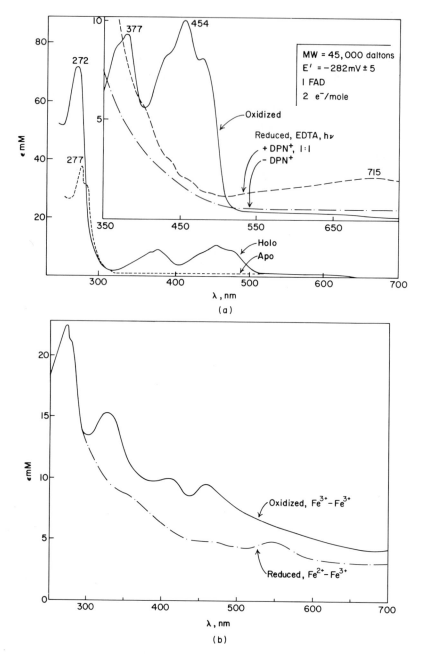

FIG. 2. Hydroxylase component absorption spectra: (a) DPNH-Putidaredoxin reductase, (b) putidaredoxin, and (c) cytochrome P450$_{cam}$.

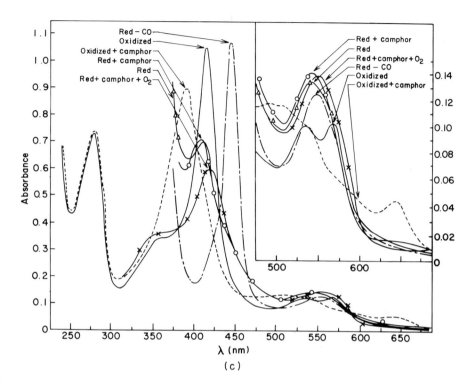

(c)

to expose an arginal residue, yields a derived protein termed Pd·des·Trp. The purified Pd·des·Trp has the absorbance spectra of the native material less that of one tryptophan in the 280-nm region and retains, though at 25-fold higher concentration, the reductant-effector role of the native material. These data will be discussed in Section IV (see Fig. 11).

*Cytochrome P450*cam (Fig. 2c) exists in four reaction states recognizable by the Soret as well as longer wavelength absorption bands. The Soret absorption is shown on expanded scale in Fig. 3. These optical spectra are sufficient for recognition of the components and their reaction states. The physical and chemical details of the state of the iron and oxygen in the redoxin and the cytochrome will appear later in this chapter.

D. Binding and Redox Regulation

Two forms of binding are critical to catalysis by multienzyme systems: (1) the acceptance or exclusion of potential substrates—and their inhibitors or competitors—and (2) the affinity among the proteins. The latter favor

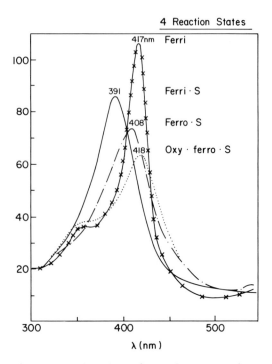

FIG. 3. Soret absorption of cytochrome P450_cam reaction states.

formation of active complexes and thus increase in the concentration of interprotein species with resultant increase in reaction rates. The cytochrome bears a central role in the P450 hydroxylase system. The carbon and the oxygen substrates are bound and undergo the reductive and oxygenative steps on the cytochrome; the redoxin, Pd, is also directly bound to this component. The flavoprotein is the DPNH reactive species. More information is needed on the protein binding and reactions between this flavoprotein and Pd·P450_cam complex.

Carbon substrate selectivity by P450_cam has been detected first by the spectral shifts as effected by the structure around the active site and second by reoxidation of DPNH, which is coupled to product release. More recently, the secondary alcohols formed as products of the hydroxylation have been found by isotope labeling to be cytochrome-bound without altering the absorption spectrum. The substrate, camphor, labeled with ^{3}H or ^{14}C shows an equimolar binding with P450 in the micromolar range; a second molecule bound at a much higher concentration may be nonspecific (76).

The search for an array of substrates for enzymes induced by growth in the presence of a specific carbon source and purified by following activity is frequently limited by the imagination of the investigator to select suitable candidates and by the uniqueness of the criteria, i.e., spectral or product release. Several reagents have been found which bind to $P450_{cam}$ and serve as competitive inhibitors; potential substrates which bind but do not react could be similarly classed.

Peterson and Griffin (88) reported a pronounced effect of cations on D-camphor binding with K^+ as the preferred ion. The K_D for camphor is in the range of 1–5 μM, depending on buffer strength and specific ion concentrations.

The flow of reductive energy to the cytochrome $P450_{cam}$ in the complete enzymic system requires the presence of the substrate, camphor, whereas the reduction of the flavoprotein by DPNH and in turn the redoxin, Pd, is substrate independent (109).

The protein–protein binding especially $Pd \cdot P450$ is described in full detail with the dynamic studies in Section IV.

III. BIOLOGY

A. Distribution

Bacterial dissimilation of hydrophobic residues has increasingly revealed oxygen reactions with direct incorporation into the substrate of either one or both atoms from a dioxygen, O_2, molecule. The monoxygenases that incorporate one atom and use two-electron reducing equivalents to form OH_2 are recognized in two general classes of reaction centers: the predominately single enzyme flavoproteins and the iron-coupled systems composed of several different protein molecules. The iron systems are now recognized in varying precision. The best understood are b-type heme proteins with redox reactive centers that catalyze changes in both the carbon and oxygen substrates. The spectral properties group them as P450 cytochromes (8). The less well-known oxygenases contain iron in a nonheme form which may not equate with iron–sulfur proteins and currently require additional study.

The bacterial cytochrome monoxygenases are recognized in several cases as catalysts of specific substrate reactions, in others merely as spectral entities by the Soret band of the ferrous-CO complexes. $P450_{cam}$, a specific monoxygenase cytochrome studied in the greatest detail, accomplishes the 5-methylene hydroxylation of D- or L-camphor and has enjoyed a precise characterization (8, 42, 47). With three components, all pure and the first, the cytochrome, in crystalline form, the physical and spectral properties of these components and of the reconstituted system have been examined and the coefficients of the heme iron substrate oxygen center has served to guide investigations of other systems. The reactions and reaction cycles were outlined in Section II. The cytochrome is central as the carbon and oxygen binding and catalytic center; the two other proteins serve redox functions, and the iron sulfide protein an additional role in the final sequence of oxygen cleavage, oxygenation and product release. The events surrounding the second, e_2, redox step are now only partially understood.

Appleby (2, 3) has reported P450-type cytochrome spectra in several symbiotic bacteroids associated with the rhizosphere. More recently, in collaboration with Peisach, Appleby (85) has examined these P450's from *Rhizobium* by EPR. A metabolic role has not been discovered; thus, the oxygenation, regulation, and cellular role remain at the spectral level as first recognized with P450 proteins in animal cells [see Boyd and Smellie (8)].

Enrichment cultures on terpenoids have yielded a diptheroid, *Mycobacterium rhodocorus*, which oxidized both D- and L-camphor with accumulation of the 6-*endo* alcohols (39). By analogy, a P450 cytochrome would be anticipated. About 20 strains with a variety of product accumulation patterns when grown in minimal salts camphor media were isolated by Heath (49) and examined by W. Toscano (private communication). Most appear to be pseudomonads of nonfluorescent acidovorous-testosteroni group. Several accumulate campholide levels reminiscent of the C5 strains used by Hedegaard and Gunsalus (50).

A cytochrome P450 was found to participate actively in the terminal hydroxylation of normal hydrocarbons to primary alcohols in a diptheroid bacterium. Octane, a preferred substrate, is converted rapidly to 1-octanol (9, 10). The reductant, DPNH, functions via a dehydrogenase flavoprotein; the system appears not to contain an iron–sulfur protein; the reductive couple to cytochrome is unclear. The equivalence of DPNH to O_2 and product conforms to a monoxygenase stoichiometry. By analogy, the substrate-O_2 steps are assumed to occur on the cytochrome (62, 118).

A second normal alkane methyl hydroxylase, the ω-hydroxylase of *Pseudomonas oleovorans*, a *Pseudomonas putida* strain (104) was isolated from an organism obtained by enrichment technique (6). A series of studies (71, 117) has reported three components, their separation and the identification of two as an FAD flavoprotein with DPNH dehydrogenase activity and an $FeCys_4$ iron sulfur protein, rubredoxin, containing two single iron active sites in a polypeptide of 19,000 molecular weight (70). The prosthetic group of the third protein, ω-hydroxylase, so far unidentified, is probably not a cytochrome. This system presents features useful in understanding the cytochrome monoxygenases; however, a full discussion is inappropriate to this chapter, by assignment and title directed to the cytochrome-containing monoxygenases.

A functionally different class of monoxygenases, possibly related in prosthetic group to the ω-hydroxylase, participates in the cleavage of carbocyclic rings adjacent to ketone groups. Camphor oxidation by the so-called 1,4- and 1,3-diketone pathways presents three cases of ring fission by monoxygenase action; cleavage of the steroid ring D to form testostolactone appears to be a fourth. Pyridine nucleotides serve as reductants via dehydrogenase flavoproteins with one or two additional components whose functional groups remain to be identified (19, 39, 40, 44). Though the ketolactonases conform to the two reduction equivalent, the one ketolactonase studied seems not to contain a cytochrome (123). Thus, like the ω-hydroxylase, their detailed analysis is not germane to the present text.

Nitrilotriacetate (NTA) oxidation furnishes still another notable example of monoxygenase stoichiometry in a *Pseudomonas* sp. Cell extracts were prepared which use DPNH as reductant and remove one acetate residue (108).

The extensive and increasing use of substituted hydrocarbons, both aromatic and aliphatic, with a variety of functional groups presents numerous "inert" reagents to the environment. The number and variety of known cytochrome monoxygenases seems certain to be expanded (27).

B. Formation

In bacteria, the oxygenase systems studied are found under substrate control. Like peripheral metabolic pathways in general, the enzymes of P450 cytochrome systems are formed in response to substrate and related compounds; that is, potential energy and carbon sources, not essential as cellular intermediates, are transformed by enzymes under inductive, or derepression, control. The physiologic and genetic studies needed to understand in detail the regulation of enzyme formation and activity have not been applied to many peripheral metabolic processes. The attention to inducer selectivity has been most exacting in the β-keto adipate pathway of aromatic cleavage (83, 104). Early work on monoxy-genase induction in the terpene series concerned the flavoprotein(s) and ketolactonase I (38, 40). Later, cytochrome P450 levels, and to a lesser extent, putidaredoxin levels, were considered (91). Genetic studies and enzyme separations showed a commonality of some camphor pathway components for D- and L-camphor. An exception is the substrate reactive enzyme of the D- and L-camphor ketolactonases (91, 109). In the D- and

L-camphor pathways the early enzymes, including the P450$_{cam}$ hydroxylase components, were found to form in response to each of the first five intermediates in the reaction pathway. Genetic investigations in particular have permitted the accumulation of intermediates and the identification of specific inducers in the aromatic (83) and D- and L-camphor oxygenation pathways (91).

Much could be gained by the optimal selection of strains and conditions for maximum enzyme formation, both in the ease of enzyme separation and in yield. Hartline and Gunsalus (48) confirmed and extended the earlier studies on inducer specificity (see Table II). Oxygen in addition to carbon substrate stimulates induction, possibly in part via energy release to drive protein biosynthesis. Many substituted bornanones induced alcohol dehydrogenases active on the 5-*exo* alcohol; few resulted in diketocamphane accumulation or supported growth.

The catabolic repression of oxygenase protein formation has been observed (36, 39, 60, 65). Succinate (60) and, in some instances, glucose (48, 50), are particularly strong repressors. Citrate and glutamate on the other hand cause minimal catabolite repression and are especially useful for enzyme induction studies with strains carrying mutant loci within the camphor oxidation pathway (48).

C. Genetics

Quite aside from the magnificent impact of genetics as a science, in practice, biochemistry flies as on a broken wing without her insight and tools. Thus, concomitantly with the exploration of the chemistry and enzymology of terpene metabolism, the authors initiated an intensive search for the genetic basis of metabolic diversity in the predominantly soil and water organisms able to recycle nature's carbon. Among the pseudomonads found to play a prominent role in the recycling of complex organic molecules are strains positive for both terpene and aromatic oxygenases —see the systematic analysis of Stanier *et al.* (104).

Starting with a single fluorescent pseudomonad, over a period of 10 years the authors have documented systems of genetic exchange by transduction (41) and by conjugation (15, 16) on both an intra- and interstrain level. While it would be premature to proclaim an understanding of the methods of generation and maintenance of biological diversity, valuable insight has been gained and at least one mechanism outlined for exchanges important to evolutionary accommodation of new structures and processes. The authors found homologies by complementation of proteins from heterologous strains with full catalytic activity (89) and were able to extend the range of genetic compatibility ("relatedness") to include

Pseudomonas aeruginosa and the saprophytic fluorescent group (17). This exchange with *P. aeruginosa*, the most homogeneous species within the fluorescent *Pseudomonas* group, furnished a genetic basis for joining newly acquired knowledge with the studies initiated earlier and continued in the hands of Holloway and his colleagues (57). The latter studies stem from the medical importance, ubiquitous distribution in hospitals, and high lysogenic frequency among *P. aeruginosa* strains.

The genetic explorations that started with the camphor-utilizing *Pseudomonas putida* strain PpG1 have gradually led to conditions for transfer by transduction of both chromosomal and plasmid genes. Host range phage mutants were selected for Stanier's primary strain and used to transfer the mandelate gene cluster to PpG1 (14). Through collaboration and residence in our laboratory of Stanier's students, the Ornstons and Mark Wheelis, and with the generous tutelage and collaboration of I. P. Crawford, transduction with the phage pf16 is now in general use (32, 54, 67, 83, 84, 91).

More recently, conjugation with transfer of chromosomal loci has been accomplished with suitable frequency for genetic analysis (97, 98). A liquid mating system permitting interrupted mating and chromosome mapping has not been perfected but seems within the realm of the possible (13).

The genetic studies have revealed plasmid coding for several peripheral metabolic pathways; the activities for the early reaction steps are lost upon curing (91); the genes are reintroduced by conjugation (13) or, in the case of smaller sequences, by transduction (20). Transfer among strains was shown to form multiplasmid-containing exconjugants (13, 20). Strains carrying single plasmids are usually stable for those characters; multiplasmid-bearing strains may show enhanced segregation rates reaching as high as 10% per generation for the nonselected phenotypes (20, 97, 98). In rare exceptions compatibility is established, apparently by recombination of two or more plasmid elements (13,20).

The genetics of the camphor pathway from the initial substrate to isobutyrate were blocked out by Rheinwald (90). Point mutants in genes coding early steps of camphor oxidation permitted Rheinwald *et al.* (91) to separate the loci controlling enzymes for early, mid, and late reactions and to examine induction selectivity. Data are shown in Table III for critical point mutants and two cured strains, /CAM[d]; the enzyme levels in camphor-induced mutants with glutamate as a secondary carbon source appear in Table IV.

Chromosomal genes control the enzymes for isobutyrate and beyond; Helen Dunn clarified their relationship (31). Redundancy was noted with camphor plasmid, /CAM, genes; the CAM plasmid was transferred to

TABLE II

COMPOUNDS TESTED AS INDUCERS OF CAMPHOR OXIDATION

Name of compound	Structural formula	Maximum differential rate[a] of synthesis			Appearance[b] in culture broths		Growth
		5-*exo*-DH	2-*endo*-DH	Activity[c] of compound with extract	Bornane-diones	Dihydroxy- or mono-hydroxy-bornanones	Maximum optical density at 2 mM concn.
2-Bornanone (D-camphor)		136–303	25–58	—	45	90	0.38
2-*endo*-Hydroxybornane		334, 418	53	53	45	135	0.520
2-*exo*-Hydroxybornane		105, 310	20	2	45	135	0.410
3-*endo*-Hydroxybornane		186	8	3	None	135	0
3-*exo*-Hydroxybornane		388	20	66	None	135	0.100
3-Bornanone		171, 486	3	—	None	None	0.175

Compound	Structure						
5-endo-Hydroxybornane[d]		90	4	9	None	135	0
5-exo-Hydroxybornane[d]		172	4	57	None	135	0
5-Bornanone		237	6	—	None	None	0
6-endo-Hydroxybornane[d]		178	42	113	None	135	0.530
6-exo-Hydroxybornane		278	21	—	None	90	0.500
6-Bornanone (L-camphor)		297	24	—	135	135	0.560
2,3-Bornanedione		99	4	—	—	—	0
2,5-Bornanedione		197, 96	18, 52	—	—	—	0.370
2,6-Bornanedione		58	0	—	—	—	0

TABLE II

COMPOUNDS TESTED AS INDUCERS OF CAMPHOR OXIDATION

Name of compound	Structural formula	Maximum differential rate[a] of synthesis		Activity[c] of compound with extract	Appearance[b] in culture broths		Growth
		5-*exo*-DH	2-*endo*-DH		Bornane-diones	Dihydroxy- or mono-hydroxy-bornanones	Maximum optical density at 2 mM concn.
3,6-Bornanedione		257	18	—	—	—	0.590
5,6-Bornanedione		57	1	—	—	—	0
2-*endo*-Hydroxy-5-bornanone		96	2	—	—	—	—
2-*endo*, 5-*exo*-Dihydroxybornane		266	0	—	—	—	—
3,4,4-Trimethy-5-carboxymethyl-2-cyclopentenone (6 mM)[e]		100	64	—	—	—	—
Bornane (3 mM)[e]		150	0	—	—	—	0

D(−)2-β-Aminobornane·HCl (6 mM)[e]		106	0	—	None	None	0
1,4-Cyclohexanedione		2–20	0.5–2	—	None	None	1.6
Glucose (10 mM)		7	—	—	—	—	—

[a] Differential rates expressed as units of enzymic activity per milligram of dry cell weight.
[b] Expressed as minutes after induction.
[c] Measured as a substrate for DPN reduction.
[d] Maximum occured between 45- and 90-min samples.
[e] Induced in cells growing on 5 mM glucose.

isobutyrate-positive *Pseudomonas* strains including two typical strains of *P. aeruginosa* (91). In this high temperature (45°) competent strain, the expression of plasmid-borne genes is temperature sensitive (13). Genetic studies have been extended to plasmids governing naphthalene oxidation (32), salicylate (12), and the octane (from *P. oleovorans*; 13, 21). These plasmids were transferred to strain PpGl to facilitate genetic analysis. The genetic studies of the mandelate-aromatic ring cleavage pathway are reviewed by Wheelis and Stanier (120) and by Ornston (83).

The genetic organization of biosynthetic pathways in *Pseudomonas putida* was studied to build experience and strains. These include tryptophan (14, 41, 54, 89), arginine (20), histidine, and methionine. Several genetic firsts met along the way have permitted an integration with the Holloway and Loutit studies on *P. aeruginosa* (58). Clarke (23) and

TABLE III

CAMPHOR MUTANT GENES AND GROWTH PHENOTYPES[a]

Stock No.[b]	Genotype	Phenotype	D-Cam	F	D	X_1	Ibu	Prp	Suc
1[c]	wt/CAM	wt/Cam+	+[c]	+	+	+	+	+	+
544	/CAM*camA101*	OH,E_A^-	−	+	+	+	+	+	+
545	/CAM*camB102*	OH,E_B^-	−	+	+	+	+	+	+
543	/CAM*camC100*	OH,E_C^-	−	+	+	+	+	+	+
552	/CAM*camD120*	FdeH^{-c}	−	−	+	+	+	+	+
553	/CAM*camD121*	FdeH$^-$	−	−	+	+	+	+	+
557[c]	/CAM*camG133*	KL,1^{-c}	−	−	−	+	+	+	+
556	/CAM*cam-132*	KL,1^-	−	−	−	+	+	+	+
566[c]	/CAM*cam-206*	Z	−	−	−	−	+	+	+
560	/CAM*cam-200*	Z	−	−	−	−	+	+	+
568	/CAM*cam-208*	Z	−	−	−	−	+	+	+
571	/CAM^d		−	−	−	−	+	+	+
572	/CAM^d		−	−	−	−	+	+	+
575	*ibu-100*/CAM		−	−	−	−	−	−	+
577	*ibu-102*/CAM		−	−	−	−	−	−	+

[a] From Rheinwald *et al.* (91).

[b] All by treating wt/CAM with nitrosoguanidine, except 571 and 572.

[c] 1, 557, 566 L-(−)-Camphor +. Trivial names of compounds and enzymes: D-Cam = (−)-camphor; F = 5-*exo*-hydroxycamphor; D = 2,5-diketocamphane; X_1 = cyclopentenone acetic acid; Ibu = isobutyrate; Prp = propionate; Suc = succinate; OH$^-$ = lack of hydroxylase enzymes E_A, E_B, E_C, etc.; KL = ketolactonase; FdeH = F dehydrogenase; and Z = reactions, enzymes unresolved.

TABLE IV

MUTATIONS AND THE INDUCTION OF ENZYMES FOR CAMPHOR
OXIDATION[a]

	Specific activity			
Strain[b]:	1	543	545	544
Phenotype:	Cam$^+$	E$_C^-$	E$_B^-$	E$_A^-$
Enzymes				
Hydroxylase				
Cytochrome P450, E$_C$	45	0	30	44
Putidaredoxin, E$_B$	4	3	0	4
Putidareductase, E$_A$	4	3	4	0
Alcohol dehydrogenase				
5-exo	18	14	10	14
Ketolactonase 1				
Oxygenase function, E$_{2,3}$	0	9	8	7
DPNH-dehydrogenase, E$_1$	75	63	67	57

[a] From Reinwald (90).

[b] Grown to late logarithmic phase (OD $\sim$ 2) in PSA,
containing 10 mM glutamate (MSG) and 10 mM camphor.
Extract prepared in phosphate buffer pH by sonic oscillation.

Richmond (92) among others have extended these genetic studies; a
monograph on *Pseudomonas* biology and genetics is in process in which
P. aeruginosa emerges as the prima ballerina (93).

D. Evolution

Oxygenases of the cytochrome P450 class have been observed as spectral
entities in all plants and animals so far examined. The metabolic functions
of most are unknown. The prime exception are for those systems known
to be induced by substrates and the "nonspecific" hepatic systems.

A striking similarity of components and regulation exists between the
P450 systems in adrenal cortex and *P. putida* (46, 64, 96). The iron–
sulfide proteins adrenodoxin and putidaredoxin are indistinguishable by
resonance spectral measurements, show regions of homology, and are
almost of equal size (106, 107, 110). The two, however, are not reciprocally
active in the complete hydroxylase systems nor is Ad$^\circ$ an effector in the
e_2 activity with the oxygenated P450$_{cam}$. The flavoproteins, TPNH versus
DPNH specific, are otherwise similar in reduction of both iron sulfide
proteins. One can scarcely demand a broader range of phylogenetic separation than an aerobic soil pseudomonad and the mammalian adrenal

cortex—the deduction is one of early origin and selective protein requirement.

Evolution in substrate variety and specificity has not been explored systematically among the P450 cytochromes. The selectivity varies over a wide range, possibly depending on the biological function in response to a role in biosynthesis versus degradation. The search for function(s) of a protein starting with spectral identity, as indicated in Sections II, A and B, is not a rational study; the range of substrate selectivity is subject to the same burden. The organismal gain from narrow versus broad selectivity is open to rationalization if not to study. In the mammal sharp specificity in the adrenal may be essential to regulation and survival; the low order in the liver is required to deal with detoxification of a great variety of insults within a limited potential for total protein synthesis (22). In the microbe, high turnover rate with rapid access to carbon and energy may require a narrow selectivity range for competitive advantage and a strict control at the repressor level to avoid synthesis of unused cellular components.

Initial experiments aimed at broadening the metabolic capacity of selected bacterial strains have met with modest success and, equally, with the counter argument that the systematics of species lines would blend if continuous variation were of common occurrence—perhaps, indeed, they do, subject to a competitive environment of natural enrichments. Only as additional workers with clear heads and sharper tools (1, 28) join in investigations of organic residue recycling will there emerge clearer pictures of nature's potential for continuous nonlimiting variation.

IV. CHEMISTRY AND PHYSICS

A. Structure

The separation and purification of the three protein components of the methylene hydroxylase from camphor-induced *Pseudomonas putida* is accomplished by relatively simple and gentle anion exchange and sizing chromatographic procedures (46, 109). The FAD-flavoprotein (reductase, fp), iron–sulfur protein (redoxin, Pd), and the cytochrome ($P450_{cam}$) are obtained in soluble, homogeneous fractions which can be recombined to yield a fully active system with monoxygenase stoichiometry.

Following a rather brief general analysis of the composition and properties of the three hydroxylase components, the resonance spectral properties of each will be discussed, followed by those dynamic properties of component and substrate interactions that reveal structural-catalytic roles.

TABLE V

CAMPHOR 5-METHYLENE HYDROXYLASE COMPONENTS[a]

Amino acid	Cytochrome E_C	Reductase E_A	Redoxin E_B	
Asp	27	25	10	14(13)[b]
Asn	9	15	4	
Thr	19	20	5	
Ser	21	18	7	
Glu	42	16	9	12(10)[b]
Gln	13	24	3	
Pro	27	18	4	
Gly	26	33	9 (8)[b]	
Ala	34	48	10 (9)[b]	
Val	24	34	15 (14)[b]	
Met	9	6	3	
Ile	24	24	6	
Leu	40	42	7 (6)[b]	
Tyr	9	6	3	
Phe	17	10	2 (1)[b]	
His	12	6	2	
Lys	13	13	3	
Trp	1	3	1	
Arg	24	24	5	
Cys/2	6	6	6	
Total	397	393	114 (106)[b]	
Free SH	6	6	4	
N-terminus	Asx	Ser	Ser	
C-terminus	Val	Ala	Gln (Trp)[b]	
MW ($\times 10^3$)	45	43.5	12.5	
Prosthetic group	Heme	1 FAD	$(FeS)_2$	

[a] From Tsai et al., (110).
[b] According to Tanaka et al. (107).

Table V summarizes the properties of the three enzymes which comprise the cytochrome $P450_{cam}$ methylene hydroxylase, including the prosthetic groups, size, composition and C- and N-terminal residues. Each protein is composed of a single polypeptide chain and a molecule of prosthetic group (110). The primary structure of the redoxin as determined by Yasunobu and colleagues (107) is reproduced in Table VI. The 106 residues differ slightly from the composition published earlier (110). Of special interest is the single tryptophan residue at the C-terminus.

The putidaredoxin, Pd, prosthetic group contains two atoms each of acid-labile iron and sulfur and is attached to the protein through 4-alkyl sulfide linkages provided by 4 of the 6 cysteine residues of the primary structure (37, 107). Holm and Phillips (see ref. 77) have introduced the convention $Fe_2S_2*Cys_4$ to describe this prosthetic group and have prepared a synthetic analog with properties approximating those of iron–sulfide redoxins. In analogy to adrenodoxin, and to the less closely related plant ferridoxins, four specific Cys residues of the six present have been suggested as iron ligands, but in the absence of X-ray data the assignments are not firm.

The symmetrical arrangement of the iron–sulfide active center in Pd and the interaction of the iron atoms are shown by the spectra, EPR (29, 110) and Mössbauer (24, 79). The redox potential of Pd (-239 mV) (121) compares favorably to the most recent value reported for adrenodoxin (-270 mV) (59) and, in contrast, is relatively high as compared to the ferridoxins of spinach, *ca.* -420 mV, and of *Clostridium pasteurianum* ($Fe_8S_8*Cys_8$), -390 mV (72). The potentials of the three components of the P450$_{cam}$ system and representative redoxins are tabulated in Table VII with the size and prosthetic group components included. Perhaps the potentials are not surprising when one considers that the green plant ferridoxins are intermediates between the light-generated reductant and reduced pyridine reductides, -320 mV, the clostridial redoxin between a reductant generated in pyruvate cleavage and energy transfer in nitrogen reduction.

P450$_{cam}$ crystals were first grown by Yu and Gunsalus (124) and recently in a form suitable for minimal X-ray study by Jacobs and Lipscomb (59a). The single molecule of ferriprotoporphyrin IX in P450$_{cam}$ appears in a relatively open and apparently flexible active site. These crystals possess the P450$_{cam}$·camphor absorption spectra and undergo reduction and formation of the ferrous·CO spectrum without resolution or shattering.

TABLE VI

PUTIDAREDOXIN SEQUENCE[a]

Ser-Lys-Val-Val-Tyr-Val-Ser-His-Asp-Gly-Thr-Arg-Arg-Gln-Leu-Asp-Val-Ala
Asp-Gly-Val-Ser-Leu-Met-Gln-Ala-Ala-Val-Ser-Asp-Gly-Ile-Tyr-Asp-Ile-Val
Gly-Asp-Cys-Gly-Gly-Ser-Ala-Ser-Cys-Ala-Thr-Cys-His-Val-Tyr-Val-Asn-Glu
Ala-Phe-Thr-Asp-Lys-Val-Pro-Ala-Ala-Asn-Glu-Arg-Glu-Ile-Gly-Met-Leu-Glu
Cys-Val-Thr-Ala-Glu-Leu-Lys-Pro-Asn-Ser-Arg-Leu-Cys-Cys-Gln-Ile-Ile-Met
Thr-Pro-Glu-Leu-Asp-Gly-Ile-Val-Val-Asp-Val-Pro-Asp-Arg-Gln-Trp

---

[a] From Tanaka *et al.* (107).

TABLE VII

REDOX POTENTIAL OF "HYDROXYLASE COMPONENTS"[a]

| Protein | E_0' (mV) | Prosthetic group | | | MW (thousands) |
		Fe	S	Cys	
Ferredoxin					
Spinach	−420	2	2	4	13
C. acidi-urici	−390	8	8	8	6
C. pasteurianum	−390	8	8	8	6
E. coli	−360	2	2	−	12
P450$_{cam}$ cytochrome	∼−270	Heme			45
Adrenodoxin	−270	2	2	4	13
Putidaredoxin	−240	2	2	4	12.5
P450$_{cam}$·S	∼−170	Heme			45
Rubredoxin					
C. pasteurianum	−57	1	0	4	6
P. oleovorans	−37	1[b]	0	4[b]	19
P. aerogenes	−60	1	0	4	5.5
P450$_{cam}^r$·S·O$_2$	∼0	Heme			45
cyt b$_5$ (rabbit)	+30	Heme			14
cyt b$_{562}$ (E. coli)	+130	Heme			12
cyt c	+230	Heme			13
HiPIP (Chromatium)	+330	4	4	4	10

[a] From Gunsalus and Lipscomb (42).
[b] Contains two active centers of this conformation.

The binding of both camphor and oxygen near the iron is shown by marked changes in the optical spectra and apparent redox potentials (42, 111, 120a). The potential shift in the ferric heme on substrate addition appears to facilitate reduction to the ferrous state. Tentative values for the redox potential of P450$_{cam}$ and P450$_{cam}$·camphor complex in Table VII are −270 and −170 mV. Both the ferrous P450$_{cam}$ and P450$_{cam}$·camphor form dioxygen adducts. The latter is more stable and has the lower apparent redox potential for the second electron addition, i.e., ca. 0 mV, whereas the substrate-free form is near +270 mV. The stability and the dynamics of the several reactions by which the oxyferrous substrate P450$_{cam}$ decays both with and without product formation are discussed in Section IV, C.

The first stage in the loss of catalytic activity by P450 is detected as a form termed P420 to indicate the position of the Soret absorption of the ferrous·CO derivative. Soret absorption near 420 nm for CO adducts is characteristic of nearly all ferrous heme proteins which accommodate a CO ligand. The first decay of P450 to P420 can be reversed by treatment with sulfhydryl reagents such as cysteine (127). This observation contributes to the speculation that the P450 has one axial sulfur ligand (68), whereas P420 may possibly contain two nitrogen ligands like myoglobin. The possibility of a sulfur ligand, first suggested by Mason (75) and Jefcoate and Gaylor (61), and by the similarity of the EPR spectra of P450 and sulfmyoglobin as well as other model complexes (103) is strengthened by these observations. The EPR spectra of P450 and P420 are not, however, significantly different, suggesting that the conversion to P420 may be more subtle.

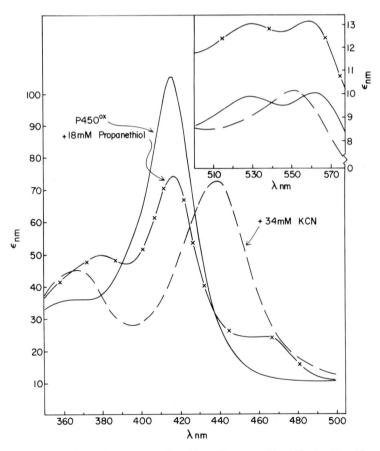

Fig. 4. P450$_{cam}$ absorption spectra: Cyanide and mercaptide adducts 50 mM potassium phosphate buffer, pH 7.0 containing 430 μg/ml ferric cytochrome, 25°.

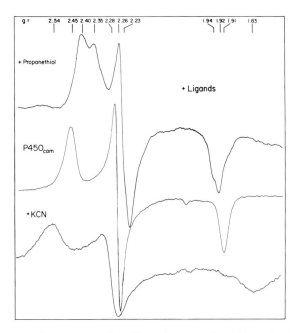

FIG. 5. P450cam EPR spectra: Cyanide and mercaptide adducts buffer as in Fig. 4. T = 79°K. Also as in Fig. 4, 19.5 mg/ml ferric cytochrome. Frequency modulation = 100 Hz; microwave = 9.106 GHz. Amplitude modulation = 16 G. Gain = 400. Microwave power = 180 mW. Varian = E-9.

B. Resonance Spectra

The presence of iron in the redox centers of both cytochrome P450cam and putidaredoxin has facilitated the use of several physical probes pointed toward characterization of the structure and dynamics of their active sites. Each protein has characteristic optical and resonance spectra which change with redox state, substrate binding, and on the addition of several sorts of ligands, e.g., N-phenyl-imidazoles, isocyanates, CO, and CN⁻. Optical absorption and EPR spectra, respectively, for the cyanide and thiol adducts are shown in Figs. 4 and 5. Both proteins have been subjected to replacement of the iron and of the chalcogenide with isotopes by chemical (113) and growth procedures (99). These changes have permitted differentiation of states and lent credence to a cyclic reaction sequence proposed by Katagiri et al. (63) and Gunsalus (37) for the camphor methylene hydroxylase.

The optical absorption spectra of the three protein components of the cytochrome P450 methylene hydroxylase between 200 and 700 nm were shown in Fig. 2 and the major maxima of the several states given in Table I.

1. Putidaredoxin

As discussed in Section II, the 2 Fe-2 S protein, putidaredoxin, is the second component of the camphor hydroxylase system of *P. putida* (37). It transfers two reducing equivalents, one at a time, from the reductase, fp, to the P450 cytochrome (see Fig. 1a). The standard potential of putidaredoxin is intermediate between the potentials of the reductant, DPNH with flavoprotein, and of the oxidant, the substrate adduct of cytochrome $P450_{cam}$. The first reduction of $P450_{cam}$, e_1 (Fig. 1a or 1b), is accompanied by the one equivalent ferrous–ferric oxidation, Pd^r to Pd^o. The Pd^r can be replaced by a number of low potential chemical systems (Fig. 1c) to form the stable ferrous cytochrome. The second reduction, e_2, of the oxygenated $P450_{cam}$ is more selective and is accompanied by product formation (for approximate potentials, see Table VII). Pd^r serves both reduction roles; the oxidized molecule, Pd^o, catalyzes a disproportionation to yield 1 mole of product per 2 moles of oxyferrous $P450_{cam}$·camphor complex (116).

Putidaredoxin shares many characteristics with other iron–sulfur proteins (115): (a) It contains an equal number, i.e., two each, of inorganic, acid-labile sulfur and of iron atoms (115); (b) under reducing conditions it shows an electron spin resonance (ESR) signal with g values, $g_\perp = 1.94$ and $g_\parallel = 2.01$ (114) which is typical for all proteins of this class; and (c) it is diamagnetic in the oxidized state. The optical spectrum of putidaredoxin resembles that of other iron–sulfur proteins. Oxidized putidaredoxin has absorption maxima at 455, 412, 328, and 276 nm, whereas the reduced protein shows a weak maximum at 540 and the original 274 nm only [see Fig. 2b or Tsibris and Woody (115)]. Replacement of the labile sulfur by selenium causes an almost uniform red shift of all the bands by 10–30 nm (113), consistent with a ligand-to-metal charge transfer assignment.

The structure of the iron–sulfur complex in putidaredoxin, as well as those structures of other 2 Fe-2 S proteins, is still unknown, but a model proposed in 1966 by Gibson *et al.* (35) has been quite successful in explaining the ESR, ENDOR, and Mössbauer data to be discussed below. According to this model, the complex contains an antiferromagnetically coupled pair of iron ions which are both ferric in the oxidized protein, while one is ferrous and the other ferric in the reduced protein. To account for the unusual g values, Gibson *et al.* further assumed that the ligand field of the ferrous ion has distorted tetrahedral symmetry. The X-ray diffraction studies of related 1 Fe, 4 Fe-4 S, and 8 Fe-8 S proteins (11, 51, 100) show that in each case the redox center contains iron that is more or less tetrahedrally coordinated to four sulfur atoms, which may be either cysteine or inorganic sulfur. These results and studies of synthetic compounds (5, 52, 77) make Gibson's hypothetical structure even more plausible.

Reduced putidaredoxin has a spin of $S = \frac{1}{2}$ (78) and, in the native form, an axially symmetric g tensor, $g_\perp = 1.94$, $g_{||} = 2.01$; the g tensor of the selenium derivative is rhombic, $g_x = 1.93$, $g_y = 1.98$, and $g_z = 2.04$ (82). The enrichment of the protein in ^{57}Fe leads to a characteristic broadening of the ESR spectrum as a result of the magnetic hyperfine interaction between the unpaired electron and the magnetic moments of the two iron nuclei. Similar experiments with ^{33}S-enriched protein and with a ^{77}Se analog show that the unpaired spin also interacts with the labile and with some organic sulfur of the amino acid residues in the primary sequence of the protein. Through a series of ENDOR and Mössbauer measurements, the model proposed by Gibson and co-workers was shown to be essentially correct (34, 79).

ENDOR experiments by Sands and co-workers (34) prove convincingly that the two iron atoms in the redox center of reduced putidaredoxin are inequivalent. The absolute values of the ^{57}Fe hyperfine coupling constants are compatible with the assignment of high-spin ferric and high-spin ferrous states, respectively, to the two iron atoms. Mössbauer studies confirm this interpretation (79). From a determination of the sign of the hyperfine coupling constants, it follows unambiguously that the spins, $S_a = 5/2$ and $S_b = 2$, of the two iron atoms couple antiferromagnetically to a total spin of $S = \frac{1}{2}$, $\vec{S} = \vec{S}_a + \vec{S}_b$, as predicted by the Gibson model. Moreover, the quadrupole interaction of the ferrous iron is compatible with an iron site of almost tetrahedral symmetry. The symmetry axis is the same as that found in adrenodoxin but differs from that of the plant ferridoxins (34).

In oxidized putidaredoxin, the two iron atoms are indistinguishable by Mössbauer spectroscopy. Both the quadrupole interaction and the isomer shift are the same as that found for the ferric iron in the reduced protein. In oxidized putidaredoxin both iron atoms are thus in the high-spin ferric state, $S_a = S_b = 5/2$, coupled to a diamagnetic ground state, $\vec{S} = \vec{S}_a + \vec{S}_b = 0$. The exchange interaction responsible for the antiferromagnetic coupling is estimated to be at least 60°K (78).

In summary, Gibson's tetrahedral model with spin coupling provides a consistent explanation of the experimental results, but it does not explain all the facts. To account for the strong covalency, the exchange interaction, the optical properties, and possibly the redox reaction, a more sophisticated model will be required.

2. Cytochrome P450cam

Oxidized $P450_{cam}$ in the camphor-free state exhibits an optical absorption spectrum characterized by a Soret band at 417 nm with a shoulder at 360 nm. The α and β bands are observed at 567 and 538 nm, respectively. Upon addition of camphor to form the oxidized substrate complex, a new state is established with the Soret band shifted to 391 nm and charge transfer bands appearing at 500 and 640 nm. Electron spin resonance investigations by Tsai *et al.* (111) gave the following results: (a) Camphor-free preparations yield g values at $g = 2.46$, 2.26, and 1.91, typical of heme iron in a low-spin ferric state; (b) in a camphor-saturated sample, a high-spin ferric signal is observed at $g = 8$, 4, and 1.8 in addition to the low-spin signal. The high-spin signal accounted for only 60% of the total ferric heme, whereas 40% remained in the low-spin ferric state. Titrations of oxidized $P450_{cam}$ with camphor have indicated that one molecule of camphor is bound per molecule of protein and that complete camphor binding results in a high-spin population of approximately 60% at 12.5°K as measured by ESR (W. Orme-Johnson, private communication). Peterson found 75% high-spin ferric material between 94° and 253°K from magnetic susceptibility measurement (87). The low temperature ESR and the high temperature susceptibility data are qualitatively confirmed by Mössbauer measurements (99). The high-spin fraction appears to increase with the temperature. It was found from the Mössbauer measurements that the spin populations can differ by about 10% for different samples. More investigations are clearly necessary before the temperature-dependent spin equilibrium can be evaluated in terms of thermodynamic parameters. It remains to be investigated how the spin equilibrium depends on the concentration, the buffer, the freezing rate and, perhaps, the history of the sample. Since each method used has its own inherent difficulties in evaluating spin populations, the different techniques should all be applied to the same sample.

The g values associated with the low-spin ferric material have been analyzed by Blumberg and Peisach (7) in the framework of ligand field theory. For a pure low-spin ferric compound the wave function of the ground state can be represented in a good approximation by a linear combination of t_{2g} orbitals. In this approximation, and assuming an orbital reduction factor $k = 1$, the g values can be expressed by two parameters describing the environment of the heme iron, the tetragonality and the rhombicity. Blumberg and Peisach have presented the results of the analysis of a series of ESR measurements on heme proteins in the form of "truth tables," in which each derivative is classified according to tetragonality and rhombicity. The g values of oxidized $P450_{cam}$ define a unique class (P type). Compounds in the same class have been prepared by addition of thiols to hemoglobin and myoglobin (7, 53). These observations suggest that the heme iron in oxidized P450$_{cam}$ is bound to a cysteine residue. Though preliminary data are available (53), much less is known regarding the chemical nature of a (possible) sixth ligand (88).

Many different external ligands can be reacted with oxidized P450 including imidazoles, sulfides, pyridine, and pyrimidine. These ligands cause the Soret band to shift a few nanometers, from 417 nm to longer wavelengths. Electron spin resonance measurements show that the g values again fall into the P-type class. Even P420, prepared by treatment of P450$_{cam}$ with 20% acetone at 25°C shows these characteristic g values (2.42, 2.25, and 1.91). There is no direct proof that the external ligands bind to the heme iron. Though all these complexes look quite similar when produced with oxidized P450$_{cam}$, quite different species are observed upon reduction. The reduced complexes fall roughly into two classes, those with Soret bands at 410–420 nm, indicating high-spin ferrous compounds, and those with Soret bands at 440–450 nm, indicating low-spin ferrous material. Some of these compounds have been investigated with Mössbauer spectroscopy confirming the spin assignments.

The observed ESR signals ($g = 8$, 4, and 1.8) on camphor-saturated samples of oxidized P450 have been evaluated in the framework of a spin Hamiltonian characteristic of high-spin ferric iron (111). The observed g values imply $E/D = 0.087$, where D and E are the tetragonal and rhombic coefficients of the spin Hamiltonian; E describes the departure of the heme iron system from tetragonal symmetry. The value of $E/D = 0.087$ represents a large departure from axial symmetry, the largest observed so far for high-spin ferric heme proteins. Recently, similar rhombic distortions have been observed for iodide and fluoride complexes of ferric chloroperoxidase (18).

The low temperature Mössbauer spectra of camphor-complexed P450$_{cam}$ clearly display the rhombic distortions also. Figure 6 shows a typical Mössbauer spectrum of P450$_{cam}$ taken at 4.2°K. As mentioned above, camphor-saturated P450$_{cam}$ exists in a spin equilibrium even at 4.2°K. Figures 6b and 6c show the decomposition of spectrum (Fig. 6a) into high- and low-spin spectra [for details, see Sharrock et al. (99)]. The spectrum at 200°K can be interpreted unambiguously as a superposition of two quadrupole doublets. The doublet with the small splitting ($\Delta E_Q = 0.79$ mm/sec) is attributed to high-spin ferric iron, the other doublet ($\Delta E_Q = 2.66$ mm/sec) to low-spin ferric iron. The analysis of the low temperature spectra agrees fairly well with information obtained from ESR measurements. Figure 6d shows that camphor-depleted P450$_{cam}$ is predominantly low-spin. It is worthwhile to note that the isotropic

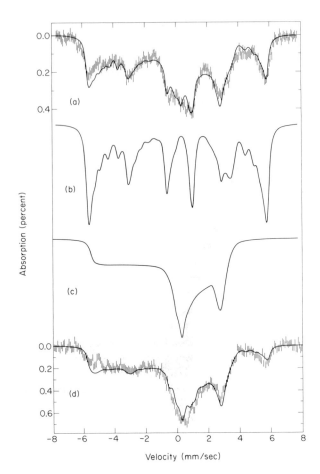

Absorption (percent)

Velocity (mm/sec)

FIG. 6. Mössbauer spectra of oxidized P450$_{cam}$: (a) camphor >3.5 mM, (d) camphor-depleted. 4.2°K, applied field $\simeq$120 G, parallel to the observed γ-rays. Solid curves are composites of the computed spectra for (b) high-spin ferric iron and (c) low-spin ferric iron. High-spin fraction 0.45 in (a) and 0.20 in (d).

hyperfine coupling constant found for the high-spin ferric component of P450$_{cam}$ is appreciably less than found for other heme proteins, indicating a large degree of covalency at the heme iron.

Preliminary X-ray work by Dickerson and co-workers (B. L. Trus, private communication) has shown that the substrate complex of P450$_{cam}$ crystallizes with lattice constants $a = 69$, $b = 64$, and $c = 97$ Å into the space group $P2_12_12$ with four molecules per unit cell. In addition, Sligar and Jacobs in our laboratories have measured the optical absorption spectrum of these crystals and have found that the initial crystallization takes place with the substrate, camphor, bound to the protein, i.e., 391 nm Soret absorption. The intact crystal can also be the reduced and the CO adduct formed without shattering the crystals, suggesting that the space group for these stable states is not different from that of the oxidized species. Absorption measurements using polarized light are currently in progress and promise to reveal crucial information concerning the electronic structure of the heme and of the nearby environment as well as the angles between the prophyrin plane and the crystallographic axis. Analogous measurements made in the near ultraviolet, coupled with fluorescence lifetime data, then yield the distances between the aromatic amino acids and the heme plane.

Reduced $P450_{cam}$ is characterized by a Soret band at 408 nm; neither the wavelength nor the magnitude is appreciably changed in adding substrate. The Mössbauer spectrum shown in Fig. 7 demonstrates unambiguously that the heme iron is high-spin ferrous. The following parameters have been found: the quadrupole splitting ΔE_Q and the isomeric shift δ relative to Fe metal are at 4.2° and 173°K, respectively, $\Delta E_Q = 2.45$ and 2.39 mm/sec, $\delta = 0.83$ and 0.77 mm/sec with uncertainties of ± 0.02 mm/sec). Thus in contrast to hemoglobin, myoglobin, reduced horseradish peroxidase, and reduced Japanese radish peroxidase *a*, the reduced $P450_{cam}$ samples do not show a pronounced temperature dependence of ΔE_Q. This difference in temperature behavior for $P450_{cam}$ suggests that excited electronic states are at higher energies than those of other heme proteins known to be high-spin ferrous. Though it seems

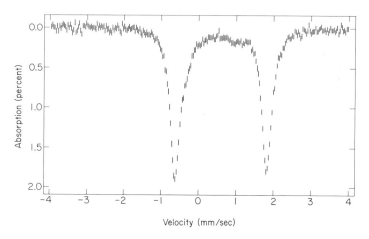

FIG. 7. Mössbauer spectrum of reduced camphor-complexed $P450_{cam}$ 4.2°K.

unlikely that the heme iron is bound to a histidine residue, it would be premature to draw a firm conclusion.

Oxygenated P450$_{cam}$ is formed on the addition of O_2 to the anaerobically reduced, camphor P450$_{cam}$ complex to produce a new state with an optical absorption spectrum very similar to that of oxyhemoglobin. Mössbauer spectroscopy on oxygenated P450$_{cam}$ gave the following results (99): (a) the spectra are quite similar to those of oxyhemoglobin, though oxy-hemoglobin displays a stronger temperature dependence of the quadrupole splitting; and (b) spectra taken in an applied magnetic field of 39 kG prove that the complex is diamagnetic like oxyhemoglobin and the quadrupole coupling constant is also negative. From these data, it can be concluded that oxygen binds in a similar way to the heme iron as in hemoglobin.

Weiss (119) proposed the Fe(III)-O_2^- model of oxygen binding to account for the diamagnetism, the acid dissociation properties, and the optical absorption spectrum of oxyhemoglobin. He noted that the optical spectrum is very similar to that of alkaline low-spin ferric hemoglobin. Spectral relationships that correspond have been pointed out for a number of heme proteins (122). Lang and Marshall (66) proposed a molecular orbital scheme providing the assignment of approximately five d electrons to the iron in order to account for the Mössbauer data. In the case of coboglobin, ESR measurements have shown that the oxy form is paramagnetic ($S = \frac{1}{2}$) and that the unpaired spin resides, for the most part, on the bound super-oxide ion (55).

Taking into account the similarities of the optical and the Mössbauer spectra of oxygenated P450$_{cam}$ and hemoglobin, there is good reason to assume that the oxygenated P450$_{cam}$ should be described as a complex of ferric iron where electron density has been transferred from the ferrous ion to the oxygen to form the superoxide anion O_2^-. It must be noted, however, that despite the close similarities between the oxygenated forms of P450$_{cam}$ and hemoglobin the heme sites of the two proteins differ in some respects as witnessed by the different temperature dependencies of the quadrupole splittings. It might be asked whether the similarities in the spectra of oxy P450 and the known oxygen carriers, myoglobin and hemoglobin, imply that they all have histidine as a fifth ligand. Recent results from Mössbauer emission spectroscopy on oxygenated heme model complexes in frozen solution show that the typical Mössbauer spectrum of oxyhemoglobin is obtained regardless of whether the axial ligand is pyridine, 1-methyl imidazole, 1,2-dimethyl imidazole, piperidine, or ethylmethylsulfide (80). It thus appears that the typical Mössbauer spectrum of oxyhemoglobin reflects essentially the iron–oxygen linkage and is rather insensitive to the ligand bond in the other axial position.

Reduced P450$_{cam}$ *complexed with CO* exhibits the Soret absorption at

446 nm, i.e., about 450 nm, from which this class of cytochrome derives the name P450. The optical spectrum indicates an unusual metal–porphyrin complex. Since complexes of reduced $P450_{cam}$ with bases like imidazole, pyrimidine, dimethylsulfide, thioxane, and tetrahydrothiophene exhibit the Soret band at long wavelengths (>440 nm) also, unusual CO binding cannot be made responsible for the peculiar spectra.

Surprisingly, the Mössbauer spectra do not show differences between the CO complexes of $P450_{cam}$ and those of hemoglobin and myoglobin. The $P450 \cdot CO$ complex is low-spin ferrous and, within the uncertainties of the Mössbauer parameters, agrees with those for hemoglobin. The agreement of the quadrupole splittings might be coincidental since the sign of the electric field gradient tensor and the asymmetry parameter have not as yet been determined.

The CO adducts of $P450_{cam}$ and Mb show similar but quite complex recombination kinetics after photodissociation by an intense light pulse. Frauenfelder and co-workers (4) have observed at least four distinct types of kinetic behavior as the temperature is lowered from 25°C to near absolute 0° (see Fig. 8). In region I, above 240°K, the reaction is second order; in region II, between 240° and 200°K, the rebinding becomes exponential and independent of CO. In region III, the reaction is either too fast or too slow to follow whereas in region IV, below 150°K, it follows a power law dependence. Myoglobin and P450 differ significantly only in the fourth or lowest temperature region, but even here only in magnitude and not in type of kinetics observed (4).

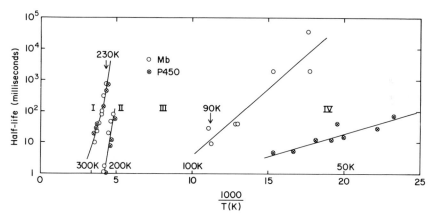

FIG. 8. Temperature dependence of CO binding to Mb and P450 heme protein 10 μM in glycerol:water (3:1 v/v); CO 1 atm. Half-lives of rebinding are CO concentration dependent in region I and independent in regions II and IV (4).

TABLE VIII

FERROUS HEME PROTEINS: CO BINDING RATE

Protein	C 5°C	15°C	25°C	ΔE_{act}
	$(k = sec^{-1} M^{-1} \times 10^{-5})$			$(eV\ mole^{-1})$
P450	27	47	84	0.38
C·P450	0.8	1.3	2.3	0.38
Mb	10	20	30	0.26
Hb	0.2	0.5	1	0.50

A first comparison of the binding rates and energies of activation of CO to the ferrous and ferrous-substrate forms of P450 are compared to those for myoglobin and hemoglobin in the region of biological temperatures in Table VIII. As the data in Table VIII show, the activation energy for ferrous $P450_{cam}$·CO recombination is independent of the presence of substrate and lies intermediate between the myoglobin and hemoglobin values. Analyses of these data as well as the complete analysis of the temperature dependence of CO binding over the range 4°K to room temperature are in progress.

3. Spectral Similarities between P450$_{cam}$ and Chloroperoxidase

In the course of the Mössbauer investigations of $P450_{cam}$ and chloroperoxidase (CPO) from *Caldariomyces fumago*, we noted that the spectra of the reduced forms of P450 and CPO are almost identical (18). The Mössbauer investigations led to a further search for spectral correlations between the two proteins. Surprisingly, the optical absorption spectrum of reduced CPO·CO complex was found to be characteristic of the cytochromes of the P450 class (18). Furthermore, the optical spectra of the nitroxide complexes of ferrous P450 and CPO are remarkably similar but differ appreciably from most other heme proteins.

Recently, Hollenberg and Hager (56) pointed out that there are also some resemblances between the ferric forms of these two proteins. There is one remark, however, regarding the g values which needs some comment. The authors state that g values at 7.44 and 4.30 have been observed for CPO and that these g values are quite similar to $P450_{cam}$. Native CPO, however, undergoes a nearly complete transition to low-spin ferric state at 4.2°K. At the present time, it is not clear whether the observed signals are associated with an impurity. It is worth noting, however, that the low-spin ferric form of CPO can be classified close to the P-type heme proteins in the "truth tables" of Blumberg and Peisach (7). As mentioned above, it was suggested that proteins in the P-type class have a cysteine residue bound to the heme iron. If this interpretation is correct, the classification of CPO close to the P-type must be coincidental since CPO is reported to contain only two cysteine residues and these are present in the form of a disulfide bridge (L. P. Hager, private communication).

Though many aspects of the similarities between $P450_{cam}$ and CPO remain to be investigated, there is good evidence for close structural and fundamental relations between the two proteins.

C. Dynamics and Component Interactions

The reaction cycle of cytochrome P450 during methylene hydroxylation, illustrated in Fig. 1, was measured in four single steps from the rate limiting values on increasing the concentration of the several reactants (43, 42).

The rates of the first and third reactions, carbon and oxygen substrate additions, are diffusion controlled and >100-fold more rapid than the reduction steps, two and four. These single equivalent transformations, called e_1 and e_2, require energy; the latter is accompanied by product release. In the reconstructed *in vitro* system driven by DPNH energy, the maximal turnover number (TON) is about 1200, corresponding to the first-order rate constant of about 17 sec^{-1} for the rate controlling step in the presence of saturating levels of reducing enzymes. The first reduction step, calculated as a first-order rate constant for limiting cytochrome, is about 35 sec^{-1}, or at least double the rate limiting velocity.

By reducing the ferric P450·camphor complex in an inert atmosphere, the ferrous complex is readily accumulated; on exposure to air, dioxygen adds to form the oxyferrous substrate cytochrome. The decay of the pure cytochrome in this state is first order with a rate constant of about 0.01 sec^{-1} at room temperature, that is, equivalent to a half-life of about 90 sec. This rate is less than 10^{-3} the product-forming rate, 17 sec^{-1}, when saturated with Pd^r. Thus, the oxyferrous camphor intermediate can be used at substrate level to resolve the fourth transformation into individual events. The e_2 reduction with product formation can, in fact, be written conveniently only as several single steps.

Figure 9 illustrates the rates of disappearance of oxyferrous camphor $P450_{cam}$ and the maximal product yields. The first k_6 is first order with zero product yield. The ferric substrate P450 is regenerated; the fate of the O_2 and one electron are unknown. The rate constants k_4' and k_4, here illustrated with Pd^o and Pd^r, are second order, and yield 0.5 and 1.0 moles, respectively, of product as limiting values. The Pd^o and Pd^r differ only in the latter furnishing e_2, whereas Pd^o acts merely as an effector of product formation and reaction rate enhancement. In this case a second

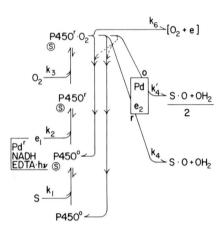

FIG. 9. Oxyferrous P450·camphor decay reactions and rate constants.

molecule of the ferrous cytochrome serves as reductant to regenerate the ferric substrate form presumably with O_2 release. The data for these deductions are reproduced in Table IX. The theory for a mixed function oxidase is two reducing equivalents per hydrocarbon and O_2, and per product, alcohol and OH_2, formed. The complete system driven by DPNH oxidation (Table IX, line 1) conforms to theory. The theory is also confirmed in the Pd-mediated product formation. Both $P450_{cam}$ and Pd are one equivalent redox carriers. With either in reduced form, the stoichiometry dictates a half mole of substrate to product conversion, with both reduced, one electron each, a full mole. As indicated earlier, cytochrome alone is not a product-forming species, though Pd^o without reducing equivalents stimulates the decay rate and induces product release.

Until recently, Pd^r was the only known reductant-effector, whereas several effectors with differing P450 affinities and maximal rates measured as K_4' were known. Now single chemical reductant effectors are known—though the apparent affinities are slight, the maximal yields approximate theory.

One could visualize the product-free reaction (Fig. 9, dashed lines) to result from a dismutation between two oxyferrous S·P450 molecules—redox state superoxide anion—to release one molecule each of dioxygen and an oxygen at the redox state of a peroxy anion. The latter on protonation would form hydrogen peroxide. A similar dismutation, without "peroxy anion" release but attack on the substrate, would account for the maximal yield of a half mole of product per mole of input substrate, as indicated in the reaction with velocity constant k_4'. The reduced species O^{2-} on protonation would form OH_2. In contrast, for the rate constant k_4 with a second electron derived from an external reductant-effector, a second molecule of the oxyferrous substrate cytochrome would not be required to form the peroxy anion. Depending on the rate determining step, this process could exceed k_4' in velocity. On the other hand, if product formation is rate limiting, the two could be equal.

An important aspect of the high activity of Pd can be attributed to a specific affinity to the cytochrome. The binding of substrate to the cytochrome modifies the apparent redox potential for addition of each of two reducing equivalents, e_1 and e_2, and also increases the stability of the oxygen adduct.

In the case of redoxin, the interaction with the oxygen active center of cytochrome P450 most definitely occurs through the formation of a stable enzyme–enzyme complex. Perhaps the most direct and unambiguous demonstration of this effector complex has come from the results of fluorescence quenching. The fluorescent dye fluoresceinisothiocyanate has been bound to P450$_{cam}$, with approximately one dye molecule per protein (101). This selectively modified cytochrome retains 100% of the catalytic activity of the native enzyme as determined by the unaltered rate of NADH oxidation and total product yield from the complete hydroxylase system. Figure 10 shows the titration of the labeled oxidized P450 with oxidized putidaredoxin. The intercept of the double reciprocal plot indicates a dissociation constant of 3 μM for the complex completely consistent with the estimates made from kinetic measurements in the complete system, the $K_m = 4.1$ μM. Extrapolation of the initial slope of

TABLE IX

PRODUCT YIELDS—SYSTEM VERSUS OXY P450·EFFECTOR

		Product	
Reactants	e^-	Theory	Expt.
DPNH + total system + S + O_2	2	1	0.95
Pd^r + P450^o + S + O_2	1	0.5	0.45
Pd^r + P450^r·S	2	1	0.90
Pd^o + P450^r·S | O_2	1	0.5	0.35
— + P450^r·S | O_2	1	0.5	0

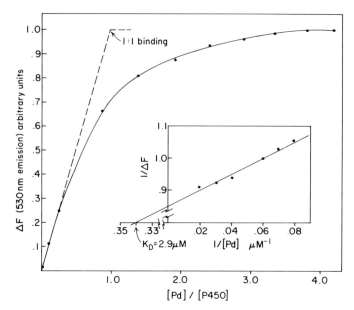

FIG. 10. Fluorescent probe of $P450_{cam} \cdot Pd^o$ binding. Fluoresceinisothiocyanate-labeled $P450_{cam} \simeq 1.4 : 1$.

the titration curve indicates that P450 saturates one-to-one with the effector protein. Since our molecular probe is present in only a ratio dye:protein of 1.4:1, however, we have information only at these points and cannot rule out the possibility of multiple Pd binding to $P450_{cam}$.

An enzymic modification of putidaredoxin has been accomplished readily because of the very fortunate circumstance in the primary structure of the last three residues of the carboxy terminus as shown in Table VI. The final sequence Arg-Gln-Trp is susceptible to carboxypeptidase A cleavage to release only the first two residues forming a derived protein with an arginyl carboxy terminus. The molecule contains but one tryptophan residue as shown both by the sequence (Table VI) and the composition (Table V). The fluorescent yield of the tryptophanyl residue is about 0.2 compared to approximately 1 for free tryptophan. Thus, the release of tryptophan on carboxypeptidase A treatment can be followed by increase in fluorescence and by the free tryptophan remaining in the solution after trichloroacetic acid precipitation of the residual protein. The parallel release of tryptophan and the loss of Pd effector activity in the complete hydroxylase assay are shown in Fig. 11. The derived protein, Pd·des·Trp has been purified by DEAE-cellulose chromatography. The absorption spectrum is identical to the native protein less one tryptophan absorption equivalent in the 280-nm region (see Fig. 2b). The derived protein is free of tryptophan fluorescence. This modified protein is a reductant for e_1 in the complete system, but as shown in Fig. 11 is much less active at the concentration employed for Pd. In the complete system, the Michaelis constant K_m for Pd is about 4.1 μM; for Pd·des·Trp the K_m is 91. An additional measurement of the affinity of Pd for P450$_{cam}$ was obtained by measuring the inhibitor constant of the cytochrome toward the carboxypeptidase cleavage of the C-terminus measured as tryptophan release. The K_I for P450$_{cam}$ is 2 μM. Thus, three independent measurements, one by fluorescence quenching on binding, one as an inhibitor constant, and the third a Michaelis constant for the saturation for P450$_{cam}$ with Pd, give values in the range of 2–4 μM.

A series of effector molecules was evaluated in comparison with Pd and Pd·des·Trp in the decay reaction for oxyferrous substrate P450. The evaluation of the first- and second-order rate constants can be accomplished by the differential equation

$$\frac{d[P450^r \cdot S \cdot O_2]_t}{dt} = -k_6[P450^r \cdot S \cdot O_2]_1 - k_4[effector][P450^r \cdot S \cdot O_2]_1 \quad (1)$$

On integration, this equation gives a straight line in double reciprocal

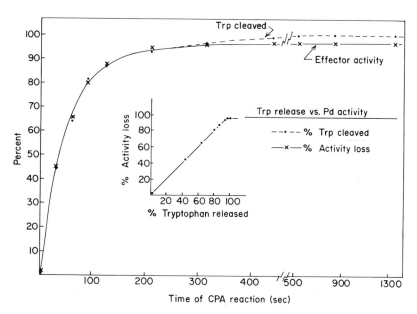

FIG. 11. Tryptophan and Pd° activity. Parallel release of tryptophan and loss of activity by Pd upon carboxypeptidase A treatment.

plots of product yield versus effector concentration. The abscissa is the intercept of the negative value of k_4'/k_6 from the equation

$$\frac{[P450^r \cdot S \cdot O_2]_t}{2P} = (1/\text{eff})(k_6/\alpha k_4') + 1/\alpha \tag{2}$$

where α is the maximal product formed from two molecules of $P450^r \cdot S \cdot O_2$ at infinite time and effector concentration. This value is represented by the ordinate. The rate constants can be evaluated from a plot of the rate of substrate disappearance as a function of effector concentration. At 0 effector, the ordinate intercept is k_6, the first-order rate constant; the slope is k_4', the second-order rate constant. These values are illustrated in two plots in Fig. 12.

The spontaneous first-order decay of oxyferrous $P450_{cam}$, in addition to the expected temperature dependence, is greatly influenced by the presence of the substrate. The data in Fig. 13 show both the temperature dependencies and the activation energies. The substrate-free oxyferrous $P450_{cam}$ decays nearly 10^2 more rapidly. Thus, k_6 defines both temperature and camphor adduct to the protein. Earlier experiments had shown that the potential for the second electron transfer is ~ 0 mV at pH 7.0, whereas

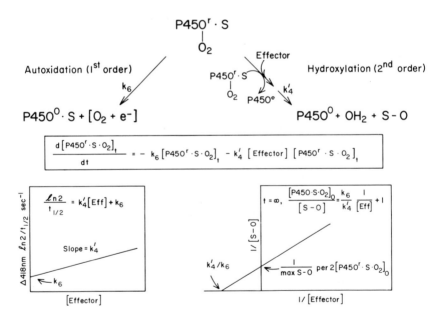

FIG. 12. First- and second-order (product) decay of oxyferrous $S \cdot P450_{cam} \cdot$ Differential equation for graphic analyses apply to the case of oxyferrous P450 camphor decay.

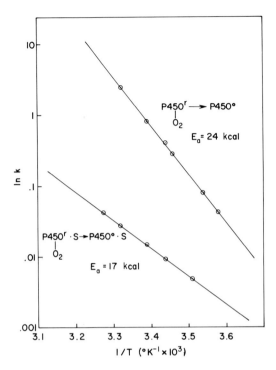

FIG. 13. Decay rates of oxycytochrome P450$_{cam}$ and substrate effect. 50 mM phosphate buffer pH 7.0, photo reduction; rates observed by absorption decrease at 418 nm.

it is considerably higher with the substrate-free P450$_{cam}$. When poised at pH 7 with redox dyes, the tentative values are 0 and $+270$ mV. In contrast, the CO adducts indicate about equal values at *ca.* 0 mV. Clearly, further experiments with chemical effectors using both oxygenated intermediates are in order. A number of hypothetical structures of the nature of oxygenated intermediates and for the reaction sequence leading to the product can be written and several can be separated by experiments.

An evaluation of effectors of product formation rates and yields is shown in Fig. 14. To evaluate the rate constant k_4' Pd° was used. The minimal value is shown by the adrenodoxin Ad° which does not mediate product formation nor modify the decay rate, i.e., does not influence k_6. Three rubredoxins are quite active with the *Peptococcus aerogenes* protein very nearly equal to Pd°. The rates measured for k_4' (Fig. 12) do not include an external reduction step, i.e., these effectors do not serve the reductant function of Pd^r. As mentioned earlier, rabbit liver cytochrome b$_5$ is also active in this system (see Fig. 14). The limiting value of k_6, 0.008 sec^{-1}

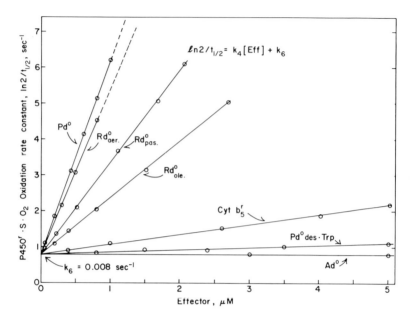

FIG. 14. Oxyferrous S·P450$_{\text{cam}}$ product forming decay rates. 50 mM phosphate buffer, pH 7.0 30 μM photoreduced oxygenated P450 camphor. Rates measured by 418 nm disappearance at 20°.

at 25° is a refinement on the approximate value 0.01. The relative values of the second-order k_4' are tabulated in Table X. They vary in M^{-1} sec^{-1} from 50,000 for Pd to 970 for the Pd·des·Trp.

To evaluate maximal product yield, the double reciprocal plots of effector versus product yields are shown in Fig. 15 for the same proteins. The ordinate intercept is the maximum product yield per 2 moles of substrate (i.e., oxyferrous P450·camphor); the negative intercept on the abscissa is the ratio of k_4'/k_6, as illustrated in Fig. 12. The extrapolated values are tabulated in the legend of Fig. 15 and summarized in Table X. The highest yields are found with cyt b$_5$ and Pd°; the lowest with Pd·des·Trp and rubredoxin from *Clostridium pasteurianum*, i.e., 0.3 of the theory. The other two rubredoxins fall midway at *ca.* 0.5 of the expected yield. The ratio of k_6 to k_4, which retains the dimension of molarity, gives small values from 0.16 μM for Pd° to 10 μM for Pd°·des·Trp or a factor of about 60. The rates vary by about 55-fold. Further evaluation of the meaning of the product yields are in progress; they may carry importance in terms of the decay route and products. As indicated earlier, the values in Figs. 14 and 15 are taken from the data of Gunsalus and Lipscomb (42) and from more recent data of Sligar *et al.* (102).

We have now shown (69) the replacement of the effector-reductant role of Pd^r with chemical reagents. The action of vicinal dithiols using the same assay as illustrated in Figs. 14 and 15 are plotted in Fig. 16, together with representative values for Pd^o and Pd^r and Rd$_{aer}$. Three scales are used on the abscissa to accommodate the large differences and association

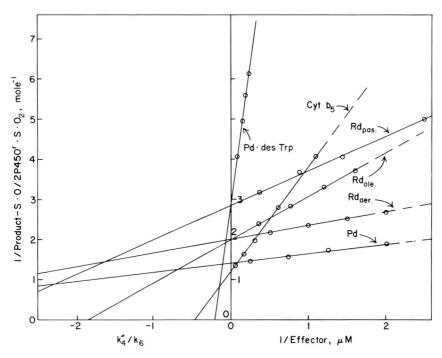

FIG. 15. Effector modulated product yield from oxyferrous S·P450$_{cam}$·product, 5-exo-OH camphor, measured by vpc at time of final zero substrate (see Fig. 14).

Hydroxylated product yields
$$1/\text{S·O} = k_6/\alpha k'_4 \cdot 1/\text{Eff} + 1/\alpha$$

	k_6/k'_4 (μM)	$\alpha = P/2e$ at ∞ Eff.
Pd^o	0.16	0.72
Rd$^o_{aer.}$	0.17	0.49
Rd$^o_{pas.}$	0.30	0.32
Rd$^o_{ole.}$	0.54	0.50
Cyt $b_5{}^r$	2.6	0.83
Pd^o·des·Trp	10	0.33

of the small chemical effectors and the more tightly bound proteins. As shown earlier in Table IX, Pd^r yields nearly two product molecules; both dithiols, lipoate and dithiothreatol, are active and the product yields equal Pd^r. The ratio of k_6 to k_4 is the millimolar region with the DTE requirement approximately 30-fold the level for lipoate. Monothiols

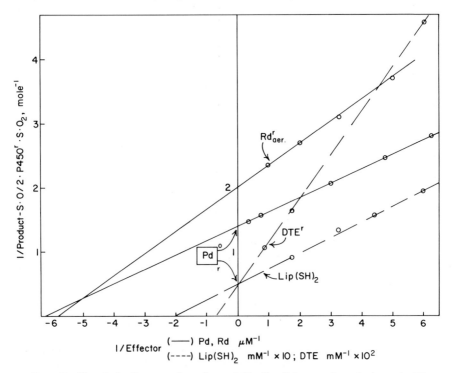

FIG. 16. Chemical effectors of product yield. Conditions and analysis as in Figs. 14 and 15.

Hydroxylated product yields

	k_6/k_4	$\alpha = P/2e$ at ∞ Eff.
Pd^o	0.16 μM	0.72
$Rd^o_{aer.}$	0.17 μM	0.49
Pd^r	—	1.9
$Lip(SH)_2$	5 mM	2
DTE^r	140 mM	2

TABLE X

REDOX EFFECTORS—PRODUCT FORMATION

Effectors	E_0' (mV)	Prosthetic group			Product	
		Fe	S*	Cys	$k_4{}^a$ (M^{-1} sec^{-1})	$Y/2e^b$ (moles)
Putidaredoxin	−240	2	2	4	51,000	0.73
Rubredoxin						
P. aerogenes	−60	1	0	4	46,000	0.5
C. pasteurianum	−57	1	0	4	27,000	0.32
P. oleovorans	−37	1	0	4^c	16,000	0.5
cyt b₅, hepatic	+30	Heme			2,600	0.82
Pd·des·Trp		2	2	4	970	0.33
Adrenodoxin	−270	2	2	4	∼0	0
Ferredoxin						
Spinach	−420	2	2	4	Nil	0
C. acidi urici	−390	8	8	8	Nil	0
cyt b₅₆₂	+130	Heme			Nil	0
HiPIP	+330	4	4	4	Nil	0

^{a} See Fig. 9.

^{b} Yield of 5-*exo*-OH camphor per two electrons from P450·S·O₂.

^{c} ×2.

have not been found active in the effector role regardless of potential. The reaction of a dithiol structure with the superoxide anion active center in the oxygenated cytochrome may be critical to the mechanism of mixed function oxidation.

V. SUMMARY

Prominent among the unrationalized enigmas uncovered during the previous discussion of this hydroxylase complex and common to other enzyme-redox systems are (1) the mechanism of reduction and electron transfer, (2) level of perception between protein and substrate, (3) energy levels and shunts, (4) nature of unstable intermediates, and (5) handles for investigation beyond the "stable" intermediates. The inseparable nature of each of these problems does not accommodate an individual rationale but rather lends itself to a more concerted explanation. Tying together on a molecular scale a step-by-step reaction scheme for the monoxygenation of the substrate, camphor, via three discrete proteins and three interposing substrates is, to say the least, an intricate task.

Peripheral shuttling of electrons via the DPNH-redoxin reductase flavoprotein and putidaredoxin to the cytochrome P450 reaction center is apparently mediated by the electromotive potential of the interacting proteins (discussed previously), the exact mechanism of which is not fully understood. The rapid and efficient passage of electrons from the flavoprotein to the iron–sulfur protein is of particular interest.

An understanding of the recognition of substrate, camphor, by cytochrome P450 is a problem basic to enzyme catalysis. Generalities drawn from hydrophobic interactions between the bicyclic monoterpene ring and aliphatic and aromatic amino acids in or near the binding pocket offer no real clues to the nature of the active site. The position of the ketone group relative to the 5-*exo* hydroxylating position and the three-dimensional conformation of camphor are possibly key factors in the specificity characteristics of the hydroxylating site. Proximity of the substrate to the iron heme plane and axial positions are not known with any degree of certainty. Current evidence suggests that the substrate lies adjacent to or on top of the O_2 and CO binding sites several Ångstroms away from the iron atom. Such a close-pocket arrangement facilitates rapid catalysis of any unstable superoxide anion intermediates without releasing them into the reaction medium. Detection of any unstable species is undoubtedly going to require sophisticated techniques carried out at reduced temperatures, now available in the liquid helium range.

Additional information concerning the nature of the axial ligands and O_2, substrate and effector binding sites must await X-ray diffraction studies on crystalline P450 currently in progress. Future insight into the mechanism of concerted hydroxylation will be gained only by concentrated physical studies on P450 and its interaction with energy transducing active centers of native and derived proteins or their replacements.

The solution of these problems represents a most challenging field for the application of the scientific knowledge and insight gleaned from a wide range of disciplines and educations. We have indeed been fortunate to have an active collaboration of biochemists, geneticists, physicists, and organic chemists. Through such interaction an understanding of the mechanisms of energy transduction and segregation in mixed function oxygenation may not be out of reach.

ACKNOWLEDGMENTS

This chapter reviews work supported in part by grants from the National Science Foundation, GB 33962Xi, and the National Institutes of Health, AM 00562, and GM 16406 and 18051.

REFERENCES

1. Alexander, M. (1973). *ASM News* **39**, 483.
2. Appleby, C. A. (1967). *Biochim. Biophys. Acta* **147**, 399.
3. Appleby, C. A. (1969). *Biochim. Biophys. Acta* **172**, 88.

4. Austin, R. H., Beeson, K., Eisenstein, L., Frauenfelder, H., Gunsalus, I. C., and Marshall, V. P. (1973). *Science* **181,** 541.
5. Averill, B. A., Herskovitz, T., Holm, R. H., and Ibers, J. A. (1973). *J. Amer. Chem. Soc.* **95** (in press).
6. Baptist, J. N., Gholson, R. K., and Coon, M. J. (1963). *Biochim. Biophys. Acta* **69,** 40.
7. Blumberg, W. E., and Peisach, J. (1971). In "Probes of Structure and Function of Macromolecules and Membranes" (B. Chance, T. Yonetani, and A. S. Mildvan eds.), Vol. 2, Part 2, p. 215. Academic Press, New York.
8. Boyd, G. S. (1972). In "Biological Hydroxylation Mechanisms" (G. S. Boyd and R. M. S. Smellie, eds.), p. 1 Academic Press, New York.
9. Cardini, G., and Jurtshuk, P. (1968). *J. Biol. Chem.* **243,** 6070.
10. Cardini, G., and Jurtshuk, P. (1970). *J. Biol. Chem.* **245,** 2789.
11. Carter, C. W., Jr., Freer, S. T., Xuong, N. H., Alden, R. A., and Kraut, J. (1971). *Cold Spring Harbor Symp. Quant. Biol.* **36,** 381–385.
12. Chakrabarty, A. M. (1972). *J. Bacteriol.* **112,** 815.
13. Chakrabarty, A. M., Chou, G., and Gunsalus, I. C. (1973). *Proc. Nat. Acad. Sci. U.S.* **70,** 1137.
14. Chakrabarty, A. M., Gunsalus, C. F., and Gunsalus, I. C. (1968). *Proc. Nat. Acad. Sci. U.S.* **60,** 168.
15. Chakrabarty, A. M., and Gunsalus, I. C. (1969). *Virology* **38,** 104.
16. Chakrabarty, A. M., and Gunsalus, I. C. (1969). *Proc. Nat. Acad. Sci. U.S.* **64,** 1217.
17. Chakrabarty, A. M., and Gunsalus, I. C. (1970). *J. Bacteriol.* **103,** 830.
18. Champion, P. M., Münck, E., Debrunner, P. G., Hollenberg, P. F., and Hager, L. P. (1973). *Biochemistry* **12,** 426.
19. Chapman, P. J., Meerman, G., and Gunsalus, I. C. (1966). *Biochemistry* **5,** 895.
20. Chou, I. N., Chakrabarty, A. M., and Gunsalus, I. C. (1973). In preparation.
21. Chou, I. N., and Gunsalus, I. C. (1973). In preparation.
22. Cinti, D. L., Grudin, R., and Orrenius, S. (1973). *Biochem. J.* **134,** 367.
23. Clarke, P. H. (1972). *J. Gen. Microbiol.* **73,** ii.
24. Cook, R., Tsibris, J. C. M., Debrunner, P., Tsai, R. L., Gunsalus, I. C., and Frauenfelder, H. (1968). *Proc. Nat. Acad. Sci. U.S.* **59,** 1045.
25. Coon, M. J., Strobel, H. W., Autor, A. P., Heidema, J., and Duppel, W. (1972). In "Biological Hydroxylation Mechanisms" (G. S. Boyd and R. M. S. Smellie, eds.), p. 45. Academic Press, New York.
26. Cushman, D. W., Tsai, R. L., and Gunsalus, I. C. (1967). *Biochem. Biophys. Res. Commun.* **26,** 577.
27. Dagley, S. (1972). In "Degradation of Synthetic Organic Molecules in the Biosphere," Proc. Conf. Nat. Acad. Sci., Washington, D. C.
28. Dagley, S. (1973). *ASM News* **39,** 616.
29. Der Vartanian, D. V., Orme-Johnson, W. H., Hansen, R. E., Beinert, H.. Tsai, R. L., Tsibris, J. C. M., Bartholomaus, R. C., and Gunsalus, I. C. (1967). *Biochem. Biophys. Res. Commun.* **26,** 569.
30. Dickerson, R. E. (1972). *Sci. Amer.* **226,** 58.
31. Dunn, H., and Gunsalus, I. C. (1973). In preparation.
32. Dunn, N. W., and Gunsalus, I. C. (1973). *J. Bacteriol.* **114,** 974.
33. Estabrook, R. W., Hildebrandt, A., Remmer, H., Schenkmon, J., Rosenthal, O., and Cooper, D. Y. (1968). *Coll. Ges. Biol. Chem.* **19,** 142.

34. Fritz, J., Anderson, R., Fee, J., Petering, D., Palmer, G., Sands, R. H., Tsibris, J. C. M., Gunsalus, I. C., Orme-Johnson, W. H., and Beinert, H. (1971). *Biochim. Biophys. Acta* **253**, 110.
35. Gibson, J. F., Hall, D. O., Thornley, J. H. M., and Whatley, F. R. (1966). *Proc. Nat. Acad. Sci. U.S.* **56**, 987.
36. Gunsalus, C. F., Chakrabarty, A. M., and Gunsalus, I. C. (1970). *Bacteriol. Proc.* p. 34 (ASM abstr.).
37. Gunsalus, I. C. (1968). *Hoppe-Seyler's Z. Physiol. Chem.* **349**, 1610.
38. Gunsalus, I. C., Bertland, A. U., II, and Jacobson, L. A. (1967). *Arch. Mikrobiol.* **59**, 113.
39. Gunsalus, I. C., Chapman, P. J., and Kuo, J. F. (1965). *Biochem. Biophys. Res. Commun.* **18**, 924.
40. Gunsalus, I. C., Conrad, H. E., Trudgill, P. W., and Jacobson, L. A. (1965). *Isr. J. Med. Sci.* **1**, 1099.
41. Gunsalus, I. C., Gunsalus, C. F., Chakrabarty, A. M., Sykes, S., and Crawford, I. P. (1968). *Genetics* **60**, 419.
42. Gunsalus, I. C., and Lipscomb, J. D. (1972). In "The Molecular Basis of Electron Transport" (J. Schultz and B. F. Cameron, eds.), p. 179. Academic Press, New York.
43. Gunsalus, I. C., Lipscomb, J. D., Marshall, V. P., Frauenfelder, H., Greenbaum, E., and Münck, E. (1972). In "Biological Hydroxylation Mechanisms" (G. S. Boyd and R. M. S. Smellie, eds.), p. 135. Academic Press, New York.
44. Gunsalus, I. C., and Marshall, V. P. (1971). *CRC Crit. Rev. Microbiol.* **1**, 291.
45. Gunsalus, I. C., Meeks, J. R., and Lipscomb, J. D. (1973). *Ann. N. Y. Acad. Sci.* **212**, 107.
46. Gunsalus, I. C., Tyson, C. A., and Lipscomb, J. D. (1973). In "Oxidases and Related Redox Systems" (T. E. King, H. S. Mason, and M. Morrison, eds.), p. 583. Univ. Park Press, Baltimore, Maryland.
47. Gunsalus, I. C., Tyson, C. A. ,Tsai, R. L., and Lipscomb, J. D. (1971). *Chem.-Biol. Interat.* **4**, 75.
48. Hartline, R. L., and Gunsalus, I. C. (1971). *J. Bacteriol.* **106**, 468.
49. Heath, H. E., and Gunsalus, I. C. (1973). *ASM Proc.* Abstract G249, p. 67.
50. Hedegaard, J., and Gunsalus, I. C. (1965). *J. Biol. Chem.* **240**, 4029.
51. Herriott, J. R., Siecker, L. C., Jensen, L. H., and Lovenberg, W. (1970). *J. Mol. Biol.* **50**, 391.
52. Herskovitz, T., Averill, B. A., Holm, R. H., Ibers, J. A., Phillips, W. D., and Weiher, J. F. (1972). *Proc. Nat. Acad. Sci. U.S.* **69**, 2437.
53. Hill, H. A. O., Röder, A., and Williams, R. J. P. (1970). *Struct. Bonding (Berlin)* **8**, 123.
54. Hoch, S. O., Roth, C. W., Crawford, I. P., and Nester, E. W. (1971). *J. Bacteriol.* **105**, 38.
55. Hoffman, B. M., and Petering, D. H. (1970). *Proc. Nat. Acad. Sci. U.S.* **67**, 637.
56. Hollenberg, P. F., and Hager, L. P. (1973). *J. Biol. Chem.* **248**, 2630.
57. Holloway, B. W. (1969). *Bacteriol. Rev.* **33**, 419.
58. Holloway, B. W. (1973). In "The Biology of *Pseudomonas*" (M. H. Richmond and P. H. Clark, eds.). In preparation.
59. Huang, J. J., and Kimura, T. (1973). *Biochemistry* **12**, 406.
60. Jacobson, L. A., Bartholomaus, R. C., and Gunsalus, I. C. (1966). *Biochem. Biophys. Res. Commun.* **24**, 955.

61. Jefcoate, C. R. E., and Gaylor, J. C. (1969). *Biochemistry* **8,** 3464.
62. Jurtshuk, P., and Cardini, G. (1971). *CRC Crit. Rev. Microbiol.* **1,** 239.
63. Katagiri, M., Gangali, B. N., and Gunsalus, I. C. (1968). *J. Biol. Chem.* **243,** 3543.
64. Katagiri, M., and Takemori, S. (1973). *Proc. Int. Congr. Biochem. 9th,* 1973 Abstracts, p. 327.
65. Kornberg, H. L. (1965). *Angew Chem., Int. Ed. Engl.* **4,** 558.
66. Lang, G., and Marshall, W. (1966). *Proc. Phys. Soc., London* **87,** 3.
67. Leidigh, B. J., and Wheelis, M. L. (1973). In preparation.
68. Lipscomb, J. D., and Gunsalus, I. C. (1973). In "Microsomes and Drug Oxidation" (R. Estabrook, J. Gillette, and K. Leibman, eds.), pp. 1–5. Williams & Wilkins, Baltimore, Maryland.
69. Lipscomb, J. D., Namtvedt, M. J., Sligar, S. L., and Gunsalus, I. C. (1973). In preparation.
70. Lode, E. T., and Coon, M. J. (1971). *J. Biol. Chem.* **246,** 791.
71. McKenna, E. J., and Coon, M. J. (1970). *J. Biol. Chem.* **245,** 3882.
72. Mahler, H. R., and Cordes, E. H. (1971). "Biological Chemistry." Harper, New York.
73. Marbach, W. (1973). Thesis, University of Illinois, Urbana.
74. Mason, H. S. (1957). *Science* **125,** 1185.
75. Mason, H. S. (1965). *Annu. Rev. Biochem.* **34,** 595.
76. Massey, V., Palmer, G., and Williams, C. H., Jr. (1966). In "Flavins and Flavoproteins" (E. C. Slater, ed.), p. 131. Amer. Elsevier, New York.
77. Mayerle, J. J., Frankel, R. B., Holm, R. H., Ibers, J. A., Phillips, W. D., and Weiker, J. R. (1973). *Proc. Nat. Acad. Sci. U.S.* **70** (in press).
78. Moleski, C., Moss, T. H., Orme-Johnson, W. H., and Tsibris, J. C. M. (1970). *Biochim. Biophys. Acta* **214,** 584.
79. Münck, E., Debrunner, P. G., Tsibris, J. C. M., and Gunsalus, I. C. (1972). *Biochemistry* **11, 855.**
80. Münck, E., Sharrock, M., Marchant, L., and Hoffman, B. M. (1973). In preparation.
81. Omura, T., Sato, R., Cooper, D. Y., Rosenthal, O., and Estabrook, R. W. (1965). *Fed. Proc., Fed. Amer. Soc. Exp. Biol.* **24,** 1181.
82. Orme-Johnson, W. H., Hansen, R. E., Beinert, H., Tsibris, J. C. M., Bartholomaus, R. C., and Gunsalus, I. C. (1968). *Proc. Nat. Acad. Sci. U.S.* **60,** 368.
83. Ornston, L. N. (1971). *Bacteriol. Rev.* **35,** 87.
84. Ornston, L. N., Ornston, M. K., and Chou, G. (1969). *Biochem. Biophys. Res. Commun.* **36,** 179.
85. Peisach, J., Appleby, C. A., and Blumberg, W. E. (1972). *Arch. Biochem. Biophys.* **150,** 725.
86. Peisach, J., and Blumberg, W. E. (1970). *Proc. Nat. Acad. Sci. U.S.* **67,** 172.
87. Peterson, J. A. (1971). *Arch. Biochem. Biophys.* **144,** 678.
88. Peterson, J. A., and Griffin, B. W. (1973). *Drug Metab. Disposition* **1,** 14.
89. Queener, S. F., and Gunsalus, I. C. (1970). *Proc. Nat. Acad. Sci. U.S.* **67,** 1225.
90. Rheinwald, J. G. (1970). M. S. Thesis, University of Illinois, Urbana.
91. Rheinwald, J. G., Chakrabarty, A. M., and Gunsalus, I. C. (1973). *Proc. Nat. Acad. Sci. U.S.* **70,** 885.
92. Richmond, M. H. (1972). *J. Gen. Microbiol.* **73,** iv.
93. Richmond, M. H., and Clarke, P. H., eds. (1973). "The Biology of *Pseudomonas*." In preparation.

94. Salemme, F. R., Freer, S. T., Xuong, N. H., Alden, R. A., and Kraut, J. (1973). *J. Biol. Chem.* **248**, 3910.

95. Sato, R., Oshino, N., Shimakata, T., Mihara, K., and Imai, Y. (1973). *Proc. Int. Congr. Biochem., 9th,* Abstract 7Sa3, p. 327.

96. Schleyer, H., Cooper, D. Y., Levin, S. S., and Rosenthal, O. (1972). In "Biological Hydroxylation Mechanisms" (G. S. Boyd and R. M. S. Smellie, eds.), p. 187. Academic Press, New York.

97. Shaham, M., Chakrabarty, A. M., and Gunsalus, I. C. (1973). *J. Bacteriol.* **114.**

98. Shaham, M., and Gunsalus, I. C. (1973). In preparation.

99. Sharrock, M., Münck, E., Debrunner, P., Lipscomb, J. D., Marshall, V. P., and Gunsalus, I. C. (1973). *Biochemistry* **12**, 258.

100. Siecker, L. C., Adman, E., and Jensen, L. H. (1972). *Nature (London)* **235**, 40.

101. Sligar, S. G., Debrunner, P. G., Lipscomb, J. D., and Gunsalus, I. C. (1973). *Int. Conr. Biochem., 9th,* Abstract 7c12, p. 339.

102. Sligar, S. G., Lipscomb, J. D., and Gunsalus, I. C. (1973). *Biochem. Biophys. Res. Commun.* (in press).

103. Smith, D. W., and Williams, R. J. P. (1970). *Struct. Bonding (Berlin)* **7**, 1.

104. Stanier, R. Y., Palleroni, N. H., and Duodoroff, M. (1966). *J. Gen. Microbiol.* **43**, 159.

105. Szent-Györgyi, A. V. (1937). In "Studies on Biological Oxidation and Some of its Catalysts," Preface, p. 3. Karl Rényi, Budapest.

106. Tanaka, M., Haniu, M., and Yasunobu, K. (1973). *J. Biol. Chem.* **248**, 1141.

107. Tanaka, M., Haniu, M., Yasunobu, K., Dus, K., and Gunsalus, I. C. (1973). *J. Biol. Chem.* (in press).

108. Tiedje, J. M., Firestone, M. K., Mason, B. B., and Warren, C. B. (1973). *Bacteriol. Proc.* p. 171.

109. Tsai, R. L. (1969). Ph.D. Thesis, University of Illinois, Urbana.

110. Tsai, R. L., Gunsalus, I. C., and Dus, K. (1971). *Biochem. Biophys. Res. Commun.* **45**, 1300.

111. Tsai, R. L., Yu, C.-A., Gunsalus, I. C., Peisach, J., Blumberg, W. E., Orme-Johnson, W. H., and Beinert, H. (1970). *Proc. Nat. Acad. Sci. U.S.* **66**, 1157.

112. Tsibris, J. C. M., Münck, E., Debrunner, P. G., Woody, R. W., Frauenfelder, H., and Gunsalus, I. C. (1970). *Mag. Resonance Biol. Syst., Proc. Int. Conf., Oxford,* p. 55.

113. Tsibris, J. C. M., Namtvedt, M. J., and Gunsalus, I. C. (1968). *Biochem. Biophys. Res. Commun.* **30**, 323.

114. Tsibris, J. C. M., Tsai, R. L., Gunsalus, I. C., Orme-Johnson, W. H., Hansen, R. E., and Beinert, H. (1968). *Proc. Nat. Acad. Sci. U.S.* **59**, 959.

115. Tsibris, J. C. M., and Woody, R. W. (1970). *Coord. Chem. Rev.* **5**, 417.

116. Tyson, C. A., Tsai, R. L., Lipscomb, J. D., and Gunsalus, I. C. (1972). *J. Biol. Chem.* **247**, 5777.

117. Ueda, T., Lode, E. T., and Coon, M. J. (1972). *J. Biol. Chem.* **247**, 2109.

118. Van der Linden, A. C., and Thijsse, G. J. E. (1965). *Advan. Enzymol.* **27**, 469.

119. Weiss, J. J. (1964). *Nature (London)* **202**, 83.

120. Wheelis, M. L., and Stanier, R. Y. (1970). *Genetics* **66**, 245.

120a. Wilson, G. S. (1900). Unpublished data.

121. Wilson, G. S., Tsibris, J. C. M., and Gunsalus, I. C. (1973). *J. Biol. Chem.* **248**, 6059.

122. Wittenberg, J. B., Wittenberg, B. A., Peisach, J., and Blumberg, W. E. (1970). *Proc. Nat. Acad. Sci. U.S.* **67**, 1846.

123. Yu, C.-A., and Gunsalus, I. C. (1969). *J. Biol. Chem.* **244,** 6149.
124. Yu, C.-A., and Gunsalus, I. C. (1970). *Biochem. Biophys. Res. Commun.* **40,** 1431.
125. Yu, C.-A., and Gunsalus, I. C. (1973). *J. Biol. Chem.* **248.**
126. Yu, C.-A., Gunsalus, I. C., Katagiri, M., Suhara, K., and Takemori, S. (1973). *J. Biol. Chem.* **248.**
127. Yu, C.-A., Gunsalus, I. C., Katagiri, M., Suhara, K., and Takemori, S. (1973). *J. Biol. Chem.* **248.**

AUTHOR INDEX

Numbers in parentheses are reference numbers and indicate that an author's work is referred to although his name is not cited in the text. Numbers in italics show the page on which the complete reference is listed.

A

Abbott, M. T., 7, 18(9), 19(78, 79, 80), 25, 27, 28, 168(4), 185(165, 166), 186 (3, 5, 141, 163, 164, 165, 183), 187 (119, 120), 188(119, 121), 189(119, 124, 166), 190(1, 2, 3, 4, 121, 133, 165), 191(3, 124, 133), 192(3, 124), 193(3, 119, 183), 194(5, 119, 183), 196(163, 164, 165), 197(119, 165), 198(119, 163, 164), 199(61, 124, 165), 205(121), 206(183), 209, 210, 212, 213, 214

Abeles, R. H., 254(20, 21, 22, 23), 256 (20, 21, 22, 23), 280

Aberhart, D. J., 58, 81

Able, E., 467(83), 476

Abraham, E. P., 90(62), 131

Acs, G., 90(32), 103(90), 119(32), 120(32), 130, 132

Adachi, K., 137(1, 47), 148(47), 152(2), 162, 164

Adam, G., 494, 531

Adams, E., 174(53), 210

Adams, G. E., 471(121), 477

Addink, R., 290(113), 298(113), 368

Adelberg, E. A., 453(1), 457(1), 474

Adman, E., 589(100), 612

Adolf, P. K., 409(9), 411(9), 448

Aerts, F. E., 387(1), 388(1), 392(1), 394 (1), 399

Agner, K., 550(147, 148, 151), 558

Ahern, D. G., 44(63), 78

Ainsworth, S., 457(34), 474

Air, G., 550(143), 557

Aisen, P., 383(120), 403

Akazawa, T., 536(11), 554

Åkeson, Å., 22(103), 28, 548(132), 557

Akhtar, M., 56(170), 57(180, 186), 67 (260), 81, 83

Akino, M., 297(143), 310(114), 329(143), 330(143), 332(143), 334(143), 335 (143), 338(143), 340(143), 341(143), 342(143), 343(143), 345(143), 351 (143), 354(143), 356(143), 357(143), 358(143), 368, 435(85), 450

Al-Adnani, M. J., 177, 209

Albonico, S. M., 44(66), 78

Alden, R. A., 563(94), 589(11), 609, 612

Alderson, T., 472(133), 477

Alexander, M., 582(1), 608

Alexander, N. M., 545(92), 550(92), 556

Alexander, P., 469(103), 476

Allen, A. O., 460(56), 477(56), 475

Allen, P. Z., 548(141), 550(141, 142), 557

Altmann, H., 473(135), 477

Altschul, A. M., 488(48), 529

Alvares, A. P., 231(161), 242

Alyea, H. N., 471(115), 476

Amiconi, G., 548(129), 557

Anagnoste, B., 395(37), 400

Anan, F., 152(14), 163

Anbar, M., 412(27), 449, 469(107), 476

Ander, S., 471(123), 477

Anderson, R., 589(34), 610

Ando, K., 493(80), 530

Ando, N., 61(206), 62(206), 82

Andreasen, A. A., 18(74), 27

Andrews, T. T., 24(107), 28

Änggård, E., 37(37), 42(37), 77

615

Coon, M. S., 411(22), *448*

Cooper, A., 526(204), *534*

Cooper, D. Y., 9(18), *26*, 59(190, 193), 61(190, 193, 195, 196, 197), 62(197), 64(236), 66(248), *81, 82, 83*, 217(26), 219(26, 43, 46), 222(79, 80, 93, 94), 227(125, 126), 230(149), 232(94), *238, 239, 240, 241, 242*, 581(96), *609, 611, 612*

Cooper, G. W., 181, *210*

Cooper, J. L., 13(42), *26*

Cooper, J. R., 215(4), *237*, 287(162), 327(163), 349(15), 360(16), *365*

Cordes, E. H., 584(72), *611*

Corey, E. J., 44(66), 54, 55(158), *78, 80, 81*, 413, *449*

Cormier, M. J., 538, *555*

Corrodi, H., 344(11), *365*

Coryell, C. D., 547(123), *557*

Cosmides, G. J., 215(16), *238*

Costa, M., 155(19, 73), *163, 164*

Coulson, A. F. W., 551, 552(160), *558*

Cowburn, D. A., 146(28), *163*

Craine, J. E., 288(17), 289(16), 348(17), *365*

Cramer, J. W., 20(86), *28*

Cramp, W. A., 472(134), *477*

Crandall, D. I., 22(98), *28*, 152(13, 14), 153(20), *163*, 201(36), *210*

Crane, F. L., 487(37, 41), *529*

Crawford, I. P., 574(41), 575(54), 580(41, 54), *610*

Cremona, T., 221(67), *239*

Creveling, C. R., 395(24), 346(24, 25, 144), *400, 403*

Criddle, R. S., 488(44), *529*

Cripps, R., 327(4), *365*

Crist, R. H., 467(82), *476*

Critchlow, J. E., 542(80), *556*

Cronholm, T., 76(323), *85*

Curzon, G., 360(48), *366*

Cushman, D. W., 222(82), 228(82), *240, 609*

Cutler, M. E., 495, 497, 499(114), 504(97), *531*

Cvetanovic, R. J., 417(36), 418(36), *449*

Czapski, G., 411(15), *448*, 460(54), 471 (54, 124, 129), 472(129, 131), *475, 477*

Czerlinski, G., 491(70), *530*

D

Dagley, S., 136(16), 137(15), 144, 150(17), *163*, 582(27), *609*

Dahm, H., 202(37, 38), 206, 207, *210*

Dahm, K., 69(282), *84*

Dairman, W., 346, *365*

Dalgleish, D., 457(41), *475*

Dallman, P. R., 235(190), *243*

Dallner, G., 219(44), 221(66, 68), 235(66, 189, 190, 191, 192), 237(215), *239, 243, 244*

Daly, J. W., 13(43), *26*, 202(39), 206(62), *210*, 299(51, 70, 71), 300(5, 72), 306 (19), 310(71), 335(53), 336(20), *365, 366, 367*, 394(55), 395(24), 396(24), *400, 401*, 418(37, 38, 39, 40, 41), 420 (37, 38, 39, 40, 41), 426(38, 61, 71), 427(61, 71), 428(61, 71), 431(71), 435(37, 38, 39, 40), 438(37, 38, 39, 40, 41), 439(37, 38, 39, 40, 41, 61), 442(37, 38, 39, 40, 41), *449, 450*

D'Angelo, G. L., 328(119), 329(120), 332(119), 338(119), 347(119), 348(119), *368*

Daniels, E. G., 44(62), *78*

Danielsson, H., 18(75), 22(75), *27*, 37(30, 31), 48(104, 105), 49, 69 (283), 73(306), 75(309, 310, 318), 76(310), *77, 79, 84, 85*, 230(150), 233(150), *242*, 445(107), *451*

Danner, D. J., 536(18), *554*

Das, M. L., 48(110), 49, 70(110), *79*, 236(199), 237(199), *243*

Dashman, T., 118(103), *132*

Davidson, N., 471(117), *477*

Davies, D. C., 215(11), 228(11), 229(11), 233(11), *238*

Davies, D. S., 215(11), 227(121), 228(11, 121), 229(11), 233(11), *238, 241*

Davies, H. C., 488(52), 490(52), 493(52), *529*

Davies, M. O., 469(105), *476*

Davison, A. J., 498(112), 499, *531*

Dawson, C. R., 373(26, 109), 378, 382(88), 383(26), 384(89), 387, 388(15, 109), 389, 390(109), *400, 402*

Dearden, M. B., 423(50), *449*

Debrunner, P. G., 566(112), 584(24, 79), 587(99), 588(99), 590(99), 591(18, 99),

Fujinga, K., 536(10), *554*
Fujisawa, H., 13(34, 35), *26*, 137(63), 138(21, 23), 140(59, 66), 141(59), 142(21, 23), 143, 144(23, 59, 71), 145(63), 148(63, 64, 66), 149(66), 151(51), 156(23, 59, 63, 66), 158(23, 59, 66), 159(23), 160, 161(24), 162(66), *163, 164*
Fujisawa, M., *133*
Fujita, T. S., 200(43, 44), 201(44), *210*
Fulco, A. J., 45, 52(142), *79, 80*
Fuller, E. C., 467(82), *476*
Funan, H., 373(102), *402*
Funatsu, M., 379(53), *401*
Furutachi, T., 63(226), *82*
Furuya, E., 145(82), 153(82), *165*

G

Gaber, B. P. 459(53), 461(76), 463(53), *475*
Gafford, R. D., 142(29), *163*
Gal, E. M., 310(38), 352(40), 353, 355, 356, *366*
Gale, P. H., 222(77), *240*
Galli, G., 57(178), *81*
Galliard, T., 36(28), 47(90), *77, 79*
Gallop, P. M., 168(54), *210*
Ganapathy, K., 380(158), *404*, 489, *530*
Gander, J. E., 151(68), 152(68), *164*
Ganguli, B. N., 9(20), *26*, 587, *611*
Ganschow, E., 236(205), *244*
Gardner, H. W., 36(27), *77*
Garfinkel, D., 216, 232(177), *238*
Garland, P., 321(159), *369*
Garssen, G. J., 36, *77*
Garvey, T. Q., 293(41), *366*, 397(41), 398(41), *401*, 440(95), *451*
Gaudette, L., 215(4, 5), *237*
Gauthier, J. J., 153, 154(25, 26), *163*
Gautschi, F., 55(161), *81*
Gaylor, J. L., 56(163, 165, 166, 167, 168, 169, 173, 174), 57(175), *81*, 206(9), *209*, 224(106), *241*, 586, *611*
Geary, P. J., 150(17), *163*
Gebert, R. A., 327(135), *368*
Geiger, P., 315(3), *365*
Gelboin, H. V., 231(164), 236(200, 201), *242, 243*
Gelehrter, T. D., 120(108), 121(108), *132*

Genge, C. A., 431(78), *450*
George, P., 407(5), 409(5), 411(5), 414(5), 417(5), 418(5), *448*, 471(116), *477*, 536(22), 537, 539, 546, 551(163), 552(163), *554, 555, 558*
Gerald, P. S., *367*
Gessa, G. L., 358(146), *368*
Gesswagner, D., 473(135), *477*
Gey, K., 310(10), 311(10), *365*
Ghislandi, V., 396(125), *403*
Gholson, R. K., 572, *609*
Ghosh, D., *132*
Giacin, J. R., 426(63, 64, 69), 427(63, 64), 432(63), 438(63), *450*
Gibbons G. F., 57(179), *81*
Gibbs, R. H., 220(58), *239*
Gibson, D. T., 150(27), *163*
Gibson, J. F., 471(126), *477*, 588, *610*
Gibson, Q. H., 220(56, 57), *239*, 260(52), 264(52), *281*, 492(74), 495, 497(100), 498(74, 106, 107, 108), 499, 500, 503, 505(143), 506, 514, *530, 531, 532*, 539, *555*
Gielen, J. E., 20(91), *28*
Gifford, E. M., 185(171), *213*
Gigon, P. L., 227(121, 122), 228(121), *241*
Gillette, J. R., 20(87), *28*, 70(297), 71(297), 72(297), *84*, 215(6, 7, 8, 11, 16), 219(48), 227(121, 122), 228(121), 229(8, 11), 231(48), 233(8, 11), 237(216), *237, 238, 239, 244*
Gilman, A. G., 236(198), *243*
Gilmour, M. V., 484(29), 485, 494(91), 499, 500(91), 509(29), 511(29), *529, 531*
Ginsberg, B., 352(40), 353(40), 355(40), 356(40), *366*
Glaumann, H., 235(191), *243*
Glaze, R. P., 521(188), *534*
Glazer, R. I., 220(52), 234(52), *239*
Gledhill, W. E., 453(8), 472(8), *474*
Glowinski, J., 362(111), *366, 368*
Gnosspelius, Y., 231(159), 236(199), 237(199), *242, 243*
Goad, L. J., 54, 57(179), 58, *80, 81*
Goddard, J., *369*
Goldberg, B., 176(55), *210*
Goldenberg, H., 229(141, 142), *242*
Goldstein, L., 120(106), *132*
Goldstein, M., 293(41), 366(75), 364(42),

SUBJECT INDEX

A

Acetabularia, thymidine metabolism in, 185

Acetanilide, monooxygenation of, 219

Acetate, as lanosterol precursor, 54

Acetomyces sp., arginine monooxygenase from, 254

Acetyl-coenzyme A (Acetyl-CoA), from 3-hydroxyanthranilic acid, 17

Acidovorus-testosteroni pseudomonads, cytochrome P450 activity in, 572

Acids, from aldehydes, by biological oxidation, 3, 4

Actinomycin D, effects on tryptophan oxygenase induction, 90, 119, 120, 122

Acyl carrier protein (ACP), in stearyl-ACP, 52

Acyl CoA:phospholipid transferases, substrates for, 32

Acyl thioesters, enzymic formation of, 4

Adrenal cortex
cytochrome P-450 in, 219, 222, 226, 232, 233
steroid hormone synthesis in, 58, 62, 63, 66
steroid monooxygenases in, 22

Adrenal cortical hormones, tryptophan oxygenase induction by, 89

Adrenochrome, from epinephrine oxidation, 228, 460

Adrenodoxin
as cytochrome P450 component, 581
isolation and properties of, 59
product formation by, 607
redox potential of, 585

Agromyces ramnosus, catalase absence in, 453

L-Alanine, as oxygenase substrate, 14–15

Alcohol, enzymic oxidation of, 10

Alcohol dehydrogenase
induction of, 574, 581
in lanosterol conversion, 56

Aldehyde dehydrogenases
in lanosterol conversion, 56
reactions catalyzed by, 4

Aldehyde oxidase, cytochrome c reduction by, 455

Aldehydes
acids from, by biological oxidation, 3, 4
from enzymic oxidation, 10

Alfalfa, lipoxygenase from, 36

Alkaloids, oxygenases in biosynthesis of, 22–23

Alkanes
carbene insertion into, 416
oxidations of, characteristics, 428, 430

1-Alkanols, from hydrocarbon hydroxylation, 51

Alkaptonuria, enzyme deficiency in, 18

Alkenes, carbene addition to, 416

Allylic peroxides, in oxygenase reaction, 408

Amides, of fatty acids, ω-oxidation of, 48

Amino acid decarboxylase, lysine monooxygenase as, 18

Amino acid hydroxylases, 285–369

Amino acid oxidase(s)
lysine monooxygenase as, 14
oxidative decarboxylases as, 441

D-Amino acid oxidase, 253–254
reaction catalyzed by, 3

L-Amino acid oxidase, 253–254

Amino acids
oxidation of, in presence of alkylamines, 15
oxygenation of, 24
oxygenases in metabolism of, 15–18, 24

651

"Catalase effect," 204

Catechol(s)
  as enzyme inhibitors, 310–311, 345
  enzymic cleavage of, 5, 137, 138
  enzymic formation of, 7
  as tryptophan metabolite, 113–114
  as tryptophan hydroxylase inhibitors, 364

Catechol dioxygenases, 136–151
  classification of, 156
  mechanism of, 443
  ring fission by, 136–137

Catechol: oxygen 1,2-oxidoreductase, *see* Pyrocatechase

Catechol: oxygen 2,3-oxidoreductase, *see* Metapyrocatechase

Catecholase activity, of tyrosinases, 436–437

Cathepsins, as phenylalanine hydroxylase activators, 323

Cell membrane, lipids vital for, 18

Cerebrocuprein, superoxide dismutase as, 457

Chemiluminescence, from dioxetane decomposition, 444

Chlorella vulgaris, fatty acid desaturation in, 52, 53

β-Chloroalanine, as lactate oxidase substrate, 254

Chlorobenzene, enzymatic oxidation of, 420

Chlorocyclizine, *N*-demethylation of, 70

Chloroiridite ions, as reducing agent, 539

β-Chlorolactate, as lactate oxidase substrate, 254

Chloroperoxidase
  comparison with cytochrome P-450$_{cam}$, 596–597
  heme in, 545, 548

p-Chlorophenylalanine, as enzyme inhibitor, 311

Choleglobin, 543

5α-Cholesta-8, 14-dien-3β-ol, biosynthesis of, 57, 58

5,7-Cholestadien-3β-ol, in cholesterol biosynthesis, 58

5β-Cholestane-3α, 7α-diol, hydroxylation of, 76

5β-Cholestane-3α, 7α, 12α-triol, hydroxylation of, 76

5α-Cholest-7-en-3β-ol, in cholesterol biosynthesis, 58

5α-Cholest-8(14)-en-3β-ol, 57–58

Cholesterol
  bile acid biosynthesis from, 73–75
  biosynthesis of, 54–58
  hydroxylation of, 75
  from pregnenolone, 61
    conversion pathway, 63–65
  side-chain cleavage of, 65
  steroid hormones from, 58–69

Cholesterol 20α-hydroperoxide, as pregnenolone precursor, 63, 64

α-Chymotrypsin, as phenylalanine hydroxylase activator, 317–323

Claisen rearrangement, in oxenoid mechanism, 434

Claviceps purpurea, ricinoleic acid biosynthesis in, 47

Clostridia, ferredoxins in, autoxidation of, 468

CO_2 assay method, for α-ketoacid dioxygenases, 176

Codeine, monooxygenation of, 219

Collagen
  hydroxylating enzyme for, 168, 174
  hydroxyprolines in, 18
  unhydroxylated, as enzyme cofactor, 203

Collagen proline hydroxylase, as early name for prolyl hydroxylase, 169

Copper
  in cytochrome c oxidase, 486, 487
    role in catalytic action, 495–500
  in dopamine β-hydroxylase, 440
  in oxygenases, 7, 8, 436
  in superoxide dismutase, 457, 458, 462–463, 464
  in L-tryptophan oxygenase, 99
  in tyrosinases, 385–386

Copper-containing oxygenases, 371–404
  dopamine β-hydroxylase, 394–398
  mechanism of, 371
  quercetinase, 398
  tyrosinases, 374–394, 436

Corn, lipoxygenase from, 36

Corpus luteum, steroid hormone synthesis in, 58, 63

Corticosterone, tryptophan oxygenase induction by, 120

D

J

Japanese radish peroxidase, heme in, 545

K

α-Ketoacid dioxygenase
p-hydroxyphenylpyruvate hydroxylase as, 168
   CO_2 assay method for, 176
   mechanism of reaction of, 205, 207–208
Ketoacids
   enzymatic decarboxylation of, 249, 253
   oxidative decarboxylation of, mechanism for, 441
α-Keto-α-aminocaproate, enzymatic formation of, 257
2-Keto fatty acids, as fatty acid intermediate, 46
α-Ketoglutarate, as oxygenase cosubstrate, 7, 19
α-Ketoglutarate-coupled dioxygenases
   γ-butyrobetaine hydroxylase as, 181–185
   5-formyluracil dioxygenase as, 192–194
   general properties of, 203–208
   5-hydroxymethyluracil dioxygenase as, 192–194
   p-hydroxyphenylpyruvate hydroxylase as, 199–203
   lysyl hydroxylase as, 178–181
   oxenoid mechanism in, 442–443
   prolyl hydroxylase as, 168–178
   pyrimidine deoxyribonucleoside 2'-hydroxylase as, 195–199
   in pyrimidine and nucleoside metabolism, 185–208
   reactions of, 168
   thymine 7-hydroxylase as, 189–192
Ketolactonases
   induction of, 581
   lack of cytochrome in, 573
Kidney, cytochrome P-450 in, 232, 233
L-Kynurenine
   enzyme induction by, 91
   as serotonin intermediate, 17
   as tryptophan metabolite, 88, 113
Kynurenine formamidase, induction of, 114, 115
Kynurenine hydroxylase, 278–279
   hormonal control of, 279

as mitochondrial marker, 278
reaction catalyzed by, 247, 278
source of, 247, 278
Kynurenine 3-monooxygenase
   distribution of, 22
   reaction catalyzed by, 17

L

Lactate monooxygenase, 250–254, 541
   isolation and purification of, 8–9, 250
   reaction catalyzed by, 208, 246, 251–252
      mechanism, 253
   sources of, 246
Lactate oxidase, see Lactate monooxygenase
Lactate oxidative decarboxylase, see Lactate monooxygenase
Lactic acid bacteria catalase absence in, 453
Lanosterol
   biosynthesis of, 54
   demethylation of, 55–56
Lauric acid
   ω-hydroxylation of, 70
Lecithin
   as enzyme inhibitor, 319–320
Lignin
   biosynthesis of, 23
Linoleic acid
   as essential fatty acid, 52
   oxygenation of, 31–33, 35, 43
      compounds formed, 36
   as ricinoleic acid precursor, 47–48
γ-Linolenic acids
   labeled, formation of, 40
Lipids
   as oxygenase substrates, 12, 15, 18, 22
Lipooxygenase(s), 30–36
   distribution and activity of, 22
   mechanism for, 445–446
   from sheep vesicular gland, 43
   from soybean, 30–32
   from wheat flour, 35–36
Lithocholic acid, hydroxylation of, 76
Liver
   cytochrome P-450 linked monooxygenase system in, 229–232
   microsomes of, cytochrome P-450 linked monooxygenase system in, 215–244

oxy form of, 553
prosthetic group of, 545
Mitochondrial membrane, cytochrome c
    interaction with, 516–528
Mixed function amine oxidase(s), 248,
    277–278, 536
    definition of, 297
    reaction catalyzed by, 247, 277–278, 552
    source of, 247
Monooxygenases
    from bacteria, 561
    catalytic action of, 87–88
    definition and activity of, 7–10
    external type, 8, 9–10
    internal type, 8–9
    list of, 8
    oxenoid mechanism for, 431–440
    reactions catalyzed by, 15–17
Monoperoxy succinic acid, source of, 205
Monophenol monooxygenase, reaction
    catalyzed by, 10
cis, cis-Muconic acid
    from catechol cleavage, 137, 139
    enzymic formation of, 5
Mung bean, lipoxygenase from, 36
Mushroom, tyrosinase in, 375–376, 378
Mycobacteria, oxygenases in, 21
Mycobacterium phlei
    fatty acid desaturation in, 52
    lactate monooxygenase from, 246, 249,
        250
    oxygenase from, 8–9
Mycobacterium phodocorus, cytochrome
    P450 activity in, 572
Mycobacterium smegmatis, lactate mono-
    oxygenase from, 246, 250
Myeloperoxidase
    hemoprotein of, 548
    inhibition by superoxide dismutase, 470
    spectral properties of, 545
Myoglobin
    artificial hemoproteins for, 548
    as cytochrome c reduction inhibitor, 456
    oxy form of, 553
    peroxidase compared to, 551
    peroxide compounds of, heme structure
        in, 546
    redox potential of, 552
Myoinositol, D-glucuronate from, 19, 441

N

NAD, in serotonin biosynthesis, 17
NADH, peroxidase-catalyzed oxidation of,
    536
NADH-rubredoxin reductase, in ω-hy-
    droxylation, 50
NADP, low levels of, in hyperthyroidism,
    17
NADPH
    in cholesterol side chain cleavage 64–65
    in 11β-hydroxylating system, 61
    in 19-hydroxylating system, 67–68
NADPH-adrenodoxin reductase, in steroid
    hydroxylation 59
NADPH-cytochrome c reductase
    in ω-hydroxylation, 48–49
    isolation of, 49
NADPH-cytochrome P-450, 59
NADPH-cytochrome P-450 reductase
    in hydroxylation, 70, 72, 73, 75, 76
    in monooxydation, 219–220, 228, 233–
        234, 235
NADPH-2,6-dichlorophenol indophenol
    reductase, 220
NADPH-ferredoxin reductase, in ω-hy-
    droxylation, 50
Naphthohydroquinone, peroxidase-cata-
    lyzed oxidation of, 536
α-Naphthoquinoline, as enzyme inhibitor,
    147
Naphthylamines, as amine oxidase sub-
    strates, 278
Naproxen, as prostaglandin synthetase
    inhibitor, 44
Neocuproine, as enzyme inhibitor, 201
Neotetrazolium, enzymic reduction of, 220
Neotetrazolium reductase, cytochrome
    P-450 system and, 233
Neurospora sp.
    oxygenases in, 19
    protocatechuate 3,4-dioxygenase from,
        142
Neurospora crassa
    pyrimidine metabolism in, 185–186
        enzymes for, 186, 187, 188, 199
    superoxide dismutase in, 466
    tyrosinase in, 375–376, 378
Niacin, tryptophan as precursor of, 90

P

Molecular Biology

An International Series of Monographs and Textbooks

Editors

BERNARD HORECKER

Roche Institute of Molecular Biology
Nutley, New Jersey

NATHAN O. KAPLAN

Department of Chemistry
University of California
At San Diego
La Jolla, California

JULIUS MARMUR

Department of Biochemistry
Albert Einstein College of Medicine
Yeshiva University
Bronx, New York

HAROLD A. SCHERAGA

Department of Chemistry
Cornell University
Ithaca, New York

WALTER W. WAINIO. The Mammalian Mitochondrial Respiratory Chain. 1970

LAWRENCE I. ROTHFIELD (Editor). Structure and Function of Biological Membranes. 1971

ALAN G. WALTON AND JOHN BLACKWELL. Biopolymers. 1973

WALTER LOVENBERG (Editor). Iron-Sulfur Proteins. Volume I, Biological Properties—1973. Volume II, Molecular Properties—1973. Volume III, Structure and Metabolic Mechanisms—1977

A. J. HOPFINGER. Conformational Properties of Macromolecules. 1973

R. D. B. FRASER AND T. P. MACRAE. Conformation in Fibrous Proteins. 1973

OSAMU HAYAISHI (Editor). Molecular Mechanisms of Oxygen Activation. 1974

FUMIO OOSAWA AND SHO ASAKURA. Thermodynamics of the Polymerization of Protein. 1975

LAWRENCE J. BERLINER (Editor). Spin Labeling: Theory and Applications. 1976

T. BLUNDELL AND L. JOHNSON. Protein Crystallography. 1976

in preparation

HERBERT WEISSBACH AND SIDNEY PESTKA (Editors). Molecular Mechanisms of Protein Biosynthesis

B 7
C 8
D 9
E 0
F 1
G 2
H 3
I 4
J 5